Materials Science Applications of Ion Beam Techniques

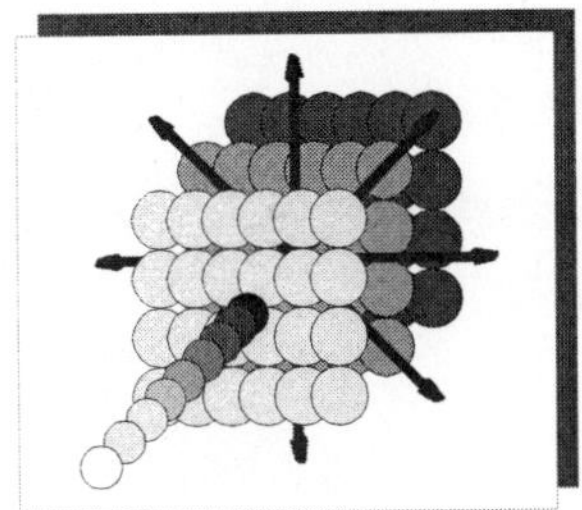

Seeheim, Germany
September 9-12, 1996

MATERIALS SCIENCE APPLICATIONS OF ION BEAM TECHNIQUES

Proceedings of the International Symposium
on Materials Science Applications of Ion Beam Techniques,
incorporating the
1st German-Australian Workshop on Ion Beam Analysis,
Seeheim, Germany, September 9-12 1996,

Editors:

A.G. Balogh and G. Walter

TRANS TECH PUBLICATIONS
Switzerland • Germany • UK • USA

Volumes 248-249 of
Materials Science Forum
ISSN 0255-5476

Distributed in the Americas by

Trans Tech Publications Ltd
c/o Enfield P & D Co, Inc
Post Box 699
Enfield, New Hampshire 03748
USA
Phone (603) 632-7377
Fax   (603) 632-5611

and worldwide by

Trans Tech Publications Ltd
Brandrain 6
CH-8707 Uetikon-Zuerich
Switzerland
Fax +41 (1) 922 10 33
e-mail: ttp@ttp.ch
http://www.ttp.ch

Printed in the United Kingdom
by Hobbs the Printers Ltd,
Totton, Hampshire SO40 3WX

Conference Organization

Local Organizing Committee:

A.G. Balogh	Darmstadt
K. Bethge	Frankfurt
J.D. Meyer	Frankfurt
G. Walter	Darmstadt

International Advisory Board:

N. Angert	Germany
J. Baglin	USA
D. Cohen	Australia
J. Gyulai	Hungary
S. Roorda	Canada
S. Klaumünzer	Germany
W. Möller	Germany
M. Nastasi	USA
J. O'Connor	Australia
K.-P. Lieb	Germany
B. Stritzker	Germany
P. Thevenard	France

Programme Committee:

A.G. Balogh	Darmstadt
H. Baumann	Frankfurt
K. Bethge	Frankfurt
D. Carstanjen	Stuttgart
H. Hahn	Darmstadt
S. Kalbitzer	Heidelberg
K. Maier	Bonn
F. Rauch	Frankfurt
W. Wesch	Jena

<u>Sponsors:</u>

- **Deutsche Forschungsgemeinschaft (DFG)**

- **Bundesministerium für Bildung, Wissenschaft, Forschung und Technologie (BMBF)**

- **Hessisches Ministerium für Wissenschaft und Kunst (HMWK)**

- **Gesellschaft für Schwerionenforschung (GSI)**

- **Technische Hochschule Darmstadt (THD)**

- **Vereinigung von Freunden der Technischen Hochschule zu Darmstadt**

- **Fachgebiet Dünne Schichten des Fachbereichs Materialwissenschaft der THD**

- **Institut für Kernphysik der Johann-Wolfgang-Goethe-Universität Frankfurt/Main**

- **Institut für Kernphysik der THD**

<u>Exhibitors:</u>

- **High Voltage Engineering Europe B.V.**

- **Goodfellow GmbH**

- **Pfeiffer Vakuum Vertriebs GmbH**

- **Canberra-Packard Instruments, Ltd.**

- **Varian Vacuum Products**

- **Campro Scientific**

Preface

„Ages, quod agis"

This proceedings volume contains selected papers that were presented at the „International Symposium on Materials Science Applications of Ion Beam Techniques" held in Seeheim, Germany, from 9 to 12 September 1996. More than 140 scientists from 22 countries attended the conference. In addition to the 46 oral contributions about 80 papers were presented as posters at the meeting, which was organised without parallel sessions. The main topics of the conference were:

- Basic Processes
- Ion Beam Modification of Materials
- Ion Beam Analysis
- New Technical Developments
- Industrial Applications.

The first particle accelerators were built in the early 30's, e.g. some 66 years ago. For a long time, these apparatus were exclusively used by nuclear physicists. In the 60's extensive development in measuring techniques took place, mainly as a consequence of the newly developed semiconductor devices. The first materials science application of an Ion Beam Analysis Method, as we would call it today, was performed in 1967 during the moon mission of the Surveyor V satellite. The moon dust was analysed using the 6.1 MeV α-particles of a ^{242}Cm source. A further strong impact came from the semiconductor industry, when the ion implantation technique has become widely accepted as the ultimate tool in Si-based device fabrication. In the last decade many materials science application experiments using ion beam techniques were reported. Modification of materials structure and materials properties by ion beams and analysis of the modified properties using ion beam techniques are widely used today in a broad spectrum of topics including amorphisation-recrystallisation studies, improvement of material properties like adhesion, hardness, wear resistance, corrosion, etc., influence of interface properties by ion beam mixing, fabrication of semiconductor devices by low and high energy ion implantation, etc. In summary, up-to-date materials science without application of ion beam techniques is no longer conceivable. Therefore, it seemed to be necessary to organise a conference, where the most important and recent aspects and results of the materials science applications of these techniques will be highlighted.

Many colleagues, participated at this symposium, assured me during and after the conference, that this meeting was a successful event. This success is credited to many people. I wish to thank all those, who have fundamentally contributed to it, in particular to the sponsors of the conference, the members of the Advisory Board, the Programme Committee and the Organising Committee and last but not least the stuff of the Conference Bureau.

„Nihil est tam perfectum,
quod non habeat defectum"

Darmstadt, Germany, March 1997

Adam Georg Balogh

Table of Contents

Basic Processes

Ion Beam Modifications

Ion Beam Analysis

Turm C
Tower
Casi
Audi

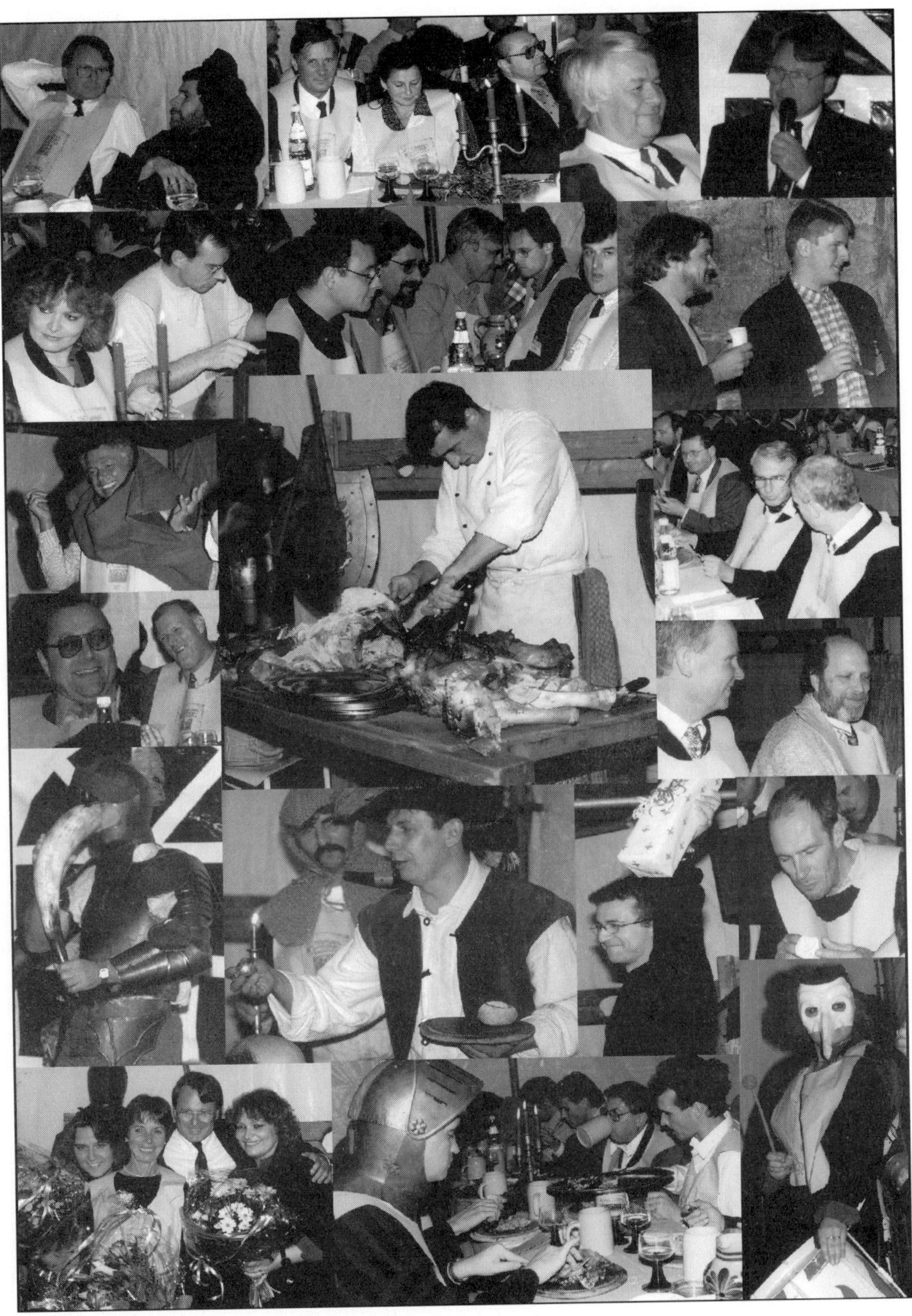

Basic Processes

Materials Science Forum Vols. 248-249 (1997) pp. 3-12
© *1997 Trans Tech Publications, Switzerland*

Ion Beam Induced Amorphization of Crystalline Solids: Mechanisms and Modeling

H. Trinkaus

Institut für Festkörperforschung, Forschungszentrum Jülich, D-52425 Jülich, Germany

Keywords: Amorphization, Amorphous Zones, Ion Tracks, Thermal Spikes, Recrystallization

Abstract There are two main mechanisms of irradiation induced amorphization of crystalline solids: (1) continuous accumulation of defects which destabilize the crystal structure at some critical level, and (2) rapid quenching of irradiation induced liquid thermal spike regions. Which mechanism occurs or dominates depends on type of material, energy and type of projectile and irradiation temperature. In connection with a brief review of the main experimental findings and their current understanding the role of thermal spikes in ion beam induced amorphization (IBIA) is emphasized. An amorphous nucleus is considered to be the result of melting and incomplete recrystallization of a thermal spike region. The evolution of a liquid/amorphous region during as well as after the thermal spike within the defective crystalline environment is controlled by crystalline/liquid (amorphous) interface kinetics. On the basis of these ideas, the dependence of IBIA on main parameters such as type of material, projectile and energy deposition is discussed.

1. Introduction

The amorphous (or glass) state is perhaps the most familiar amongst the metastable non-equilibrium solid states. An amorphous solid is frequently considered to be a frozen liquid since both are characterized by the lack of long range order. Otherwise, the amorphous state is not uniquely defined and for one material a whole spectrum of states is conceivable differing for instance in the degree of short range order. Amorphization can be achieved by various techniques [1] such as rapid melt quenching, mechanical alloying, hydrogenation, interdiffusion reactions or irradiation.

Radiation and particularly ion beam induced amorphization (RIA, IBIA) of surface near layers has become a widely used technique in which the amorphization process can be rather well controlled by the relevant irradiation parameters. For a given material, the occurrence of RIA and its evolution depends on the type and energy of the projectile (electrons, light or heavy ions, neutrons), dose rate, dose (up to the "amorphization dose") and temperature (below the "critical temperature" for amorphization) [2-5].

An energetic particle penetrating a solid slows down by two different types of interactions: quasi-elastic collisions characterized by an "nuclear stopping power", S_n, and electronic excitations associated with ionizations characterized by an "electronic stopping power", S_e [6]. S_n and S_e attain maximum values of the order of 100 eV/nm (for light ions in low Z materials) up to 10 keV/nm and 100 keV/nm (for heavy ions in high Z materials) at ion energies in the keV to MeV and MeV to GeV range and are thus dominate at low and high energies, respectively. For recoils with energies above the displacement threshold, elastic collisions result in displacements, whereas electronic excitations can generate lattice defects only indirectly via a transfer of the energy from the electronic into the lattice system. Therefore, the sensitivity of a material to amorphization will not only depend on the amount of energy deposition but also on its type [4].

For nuclear energy deposition, two main mechanisms of radiation induced amorphization have been considered: (1) continuous defect accumulation [7] resulting in an elastic [8,9] or thermodynamic [1-3] instability of the crystal structure, corresponding to the new routes to amorphization such as mechanical alloying, hydrogenation or interdiffusion reactions, and (2) local melting and subsequent rapid quenching of the liquid state within an irradiation induced thermal spike region [10-12] corresponding to the "classical" route to amorphization by rapid melt quenching. Which of the two mechanisms occurs or dominates depends mainly on the type of material and the type and energy of the projectile particle. A crucial difference between both mechanisms is that, in the first case, amorphization is quasi-homogeneous while in the latter it is generically inhomogeneous. For electronic energy deposition, the thermal spike mechanism [12] seems now being increasingly favoured above the mechanism of Coulomb explosion iniated by ionization [13] while global lattice destabilisation by defect accumulation has not been considered in this case.

There are mechanisms between and combinations of these two main mechanisms which should be considered: Between the two cases, for instance, are contributions from locally enhanced defect concentrations arising from displacement cascades which do not result in pronounced thermal spikes (light ions in low Z materials). On the other hand, cascades or thermal spikes may nucleate small amorphous zones within a defective crystalline environment [14]. This would represent an example for a combined effect of two different types of damage which are generally not specified in existing phenomenological damage overlap models [15,16].

A full understanding of IBIA should include an understanding of the phenomena occurring around the upper temperature limit where amorphization occurs heterogeneously at interfaces ("ion beam induced interfacial amorphization", IBIIA) [17] and the balance between amorphization and crystallization ("ion beam induced epitaxial recrystallization", IBIEC) [18] depends sensitively on the irradiation parameters. Models of these phenomena are mostly based on single point defect kinetics and contributions of cascades or thermal spikes are considered only in a rather crude phenomenological way [19,20].

Concluding this short summary of mechanisms and modeling we may say that there is still no fully consistent picture of IBIA in which both aspects, the build-up of a quasi-homogeneous distribution of lattice defects and the formation of local defect structures (amorphous zones) in cascades or thermal spike regions as well as the kinetic interaction between both constituents are taken into account.

In the present contribution, first a brief review of the main experimental findings and their current understanding is presented. Then, the role of cascades and thermal spikes in IBIA is discussed. An amorphous nucleus is suggested to be the result of melting and incomplete recrystallization in the thermal spike region and the overall amorphization is proposed to be due the production and evolution of such nuclei within the defective crystalline environment. For a description of these processes, the kinetics of the crystalline/liquid (amorphous) interface must be considered. The dependences of IBIA on the main parameters such as type of material, projectile and energy deposition are discussed in this theoretical framework.

2. Radiation induced amorphization (RIA) - main experimental findings

The amorphizability of a crystalline solid by irradiation depends on the type of material (elements/compounds; metals/semiconductors/insulators), type of projectile particle (electrons, light and heavy ions, neutrons), projectile energy and type of energy deposition (nuclear, electronic), dose rate, dose and temperature.

Type of material: (1) No RIA occurs pure metals and metallic solid solutions;
(2) RIA is possible in intermetallic compounds and semiconductors;
(3) RIA is easy in insulators.

Generally, RIA is associated with (or preceded by) disordering (RID) but some short range order may remain. Among the intermetallic compounds which represent the best investigated class of materials [2] two subclasses may be distinguished [1,5]. (1) "RID alloys" where only radiation induced disordering takes place (Cu_3Au, Ni_3Mn, Ni_3Al, Cu_3Pd ...). In these alloys the disordering temperature is below the melting temperature, $T_{dis} < T_m$. (2) "RIA alloys" where amorphization occurs at sufficiently low temperatures (NiAl, FeTi, CuTi, $NiAl_3$, MoSi ...). These alloys remain ordered up to T_m. There are intermediate cases ("RID-RIA alloys") where first disordering and subsequently amorphization takes place, for instance in Zr_3Al and NiTi under ion bombardement at low temperatures (Zr_3Al, ...). Also these alloys remain ordered up to T_m.

Type of projectile particle: RIA is easier with (fast) neutrons and heavy ions than with (MeV)-electrons and light ions. In materials where RIA is possible with electrons it is also possible with ions but this need not be true vice versa. Thus, Si can be amorphized by neutron and ion but not by electron irradiation. RIA occurs quasi-homogeneously under electron and light ion irradiations but inhomogeneously under neutron and heavy ion irradiations (amorphous zones, ion tracks).

Particle energy and type of energy deposition: In the "nuclear regime" (ion energies in the keV to MeV range) the susceptibility of a given material to RIA generally increases with increasing average transferred recoil energy, E_{rec}, starting from a low level for $E_{rec} \leq 1$ keV. In the "electronic regime" (ion energies in the MeV to GeV range [4]) RIA/IBIA is possible only above a threshold value, S_e^{th}, of the electronic stopping power. S_e^{th} depends on the type of materials; it is of the order of a few keV/nm for insulators and of the order of 10 to a few 10 keV/nm for metallic systems. In materials where RIA is possible with electronic excitations it is also possible with elastic collisions but not necessarily vice versa. Thus Si can be amorphized by elastic collisions transferring sufficiently high recoil energies but neither by (MeV) electron irradiation nor by ion beam induced electronic excitations.

Dose rate and dose: The amorphization efficiency is independent of dose rate at very low temperatures but increases with dose rate at temperature close to the critical temperature for amorphiziation which itself, even if only weakly, increases with dose rate. The amorphous volume fraction, x_a, increases with dose until complete amorphization is reached at the "amorphization dose" (if no fractional saturation value is reached). The dose dependence of x_a varies between an exponential saturation behaviour with an initially linear increase of x_a with dose and a sudden change after a slow initial increase frequently characterized by a strongly nonlinear, high power law dependence. For IBIA by electronic excitations, the dose dependence is close to the first type [21,22]. For RIA by elastic collisions of ions, the initial shape of the log x_a vs. log dose curve varies between 1 (heavy ions in medium to high Z materials) to 4 (light ions or low Z materials) [14,23,24] and is generally above 3 for electron irradiations (see Fig. 1).

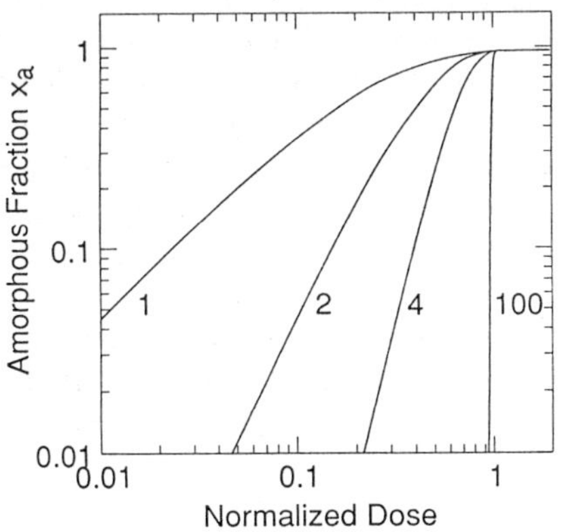

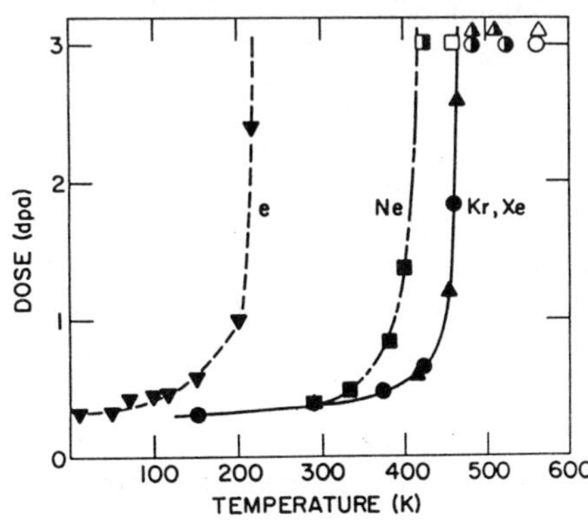

Fig. 1 (left): Log-log plot of amorphous volume fraction x_a vs. dose normalized to dose at $x_a = 99$ % for four initial slopes (exponents).

Fig. 2 (right): Amorphization dose ($x_a = 90$ %) vs. temperature for CuTi (from [23]).

Temperature: RIA disappears in a rather narrow temperature interval. The critical temperature where the amorphization dose diverges depends on the type of projectile particle. For ions (IBIA), it is substantially higher than for electrons and generally increases with increasing ion mass (see Fig. 2). Close to the critical temperature, IBIA occurs heterogeneously at interfaces with a dose rate dependent efficiency.

3. Lattice destabilization by defect accumulation

A continuous (homogeneous) accumulation of chemical or structural defects originating from elastic collisions is associated with a storage of energy [2] and may lead to elastic softening or even elastic instability [11,12]. RIA may be described as melting of a metastable defective crystal. A reduction of the transition ("melting") temperature to the rather low values required for substantial defect accumulation implies that the defect induced increase in the enthalpy of the crystalline state, ΔH_c^*, is of the order or exceeds the enthalpy difference between the amorphous and the equilibrium crystal phase ΔH_{ac} (which is equal to the heat of fusion ΔH_f at the equilibrium melting temperature T_m), $\Delta H_c^* \geq \Delta H_{ac}$. For negligible defect interaction where individual defects contribute independently, each with H_d, to $\Delta H_c^* = c H_d$ this requirement may be expressed as a condition for the atomic defect concentration, $c \geq \Delta H_c^*/H_d$.

Using the estimates $\Delta H_{ac} \approx \Delta H_f \approx kT_m$ and $H_{FP} = 50\ kT_m$ we find as condition for RIA by single Frenkel pair accumulation $c_{FP} \geq 2\ \%$. This estimate yields also a lower limit of 0.02 dpa for the initiation of RIA. The maximum value of c_{FP} attainable at low T is limited, however, by the size of the recombination volume, V_{rec}, within which spontaneous recombination occurs, $c_{FP} \leq \Omega/V_{rec}$, where Ω is the atomic volume. For pure metals and simple metallic solute solutions V_{rec} amounts to hundreds of Ω so that $c_{FP} < 1\ \%$. Therefore these materials cannot be amorphized by single Frenkel pair accumulation produced, for instance, by electron irradiation. Higher values of c_{FP} can probably be reached for covalent materials where V_{rec} is expected to be smaller or where a barrier against recombination exists as for instance in the case of Si [25]. Inspite of this, Si cannot be amorphized by electron irradiation [26] which may be due to a nucleation barrier or another kinetic constraint against the transformation. RIA by single Frenkel pair accumulation is restricted to the low temperature range where at least one of the constituents, vacancies or interstitials is virtually immobile.

In intermetallic compounds, the maximum storable energy is increased in comparison to that in pure metals and simple alloys probably by both a reduced recombination volume and chemical in addition to structural disorder. Thus, the enthalpy difference between the disordered and the ordered state, ΔH_{dis}, must be considered in addition to the Frenkel pair contribution. In fact, the relation between ΔH_{dis} and ΔH_{ac} seems to be crucial for the qualitatively different behaviour of the RID and RIA alloys introduced above. In the case of RID alloys, $\Delta H_{ac} > \Delta H_{dis}$ meaning $T_{dis} < T_m$ for $\Delta H_{ac} \approx \Delta H_f$, and the opposite for the RIA alloys. The behaviour of the "RID - RIA alloys" cannot be understood by such a crude scheme (see below).

A useful measure for the enthalpy stored in the lattice in the form of chemical and structural disorder is provided by the mean-square static atomic displacements which, in addition, correlate well with the softening of the elastic shear moduli. These correlations suggest to generalize Lindemann's melting criterion to RIA [9].

4. Melting and quenching of thermal spike regions

There is now ample evidence for melting in displacement cascades and in electronically highly excited cylindrical regions around ion tracks. Thus, the high ion beam mixing efficiency cannot be understood in terms of ballistic two body collisions but may well be attributed to atomic mixing in a liquid. MD-studies of displacement cascades in pure metals [27] clearly show structural disordering as in melting during the hot phase and recrystallization during the cooling phase of the spike. The occurrence of thermal spikes has been questioned for materials with $Z < 20$ [28]. But even for Si, MD simulations of cascades [29] show structural disordering occurring at average Si atom energies sufficient for melting. Ion beam induced creep of vitreous SiO_2 [30] may be quantitatively described by stress relaxation in thermal spike regions [31] and thus indicate the operation of thermal spikes even for this low Z material. Clear evidence for melting by electronic excitation along ion tracks in ceramics is provided by high resolution electron microscopy which exhibit well defined amorphous-crystalline interfaces [32,33].

The basic assumption in thermal spike models is that the energy contained in atomic or electronic collision cascades is randomized within short periods of 100 to 500 and 1 to 10 fs, respectively, resulting in local melting which is generally not (completely) reversed by recrystallization during the subsequent cooling down phase. In the case of electronic excitations, the energy must first be transferred from the electronic to the atomic system which occurs typically between 0.1 to 5 ps before melting can take place. In the simplest "direct impact" approximation, amorphization is then the result of the superposition of individual amorphous regions. The size of the molten and the resulting amorphous zones reaches values of up to about 10 nm. Radiation induced thermal spikes are characterized by extremely rapid quenching rates of up to 10^{15} K/s.

The evolution of an individual thermal spike is controlled by the production of heat within one of the two subsystems, its transfer to the other subsystem and its diffusional spreading. This may be described by two coupled heat transport equation for the atomic (a) and the electronic subsystem (e), respectively [12]

$$C_{a,e} \partial T_{a,e} / \partial t = q_{a,e}(\mathbf{r}, t) - g(T_{a,e} - T_{e,a}) + \nabla K_{a,e} \nabla_{a,e} T_{a,e}, \tag{1}$$

where $q_{a,e}$ are the energy deposition densities, $T_{a,e}$ are the temperatures, $C_{a,e}$ are the specific heats, g is the coupling coefficient (electron phonon coupling) and $K_{a,e} = D_{a,e} C_{a,e}$ are the thermal conductivities ($D_{a,e}$: thermal diffusion coefficients) in the two subsystems. The initial and boundary conditions are that the initial and distant temperatures are equal to the ambient temperature. An approximate analytical treatment of the problem is complicated by the non-Gaussian form and ion velocity dependence of q_e (r,t), and the temperature dependence of C_e, D_e and g. Some important qualitative conclusions can, however, be drawn directly. In the case of electronic excitations, for instance, the thermal spike effect is overestimated when assuming a spatially Gaussian energy deposition which is more concentrated than the real one. Since melting occurs in the atomic system the role of the electron-phonon coupling coefficient g is different in thermal spikes induced by collision cascades or by electronic excitations, particularly in metallic systems. Thus, for collision cascades induced spikes, the quenching rate during cooling increases with increasing g which is favourable for IBIA. For spikes induced by electronic excitations, the maximum temperature in the atomic system (important for the threshold S_e^{th} for amorphization) increases with increasing g.

The simplest approach to the problem is to assume that, (1) the temperature distribution in the atomic system is Gaussian, characterized by a given initial width related to the mean heat diffusion length at equilibration of the two subsystems, and the maximum temperature determined by the total deposited energy, (2) the whole region enclosed by the melting isotherm, $r_m = r(T = T_m)$, melts but no recrystallization occurs upon cooling down, (3) the latent heat of fusion is negligible in the total enthalpy invested into the thermal spike. These simplifications clearly result in upper bound estimates for size of the amorphous zones or ion tracks and for the amorphization efficiency. For sufficiently high energy deposition $Q = E_{cas}$ into a spherical cascade ($d = 3$) or $Q = S_e$ into a cylindrical ion track ($d = 2$), respectively, where r_m expands initially, the maximum r_m is given by

$$r_m \, (max) = (d/2\pi e)^{1/2} \, (Q/\rho C \Delta T_m)^{1/d} \quad , \qquad (2)$$

where $d = 2$ or 3 is the dimension of region, ρ is the mass density, C is the total specific heat per unit mass and $\Delta T_m = T_m - T_o$ is the difference between the melting and the ambient temperature. Assuming in addition that the whole energy lost by the ions is deposited in thermal spikes, one obtains an upper band estimate corresponding to the "direct impact" approximation for the amorphous volume fraction

$$x_a \leq 1 - exp \, (- \, \alpha \, \phi \, S/\rho C \, \Delta T_m) \quad , \qquad (3)$$

where $\alpha = (d/2e)^{d/2} / (d/2)! \approx 1/3$ is a numerical factor and ϕ is the ion fluence. With $\rho C \Omega \approx 3k$, and $\Delta T_m \approx 1000$ K, Eq. (3) corresponds to an average energy to be deposited per atom to induce amorphization of the order of 1 eV which is in order of magnitude agreement with values found for low temperature IBIA of well amorphizable systems.

For many systems, however, experimental studies of IBIA indicate that melting in thermal spikes is incomplete and/or (partial) recrystallisation takes place. First of all, there is general agreement that in pure metals and metallic solid solutions, liquid thermal spike cores recrystallize upon cooling down. But there are hints at incomplete melting/partial recrystallization in other systems too. An analysis of ion track radii in insulators in dependence upon S_e in the framework of the above described simple approach [34], for instance, yielded less than 20 % for the apparent fraction of electronic energy deposited in Gaussian thermal spikes which was used as a fit parameter together with the initial width of the distribution. Such low values cannot be explained by deviations of the initial spatial energy distribution from a Gaussian one which would at most allow about 50 % and also not by an increase of the melting temperature due to pressure in the hot spike core which would in general allow an additional reduction of only less than 10 %. Is the remaining discrepancy due to superheating, incomplete melting and/or partial recrystallisation? In fact, a conclusion in this direction was drawn from a study of ion tracks in high T_c superconductors [32]. The joint amorphous/crystalline interfaces of contiguous tracks in the latter material class frequently show the characteristic features [33] expected as the result of interfacial stress effects in the motion of a crystal liquid interface (see Fig. 3). Recrystallization of ion tracks have also been found in rare earth/Co alloys which results in incomplete amorphization in the case of YCo_2 [22].

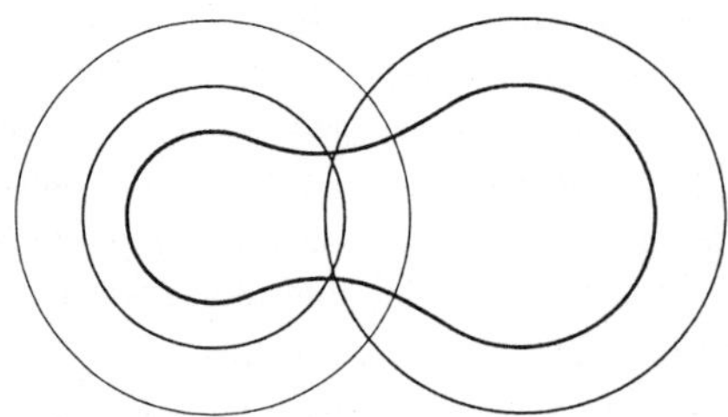

Fig. 3: Interaction of sequential ion tracks (schematic). Lines of increasing thickness indicate interfaces at progressive time steps. A partially recrystallized first track (left) recrystallizes further when touched by a subsequent track (right). The contraction of the interface is reduced in the neck joining the tracks.

In general, it is, however, not clear whether recrystallization occurs in the quenching phase of the thermal spike or, during some transient period, at ambient temperature which in most cases is far below the crystallization or even far below the glass temperature. This would imply that the amorphous state formed in thermal spikes is much less stable than that formed in classical melt quenching. In fact, swift heavy ion induced plastic deformation of amorphous solids [34] which has been recently attributed to shear stress relaxation in thermal spike regions [35] shows a decrease with increasing irradiation temperature which indicates post spike stress relaxation at temperatures far below the glass temperature, for instance below room temperature in a-SiO_2.

For a clarification of these problems, melting in a thermal spike region as well as recrystallization during and after cooling down must be considered. Melting may be initiated by an elastic instability or by the formation of liquid nuclei. Both is easier in displacement cascades where disorder is a generic feature than in thermal spikes induced by electronic excitations. In the latter case a high superheating is required (perhaps up to about $2T_m$) where liquid nucleation can occur virtually spontaneously and thus sufficiently fast within the short available time of the order of ps. This is one of the reasons for the general trend that elastic collisions are more efficient for IBIA than electronic excitations.

Once a liquid nucleus is formed melting and recrystallization are controlled by interfacial kinetics. The liquid/solid interface motion is driven by differences in the chemical potentials of the two phases at the interface. Within a linear non-equilibrium approximation the interfacial melting/recrystallization velocity (here for simplicity for a monocomponent system) may be written as

$$v_i = - M^* (\Delta\mu^* + \gamma^* \kappa V^*) < v_s \; . \tag{4}$$

Here, M^* is the interface mobility, $\Delta\mu^* = - \Delta S^* \Delta T^* + \Delta V^* \Delta p$ is the chemical potential difference between the generally perturbed liquid (amorphous) and crystalline phases where ΔS^* is the corresponding entropy difference (entropy of fusion in the undisturbed case), $\Delta T^* = T - T_m^*$ is the overheating/undercooling with respect to the effective melting temperature T_m^*, ΔV^* is the specific volume change upon melting and Δp is the interval pressure in the thermal spike. In the interfacial term $\gamma^* \kappa V^*$, γ^* is the specific interface free energy, $\kappa = (d-1)/r$ is the curvature (r: radius) and V^* is the specific volume in the liquid (amorphous) phase. The asterisk indicates radiation induced perturbations of the phases. It may be useful to split the chemical potential difference between the perturbed phases into the corresponding difference between the unperturbed phases, $\Delta\mu = - \Delta S \Delta T + \Delta V \Delta p$, and the difference in the radiation induced changes $\delta\mu^*$ which can be crucial for a shift of the melting temperature. The ultimate upper limit of v_i is set by the (longitudinal) sound velocity v_s.

The interface mobility which is the most uncertain quantity in Eq. (4) may be expressed by an effective diffusion coefficient D^* as [37]

$$M^* = D^* / \ell\, kT \quad \text{with} \quad D_b = kT / 6\pi\, a\, \eta^* \le D^* \le D_a^{th} \;\; , \tag{5}$$

where ℓ is a characteristic length of the order of the atomic distance a for M^* controlled by bulk diffusion in the unperturbed liquid (amorphous) phase or for defect reaction limited M^* on one hand, and of the order of the mean defect diffusion length for M^* controlled by radiation induced defect diffusion (radiation enhanced diffusion) on the other hand. Lower and upper limits for D^* are defined by the bulk diffusion coefficient D_b which may be expressed via the Einstein-Stokes relation by the viscosity η^*, and the thermal diffusion coefficient D_a^{th} in the atomic subsystem, respectively. In thermal spikes, v_i is expected to be bulk diffusion controlled for complex structures, $D^* \approx D_b$, but may be "impingement controlled" in simple structures, $D^* \approx D_a^{th}$, where it may reach values even close to v_s. For low irradiation temperatures, the heat flux induced by the consumption and production of latent heat upon melting and recrystallization represents only a fraction of the total heat flow through the interface for the narrow and fast thermal spikes considered here and may therefore be neglected in estimates of melting and recrystallization.

Equation (4) combines thermodynamic and kinetic aspects of RIA (at least conceptionally). Thus, the thermodynamic conditions discussed above are defined by $\Delta\mu^* \leq 0$ implying $v_i \geq 0$ for $\kappa = 0$. Inspite of uncertainties, Eq. (4) represents a useful base for understanding characteristic features of IBIA. First, it is clear without detailed calculations that interface velocities required for significant recrystallization must be very high 1nm/10ps, i.e. 100 m/s, values which are indeed reached in simple structures such as pure metals and metallic solid solutions. A more quantitative condition to be imposed on M* to achieve complete recrystallization may be derived by approximately integrating Eq. (4) [37] similarly as done in ref. 38. A combination of thermodynamic and kinetic aspects as in Eq. (4) also provides a framework for understanding the physical nature of "damage overlap" effects in IBIA. Thus, for instance, the role of disordering (RID) preceding amorphization in "RID-RIA alloys" mentioned above becomes clear. In the initial stage of RID, a liquid spike region embedded in an ordered crystalline environment is just able to recrystallize completely to a disordered crystalline zone whereas this is no longer possible in the subsequent stage of IBIA because of the larger extension of the liquid region due to a reduction of the melting temperature by RID.

The monocomponent version of Eq. (4) is not sufficient to fully describe the liquid-solid interface motion in multicomponent systems. In the latter case, the simplest approach to the problem would be to assume that the slowest component controls the interfacial motion according to Eq. (4). An interesting question is how melting and partial recrystallization can occur in complex systems such as magnetic insulators or high T_c superconductors. First of all, in the case of electronic excitation, substantial superheating during the short heating stage may occur in a perfect crystalline region due to the limited speed of melting kinetics (although melting may occur by the motion of an existing amorphous crystalline interface). Secondly, the destabilization of the lattice may be due to the fast components of the system bypassing the slow components and thus generating a specific amorphous state where the slow components remember their positions in the lattice. In high T_c superconductors, for instance, the typical layered structure could be partially preserved during "melting". Thirdly, in the strongly superheated spike core, the system would be driven into a highly excited metastable liquid/amorphous state characterized by a high effective diffusivity / low viscosity allowing partial recrystallization during and after spike cooling.

5. Quenched thermal spike regions in defective crystalline environment

An amorphous zone remaining after incomplete recrystallization forms a nucleus for overall amorphization if the ambient temperature is below the effective temperature defined by the radiation induced defect concentration. Under this condition, such a nucleus may be sub- or supercritical, i.e. it may shrink or grow, $v_i < 0$, $v_i > 0$, depending on its size. In the latter case, the nucleation rate within crystalline regions is constant for constant ion flux (a situation which can hardly be realized in common first order phase transitions in solids). Simultaneous nucleation and growth of amorphous zones result in an overall amorphization rate which increases initially with a power of dose between 3 and 4 [14] (depending on the type of growth kinetics) as frequently found in low temperature IBIA [14,24]. At higher temperatures, where an individual zone is subcritical, zone touching may still result in the formation of supercritical nuclei with a dose rate dependent rate (see Fig. 4).

The steady state defect concentration under irradiation decreases substantially close to the critical temperature for amorphization, T_c, where both interstitials and vacancies are mobile. In this temperature range, absorption of defects rather results in shrinkage than in growth of amorphous zones. An existing interface may, however, prolong the lifetime of zones located contingent to it by reducing the interfacial tension effect on them and the flux of point defects to them. Heterogeneous amorphization at interface [17] may be attributed to these effects.

In existing models of ion beam induced amorphization and recrystallization for this temperature range [19,20], the role of amorphous zone formation in collision cascades and thermal spikes is certainly underestimated.

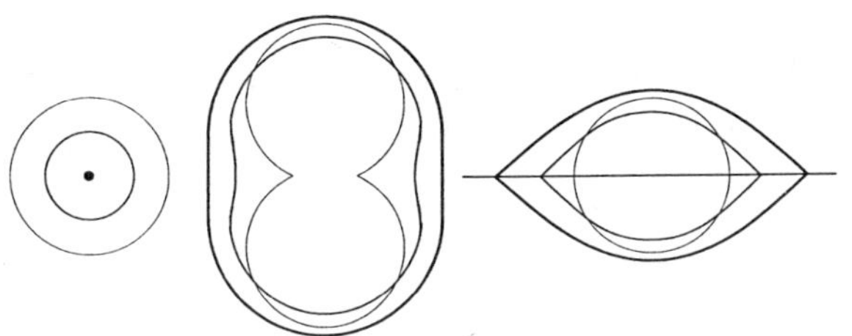

Fig. 4: Interfacial effects in local recrystallization. While a single amorphous zone is subcritical and disappears in a certain radiation environment, two contiguous zones and one zone on a inter-face may form growing supercritical nuclei.

References

[1]　W.L. Johnson, Progr. Mater. Sci. 30, 81 (1986).
[2]　P.R. Okamoto and M. Meshii, Science of Advanced Materials, 1988 ASM Materials Science Seminar, ASM International, Materials Park, Ohio (1989).
[3]　M. Nastasi and J.W. Mayer, Mat. Sci. Rep. 6, 1 (1991).
[4]　E. Balanzat, Rad. Effects & Defects in Solids 126, 97 (1993).
[5]　W. Schilling and H. Ullmaier, Mat. Sci & Techn. 10 B, VCH, Weinheim (1994).
[6]　J.F. Ziegler, J.P. Biersack and U. Littmark, "The Stopping and Range of Ions in Solids", Pergamon Press, New York (1985).
[7]　D.F. Pedraza and L.K. Mansur, Nucl. Instr. & Meth. B 16, 203 (1986).
[8]　L.E. Rehn, P.R. Okamoto, J. Pearson, R. Bhadra and M. Grimsditch, Phys. Rev. Lett. 59, 2987 (1987).
[9]　N.Q. Lam and P.R. Okamoto, Surface and Coating Tech. 65, 7 (1994).
[10]　F. Seitz and J.S. Koehler, Solid State Phys. 2, 305 (1956).
[11]　F.F. Morehead, and B.L. Crowder, Rad. Effects 6, 27 (1970).
[12]　Z.G. Wang, Ch. Dufour, E. Paumier and M. Toulemond, J. Phys.: Cond. Mat. 6, 6733 (1994).
[13]　R.L. Fleischer, P.B. Price and R.M. Walker, "Nuclear Tracks in Solids", Univ. Calif. Press, Berkeley (1975).
[14]　S.U. Campisano, S. Coffa, V. Raineri, F. Priolo and E. Rimini, Nucl. Instr. & Meth. B 80/81, 514 (1993).
[15]　J.F. Gibbons, Proc. IEEE 60, 1062 (1972).
[16]　J.R. Dennis and E.B. Hale, J. Appl. Phys. 49, 1119 (1978).
[17]　H.A. Atwater, J.S. Im and W.L. Brown, Nucl. Instr. & Meth. B 59/60, 386 (1991).
[18]　R.G. Elliman, J.S. Williams, W.L. Brown, A. Leiberich, D.M. Maher and R.V. Knoell, Nucl. Instr. & Meth. in Phys. Res. B 19/20, 435 (1987).
[19]　K.A. Jackson, J. Mater. Res. 3, 1218 (1988).
[20]　V. Heera, T. Henkel, R. Kögler and W. Skorupa, Phys. Rev. B 52, 15776 (1995).
[21]　V. Chailley, E. Dooryhée and M. Levalois, Nucl. Instr. & Meth. B 107, 199 (1996).
[22]　M. Ghidini, J.P. Nozières, D. Givord, A. Liénard, B. Gervais, E. Dooryhée, Nucl. Instr. & Meth. B 107, 344 (1996).
[23]　J. Koike, P. R. Okamoto and L.E. Rehn, J. Mater. Res. 4, 1143 (1984)
[24]　W. Bolse, J. Conrad, F. Harbsmeier, M. Borowski, this conference
[25]　H. Zillgen and P. Ehrhart, IBMM-96, to appear in Nucl. Instr. & Methods B.
[26]　D.N. Seidman, R.S. Averback, P.R. Okamoto and A.C. Baily, Phys. Rev. Lett. 58, 900 (1987).

[27] T. Diaz de la Rubia, R.S. Averback, R. Benedek and I.M. Robertson, Rad. Eff. and Def. Sol. 113, 39 (1990).

[28] Y.-T. Cheng, Mat. Sci. Rep. 5, 45 (1990).

[29] M.J. Caturla, T. Diaz de la Rubia, G.H. Gilmer, Nucl. Instr. & Meth. B 106, 1 (1996).

[30] E. Snoeks, T. Weber, A. Cacciato and A. Polman, J. Appl. Phys. 78, 4723 (1995).

[31] H. Trinkaus, J. Nucl. Mater. 223, 196 (1995).

[32] Y. Zhu, Z.X. Cai, R.C. Budhani, M. Suenaga and D.O. Welch, Phys. Rev. B 48, 6436 (1993).

[33] Ch. Simon, J. Provost, D. Grould, V. Hardy, A. Wahl, Ch. Goupil and A. Ruyter, Nucl. Instr. and Meth. in Phys. Res. B 107, 384 (1996).

[34] G. Szenes, Phys. Rev. B51, 8026 (1995).

[35] A. Benyagoub and S. Klaumünzer, Rad. Effects & Def. Sol. 126, 105 (1993).

[36] H. Trinkaus and A.I. Ryazanov, Phys. Rev. Lett. 74, 5072 (1995).

[37] H. Trinkaus, to be published.

[38] C. Abromeit and H. Wollenberger, J. Nucl. Mat. 191-194, 1092 (1992).

Materials Science Forum Vols. 248-249 (1997) pp. 13-20
© *1997 Trans Tech Publications, Switzerland*

Surface Effect on the Nature of Damage Production in Ion-Irradiated Solids: A Molecular Dynamics Investigation

M. Ghaly and R.S. Averback

Department of Materials Science and Engineering, University of Illinois at Urbana-Champaign,
Urbana, IL 61801, USA

Keywords: Surface Damage, Thermal Spikes, Sputtering, Molecular Dynamics

ABSTRACT

Molecular dynamics computer simulations have been used to investigate damage formation at surfaces due to energetic, single ion impacts. Three damage regimes are identified. For very dilute cascades, where the density of damage energy near the surface is small, linear cascade models are adequate. For most ion impacts typical of ion implantation, however, collective behavior is pronounced. In this regime, locally high temperatures and pressures developed in the cascade strongly influence both the amount and the nature of the damage that is produced. In this case, hot liquid metal in the center of the cascade convectively flows onto the surface, creating many adatoms and damage below the surface in the form of vacancy loops. At still higher energy densities, a third regime characterized as a microexplosion is found. Here, the surface ruptures owing to the high pressures, atoms and clusters of atoms spew out into the vacuum in large numbers. Comparisons with available experimental investigations will be presented

1. INTRODUCTION

Nearly forty years ago, Vineyard and co-workers [1] first realized the vast potential of molecular dynamics (MD) computer simulations for treating problems in radiation damage, but it has only been in the past decade with the advances in computational facilities and development of realistic potentials that this vision has become a practical means to solve the many body problems of energetic displacement cascades. Progress in this field has been highlighted in several review articles [2]. With exception of work on sputtering, however, nearly all MD simulations of cascades have focused on damage production in bulk crystals, examining such phenomena as defect production, defect agglomeration, and atomic mixing. The work described here focuses on changes which occur near the surface of a crystal and is relevant for ion implantation, ion sputtering, and beam enhanced deposition.

Although one effect of surfaces on damage production, viz. sputtering, has been extensively investigated by experimental, theoretical, and computer simulation methods, most work on this topic has been concerned only with the atoms leaving the surface and has not considered the remnant damage in the surface region. Other investigations on surface damage have examined the topographical development of surfaces during prolonged bombardment at elevated temperatures [3], an important topic for understanding growth of thin films, but this work, too, with a few notable exceptions [4], have neglected the topographical changes that derive from individual ion impacts. Two recently published papers, however, one computer simulation, using molecular dynamics (MD) [5], and the other experimental, using scanning tunneling microscopy (STM) [6], have illustrated that the morphological changes stemming from single impacts can indeed be profound. To illustrate this point, we note that the average sputtering yield of a 10 keV Au atom implanted into Au is ≈ 10. Past theories would suggest that the surface layer would correspondingly contain ≈ 10 vacancies. The MD simulations of a 10 keV event in Au showed, in contrast, that in addition to these some 10 sputtered atoms, another 550 atoms were transported to the surface where they formed a mound of adatoms. This is illustrated in Fig.1. As a consequence of this flow, moreover, a dislocation loop with vacancy character was created just below the surface to accommodate the missing atoms. From another perspective, a 10 keV cascade event in Au creates ≈ 20 Frenkel pairs in the crystal interior. This type of cascade event is illustrated in Fig. 2. Figs. 1 and 2 thus show how a cascade could be dramatically influenced by the presence of a surface.

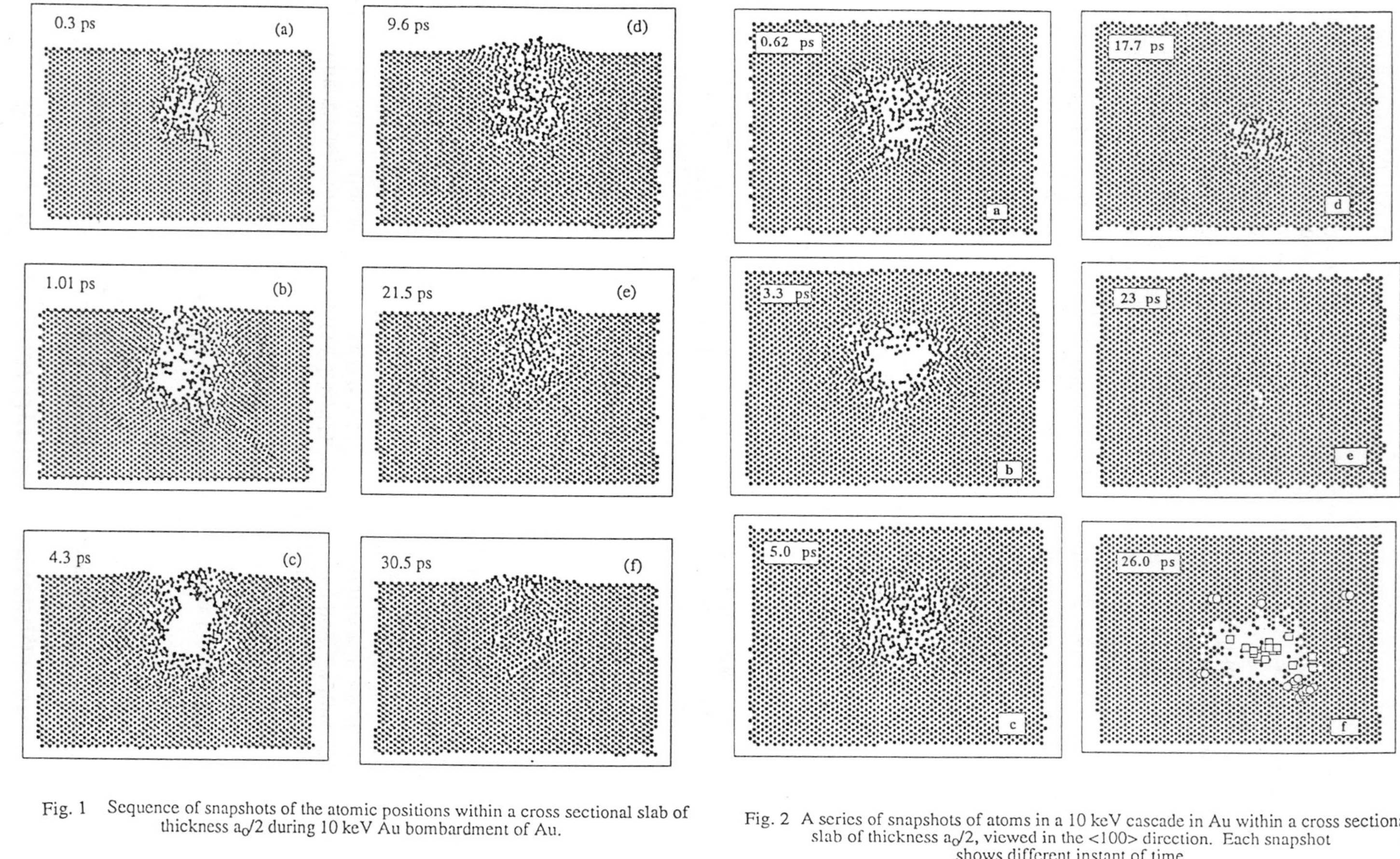

Fig. 1   Sequence of snapshots of the atomic positions within a cross sectional slab of thickness $a_0/2$ during 10 keV Au bombardment of Au.

Fig. 2   A series of snapshots of atoms in a 10 keV cascade in Au within a cross sectional slab of thickness $a_0/2$, viewed in the <100> direction. Each snapshot shows different instant of time.

2. COMPUTATIONAL METHOD

The MD simulations described here were performed using the codes MOLDYCASK which is described in the literature [7]. This code was adapted, however, to treat surfaces. Embedded Atom Method (EAM) potentials [8] were employed for all of the simulations on metals . The potentials were modified at short range to merge smoothly with the Universal potential described by Ziegler et al. [9]. It is difficult to estimate how well these potentials represent their real counterparts for the phenomena described here, since atomic collisions occur far from equilibrium while the EAM potentials are fitted and tested close to equilibrium. Preliminary studies on Cu comparing calculations of energies at close separations by pseudopotentials methods and modified EAM potentials, however, do show good agreement [10]. Similarly good agreement was found on comparing our splined Au potential to a potential obtained using the tight-binding code DMol [11], the difference being a few percent. For the present work on surface interactions, it should be noted that: the surface energies predicted by EAM appear somewhat too low for most metals; the melting point is nearly correct for Cu, but 300 K too low for Au and 450 K too low for Pt; the point defect properties and activation enthalpies for self diffusion are about correct for these same metals [12]. Possible coupling between electron and phonon systems, which tends to quench the thermal spike, is neglected in the simulations since its importance is unknown, and it is, in any case, not yet possible to reliably include such coupling in MD codes. The large size of the computational cell, $\approx 2.5 \times 10^5$ movable atoms, and damping at the boundary layers, prevent serious errors associated with energy dissipation.

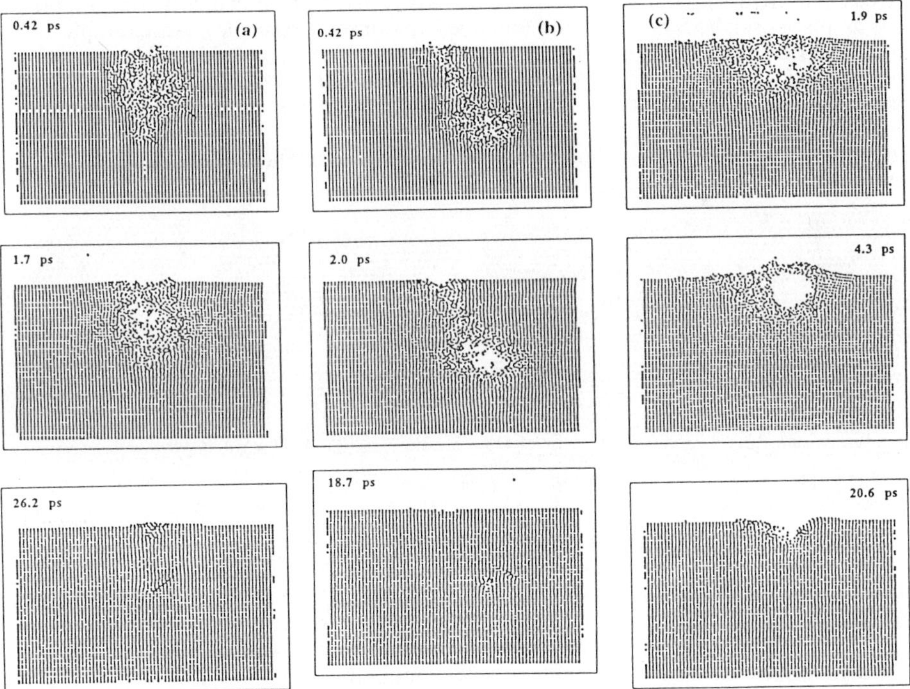

Fig. 3 Locations of atoms with a cross sectional slab, One atomic plane thick in Pt, viewed in the <100> direction, during implantation of a 10 keV Pt atom.

One difficulty of using MD to study ion implantation damage is the random nature of atomic collisions in solids; the smallest variation in the ion's angle of incidence, point of contact on the surface, or time of impact (owing to lattice vibrations) can have enormous, and unpredictable consequences. This is illustrated in Figs. 3a-3c where the impacts of three 10 keV Pt atoms on Pt (100) are illustrated. Extensive damage is created in each of the three events, but depending on how deeply the projectile had penetrated into the target, the amount and qualitative features of the surface damage can vary greatly. In the event shown in Fig. 3(a), for example, 22 adatoms were created, whereas 537 adatoms were created in the event of Fig. 3(c). Atomic mixing and sputtering are also vastly different for these three events. For this reason, it is necessary to simulate many events to provide the necessary statistics required for a detailed comparison with experiments. On the other hand, examination of just a few events often provides a rather good picture of the physical mechanisms taking place, although with just few events, rarely occurring situations, which may, in fact, dominate the surface morphology during implantation, may be missed. Nevertheless, both directions were followed in this investigation; some case studies were performed using several events to obtain needed statistics for comparison with experiments, while other simulations were performed as a survey of possible behavior.

3. Damage Production at Surfaces

Our simulation studies show that surface damage arising from single ion impacts can be classified quite similarly to the scheme devised by Sigmund for sputtering [13]. For damage energy deposition distribution typical of light-ion implantation, single knock-on events take place which lead to the production of isolated vacancies in the surface through sputtering events, and to adatoms stemming from recoil events below the surface. For dilute cascades, a linear cascade mechanism operates by which many isolated, non-interacting displacement events, as just described, occur in the localized region of a cascade. An example of such a linear cascade is depicted in Fig. 4 in which an MD simulation of a 4.5 keV Ne impact on (111) Pt is illustrated. Several isolated adatoms surround a group of vacancies in the center of the cascade. This image is remarkably similar to the structure of a linear cascade created in the bulk of a material for which vacancies are centralized in the cascade core and are surrounded by isolated interstitial atoms. Here, however, adatoms replace interstitials and the geometry is two dimensional.

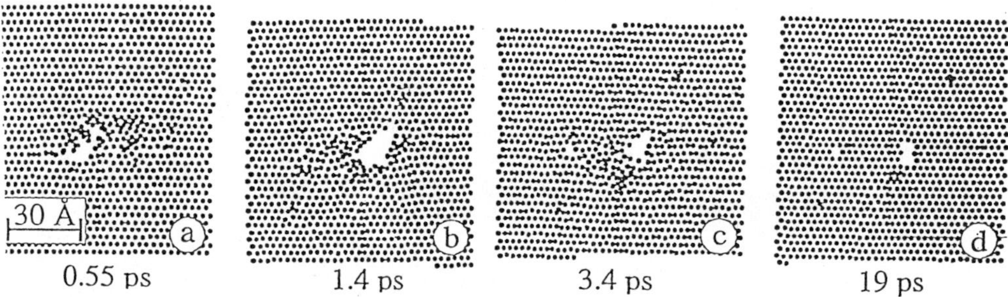

Fig. 4 Top view of Pt (111) surface during MD simulation of the impact of 4.5 keV Xe.

A second type of surface event is the thermal spike. Unlike sputtering, for which thermal spikes are of little significance in all but the most dense cascades, surface damage in metals is dominated by this mechanism. Examples of thermal spike behavior are illustrated in Fig. 1 and Fig. 3b which show 10 keV MD events of self-atom bombardment of Au and Pt, respectively. Many of the features typical of a 10 keV cascade in the bulk are also identified here, e.g., cavitation, local melting, and resolidification. In the bulk, however, the high pressure in the cascade is contained by the surrounding elastic matrix, and it appears that the pressures are not quite sufficient to cause plastic deformation by loop punching [5]. Mass transport is not, however, constrained at the surface, since there only surface tension impedes the viscous flow of hot liquid. This is clearly observed in Fig. 1. The flow in such cases is irreversible, as once the cascade cools, the atoms which have spilled onto the surface freeze in place. As a consequence, a ridge of atoms surround the site of impact, while the lattice below the surface becomes highly depleted of atoms. Some of the atoms which had initially flowed to the surface can relax back to fill some of these vacant sites. This might leave a crater in the center of the impact zone, although such a crater was not observed in Fig. 1. The residual depleted zone below the surface, moreover, often condenses to form a dislocation loop at some depth in the sample, usually near the center of the spike..

The flow of atoms onto the surface is driven by the high pressure in the cascade. The pressure, itself is generated by the rapid expansion of the metal in the hot cascade region, owing to thermal expansion and the volume change on melting. In metals, volume change is typically $\approx$ 15%. Semiconductors like Si and Ge, on the other hand, have negative changes of volume on melting (-10% and -5%, respectively) and smaller coefficients of thermal expansion, such that their total volume changes just above their melting points are negative. Consequently, there is no net driving force in these materials for mass to flow to the surface. Instead of the expected mass flow to the surface, atoms in these materials should be sucked down into the cascade volume [14, 15].

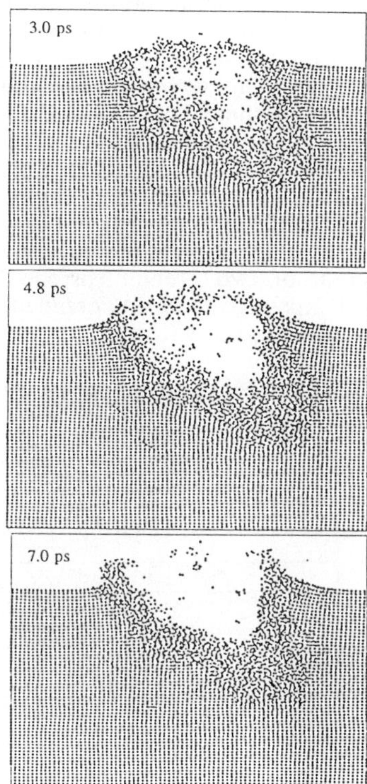

Fig.5 Cascade evolution during 20 keV Au self impact.

The last type of surface damage event is a microexplosion. An example from MD is illustrated in Fig. 5 for a 20 keV self impact of Au. A microexplosion is also in Fig. 3c for 10 keV self impact of Pt. These events occur when the energy density and pressure in the near surface layer are so high that the surface either instantaneously vaporizes or ruptures. In the event shown, the center of energy density was located well below the surface, but the high pressure in the cascade was sufficient to rupture the thin crystalline layer above it. The sputtering associated with this event, which would fall into the category of thermal spike sputtering in the Sigmund scheme, is seen to be quite different from that previously imagined. Rather than being thermal evaporation from a flat, hot surface, thermal spike sputtering is seen here in this MD event to be an explosion at the surface where both single atoms and clusters of atoms are spewed into the vacuum. Evaporation of atoms comes from both the original surface and the newly created surface in the crater. For this reason, some sputtered particles evolve from locations below the original surface.

The pictures of surface damage presented above were deduced from our MD simulations, and it is important to determine whether these pictures are valid in both qualitative and quantitative respects. Unlike damage creation in the bulk, which is extremely difficult to examine experimentally with the needed atomic resolution, surface damage is conducive to investigation by surface probes such as scanning tunneling microscopy (STM). Indeed, such studies have been begun for this purpose using Pt, Au, and Ge as model systems [6, 16]. Consider first the linear cascade regime exemplified above by MD simulations of 4.5 keV Ne bombardment of Pt. STM experiments were performed, in situ, at 40 K for this same case. Like the simulations, isolated adatoms are found several atomic distances away from a central location of the vacancies. Part of the results of this work are collected in Table 1 along with the results of the simulations. Included in these results are the variances of the radial distribution of adatoms about their centers of mass. Agreement between experiment and simulation is good, although this will be discussed more thoroughly, below. One point of some interest is how the adatoms arrived at their final locations. STM images, of course, do not answer this question, but MD does. In some cases, the adatoms are produced by an atomic replacements originating just below the surface near the adatom site. In one case, however, the adatom was ejected near the point of ion impact, but without sufficient velocity normal to the surface to overcome the surface binding energy. This atom followed a low trajectory over the surface to its final location.

An example of thermal spike behavior is illustrated by 4.5 keV Xe bombardment of Pt as shown in Fig 6. Data for this case are also listed in Table 1. Here, like the MD simulations, the surface damage consists of clusters of adatoms surrounding a vacancy crater near the point of ion impact. As expected, the variance of the radial distribution is smaller in this case than it is for the 4.5 keV Ne bombardment, since the cascade volume is more compact. It is interesting to note that the number of adatoms is not equal for 4.5 keV Ne and Xe bombardments, even though the deposited damage energy is nearly the same for the two cases. The number of adatoms is a factor of 4 greater for the Xe bombardment, which incidentally is just the opposite trend that has been shown for defect production in the bulk, i.e., the damage efficiency decreases in the bulk with increasing ion mass. One explanation for this observation is that the damage created by Ne bombardment is deposited deeper in the specimen. This is indeed true, and it is reflected by the lower sputtering yield for the Ne bombardment. But even if we normalize the number of adatoms produced by each type of ion bombardment by the sputtering yield, and thus account for the different damage energy deposition at the surface, we find that the normalized yield of adatoms is still a factor of two greater for the Xe bombardment. This is observed in both the simulations and experiments. We conclude that enhanced adatom production is a consequence of the thermal spike and mass flow.

Table 1. Adatoms and sputtering yield obtained from both STM and MD Simulations for Ne and Xe (4.5 keV) bombardment of Pt (111) surfaces.

Ion	$Y_{ad} \pm \sigma_{ad}$	Y_s
Ne (4.5 keV) Exp.	4 ± 2	2.6
Ne (4.5 keV) Sim.	6.5 ± 25	2.6
Xe (4.5 keV) Exp.	34 (54 $\pm$ 26)	9
Xe (4.5 keV) Sim.	70 ± 40	10

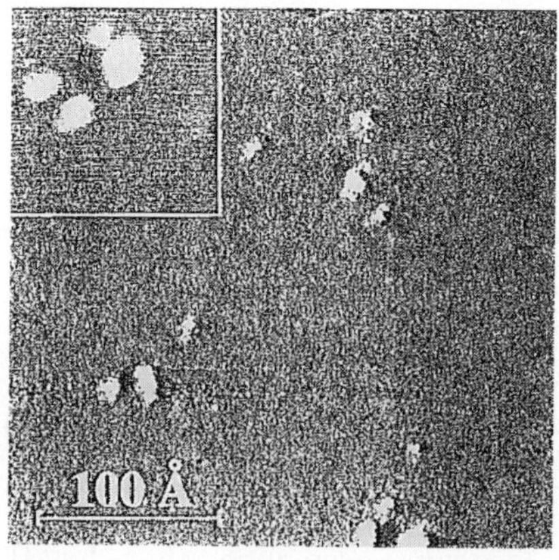

Fig. 6 STM image of a P (111) surface following its impact with Xe.

Despite the good agreement between simulation and experiment, at least in qualitative terms, Table 1 does show that the calculated number of adatoms is a factor of two greater than the observed number. This discrepancy appears too large to be explained entirely by statistical uncertainties, although it might be. Another possibility is a deficiency in the interatomic potential. Recall, the melting temperature predicted by the EAM potential is 450 K too low, and this may provide additional time for mass to flow before freezing. In addition, relaxation of adatoms back into the crater may be affected by the surface potential, or it may just require a times longer than MD times. Partial quenching of the thermal spike by electron-phonon coupling is a final possible source of error. Further work will be necessary to clarify this point. It is, of course, vital to evaluate the quantitative reliability of the simulations; it is heartening that already, with no special efforts at improving the interatomic potentials, quantitative predictions are within a factor of two of experimental values.

Finally, verification of the prediction of microexplosions has been reported in both Au and Ge. Merkel and Jäger have shown by transmission electron microscopy that self bombardment of Au can lead to very large craters in the surface [17]. The probability for these events was observed to be small. These experiments, however, were performed at room temperature so that surface diffusion may have reduced the actual number. Moreover, the limited resolution for imaging small craters in Au may have resulted in a large underestimation of their true number. Too few simulation events on Au were performed to estimate the probability for microexplosion events and so no detailed comparison is possible. Similarly, Bellon et al. observed large craters typical of microexplosions in Ge [15]. Again the probability for such events using 20 keV Ga ions was only $\approx 0.1\%$. Three simulation events were performed on Ge but without evidence of microexplosions; however, if the true yield is indeed 0.1%, our failure to observe them in three simulation events is expected.

4. SUMMARY AND CONCLUSIONS

The interaction of energetic ions with solid surfaces has been shown here to give rise to a number of different possible damage structures. These could mostly be explained on the basis of a simple model involving local melting and viscous flow. When the volume change on melting is positive, pressure builds in the molten cascade and forces atoms onto the surface. When the volume of melting is negative, there is a tendency for atoms to be drawn into the bulk and away from the surface. The difficulty in accommodating extra atoms in the crystal interior, however, prevents this inward flow from being a large effect, unless the material can easily deform. A third phenomenon, microexplosions, was also elucidated in this work and shown to be important for Au and Ge. It was also demonstrated that a combined study involving STM and MD simulation offers a unique opportunity to evaluate the accuracy of MD simulations. This is especially important since a significant effort is currently in progress to use simulation, rather than experiment, to predict the response of a material to prolonged irradiation.

ACKNOWLEDGMENTS

The authors are grateful to Dr. Th. Michely for the use of Fig. 6 and to Drs. M Morgenstern and Michely for sharing their as yet unpublished STM results. The research at the University of Illinois was supported in part by the U.S. Dept. of Energy, Basic Energy Sciences, grant DEFG02-91-ER45439 and the National Science Foundation, grant NSF DMR-9632252. The authors are also grateful for grants of computing time from the National Center for Supercomputing Applications at U.I. and the National Energy Research Supercomputer Center (NERSC), and the Materials Research Laboratory Center for Computation at the University of Illinois.

REFERENCES

1. J.B. Gibson, A.N. Goland, M. Milgram, and G.H. Vineyard, Phys. Rev. <u>120</u>, 1229.
2. R.S. Averback, T. Diaz de la Rubia, H. Hsieh. and R. Benedek, Nucl. Instr. and Meth. B <u>59/60</u> 709 (1991); D.J. Bacon and T. Diaz de la Rubia, J. Nucl. Mater. <u>216</u>, 275 (1994).

3.   See e.g., E. Chason, T.M. Mayer, B.K. Kellerman, D.T. McIlroy, and A.J. Howard, Phys. Rev. Lett. $\underline{72}$, 3040 (1994).

4.   See e.g., I.H. Wilson, N.J. Zheng, U. Knipping, and I.S.T. Tsong, Phys. Rev. $\underline{38}$, 8444 (1988).

5.   M. Ghaly and R.S. Averback, Phys. Rev. Lett. $\underline{72}$, 364 (1994).

6.   C. Teichert, M. Hohage, Th. Michely and G. Comsa, Phys. Rev. Lett.,$\underline{72}$, 1682 (1994).

7.   MOLDYCASK is a derivative of MOLDY, see M.W. Finnis, Atomic Energy Authority Report No. AERE-R-13182, (1988).

8.   S.M. Foiles, M.I. Baskes, and M.S. Daw, Phys. Rev. B $\underline{33}$, 7983 (1986).

9.   J. Ziegler, J.P. Biersack, and U. Littmark, in Stopping and Range of Ions in Solids, (Pergamon Press, New York, 1985) Vol. 1.

10. M. Ghaly, A. Smith, R. Benedek, and R.S. Averback, unpublished.

11. DMol is a trade mark of Bio Sym. Inc, San Diego, California, USA.

12. M.S. Daw, S.M. Foiles and M.I. Baskes, Mats. Sci. Rep. $\underline{9}$ 251 (1993).

13. P. Sigmund, in Sputtering by Particle Bombardment, ed. R. Behrisch, (Springer-Verlag, Berlin, 1981) Chapter2.

14. M.J. Caturla and L. Marques, and T. Diaz de la Rubia, unpublished.

15. P. Bellon, S. J. Chey, J.E. Van Nostrand, M. Ghaly, D.G. Cahill, and R.S. Averback, Surface Science, $\underline{339}$, 135 (1995).

16. M. Morgenstern and T. Michely, unpublished; M. Morgenstern, Ph. D. dissertation, KFA, Jülich (1996).

17. K.L. Merkle and W. Jäger, Philos Mag. A. $\underline{44}$, 741 (1981).

Materials Science Forum Vols. 248-249 (1997) pp. 21-31
© 1997 Trans Tech Publications, Switzerland

The Thermal Spike Model: A Possible Way to Describe the Effects Induced in Y_2O_3 by Swift Heavy Ion Irradiations

Ch. Dufour[1], S. Hemon[2], F. Gourbilleau[1], E. Paumier[1,2] and E. Dooryhee[2,3]

[1] LERMAT, URA-CNRS 1317, ISMRA, 6 Bd Mal Juin, F-14050 Caen Cedex, France

[2] CIRIL, CEA-CNRS, rue Claude Bloch, BP 5133, F-14040 Caen Cedex, France

[3] present address: E.S.R.F., av. des Martyrs, BP 220, F-38043 Grenoble Cedex, France

__Keywords:__ High Energy Irradiation, Thermal Spike, Heavy Ions, Electronic Energy Loss, Phase Transformation, Y_2O_3

Abstract

Micrometric powders of yttrium oxide have been irradiated with swift heavy ions. Due to the huge energy deposited by the 4.5MeV/u Lead ions on the target electrons, a phase transformation from cubic to monoclinic structure has been observed by X-ray diffraction. The irradiation with 11.5MeV/u Sulphur ions did not lead to such a transformation. An interpretation is given through the thermal spike model which gives the radius of a cylindrical track of monoclinic Y_2O_3 as a function of the electronic energy loss of the incident ion.

Introduction.

The interaction between swift heavy ions and bulk materials leads to specific effects due to the huge energy (S_e) deposited on the target electrons by the incident ions. These effects have now been studied for a long time in numerous materials such as metals [1, 2], semiconductors [3], insulators [2, 4, 5], metallic alloys in their crystalline [6] or amorphous phase [7]. The thermal spike model has been used in all those materials to interpret the sensitivity of materials to ion irradiations [8, 9]. Whatever its final nature, the so called 'track' is supposed to be formed after a rapid quenching of a cylindrical zone that has been molten along the ion path. Provided that the main thermodynamic properties of the material are known, this model can predict whether a given material can exhibit tracks or not. It can also predict the size of the tracks and the threshold value of S_e above which such tracks appear in agreement with experimental values.

The role of the electrons excited along the ion path may be all the more important as the size of the target material is close to the diffusion length of the electrons so that the energy density given by the incident ion can reach very high values. That idea leads to irradiate micro- and nanocrystalline powders by swift heavy ions. The confinement of the excited electrons is expected leading to the enhancement of the S_e induced effects. Two powders of oxides (micrometric Y_2O_3 and nanometric SnO_2) have been irradiated to study such effects [10]. In this paper we are reporting details of the experiments performed on Y_2O_3 and the correlations made with theoretical predictions.

Although the agreement between the thermal calculations and the experimental results has been good, the accuracy of such calculations was debated by several authors in the case of materials that undergo phase transformation under pressure [11, 12]. According to their analysis, the temperature reached along the ion path is far below the melting temperature. Therefore they concluded that the tracks observed are due to elastic shock waves. In this paper, it will be shown that the temperature determined using the thermal spike is in good agreement with the observed phase transition. The existence of very high pressure is not in contradiction with this model.

Irradiation of micrometric Y_2O_3 powders.

At room temperature, the crystallographic phase of micrometric Y_2O_3 is cubic (C). Two other phases are known to appear at a high pressure and at a temperature near the melting point [13-

15] or under shock compressions [16]. Such phases have been observed in nanometric powders prepared by the technique of ball milling that provokes a very high pressure in the grains. Both phases are monoclinic. Their cell parameters are very similar which made it difficult to separate them from each other. In this paper, the monoclinic phase will be characterized by the three peaks (L1,L2 and L3) taken from ref. [15]. They correspond respectively to the values of interplanar d-spacings: 3.09Å, 2.93Å and 2.77Å.

The powders - available from Sigma-Aldrich GmbH (Germany)- have been pressed and then put in the XRD setup (CHEXPIR) located at the end of the medium energy beam line of GANIL (Caen, France). The mean grain diameter was around 2μm. The XRD powder diffraction patterns were performed *in situ* during the irradiation experiments made at room temperature. The evolution of the phase composition Y_2O_3 was observed as a function of the ion fluence (i.e. the number of ions per cm^2).

Irradiation with sulfur ions.

The energy of the $^{36}S^{15+}$ ions was 11.5 MeV/u corresponding to an electronic energy loss S_e of 2.7 keV nm^{-1}. No phase transformation has been observed during the irradiation up to a maximum fluence of 6 10^{13} ions cm^{-2}. The XRD pattern remained unmodified.

Irradiation with lead ions.

The energy of the $^{208}Pb^{44+}$ ions was 4.5 MeV/u corresponding to an electronic energy loss S_e of 38.2 keV nm^{-1}. Figure 1 shows compares the XRD spectra of the virgin and irradiated sample. It can clearly be seen that the crystalline cubic phase gradually disappears whereas the monoclinic phase shows up for fluences larger than 10^{11} Pb cm^2 whereby the intensities of the corresponding peaks (L1, L2 and L3) increase with increasing fluence.

For the analysis, we assume that only the monoclinic (B) and the cubic (C) phases are present. We deduced the volumic fractions (f_B and f_C) of each phase which are proportional to the corresponding B and C peak intensities. To study the evolution of f_C the three peaks (L1,L2 and L3) were considered:

$$f_C = (1/3) \ \Sigma_{i=1,3} \ [I_i(\Phi t) / I_i(\Phi t = 0)]$$

The evolution of the fraction f_C is plotted in figure 2 as a function of fluence Φt.

Assuming that each ion passing through the matter creates a cylindrical zone of monoclinic Y_2O_3, we define the cross section σ_B of this cylinder using the following law:

$$d \, f_B / d \, \Phi t = \sigma_B \, (1 - f_B)$$

Given that $f_C + f_B = 1$, the fraction of cubic phase hence follows the law:

$$f_C = \exp(-\sigma_B \, \Phi t)$$

The value of σ_B was deduced from an exponential fit of the data in fig. 2 ($\sigma_B = 2.68 \ 10^{-13} \ cm^2$) which corresponds to a cylindrical track of monoclinic Y_2O_3 with a radius of 2.9 nm.

Thermal spike calculations.

Toulemonde et al. used the thermal spike model to interpret the track radii observed in insulators after swift heavy ion irradiations [5, 9]. They show that the key parameter driving the response of an insulator to ion irradiation is the mean energy diffusion length λ of the target electrons. Numerous experiments allowed the correlation between λ and the band gap energy E_g. Using a commonly accepted value of 5eV for the band gap in Y_2O_3[17], the value of λ is expected to be 6 nm. The equations that describe the energy flows in the target are written in a cylindrical geometry, with the z-axis being the ion path and r the radius of the cylindrical shell of a thickness dr on which the energy balance is written per unit time and unit volume:

$$C_e \frac{\partial T_e}{\partial t} = \frac{1}{r} \frac{\partial \left(r K_e \frac{\partial T_e}{\partial r} \right)}{\partial r} - g(T_e - T_l) + A(r,t)$$

$$C_l \frac{\partial T_l}{\partial t} = \frac{1}{r} \frac{\partial \left(r K_l \frac{\partial T_l}{\partial r} \right)}{\partial r} + g(T_e - T_l)$$

C_i is the specific heat (J cm^{-3} K^{-1}), K_i the thermal conductivity(W cm^{-1} K^{-1}), the index *i* must be replaced by *e* for the electrons and *l* for the lattice. A(r,t) (in Wcm^{-3}) is the energy transferred by the incident ion to the target electrons and may be written as: A(r,t) = S_e exp (-t / τ) f(r). f(r) is an analytical function derived from Monte Carlo calculations that describe the radial distribution of the energy on the target electrons. τ is a characteristic diffusion time linked to λ by: $\lambda^2 = D_e \, \tau$ where D_e is the electronic thermal diffusivity taken equal to 2 cm^2 s^{-1}[5]. g(T_e-T_l) is the energy exchanged between the target electrons and the atomic lattice by electron-phonon coupling. g is related to τ by: τ = C_e / g. Solving both equations numerically, the temperatures T_e(r,t) and T_l(r,t) are calculated as a function of the radial distance from the ion trajectory. The track is supposed to be the cylindrical zone in which the maximum temperature has surpassed the melting temperature.

We performed calculations of track radii versus the electronic energy loss S_e of the incident ion using the thermodynamic parameters of Y_2O_3 as reported in tables 1a, 1b and 1c. In figure 3, we compare the calculated track radii as a function of S_e with two experimental points. The Sulfur ions do not create tracks because their S_e value is lower that the theoretical threshold value S_{eth} (5 keV nm^{-1} < S_{eth} < 15 keV nm^{-1}). On the contrary, the value of S_e for the lead ions is higher than S_{eth} so that tracks can be created. Moreover the experimental value (2.9nm) is of the same order of magnitude as the theoretical value (4.5 nm) for S_e = 38 keV nm^{-1}. Within the experimental errors the agreement can be considered as significant.

Discussion.

If ion tracks correspond to a cylinder of molten matter, the following two points have to be confirmed:
i) the way to calculate the temperature is correct even if high pressures appear in the matter.
ii) the pressure increase induced by an increase of the temperature along the ion path is sufficient to induce a phase transformation from cubic to monoclinic.

Concerning the first point, we must recall that, up to now, the energy variation dQ/dt per unit time of a unit volume of the target has been considered as C dT/dt. In fact, if there is a pressure variation dP, dQ should be replaced by dQ = C dT + h dP where h can easily be found, using classical thermodynamics.
A first step leads to h = $-\alpha$ V T, where α is the thermal expansion coefficient, V the volume and T the temperature. The next step consists in using the thermal compressibility coefficient β = 1/P *dP/dT* and express dP as a function of dT: dP = β P dT. In the last step, using the general relationship between α, β and the adiabatic compressibility χ (α = $-\beta\chi$P), h dP can be written as h dP = α^2/χ V T dT corresponding to h dP = α^2/χ T dT per unit volume. We can now compare the energy amount due to the temperature increase (C dT) and due to the pressure increase (h dP). These quantities are expressed in J m^{-3} whereby values for $\alpha \approx$ 5 10^{-6} K^{-1} and $\chi \approx$ 10^{-10} Pa^{-1} for temperatures of about 1000K are assumed to be realistic.

CdT* = 10^6 *dT and **h *dP*** = 250 *dT* , where *dT* is the elementary temperature increase.

Therefore we conclude that the contribution due to a pressure increase may be neglected. From this point of view ,our calculations are still valid.

The second point is treated, considering that the coefficients α and χ are independent on the temperature and on the pressure. The relationship between dP and dT has been developed above: $dP = \beta P\, dT = \alpha / \chi\, dT$. With the numerical values taken above, we can write the following approximation: $dP \approx 5\ 10^4\, dT$ where dP is expressed in Pa and dT in K.

In the molten zone, the temperature increase is at least 2200K which corresponds to a pressure increase of 10^8 Pa. Although this value must be considered as an order of magnitude, it is compatible with the very high pressure required for a transition from the cubic to the monoclinic phase.

Excepting that the calculations take into account the latent heat of fusion, the study of what happends during the melting process (i.e. the mechanism that could allow the formation of a liquid phase in a solid phase) has not yet been investigated. The aim of the present experimental program is to manage to work on powders whose mean grain size should be close to the size of the ion track.

References.

[1] Dunlop A. and Lesueur D. *Rad. Eff. Def. Sol.* (1993) **126**, 123

[2] Balanzat E. *Rad. Eff. Def. Sol.* (1993) **126**, 97

[3] Levalois M., Bogdanski P. and Toulemonde M. *Nucl. Instr. Meth.* (1992) **B63**, 14

[4] Matzke H. *Rad. Eff.* (1983) **64**, 3

[5] Toulemonde M., Bouffard S. and Studer F. *Nucl. Instr. Meth.* (1994) **B91**, 108

[6] Audouard A., Balanzat E., Fuchs G., Jousset J.C., Lesueur D., Thomé L.
     Europhys. Lett. (1987) **5**, 241

[7] Barbu A., Dunlop A., Lesueur D. and Averback R.S.
     Europhys. Lett. (1991) **15**, 3713

[8] Wang Z.G., Dufour Ch., Paumier E. and Toulemonde M.
     J. Phys. :Condensed Matter (1994) **6**, 6733 and (1995) **7**, 2525

[9] Meftah A., Brisard F., Costantini J.M., Dooryhee E., Hage-Ali M., Hervieu M.Y.,
     Stoquert J.P. Studer F. and Toulemonde M. *Phys. Rev.* (1994) **B49**, 12457

[10] Gourbilleau F. et al E.M.R.S. '96 Conference
      to be published in Nucl.Instr.Meth. in Phys. Res. B

[11] Dammak H., Dunlop A., Lesueur D., Brunelle A., Della-Negra S. and Le Beyec Y.
      Phys. Rev. Lett. (1995) **74**, 1135

[12] Dammak H., Lesueur D., Dunlop A., Legrand P. and Morillo J.
      Rad. Eff. Def. Sol. (1993) **126**, 111

[13] Hoekstra H.R., Gingerich K.A., *Inorg. Chem.* (1964) **3**, 711

[14] Hoekstra H.R., *Inorg. Chem.* (1966) **5**, 754

[15] Atou T., Kusada K., Fukuova K., Kikuchi M., Syono Y. J.
      Solid State Chem. (1990) **89**, 378

[16]   German V.N., Podurets A.M., Tarasova L.A. *Neorg. Mater.* (1982) **18**, 1736

[17]   Jollet F. *Phys. Rev. B: Cond. Matt.* (1990) **42**, 7587

Table 1a   Specific heat in J g-1 K^{-1} as a function of temperature.

T(K)	specific heat C (J g-1 K^{-1})
50 < T < 300	0.2
300< T < 2500	0.48
T > 2500	0.86

Table 1b   Thermal Conductivity K (W cm^{-1} K^{-1}) as a function of temperature

T(K)	K(W cm^{-1} K^{-1})
80 < T < 2500	0.5
T > 2500	0.01

Table 1c   Physical parameters of Y_2O_{3+}.

Latent heat of fusion ΔH_F (J g^{-1})	464
Latent heat of vaporization ΔH_V (J g^{-1})	4000
Melting temperature (K)	2500
Vaporizing temperature (K)	20000
Ionizing potential (eV)	10
Molar mass (g)	45.2
specific mass (solid) (g cm^{-3})	5.03
specific mass (liquid) (g cm^{-3})	4.9

Figure captions

Figure 1

X Ray Diffraction patterns of the micrometric Y_2O_3 powder before irradiation (lower part) and after irradiation with lead ions ($4.2\ 10^{12}$ Pb cm^{-2}) (upper part).

Figure 2

Fraction f_C of the cubic phase versus the ion fluence Φt (Pb cm^{-2}). The continuous line follows the exponential law : $f_C = \exp(-\sigma_B\ \Phi t)$ with $\sigma_B = 2.68\ 10^{-13}$cm^2.

Figure 3

Calculated (black squares) and experimental (crosses) track radii in micrometric Y_2O_3 powders versus the electronic energy loss S_e (keV nm^{-1}). The calculation has been performed for an incident ion with an energy of 4.5 MeV/u. The two experimental points correspond respectively to the sulfur ($S_e = 2.7$ keV nm^{-1}) and lead ($S_e = 38.2$ keV nm^{-1}) irradiations.

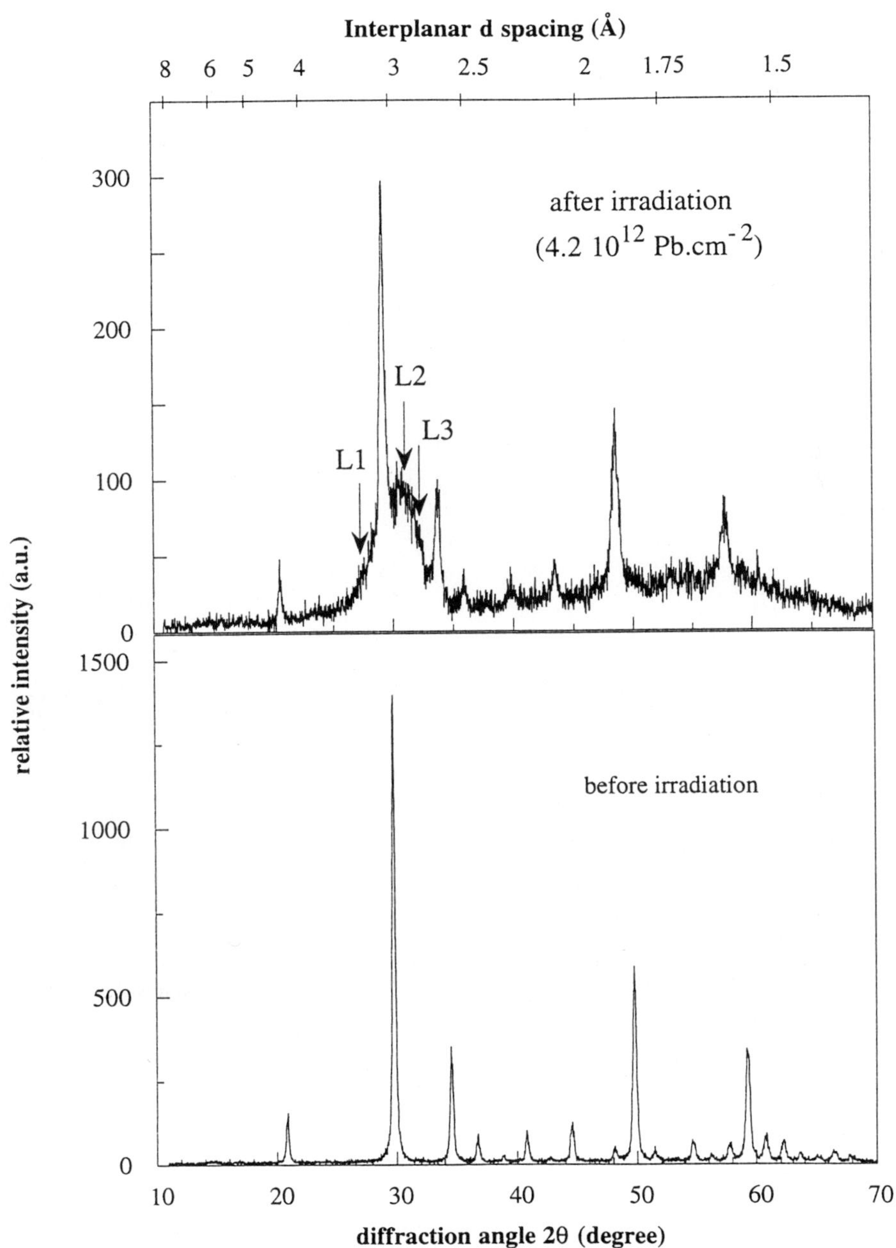

Interplanar d spacing (Å)
8 6 5 4 3 2.5 2 1.75 1.5
300
200
100
0
after irradiation
(4.2 10^12 Pb.cm^-2)
L1
L2
L3
1500
1000
500
0
before irradiation
relative intensity (a.u.)
10 20 30 40 50 60 70
diffraction angle 2θ (degree)

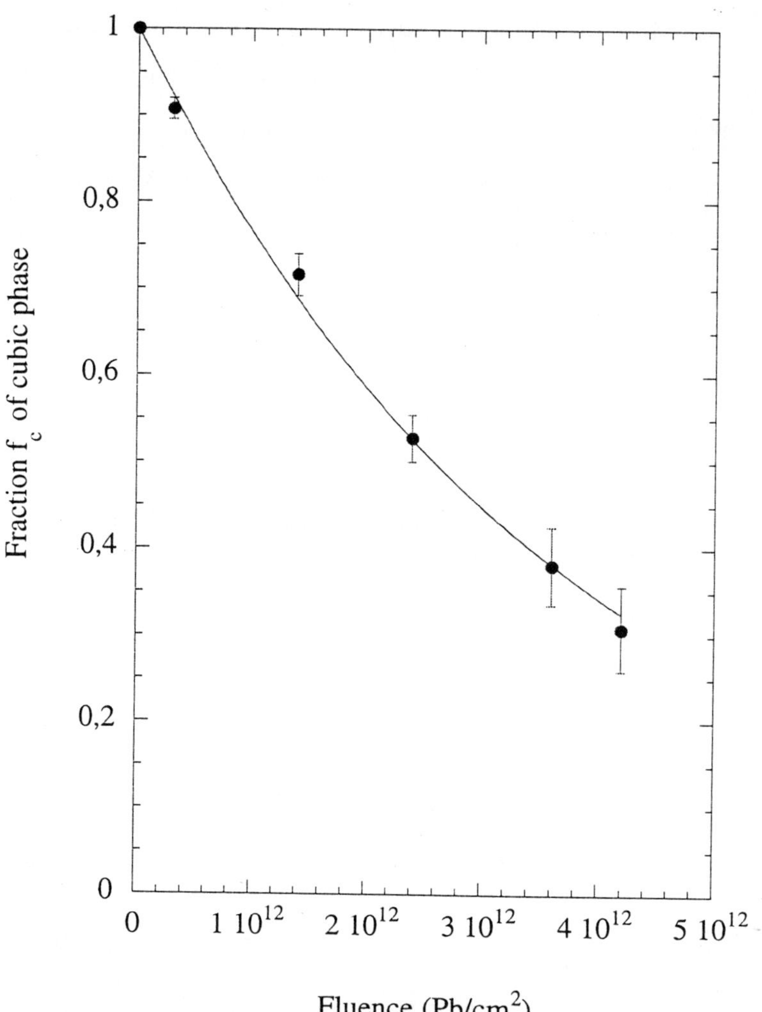
1
0,8
0,6
0,4
0,2
0
Fraction f_c of cubic phase
0
1 10^12
2 10^12
3 10^12
4 10^12
5 10^12
Fluence (Pb/cm^2)

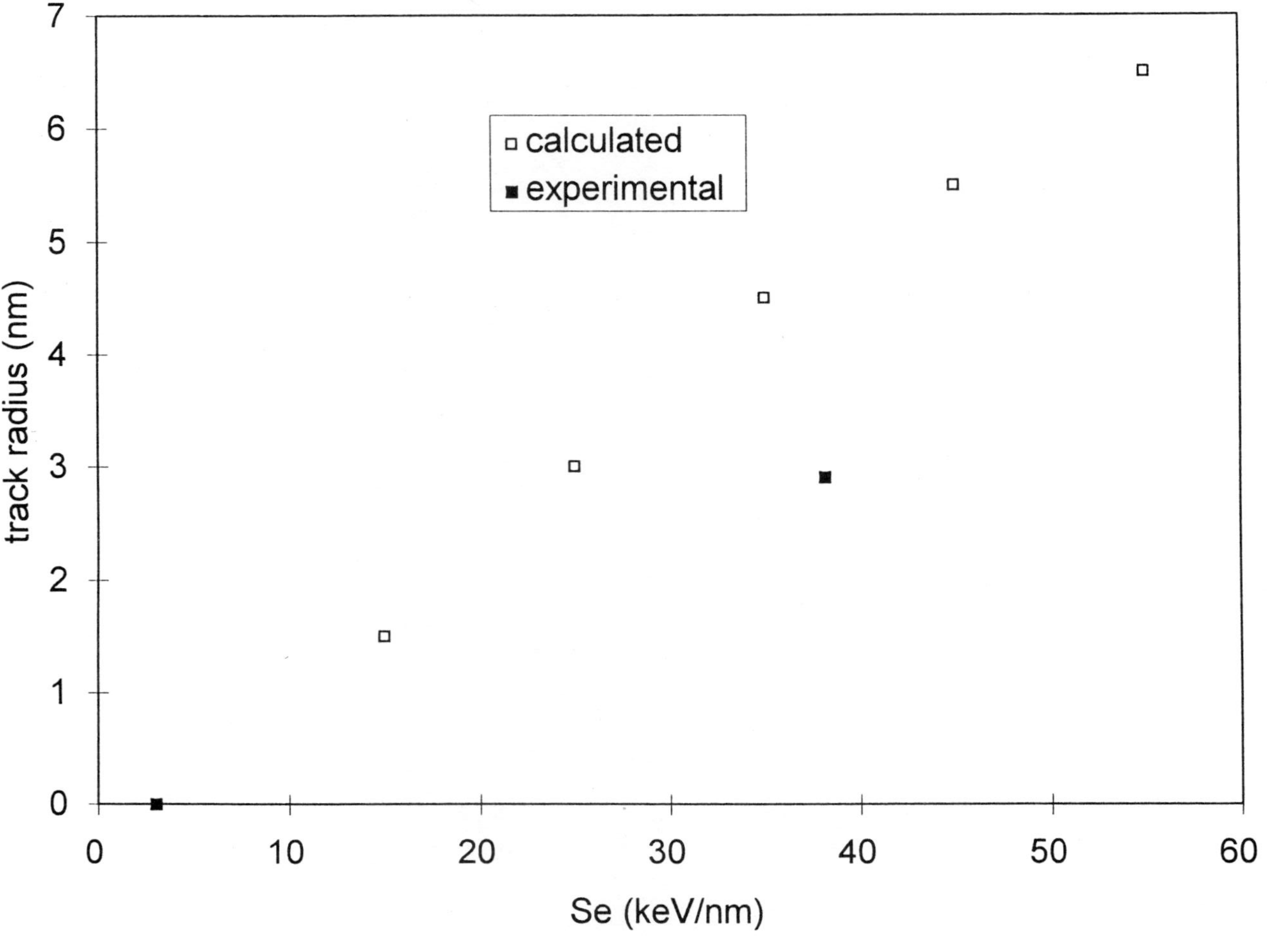
calculated
experimental
track radius (nm)
Se (keV/nm)

Materials Science Forum Vols. 248-249 (1997) pp. 33-39
© 1997 Trans Tech Publications, Switzerland

Stopping Power and Range Relations for Low and High Z Ions in Solids: A Critical Analysis

H.S. Virk and G.S. Randhawa

Department of Physics, Guru Nanak Dev University, Amritsar-143005, India

Keywords: Heavy Ions, Stopping Power, Range-Energy, TRIM Code, SSNTDs

Abstract

A critical analysis of various stopping power and range formulations has been made by comparing the calculated stopping power and range values with corresponding experimental values for different low Z $(1 \leq Z \leq 8)$ and high Z projectiles $(54 \leq Z \leq 92)$ in different targets, e.g. Be, C, Al, Au, Pb, CR-39, Lexan, Mylar, LR-115, CH, (CH)n , TRIFOL-TN, etc. at various low and high energies. A comparative study has been made by taking into consideration different target and projectile combinations, e.g., heavy ion-light target, light ion-heavy target and light ion -light target etc.. Overall the Ziegler et al. (1985) formulation [1] (TRIM-95) provides the best agreement with the experimental results for all projectile and target combinations except for heavy ion-light target combination where it underestimates the stopping power data and overestimates the range data in the range, 2-50 MeV/u. Mukherjee and Nayak (1979) formulation [2] totally fails at relativistic and low energies of the projectile, irrespective of the projectile-target combination. Northcliffe and Schilling (1970) formulation [3] does not show any particular trend. Benton and Henke (1969) formulation [4] gives good agreement between experimental and theoretical data within the range of experimental errors.

Introduction

The theoretical and experimental investigations about the penetration of charged particles in matter played a very important role in the development of modern physics. Research in many fields, e.g. astrophysics, biophysics, nuclear physics, material science etc. relies on the experimental techniques utilizing accelerated charged particles and subsequent interaction process involved therein [5]. Particle track detectors (cloud chambers, bubble chambers, nuclear emulsions, etc.) have been directly responsible for the discovery of most of the known elementary particles. Many review articles and new formulations have been written on the subject of charged particle penetration in matter [1-10].

In the present work, a comparative study of different stopping power and range formulations, e.g., and Henke (1969), Northcliffe and Schilling (1970), Mukherjee and Nayak (1979) and Ziegler et al. (1985), has been made by comparing the calculated stopping power and range values adopting these formulations with the corresponding experimental stopping power and range data for different projectile and target combinations at various energies. The Northcliffe and Schilling data tables have been used to calculate the stopping power and range data for all heavy ions in the energy range 0.0125 to 12.0 MeV/u in different elemental and compound targets (using Bragg additivity rule). The computer code TRIM-95 developed by Ziegler (1995) [10] is used for the stopping power and range calculations based on the Ziegler et al. formulations [1]. The experimental stopping power and range data of references [11-31] have been used in the present study in order to check the validity of various stopping power and range formulations.

Results and Discussion

(a) Heavy ion -light target combination

The experimental stopping power data [11-13] for different heavy ions, viz. U, Pb, Au and Xe in light targets like C, Be, Al, Mylar, Formvar, Hostaphan, $(CH)_n$, etc. have been compared with the theoretically computed data. It is noted that Ziegler et al. formulation strongly underestimates the stopping power data in comparison to experimental data. In some cases, like ^{238}U ion in Be and Formvar in the energy range 3.5 MeV/u to 5.0 MeV/u this deviation is upto 40 percent. Mukherjee and Nayak formulation gives satisfactory results in this range of energy. An interesting point to be noted here is that for Au ion in $(CH)_n$ target, as the energy of the incident ion increases from 15.0 MeV/u to 150.0 MeV/u, Ziegler et al. formulation starts giving better results in comparison to Mukherjee & Nayak formulation (fig.1). For this ion-target combination, below 50.0 MeV/u, Ziegler et al. formulation underestimates the experimental data and above this energy it gives results within experimental errors. Mukherjee and Nayak formulation starts underestimating the experimental data upto 15 percent above 75.0 MeV/u. The results predicted by Benton and Henke and Northcliffe and Schilling formulations do not show any particular trend. As the range is a cummulative effect of stopping power which can be mathematically written as

$$R(E) = \int_0^E (dE / dx)^{-1} dE \tag{Eq.1}$$

TRIM programme for heavy ion-light target combination overestimates the range values in comparison to the experimental values[14]. A similar study has been made on the range data [15] for ^{197}Au in CH_n in the energy range 12.26 MeV/u to 150.06 MeV/u and a similar behaviour is observed. Ziegler et al. formulation upto 50 MeV/u energy overestimates the range data (underestimates the stopping power data) and above this energy gives satisfactory results. Mukherjee and Nayak formulation shows an opposite behaviour to Ziegler et al. (fig. 1). Similar results are depicted from the range data [14,16-21] for heavy ions U, Bi, Pb, Au, La, and Xe in light targets like CR-39, Lexan, Hostaphan, TRIFOL-TN, etc. which have C, H, O and N as their main constituents. Mukherjee and Nayak formulation gives reasonable agreement for these combinations except for the Au (900 MeV/u) ion in C, CH, CH_2 and Al targets where it totally fails to estimate the range values. As the energy of the U ion decreases from 16.5 MeV/u to low values in different light targets like CR-39, Lexan, Hostaphan, TRIFOL-TN, the computed values by TRIM code start giving agreement to the experimental values (fig.2). At relativistic energy of Au ion in C, CH, CH_2, and Al target, Ziegler et al. formulation gives good agreement to the experimental values. Benton & Henke formulation gives satisfactory results with a percentage deviation of ± 8 percent in the given energy range.

(b) Light ion-heavy target combination

A comparison between theoretical and experimental stopping power data [22] has been made for light ions, He and Li, in heavy target Au in the energy range 1.0 to 4.0 MeV/u. Benton & Henke and Ziegler et al. formulations show an identical behaviour and give an agreement to the experimental data within a deviation of ± 4 percent. For this combination Mukherjee and Nayak formulation overestimates the experimental data in the given energy range. However, Northcliffe and Schilling formulation underestimates the experimental data.

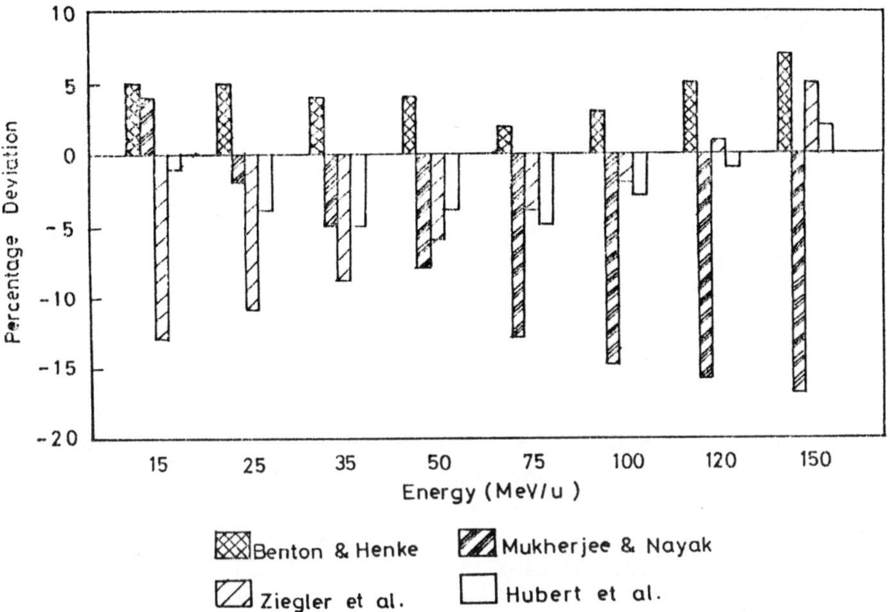

Fig. 1. Percentage deviation of theoretical calculated stopping power data from the experimental data for Au ion at different energies in $(CH)_n$ target.

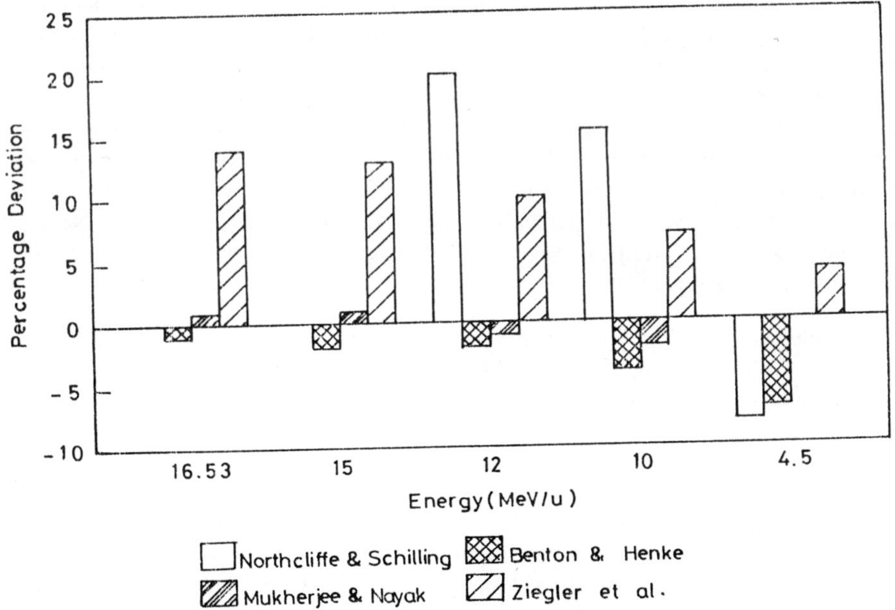

Fig. 2. Percentage deviation of theoretical calculated stopping power data from the experimental data for U ion at different energies in CR-39 target.

(c) Light ion-light target combination

The stopping power data [23] for light ions, viz. H (0.695 to 3.23 MeV/u), He (0.303 to 2.631 MeV/u), Li (0.513 to 2.446 MeV/u), B (0.547 to 1.884), C (0.53 to 1.919 MeV/u), N (0.536 to 1.649 MeV/u) and O (0.663 to 1.77 MeV/u) ions in the light target, LR-115 is analysed. For H ion, the Ziegler et al. and Benton & Henke formulations give the theoretical results in agreement to the experimental stopping power data within a deviation of ±5 percent. Mukherjee and Nayak formulation underestimates the experimental data between 10 to 30 percent. In the case of He ion projectile, Mukherjee and Nayak formulation underestimates the results upto 40 percent, where as Ziegler et al. and Benton & Henke formulations yield satisfactory results (fig. 3). A similar trend is noted for the Li, B, C, N, and O ions in the LR-115 target where Mukherjee and Nayak formulation underestimates the experimental data upto 30 percent and Ziegler et al. and Benton & Henke predict the results within a percentage deviation of ±5 percent. Hence it   can be concluded that for different light ions, Ziegler et al. and Benton and Henke formulations show agreement with the experimental results within ±5 percent deviation. Northcliffe and Schilling formulation does not show any particular trend. Mukherjee and Nayak formulation predicts results underestimating the stopping power data for these ions in LR-115 target upto ±40 percent. The stopping power data [24] for H (0.328   to 2.823 MeV/u), He (0.248   to 2.544 MeV/u), Li (0.372 to 2.956 MeV/u), B(0.483 to 1.896 MeV/u), C (0.565 to 2.106 MeV/u), N (0.569 to 1.675 MeV/u), O (0.561 to 1.782 MeV/u), Na (0.556 to 0.863), Al (0.646 to 0.828) and Si (0.665   to 0.771 MeV/u) in light target, CR-39 is analysed by using all the four formulations. Mukherjee and Nayak formulation underestimates the experimental data between 12% to 22% upto 0.84 MeV/u energy of H projectile; above this energy it underestimates below 10 percent. The results predicted by Ziegler et al. and Benton and Henke formulations are deviated upto a maximum of ±6% from the experimental results. For the He and Li ion, in the given energy range, Mukherjee and Nayak formulation underestimates the data between 10 to 30 percent. TRIM calculations based on Ziegler et al. and Benton and Henke formulation predict the results within a deviation of ±6 percent from the experimental ones. For B, C, N, O, Na, Al and Si ions having energy below 1.0 MeV/u, the Mukherjee and Nayak formulation provides results underestimating the experimental results between 10 to 35 percent; above this energy, the results are underestimated below 10 percent. Ziegler et al. formulation predicts results for the B, C, N and O ions in CR-39 target within a percentage deviation of ±5 percent and for Na, Al and Si ions, it underestimates the   stopping power data between 5 to 9 percent. Benton and Henke formulation shows an almost similar trend.

A comparison between theoretical and experimental range data [25,26] for C, N, O, Ne, Mg, Al, P, Ar, Fe, and Cu ions having energy 9.6 MeV/u in CR-39 target is made. Ziegler et al. and Benton & Henke formulations give theoretical range values which are in agreement to the experimental values (maximum deviation ±8 percent). Mukherjee and Nayak formulation overestimates the data upto 15 percent. Northcliffe and Schilling data tables overestimate the experimental range and this deviation increases with the increase of Z of the ion. The experimental stopping power data [27-30] for He (0.38 to 1.74 MeV/u), N (0.62 to 1.21 MeV/u) and O (0.50 to 93.10 MeV/u) ions in Mylar target (light ion-light target) have been compared with the theoretical stopping power data. For the He ion, in the given energy range, Benton and Henke and Ziegler et al. formulations predict results which have a maximum deviation of ±5 percent from the experimental ones. Mukherjee and Nayak formulation underestimates the results upto 20 percent. For the N ion, all these four formulations show an identical behaviour as for the He ion in Mylar target in the energy range 0.62 MeV/u to 1.21 MeV/u. In case of O ion, Northcliffe and Schilling formulation overestimates the experimental results   by about 20 percent. Ziegler et al.

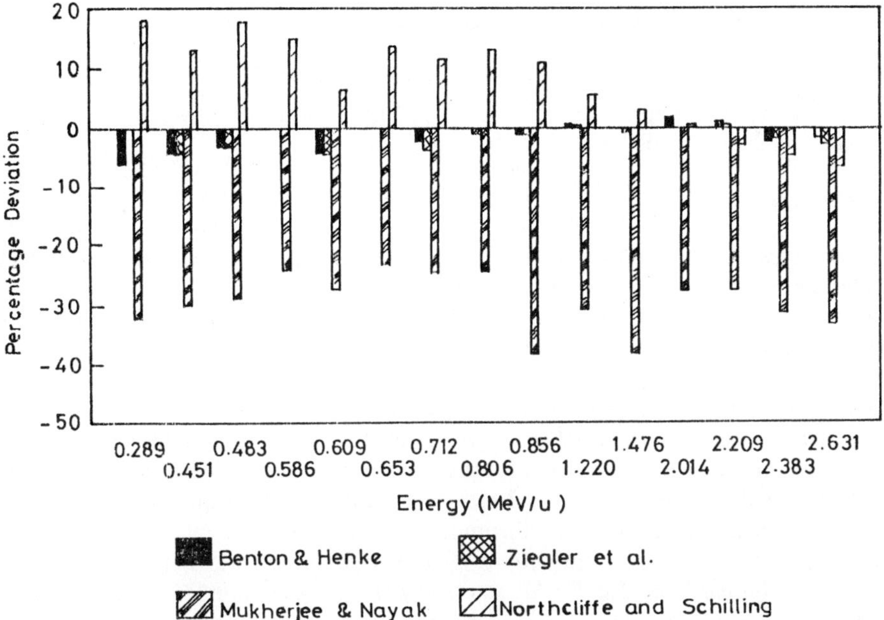

Fig. 3. Percentage deviation of theoretical calculated stopping power data from the experimental data for He ion at different energies in LR-115 target.

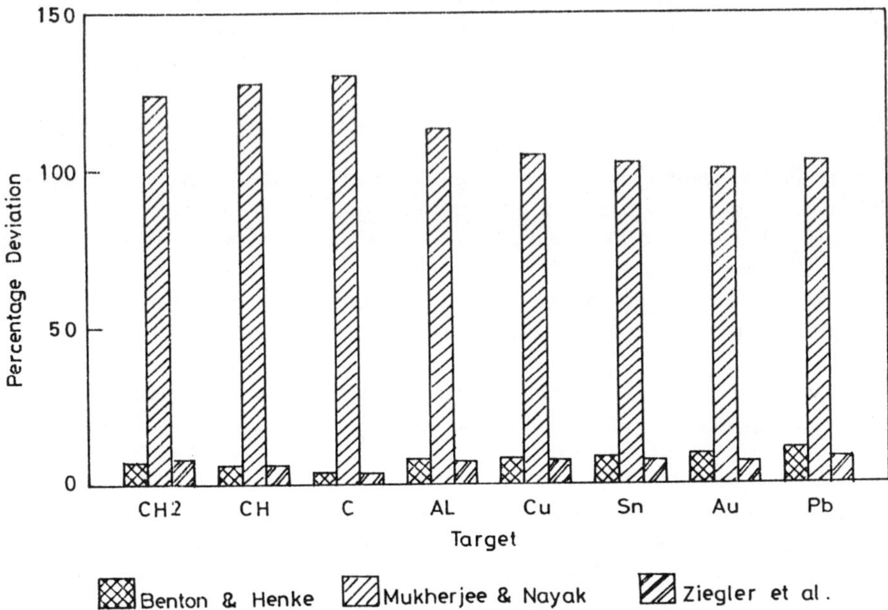

Fig. 4. Percentage deviation of theoretical calculated range data from the experimental data for Au ion having energy 900 MeV/u in different targets.

formulation yields   results which are in best agreement to the experimental results (a maximum deviation of ±3 percent). Benton and Henke formulation overestimates the experimental results upto 8 percent. Mukherjee and Nayak formulation underestimates the experimental results upto 20 percent in the low energy range (upto 1.0 MeV/u) of the O ion in the Mylar target . Above 50 MeV/u, it again underestimates the results by 10 percent.

(d) Relativistic case

The experimental stopping power data [13] for Au ion (75.0 to 150.0 Mev/u) in (CH)$_n$ and Au targets have been analysed (heavy ion- light target and heavy ion- heavy target). The theoretical results predicted by Benton & Henke and Ziegler et al. formulations are in agreement to the experimental results. Mukherjee and Nayak formulation underestimates the stopping power data upto 20 percent. The experimental range data [21] for Au ion having energy 900 MeV/u in different targets like CH, CH$_2$, C, Al, Cu, Sn, Au and Pb have been compared with the theoretical range data. Mukherjee and Nayak formulation overestimates the experimental range data by more than 100 percent, irrespective of the ion- target combination (fig. 4). Benton   and   Henke and Ziegler et al. formulations predict the   results   in   agreement to the experimental ones. A similar comparison has been made between the theoretical and the experimental stopping power data [31]for relativistic heavy ions, viz., O (690.0 MeV/u), Ar (985.0 MeV/u), Kr (900.0 MeV/u) and Xe (780.0 MeV/u) in different light (Be, C, Al) and heavy (Pb) targets. For O ion in light targets (Be, C),   Mukherjee and Nayak forumulation underestimates the experimental data by more than 60 percent. A similar result is obtained for O(690.0 MeV/u) ion in Pb. Ziegler et al. and Benton and Henke formulations predict results those are in good agreement to the experimental data. A similar behaviour is observed for Ar ion in C, Al, Cu and Pb for all three formulations. For Xe ion in C, Al, Cu and Pb targets, Mukherjee and Nayak formulation predicts results   underestimating   the experimental ones by more than 60 percent. Ziegler et al. formulation underestimates the experimental data by 6 percent and Benton and Henke formulation underestimates the data upto 10 percent.

Conclusions

(1) Ziegler et al. formulation is in best agreement with the experimental results within a percentage deviation of ±5 percent for all light ion-light target and light   ion-heavy target combinations. For heavy projectile and light target combination this formulation strongly underestimates the stopping power data (overestimating the range data) in the limited range of energy of the projectile. At low (<2.0 MeV/u) and high energies (>50 MeV/u)of the projectile this formulation gives the reasonable agreement to the experimental results for the heavy ion-light target combination.

(2) Benton and Henke formulation predicts the theoretical results with a maximum deviation of 10 percent from the experimental ones for all combinations. After Ziegler et al. formulation, Benton and Henke formulation provides the stopping power and range data which are in best agreement to the experimental results.

(3) Northcliffe and Schilling data tables do not show any particular trend for all the ion-target combinations in energy range under study.

(4) Mukherjee and Nayak formulation totally fails to predict the results in the relativistic energy range of the projectile irrespective of ion-target combination. For light ion-light target combination it underestimates the experimental results even upto 40 percent in the range of energy under study. This formulation shows agreement to experimental results only at limited points in case of heavy ion-light target combination.

References

[1] J.F. Ziegler, J.P. Biersack and U. Littmark , The stopping power and range of ions in solids, Vol. 1 , Pergamon Press, New York (1985).
[2] S. Mukherjee and A.K. Nayak, Nucl. Instrum. and Meth. **159**, 421 (1979).
[3] L.C. Northcliffe and R.F. Schilling , Atomic Data and Nucl. Data Tables A7, 233 (1970).
[4] E.V. Benton and R.P. Henke, Nucl. Instrum. and Meth. **67**, 87 (1969).
[5] S.P. Ahlen , Rev. Mod. Phys. **52**, 121 (1980).
[6] H.A. Bethe, in Handbuch der Physik, 24/1, 2nd edition, Edited by H. Geiger and K. Scheel (Springer, Berlin) 273 (1933).
[7] N. Bohr, K. Dan. Vidensik Selsk. Mat.-Fys. Medd. 18, 8 (1948).
[8] E.A. Uehling, Ann. Rev. Nucl. Sci. **4**, 315 (1954).
[9] U. Fano, Ann. Rev. Nucl. Sci. **18**, 8 (1948).
[10] J.F. Ziegler, TRIM-95: The Transport of ions in matter, IBM, Research, 28-0, Yorktown, NY 10598, U.S.A.(1995).
[11] R. Bimbot, D. Gardes, H. Geissel, T. Kithara, P. Armbruster, A. Fleury and F. Hubert, Nucl. Instrum. and Meth. **174**, 231 (1980).
[12] H. Gauvin , R. Bimbot, J. Herault, B. Kubica, R. Anne, G. Bastin and F. Hubert, Nucl. Instrum. and Meth. **B47**, 339 (1990).
[13]H.H. Heckman, H.R. Bowman, Y.J. Karant, J.O. Rasmussen , A.I. Warwick and Z. Z. Xu , Phys. Rev. **36A**, 3654 (1987).
[14] G.S. Randhawa, S.K. Sharma and H.S. Virk, Nucl. Instrum. and Meth. **108B**,7(1995).
[15] W. Schimmerling and K.G. Vosburgh , Phys. Rev., **B7**, 2895 (1973).
[16] Zs. Kocsis and R. Brandt, Nucl. Tracks and Radiat. Meas. **22**, 215 (1993).
[17] Y. Laichter, H. Geissel , M. Schadel and P. Armbruster, Phys. ReV. **26A**,1915 (1982).
[18] J. Raju and K.K. Dwivedi, Nucl. Tracks Radiat. Meas. **22**, 149(1993)
[19] A Saxena and K.K. Dwivedi, Nucl. Tracks Radiat. Meas. **15**,111 (1988).
[20] A Saxena and K.K. Dwivedi, Nucl. Tracks Radiat. Meas.**17**, 447(1990).
[21] C.J. Waddington, D.J. Fixen, H.J. Crawford, P.J. Lindstrom and H.H. Heckman, Phys. Rev. **34A**, 3700 (1986).
[22]H.H. Andersen , J.F. Bak, H. Knudsen and B.R. Nielsen, Phys. Rev. **16A**, 1929 (1977).
[23] E. Rauhala, J. Raisanen, Zs. Fulop, A.Z. Kiss and I. Hunyadi, Nucl. Tracks Radiat. Meas. **20**, 611(1992).
[24] J. Raisanen, E. Rauhala, Zs. Fulop, A.Z. Kiss, E. Somorjai and I. Hunyadi , Radiat. Meas. **23**, 749 (1994).
[25]S.A.R. Al-Najjar, R.K.Bull and S.A. Durrani , Proceedings of 11th International Conference on SSNTDs at Bristol, 167(1981).
[26] P.H. Fowler, S. Amin, V.M. Clapham and D.L. Henshaw, Proceedings of the 10th International Conference on solid state nuclear track detectors, Lyon, 239(1979).
[27]H. Gauvin, R. Bimbot, J. Herault, B. Kubica , R. Anne, G. Bastin and F. Hubert, Nucl. Instrum. and Meth. **B28**,191 (1987).
[28] A.Z. Kiss, E. Somorjai, J. Raisanen and E. Rauhala, Nucl. Instrum. and Meth. **B39**, 15 (1989).
[29] J. Raisanen and E. Rauhala, Phys. Rev. **35B**, 1426 (1987).
[30] J. Raisanen and E. Rauhala , Phys. Rev. **36B**, 9776 (1987).
[31]C. Scheidenberger, H. Geissel, H.H. Mikkelsen, F. Nickel, T. Brohm, H. Folger, H. Irnich., A. Magel, M.F. Mohar, G. Munzenberg, M. Pfutzner, E. Roeckl, I. Schall, Schardt,K.H. Schmidt, W. Schwab, M. Steiner, Th. Stohlker, K. Summerer, D.J. Vieira, B. Voss and M. Weber, Phys. Rev. Lett. **73**, 50 (1994).

Materials Science Forum Vols. 248-249 (1997) pp. 41-48
© *1997 Trans Tech Publications, Switzerland*

Modeling of Ion Implantation and Diffusion in Si

M.-J. Caturla, T. Diaz de la Rubia and P.J. Bedrossian

Lawrence Livermore National Laboratory, Livermore CA-94550 USA

Keywords: Ion Implantation, Diffusion, Molecular Dynamics, Monte Carlo

Abstract

Classical molecular dynamics simulations are used to study the damage produced during the implantation of semiconductors with different ion masses and energies between 1-25 keV. The time scale for these type of simulations is only on the order of ns, and therefore problems like transient enhanced diffusion of dopants or formation of extended defects can not be studied with these models. Monte Carlo simulations, including as input the results obtained from molecular dynamics calculations, are used to extend the simulation time, and in particular, to study processes like ion implantation and defects diffusion in semiconductors. As an example we show results for diffusion of the damage produced by implantation of silicon with 5 keV Xe ions at low doses. The results of the simulations are compared with experiments in order to validate the model.

Introduction

Predictive models for IC technologies require of a good understanding of the processes involved during ion implantation and thermal annealing of semiconductors [1]. The damage produced during ion implantation can play an important role on the diffusion of dopants during high temperature treatments. In particular, one of the key problems in the development of ultra-shallow junctions is transient enhanced diffusion [2,3,4] of dopants during the activation stage. The defects produced during implantation interact with the dopants and induce their migration over long distances, changing the initial implantation profile. The industry trend towards a decreasing size of semiconductor devices requires predictive models. Consequently, there is a strong need for a physically based description of the damage level in the crystal lattice after implantation, as well as understanding the role of this damage in the subsequent diffusion of the dopants.

We can summarize the basic mechanisms in ion beam processes as damage production by individual ions, defect diffusion and damage accumulation. Despite the important effort devoted over the last few years to elucidate some of these problems, many unknowns still remain. One of the reasons for this lack of understanding is the many parameters involved in these processes, like ion mass, energy, dose, dose rate, temperature of the lattice and impurity levels or traps. Regarding damage accumulation, it is known experimentally that at sufficiently high doses ion implantation can induce amorphization [5]. The critical dose to reach amorphization depends on the ion mass, the temperature of the substrate and the ion flux [6,7,8]. Several models have been proposed to explain the amorphization mechanism, but none can give a complete description of all the experimental data available. One of the main difficulties is the dose rate dependence of amorphization, which is not taken into account in most of these models.

On the other hand, the diffusion kinetics of defects in silicon is also a controversial issue. For example, values for the migration energies of Vacancies (V) and Interstitials (I), as well as binding energies of clusters are not known experimentally, and discrepancies for interstitial diffusivity in Si as large as 9 orders of magnitude at 700°C exist in the literature [9,10]. Recent molecular dynamics [11] and tight binding [12] simulations indicate lower migration energies for vacancies than for interstitials. The molecular dynamics simulations [11] also indicate lower binding energies for vacancy clusters with respect to interstitials. This poor understanding of the processes involved during semiconductor doping represents a major drawback in the development of predictive models. For a complete description of all these processes it is necessary to continue with the simulation efforts as well as to develop new models in order to compare directly with the experimental results.

Molecular dynamics simulations have the advantage of giving a clear description of the atomistic processes occurring in the system during ion bombardment, since they describe the motion of all the atoms in the crystal [13]. However, they also involve large computational times. Even with the capabilities of parallel machines, only time scales on the order of nanoseconds can be achieved for the simulation of ion implantation. Processes like diffusion of defects during high temperature anneal, such as in semiconductor doping, or amorphization processes, that depend on the ion flux can not be modeled using exclusively molecular dynamics. Recent simulations by M. Jaraiz et al. [14] have shown that a good approach to the simulation of ion implantation and annealing at energies on the order of tens of keV, consists on coupling between binary collision approximation models, like MARLOWE, to Monte Carlo simulations. Following that approach, we use Monte Carlo simulations to study the annealing of defects at high temperatures, and extend the simulation time and length scale. In our case, the damage produced by the implantation, as well as the binding energies of the clusters of defects and the migration energies are obtained from molecular dynamics. This coupling between molecular dynamics and Monte Carlo allows us to simulate times on the order of seconds or longer, and therefore, validate our simulations by direct comparison with experiments.

In the next section we explain the molecular dynamics calculations performed in this work and how we use this data as input for the Monte Carlo simulation. The following section is a brief description of the Monte Carlo model, as well as the parameters used in these calculations. We finish with an example on how this tool can be applied to a real problem and compare with experimental results.

Molecular dynamics simulation: source for diffusion modeling

The primary damage state induced by ion implantation is obtained using classical molecular dynamics calculations. We use the Stillinger Weber potential [15] to describe the interaction between the Si-Si atoms in the lattice. For the short range interaction the Stillinger Weber potential was modified by connecting the two body term to the Universal potential [16], as described by Gärtner et al [17]. The interaction between the energetic ions and the silicon atoms is also described by the Universal potential. In order to account for inelastic energy losses, we use the Lindhard model [18] for those atoms with energies higher than 1 eV. The cells used in these simulations contain on the order of 10^6 atoms. The boundary layers of the cell are connected to a thermostat in order to dissipate the energy deposited by the ion. The simulations were performed in a CRAY T3D MPP parallel computer at LLNL. The code employs the PVM message passing library [19] for the communication between processors.

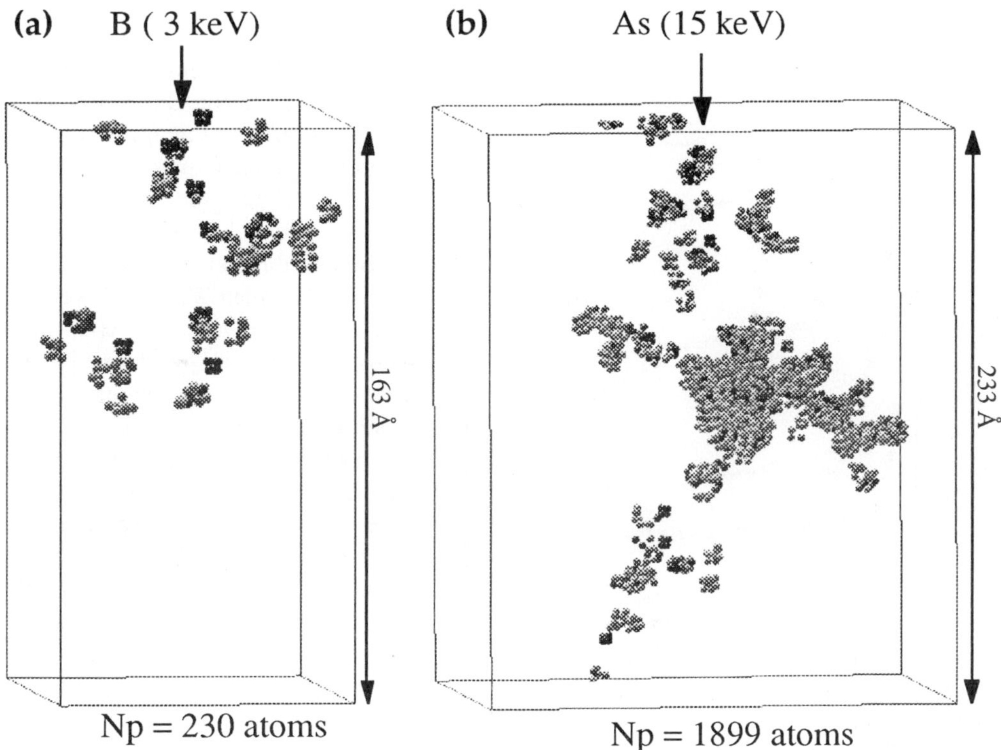

Figure 1. Molecular dynamics simulation of ion implantation in Si for (a) 3 keV B irradiation and (b) 15 keV As irradiation. Configuration of atoms with potential energy > 0.2eV with respect to the ground state, 10ps after the initial collision.

In figure 1 we show the damage produced by a 3 keV B ion, figure 1(a), as compared to a 15 keV As implantation, figure 1(b), 10ps after the implantation. These cases represent typical energies used in the production of shallow junctions, and correspond to the same range of the ions, approximately 130 Å. In this figure we show only those atoms that have a potential energy higher than 0.2 eV with respect to the ground state of the perfect crystal. The grays represent the atomic level stress of the atoms, with dark grays being atoms under tension (and therefore, related to atoms surrounding a vacancy) and light grays representing those atoms under compression (mainly interstitials). It is clear from this picture that the morphology of the damage produced by these two ions is very different. While light ions only produce small clusters of defects, or isolated defects, heavier ions, like As, will produce large disordered zones. These zones have been reported earlier [20,21] and have been proved to be amorphous in structure. This is also in agreement with experimental results for heavy ion implantation at low doses [22]. From these results we can conclude that the damage induced during doping of a semiconductor will depend on the type of dopant, and its morphology can range from small clusters of defects to localized amorphous pockets. All these effects have to be taken into account to model the later evolution of the defects during high temperature annealing.

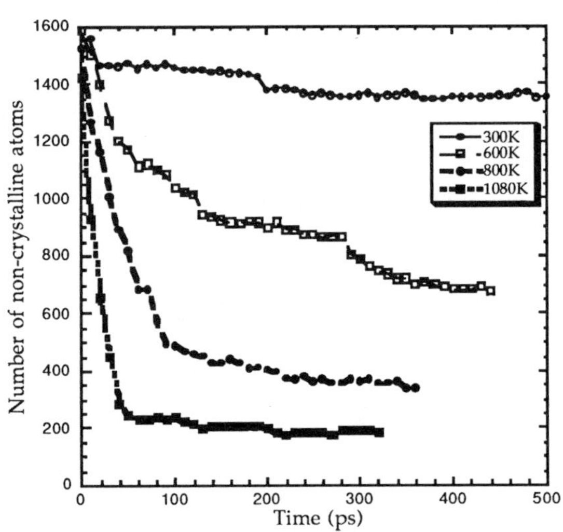

Figure 2. Annealing of the damage produced by 15 keV As implantation in Si. Number of non-crystalline atoms as a function of annealing time, for different temperatures.

To properly understand damage accumulation during implantation, the stability at different temperatures of the amorphous pockets produced by heavy ion bombardment must be understood. We have simulated the damage induced by bombardment with As and Pt at energies between 5 and 25 keV, and studied the annealing kinetics of the amorphous pockets produced by these ions at different temperatures. In figure 2 we present results for the number of amorphous atoms as a function of annealing time and temperature for the case of damage produced by bombardment with 15 keV As ions. Observe that these curves do not follow an exponential behavior, but instead there is an initial stage when recrystallization occurs in a very short annealing time. Later the recrystallization occurs in a series of plateaus and steps. This behavior is in agreement with the recrystallization models proposed by Spaepen and Tumbull [23], and modified by Aziz [24] and Williams [25] for a planar amorphous-crystalline interface. According to these models recrystallization occurs by the production and migration of dangling bonds at the interface. In our case, the irregular shape of the amorphous pockets provides the nucleation sites for recrystallization, as pointed out by Priolo [26], and recrystallization can occur at temperatures lower than for the case of a planar interface.

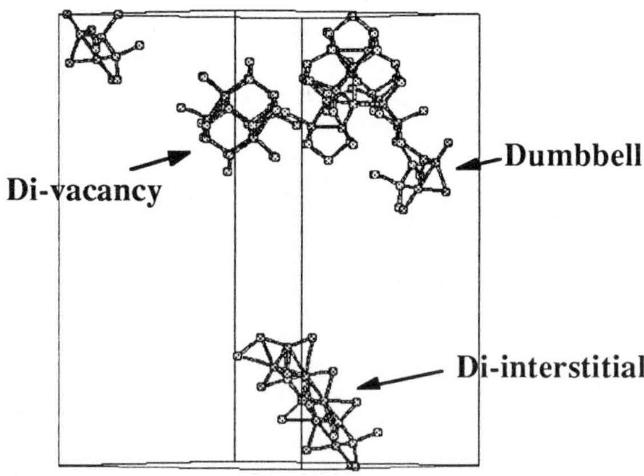

Figure 3. Defects at the Si lattice after 0.5 ns anneal at 800K. Only point defects and small clusters of defects are present.

Considering the results from figure 2, at high temperatures the damage anneals in a very short time; after the first few picoseconds only point defects and small clusters of defects are left in the lattice. In figure 3 we show the defects present in the crystal after 0.5 ns anneal of the damage at 800K. From these results we can conclude that at high temperatures, like the ones used for the activation of the dopants in a semiconductor, only point defects and small clusters will be present, and no local amorphization is left. Therefore, treatment of defect diffusion at high temperature after long times can be done using simple kinetic Monte Carlo simulations.

Modeling of defect diffusion: Monte Carlo simulations

Two types of defects are considered in the simulation, vacancies and interstitials. The initial positions of these defects are obtained from the molecular dynamics simulations described above. The values for migration energies and binding energies of clusters of different sizes and are also obtained from molecular dynamics using the Stillinger Weber potential. These values are shown in Table I. The pre-factor for the self-interstitials has been obtained from fitting kinetic Monte Carlo results to the diffusion of boron in silicon [27] and using *ab initio* results for the energetics of the boron-silicon interaction [28]. Clusters larger than 5 have binding energies fitted to a function of the form: $Eb_V(N) = 3.6 - 4.9(n^{2/3} - (n-1)^{2/3})$ eV and $Eb_I(n) = 2.5 - 2.17(n^{1/2} - (n-1)^{1/2})$ eV, for V and I of size n respectively. The program is described in detail in ref. [14].

As an example of the capabilities of this simulation tool we consider the annealing at different temperatures of the damage produced by bombardment with 5 keV Xe ions at a dose of 1.5×10^{13} ions/cm^2. The final dose is obtained by accumulating the damage induced by individual ions at room temperature. In this case we consider that there is no relaxation of the damage between cascades, as justified by the calculations described above. The kinetic-Monte Carlo simulation includes a total of 152 ion trajectories in a box 0.32 μmx0.32μmx5μm.

Table I. Parameters used for the Monte Carlo simulation. Values for diffusivities and binding energies from classical molecular dynamics calculations [11]. The pre-factor for interstitials is obtained from fitting to boron diffusion experiments [27] and *ab initio* calculations [28].

	DIFFUSIVITY (cm^2/s)	BINDING ENERGY SIZE 2 (eV)	BINDING ENERGY SIZE 3 (eV)	BINDING ENERGY SIZE 4 (eV)	BINDING ENERGY SIZE 5 (eV)
VACANCY	0.001EXP(-0.43/KT)	0.62	0.78	1.2	1.82
INTERSTITIAL	5.EXP(-0.9/KT)	1.6	2.25	1.29	2.29

First we study the evolution of the damage at low temperature, 350°C, during a total annealing time of 4 hours. In figure 4 we show the number of free and clustered vacancies and interstitials in the bulk, as well as the number of vacancies and interstitials reaching the surface as a function of annealing time. During the initial stages of the anneal, when free defects are still present in the lattice, the diffusion is governed by the individual migration energies of vacancies and interstitials. With the parameters considered in this simulation the diffusivities for V and I at this temperature are very similar. Therefore these defects migrate and annihilate, recombine with the surface or form clusters of defects. After 10^{-7} s only clusters of defects remain in the bulk. Then diffusion is governed by the binding energies of the different clusters. In our case, the binding energies of vacancy clusters are always smaller than those of the interstitials.

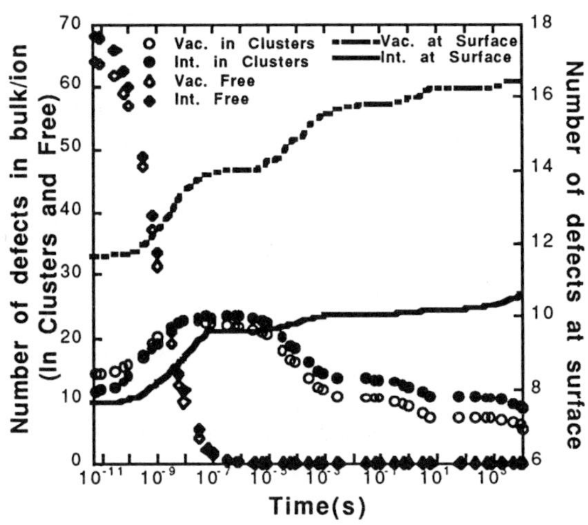

Figure 4. Kinetic Monte Carlo results for the annealing of the damage produced by 5 keV Xe implantation at 1.5×10^{13} ions/cm^2. Number of V and I in clusters and free (left hand side of the plot), as well as the number of defects at the surface (right hand side) as a function of the annealing time. Temperature of annealing is 350°C.

Therefore, the small vacancy clusters start releasing defects, and some of those will reach the surface. As a consequence the total number of vacancies reaching the surface after 4 hours of anneal is higher than the number of interstitials. Also the vacancy clusters increase in size, with an average cluster size of 6 for vacancies while the interstitial average cluster size is less than 3. The sputtering yield obtained from the molecular dynamics simulations is of 2.25 atoms/ion consistent with experimental results [29]. After the 4 hours anneal at 350°C the excess number of vacancies at the surface is ~ 6 V/ion, and therefore, much higher than the sputtering yield.

In figure 5 we show the results of a later annealing of the damage at higher temperature, 550°C. In this case both vacancy and interstitial clusters dissociate. All the vacancies present in the bulk quickly disappear either on the surface or recombined with interstitials. Some of the interstitials also reach the surface, decreasing the excess number of vacancies per ion at the surface.

These results agree with experimental observations on Si (111)-7x7 [30]. In these experiments a depopulation of the adatom layer is observed under annealing at 350°C, that could be explained as an arrival of vacancies to the surface from the bulk. The density of vacancies in the experiment is 1.2×10^{14} V/cm^2, in reasonable agreement with the 0.8×10^{14} V/cm^2 from our simulation. Annealing at 550°C induces repopulation of the surface layer, that considering our model, is due to the arrival of interstitials that evaporate from clusters in the bulk. In the experiment the density of interstitials at the surface after 2 minutes anneal is of $\sim 7 \times 10^{13}$ I/cm^2 as compared to $\sim 3 \times 10^{13}$ I/cm^2 obtained in the simulation.

Conclusions

In summary, a combination of molecular dynamics and kinetic Monte Carlo simulations has been used to study ion implantation and diffusion in silicon. The time scale achieved with this type of simulations allows for direct comparison with experiments. In our case we apply the model to study the diffusion of defects produced during the irradiation of Si with 5 keV Xe ions, and the later diffusion at different temperatures. It is observed that, depending on the temperature, different defects reach the surface. At low temperatures an excess number of vacancies arrive to the surface. At higher temperatures, interstitial clusters dissolve decreasing the excess of vacancies at the surface. These results can explain the experimental observation of depopulation and repopulation of the Si (111)-7x7 surface at different annealing temperatures [30].

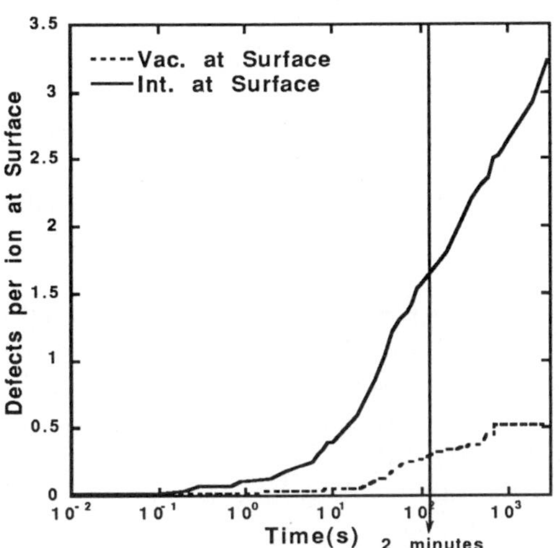

Figure 5. Number of defects per ion recombined at the surface as a function of the annealing time. The annealing temperature is 550°C.

Acknowledgments

This work was performed under the auspices of the U.S. Department of Energy by Lawrence Livermore National Laboratory under Contract No. W-7405-Eng-48. M. Jaraiz and G.H. Gilmer of Lucent Technologies provided theMonte Carlo code. We thank L. Pelaz for useful discussions.

References

[1] J.D. Plummer, P.G. Griffin, Nucl. Instrum. Meth. **B102** (1995) 160

[2] T. Motooka and O.W. Holland, Apply. Phys. Lett. **61** (1992) 3005

[3] K.S. Jones, H.G. Robinson, J. Listebarger, J. Chen, J. Liu, B. Herner, H. Park, M.E. Law, D. Sieloff, J.A. Slinkman, Nucl. Instrum. and Methods, **B96** (1995) 196

[4] D.J. Eaglesham, P. A. Stock, H-J Gossman, and J.M. Poate, Appl. Phys. Lett. **65** (1994) 2305

[5] J.S. Williams, Mater. Res. Soc. Bull. **17**, 47 (1992)

[6] R.G. Elliman, J. Linros, and W.L. Brown, Mater. Res. Soc. Synp. Proc. **100** (1988) p. 60

[7] P.J. Schultz, C. Jagadish, M. C. Ridgeway, R. G. Elliman and J.S. Williams, Phys. Rev. **B44** (1991) 9118

[8] R.D. Goldberg, R.G. Elliman, and J. S. Williams, Nucl. Intrum. Meth. **B80/81** (1993) 596

[9] H-J Gossmann, P.A. Stock, D.J. Eaglesham,. G. H. Gilmer and J.M. Poate, in *Process Physics and Modeling in Semiconductor Technology*, edited by G.R. Srinivasan, D.S. Murthy and S.T. Dunham (Electrochemical Society, Pennington, New Jersey, 1996)

[10] U. Gösele, A. Plößl and T.,Y. Tan in *Process Physics and Modeling in Semiconductor Technology*, edited by G.R. Srinivasan, C.S. Murthy and S.T. Dunham (Electrochemical Society, Pennington, New Jersey, 1996) p.309

[11] G.H. Gilmer, T.Diaz de la Rubia, D. Stock and M. Jaraiz, Nucl. Instrum. and Methods **B102** (1995) 247

[12] M. Tang, L. Colombo, J. Zhu, T. Diaz de la Rubia, Phys. Rev. B (accepted for publication)

[13] M.P. Allen and D.J. Tildesley in *Computer simulation of liquids*, Oxford science publications (1987)

[14] M. Jaraiz, G.H. Gilmer, J.M Poate, T. Diaz de la Rubia, Appl. Phys. Lett. **68** (1996) 409

[15] F.H. Stillinger and T.A. Weber, Phys. Rev. **B31** (1985) 5262

[16] J.F. Ziegler, J.P. Biersack, U. Littmark, *The stopping and Range of Ions in Solids*, Vol.1 of *The Stopping and Range of Ions in Matter*, ed. J.F. Ziegler (Pergamon, New York, 1985) p. 25 ff

[17] K. Gärtner, Nucl. Instrum. and Methods, **B102** (1995) 183

[18] J. Lindhard and M. Sharff, Phys. Rev. **124** (1961) 128

[19] *PVM3 Users Guide and Reference Manual.* ORNL/TM-12187, May 1993

[20] T. Diaz de la Rubia, G.H. Gilmer, Phys. Rev. Lett. **74** (1995) 2507

[21] M-J Caturla, L. A. Marques, T. Diaz de la Rubia and G.H. Gilmer, Phys. Rev. B (accepted for publication)

[22] J. Narayan, O.S. Oen, D. Fathy and O.W. Holland, Materrials Letters **3** (1985) 67

[23] F. Spaepen, D. Turnbull, *Laser Solid Interactions and Laser Processing*, ed. S.D. Ferris, H.J. Leamy, J.M. Poate, Mat. Res. Soc., Boston, 1978

[24] G.-W. Lu, E. Nygren, M.J. Aziz, J. Appl. Phys. **70** (1991) 5323

[25] J.S. Williams, R.G. Elliman, Phys. Rev. Lett. **51** (1983) 1069

[26] F. Priolo, A. Battaglia, R. Nicotra, Appl. Phys. Lett. **57** (1990) 768

[27] L. Pelaz and G.H. Gilmer, private communication

[28] J. Zhu, T. Diaz de la Rubia, L. Yang, C. Mailhiot and G. H. Gilmer, Phys. Rev. B 1996

[29] P.C. Zalm, J. Appl. Phys. **54** (1983) 2660

[30] P. Bedrossian, M-J Caturla, T. Diaz de la Rubia, Appl. Phys. Lett (accepted for publication).

Materials Science Forum Vols. 248-249 (1997) pp. 49-52
© *1997 Trans Tech Publications, Switzerland*

Crystal-Grid Investigation of Atomic Collision Cascades in Ionic Compounds

M. Jentschel[1,2], K.H. Heinig[1], H.G. Börner[2] and C. Doll[2]

[1] Institute of Ion Beam Physics and Materials Research,
FZ-Rossendorf, D-01314 Dresden, Germany

[2] Institut Laue-Langevin, BP 156, F-38042 Grenoble Cedex 9, France

Keywords: Atomic Collisions, Molecular Dynamics, Nuclear Lifetimes

Abstract: In the Crystal-GRID technique γ-lineshapes Doppler broadened due to the recoil motion of atoms in the bulk of solid state targets are measured. The recoils were induced due to the emission of a primary γ-ray after thermal neutron capture. This technique offers the unique opportunity to obtain direct experimental information on collision processes in solids at kinetic energies of several hundreds of eV. The experimental results of recoiling ^{36}Cl atoms in NaCl single crystals are presented and compared to MD-simulations. It is shown that the Coulomb interaction plays an essential role in a correct description of the slowing down process.

1. The Experimental Technique

In the **Gamma Ray Induced Doppler** broadening (GRID) technique [1] the following effect is used: Atoms placed in a thermal neutron flux undergo neutron capture reactions. These reactions produce excited nuclei, which decay via a γ-ray cascade. Each emitted γ-quant induces due to the momentum conservation a recoil to the nucleus. Consequently, the energy of subsequently emitted γ-rays in the cascade is Doppler shifted if they are observed in a laboratory system. In general, as the initial recoil directions are isotropically distributed, a Doppler broadened line profile will be obtained. It should be mentioned that high energetic primary γ-rays induce recoils with kinetic energies of several hundreds of eV leading to a Doppler broadening of several hundreds of eV. The effect depends in general on two time scales: i) The lifetime of the nuclear state, which was populated by the recoiling γ-ray and depopulated by the measured γ-ray. ii) The time, which is needed for the recoiling atom to slow down due to interaction with its surrounding atoms. In the case that both timescales are comparable the Doppler broadening can be experimentally resolved, using ultra high resolving γ-spectrometers.

As the measured line shapes depend on two timescales, the knowledge of one of them would be required for the study of the second one. The main purpose of the GRID technique was actually the measurement of nuclear state lifetimes, calculating the slowing down by computer simulations. However, the investigation of atomic collisions with several hundreds of eV kinetic energy is of interest for a number of important ion beam techniques, and the optimization of the GRID-technique for this purpose would allow to obtain very direct information. A decisive step in this direction is the idea to carry out GRID-experiments with single crystalline targets (Crystal-GRID). It has been predicted that in this case a characteristic structure on the Doppler-broadened lineshapes will be obtained, which changes for different alignments of the target crystal with respect to the spectrometer axis. As this enables the accumulation of information on the same microscopic processes it can be expected that it will be possible to extract both timescales simultaneously from GRID-measurements. Using MD-simulations for the calculation of the slowing down it would thus enable the study of interatomic potentials.

2. Crystal-GRID Experiments with NaCl

In order to test the applicability of Crystal-GRID experiments for the investigation of atomic collision cascades, experiments with single crystals of NaCl oriented in $< 100 >$ and $< 110 >$ were carried out. The ^{36}Cl isotope created in these crystals after thermal neutron capture is recoiling with a kinetic energy of 468eV due to deexcitation from the capture state to the 2864keV state. The distribution of the recoil energies is nearly monochromatic and the nuclear lifetime of the 2864keV state is expected to be about 20fs. Thus, the 2864keV transition depopulating this level to the ground state should be strongly Doppler broadened and well suited to show the sensitivity of the experiment.

A first step in the experiment was the determination of the instrumental response function, which can be done with high accuracy. As the energetic resolution of the spectrometer is achieved by diffracting the γ-rays on two flat Si-crystals following Braggs law an improvement of the resolution with an increasing reflection order may be expected. On the other hand the diffraction probability of the γ-rays is decreasing with higher order, which means loss in statistics. The 2864keV line was scanned in first and second order of reflection for both target sets. The experimental results are shown in Fig.1

As can be seen there, the predicted structure on the lineshape and its dependence on the relative alignment of the target crystal with respect to the spectrometer axis could be verified experimentally. However, only in the second order of reflection are details of the structure well resolved, but with large statistical errors. For a further application of this experimental technique the luminosity of the spectrometer will have to be optimized.

The slowing down of the recoiling atoms was simulated using MD-simulations. Three types of simulations were performed. First pure repulsive screened Coulomb potentials were used (in this paper the KrC potential [2]). Since the simulation cell in MD-simulation with such potentials is not stable, the interaction of the atoms was only "switched on" in a sphere of 7 $\overset{\circ}{A}$ around the recoiling (excited) atom. The accuracy of this kind of simulation is somewhere in-between the well known Binary Collision Approximation and full MD-simulations. In the second approach only the recoiling atom was assumed to interact with the other atoms via a pure repulsive potential, whereas all other atoms interact via an attractive equilibrium potential, (Karasawa et al. [3]), connected to the same repulsive potential by a smooth step function. The connection distance and smoothness were chosen such, that on the one hand at interaction energies above 5eV the repulsive potential was valid and on the other hand the equilibrium properties (elastic constants, cohesive energy) were not disturbed. In the third approach, for the interaction of all atoms, the connected potential was taken. The long ranging ionic forces in the second and the third approach were treated using the Ewald technique. The size of the simulation cell was $5 \times 5 \times 5$ elementary cells (1000 atoms) in the case of the connected potentials and $7 \times 7 \times 7$ in the case of the pure repulsive potentials. The number of simulated recoil events was 1000 and each of them was followed by 125fs.

On the basis of such simulations one can calculate theoretical lineshapes for arbitrary

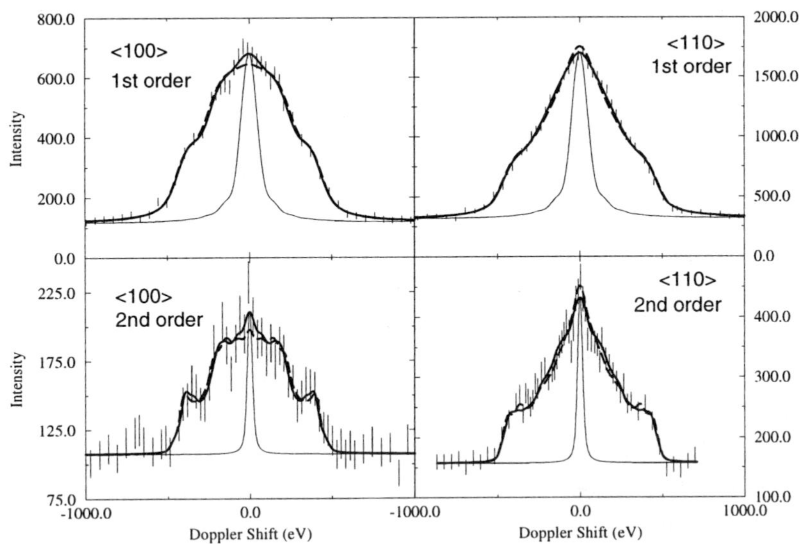

Figure 1: Experimental results of the 2864keV transition are compared to lineshapes calculated with MD-simulations. The thick dashed line results from a simulation with a pure repulsive KrC potential following the first approach. The thick solid line corresponds to a simulation following the third approach, where the KrC-potential was smoothly connected to the Karaswa potential. The inner lineshapes are the instrumental response functions.

assumed lifetimes. Folding these theoretical lineshapes with the measured instrumental response function, the direct analogon to the experimental lineshape is obtained. As the lifetime has entered in the lineshape calculation as a free parameter, it can be adjusted by fitting the lineshapes to the experiment, using a least squares fitting routine. In such a way, for each simulation, which was carried out a lifetime value τ and a number χ^2 depending on the agreement/disagreement of the theoretical and experimental lineshape are obtained.

3. Results and Discussion

For the comparison of the simulations with the experiments, the following procedure was chosen: The lineshapes were fitted as correlated fits to the data of both reflection orders of one crystal direction. Thus, two independent results for the lifetime are obtained. After this, the lineshapes were fitted as correlated fits to all data simultaneously, which results in a mean value for the lifetime. The lineshapes corresponding to the latter procedure are displayed in Fig.1 together with the experimental data. The τ and χ^2 values of all fits are listed in Table 1.

The τ-values, obtained from the independent fits of the theoretical lineshapes of the first approach to the experimental data of the two crystal directions, show a reasonable difference of 4.4fs. As the error indicates, this deviation can not be explained by the statistics. However, as the only difference in-between these two measurements is a different alignment

Simulation		$< 100 >$	$< 110 >$	$< 100 > + < 110 >$
KrC	τ	$27.4^{0.8}_{0.9}$	$23.0^{0.5}_{0.5}$	$24.1^{0.4}_{0.4}$
	χ^2	1.1683	1.0937	1.1268
Kar&KrC	τ	$21.4^{0.7}_{0.7}$	$20.2^{0.4}_{0.4}$	$20.5^{0.4}_{0.4}$
	χ^2	1.1622	1.0715	1.1012
Kar+KrC	τ	$21.2^{0.8}_{0.8}$	$20.8^{0.4}_{0.4}$	$20.9^{0.4}_{0.4}$
	χ^2	1.1496	1.0713	1.1002

Table 1: Summary results of the lineshape fits based on different MD-simulations to the experimental data of the 2864keV line of ^{36}Cl. As can be seen the best agreement can be obtained from the simulation with the smoothly connected potential (Kar+KrC).

of the target with respect to the spectrometer, the obtained lifetime values have to be the same! For this reason it can be concluded that the first simulation approach is not suited to describe correctly atomic collision cascades with $\sim$500eV kinetic energy in ionic compounds. This conclusion is also supported by the bad χ^2-value. In the second approach, where an equilibrium interaction of the lattice atoms (not the projectile) is added, whereas the repulsive interaction with the projectile is the same as in the first approach, the difference of the lifetime values is remarkably smaller, but the values are still not overlapping. Also the χ^2-value is remarkably better. This indicates that the "elastic" reaction of the lattice is of essential importance for the correct description of the collision dynamics. Only in the third approach, which uses the connected potentials, the lifetime values overlap and also the best χ^2-value is obtained. Comparing this approach to the second one, the only difference is the additional longranging Coulomb interaction of the projectile with the lattice atoms. This suggests that the recoiling atom still remains to be ionized after the recoil.

In conclusion one can state that the Crystal-GRID method is very sensitive to details of the slowing down process, and provides very direct information on atomic motion in the bulk of solid state targets. It can be used to test interatomic potentials by comparing the experimental data with MD-simulations. Thus it has been shown that in the case of collision cascades in ionic compounds, the Coulomb interaction plays an essential role. Therefore, simple models such as the Binary Collision Approximation and simplified MD-simulations lead to a wrong description of the slowing down process.

4. Acknowledgements

This work was financially supported by the Sächsisches Ministerium für Wissenschaft und Kunst.

References

[1] H.G. Börner and J.Jolie, J. Phys. G: Nucl. Part. Phys. 19 (1993) 217-248

[2] W.D. Wilson, L.G. Haggmark and J.P. Biersack, Phys. Rev. 15 (1977) 2458

[3] N. Karasawa and W.A. Goddard III J. Phys. Chem. 93 (1989) 7320-7327

Materials Science Forum Vols. 248-249 (1997) pp. 53-56
© 1997 Trans Tech Publications, Switzerland

Thermal Spike Description of the Damage Creation in $Y_3Al_5O_{12}$ Induced by Swift Heavy Ions

A. Meftah[1], M. Djebara[1], N. Khalfaoui[1], J.P. Stoquert[2], F. Studer[3] and M. Toulemonde[4]

[1] USTHB, Institut de Physique, BP 32, El-Alia, Bab-Ezzouar, 16111 Algiers, Algeria

[2] CRN, Groupe Phase, F-67037 Strasbourg Cedex, France

[3] CRISMAT, Université de Caen, F-14032 Caen Cedex, France

[4] CIRIL, BP 5133, F-14040 Caen Cedex, France

Keywords: Swift Heavy Ions, Electronic Stopping Power, Latent Track, Thermal Spike Model

Abstract : Single crystals of $Y_3Al_5O_{12}$ have been irradiated with 104 MeV ^{52}Cr, 282 MeV ^{86}Kr, 278 MeV ^{128}Te and 863 MeV ^{208}Pb ions. The damage cross section is extracted using Rutherford backscattering spectroscopy in channeling geometry with a 2-MeV ^{4}He beam. The thermal spike model is applied to calculate track radii assuming that the observed damage is due to a rapid quench of a molten liquid phase along the ion path. By fitting the latent track radii versus the electronic stopping power for different beam energies, a value of 4.6 ± 0.3 nm is extracted for the mean diffusion length λ of the energy deposited on the electrons. The available values of λ determined in the same conditions for radiolysis resistant insulators SiO_2 quartz, $BaFe_{12}O_{19}$ and $Y_3Fe_5O_{12}$ indicate that λ should decreases when the band gap energy increases.

1. Introduction

Swift heavy ions (energy ≥ 100 keV/amu) penetrating into matter lose their energy (a few keV/Å) mainly via ionization and electronic excitation processes. In most solids, a latent track is produced along the ion path resulting from collective electronic excitations. Several physical characterizations made it possible to have a coherent experimental results on tracks in different materials [1-6]. However the damage mechanism is not well understood. Several theoretical models have been proposed to explain this mechanism since the beginning of 1950, but this question still remains controversial. Among these, the Coulomb spike [7] and the thermal spike [3,8-9] models are most often quoted. In the Coulomb spike model it is assumed that the incident ion creates along its trajectory a region of highly ionized matter. The strong repulsive electrostatic forces give rise to a violent explosion leading to the creation of important defects within a localized region of the lattice. In the thermal spike model, it is supposed that the temperature increases along the ion path and reaches its maximum value after the melting process. The rapid quenching of the entire cylinder of the melted matter lead to the formation of an amorphous structure. In both models the key parameter is the electrical resistivity. This explained that most insulators are sensitive in contrast to metallic materials. Since the beginning of the 80 when heavy ion accelerators became available, it is possible to study ion induced effects in a wide variety of materials (metal, semiconductor, insulator) and to realize irradiations at extremely high electronic stopping power dE/dx. Numerous new experimental results lead to the conclusion that the electronic structure of all types of materials is obviously not the only important parameter responsible for swift heavy ions induced effects. Several recent experiments [4-6] have shown that the electronic stopping power is efficient to create damage in crystalline metals, amorphous metals and semi-conductors. Sigrist and Balzer [10] have shown later on that the electronic stopping power threshold for chemical etching of tracks in various insulators cannot be scaled by the parameters governing the Coulomb spike, on the contrary, a better correlation appears when considering the thermal conductivity of the insulators. Also, the electronic sputtering [11-12] was interpreted on the basis of a thermal process. A recent development [13-14] of the thermal spike model allows a relatively good description of the observed effects in radiolysis resistant oxides SiO_2, $BaFe_{12}O_{19}$ and $Y_3Fe_5O_{12}$. The present paper aims at extending this model to the garnet $Y_3Al_5O_{12}$.

2. Experimental

Single crystals of $Y_3Al_5O_{12}$ have been irradiated with 104 MeV [52]Cr, 282 MeV [86]Kr, 278 MeV [128]Te and 863 MeV [208]Pb ions. The irradiations were performed at the GANIL accelerator in Caen using the medium-energy beam line [15]. A degrador was used whenever possible for covering different values of dE/dx in the same irradiation. The beam flux was between 10^8 to 10^9 p/s.cm^2 depending on the incident ion in order to avoid sample heating. The maximum fluence was between 10^{13} and 10^{14} p/cm^2. The dE/dx values were calculated using the TRIM93 code[16].

Rutherford backscattering spectrometry in channeling (RBS-C) geometry was performed as described in a previous paper [3]. All the samples were analyzed with a 2 MeV [4]He beam at the 4-MV Van de Graaff accelerator at the Centre de Recherches Nucléaires of Strasbourg. The damage cross section A can be extracted by comparing the scattering spectra of unirradiated and irradiated samples [3]. It defines the cross section of a cylinder in which is created an amorphous phase. The effective radius R_e of the cylindrical track is deduced from the relation $A = \pi R_e^2$. It has been shown earlier that the radius determined by high-resolution electron microscopy [2] is directly correlated to the damage cross section deduced from RBS-C for values of A bigger than 3×10^{-13} cm^2. The experimental results of the irradiation performed on $Y_3 Al_5 O_{12}$ are reported in table 1.

Table 1 :

Ion	Energy (MeV/a.m.u.)	dE/dx (keV/nm)	A (cm^2)	R_e (nm)
[52]Cr	2	11	$(2.53\pm0.38)10^{-14}$	0.9 ± 0.1
[86]Kr	3.28	17	$(1.42\pm0.28)10^{-13}$	2.1 ± 0.2
[128]Te	1.24	21.5	$(4.30\pm0.70)10^{-13}$	3.7 ± 0.3
[128]Te	2.17	24.5	$(4.24\pm0.60)10^{-13}$	3.7 ± 0.3
[208]Pb	1.44	32	$(8.22\pm1.40)10^{-13}$	5.1 ± 0.4
[208]Pb	4.15	39.5	$(8.64\pm2.87)10^{-13}$	5.2 ± 0.9

3. The thermal spike description

The thermal spike model is mathematically described by two coupled equations [17] governing the energy diffusion on the electron and lattice subsystem. As radiation defects created in materials by swift ions are cylindrical, these equations are expressed in cylindrical geometry:

$$C_e \frac{\partial T_e}{\partial t} = \nabla(K_e \nabla T_e) - g (T_e - T) + B(r,t) \qquad (1)$$

$$\rho\, C(T) \frac{\partial T}{\partial t} = \nabla(K(T) \nabla T) + g (T_e - T) \qquad (2)$$

Where T_e, T, C_e, C(T) and K_e, K(T) are the temperature, the specific heat and the thermal conductivity for the electronic and atomic systems respectively. ρ is the specific mass of the lattice and g is the electron-phonon coupling constant. B(r,t) is the energy density per unit time deposited by the incident ion to the electronic system at radius r and time t. The influence of the ion velocity on irradiation effect in the electronic energy loss regime has clearly been established in insulators [18]. In order to take into account this effect, the radial shape of the initial energy deposition D(r) is given by the distribution of Waligorski et al. [19]. As previously [20], it is assumed that the energy deposition time τ is 10^{-15} s, corresponding to the time necessary to slow down the δ rays electrons[19]. In the present case we assume that $B(r,t) = \beta\, D(r)\, \alpha\, e^{-\alpha t}$ where $\alpha = 1/\tau$ and β is chosen so that

$$\int_{t=0}^{\infty} \int_{r=0}^{r=r_m} B(r,t)\, 2\pi r\, dr\, dt = \frac{dE}{dx} \qquad (3)$$

r_m being the maximum projected range for electrons perpendicularly to the ion path. As previously tested [20] a variation of τ by a factor of 5 has no influence on the results. The electronic thermal diffusivity is defined by the relation $D_e = K_e/C_e$. The electron phonon coupling constant g is linked to the mean electron-phonon interaction time τ_a by the relation $\tau_a = C_e/g$ [17] and to the mean diffusion length λ of the deposited energy by the relation $\lambda^2 = D_e \tau_a$. As suggested by Baranov et al. [11] hot electrons in the conduction band of an insulator will behave like hot electrons in a metal. Consequently in our calculations the diffusivity of the hot electrons and their specific heat are considered as constant [21] : $D_e \sim 2$ cm^2 s^{-1} and $C_e \sim 1$ J cm^{-3} K^{-1}.

In the present calculation, λ will be considered as the free parameter. Since D_e and C_e are constant, this parameter is directly related to the electron-phonon coupling constant. In order to take into account the thermodynamical parameters of the lattice, including the transition from the solid to the liquid state, both equations (1) and (2) are numerically solved. The following macroscopic thermodynamical paramters for $Y_3Al_5O_{12}$ were used : solid and liquid volumic masses respectively (4.6 g/cm^3) and (4.3 g/cm^3), specific heat (0.84 J/g K), melting temperature (2223 K) and latent heat of fusion (702 J/g). The dependence of the thermal conductivity on the temperature T is given by 39/T (Watt/K cm).

In the thermal spike model, the radius of the track corresponds to the radius of the molten cylinder. The comparison between theory and experiment is limited to track radii larger than 3 nm when the track is continuous [3]. For each radius determined from the RBS-C experiment, a λ_i value is extracted that best reproduced the experimental data point. Considering all data, a mean value of $\lambda = (4.6 \pm 0.3)$ nm is deduced. The radius evolution of the track versus dE/dx, as extracted from the model calculation for the extreme values of λ and ion energy, is reported in fig. 1. This shows that the thermal spike calculation qualitatively describe the evolution of R_e as function of the energy loss.

The same calculations have earlier been performed for $BaFe_{12}O_{19}$ [13], $Y_3Fe_5O_{12}$ [14], and SiO_2 quartz [13]. For each material, a mean value of λ is extracted by fitting all the experimental results. Table 2 summarizes the calculated values of λ and the band gap energy E_g. One can clearly see that λ decreases for increasing E_g.

Table 2

material	E_g (eV)	λ (nm)
$BaFe_{12}O_{19}$	1	8.2 ± 1.3
$Y_3Fe_5O_{12}$	2.8	$< 6.3 \pm 0.3$
$Y_3Al_5O_{12}$	6.3	4.6 ± 0.3
SiO_2	12	4.0 ± 0.3

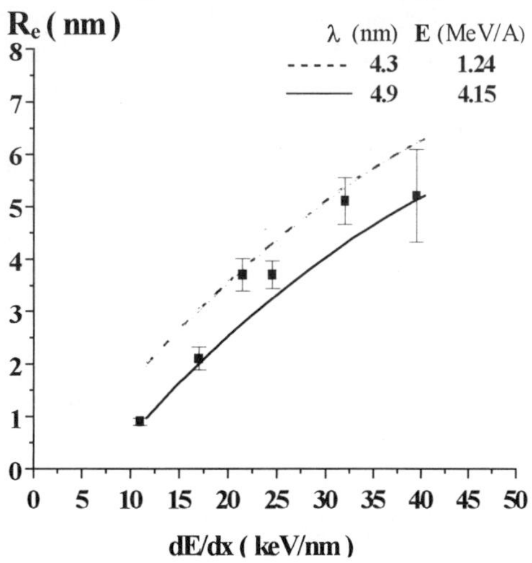

Figure 1 : Experimental data of track radius R_e versus dE/dx for $Y_3Al_5O_{12}$. The two lines give the calculated radius evolution of the track versus dE/dx for the extreme values of λ and the ion energy.

This result confirms the conclusion made by Toulemonde et al. [13] and which found theoretical support in the description of Katin et al. [22] and Baranov et al. [11] : λ in solid insulators should decrease when the band gap energy increases.

4. Conclusion

Damage cross section A and latent track radii R_e have been determined for $Y_3Al_5O_{12}$. The thermal spike model is applied in order to get the evolution of the track radii versus the electronic stopping power dE/dx in this garnet. Only one fitting parameter is used : the mean diffusion length λ of the energy deposited on the electrons. A mean value of $\lambda = (4.6 \pm 0.3)$ nm is deduced. The radius evolution of the track versus dE/dx for the extreme values of λ and beam energy is in good agreement with the experimental results. We have correlated the mean diffusion length λ of the deposited energy of several oxides materials ($BaFe_{12}O_{19}$, $Y_3Fe_5O_{12}$, $Y_3Al_5O_{12}$ and SiO_2 quartz) to the band gap energy. This work confirms the conclusion that for radiolysis resistant insulators λ decreases as band gap energy E_g increases. Whether this dependence is also true for non amorphizable materials and for radiolysis sensitive materials is still an open question.

References

[1] M. P. A. Viegers, J. Electron Microsc. **2**, 187 (1982).

[2] M. Toulemonde and F. Studer , Phil. Mag. A **58**, 799 (1988).

[3] A. Meftah, F. Brisard, J.M. Costantini, E. Dooryhee, M. Hage-Ali, M. Hervieu, J.P. Stoquert, F. Studer, and M. Toulemonde, Phys. Rev. B **49**, 12457 (1994).

[4] M. Toulemonde, S. Bouffard, and F. Studer, Nucl. Instrum. Meth. B **91**, 108 (1994).

[5] S.M.M. Ramos, B. Canut, M. Ambri, C. Clement, E. Dooryhee, M. Pitavale, P. Thevenard, and M. Toulemonde, Nucl. Instr. Meth. B **107**, 254 (1996).

[6] Ch Dufour, Z.G. Wang, M. Levalois, P. Marie, E. Paumier, F. Pawlak , and M. Toulemonde, Nucl. Instr. Meth. B **107**, 218 (1996).

[7] R.L. Fleischer, P.B. Price, and R.M. Walker, Nuclear Track in solids, Univ. of California Press, 1975.

[8] L. T. Chadderton and I. Mc C. Torrens, Fission damage in crystals (Ed. Methuen &Co. ltd 1969) 113 and 190.

[9] M. Toulemonde, C. Dufour, and E. Paumier, Phys. Rev. B **46**, 14362 (1992).

[10] A. Sigrist and R. Balzer, Helv. Phys. Act. **50**, 49 (1977) and Rad. eff. **34**, 75 (1977).

[11] I.A. Baranov, Yu V. Martinenko, S.O. Tsepelevitch, and Yu N. Yavlinskii, Sov. Phys. Usp. **31**, 1015 (1988).

[12] R. E. Johnson, B.U.R. Sundquist, A. Hedin, and D. Feyn., Phys. Rev. B **46**, 14362 (1992).

[13] M. Toulemonde J.M. Costantini, C. Dufour, A. Meftah, E. Paumier, and F. Studer, Nucl. Instr. Meth. B **116**, 37 (1996).

[14] A. Meftah, J. M. Constantini, M. Djebara, N. Khalfaoui, J. P. Stoquert, F. Studer, and M. Toulemonde, to be published.

[15] S. Bouffard, J. Dural, F. Levesque, and J.M. Ramillon, Ann. Phys. **14**, 385 (1989).

[16] J.P. Biersack and L.G. Haggmark, Nucl. Instr. Meth. **174**, 257 (1980).

[17] I.M. Lifshitz, M.I. Kaganov, and L.V. Tanatarov, J. Nucl. Energ. A **12**, 69 (1960).

[18] A. Meftah, F. Brisard, J.M. Costantini, M. Hage-Ali, J.P. Stoquert, F. Studer, and M. Toulemonde, Phys. Rev. B **48**, 920 (1993).

[19] M.P.R. Waligorski, R.N. Hamm, and R. Katz, Nucl. Tracks Radiat. Meas. **11**, 309 (1986).

[20] C. Dufour, B. Lesellier De Chezelles, V. Delignon, M. Toulemonde, and E. Paumier, Modifications induced by irradiation in glasses, edited by P. Mazzoldi (1992) Elsevier Science Publishers B.V. p. 61.

[21] N.W. Ashkroft and N.D. Mermin 1976, Solid State Physics (New York : Holt, Reinhart and Winston).

[22] V.V. Katin, Yu V. Martinenko and Yu N. Yavlinskii, Sov. Techn. Phys. Lett. 13 (1987) 276.

Materials Science Forum Vols. 248-249 (1997) pp. 57-60
© *1997 Trans Tech Publications, Switzerland*

Track Formation in Metals under Swift Heavy Ion Bombardment

Yu.N. Yavlinskii

Russian Research Centre "Kurchatov Institute", Nuclear Fusion Institute, 123182 Moscow, Russia

Keywords: Heavy Multicharged Ions, Excited Electrons, Tracks, Point Defects, Melting

Abstract. Swift heavy ions passing the crystalline lattice can create a cylindrical damage zone in the central track core. In metals this damage zone leads to various thermal phase transitions (melting). The condition of the track formation in metals is considered. Under heavy ion irradiation point defects in metals are created by elastic collisions whereas thermal phase transitions in the track are induced by the interaction of the electron gas (plasma state) with the atoms of the lattice. The calculation of the ion induced temperature in ideal cristalline iron is noticeable below the melting point and can not induced a track damage zone due to a phase transition solid state $\leftrightarrow$ liquid. Therefore, the experimentally observed ion tracks in metals are produced by radiation stimulated clustering of intrinsic lattice defects existing before irradiation.

The slowing down of a swift multicharged ion (MCI) in solids results from the energy loss due to inelastic collisions with electrons and elastic collisions with nuclei of target atoms. The elastic collisions dominate for small ion velocity $V \leq V_0$, V_F ($V_0 = e^2/\hbar$ is the Bohr velocity, V_F is the Fermi velocity for metals), while for $V \geq V_0$, V_F the slowing down results mainly from inelastic collisions (ionization and excitation of target electrons). Usually, the fast particle loses its energy due to ionization processes, and in the end of its path (when its velocity becomes less than V_0) the particle slows down due to elastic collisions. Practically almost all energy leads to excited electrons. The efficiency of the electron energy transfer to the solid lattice determines the heating of target atoms. The relaxation time of the excited electron subsystem (overheated electron gas) in metals is of some orders of magnitude smaller than the typical time of energy transfer from electrons to the lattice [1,2]. Therefore, the condition of the track damage formation depends on the crystal structure of the target and on its electrical characteristics. The appreciable lattice heating of a fine-grained metal [3] and of an insulator [4] up to melting was pointed out previously.

In a dielectric target usually there is created a track damage, which is the melting trace in the core. In ionic crystals the track can include a metallic phase (colloid) [5]. Such macroscopic track damage was observed in mica irradiated with uranium fission fragments [6] (i.e., heavy ions with the energy $E \sim 100 MeV$, the mass $M \sim 100au$, and the initial effective charge $Z_{eff} \sim 20$). In dielectric materials the cooling of the excited electrons is determined by the low temperature ionization wave [7], and the excited region enlarges very slowly. Really all the electron energy is transferred to the target atoms. The temperature of crystal lattice in the track core exceeds essentially the melting point, and the macroscopic track damage is formed.

In metal targets, the cooling of the excited electrons is determined by the nonlinear temperature waves [8]. As it will be shown below, the radius of the excited electron gas region is enlarged so much, that the electron temperature drops quickly, and the transfer of the electron energy to lattice leads to an insignificant rise of the lattice temperature if the crystal size is not very small. In the fine-grained metal crystals the electron heat appears to be locked, and the energy transfer to lattice is more effective. The above mentioned investigations [6] did not reveal the tracks under fission fragment bombardment of steel with the grain size of $\sim 20\mu m$. However, the change of the crystal structure and the melting trace are observed near the grain boundary. Nevertheless, the excitation density by irradiation with MCI is several times larger than that of fission fragments. Therefore new attempts [9,10] were undertaken to estimate the track formation in solids.

To explain the origin of the tracks (i.e., macroscopic damage zones), first of all, it is necessary to know the nature and the characteristics of the structure of the metallic target. Lattice atoms get energy from the excited electrons as a result of electron-phonon interaction. As it was shown in [8], the final lattice temperature of amorphous targets exceeds in $5 \div 10$ times the final lattice temperature of ideal single crystals. The single crystal has a periodic potential, which oscilates with the lattice constant. Electrons in single crystals are the Bloch electrons. If such electron moves in the periodic potential of the crystal, its wave function is adjusted to the lattice period, the electron wave "streamlines" the lattice nuclei and scatters weakly due to the thermal vibrations of atoms. Therefore the scattering cross-section of Bloch electrons on the lattice is much less than the corresponding cross-section in amorphous targets where the periodic potential is absent, and binary encounter of electrons and lattice atoms takes place. Just for this reason the energy transfer from excited electrons to the lattice is more effective in amorphous targets than in single crystals. Any distortion of the periodic potential in single crystals (point defects, grain boundaries, dislocations, infringements of stoichiometry in alloys, etc.) can lead to a local increase of the lattice temperature. In this case the interrupted track damage can be expected.

If the target is not a single crystal, the formation of the track damage or of appreciable infringements of the lattice is evident. The increase of the point defect density after irradiation with uranium ions in rolled and annealed iron ribbon [9] (which is not single crystal) can be interpreted as a result of a strong increase of the target temperature exceeding the melting point.

Below the track damage creation in an ideal metallic crystal by irradiation with uranium ions of $\sim 10 MeV/amu$ will be considered.

The outline of physical processes can be described as following Ref.[8]. The excited electrons acquire the temperature T, because the Maxwellian distribution is developed during the time of electron-electron collisions ($\sim 10^{-16}s$). According to the Saha formula the ionization equilibrium of Maxwellian electrons with the lattice ions is developed during the typical time of ionization ($\sim 10^{-15}s$). The lattice ions acquire an average charge which can be approximated by expression:

$$z(T) = \frac{\sum\limits_{k} C_k \cdot k}{\sum\limits_{k} C_k} = \kappa \cdot T, \tag{1}$$

where C_k is density of electrons ionized from k-th level, $n(T) = N \cdot z(T)$, $n(T)$, and N is the total electron and atomic density. Therefore the total energy of electron subsystem is the sum of "kinetic" energy $(3/2)z(T) \cdot T$ and "potential" energy (spent by ionization) $U(T)$ which can also be approximated by

$$U(T) = \sum\limits_{k} C_k \sum\limits_{n=2}^{k} I_n = \mu \cdot T^2, \tag{2}$$

where I_k is k-th ionization potential. Direct calculations of $z(T)$ and $U(T)$ for Au and Fe was carried out in [8]. Each the metal has its own values κ and μ which are determined by the consecutive ionization of atom electrons; κ and μ for different metals have a small variation.

The electron relaxation is determined by electron heat conductivity and electron-phonon interaction. As mention above the typical cooling time of the excited electrons due to the electron heat conductivity is much less than the typical time of the electron energy transfer to lattice atoms. Therefore, these two processes can be considered separately. At first the problem of electron cooling (as a result of the increasing of the excited electron region) will be solved, and then the function of $T(r,t)$ is used for the calculation of the lattice temperature. The dynamics of the initial cooling of the hot region is described by the heat conduction equation. The electronic thermal diffusivity $\chi(T)$ ($\chi(T)z(T)$ is the heat conductivity per atom) calculated in [11] is approximated satisfactorily in the range $\varepsilon_F \leq T \leq T_0$ as

$$\chi = \gamma \cdot T^{3/2}, \tag{3}$$

where $\gamma \simeq 3/\varepsilon_F^{3/2}$, ε_F is the Fermi energy, and the initial temperature of the excited electrons T_0 is equal to

$$T_0 = \sqrt{[\pi(1.5\kappa + \mu)r_0^2 N]^{-1} \cdot (dE/dx)}, \tag{4}$$

where $r_0 = 10\text{Å}$ is the initial radius of the cylindrical region of the excited electrons, and dE/dx is the energy loss.

As far as the thermal diffusivity $\chi(T)$ Eq.3 is the function of the electron temperature, the heat conduction equation appears to be nonlinear:

$$\frac{\partial T^2(r,t)}{\partial t} = b\frac{1}{r}\frac{\partial}{\partial r}\left(T^{3/2}(r,t)r\frac{\partial T^2(r,t)}{\partial r}\right). \tag{5}$$

The Eq.5 was solved analytically [8], and the so-called self-similar solution is a nonlinear temperature wave, which can be written in variables $(r,\ t)$ as

$$T(r,\ t) = T_0 \cdot (t_0/t)^{2/7}[1 - (t_0/t)^{4/7} \cdot (r/r_0)^2]^{2/3}, \tag{6}$$

where $t_0 = (3/28)(r_0^2/bT_0^{3/2})$ is a parameter of the self-similar solution development, and $b = \gamma\kappa/(3\kappa + 2\mu)$. The parameter t_0 for iron is equal to $\sim 10^{-15}s$ that corresponds with initial condition of the task as far as the time is counted from $t = t_0$. The space dependence of electron temperature $T(r,\ t)$ is almost a right-angled step moving in the direction of variable r with increase of the time t. The height of the step is decreased simultaneously as $(t_0/t)^{2/7}$. The drop of temperature $T(r,\ t)$ at $r = 0$ happens more slowly than in the case of linear heat conduction according to $T(t) \propto 1/\sqrt{t}$ [1].

Following [8] the lattice temperature Θ_L is equal to

$$\Theta_L = \sqrt{\frac{100}{3}\frac{m}{M_2}\frac{\sigma}{\hbar N^{1/3}\gamma}\frac{dE}{dx}\Theta_0 + \Theta_0^2}, \tag{7}$$

where σ is the cross-section of electron-isolated atom interaction [1], M_2 is the target atom mass. Using Eq.7, a universal condition of thermal track damage formation in single crystals can be obtained for usual and refractory metals with $Z_2 \geq 20$. Among the applied projectiles, the uranium MCI creates the largest damage in solids and all other ions will have a less efficient radiation damage effect. The energy loss dependence on the energy of the ion is a curve having a maximum, which is distinguished for targets with different nucleus charge Z_2. Calculations lead to the following approximation

$$(dE/dx)_m \cong 0.72 \cdot \frac{M_2 N}{\sqrt{Z_2}}, \tag{8}$$

which describes with a good accuracy the maxima of (dE/dx) from the results of [12]. The coefficient 0.72 has a dimension of cm^4/s^2.

After inserting Eq.8 into Eq.9 the lattice temperature is equal to

$$\Theta_L \cong 9.13 \cdot 10^{-8}\frac{N^{1/3}\Theta_0^{1/2}}{\alpha Z_2^{1/4}}, \tag{9}$$

where Θ_0 is the initial temperature of the lattice, i.e., the temperature, at which irradiation is performed. Usually, Θ_0 is chosen to be small to freeze the lattice changes and to exclude annealing. In [9] the value of $\Theta_0 = 1.3 \cdot 10^{-3} eV$ that corresponds to a temperature of 15K. The dimensionless quantity $\alpha = e^2/\hbar V_F \geq 1$ is a frequently used parameter of metals, which for most metals vary by a factor of 1.5. For example, for iron $\alpha_{Fe} = 1.1$, and for gold $\alpha_{Au} = 1.5$. The numerical factor $9.13 \cdot 10^{-8}$ has a dimension of $(eV)^{1/2} \cdot cm$. The final temperature Θ_L has a very weak dependence on Z_2.

The Eq.9 determines the upper limit of the lattice temperature. Therefore, if the obtained value of Θ_L is less than the melting point, the formation of a melted track damage zone is impossible. Such condition is satisfied for most metals with the exception of alkali and some fusible metals. For example, for iron with the atomic density $N = 8.5 \cdot 10^{22} cm^{-3}$, $\alpha = 1.1$, $Z_2 = 26, and\ \Theta_0 = 1.3 \cdot 10^{-3} eV$ this corresponds to $\Theta_L \cong 0.06 eV$. The estimated value of the lattice temperature is 660K which is much smaller than the melting point. The increase of Z_2 will entail a weak drop of the lattice temperature while other parameters have little changes.

There exists also another mechanism of lattice heat creation. The MCI creates point defects along its trajectory. As it was shown in [8] the electron energy transfer to point defect rises for $5 \div 10$ times of that in an ideal crystal. If the point defect density is so large that it is near to the atomic density, melting can be possible. However this situation can be realized only at a projectile fluence higher than $10^{16} cm^{-2}$, because the average impact parameter of swift MCI is of the order (or less) of the atomic size, and the MCI produces point defects only in the track core (trajectory line). At present such experiments are not known.

A correct explanation of the relaxation processes in metals under swift MCI irradiation is possible only if the target is an ideal or quasi-ideal crystal. Each experiment with fine-grained, amorphous, or complex targets requires an individual interpretation and analysis.

References

[1] Yu.V.Martynenko, Yu.N.Yavlinskii, Sov. At. Energy **62** , 93 (1987).

[2] Yu.N.Yavlinskii, Nucl. Instrum. & Methods in Physics Research B**90**, 96 (1994).

[3] Yu.V.Martynenko, Yu.N.Yavlinskii, Poverkhnost No.6, 5 (1988).

[4] V.V.Katin, Yu.V.Martynenko, Yu.N.Yavlinskii, Zh. Tech. Fiz. **59**, 88 (1989).

[5] K.Schwartz, Nucl. Instrum. & Methods in Physics Research B**107**, 128 (1996)

[6] I.M.Lifshitz, M.I.Kaganov, L.V.Tanatarov, At.Energ. **6**, 391 (1959).

[7] V.V.Katin, Yu.V.Martynenko, Yu.N.Yavlinskii, Zh. Tech. Fiz. **61**, 68 (1991).

[8] Yu.N.Yavlinskii, Nucl. Instrum. & Methods in Physics Research B**115**, 594 (1996).

[9] A.Dunlop, D.Lesueur, P.Legrand, H.Dammak, J.Dural, Nucl. Instrum. & Methods in Physics Research B**90**, 330 (1994).

[10] Z.G.Wang, Ch.Dufour, E.Paumier, M.Toulemonde, J.Phys.: Condens. Matter **6**, 6733 (1994).

[11] Yu.V.Martynenko, Yu.N.Yavlinskii, Sov. Phys. Dokl. **28**, 391, (1983).

[12] F.Hubert, R.Bimbot, and H.Gauvin, At.Data & Nuclear Data Tabl. **46**, 1 (1990).

Materials Science Forum Vols. 248-249 (1997) pp. 61-64
© *1997 Trans Tech Publications, Switzerland*

Diffusion in Ion Irradiated Nickel: Determination of Atomic Mixing, Sink Strength and Fraction of Freely Migrating Defects

P. Fielitz, M.-P. Macht, V. Naundorf and H. Wollenberger

Hahn-Meitner-Institut Berlin, Glienickerstr. 100, D-14109 Berlin, Germany

Keywords: Ion Irradiation, Atomic Mixing, Radiation-Enhanced Diffusion, Freely Migrating Defects, Effective Sink Strength

Abstract

Diffusion coefficients under heavy ion irradiation at 77 K, 800 K and 950 K were measured in nickel using specimens with several inserted thin layers of ^{63}Ni or Co tracer in different depth. At 77 K the diffusion coefficient follows closely the depth dependence of the calculated displacement rate indicating that athermal atomic mixing is responsible for the observed transport at this temperature. Considerable self-diffusion is observed at 800 K and 950 K even far from the irradiated surface in a region were the displacement rate is negligible indicating long range diffusion of mobile radiation-induced point defects. The fit of a diffusion model for these defects to the observed depth dependence of the radiation-enhanced self-diffusion coefficients yielded effective values of the sink concentration and the fraction of freely migrating defects. In the non-damaged region of the specimens the sink concentration amounts to less then 4×10^{-7}, while in the damaged region near the surface it is of the order of 10^{-6} to 10^{-5}. The fraction of freely migrating defects is estimated to be between 0.7% and 7%, depending on the kind of irradiation.

1. Introduction

Energetic ions are frequently used in order to study irradiation effects in structural materials of fission and fusion reactors. These irradiation effects include microstructural changes of the material, e.g. void formation or phase transformations. The microstructural changes originate essentially from the radiation-induced atom transport [1]. Hence, one of the main objects of radiation damage modeling is the proper understanding of the relation between the defect production process and the long-range atom transport.

The interpretation of transport effects caused by cascade forming ion irradiation is considerably impeded by two specific properties of this type of radiation damage. These are i) simultaneous production of mobile point defects and of an immobile microstructure, which acts as defect sink, and ii) damage production restricted to a volume near to the surface. The first property has been discussed in detail for the steady state conditions [2,3]. This discussion, however, did not consider the possible influence of the surface and the inhomogeneous damage distribution on atom transport, mainly because experimental data were not available. The present paper reports first results of spatially resolved transport measurements at low and high temperatures near the surface of nickel irradiated with heavy ions.

2. Experimental Procedure

The investigation was performed using two types of nickel multilayer specimens. One was grown on a sapphire substrate 12×12 mm^2 wide by repeating five times an alternate vapor deposition of about 50 nm nickel with a purity of 5N and sputter deposition of about 1 nm of cobalt. Before layer deposition the substrate was covered with about 300 nm of nickel. These specimens were used for the investigation of atomic mixing at 77 K. Since Ni and Co are neighbors in the

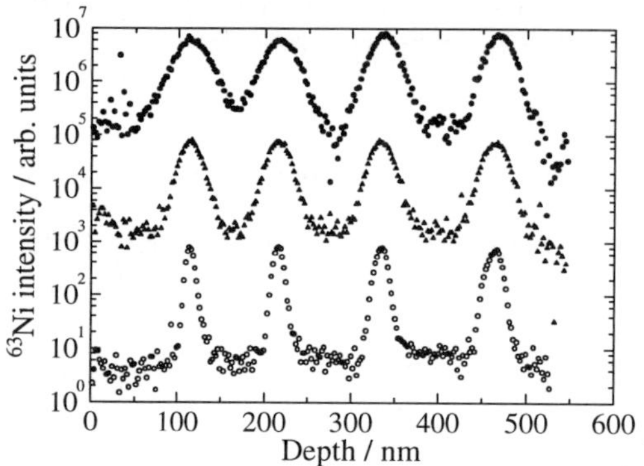

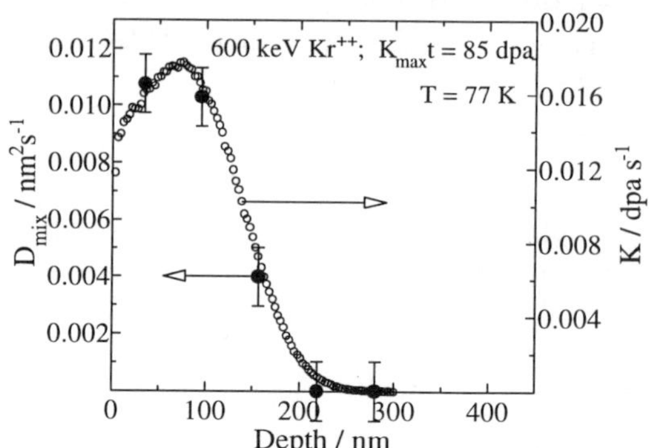

Fig. 1 ^{63}Ni-Intensity of a Ni/^{63}Ni layer structure as measured by SIMS. Open circles: as prepared; closed triangles: 950 K; closed circles: 950 K, 300 keV Ne$^+$.

Fig. 2 Depth dependence of the mixing diffusion coefficient D_{mix} (data points with error bars) and of the calculated displacement rate K (open circles).

periodic table and their diffusion behavior is quite similar [4] atomic mixing of Co in nickel should represent that of the nickel matrix itself [5]. For irradiations at 800 K and 950 K another type of specimens was used. These were grown on a well-polished Ni single crystal with a diameter of 12 mm cut about 5° off the (112) zone axis. Four layers about 1 nm thick of the ^{63}Ni tracer were sputter deposited between nickel layers about 100 nm thick. Electron microscopic investigation showed that the specimens grown on the sapphire substrate were polycrystalline with a grain size of about 50 nm, while the layer structure grown on the single crystal had the orientation of the base crystal.

The specimens were partly irradiated on an area of about 0.2 cm^2 through an appropriate aperture. Atom transport was measured through the broadening of the thin tracer layers by sputter sectioning in combination with secondary ion mass spectrometry (SIMS). Details of this measuring technique and a discussion of its depth resolution are given elsewhere [6,7]. As an example Fig. 1 presents the concentration depth profiles of ^{63}Ni as measured by SIMS before (open points) and after 300 keV Ne$^+$ irradiation at 950 K (closed points) at one and the same specimen. While the measurement performed on the as-prepared specimen before irradiation demonstrates the depth resolution of the SIMS technique [7], the broadening observed after the experiment in the non-irradiated area of the specimen (closed triangles) indicates thermal diffusion, and that observed in the irradiated area of the specimen (closed circles) shows the combined effect of three contributions, atomic mixing, thermal and radiation-enhanced diffusion. In terms of the effective diffusion coefficient under irradiation, D_{irr}, it is assumed that $D_{irr}=D_{mix}+D_{therm}+D_{rad}$ is valid.

Effective diffusion coefficients were derived from the variance of each of the layers in the different depths by fitting a Gaussian curve to the respective peak. If σ^2 is the variance of the Gaussian curve obtained in the irradiated area and σ_0^2 that in the non-irradiated area of the same specimen, then $D_{mix}+D_{rad}=(\sigma^2-\sigma_0^2)/2t$, where t is the irradiation time.

3. Results and discussion

Fig. 2 presents the depth dependent diffusion coefficients derived from a 600 keV Kr^{++} ion irradiation at 77 K together with the corresponding depth dependent displacement rate K(x) (given in units of displacements per atom and per second, dpa/s), which was calculated by means of the binary collision code TRIM [8] using a displacement threshold of 40 eV [9]. Since at low temperatures thermal mobility of the defects is strongly reduced, this measurement yields the diffusion coefficient of atomic mixing, D_{mix}. The coincidence of D_{mix} and K in Fig. 2 is obvious. The average value of the mixing efficiency $D_{mix}/K=0.67\pm0.12$ nm^2/dpa is derived within the first 150 nm below the surface. The magnitude of the mixing efficiency indicates that thermal spike mixing [5] is responsible for the observed effect. Beyond a depth of about 200 nm, i. e outside the damaged region, no broadening of the layers could be resolved. For irradiation with 300 keV Ar$^+$ and Ne$^+$ ions on Ni at 77 K the mixing efficiency was 0.56±0.09 nm^2/dpa and 0.42±0.08 nm^2/dpa, respectively.

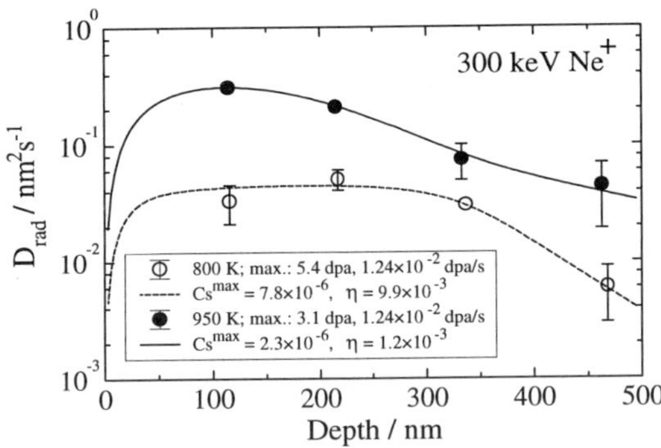

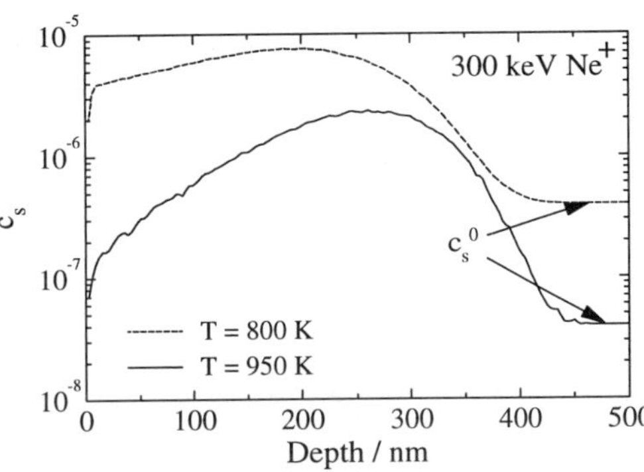

Fig. 3 Depth dependence of the radiation enhanced diffusion coefficient (data points with error bars) for irradiation with 300 keV Ne$^+$ at two temperatures. Solid and dashed lines are fits of Eq. 1 of the text.

Fig. 4 Depth dependence of the effective concentration of sinks used for a fit of Eq. 1 to the data shown in Fig. 3.

The depth dependence of the diffusion coefficient under 300 keV Ne$^+$ irradiation at 800 K and 950 K is presented in Fig. 3. Considerable transport of the nickel tracer is observed beyond 450 nm, i. e. in a region where the displacement rate is negligible small. The obvious reason for this result is the thermally activated mobility of the freely migrating defects which at the temperature of irradiation can escape the production region by long range diffusion.

The radiation-enhanced diffusion is essentially caused by the enhanced concentration $c_{i,v}$ of radiation-induced and freely migrating single interstitials and vacancies (subscripts i and v, respectively). The corresponding diffusion coefficient of the radiation-enhanced self-diffusion is $D_{rad}=f_v D_v c_v + f_i D_i c_i$, where $D_{i,v}$ are the diffusion coefficients of the defects and $f_{i,v}$ are the correlation factors. According to refs. [1,10] D_{rad} can be approximately derived from a diffusion equation for the point defects, if steady state is assumed, and different production rates of freely migrating vacancies and interstitials as well as an annihilation bias for these defects at sinks are neglected. If it is moreover assumed that the production rate of freely migrating defects, $\eta K(x)$, is proportional to the depth dependent displacement rate $K(x)$, then the spatial dependence of $D_{rad}(x)/(f_i+f_v)=C(x)$ is given by the diffusion equation [1]

$$\eta K(x) - \frac{K_{iv}}{D_i D_v} C^2 - \frac{K_{vs}}{D_v} c_s(x) C + \frac{\partial^2 C}{\partial x^2} = 0. \tag{1}$$

with K_{iv} and K_{vs} the rate constants for recombination and annihilation of vacancies at sinks, respectively [1], and $c_s(x)$ the depth dependent sink concentration in atomic units. The influence of point defect annihilation at the surface is taken into account by the boundary condition of $C(x=0)=0$, which is also assumed in very large distance from the irradiated surface.

Eq. (1) is approximately solved by an exponential decay of the diffusion coefficient for negligible defect production and a constant sink concentration as prevails beyond the damaged region in larger depths. An exponential slope of the diffusion coefficient was indeed observed beyond a depth of about 200 nm in 600 keV Kr^{++} irradiations at 800 K [10]. The corresponding

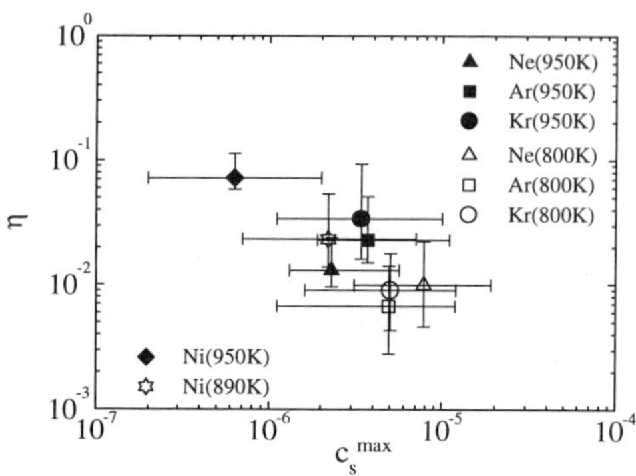

Fig. 5. Fraction of freely migrating defects η and maximum effective concentration of sinks c_s^{max} for irradiation with different ions at different temperatures.

decay length is proportional to $(c_s^0)^{-1/2}$ and yielded a sink concentration of 2×10^{-7} which may be regarded as typical for the Ni single crystals.

In the region of defect production a considerable higher effective concentration of sinks is expected. These are produced in the form of immobile defect clusters and dislocation loops either directly as cascade remnants or by defect reactions which in the case of noble gas irradiation may include reactions with the implanted gas. The depth dependence of the sinks is only incompletely known. Electron microscopic investigations of dislocation loops in He$^+$ ion irradiated Ni indicate that the centre of the loop density shifts into larger depth for irradiations performed at higher temperatures [11]. Since at lower temperature (e. g. 800 K in Ni) these defect clusters and loops are reasonably stable the respective sink distribution can be assumed to be produced as a fixed fraction of the displacement rate. At higher temperatures (e. g. 950 K in Ni) defect clusters which are not stabilized by the implanted gas atoms can dissolve thermally and the sink distribution is expected to resemble the implantation profile of the irradiating noble gas. Fig. 4 presents the sink distribution as used for a fit of Eq. 1 to the data shown in Fig. 3.

The parameters, maximum sink concentration, c_s^{max}, and fraction of freely migrating defects, η, in Eq. 1 were determined by a numerical fit procedure. Fig. 5 shows the optimum fit parameters for 300 keV Ne$^+$, Ar$^+$ and Ni$^+$, and for 600 keV Kr^{++} irradiation at different temperatures. As can be seen in particular for the result of the self-ion irradiation at 950 K the uncertainty of the derived sink concentration is much larger than that of the fraction of freely migrating defects. This indicates that the actual value of the sink concentration is of minor importance for the observed diffusion coefficients near the surface. This behavior can be understood by inspection of Eq. 1 which shows that a large diffusion term can indeed suppress the effect of recombination and sinks [10]. The data presented in Fig. 5 suggest a tendency of an increasing fraction of freely migrating defects and decreasing sink concentration at higher irradiation temperatures.

References

[1]   R. Sizmann, J. Nucl. Mater. **69+70**, 386 (1978).

[2]   H. Trinkaus, V. Naundorf, B. N. Singh, and C. H. Woo, J. Nucl. Mater. **210**, 244 (1994).

[3]   V. Naundorf and H. Wollenberger, J. Nucl. Mater. **212-215**, 226 (1994).

[4]   G. Neumann, in Diffusion in Solid Metals and Alloys, H. Mehrer, ed., Landold-Börnstein, New Series **III/26**, p. 183 (Springer, Berlin, 1990).

[5]   Y.-T. Cheng, Mater. Sci. Rep. **5**, 47 (1990).

[6]   A. Müller, V. Naundorf, and M.-P. Macht, J. Appl. Phys. **64**, 3445 (1988).

[7]   M.-P. Macht, R. Willeke, and V. Naundorf, Nucl. Instr. Meth. Phys. Res. **B43**, 507 (1989).

[8]   J. P. Biersack and L. G. Haggmark, Nucl. Instr. Meth. **174**, 257 (1980).

[9]   ASTM/E 521-89, Standard Practice for Neutron Radiation Damage Simulation by Charged-Particle Irradiation. Philadelphia. PA: Am. Soc. Test. Mat., 167 (1989).

[10]   P. Fielitz, M.-P. Macht, V. Naundorf, and H. Wollenberger, Appl. Phys. Lett. **69**, 331 (1996).

[11]   K. Niwase, T. Ezawa, T. Tanabe, M. Kiritani and F. E. Fujita, J. Nucl. Mater. **203**, 56 (1993).

Ion Beam Modifications

Materials Science Forum Vols. 248-249 (1997) pp. 67-72
© 1997 Trans Tech Publications, Switzerland

Ion-Beam Modifiction of Semiconductors and Related Electronic Materials

R.G. Elliman

Electronic Materials Engineering Department, Research School of Physical Sciences and Engineering, Institute of Advanced Study, Australian National University, Australia

Keywords: Ion-Implantation, Semiconductors, Electronic Materials, Optoelectronics

Abstract

This paper reviews the ion-implantation research program at the Australian National University. The emphasis of the research is on the modification and analysis of semiconductors and related electronic and optoelectronic materials, and includes both device and materials based studies. These are discussed under the headings: a) solid-phase epitaxy, b) defect engineering, doping and gettering, c) the synthesis of alloys and compounds, and d) optical materials.

Introduction

Ion-implantation is one of the key technologies in modern semiconductor-device fabrication, where the most routine application is that of doping Si-based integrated circuits, which typically involves implant energies less than 200 keV. However, over recent years, an increasing number of novel applications have emerged, many involving high-energy (MeV) ion-implantation [1]. These include the fabrication of advanced Si-based devices [1] as well as the fabrication of advanced electronic and optoelectronic devices in III-V compound semiconductors [1]. In addition to these important applications, ion-implantation is also a very effective tool for studying the physics of materials. It enables controlled concentrations of impurities or defects to be added to materials and, as such, can be used to make unique material structures for subsequent studies. The Electronic Materials Engineering Department at the Australian National University (ANU) undertakes both device-related and materials-related research. This ranges from point defect studies, requiring ion fluences less than 1×10^7 ions.cm^{-2} [2], to the formation of buried compound and alloy layers such as SiO_2 [3] and GeSi [4], respectively, requiring doses up to 1×10^{18} ions.cm^{-2}. It also involves a broad range of implant energies, which range from 15 keV to greater than 10 MeV. A selection of these applications is discussed in this paper.

Experimental

The studies outlined below were conducted at ANU using an NEC 5SDH-4 tandem accelerator with a SNICS negative-ion source. The machine has a maximum terminal voltage of 1.7 MV and in normal operation employs N_2 stripping gas to charge exchange ions as they pass through the terminal. For low energy ions, less than 100 keV, the accelerator is operated in an accel-decel mode in which the stripping gas is turned off and a terminal voltage of ~300 kV is employed. Negative ions from the injector are therefore accelerated and decelerated as they pass through the tandem, exiting with the injection energy. This mode of operation provides greater transmission of low energy negative ions.

Results and Discussion
ANU research activities are discussed under the following broad headings: a) solid-phase epitaxy, b) defect engineering, doping and gettering, c) ion-beam synthesis of compounds and alloys, and d) optical materials. It is not possible to provide detailed discussion in a review of this nature nor is it possible to cover all aspects of the research undertaken by the ANU group. Rather, a selection of topics is presented to illustrate the range of research interests. For further detail the reader is referred to the cited papers.

a) Solid-Phase Epitaxy
The solid-phase epitaxial crystallisation (SPEC) of amorphous semiconductor layers is an interesting and technologically important process which has been studied for over twenty years [5-7]. Our group has contributed to the understanding of SPEC, and especially to the understanding of doping effects [8] and ion-beam induced SPEC [9]. More recent studies have concentrated on SPEC in GaAs [10, 11] and in GeSi alloy layers, including GeSi/Si strained-layer heterostructures [4, 12] and strain-relaxed GeSi layers [13, 14].

SPEC in Si and Ge is thermally activated, being well characterised by a single activation energy of 2.7 eV and 2.0 eV for Si and Ge, respectively. The crystallisation rate is also much faster in Ge than in Si. Bulk GeSi alloys might therefore be expected to have a crystallisation rate which increases with increasing Ge content and an activation energy which varies monotonically from 2.7 eV to 2.0 eV as the alloy changes from pure Si to pure Ge. Our studies show that although the crystallisation rate increases with increasing Ge concentration, the activation energy for crystallisation does not vary monotonically but increases from 2.7 eV to a maximum of 2.8 eV for a Ge concentration of 20-30% before decreasing towards 2.0 eV. The activation energy and pre-exponential factor for the crystallisation velocity are shown in Figs. 1 and 2. Two models have been proposed to account for this behaviour: a microscopic strain model [14], and a model based on the probability of bond-breaking [15]. In the former, the energy required for bond-breaking, a major component of the activation energy, is assumed to be increased by the microscopic strain associated with bonding between atoms of different bond length, whilst in the latter model, the increase in the activation energy is viewed as an artefact arising from the temperature and concentration dependence of the crystallisation velocity. Further investigation is required to resolve this issue.

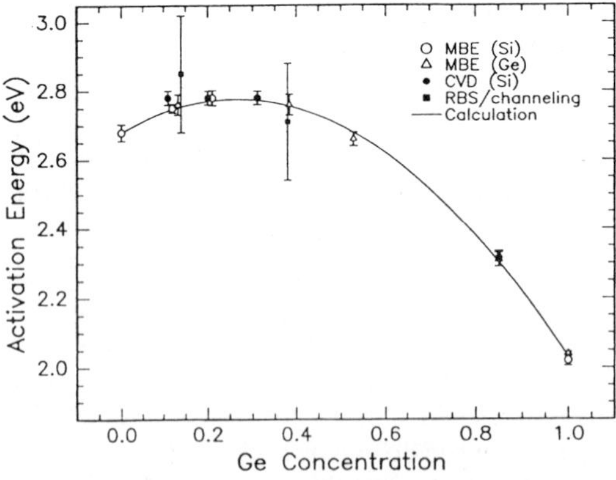

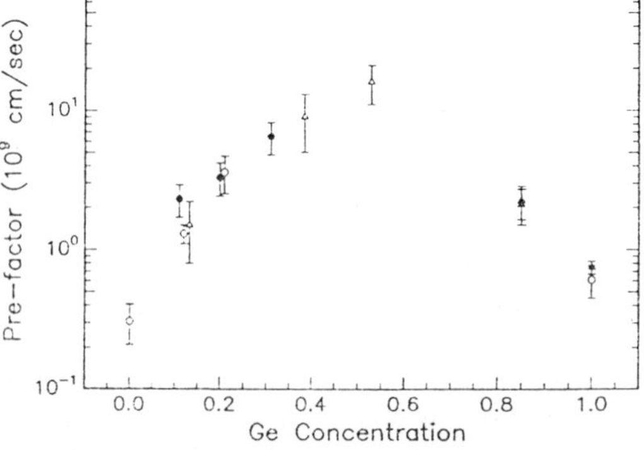

Fig. 1: Activation energy for solid-phase epitaxial crystallisation velocity of GeSi alloys as a function of Ge fraction [14].

Fig. 2: Pre-exponential factor for solid-phase epitaxial crystallisation velocity of GeSi alloy layers as a function of Ge fraction [14].

b) Defect Engineering, Doping and Gettering

Recent studies in this area include: the effect of ion-implantation damage on strain relief in GeSi/Si strained layer heterostructure [16]; transient-enhanced diffusion of B in ion-implanted Si [17, 18]; point-defects and their interactions in Si [2, 19]; electrical isolation [20, 21], doping [22, 23] and quantum-well mixing [24-25] in III-V compound semiconductor structures; and impurity gettering to voids in Si [26-27].

Gettering is used to remove impurities from the device region of Si wafers. This is usually achieved by precipitating SiO_2 within the wafer, by heavily doping the backside of the wafer, or by introducing defects into selected regions of the wafer. Recent work has shown that the cavities produced by hydrogen implantation of Si are effective at gettering metallic impurities such as Au and Cu [16, 27].

Cavities are typically formed by implanting with 100 keV H ions to a fluence of $3 \times 10^{16} cm^{-2}$ and then annealing at 850°C for 1 hour. Hydrogen is released from the cavities during this anneal, leaving a band of faceted cavities at a depth of ~1 μm. The effectiveness of these cavities as gettering sites is illustrated in Figs. 3 and 4 which show Au preferentially trapped at the voids. Au was implanted into samples containing cavities and moved from the surface to the cavities during subsequent annealing at 850°C for 1 hour. Detailed analysis shows that the Au initially adsorbs on the internal surfaces of the cavities and forms the bulk Au phase only after the internal surfaces are saturated.

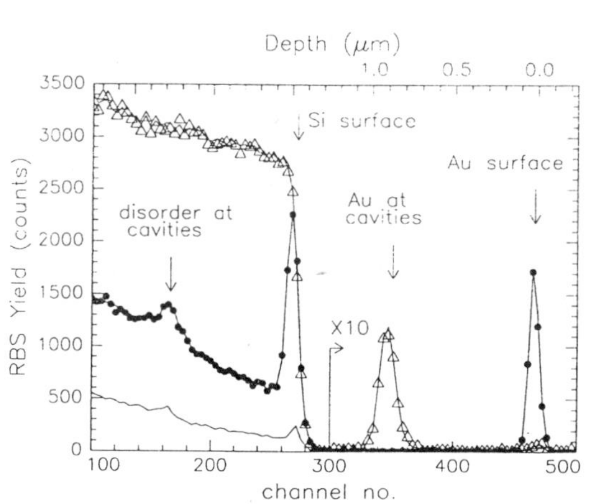

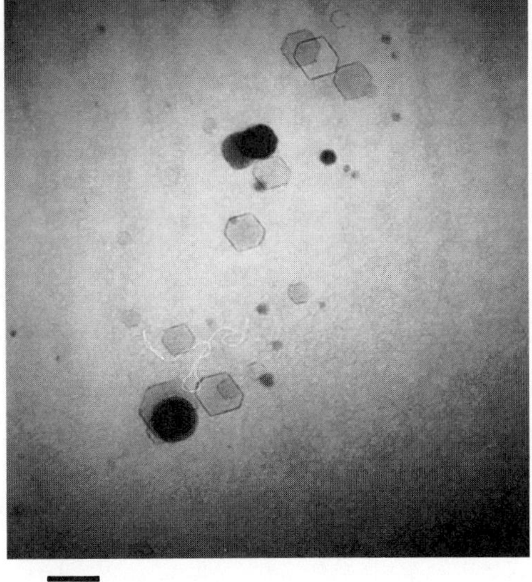

200 Å

Fig. 3: Rutherford backscattering spectra showing the trapping of Au at hydrogen induced cavities. Channelled spectra are shown for samples as-implanted (solid circles) and after annealing at 850°C for 1 hr (Si-only) (solid line). The random spectrum is after annealing at 850°C for 1 hr (open triangles).

Fig. 4: Transmission electron micrograph of Au trapped at cavities and precipitates in Si after annealing to 850°C for 1 hr. The contrast around the faceted cavities arises from absorbed Au. The dark objects are Au precipitates

c) Compound and Alloy formation

High dose ion-implantation can be used to form alloys and compounds. The ANU group has studied the synthesis of SiO_2 layers [3, 28] and Ge_xSi_{1-x} strained-layers [4, 12, 29]. SiO_2 layers were produced by the SIMOX method using MeV O ions to minimise the damage to the overlying Si layer, whilst Ge_xSi_{1-x} strained layers were produced by ion-implantation and solid-phase epitaxy. The latter is chosen as an example.

Fully strained GeSi alloy layers can be fabricated in Si by Ge implantation and annealing if the Ge concentration is kept below a critical value [12]. To test the usefulness of such alloys, a series of test devices was fabricated and their performance compared to identical devices without the GeSi alloy layer [29]. Figure 5 shows a schematic of a field-effect transistor employed for this comparison. The channel mobility was determined for a range of implant conditions and is plotted in Fig. 6. This showed that Ge implantation could improve the mobility, and hence the speed of the FETs by up to 40%. However, if the Ge concentration exceeded the critical value the mobility decreased due to the presence of strain relieving dislocations.

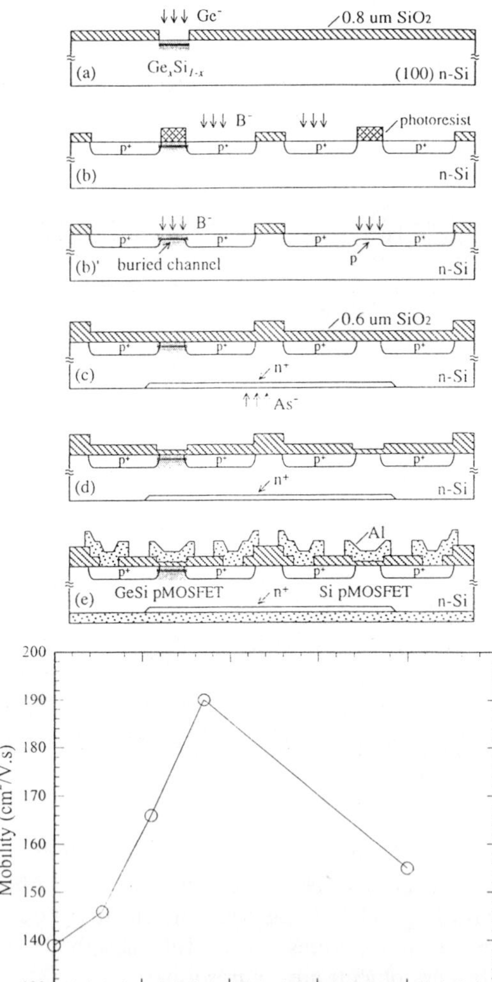

Fig. 5: Schematic representation of the fabrication of GeSi-channel p-MOSFETs by Ge implantation. a) Ge implantation for GeSi alloy formation, b) source/drain p$^+$ doping, c) p-doping for buried-channel transistors, SiO_2 deposition and backside n$^+$ doping, d) gate opening and thermal oxide growth, and e) opening of contact windows and metallisation.

Fig. 6: Hole mobilities as a function of as-implanted Ge concentration determined from transconductance measurements.

d) Optical Materials

Studies of optical materials include: the formation of optical waveguides [30-31], the fabrication of devices, including quantum wire lasers [25], and the study of non-linear optical properties of metallic and semiconducting nanocrystals [32].

Optical waveguides have been produced in a range of materials by ion-implantation [33]. However, a detailed understanding of the response of these materials to ion-irradiation and subsequent annealing is still lacking. Three processes have been identified in ion-irradiated quartz: densification, viscous flow and anisotropic deformation. Densification is thought to be responsible for the increase in refractive index and is therefore an important consequence of irradiation. Fig. 7 shows the effect of annealing on compaction and light absorption. Another important response which has recently been observed is cracking [31]. This occurs when samples, irradiated at low temperatures to fluences which maximise the internal stress within the film, are warmed to room temperature. Fig. 8 shows the consequences of such irradiation.

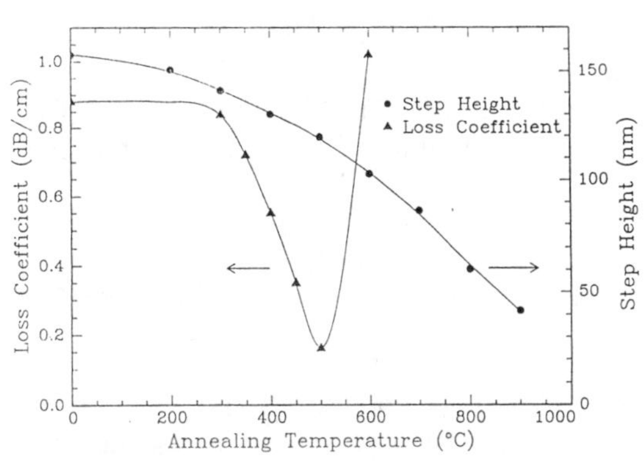

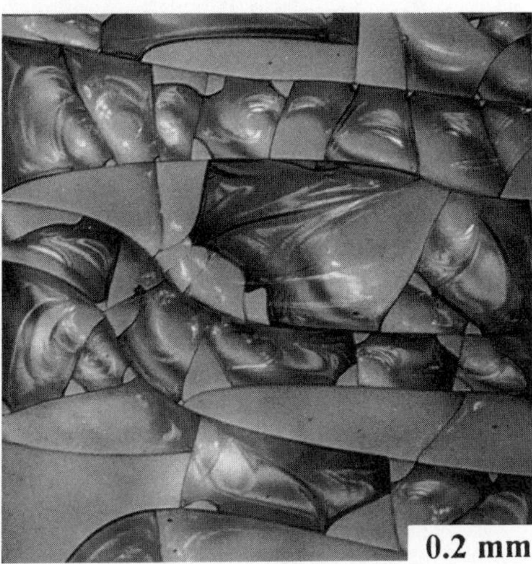

Fig. 7: Step height and loss coefficient (1.3 μm) as a function of annealing temperature for quartz implanted at -196°C with 5 MeV Si ions to a fluence of 2×10^{15} cm^{-2}.

Fig. 8: Optical micrograph of quartz after implantation at -196°C with 5 MeV C ions to a fluence of 8×10^{13} C.cm^{-2} and warmed to RT.

Conclusions

A brief summary has been presented of ion-implantation based research projects conducted at ANU using an NEC 5SDH Tandem accelerator. The emphasis of the research is on electronic and optoelectronic materials and devices, and examples were selected from the following areas: a) solid-phase epitaxy, b) defect engineering, doping and gettering, c) ion-beam synthesis of alloys and compounds, and d) optical materials.

Acknowledgments

The results presented in this paper represent the activities of the Electronic Materials Engineering Department. It is a pleasure to acknowledge the contributions of Prof. J.S. Williams, Dr. M.C. Ridgway, Dr. C. Jagadish, and students Mr. W.C. Wong, Mr. J. Glasko, Ms. J. Wong-Leung, Ms. A. Dowd, Mr. H. Tan, Ms. S. Fatima, Mr. K. Belay and Mr. C. Johnson.

References

1.   see for example, the proceeding of the international conference series on Ion Implantation Technology, published in Nuclear Instruments and Methods.
2.   J. Lalita, C. Jagadish, and B.G. Svensson, *Nucl. Instr. Meth.* **B106**, 183 (1995).
3.   S.L. Ellingboe and M.C. Ridgway, *J. Appl. Phys.* **75**, 4939 (1994).
4.   R.G.Elliman, W.C.Wong and P.Kringhoj, *Mat.Res.Soc.Symp.Proc.* **321**, 375-385 (1994).
5.   G.L. Olson and J.A. Roth. *Mat. Sci. Rep.* 3, 1 (1988).
6.   G.L. Olson and J.A. Roth. Handbook of Crystal Growth 3, Elsevier, NY (1994)
7.   O. Hellman, *Mat. Sci. & Eng.*, **R16**, 1 (1996).
8.   J.S. Williams and R.G. Elliman, *Phys. Rev. Lett.*, **51**, 1069 (1983).
9.   J. Linnros, R.G. Elliman and W.L. Brown, *J. Mat. Res.*, **3**, 1208 (1988).
10.   K.B. Belay, D.J. Llewellyn and M.C. Ridgway, *Mat. Res. Soc. Symp. Proc.* **398**, 393 (1996).
11.   K.B. Belay, D.J. Llewellyn and M.C. Ridgway, *Appl. Pjys. Lett.* (In Press) (1996).
12.   R.G. Elliman and W.C. Wong, *Mat. Sci. Forum*, **143-147**, 507 (1994)
13.   P.Kringhoj and R.G. Elliman, *Phys. Rev. Lett.*, **73**, 858 (1994).
14.   P. Kringhoj, R.G. Elliman, M. Fyhn, S.Y. Shiryaev and A. Nylandsted-Larsen, *Nucl. Instr. Meth.* B106, 346 (1995).
15.   T.E. Haynes, M.J. Antonell, C. Archie Lee and K.S. Jones, *Phys. Rev.* **B51**, 7762 (1995)
16.   P. Kringhoj, J.M.Glasko and R.G.Elliman, *Nucl. Instr. Meth.* **B96**, 276 (1995).
17.   K.S. Jones, R.G. Elliman, M. Petravic and P. Kringhoj, , *Appl. Phys. Lett.*, **68**, 3111 (1996)
18   K.S. Jones, V. Krishnamoorthy, L.H. Zhang, M. Law, D.S. Simons, P. H. Chi, L. Rubin, and R.G. Elliman, *Appl. Phys. Lett.* **68**, 2672 (1996).
19.   H.H. Tan, J.S. Williams and C. Jagadish, *J. Appl. Phys.* **78**, 1481 (1995).
20.   M.C. Ridgway, R.G. Elliman and N. Hauser, *Nucl. Instr. Meth* **B80/81**, 636 (1993)
21.   M.C. Ridgway, S.L. Ellingboe, R.G. Elliman and J.S. Williams, *Nucl. Instr. Meth.*, **B89**, 290 (1994)
22.   M.C. Ridgway and P. Kringhoj, *J. Appl. Phys.*, **77**, 2375 (1995).
23.   M.C. Ridgway, P. Kringhoj and C.M. Johnson, *J. Appl. Phys.*, **79**, 7545 (1996)
24.   H.H. Tan, J.S. Williams, C. Jagadish, P.T. Burke and M. Gal, *Appl. Phys. Lett.*, **68**, 2401 (1996)
25.   H.H. Tan, F. Karouta, J.S. Williams and C. Jagadish, Submitted to *Appl. Phys. Lett.* (1996)
26.   J. Wong-Leung, C.E. Ascheron, M. Petravic, R.G. Elliman and J.S. Williams, *Appl. Phys. Lett.*, **66**, 1231 (1995)
27.   J. Wong-Leung, J.S. Williams and E. Nygren, *Nucl. Instr. Meth.*, **B106**, 424 (1995)
28.   S.L. Ellingboe and M.C. Ridgway, *Nucl. Instr. Meth.*, **B106**, 409 (1995).
29.   H. Jiang and R.G. Elliman, *IEEE Trans. Electron Devices*, **45**, 97 (1996).
30.   P.W. Leech, M.C. Ridgway, M. Faith and O. Leistiko, *Elect. Lett.*, **32**, 1000 (1996)
31.   C.M. Johnson and M.C. Ridgway, private communication.
32.   R.G. Elliman, B. Luther-Davies, M. Samoc and A. Dowd, To be presented at the 1996 Fall Meeting of the Materials Research Society.
33.   P.D. Townsend, P.J. Chandler and L. Zhang, '*Optical Effects of Ion Implantation*', Cambridge University Press, Cambridge, (1994).

Materials Science Forum Vols. 248-249 (1997) pp. 73-78
© *1997 Trans Tech Publications, Switzerland*

Ion Beam Induced Epitaxial Crystallization of Buried SiC Layers in Silicon

J.K.N. Lindner, K. Volz and B.Stritzker

Universität Augsburg, Institut für Physik, D-86135 Augsburg, Germany

Keywords: Ion Implantation, Ion Beam Induced Crystallization, IBIEC, IBIIA, Amorphization, SiC, Silicon, Buried Layers, Epitaxy, TEM

Abstract

The ion beam induced crystallization of buried amorphous SiC layers in silicon is studied by cross-sectional transmission electron microscopy as a function of dose, dose rate, target temperature and substrate orientation. For this purpose homogeneous, crystalline, buried SiC layers were formed in (100) and (111)Si by ion beam synthesis. The resulting Si/SiC/Si layer systems were irradiated with 2 MeV Si^+ ions at 300 K to amorphize the buried carbide layer without changing its chemical composition or damaging the Si top layer seriously. The amorphization process as well as the amorphous state of the SiC are considered in detail. In order to achieve ion beam induced epitaxial crystallization (IBIEC), layer systems were irradiated with 800 keV Si^+ ions at 320 and 600 °C. It is demonstrated that IBIEC works well on buried SiC layers and results in a single phase heteroepitaxial recrystallization at target temperatures considerably below those necessary for thermal annealing. The IBIEC process starts from both SiC/Si interfaces with different crystallization rates. Depending on irradiation conditions, it may be accompanied by heterogenous nucleation of poly-SiC as well as interfacial layer-by-layer amorphization.

1. Introduction

Numerous superior physical properties make the wide-band gap semiconductor silicon carbide (SiC) a promising material for future high-temperature, high-frequency and high-power electronic devices [1]. In addition, the possibility to achieve high p- and n-type dopand activation and to form a thermal oxide make this material technologically attractive. Since doping by diffusion is limited by the slow diffusivities of dopands, ion implantation is necessary for the formation of laterally structured p/n junctions. Ion implantation, however, may result in amorphization. For 6H-SiC it has been demonstrated [2], that even up to temperatures of 1700 °C it is not possible to recrystallize amorphous SiC epitaxially without polytype transformation. Therefore it is necessary to have an alternative low temperature technique which enables single phase epitaxial crystallization of SiC. Epitaxial crystallization has been shown to take place in silicon and other materials at considerably lower target temperatures than necessary for thermal annealing when performed under ion irradiation (e.g. [3]). IBIEC experiments on 6H-SiC bulk material have been carried out by Heera et al. [4,5] using 200 keV Ge implantation for amorphization and 300 keV Si irradiation for crystallization. Annealing of the end-of-range damage was obtained at temperatures as low as 300° C. IBIEC has been reported to take place and result in a multilayer structure including a columnary grown epitaxial 6H-SiC region and polycrystalline 3C-SiC.

In this work the ion beam induced amorphization and epitaxial crystallization of buried 3C-SiC layers formed by ion beam synthesis in silicon is studied. It will be demonstrated that IBIEC can be applied to achieve poly-type free epitaxial crystallization of 3C-SiC at 600° C.

2. Experimental

c-Si/c-SiC/c-Si layer structures were fabricated by implantation of 9×10^{17} cm^{-2} 180 keV C^+ ions in Si(111) and Si(100) under beam heating conditions (330-440° C) and subsequent annealing in argon atmosphere for 5 h at 1250 °C. A detailed description of the ion beam synthesis of SiC layers has been given earlier [6,7]. In order to amorphize the 170 nm thick crystalline silicon carbide (c-SiC) layer without turning the 300 nm thick crystalline Si (c-Si) top layer completely amorphous, 2 MeV Si^{2+} ions were implanted at a constant target temperature of 300 K in a dose range from 1 to 37 x 10^{15} Si/cm^2 at a current density of approximately 5 μA/cm^2. Due to the projected range of approximately 2 μm for 2 MeV Si ions, the Si concentration in the SiC layer increases by less than 0.01%.

Prior to IBIEC experiments, samples were annealed for 2 h at 550° C in a flowing argon ambient, in order to achieve epitaxial recrystallization of any amorphous silicon (a-Si) present after Si irradiation. At 550° C recrystallization of amorphous SiC (a-SiC) can be excluded and the c-Si/a-SiC/c-Si starting layer structure for IBIEC experiments is obtained. For ion beam induced crystallization, layer systems were irradiated at constant temperatures of 320 and 600 °C with 800 keV Si$^+$ ions in a dose interval from 4 to 10 x 10^{16} Si/cm^2 at constant current densities between 0.25 and 1.0 μA/cm^2.

3. Results and discussion

3.1 Structure of ion beam synthesized c-Si/c-SiC/c-Si layer systems

The ion beam synthesized c-Si/c-SiC/c-Si layer systems used as a starting material consist of a 170 nm thick 3C-SiC layer, which is covered with a 300 nm thick c-Si top layer and a surface oxide film. In this work samples were used which did not exceed a target temperature of 390° C during C$^+$ implantation and consequently [6,7] showed a portion of poly-SiC mainly near the upper Si/SiC interface, while the lower part of the SiC layer was completely aligned with the c-Si bulk.

3.2 Amorphization

The layer structure obtained after additional 2 MeV Si$^+$ implantation is diplayed in Fig. 1. For doses between 1 x 10^{15} and 1.1 x 10^{16} Si/cm^2 (Fig. 1(a)) a buried amorphous silicon layer is formed within the c-Si bulk. The lower a/c interface is located at a nearly constant depth of 2 μm while the upper c/a interface moves towards the buried SiC layer with increasing dose. For a dose of 1.3 x 10^{16} Si/cm^2 amorphization of the buried SiC layer and of isolated SiC precipitates in the Si top layer is reached (Fig. 1(b)). A further dose increase causes the a/c interface to proceed further in the Si top layer (Fig. 1(c)).

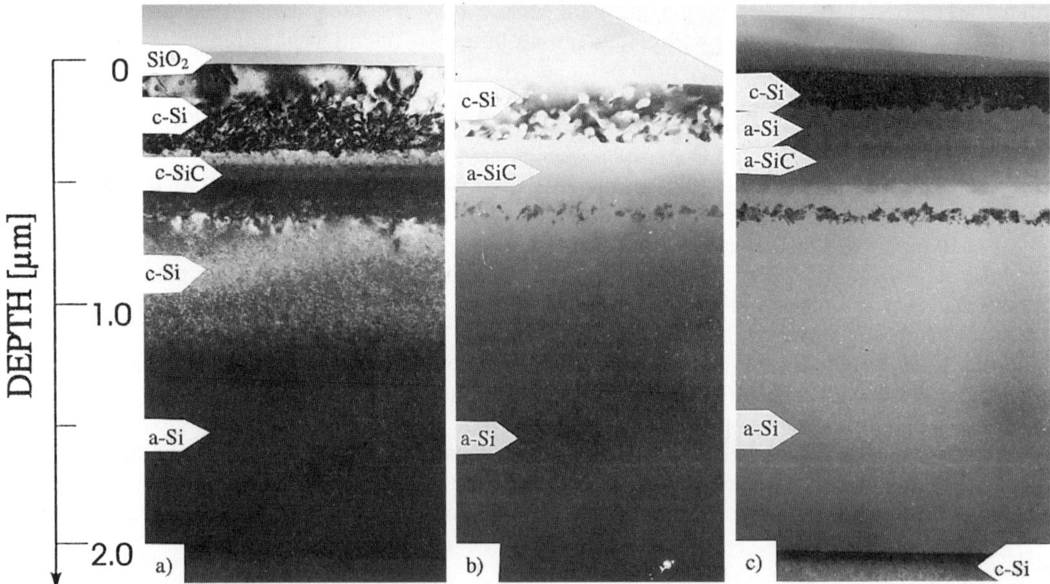

Fig. 1: Bright field overview micrographs of the layer structures obtained after 2 MeV Si irradiation with different doses: a) 2.0 x 10^{15} , b) 1.3x 10^{16} , c) 3.7 x 10^{16} Si/cm^2

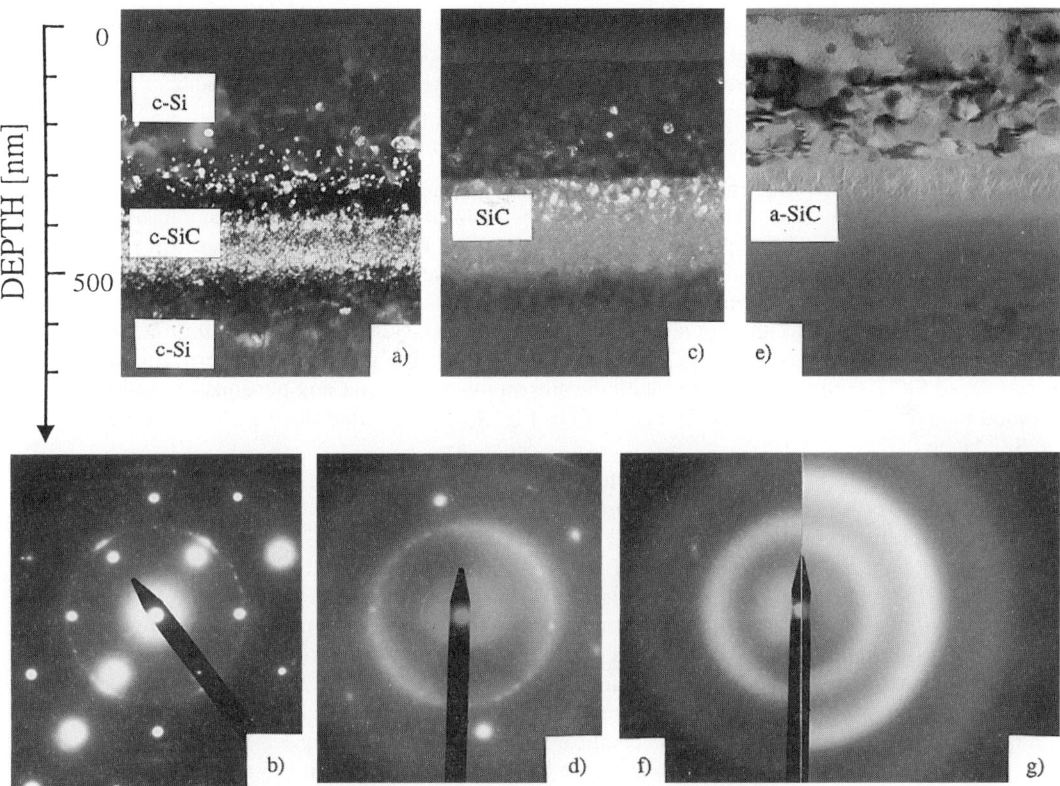

Fig. 2: Defect accumulation in the buried SiC layers after 2 MeV Si irradiation with doses of
D=2.0 x 10^{15} Si/cm^2: a) SiC $_{g=(222)}$ dark field micrograph, b) SAD near [011],
D=7.0 x 10^{15} Si/cm^2: c) SiC $_{g=(222)}$ dark field micrograph, d) SAD near [114],
D=1.3 x 10^{16} Si/cm^2: e) bright field micrograph, f) SAD of a-SiC, g) SAD of a-Si

The defect accumulation within the buried SiC layers during irradiation is considered in Fig. 2. For irradiation doses up to 2 x 10^{15} Si/cm^2 (Fig. 2(a-b)) SiC dark field images show the same characteristic grainy contrast as unirradiated SiC layers and TEM diffraction patterns do not display any amorphous portions, i.e. the damage is below the detection limit. Increasing the dose to 7 x 10^{15} Si/cm^2 (Fig. 2(c-d)) leads to more faded contrasts in the SiC layer, indicating a stronger damage, and the selected area diffraction (SAD) pattern of the initial state gets superimposed with the broad rings of amorphous SiC. A further dose increase to 1.3 x 10^{16} Si/cm^2 results in the complete amorphization of the buried SiC layer (Fig. 2(e-f)), as concluded from TEM tilting experiments and the disappearence of crystalline spots in SAD patterns. From an evaluation of SAD patterns of the a-SiC layer and the buried a-Si layer, displayed in Figs. 2(f) and 2(g), it is concluded that at the irradiation conditions chosen, the SiC layer became structurally amorphous under preservation of the local chemical order.

With a critical dose for amorphization between 1.1 and 1.3 x 10^{16} Si/cm^2 and the mean TRIM [8] nuclear stopping power of 2 MeV Si ions in the buried SiC layer, a critical energy density e_c for the amorphization of buried 3C-SiC layers at 300 K of (1.1 ± 0.1) x 10^{25} eV/cm^3 is calculated. This value is considerably higher than that of 2.0 x 10^{24} eV/cm^3 [9] for the room temperature amorphization of bulk 6H-SiC. This is astonishing since according to computer simulations [10] the cubic SiC should be less stable than the hexagonal polytype. The higher value is most probably due to the fact that amorphization of the near surface SiC layer is performed with the upper wing of the nuclear stopping power profile of the 2 MeV ions and that it occurs when the adjacent Si substrate is already amorphized.

For the IBIEC experiments the c-Si/a-SiC/a-Si/c-Si layer systems obtained with a dose of 1.3 x 10^{16} Si/cm^2 were transformed into c-Si/a-SiC/c-Si layer systems by epitaxial crystallization of a-Si regions in a 2 h thermal anneal at 550 °C. At this temperature a-SiC stays amorphous.

3.3 Ion beam induced crystallisation of buried SiC layers

Fig. 3 (a-c) shows XTEM micrographs and the electron diffraction pattern (d) of a (111) layer system, which has been irradiated at 600° C with 800 keV Si$^+$ ions at a dose of 8 x 10^{16} Si/cm^2 and a current density of 0.25 µA/cm^2. The bright field micrograph (BF) in Fig. 3(a) gives an overview of the whole layer structure: the 300 nm c-Si top layer, the 170 nm SiC layer and the c-Si bulk. The a-SiC layer has epitaxially recrystallized from both interfaces with the Si bulk and the top layer (Fig. 3(c)), leading to an upper epitaxial part of 90 nm aligned 3C-SiC, a 20 nm thick central amorphous region, and a lower 60 nm thick epitaxial 3C-SiC zone above the c-Si bulk. The diffraction pattern shows that the recrystallized SiC is completely aligned with the silicon matrix without any portions of random or twin oriented material. The Si dark field image (DF) in Fig. 3(b) proves that there are no c-Si inclusions in the SiC layer.

The dependence of the layer structure after Si ion beam treatment on the dose, dose rate, temperature, and crystal orientation is summarized in Fig. 4.

For the same irradiation conditions (D = 8 x 10^{16} Si/cm^2; T = 600° C; j = 0.25 µA/cm^2) as in Fig. 3, but a (100) *crystal orientation* instead of (111), the SiC layer recrystallizes over the whole 170 nm instead of forming two interfacial zones (Fig. 4(a)). There is, however, a considerable fraction of SiC at random orientation (Fig. 4(b)), which can also be seen from the electron diffraction pattern in Fig. 4(c). These observations may be explained with a higher recrystallization velocity of (100) oriented material compared to (111)SiC as observed for Si [3]. A high recrystallization velocity may lead to the occupation of anti-site positions and consequently to deviations from the (111) stacking sequence of Si and C layers in this composite material. Therefore misoriented crystal regions occur.

An increase of the *current density* from 0.25 µA/cm^2 (Fig. 3) to 0.5 µA/cm^2 (Fig. 4(d-f)) at otherwise unchanged irradiation conditions ((111) material; D = 8 x 10^{16} Si/cm^2; T = 600° C) leads to a reduced amount of epitaxially recrystallized carbide (Fig. 4(d)) and a detectable portion of random and twin oriented material in the SiC layer (Fig. 4(e)). The presence of twin oriented SiC is also visible in the SAD pattern of Fig. 4(f). This behaviour has also to be referred to the decreased time available for an ordered layer-by-layer crystallization of epitaxial SiC.

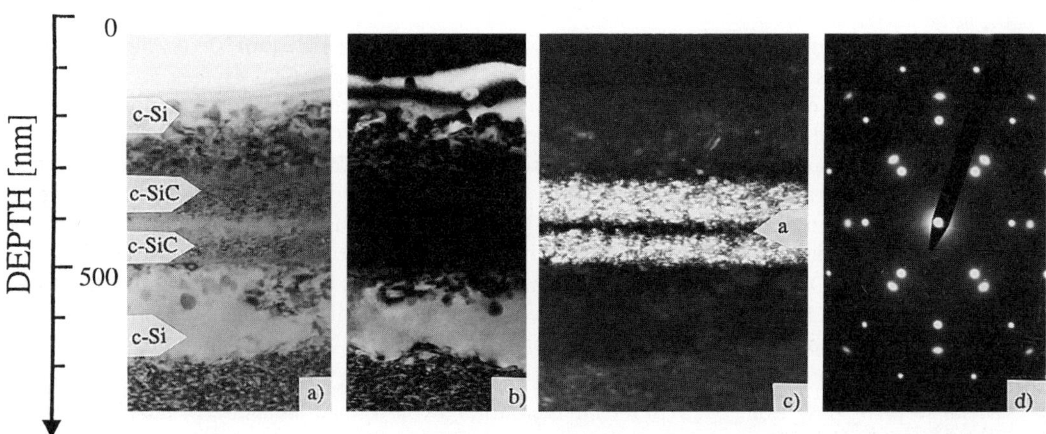

Fig. 3:
a) bright field, b) Si$_{g=(220)}$ dark field, c) SiC$_{g=(220)}$ dark field micrograph and d) [011] SAD pattern of a (111) layer system, recrystallized by ion irradiation at 600° C (D = 8 x 10^{16} Si/cm^2; j = 0.25 µA/cm^2).

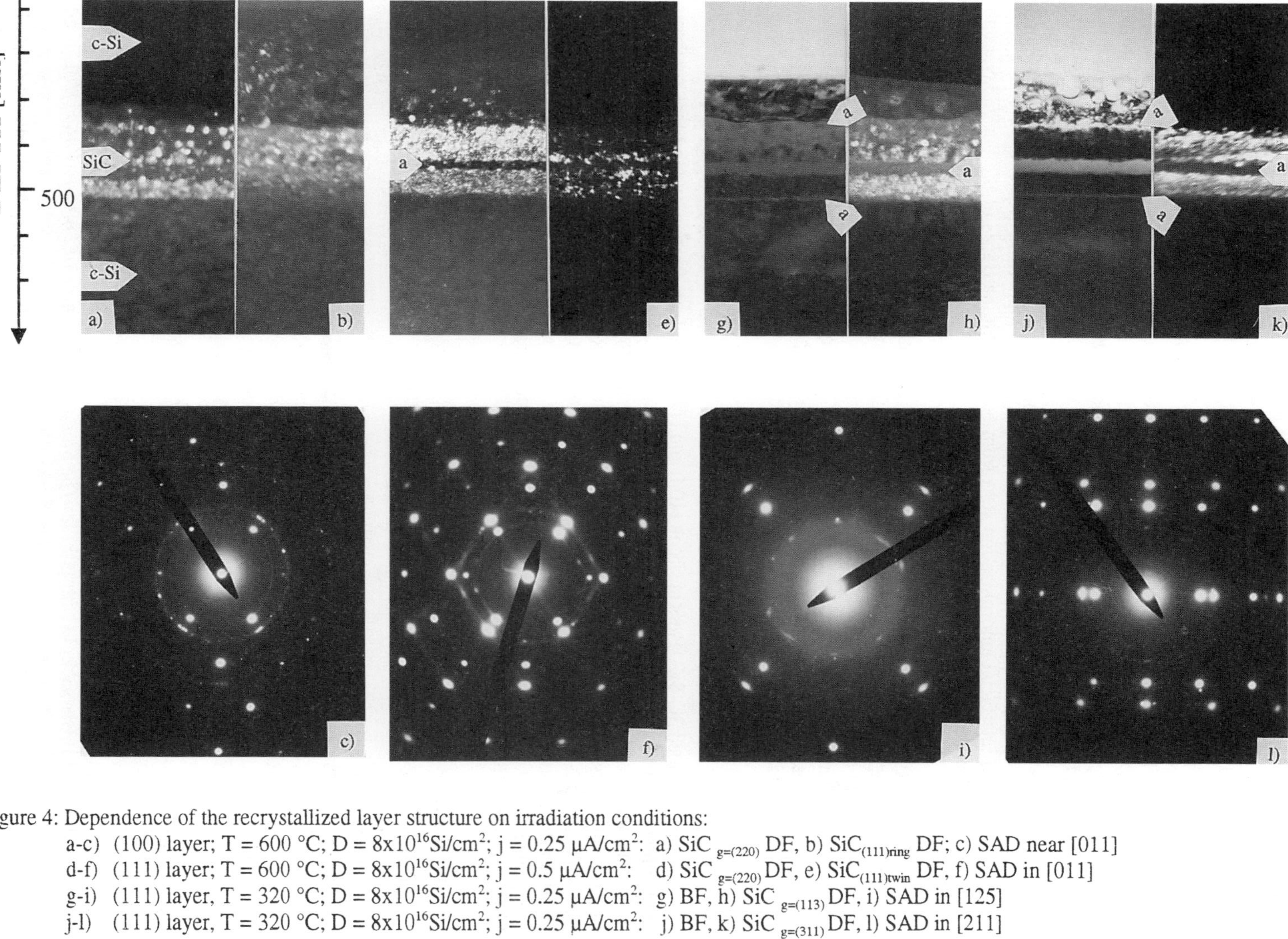

Figure 4: Dependence of the recrystallized layer structure on irradiation conditions:

a-c) (100) layer; T = 600 °C; D = 8×10^{16}Si/cm^2; j = 0.25 µA/cm^2: a) SiC $_{g=(220)}$ DF, b) SiC$_{(111)ring}$ DF; c) SAD near [011]

d-f) (111) layer; T = 600 °C; D = 8×10^{16}Si/cm^2; j = 0.5 µA/cm^2: d) SiC $_{g=(220)}$ DF, e) SiC$_{(111)twin}$ DF, f) SAD in [011]

g-i) (111) layer, T = 320 °C; D = 8×10^{16}Si/cm^2; j = 0.25 µA/cm^2: g) BF, h) SiC $_{g=(113)}$ DF, i) SAD in [125]

j-l) (111) layer, T = 320 °C; D = 8×10^{16}Si/cm^2; j = 0.25 µA/cm^2: j) BF, k) SiC $_{g=(311)}$ DF, l) SAD in [211]

A reduction of the implantation *temperature* from 600° C (Fig. 3) to 320° C (Fig. 4(g-i)) ((111) material; $D = 8 \times 10^{16}$ Si/cm^2; $j = 0.25$ μA/cm^2) results in a subdivision of the SiC layer into five zones. The recrystallized SiC parts are now embedded in amorphous SiC layers (Fig. 4(g-h)), which contain isolated SiC crystallites at random orientation. This is attributed to the fact that at lower target temperatures less defects contribute to recrystallization, resulting in ion beam induced interfacial amorphization (IBIIA) at the expense of already epitaxially recrystallized SiC regions. Thus, the temperature of 320° C is expected to be near the IBIEC reversal temperature of 3C-SiC. Due to ion beam induced polynucleation randomly oriented SiC precipitates are formed in the a-SiC layer. The large amount of amorphous material leads to the broad rings of a-SiC in the diffraction pattern in Fig. 4(i).

Reducing the *dose* by a factor of two ((111) material; $T = 320°$ C; $j = 0.25$ μA/cm^2) leads to smaller outer and slightly thicker central a-SiC zones (Fig. 4(j-k)), supporting the conclusion that both crystallisation and interfacial amorphization start from both SiC/Si interfaces of the buried layer and proceed with increasing dose towards the centre of the SiC layer. According to the electron diffraction pattern in Fig. 4(l) the recrystallized SiC is well aligned with the Si matrix.

The recrystallization of the buried layer starts at both interfaces and is generally more pronounced at the upper interface. Since the nuclear stopping power within the Si top layer is smaller than that in the Si bulk next to the buried SiC layer, one would expect that recrystallization is stronger at the lower interface, if the defect responsible for IBIEC is generated far away from the interfaces. Since the opposite is observed, one may conclude that this defect is generated directly at the amorphous/crystalline interface.

4. Summary

Homogeneous, buried 3C-SiC layers in silicon (100) and (111) have been formed by ion beam synthesis and subsequently amorphized by 2 MeV Si$^+$ irradiation at 300 K. The c-Si/a-SiC/c-Si layer structures have been irradiated with 800 keV Si$^+$ ions at different target temperatures, doses and current densities in order to achieve IBIEC. Cross-sectional TEM investigations give evidence that IBIEC works well for c-Si/a-SiC/c-Si structures and leads to heteroepitaxial silicon carbide layers. The epitaxial crystallization starts from both SiC/Si interfaces. Depending on irradiation conditions, in addition to the ion beam induced epitaxial crystallization ion beam induced polynucleation and interfacial amorphization may occur, resulting in complex multilayer structures.

Acknowledgement

The authors are grateful to W. Brückner and D. Donimierski for technical assistance.

References

[1] H.Morkoç, S.Strite, G.B.Gao, M.E.Lin, B.Sverdlov, M.Burns, J. Appl. Phys. 76 (1994) 1363.

[2] W.Wesch, A.Heft, J.Heindl, H.P.Strunk, T.Bachmann, E.Glaser, E.Wendler, accepted for publ.

[3] F.Priolo, E.Rimini, Mater. Sci. Rep. 5 No. 7,8 (1990) 319.

[4] V.Heera, J.Stoemenos, R.Kögler, W.Skorupa, J. Appl. Phys. 77 (1995) 2999.

[5] V.Heera, R.Kögler, W.Skorupa, J.Stoemenos, Inst. Phys. Conf. Ser. No 142 (IOP Publishing Ltd., Bristol, 1996) p. 533.

[6] J.K.N.Lindner, K.Volz, B.Stritzker, Inst. Phys. Conf. Ser. No 142 (IOP Publishing Ltd., Bristol, 1996) p. 145.

[7] J.K.N.Lindner, K.Volz, U.Preckwinkel, B.Götz, A.Frohnwieser, B.Rauschenbach, B.Stritzker, Mater. Chem. and Phys. 1996, in press.

[8] J.F.Ziegler, J.P.Biersack, U.Littmark, in: *The Stopping and Range of Ions in Matter*, Vol. 1, ed. by J.F. Ziegler, (Pergamon Press, New York, 1985).

[9] J.A.Spitznagel, S.Wood, W.J.Choyke, N.J.Doyle, J.Bradshaw, S.G.Fishman, Nucl. Instr. and Meth. B16 (1986) 237.

[10] V.Heine, C.Cheng, R.J.Needs, Materials Science and Engineering B11 (1992) 55.

Materials Science Forum Vols. 248-249 (1997) pp. 79-86
© *1997 Trans Tech Publications, Switzerland*

Ion Beam Induced Epitaxial Regrowth and Interfacial Amorphization of Compound Semiconductors

E. Glaser, T. Fehlhaber, R. Schulz and T. Bachmann

Friedrich-Schiller-Universität Jena, Institut für Festkörperphysik,
Max-Wien-Platz 1, D-07743 Jena, Germany

Keywords: Ion Implantation, Ion Beam Induced Crystallization, Epitaxial Regrowth, Amorphization

ABSTRACT

Ion beam induced epitaxial crystallization (IBIEC) and interfacial amorphization (IBIIA) are investigated in preamorphized layers of (100)-GaAs, InAs, GaP and InP using energetic (MeV) ions. The IBIEC rate is shown to be controlled by point defect diffusion towards the a/c-interface and additionally modified by the interface structure. The suppression of microtwin and stacking fault formation during IBIEC is explained by the fact that the ion beam modifies the orientation dependence of the crystallization kinetics avoiding the disintegration and (111)-faceting of the (100)-interface. Reversal temperatures for IBIEC $\leftrightarrow$ IBIIA transitions are determined as a function of the ion dose rate and the density of displacements. The capture of diffusing defects by crystallites growing in the amorphous layers is considered to be responsible for the stopping of the IBIEC interface above critical temperatures and doses. The limited temperature ranges for undisturbed IBIEC and IBIIA in III/V-compounds are explained by low nucleation barriers and high growth rates both of crystallites and of amorphous zones.

1. INTRODUCTION

Ion implantation has become a well established technique for doping and modifying surface properties of semiconducting materials. Annealing processes necessary to remove the produced radiation damage in the case of III-V-compound semiconductors because of their binary nature and low dissociation temperatures are much more complicated than for elemental semiconductors (Si, Ge). After thermal solid phase crystallization (SPE) of amorphous III/V-layers from an adjacent crystalline substrate high densities of planar defects (microtwins and stacking faults) remain [1]. High temperature annealing necessary to remove these residual defects causes surface dissociation of the compounds. Ion beam induced crystallization based on athermal growth processes is a promising alternative because it proceeds at temperatures lower than that necessary for thermal SPE and the formation of planar defects can be avoided. In this way monocrystalline layers are obtained.

The process of ion beam induced epitaxial crystallization (IBIEC) is a growth process far from thermodynamic equilibrium and induced by high energy ion irradiation through an amorphous/crystalline (a/c–) interface. If the temperature during irradiation is decreased below a critical value (reversal temperature T_R) the interface motion reverses and planar interfacial amorphization (IBIIA) occurs instead [2]. The processes of IBIEC and IBIIA are extensively investigated in silicon [3,4,5]. The microscopic mechanisms controlling these rather complex growth processes are not yet completely understood. It is assumed that due to generation of point defects in the amorphous and the crystalline phase both crystallization and amorphization mechanisms occur simultaneously at the a/c-interface and that the net motion of the interface is a consequence of the balance between the two mechanisms [6]. As a consequence of the interaction with the defects the a/c-interface itself becomes roughened [4]. But, there is a controversy whether the kinetics of IBIEC or IBIIA is controlled by the structure of the a/c-interface (reaction limited) [6] or by radiation enhanced diffusion of point defects towards the interface (diffusion limited) [5].

The kinetics of IBIEC in III/V-compounds additionally is substantially affected by random nucleation and growth (RNG) of crystallites competing with epitaxial regrowth [7,8]. On the other hand the observation of interfacial amorphization effects in compounds is difficult because of the simultaneously occurring RNG of amorphous zones [7]. In this paper the phenomena of IBIEC and IBIIA are studied in preamorphized layers of (100)-GaAs, InAs, GaP and InP using MeV-ions. Predominantly new results are presented and discussed together with some results already published by our group. The dose and dose rate dependences of the IBIEC-rate are explained by the diffusion model of IBIEC. The suppression of twin and stacking fault formation during IBIEC in III/V-compounds is explained by the orientation dependence of the growth kinetics essentially differing from that during SPE. In the different substrates the kinetics of IBIEC $\leftrightarrow$ IBIIA - transitions is studied as a function of temperature T, nuclear energy deposition ν (corresponding to a linear density of displacements), and ion dose rate j. The reversal temperatures T_R show characteristic dependences $T_R(j,\nu)$ in agreement with that for Si and Ge. But, contrary to Si and Ge the temperature ranges $T > T_R$ and $T < T_R$ where undisturbed IBIEC and IBIIA can be observed turn out to be limited by RNG of crystallites and amorphous zones, respectively. The disturbance of the IBIEC above critical temperatures and doses is explained by a model that considers the capture of migrating defects promoting the growth of crystallites in the amorphous layers.

2. EXPERIMENTAL

To form amorphous layers with thicknesses of about 250 nm (InP, GaP), 140 nm (GaAs, Ge) and 100 nm (InAs) (100)-semi-insulating InP and GaAs, low-doped GaP and InAs as well as Ge wafers were implanted at liquid nitrogen temperature (LNT) with 120 keV Si^+, 1×10^{15} cm^{-2} (InP), 300 keV Se^+, 5×10^{14} cm^{-2} (GaP), 100 keV Se^+, 5×10^{14} cm^{-2} (InAs) and 50 keV N^+, 1×10^{15} cm^{-2} (GaAs, Ge). The wafers were then irradiated with 2.5 MeV Kr^+, 1.6 or 1.0 MeV Ar^+, and 500 keV N^+, respectively, with dose rates varying from 1×10^{11} to 1.6×10^{12} $cm^{-2}s^{-1}$. The samples were mounted on a temperature controlled sample holder and maintained at a constant temperature in the range between -10°C and +300°C. The position of the a/c-interface and the defect generation by the MeV-ion irradiation were measured by means of RBS/C using 1.4 MeV He^+ ions and a backscattering angle of 170° and 120°, respectively. Results of cross-section TEM (not shown) are considered additionally.

3. RESULTS AND DISCUSSION

Results of IBIEC in InP after irradiation of 1.6 MeV Ar^+ at different temperatures are shown in Fig. 1. The thickness of the layer recrystallized by 5×10^{15} cm^{-2} at 170°C amounts to $\approx$ 120 nm. But, the results show that this thickness cannot be increased if the dose and the temperature are increased (200°C, 1×10^{16} cm^{-2}). Obviously the a/c-interface is stopped at this position. Considering the spectra for the 'thermal effect' during irradiation (unirradiated regions) one can see that at 200°C the whole amorphous layer has already been thermally crystallized by SPE whereas at 170°C the rate of thermal regrowth is low (thickness of the recrystallized layer $\approx$ 25 nm). TEM cross section micrographs of these layers [7] showed that after thermal regrowth at 200°C only $\approx$ 30 nm of good crystal growth is observed followed by a layer containing a high density of stacking faults and twin lamellae. Contrary, IBIEC at 170°C or 200°C produces a 120 nm thick monocrystalline layer free of planar defects. The residual surface amorphous layer not yet recrystallized is transformed to a polycrystalline layer. Whereas the a/c-interface after IBIEC is still sharp and plane [7] the plane (100)-interface after thermal SPE is disintegrated and a faceted interface is formed [9]. Similar results are obtained also for the other III/V-compounds.

These observations mean that both in the temperature range of thermal SPE (200°C for InP) and at temperatures below that necessary for SPE (170°C for InP) the onset of twinned regrowth is suppressed by the ion beam (IBIEC). Thus, the suppression of twin and stacking fault formation cannot be ex-

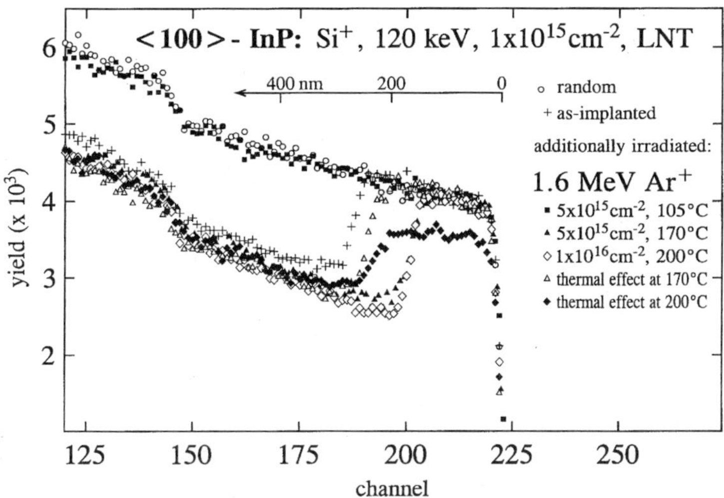

Fig.1. RBS/C spectra of amorphized InP irradiated with different doses of 1.6 MeV Ar⁺ at different temperatures (105, 170 and 200°C). The effects of thermal regrowth at 170°C (during the irradiation time 1h) and at 200°C are also shown.

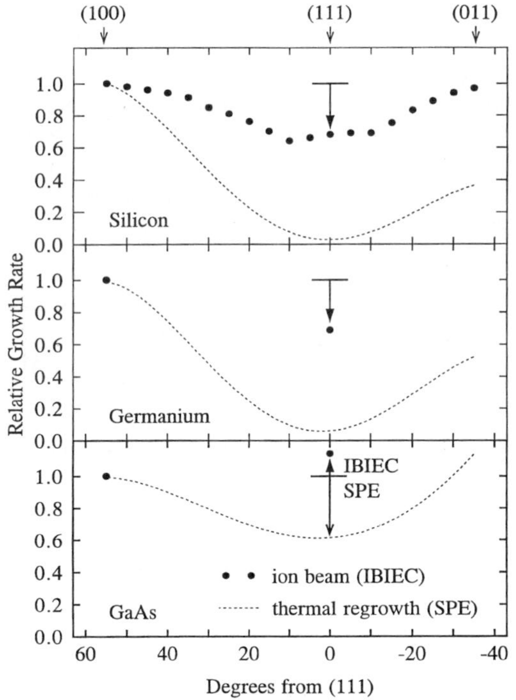

Fig.2. Regrowth rates in silicon, germanium, and GaAs, relative to that for (100) for IBIEC and SPE at orientations along (100), (111), and (011).

plained by subthreshold temperature. An explanation can be given by the results represented in Fig.2. For silicon, germanium and GaAs the relative growth rates during IBIEC and SPE are depicted for different growth directions. The SPE data are fitted using results from Refs. [10-12]. The IBIEC data for Si are taken from Ref. [4], for Ge from Ref. [13], and for GaAs from Ref. [14]. The curves for thermal growth (SPE) in Fig. 2 show that in all materials considered the lowest growth rates are measured in ⟨111⟩-direction. Therefore the nonplanar interfaces formed in III/V-compounds due to local non-stoichiometries or dopants [1] lead to the formation of (111)-facets. Only on these (111)-facets microtwins are assumed to be formed by the attachment of three-atoms-clusters twisted by an angle of 180°; on (100)-faces twin formation is impossible because (100)-growth proceeds by an attachment of single atoms [15].

Thus, there are two possibilities to explain the suppression of twin nucleation during IBIEC of (100)-compound layers: (a) the formation of (111)-facets is avoided or (b) the (111)-facets formed are roughened by the IBIEC process so that nucleation of three-atoms-clusters is no longer necessary for (111)-regrowth. A

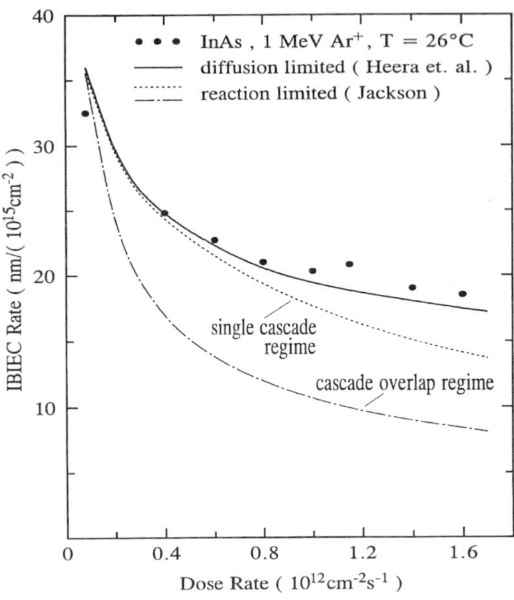

Fig.3. IBIEC rate vs dose rate for amorphized InAs irradiated with 1MeV Ar$^+$ at 26°C (—— model of diffusion-limited kinetics of IBIEC [5]; ----, — · — model of reaction-limited kinetics [6]).

roughening of the interface indeed has been demonstrated by Custer et al. [4] using Monte-Carlo-Simulation of the IBIEC process in Si. But, our investigations on ion beam induced growth of crystallites in III/V-compounds have shown [9] that twin nucleation on (111)-faces is not suppressed by such a roughening effect. So the only explanation for the suppression of twin nucleation is the suppression of (111)-faceting during IBIEC. This can be understood comparing the IBIEC rates for (111) and (100) in Fig. 2. In the case of GaAs the IBIEC-rate in (111) is higher than in (100). This behaviour is opposite to that during thermal regrowth and also to that during IBIEC of Si and Ge. As also during IBIEC the a/c-interface should be bound by the slowest-growing planes a plane (100)-interface in III/V-compounds is conserved and (111)-faceting is avoided. In this way the nucleation of planar defects is suppressed.

One possibility to give some insight into the microscopic mechanisms controlling IBIEC is to study the dose rate dependence of the IBIEC rate [5]. In Fig. 3 such a dependence is shown for InAs irradiated with 1 MeV Ar$^+$ at 26°C. The different curves correspond to different models of the IBIEC mechanism considering the kinetics of IBIEC to be limited by a (long range) <u>diffusion</u> of point defect towards the a/c-interface [5] or by point defect <u>reactions</u> at the interface (in the single cascade or the cascade overlap regime) [6]. Both models yield a decrease of the IBIEC rate with increasing dose rate but the best agreement with the experimental data is obtained using the diffusion model of IBIEC. In this model the increased annihilation of point defects decreases their diffusion length with increasing

dose rate and thus the amount of defects arriving at the interface is decreased. In this way Heera et al. obtained a decrease of the IBIEC rate with j$^{-1/4}$ [5] (solid line in Fig. 3) which is in good agreement with the experimental results of this work. Also for silicon a good agreement with the diffusion model has been demonstrated [5]. From these results one can conclude that IBIEC is caused by a diffusion of point defects towards the a/c-interface, where the IBIEC rate is also controlled by the interface structure (see Fig. 2). This conclusion is in agreement with the model of Custer et al. [4].

To study the kinetics of IBIEC ↔ IBIIA transitions in the different substrates the irradiation temperature was decreased and additionally the

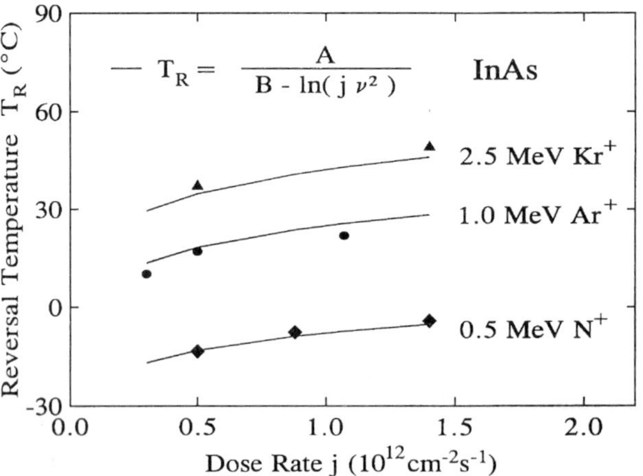

Fig.4. Reversal temperature T_R vs dose rate j for irradiation of InAs with different ions (i.e. different densities of displacements). The curves are calculated using the empirical equation shown in the figure.

density of displacements ν (via the ion mass) and the dose rate j were varied. For fixed density of displacements and dose rate (1.0 MeV Ar⁺, 1.4×10^{12} cm⁻²s⁻¹) the reversal temperatures measured are summarized in Table 1. These values are in good agreement with the critical temperatures, below which stable damage is generated in the corresponding crystalline layers (not shown here). For a given substrate (InAs) the values of T_R estimated for different j and ν (2.5 MeV Kr⁺, 1.0 MeV Ar⁺, and 0.5 MeV N⁺) are shown in Fig. 4. These experimental values are fitted to an empirical equation $T_R[K] = A/(B-\ln(j[cm^{-2}s^{-1}]\cdot\nu^2[Å^{-2}]))$ (solid lines). ν was calculated by TRIM and the constants are estimated as A=9060 and B=59.1 in this case. In earlier work it was also demonstrated for Si [16] and Ge [13] that the parameter dependences $T_R(j,\nu)$ can be described well by this parameter equation.

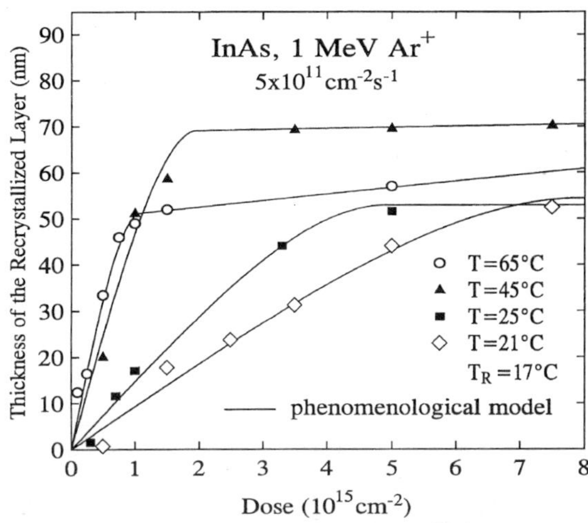

Fig.5. Thickness of InAs layers recrystallized with 1 MeV Ar⁺ at different temperatures vs ion dose. The curves are calculated by a phenomenological model described in the text.

To explain the stopping of the IBIEC process demonstrated in Fig.1 for InP (but characteristic for all III/V-compounds) the dose dependences of the thickness of the recrystallized InAs layers for different temperatures are shown in Fig. 5. As can be seen at low doses the slope of the curves (i.e. the IBIEC rate) increases with increasing temperature which is in qualitative agreement with the behaviour of Si [2]. But, for higher doses only at low temperatures (T = 21K, $T–T_R$ = 4K) a nearly linear dose dependence is observed. With increasing temperature the critical dose where a substantial deviation from the linear dependence is found becomes smaller and smaller. The decrease of the slope of the curves above these critical doses demonstrates that the IBIEC rate drastically decreases and finally the interface is stopped. Investigations by TEM have shown, that at the temperatures where nonlinear dose

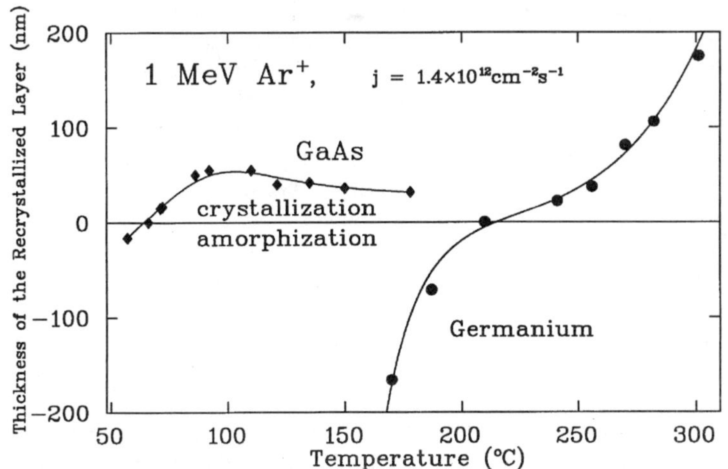

Fig.6. Thickness of GaAs and Ge layers recrystallized with 1 MeV Ar⁺ (dose 5x10¹⁵ cm⁻²) vs temperature.

dependences of the recrystallized layer thickness are found random nucleation and growth (RNG) of crystallites occurs [7,9] and that the size and the density of the crystallites increase with dose [17] and with temperature [8]. The solid lines in Fig. 5 are the result of a phenomenological model based on the diffusion model of IBIEC. In this model both the epitaxial regrowth and the growth of the crystallites are explained by a diffusion of point defects and their capture at the corresponding a/c-interfaces. As the capture of migrating defects by the surface of the growing crystallites decreases the defect flux to the plane a/c-interface the rate of epitaxial regrowth is decreased simultaneously to the growth of the crystallites and the nucleation of new grains. By this model the measured dose dependences shown in Fig. 5 are described well indicating that the proposed stopping mechanism of the IBIEC process should be realistic.

These RNG processes competing with epitaxial regrowth also lead to substantial changes of the temperature dependence of the thickness of the recrystallized layer compared with that for Si or Ge. In Fig. 6 these dependences for Ge and GaAs are compared. For the elemental semiconductor Ge where nucleation and growth of crystallites is not observed the layer thickness increases progressively with increasing temperature. This temperature dependence can be described by an Arrhenius-dependence [13]. Due to the RNG of crystallites in GaAs above a critical temperature of about 80°C the slope of the curve decreases again and at T > 100°C a decrease of the thickness of the recrystallized layer is observed. It was shown that during irradiation of amorphous GaAs with 1.0 MeV Ar$^+$ at 80°C crystallites are not yet formed [8] and a nearly linear dose

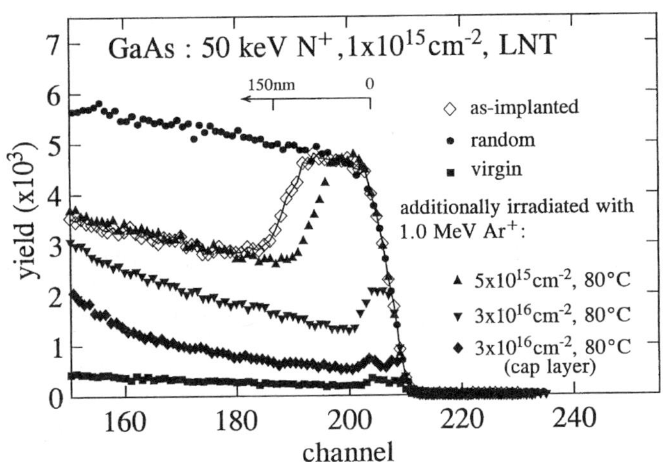

Fig.7. RBS/C spectra of amorphized GaAs irradiated with different doses of 1 MeV Ar$^+$ at 80°C.

dependence of the layer thickness was measured [17]. But, at T > 100°C the nucleation process starts explaining the stopping of the interface motion at these temperatures as demonstrated in Fig. 6. The results of Fig. 6 show that the optimum temperature range for IBIEC in GaAs not disturbed by RNG of crystallites is 65°C <T< 90°C if 1.0 MeV Ar$^+$ ions are used. The backscattering spectra of Fig. 7 indicate that an IBIEC experiment performed within this temperature range (at 80°C) yields complete epitaxial recrystallization of a 150 nm thick amorphous GaAs layer without stopping of the a/c-interface. To obtain a minimum density of residual defects near the surface of the regrown layer an aluminum cap layer was used. Another possibility to achieve epitaxial regrowth of thick amorphous GaAs layers even at T > 100°C is discussed in Ref. [18]. As the interface at these temperatures is stopped after regrowth of about 50 nm (see Fig. 6) a multiple-step IBIEC process: recrystallization → reamorphization of the formed polycrystalline layer → recrystallization of this layer (and so on) has to be performed.

Whereas the IBIEC process <u>above</u> critical temperatures is disturbed by RNG of <u>crystallites</u>, the IBIIA process <u>below</u> critical temperatures is disturbed by RNG of <u>amorphous zones</u> [18]. Below these temperatures the irradiated layer is totally amorphized by accumulation of the amorphous zones and IBIIA is no longer detectable. Thus, only limited temperature ranges ΔT^a_{RNG} below T_R and ΔT^c_{RNG} above T_R exist for each substrate where IBIIA and IBIEC, not disturbed by RNG of amorphous zones and crystallites, respectively, can be observed. In Table 2 these temperature ranges are summarized for different substrates together with the rates of crystallization (IBIEC) or amorphization (IBIIA). The largest widths of these temperature ranges are observed for Si, the smallest for InAs. The narrowest temperature ranges for undisturbed IBIEC and IBIIA are correlated with the highest IBIEC and IBIIA rates. This can be understood by the fact that the crystallites and the amorphous zones also grow by the IBIEC and IBIIA mechanisms, respectively. But, the large differences between the elemental and the compound semiconductors show in addition that the nucleation barriers for crystallites or amorphous zones in III/V-compounds are much lower than in Si and Ge.

4. CONCLUSIONS

The rate of IBIEC is controlled by the diffusion of ion beam generated point defects towards the a/c-interface additionally modified by the structure of the interface. During IBIEC the formation of micro-twins and stacking faults which are characteristic residual defects after thermal SPE is suppressed and monocrystalline layers can be formed. This is caused by the fact that, due to a modification of the growth kinetics by the ion beam, the disintegration and (111)-faceting of the (100)-interface is avoided. The reversal temperatures T_R for the different substrates are correlated with critical temperatures, below which stable damage is generated in the crystallized layers. Consequently, during IBIEC ($T > T_R$) the defect production by the ion beam is substantially reduced. The dependence of T_R on the dose rate j and on the density of displacements v per ion is in agreement with that for Si and Ge supporting conclusions derived from the model of simultaneous crystallization and amorphization mechanisms at the interface. At high values of j and v the IBIEC rate is decreased due to increased defect annihilation and the IBIIA rate is increased due to an increased probability of defect agglomeration at the interface. The capture of diffusing defects by crystallites growing within the amorphous layers give rise to a stopping of the IBIEC interface above critical temperatures and doses. Both for undisturbed IBIEC and undisturbed IBIIA the temperature ranges are limited because of the low nucleation barriers and high growth rates both for crystallites and for amorphous zones in III/V-compounds.

Substrate	T_R (°C)
Ge	215
Si	140[*]
InP	135
GaP	120
GaAs	65
InAs	17

[*] for 1.5 MeV Ar⁺, taken from Ref. [2]

Table 1.   Reversal temperatures T_R for IBIEC ↔ IBIIA transitions (1.0 MeV Ar⁺, j = 1.4x10¹² cm⁻² s⁻¹)

Substrate	T_R (°C)	$R_{IBIEC}(T)^{*)}$ (nm/10^{16} cm^{-2})	$R_{IBIIA}(T)^{**)}$ (nm/10^{16} cm^{-2})	ΔT^c_{RNG} (K)	ΔT^a_{RNG} (K)
Si	190	5	3	350	≥ 110
Ge	215	15	13		≥ 55
GaAs	65	37	22	25	≥ 7
InAs	17	135	130	15	≥ 3

$^{*)}$ $T - T_R = 7$ K ; $^{**)}$ $T_R - T = 5$ K

Table 2. IBIEC and IBIIA rates, $R_{IBIEC}(T)$ and $R_{IBIIA}(T)$, at fixed temperature differences from T_R and temperature ranges ΔT^c_{RNG} and ΔT^a_{RNG} for undisturbed IBIEC and IBIIA, respectively. The irradiations were performed with 1.0 MeV Ar$^+$, in the case of Si [19] with 680 keV Kr^{++}.

REFERENCES

[1] D.K. Sadana, Nucl. Instr. and Meth. **B7/8**, 375 (1985).

[2] J. Linnros, R.G. Elliman, and W.L. Brown, J. Mater. Res. **3**, 1208 (1988).

[3] F. Priolo and E. Rimini, Mater. Sci. Rep. **5**, 319 (1990) and references therein.

[4] J.S. Custer, A. Battaglia, M. Saggio, and F. Priolo, Phys. Rev. Lett. **69**, 780 (1992).

[5] V. Heera, T. Henkel, R. Kögler, and W. Skorupa, Phys. Rev. B **52**, 15776 (1995) and references therein.

[6] H.A. Jackson, J. Mater. Res. **3**, 1218 (1988).

[7] E. Glaser, T. Bachmann, R. Schulz, S. Schippel, and U. Richter, Nucl. Instr. and Meth. B **106**, 281 (1995).

[8] T. Bachmann, E. Glaser, R. Schulz, U. Kaiser, and G. Sáfrán, Nucl. Instr. and Meth. B **113**, 214 (1996).

[9] E. Glaser. P. Gaiduk, U. Kaiser, R. Schulz, and T. Bachmann, to be published.

[10] L. Csepregi, E.F. Kennedy, J.W. Mayer, and T.W. Sigmon, J. Appl. Phys. **49**, 3906 (1978).

[11] L.Csepregi, R.P. Küllen, J.W. Mayer, and T.W. Sigmon, Solid State Commun. **21**, 1019 (1977).

[12] C.Licoppe, Y.J. Nissim, P. Henoc, Appl. Phys. Lett. **48**, 1441 (1986).

[13] T. Bachmann, R. Schulz, E. Glaser, U. Richter, and S. Schippel, Nucl. Instr. and Meth. B **106**, 350 (1995).

[14] S.T. Johnson, R.G. Elliman, and J.S. Williams, Nucl. Instr. and Meth. B **39**, 449 (1989).

[15] R. Drosd and J.Washburn, J. Appl. Phys. **53**, 397 (1982).

[16] R. Kögler, V.Heera, W. Skorupa, E. Glaser, T. Bachmann, and D.Rück, Nucl. Instr. and Meth. B **80/81**, 556 (1993).

[17] R. Schulz, T. Bachmann, E. Glaser, and P.Gaiduk, Nucl. Instr. and Meth. B, in press.

[18] E. Glaser, T. Bachmann, R. Schulz, and U. Richter, Proc. Int. Conf. Ion Beam Modification of Materials, Canberra, Elsevier 1996, 882.

[19] A. Battaglia, F. Priolo, E. Rimini, Appl. Surf. Sci. **56-58**, 577 (1992).

Materials Science Forum Vols. 248-249 (1997) pp. 87-93
© *1997 Trans Tech Publications, Switzerland*

Effects of Ion Irradiation on Ferromagnetic Thin Films

J.E.E. Baglin, M.H. Tabacniks, R. Fontana, A.J. Kellock and T.T. Bardin

IBM Almaden Research Center, 650 Harry Rd., San Jose, CA 95120, USA

Keywords: Permalloy Films, Ion Implantation, Radiation Damage, Magnetic Properties

Abstract

The effects of ion beam irradiation on the magnetic properties of thin Permalloy films have been examined experimentally, using the ion species He^+, Ar^+, Xe^+, C^+, N^+, O^+ and Si^+, at energies chosen either to implant within the film or to produce kinetic effects in the film during transmission. Hysteresis loop parameters, magnetoresistance measurements and XRD were used to characterize the samples before and after exposure. The principal results seen were a loss of uniaxial anisotropy and an increase of coercivity, being attributed to collisional damage of the Permalloy lattice structure and the introduction of extended defects. Effects of film stress and implanted impurities were also identified.

INTRODUCTION

Applications of thin ferromagnetic films for sensors and media for magnetic disk storage of information demand that the magnetic properties of those films be carefully tailored and controlled. Traditionally, control lies in such variables as the composition of the film, its thickness, and uniaxial anisotropy induced when the film is deposited in a magnetic field. In an attempt to gain further control of such properties, the present experiment investigates the physical effects of ion implantation or ion irradiation on the magnetic properties of thin films of Permalloy, having approximate composition $Ni_{80}Fe_{20}$. Earlier studies of radiation effects on magnetic alloys, e.g. [1-4] centered mainly on neutron irradiation of bulk materials. In this experiment, Permalloy films were irradiated with a variety of ion species, energies and doses, selected to help identify the specific contributions to magnetic effects made by ionization (electronic) energy deposition by the ions, collisional loss of ion energy (displacement damage to the NiFe lattice), and implantation of various atomic species in the film, some inert and some capable of bonding with Ni or Fe.

EXPERIMENT

A set of identical samples of NiFe (150 nm) on glass substrates was deposited by DC magnetron sputtering, in the presence of a magnetic field to establish in-plane magnetic anisotropy in each film. The as-deposited properties of each film were then measured, including hysteresis loops for both easy and hard axes, the magnetoresistance metric $\Delta R/R$, sheet resistivity ρ and x-ray diffraction data (θ-2θ and rocking curve).

Each sample was then subjected to irradiation at room temperature, at normal incidence to the sample surface, under one of the sets of ion beam conditions detailed in Table I.

Table I.   Irradiation conditions, and corresponding calculated [5] components of total energy deposition within the Permalloy film due to electronic (ΔE_e) and collisional ($\Delta E_{coll.}$) processes.

Run/ /Sample Number	Irradiation	Nuclear energy deposited in film, ΔE_{coll} (arb. units)	Electronic energy deposited in film, ΔE_e (arb. units)	Concentration of implanted species (at.%)	Comments
1	2E16 He$^+$ (2 MeV)	0.004	4.0	--	Electronic ΔE
2	4E16 He$^+$ (2 MeV)	0.008	8.0	--	Electronic ΔE
3	1.9E16 C$^+$ (3 energies)	0.6	2.0	1.4	Active implants
4	2.1E16 N$^+$ (3 energies)	0.9	2.0	1.4	Active implants
5	2.4E16 O$^+$ (3 energies)	1.3	2.0	1.7	Active implants
6	1.9E16 Si$^+$ (4 energies)	3.1	2.0	1.3	Active implants
7	1.3E16 Ar$^+$ (4 energies)	3.8	2.0	0.9	Inert implants
8	2.6E16 Ar$^+$ (4 energies)	7.6	4.0	1.8	Inert implants
9	0.9E16 Xe$^+$ (5 energies)	23.8	2.0	0.6	Inert implants
10	1.8E16 Xe$^+$ (5 energies)	47.6	4.0	1.2	Inert implants
11	0.5E16 Xe$^+$ (1.5 MeV)	10.0	2.0	--	Collisional ΔE
12	1.0E16 Xe$^+$ (1.5 MeV)	20.0	4.0	--	Collisional ΔE

Each sample was then re-tested. A selection of samples was subsequently annealed in a helium tube furnace at 300°C for 4 h., and re-tested.

Irradiation conditions

The series of ion beam conditions shown in Table I was planned on the basis of TRIM'95 [5] simulations for the total energy deposited within the NiFe film by nuclear (collisional) processes (Col.3) and by electronic (ionization) processes (Col.4). In runs 1, 2, 11 and 12, energies were chosen to place the ion ranges deep in the substrate, with little residual implantation in the NiFe. In runs 3-10, multiple energies were used to produce a somewhat uniform concentration of implanted species (Col.5) throughout the film.

Thus, in runs 1 and 2, 2 MeV He$^+$ ions (at two doses) produce only electronic energy deposition in the NiFe. By contrast, in runs 11 and 12, 1.5 MeV Xe$^+$ (2 doses) produces a similar electronic energy loss, but also a very large collisional loss, which must be responsible for any difference in magnetic properties between these samples and those of runs 1 and 2.

The effect of implanting about 1 atomic percent of inert species (Ar$^+$ or Xe$^+$) is superimposed on collisional damage effects in runs 7-10; differences in magnetic performance compared with transmission runs 11 and 12 may result from strain introduced by the inert implanted species.

Finally, in runs 3-6, ion species with the possibility to form chemical bonds with Fe or Ni were implanted, to test the effect of such lattice impurities.

RESULTS

Hysteresis loops

Fig. 1 shows the reference loops (both hard and easy axes superimposed) for a typical as-deposited NiFe sample. Fig. 2 shows loops for samples 1, 2, 11 and 12 following transmission of MeV He$^+$ and Xe$^+$ ions. No significant change resulted from purely electronic energy deposition, (He$^+$), while a substantial change resulted from Xe$^+$ irradiation. Evidently, this is the result of collisional energy $\Delta E_{coll.}$ deposited in the NiFe by the passage of Xe$^+$ ions. The most obvious consequence was the loss of most of the original uniaxial magnetic anisotropy in the Py film, together with increased coercivity and a loss of squareness in the loop. A few percent loss of saturation magnetization M_s was also observed (after allowing for the effects of sputtering loss of film thickness).

Fig. 3 shows hysteresis loops from samples implanted with Ar or Xe, and otherwise experiencing $\Delta E_{coll.}$ similar to that of samples 11 and 12. A comparison of the

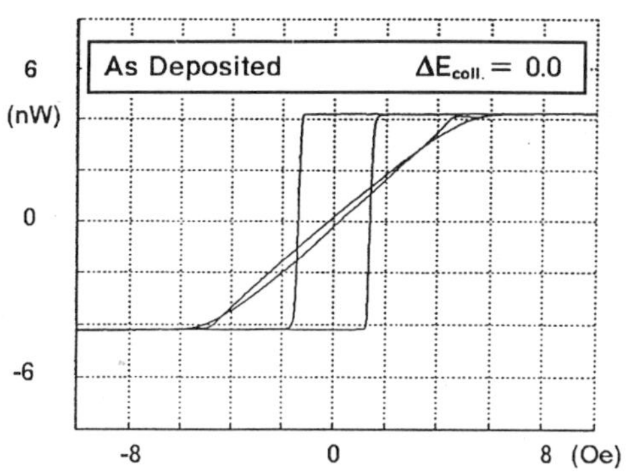

Fig. 1. Hysteresis loops (easy axis and hard axis) for typical as-deposited samples of Permalloy.

loops for samples 12 (Xe$^+$ transmission) and 9 (Xe$^+$ implant), both runs having similar $\Delta E_{coll.}$, shows the implanted sample to have suffered much greater changes. In a similar way, the Ar-implanted sample of run 8 ($\Delta E_{coll.}$ only ~ 7.6 arb. units) showed larger change than did the transmission Xe sample (no.11) having $\Delta E_{coll.} \simeq 10$ units.

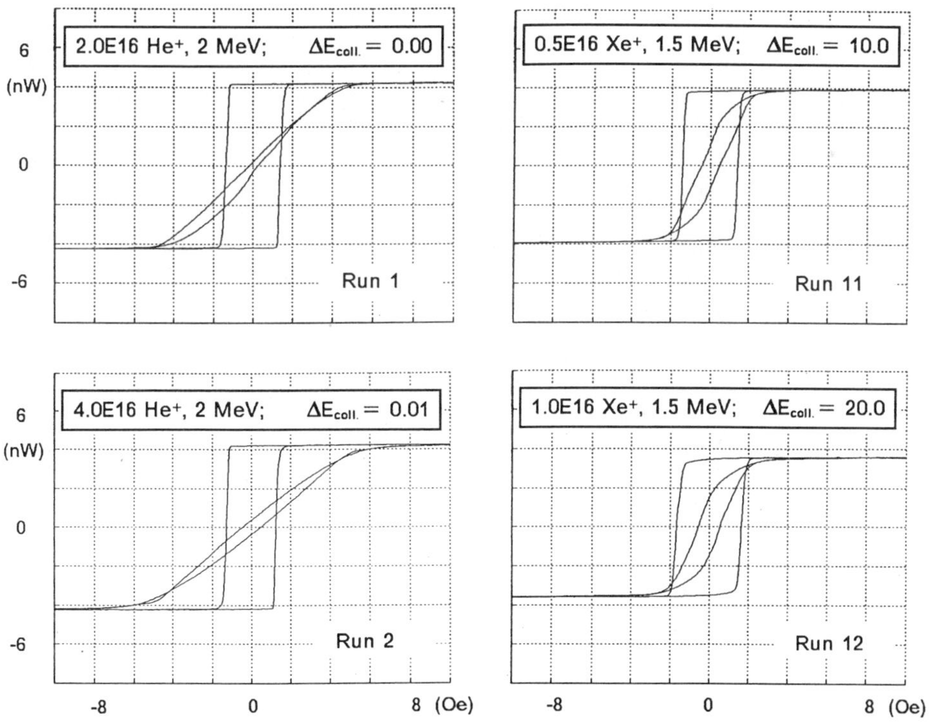

Fig. 2. Hysteresis loops for samples after irradiation with MeV He$^+$ or Xe$^+$ ions, transmitted through the film (Runs 1,2,11,12). For each irradiation, the ion species, energy and dose are shown, together with the relative metric $\Delta E_{coll.}$ describing the total collisional energy delivered within the film by the ion beam.

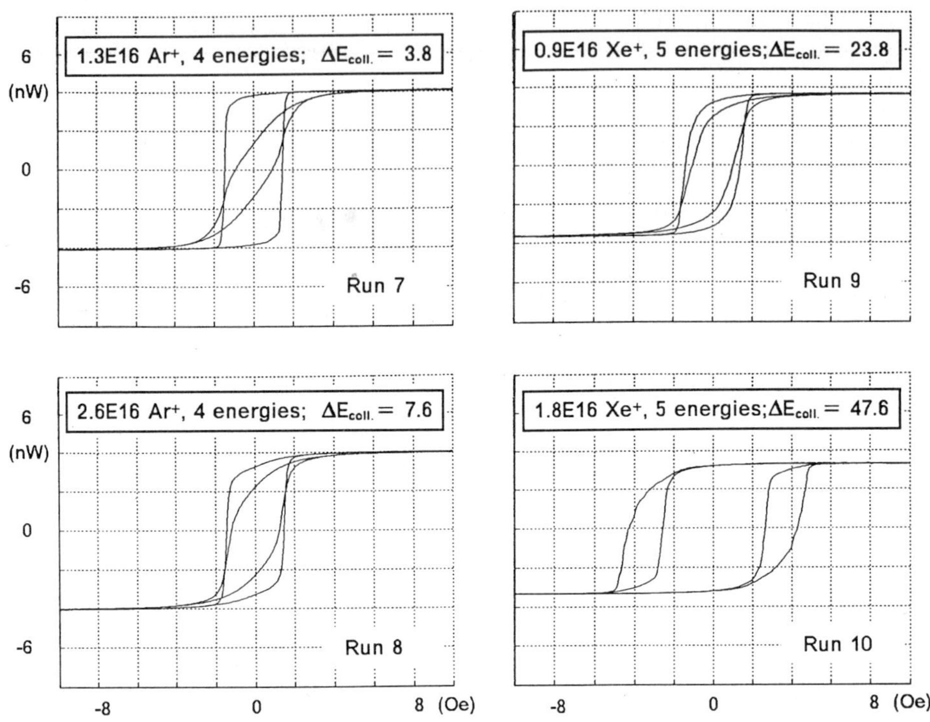

Fig. 3 Hysteresis loops for samples <u>implanted</u> with Ar or Xe, using multiple energies to improve uniformity of ΔE_{coll} with depth in the film. (Runs 7,8,9,10)

It is interesting to note here that the subsequent thermal annealing treatment had only minor effects (sometimes squaring the corners of loops) for all samples except the Ar implants. For the Ar samples, heating converted their loops to shapes very similar to those shown by Xe^+ transmission runs at the same ΔE_{coll}. This would be consistent with escape of the argon from the lattice, and gives some indication of the magnitude of the changes produced by lattice strain.

Fig. 4 shows loops from samples implanted with C^+, N^+, O^+ or Si^+. Again, the main effects are loss of magnetic anisotropy and increased coercivity. Surprisingly, N^+ had a larger effect than the other species, being comparable to that of inert implants with a higher ΔE_{coll}. This implies a chemical bonding effect within the NiFe lattice, being remarkably sensitive to the small (~ 1 at. %) doping level.

Reduced magnetic moment

Fig. 5 displays the progressive loss of "effective magnetic thickness" (a measure of M_s) observed, as a function of ΔE_{coll}. Corrections for sputtering loss of film thickness have been made. The loss of M_s is systematic, but relatively small, consistent with a disordering or segregation process in the NiFe, such as that reported by Riviere at al. [6].

Magnetoresistance ($\Delta R/R$) and sheet resistivity (ρ)

Measurements of $\Delta R/R$ and ρ were made for each sample, and Fig. 6 shows $\Delta R/R$ as a function of ΔE_{coll}. No change was found for the He^+ irradiated samples. $\Delta R/R$ decreased linearly with ΔE_{coll} for all the samples irradiated or implanted with Xe^+ ions, changing

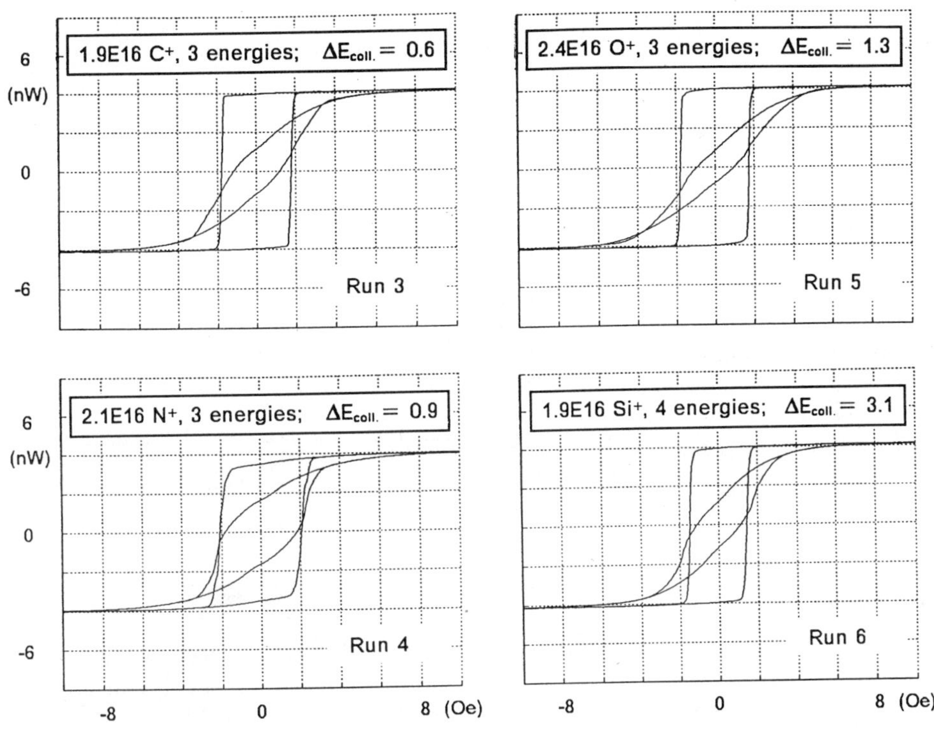

Fig. 4. Hysteresis loops for samples implanted with chemically <u>active species,</u> using multiple energies to obtain uniform concentration of added species at a level of about 1-2 atomic percent. (Runs 3,4,5,6).

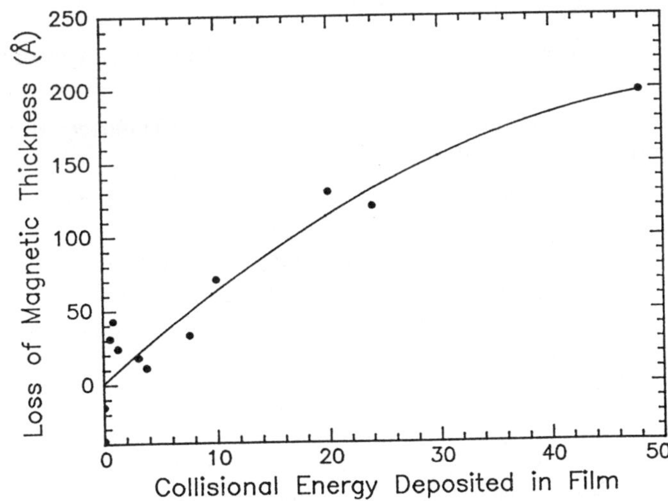

Fig. 5. Loss of "effective magnetic thickness", M_s. (Original $M_s = 1500$Å.)

from its initial value of 4.2 % to 2.3% for the highest $\Delta E_{coll.}$. However, the implants of active species or argon produced comparably large reductions in $\Delta R/R$, even though they had much smaller values of $\Delta E_{coll.}$. This effect is not yet understood. The corresponding resistivity values increased by approx. 30% for the highest value of $\Delta E_{coll.}$.

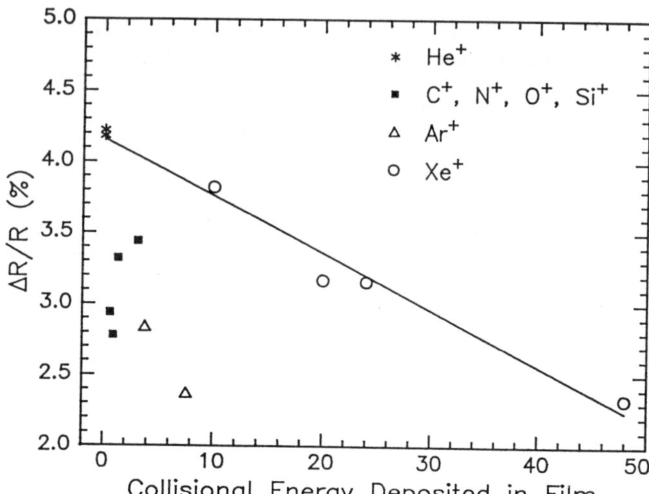

Fig. 6. Magnetoresistance ($\Delta R/R$), plotted as a function of $\Delta E_{coll.}$, for 150 nm NiFe films.

X-ray diffraction data

Standard (θ-2θ) scans and rocking curves (Cu-K$_\alpha$) for all samples, both before and after irradiation showed only sharp (111) lines corresponding to a NiFe d-spacing of 0.2046 nm. The deduced spread of fiber texture ($\sim 3.5°$ fwhm) was not affected by irradiation. Line widths indicated a grain size of about 25 nm initially, increasing to ~ 30 nm for samples with large $\Delta E_{coll.}$. Diffraction intensities were higher after irradiation, implying some ion-induced ordering at the micro-crystalline scale.

DISCUSSION AND CONCLUSIONS

The observations lead to some general inferences about ion beam effects on Permalloy films:

(a) The magnetic response is negligibly affected by electronic energy deposition, such as that of electrons or energetic light ions.

(b) Heavy ions, with large collisional energy transfer, can cause loss of uniaxial magnetic anisotropy (in-plane), greatly increased coercivity, and a small decrease in M_s. It would be expected that NiFe would undergo disordering of the aligned Fe-Fe pairs in the lattice that are believed to be primarily responsible for the initial anisotropy [2,4]. Also, for heavy ions at high doses, it is possible [6,7] that some segregation of Fe to the grain boundaries of NiFe may be induced, leaving a Ni-rich alloy with reduced M_s (see ref. 2, p. 526), and coercivity raised due to such precipitates or to other extended defects that serve to inhibit domain wall mobility.

(c) Implantation of heavy inert species in NiFe can introduce compressive stress in the film. Such stress may be the cause of the observed rise in coercivity attributable to the presence of the implanted species. This component seems to be reversible if the contaminant species is removed.

(d) Implantation of about 1 at. % of a chemically active species produced similar effects, but at a lower $\Delta E_{coll.}$ than for Ar⁺ or Xe⁺. This may indicate another stress effect. Of several ions tested, N^+ had the greatest effect on the hysteresis loop, for reasons yet to be discovered.

In summary, although much remains to be understood regarding the specific physical interactions responsible for the magnetic property changes observed, it is clear that ion beams do have the capacity to serve as a tool for controlling some of the performance properties of Permalloy films for use in technological applications.

REFERENCES

1. A.I. Schindler and C.M. Williams, J. Appl. Phys. **35,** 877 (1964)

2. C.W. Chen, "Magnetism and Metallurgy of Soft Magnetic Materials", Chapter 7, Dover Publications, Inc., New York (1985)

3. A.I. Schindler, R.H. Kernohan and J. Weertman, J. Appl. Phys. **35,** 2640 (1964)

4. L. Neel, J. Pauleve, R. Pauthenet, J. Laughier and D. Dautreppe, J. Appl. Phys. **35,** 873 (1964)

5. J.F. Ziegler, J.P. Biersack and U. Littmark, "The Stopping and Ranges of Ions in Solids", Pergamon, Oxford (1986); also: J.P Biersack and L.G. Haggmark, Nucl. Instrum. and Methods **174,** 257 (1980)

6. J.P. Riviere, P. Bouillard and J.P. Eymery, Phys. Lett. A **138,** 223 (1989)

7. D. Kurowski, R. Meckenstock, K. Brand and J. Pelzl, J. Mag. and Mag. Materials **148,** 99 (1995)

Materials Science Forum Vols. 248-249 (1997) pp. 95-100
© *1997 Trans Tech Publications, Switzerland*

Special Surface Structures in Surface Alloy Growth

D.J. O'Connor[1], J. Yao[2] and Y.G. Shen[1]

[1] Department of Physics, University of Newcastle, Callaghan, NSW 2308, Australia

[2] Department of Chemistry, University of Houston, Texas, USA

Keywords: Surface Alloys, Cu(001), Pd(001), Surface Structure, Ultrathin Films, Interfaces, Low Energy Ion Scattering

Abstract

Growth of surface alloys offers a crystal system freedom to adopt structures which may not be available to the bulk material. To investigate how this freedom is exercised an investigation of the growth of Pd on Cu(001) and Al on Pd(001) has been performed. These surfaces develop a p4g structure which has as a principal driving force the size mismatch between the elements. Proof of the role of driving force will be given by careful codeposition studies. LEED analysis of the two alloys reveals a basic similarity of structure, however ion scattering measurements reveal fundamental differences in the arrangement of surface layer atoms.

Introduction

Atomic size usually only plays a secondary role in the determination of bulk properties of metallic alloys. One example is the combination of Ni and Al to form Ni_3Al. Al has an internuclear spacing which is 15% greater than that of Ni however when combined with Ni as Ni_3Al the internuclear spacing increases by less than 2% and as NiAl by less than 0.5% (although NiAl does have a different structure). At the surface however, the size of the individual components has an important role in the structure adopted by the crystal.

There has been considerable interest in the growth modes of metal-on-metal systems both for what it can reveal about the underlying processes as well as the technological applications to producing surfaces which have valuable chemical, electronic and magnetic properties. Previous studies of the Pd/Cu(001) system (see reference lists in [1-4]) have revealed that there is evidence for observing the role played by atomic size effects in structures. The Pd atom is 7.1% larger than the Cu atom in the pure metallic state which leads to the introduction of strain in the surface region when a thin alloy layer if formed. Initially the surface forms a c(2x2) structure which reflects mixed layer bulk termination of the Cu_3Pd crystal. This surface and the more complex p4g structure have received considerable attention with over 20 studies published so far.

In this study we have extended the initial work on Pd/Cu(001) by including codeposition with Pt and Au to assess the role of atomic size in the determination of structure. The study then extended into the case of Al/Pd(001) to determine to what extent the conclusions from the Pd/Cu(001) study could be generalised.

Pd/Cu(001)

There have been three detailed studies of this system reported recently [2-4] and despite an impressive array of analysis techniques, there are three conflicting sets of conclusions on the true nature of the p4g structure for Pd/Cu(001). In general there is agreement on the fact that 0.5ML Pd deposition onto a clean Cu(001) surface leads to the formation of a c(2x2) surface formation which is the structure of the mixed layer termination of $Cu_3Pd(001)$. Furthermore it is agreed that a total deposition of approximately 1.0 ML of Pd leads to a structure which involves two atomic layers and in LEED displays a p4g periodicity. Where they disagree is the actual structure of the p4g surface. It would be a too simple to state that the size difference between Cu and Pd is sufficient to ensure that the second layer forms a c(2x2) surface which splits the fourfold sites and allows the atoms on the surface to adopt a set of rotations leading to the p4g structure. This argument would only apply if the surface was 100% Cu or Pd and all measurements so far agree that the composition of the surface of the p4g structure is mixed. The most interesting results presented recently have used different techniques to address this issue and from these we have three different interpretations.

In the study by Pope [3] the techniques of MEIS, LEED and H_2 and CO thermal desorption techniques were used. The results of these techniques were compared to molecular dynamics simulation using the embedded atom method. The conclusion arrived at from LEED analysis was that the p4g consists of 80% of the surface covered by Pd in displaced lattice sites above a second layer which has a structure of c(2x2) comprising 50% Pd. The remaining 20% of the surface was covered by a c(2x2) structure. This structure has a Pd concentration which is not in agreement with the measured Pd composition of the surface layer using MEIS and TDS so it was also claimed that there were Cu rich areas on the surface over the c(2x2) layer. From MEIS is was determined that 1.1±0.06 ML of Pd was required to achieve this structure.

In the study by Murray [4] using STM and RBS it was found that the formation of the first layer of c(2x2) mixed layer was accompanied by the growth of Cu islands created from the Cu atoms displaced by the Pd. Further deposition of Pd results in the appearance of a p(2x2) structure on the islands. It was concluded that where there were two mixed layers (Pd,Cu) the structure consisted of a p(2x2) layer over a p4g second layer. This second layer was observed to be that developing from the Cu overlayer and in the regions where there was only one atomic layer mixed, it had the c(2x2) structure. Formation of this structure required 1.3 ML of Pd as measured by RBS.

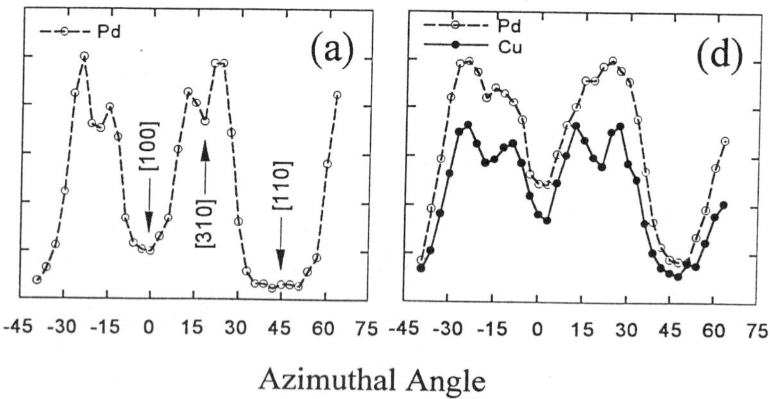

Azimuthal Angle

Fig 1 The azimuthal distribution of 1 keV Li^+ scattered off a) clean Cu(100) and b) 1ML Pd/Cu(100). For a scattering angle of 110° and an incidence angle of the ion beam to the surface of 10°.

In the study by Yao [1,2] using LEIS and LEED, a sensitive measure of the surface layers composition and structure were carefully performed. Under conditions of best p4g LEED pattern, the composition of the first two layers was found to be 47% Pd in the first layer, and 42% Pd in the second layer. The structure of the outermost atomic layer was measured using shadowing and blocking, and nearest neighbour correlation was tested with double scattering of low energy alkali ions. From these measurements there were a number of important observations which impact on the conclusions of the other studies. The first is that the azimuthal scans of Pd and Cu (fig 1) reveal that both atoms experience very similar environments in their nearest neighbours at all coverages up to and including that required for the p4g structure.

There is a reduction in the depth of the shadowing minima of the azimuthal scans, implying either that there is an increase in the atomic spacing on the surface or a reduction in the order of the atomic rows. Polar scans along the low index directions of the p4g surface reveal that the critical angle does not significantly change from that of the clean Cu(001) surface (fig 2). This is in disagreement with the model of Murray [4] which would require a doubling of spacing along these directions and hence a significantly lower critical angle.

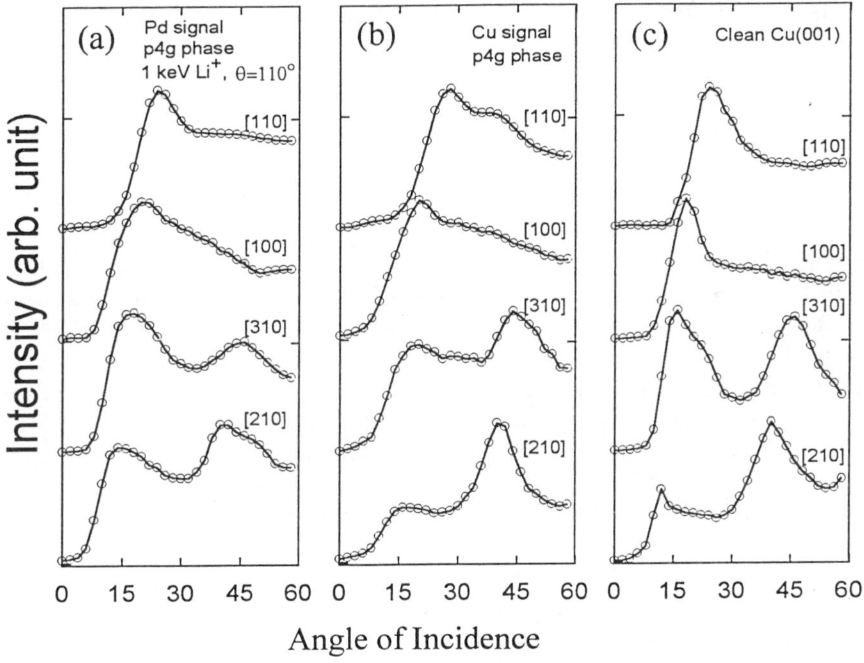

Fig 2. Intensities of 1keV Li$^+$ scattered off clean Cu(001) and p4g Pd/Cu(100) as a function of incident angle at a scattering angle of 110° along the main surrface crystal directions. The angle of the initial rise in yield is a measure of the interatomic spacing between surface atoms along a direction.

The second significant finding came from an analysis of the double scattering feature which measures near neighbour positions. The azimuthal dependence of the double scattering was much broader for the p4g surface than was the case for the clean Cu(001) surface. This was interpreted as being a consequence of the rotation of the atoms about the fourfold hollow sites on the surface further supporting a surface layer p4g structure. Comparison to computer simulation provided an estimate [1] that the surface atoms had moved laterally [1] by 0.25±0.07Å.

The conclusion of this study [1] was that the second layer had a c(2x2) structure which acted as a driving force for the surface p4g. The first layer was a randomly mixed Cu-Pd layer with Pd clusters (and perhaps Cu clusters) forming the p4g structure across the surface. This is not in agreement with the p(2x2) structure observed with STM so we are now investigating overlayer structures which have a p4g structure but would appear to be a p(2x2) in STM.

This model was further tested by the use of codeposition of Au and Pt to see if a structure could be tested under ideal controlled conditions. The motivation for this arose from the structures formed from other adsorbates with similar lattice mismatches. A comparison of the lattice mismatch to Cu is given below.

Table One

Comparison of adsorbate lattice parameters with respect to the Cu substrate.

Element	Lattice Parameter	Size Mismatch
Cu	3.61 Å	-
Pd	3.89 Å	7.8%
Pt	3.92 Å	8.6%
Au	4.08 Å	13.0%

From this table, Pd and Pt have similar mismatches and Au is much larger than Cu. The significant difference between Pd and Pt is that Pd formed the c(2x2) and p4g structures at room temperature. For Pt, it was necessary to anneal the sample to over 150°C to give the Pt atoms sufficient mobility to establish a c(2x2) surface layer. This activation barrier and the similar size difference between Cu and both Pd and Pt have been used to test the conclusions made about one of the proposed structures [1]. After forming a Pd c(2x2) Cu(001) surface, 0.5ML of Pt was deposited which resulted in the formation of a weak p4g LEED pattern. The same pattern is formed if Pd is deposited on Cu(001) c(2x2)-Pt surface. The use of Pt in both cases locks in the structures by ruling out significant mobility by diffusion.

If however Au is deposited rather than Pt on a Cu(001) c(2x2)-Pd surface then a p(2x2) surface is formed. Thus the additional size of Au is sufficient to ensure that it cannot take up the structure formed by the smaller atoms.

Al/Pd(001)

It has been previously reported that the Al deposition on Pd(001) leads to a p4g structure [5] and the size mismatch between these two elements is 4.1%. This system has been studied with LEIS (using Li^+ and He^+ projectiles) and LEED to establish the nature of the surface structure.

When the Al was deposited at room temperature, it formed an overlayer shadowing virtually all the the Pd substrate. Annealing to 450°C for 5 minutes resulted in the Al signal dropping by 15% while the Pd signal increased however there were no ordered spots in the LEED pattern and a high background. A further anneal to 650°C for 5 minutes resulted in the disappearance of all Al signal from the surface layer leaving a pure Pd surface which had a p4g structure. Measurements of the composition of the second atomic layer reveal some concentration of Al though at this stage we are not in a position to determine the Al composition accurately. It can be shown that the second layer is not 100%Al as the p4g structure can be obtained if only 0.5ML of Al is deposited.

The most interesting change is in the azimuthal scan for Pd from clean to p4g Al/Pd(001). There is clearly a shift in the principal directions of the surface in a fashion which is very different to that of the Pd/Cu(001) p4g surface. From fig 3 it is clear that the low index directions in the [100] disappears to make way for a prominent [310] direction feature, and the [110] dip reduces considerably in magnitude. This is evidence of a major rearrangement of atoms on the surface.

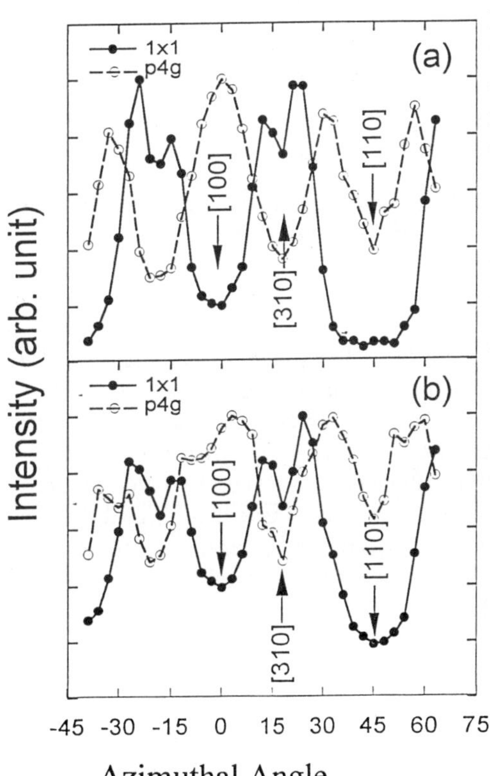

Fig 3 Azimuthal scans from clean Pd and Pd(100) p4g-Al surfaces for a)1 keV Li$^+$ and b) 1keV He$^+$ ions. The shift in the minima reflect a change in the relative location of near neighbours in the surface plane. The scattering conditions were a scattering angle of 90° and an incidence angle of the ion beam to the surface of 10°.

The structural model proposed for this surface is more ideal than that observed for the Pd/Cu(001) p4g surface. A c(2x2) layer is formed in the second layer which is mixed composition of Al and Pd. The surface layer is purely Pd which now undergoes the 'clock reconstruction' which leads to the p4g structure. This rotation involves a much larger lateral shift than that seen for Pd p4g /Cu(100) and it leads to a new set of low index directions in the surface. These correspond to a strong [310] direction with a second, less prominent direction along the [110] direction as revealed in fig two. To test this model, the experimental results were compared to a computer simulation for different values of the lateral displacement (see fig two).

From this comparison it is clear that the surface Pd atoms move off the ideal four-fold hollow position by 0.5±0.1Å.

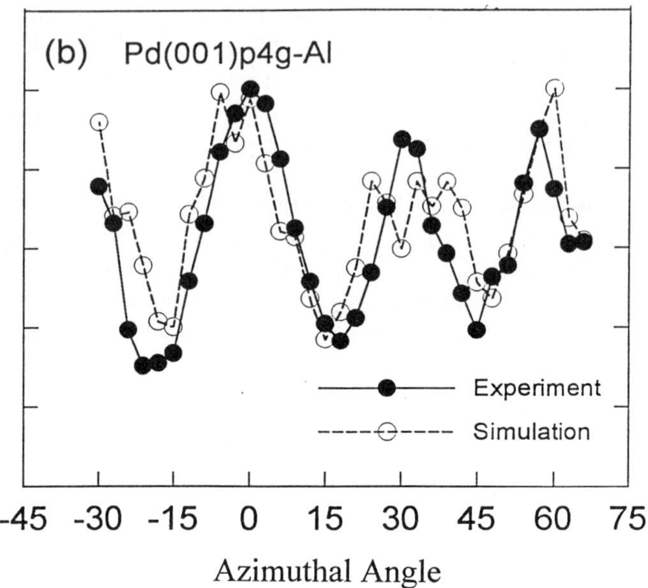

Fig 4 Comparison of the simulated azimuthal distributions with experimental results for 1 keV Li^+ scattered off the Pd(001) p4g-Al surface. The simulation was performed for surface atoms having a lateral shift of 0.5Å.

Conclusions

Both Pd/Cu(001) and Al/Pd(001) surfaces have developed a p4g structure which may have its origins in the size mismatch between the deposited atoms and the substrate. While a simple model can be introduced which establishes the second layer to be mixed (~50%) to form a c(2x2) structure, and the overlayer then is forced to take up the clock reconstruction leading to the p4g. This model breaks down for Pd/Cu(001) as there is a mixed surface layer which will not necessarily be forced to adopt the ideal p4g structure. This has lead to three different interpretations of the p4g structure for Pd/Cu(001) surface from three independent studies. Not all potential models have yet been canvassed however work is continuing to find a model which satisfies the observations from range of techniques

On the Al/Pd(001) it would appear that the ideal precursor is formed with a mixed second layer forming a c(2x2) structure and the pure Pd overlayer adopts the p4g structure as expected. While this solution is a clean conclusion, there still hangs the uncertainty of the exact nature of the concentration of Al in the second layer which may be less than 50% as assumed from the p4g structure.

References

[1] J. Yao, PhD Thesis, University of Newcastle, Australia, 1966

[2] J. Yao, Y.G. Shen, D.J. O'Connor and B.V. King, Surf. Sci. 359, 65 (1996)

[3] T.D. Pope, M. Vos, H.T. Tang, H. Griffiths, I.V. Mitchell, P.R. Norton, W. Liu, Y.S. Li, K.A.R. Mitchell, Z-J. Tian, J.E. Black, Surf Sci, 337, 79 (1996)

[4] P.W. Murray, I. Stensgaard, E. Laesgaard and F. Besenbacher, Surf Sci., in press.

[5] H. Onoshi, T. Aruga and Y. Iwasawa, Surf. Sci. 283, 213 (1993)

Materials Science Forum Vols. 248-249 (1997) pp. 101-106
© *1997 Trans Tech Publications, Switzerland*

Charge Carrier Lifetime Modification in Silicon by High Energy H⁺, He⁺ Ion Implantation

N.Q. Khánh[1], P. Tüttő[3], E.N. Jároli[1], O. Buiu[4], L.P. Biró[1], F. Pászti[2],
T. Mohácsy[1], C. Kovacsics[3], A. Manuaba[2] and J. Gyulai[1]

[1] KFKI-ATKI Res. Inst. for Materials Science, P.O.Box 49, H-1525 Budapest, Hungary

[2] KFKI-RMKI Res. Inst. for Particle and Nucl. Phys., P.O.Box 49, H-1525 Budapest, Hungary

[3] SEMILAB, Semiconductor Physics Laboratory, Inc., P.O.Box 18, H-1327 Budapest, Hungary

[4] Permanent Address: Inst. of Microtechnology, P.O.Box 38-160, RO-72225 Bucharest, Romania

Keywords: Charge Carrier Lifetime, Lifetime Tailoring, Ion Implantation, Radiation Damage, Recombination Activity, Microwave Photoconductive Decay (μ-PCD), Excess Charge Pocket

Abstract

MeV H⁺ or He⁺ was implanted into CZ-Si to set carrier lifetime with the aim of customization of power devices. Commercial Microwave Photoconductive Decay (μ-PCD, Semilab, Inc.) equipment was intended to use for wafer characterization. Realistic evaluation of μ-PCD data required a model to handle cases when the probing carrier pocket does not match the desired defect distribution. Parameters of this multilayer model were extracted from vacancy distributions using the TRIM code.

Introduction

To improve switching properties of bipolar devices, the lifetime of charge carriers has to be adjusted (reduced) to the required value. For power devices, today's industry practices introduce recombination centers by noble metal diffusion, these species (Au, Pt, Pd, etc.) are often quoted as "lifetime killers". Noble metals are interstitial, thus very fast diffusants, therefore, it is rather difficult to control their final distribution. Even so, to achieve homogeneous distribution over the whole wafer thickness, extreme long diffusion times are needed.

Recently, high energy implantation of light ions was proposed to produce recombination centers [1-4]. This technique, if feasible, adds a new horizon, namely, it allows to set the desired lifetime in the last step of the front end technology. i.e., the reduction of lifetime by ion beam introduced recombination centers can be done on fully processed wafers. This allows a single standard technology for the production to run and "customizing" can be done according to the customers' orders. It clear that this will result in substantial cost reduction.

Defects produced by light ions are mostly point defects (divacancies) and have annealing temperatures above 250°C. This is about the same temperature beyond which also noble metals may move. Thus, the technique looks promising.

Another advantage stems from the fact that the required ion doses are "easy" task for ion implantation as regular doses range from 10^9 to 10^{11} ions/cm². Energies (i.e., penetration depth, R_p) should be matched to the special device and to existing thin film structures on the wafer with compositions and thicknesses. For wafers containing typical thyristors, e.g., the proper energy is about 3-4 MeV for protons.

In terms of lifetime tailoring, recent investigations concentrated on proton implantation where lifetime was measured on test devices [5,6] or by free carrier absorption technique [7]. However, to apply ion implantation related lifetime tailoring for other purposes (e.g., latchup protection for CMOS ICs) where different depths of the recombination layer, as well as various lifetime modifications are required, further studies of the ion implantation parameters (species, energy, dose) is needed.

In this paper, we report on the results of modified lifetime by high energy light ion (H^+, He^+) implantations on unstructured wafers as measured with Microwave Photoconductive Decay (μ-PCD) method. The μ-PCD, as a routine technique in the fabrication, allows fast mapping of lifetime distribution without any prefabricated structures. One goal of the present studies was to find conditions, under which the use of commercial equipment for this purpose is feasible and proper.

Experimental

In our experiments, H^+ or He^+ was implanted into 4", n-type, <100> oriented 4 - 7.5 Ωcm Czochralski (CZ) Si wafers of 540 μm thickness using a 5 MeV Van de Graaff accelerator. Implantation was carried out in an end-station with parallel mechanical scanning and in deliberate off-channel directions. The implantation energies were, 1, 2.5, and 4 MeV. The wafers were implanted with doses ranging from 3×10^{10} to 1×10^{12} ions/cm^2. After implantation, the carrier lifetime was measured by Microwave Photoconductive Decay (μ-PCD) technique using WT - 85 type equipment at SEMILAB, Inc. (Budapest, Hungary), which allows fast mapping of lifetime distribution over the full wafer. The native oxide layer was removed by HF dipping then both sides of the wafers were passivated by an iodine solution. For each dose, the average lifetime value over an area of about 2 - 3 cm^2 (average of about 800 - 1200 data) was taken in order to suppress noise.

Results and discussion

A typical result of lifetime modification for different energies and ions are shown as a function of dose in Fig.1. As the dose increases, the measured lifetime decreases till saturation at the highest doses investigated. For a given dose, the higher the implantation energy the more lifetime degradation was detected. These tendencies can be observed both in the case of H^+ and He^+ implantations.

However, the decrease is less intensive for He^+ implantation in spite of the expectation that by creating more damage in crystalline silicon with the same dose and energy compared to H^+. The lifetime dependence on implantation dose is clearer if we draw the recombination activity, i.e. the reciprocal of lifetime, as a function of dose.

In Fig. 2, a good linear relation between recombination activity and dose can be observed for the case of 4 MeV H^+ implantation while the other curves show saturation with increasing dose. For 4 MeV H^+ implantation, a slope of $(0.21\pm0.001)\times10^{-11}$ cm^2/μs was found.

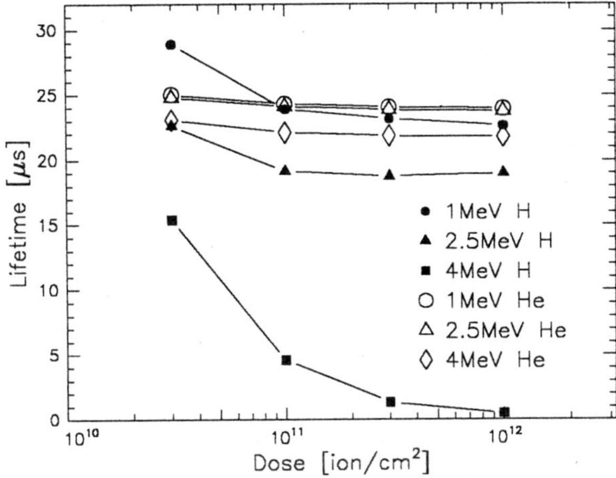

Fig.1. Measured lifetime of charge carriers in n-type CZ-silicon after irradiation with H^- or He^+ ions, as a function of the dose by μ-PCD method.

The above μ-PCD results can be explained by the relation of the mean projected range (R_p), and the light penetration of μ-PCD. In our μ-PCD measurement, a laser pulse with a wavelength of 904 nm generates excess carriers to a penetration depth of about 27 μm. The excitation pulse length is 0.2 μs, and there is a considerable spreading of the excess charge pocket even during this short time. The effective depth of the excited charge carriers at the end of the pulse is about 31 μm for holes and 39 μm for electrons.

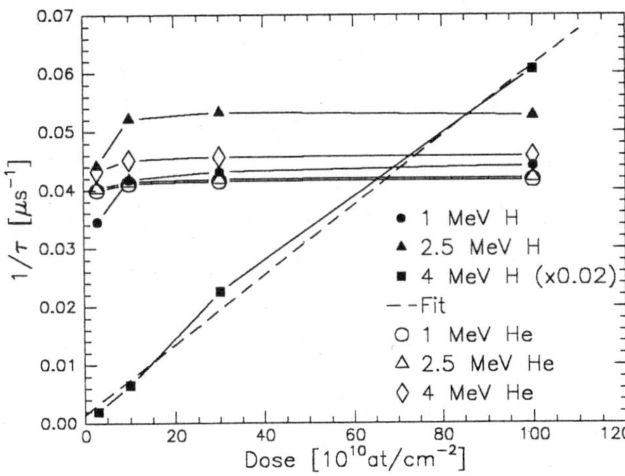

Fig. 2. Reciprocal of the ion implantation modified lifetimes (recombination activity) measured by μ-PCD as a function of dose

The second step of the measurement is the detection of microwave reflection on generated charge carriers. The measured lifetime is influenced by competitive processes, like recombination activity and the diffusion time of the excess carriers to the recombination centers. Thus, with good approximation,

$$\frac{1}{\tau_{meas}} = \frac{1}{\tau_{bulk}} + \frac{1}{\tau_{rec} + \tau_{diff}} \quad , \tag{1}$$

where τ_{meas} and τ_{bulk} are the measured lifetime (effective lifetime) and bulk lifetime, $1/\tau_{rec}$ is the recombination activity at the recombination centers (e.g., surface, crystal damage), and τ_{diff} is the diffusion time.

where τ_{meas} and τ_{bulk} are the measured lifetime (effective lifetime) and bulk lifetime, $1/\tau_{rec}$ is the recombination activity at the recombination centers (e.g., surface, crystal damage), and τ_{diff} is the diffusion time.

Usually, μ-PCD method using surface passivation ($S_{surf} = 1/\tau_{rec} \sim 0$) gives bulk lifetime. In the case of implanted samples the crystal defects act as recombination centers. According to eq. 1, the better the damage region coincides with the excess carrier pocket, the less effective the diffusion time is on τ_{meas}. The size of the excess carrier pocket can be approximated using the effective lifetime:

$$L_{diff} = \sqrt{L_{light}^2 + D_{diff}\left(\tau_{meas} + \frac{t_{light}}{2}\right)} \quad , \tag{2}$$

where L_{diff} is the size of the carrier pocket, L_{light} is the laser penetration depth (in our case 27 μm), D_{diff} is the diffusion constant ($D = 12.5$ cm^2/s for n type silicon) and t_{light} is the width of the exciting laser pulse.

The values of L_{diff} are approximately 142, 171, and 192 μm for 4, 2.5, and 1 MeV H$^+$ implantation, respectively. Comparing these values with R_p-s calculated by TRIM code [8] (148, 68, and 16 μm), one can see that the contribution of diffusion in bulk silicon is negligible at 4 MeV H$^+$ implantation because the whole carrier pocket is trapped in the damaged region. Therefore, as the dose (i.e. the damage) increases, the recombination activity proportionally increases resulting in well-defined linearity, as was shown in Fig. 2. The effect of diffusion is becoming more pronounced even dominant for lower energies, in our case at 2.5 MeV H$^+$ and 1 MeV H$^+$ implantation. Similarly, contribution of this diffusion process is dominant for all studied cases of the He$^+$ implantation. Here, e.g., for 4 MeV, the R_p is only 17.3 μm compared to the L_{diff} being 168 μm. For the cases where $L_{diff} \geq R_p$, only the upper falling edge of the excited charge pocket will sense the recombination activity of the implanted region ("quasi-surface recombination"). At low doses, the lifetime degradation due to implantation is still observable but at higher dose the lifetime in the damaged layer becomes so short that, as we mentioned above, the domination of diffusion process (τ_{diff}) makes the measured lifetime saturated.

To extract useful data for these cases, too, a "three layer" model is presented. As a first step, the thicknesses of the layers and the ratio (k) of the average vacancy densities of one and two layers are fitted to the vacancy distribution calculated by TRIM (Fig.3). With these data, the in-depth distribution of the recombination lifetime can be expressed as follows :

$$\frac{1}{\tau_1} = \frac{1}{\tau_{bulk}} + K \times Dose \tag{3}$$

$$\frac{1}{\tau_2} = \frac{1}{\tau_{bulk}} + K \times k \times Dose \quad , \tag{4}$$

where d_1, d_2 and k are determined from the simulation data, while the proportionality factor K is to be determined independently. Since all measurements were performed by chemically passivated surfaces, the excess charge distribution will have zero derivative at the surface. The solution can be written in the form of trigonometric and hyperbolic functions multiplied by $\exp(-t/\tau)$, characterizing the exponential decay of the particular eigenfunction.

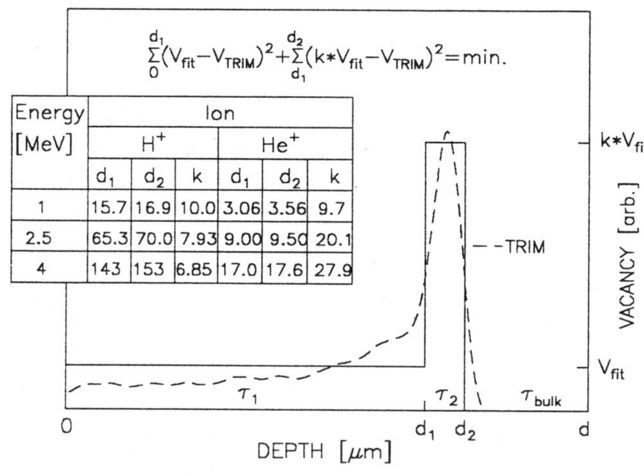

Energy	Ion					
[MeV]	H+			He+		
	d_1	d_2	k	d_1	d_2	k
1	15.7	16.9	10.0	3.06	3.56	9.7
2.5	65.3	70.0	7.93	9.00	9.50	20.1
4	143	153	6.85	17.0	17.6	27.9

Fig.3. Generation of layers for the model using the vacancy distribution from TRIM code

In the top layer, $0<x<d_1$:

$$\Delta n(x,t) = a_1 ch(q_1 x)e^{-t/\tau} \qquad \text{with:} \quad q_1 = \sqrt{\frac{1}{D}\left(\frac{1}{\tau_1}-\frac{1}{\tau}\right)} \qquad \text{if } \tau>\tau_1 \qquad (5)$$

$$\Delta n(x,t) = a_1 \cos(q_1 x)e^{-t/\tau} \qquad \text{with:} \quad q_1 = \sqrt{\frac{1}{D}\left(\frac{1}{\tau}-\frac{1}{\tau_1}\right)} \qquad \text{if } \tau<\tau_1 \qquad (6)$$

In the second layer, $d_1<x<d_2$:

$$\Delta n(x,t) = \left(a_2 ch(q_2 x)+a_3 ch(q_2 x)\right)e^{-t/\tau} \quad \text{with:} \quad q_2 = \sqrt{\frac{1}{D}\left(\frac{1}{\tau_2}-\frac{1}{\tau}\right)} \qquad (7)$$

In the undamaged layer, $d_2<x<d$:

$$\Delta n(x,t) = a_4 \cos(q_3 x)e^{-t/\tau} \qquad \text{with:} \quad q_3 = \sqrt{\frac{1}{D}\left(\frac{1}{\tau}-\frac{1}{\tau_{bulk}}\right)} \qquad (8)$$

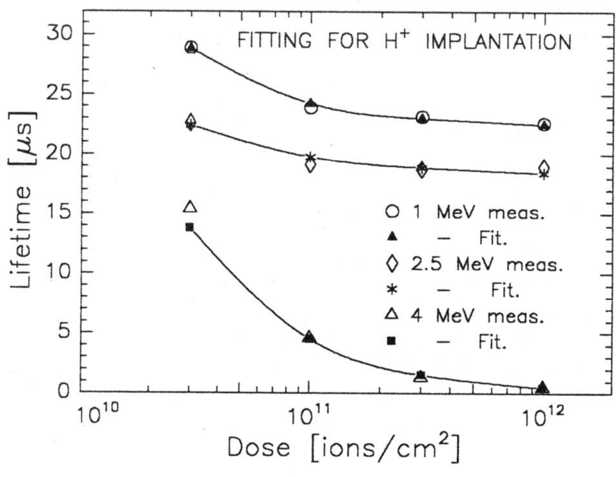

Fig.4. Comparison of measured lifetime data with values from the three layer model.

As an example, measured values and fitting by the three layer model is shown on Fig.4 (see previous page).

On the figure, direct μ-PCD lifetime data are presented as a function of dose following H^+ implantation with various energies. Complete simulation of the measurement process confirmed that the accuracy of the data received for the implantation induced recombination activity is about $\pm 10\%$ for the hydrogen implantation, while it is about $\pm 20\%$ for the helium implantation. Fitting results are shown in the Table:

Ion, Energy (MeV)	H^+, 1	H^+, 2.5	H^+, 4	He^+, 1	He^+, 2.5	He^+, 4
K $[10^{-11} cm^2/\mu s]$	1.71	0.787	0.211	33.8	15.3	5.91
K.k $[10^{-11} cm^2/\mu s]$	17.1	6.24	1.45	328	308	165

Conclusions

It was shown that to characterize ion beam modified lifetime in silicon for power device applications can be done by the μ-PCD technique. The measurement is directly applicable in cases, where the optically generated diffusing carrier pocket fully coincides with the existing defect distribution. This was the case in our experiments for 4 MeV H^+ implantation. In the case of lower proton energies or heavier ions (He^+), the measured lifetime is influenced by diffusion process. To handle this case, a three-layer model is presented using a realistic physical picture where recombination activities can be extracted from measured lifetimes.

Acknowledgements

The work was financed in major part by the European Union through CIPA-CT93-0209 "CaLif" project, additional financing was supported by OTKA grant T017344 (Hungary), both contributions are kindly appreciated.

References

[1] D.C.Sawko and Bartko, IEEE Trans. Nucl. Sci. **NS-30** (1983)1756.
[2] A.Mogro-Campero, R.P.Love, M.F.Chang, and R.F.Dyer, IEEE Trans. Elec. Dev. **ED-33** (1986)1667.
[3] A.Hallén and M.Bakowski, Solid-State Electron. **23** (1989) 1033.
[4] I.Kohno, Nucl. Inst. Meth. **B37/38** (1989) 739.
[5] A.Mogro-Campero and R.P.Love, J. Electrochem. Soc. **131** (1994) 2679.
[6] M.W.Hüppi, J. Appl. Phys. **68** (1990) 2702.
[7] J.Linnros, P.Norlin, and A.Hallén, IEEE Trans. Elect. Dev. **40** (1993) 2065.
[8] J.P. Biersack and L.J.Haggmark, Nucl. Inst. Meth. **174** (1980) 257.

Materials Science Forum Vols. 248-249 (1997) pp. 107-112
© *1997 Trans Tech Publications, Switzerland*

Range Parameters of Aluminium Implanted Targets

M. Hayes, E. Friedland and T. Hauser

Department of Physics, University of Pretoria, Pretoria 0002, South Africa

Keywords: Implantation Profiles, Nuclear Reaction Analysis, Projected Range, Range Straggling, Skewness, Kurtosis

Abstract. Range parameters for aluminium ions implanted at 150 keV into metals with atomic numbers in the $12 \leq Z_2 \leq 42$ region were analysed using a nuclear resonance reaction technique. Depth profiles were determined by detecting the 10.76 MeV photons from the 12.54 MeV to 1.78 MeV exited level transition at the 0.992 MeV resonance of the $^{27}\text{Al}(p,\gamma)^{28}\text{Si}$ reaction. The results of a four moment analysis are compared with TRIM-calculations using the 1991 and 1995 codes after correcting for proton energy straggling and beam width. The experimental projected ranges are in all cases slightly smaller than expected, while the higher moments indicate an almost gaussian shape in reasonable agreement with computed distributions.

1. Introduction

Range parameters of implanted ions are of considerable theoretical interest and of importance for technological applications in metallurgy and micro-electronics. Therefore many theoretical and experimental investigations of depth profiles are found in the literature. Most experimental results are reported for implantations of heavier ions into lighter target materials, as these can easily be analysed by RBS methods [1-9]. Modern range calculations such as those employed in the well known TRIM code [10] generally provide good agreement with experimental depth profiles for a wide range of ion-target combinations and implantation energies, although significant discrepancies have been reported in some cases. Especially for silicon targets Z_1 oscillations, indicating the importance of shell structure effects in electronic stopping, have been extensively studied [1,3,5]. Large range deviations were also reported for heavy ion implantations into boron, beryllium and carbon films [7], which were interpreted as inelastic interaction effects during atomic collision events.

Relatively few experimental range parameters for light ion implantations are found in the literature [9,11-14]. Good agreement between TRIM predictions and experimentally determined projected ranges are reported for carbon implants into silicon, gallium arsenide and stainless steel [11-13], while a severe deviation from the predicted skewness was observed for silicon [11]. Experimental depth profiles for aluminium implantations into silicon, gallium arsenide and stainless steel show reasonable agreement with theoretical predictions [13,14], while deviations are reported for carbon and aluminium implantations into magnesium targets [12,14].

In order to enlarge the limited pool of light ion implantation data, we report in this contribution on range parameters for aluminium implants into titanium, copper, zirconium, niobium and molybdenum; furthermore the depth profile for aluminium in magnesium was re-measured, as some doubts exist on the integrity of the previously used sample.

2. Experimental

Polycrystalline titanium, zirconium, niobium and molybdenum as well as <111> copper and <0001> magnesium single crystals were implanted with 150 keV $^{27}\text{Al}^+$ ions at room temperature.

The samples were implanted with fluences of 5×10^{16} ions cm^{-2} and dose rates were kept in all cases at 10^{13} ions cm^{-2} s^{-1} to prevent excessive heating of the targets. In order to limit possible channeling effects the copper and magnesium targets were tilted 7° relative to the surface normal. Depth profiles were obtained by making use of the ^{27}Al(p,γ)^{28}Si resonance at 0.992 MeV. This very strong and narrow resonance has a width of only 100 eV [15], making it ideally suited for depth profiling. The highly excited level in ^{28}Si decays with a probability of 78% to its first excited state at 1.78 MeV by emission of a 10.76 MeV photon. Energy discrimination is not very critical at this high γ-ray energy as the background is rather low. However, the counting efficiency of the detector is also low at this energy. In order to optimise the signal-to-noise ratio, the energy window was set between 9.5 and 11.0 MeV to detect the photo peak together with its two escape peaks.

Measurements were performed at the 2.5 MV van de Graaff accelerator of the University of Pretoria. The detection system consists of a 5 cm intrinsic Ge-diode and a 5-inch NaI scintillation detector. The accelerator is equipped with an automatic energy scanning system [16] with channel widths much smaller than the beam spread of 1 keV. An analysing beam current of approximately 500 nA was used for the measurements with a beam spot diameter of 4 mm. Targets were kept at room temperature during the analysis.

3. Data analysis and results

The surface position of the samples was determined by measuring yield curves for an aluminium single crystal. A typical yield curve with a channel width of 0.5 keV is illustrated in Fig. 1. This measurement was also used to determine the instrumental energy resolution, which was found to be 1 keV. However, the instrumental resolution plays only a minor role at depths

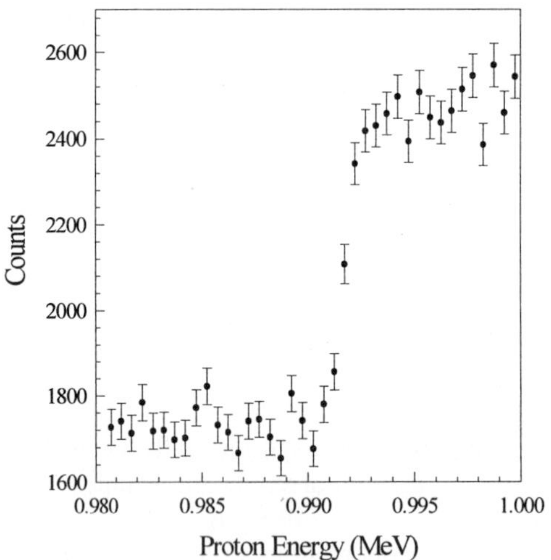

Fig. 1: Yield curve for an aluminium single crystal as a function of incident proton energy .

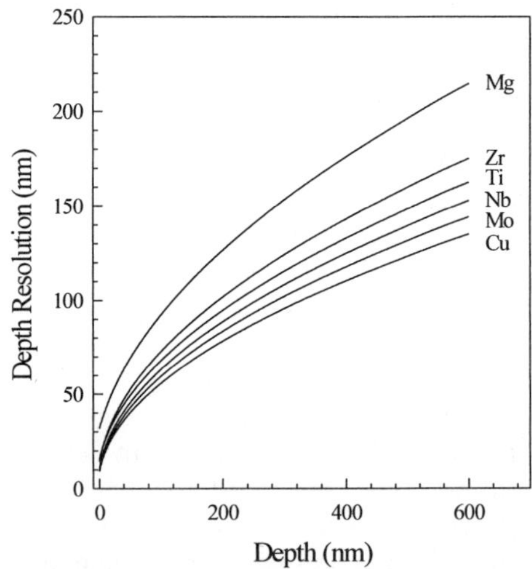

Fig. 2: Depth resolution (FWHM) for the ^{27}Al(p,γ)^{28}Si reaction in magnesium, copper, zirconium, niobium and molybdenum as a function of depth.

corresponding to the projected ranges, where energy straggling becomes the dominant factor. As beam broadening due to straggling depends linearly and energy loss quadratically on Z_1, the importance of straggling increases with decreasing charge of the probing particle. In depth profiling with protons straggling is a major effect and must be taken into account. Its importance is evident from Fig. 2, which shows the depth resolution functions for the different materials analysed. These functions are obtained by assuming Bohr straggling, which should be adequate at a proton energy of 0.992 MeV. The effect of the beam width, which is folded into the resolution function by quadratic summation, is only important right at the surface. The depth scale is obtained by using the stopping power values of reference [17].

For data analysis the effect of the resolution function is taken into account by using a deconvolution algorithm described in reference [18]. Corrected experimental depth profiles are compared with TRIM simulations obtained from computing range distributions for 10^5 ion trajectories. The projected range $R^{(1)}$, range straggling $R^{(2)}$, skewness $R^{(3)}$ and kurtosis $R^{(4)}$ values were calculated according to the definitions used in versions 91.14 and 95.06 of the TRIM code. Errors in the range moments are calculated from asymptotic expressions of the mean square deviations in the moments $R^{(i)}$, which are treated as parameters of a regression function describing the ion distribution. The statistical uncertainty of the position of the surface is included in the error quoted for the first moment.

3.1 Magnesium

The experimental depth profile together with theoretical predictions are shown in Fig. 3(a) and the moments of the depth distribution are listed in Table 1. The experimental projected range is approximately 6% smaller than the theoretical value, which is just outside the quoted experimental error. However, the shape of the distribution, which is nearly gaussian ($R^{(3)}$=0, $R^{(4)}$=3), is in reasonable agreement with the theoretical predictions. The largest deviation exhibits the third moment. TRIM predicts an asymmetric distribution with a negative skewness, but because of the relatively large statistical error of the third moment this difference is insignificant. The results are in disagreement with those of reference [14], where much larger values than expected from theory were found for the projected range and range straggling. We suspect that these discrepancies are due to an implantation error at the previous measurement.

3.2 Titanium

In Fig. 3(b) the results for implantation into titanium are depicted. Again a significantly smaller projected range is observed than expected from both TRIM simulations, while the higher moments are in reasonable agreement. By comparing the distribution moments listed in Table 1, the experimental projected range is about 11% lower than expected, which represents almost four times the statistical deviation. The three higher moments agree within the relatively large statistical uncertainties with both TRIM predictions and are consistent with a normal distribution.

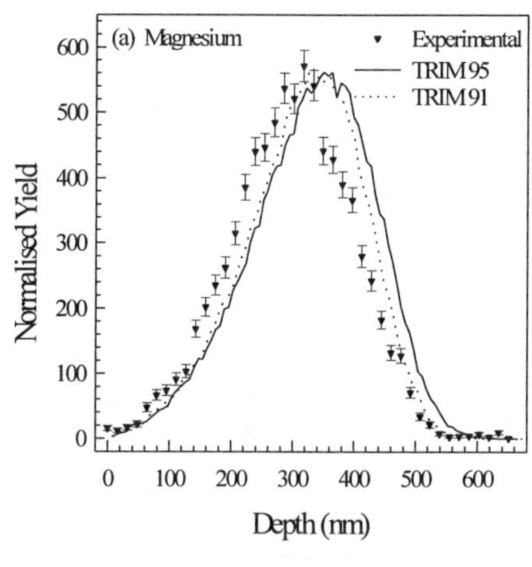

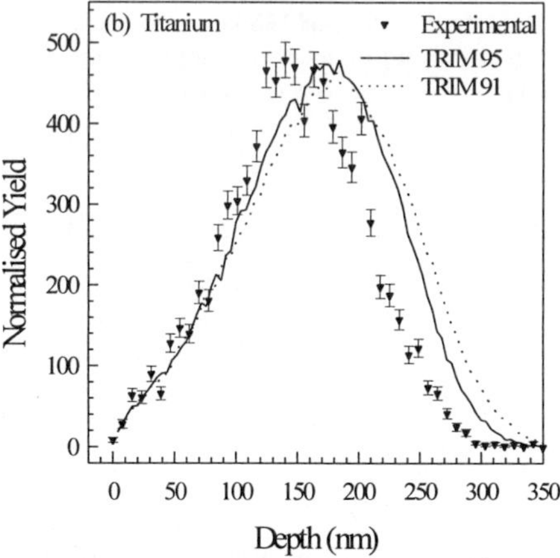

Fig. 3: Depth profiles of 150 keV ^{27}Al$^+$ ions implanted into (a) magnesium and (b) titanium at room temperature compared with TRIM predictions.

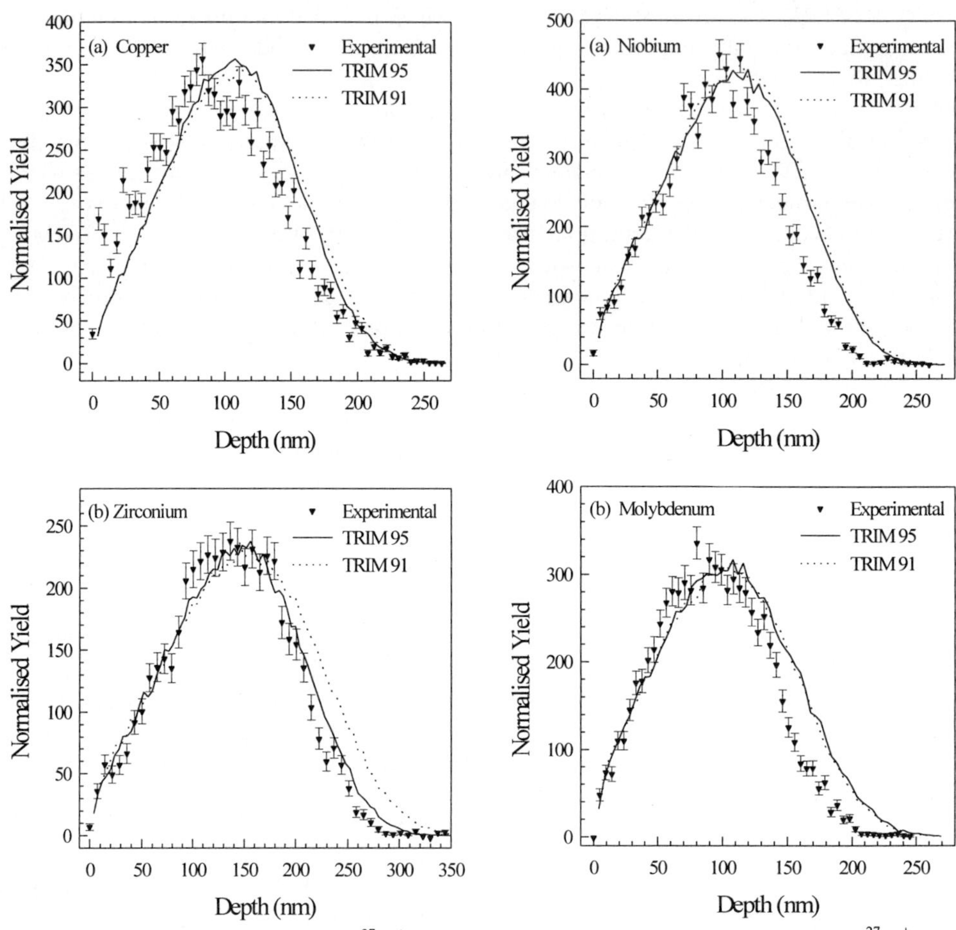

Fig. 4: Depth profiles of 150 keV $^{27}\mathrm{Al}^+$ ions implanted into (a) copper and (b) zirconium at room temperature compared with TRIM predictions

Fig. 5: Depth profiles of 150 keV $^{27}\mathrm{Al}^+$ ions implanted into (a) niobium (b) molybdenum at room temperature compared with TRIM predictions

3.3 Copper

The experimental and theoretical depth distributions for implantation into copper are shown in Fig. 4 (a). Also in this case a slightly lower projected range than expected from the TRIM simulations is obtained. Comparison of the results for the first moment as listed in Table 1 reveals a shift of almost 7% from the theoretical results which corresponds to approximately twice the statistical error. The deviations of the three higher moments from both TRIM estimates are all within the expected statistical errors and are again consistent with a normal distribution.

3.4 Zirconium

As is evident from Fig. 4(b) and Table 1, the two TRIM versions predict significantly different projected ranges for zirconium, which is not the case for the other target materials. The experimental results are in excellent agreement with TRIM 95 while TRIM 91, predicts a first range moment which is higher than experimentally observed by about four times the statistical error. The experimental third and fourth moments are again within statistical uncertainties in agreement with both a normal distribution and theoretical predictions.

3.5 Niobium

The experimental depth distribution for niobium together with both TRIM estimates is shown in Fig. 5(a). Comparing the results for the projected range in Table 1, one finds again that the experimental value is approximately 8% lower than the TRIM results, which is almost three times outside the quoted experimental error. The three higher moments agree within statistical uncertainties with both TRIM predictions but are also in agreement with a normal distribution.

3.6 Molybdenum

From Fig. 5(b) it is obvious, that also for molybdenum the experimentally determined first moment differs significantly from the TRIM estimates. The results listed in Table 1 indicate a 9% lower projected range which is almost five times outside the estimated statistical error. As far as the higher moments are concerned, the relatively large statistical errors make it again impossible to discriminate between the TRIM estimates and a gaussian distribution.

Table 1: Experimental range moments for 150 keV $^{27}Al^+$ ions implanted into samples at room temperature compared with TRIM 91 (T'91) and TRIM 95 (T'95) simulations. The quoted errors are 2σ values.

Sample	$R^{(1)}$ (nm)			$R^{(2)}$ (nm)			$R^{(3)}$			$R^{(4)}$		
	Exp.	T'91	T'95	Exp.	T'91	T'95	Exp.	T'91	T'95	Exp.	T'91	T'95
Mg	301±11	314	326	104±19	95	99	-0.11±0.44	-0.35	-0.33	2.82±0.48	2.85	2.84
Si	244±12	242	255	67±14	77	83	0.02±0.17	-0.30	-0.30	2.77±0.47	2.80	2.77
Ti	148±5	170	163	57±19	66	62	-0.06±0.33	-0.11	-0.15	2.72±0.99	2.56	2.60
S. Steel	104±6	105	105	40±9	45	44	0.04±0.31	0.06	-0.02	2.70±0.43	2.51	2.54
Cu	97±3	105	103	48±13	48	46	0.24±0.23	0.12	0.10	2.59±0.70	2.53	2.54
GaAs	189±9	170	172	69±12	76	83	-0.04±0.12	0.05	0.05	2.55±0.33	2.50	2.51
Zr	135±3	147	138	58±16	67	61	0.01±0.22	0.09	0.03	2.56±0.70	2.47	2.45
Nb	98±3	107	106	43±13	48	48	0.09±0.29	0.07	0.06	2.66±0.94	2.46	2.46
Mo	92±2	101	102	42±11	49	49	0.21±0.22	0.18	0.20	2.59±0.71	2.51	2.53

4. Conclusions

With the possible exception of zirconium the experimentally determined projected ranges for aluminium implants are between 6 and 11% lower than predicted for the metals investigated in this study. For comparison the results of reference [14] for silicon, stainless steel and gallium arsenide are also included. For silicon and stainless steel the projected ranges agree quite satisfactorily with the theoretical predictions, while the result for gallium arsenide is approximately 10% higher; this is twice the expected statistical error. A similar result is reported for fluorine implants in GaAs at energies below 100 keV, but good agreement with TRIM is found at higher energies [19]. The experimental shape agrees in all cases within statistical uncertainty with a gaussian distribution. However, in view of the relatively large statistical errors for the third and fourth moments, agreement with the TRIM estimates is not excluded. The only exception is silicon for which the predicted skewed distribution cannot be accommodated within the relatively small experimental error of the third range moment.

Acknowledgements

The authors would like to thank Professor N.A.S. Crowther for the statistical error analysis, the Wits-CSIR Schonland Research Centre for Nuclear Sciences for the implantations and the Foundation for Research Development for partially funding this work.

References

[1] F. Besenbacher, J. Bøttinger, T. Laursen, P. Loftager and W. Möller, Nucl. Instr. and Meth. **170**, 183 (1980).

[2] S. Kalbitzer and H. Oetzmann, Radiat. Eff. **47**, 57 (1980).

[3] J. Bertholt and S. Kalbitzer, Phys. Lett. **A 91**, 37 (1982).

[4] M. Behar, P.F.P. Fichtner, C.A. Olivieri, J.P. de Souza, F.C. Zawislak and J.P. Biersack, Nucl. Instr. and Meth. **B 6**, 453 (1985).

[5] P.F.P. Fichtner, M. Behar, C.A. Olivieri, R.P. Livi, J.P. de Souza, F.C. Zawislak and J.P. Biersack, Nucl. Instr. and Meth. **B 28**, 481 (1987).

[6] P.L. Grande, M. Behar, J.P. Biersack and F.C. Zawislak, Nucl. Instr. and Meth. **B 45**, 689 (1990).

[7] P.L. Grande, F.C. Zawislak, D. Fink and M. Behar, Nucl. Instr. and Meth. **B 61**, 282 (1991).

[8] P.F.P. Fichtner, M. Behar, D. Fink, P. Goppelt and P.L. Grande, Nucl. Instr. and Meth. **B 64**, 668 (1992).

[9] P. Oberschachtsiek, V. Schüle, R. Günzler, M. Weiser and S. Kalbitzer, Nucl. Instr. and Meth. **B 45**, 20 (1990).

[10] J.F. Ziegler, J.P. Biersack and U. Littmark, in: Stopping and Ranges of Ions in Solids, **vol. 1**, ed. J.F. Ziegler (Pergamon, New York, 1985).

[11] E. Friedland, M. Hayes, S. Kalbitzer and P. Oberschachtsiek, Nucl. Instr. and Meth. **B 85**, 272 (1994).

[12] M. Hayes, in: Determination of Implantation Profiles by (p,γ)-Resonance Reactions, Thesis, University of Pretoria, 78 (1996).

[13] R. Paltemaa, J. Räisänen, M. Hautala and A. Anttila, Nucl. Instr. and Meth. **218**, 785 (1983).

[14] M. Hayes, E. Friedland, S. Kalbitzer, B. Hartman and S. Fabian, in: Ion Beam Modification of Materials, 9th International Conference, ed. J.S. Williams, R.G. Elliman and M.C. Ridgway (Elsevier, 1996).

[15] M.A. Meyer and N.S. Wolmarans, Nuclear Physics **A 136**, 663 (1969).

[16] G. Amsel, E. d'Artemare and E. Girard, Nucl. Instr. and Meth. **191**, 189 (1981).

[17] H.H. Anderson and J.F. Ziegler, in: The Stopping and Ranges of Ions in Matter, vol. 3, ed. J.F. Ziegler (Pergamon, New York, 1977).

[18] E. Friedland, Nucl. Instr. and Meth. **150**, 301 (1978).

[19] M.R. Herberts, P.F.P. Fichtner and M. Behar, Nucl. Instr. and Meth. **B111**, 12 (1996).

Materials Science Forum Vols. 248-249 (1997) pp. 113-118
© *1997 Trans Tech Publications, Switzerland*

Identification of Bandgap States in Semiconductors by Transmutation of Implanted Radioactive Tracers

N. Achtziger

Institut für Festkörperphysik, Universität Jena, Max-Wien-Platz 1, D-07743 Jena, Germany

Keywords: Radioactive Isotopes, Bandgap Levels, Radiotracer, DLTS

Abstract: Bandgap levels in semiconductors are frequently detected and characterized by spectroscopic techniques that reveal their electronic or optical properties, but not the chemical identity of the observed defect. This problem is overcome by using radioactive isotopes to create bandgap states and by repeating the measurements during the elemental transmutation: Defects containing the parent isotope of the decay will vanish on a well defined timescale given by the nuclear half life and correspondingly daughter element related defects will arise. Thus a definite chemical identification of the observed bandgap states is possible.

The key to this kind of experiments is the doping with radioactive isotopes. The paper compares the techniques used up to now and presents some experiments by various authors to illustrate this kind of 'radiotracer spectroscopy'. In greater detail, radiotracer experiments on recoil implanted radioisotopes (^{75}Se, ^{48}V, ^{51}Cr) in silicon and SiC are described. The bandgap levels of these elements are identified.

Introduction

Radioactive Isotopes are being used in solid state physics for nuclear probe techniques [1] or tracer diffusion. In these cases, the nuclear radiation is detected and spectroscopic or spatial information is derived. Recently, a new application of radiotracers in conventional spectroscopic techniques of semiconductor physics came up. In this case, the emitted radiation is ignored. Instead, the elemental transmutation during the decay is the key point: The known decrease or increase of the parent or daughter isotope's concentration is correlated with the time evolution of subsequently measured spectra and element-specific properties can be identified if decreasing or increasing signals are detected. The expected concentration versus time dependence is shown in Fig.1.

Many important properties of semiconductors like conductivity type and magnitude or relaxation times are determined by impurities, either intentionally incorporated 'dopants' or unavoidable impurities. The knowledge of impurity related bandgap states is therefore a key to the understanding of a given material or to the tailoring of material properties for specific device needs.

The paper is organized as follows: chapter 2 describes the implantation of radioisotopes. Chapter 3 presents some examples of recent radiotracer experiments by various authors and compares the spectroscopic techniques used up to now. Chapter 4 describes in greater detail DLTS experiments (Deep Level Transient Spectroscopy) of the author.

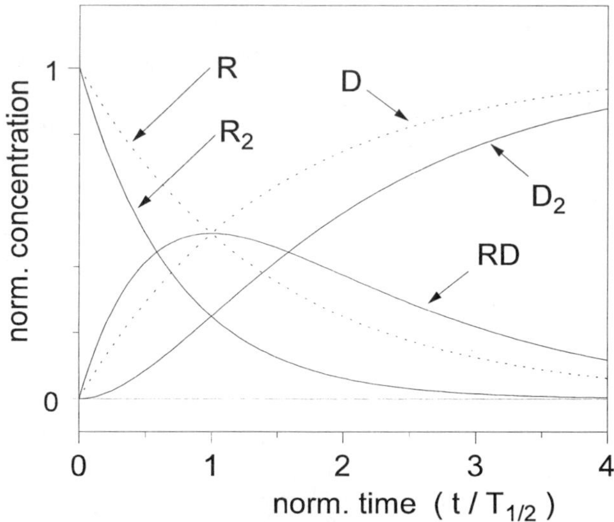

Fig.1: Concentration versus time dependence of a radioisotope R, its daughter element D and of pair defects originating from an initially formed R$_2$ pair. The curve of the intermediate mixed pair RD is based on the assumption that the two atoms in the initial R$_2$ pair are equivalent [2,3].

2. Radioactive Implantation: Requirements and Techniques

To use radioisotopes as a tracer for 'non-nuclear' spectroscopy, the following requirements to the doping process are typical:

(i) isotopic purity,
(ii) chemical purity,
(iii) total number of radioisotopes in the range of 10^9 to 10^{11} per sample,
(iv) no excessive loss of radioactivity during doping.

Requirement (iv) is completely absent in usual, non-radioactive doping procedures, where the amount of initially needed material is free of any constraints. Consequently, the transmission of the whole doping process has to be optimized which in turn favors contaminations. Requirements (i) and (ii) are more stringent here than for the other radioisotope applications like nuclear probe techniques or tracer diffusion. In those cases, the measurement is based on the nuclear radiation and is therefore not disturbed by any other stable isotopes. The techniques discussed here, however, are sensitive to any implanted atom regardless of its radioactivity.

The following implantation techniques shall be discussed:
- 'off-line', i.e. a standard ion implanter with radioactive material in the ion source
- 'on-line' isotope production, mass separation and implantation
- recoil implantation

The 'off-line' implantation generally requires chemical processing or handling of radioactive material. Because of the extremely small amount of matter to be handled, severe contamination problems exist. The radioactivity has to be produced in a separated, preceding process. This sequence restricts the minimum half life to be used.

An 'on-line' implantation facility is operating as the ISOLDE experiment [4,5] at CERN (Switzerland). A high energy proton beam (1 GeV) induces nuclear reactions in a thick target. Radioisotopes diffuse into an ion source and are then accelerated, mass-separated and implanted using a conventional beam-line. As the whole process happens within milliseconds, a short lived β-decay-precursor of the desired isotope can be implanted initially that can be either diffused (e.g. noble gases) or ionized (e.g. alkali elements) very selectively. Thus, for many isotopes a perfectly clean radioactive doping can be achieved. By varying both target and ion source, 68 elements are available at ISOLDE [5] at present. Restrictions mainly apply to non-volatile elements, e.g. transition metals. The technical effort is immense.

The recoil implantation setup is comparatively simple: a primary beam (e.g. 11 MeV protons in the case of 3d transition metals) produces the desired radioisotope via a nuclear reaction when passing a thin foil. The reaction products are kicked out of the foil and are directly implanted into samples mounted off-axis to the primary beam. The implantation is not chemically pure, because atoms of the foil (and eventually products of competing nuclear reactions) are implanted simultaneously with fluences in the same order of magnitude in favorite cases. The technique works best for series of non-volatile elements in the periodic table and is thus complementary to the techniques discussed above. The implantation of several 3d-transition metals via (p,n) or (α,n) reactions is described in Ref.[6].

Generally, the useful half lives are restricted towards short times by the time needed for implantation, annealing, further sample preparation (e.g. electrical contacts) and the measurement itself. The shortest time used up to now is 3.08 h (recoil implanted ^{45}Ti) for a DLTS experiment [7] which is practically the lower limit.

An essential point in the interpretation of any daughter element related phenomena refers to the question, whether the transmutation induces any other effects in addition to changing the element. Because of the emitted radiation quanta, there is always a recoil energy transferred to the daughter atom and generally its atomic shell is excited immediately after the nuclear transmutation. Both effects may give rise to decay induced lattice site changes or to the production of intrinsic defects [8]. A prediction about the possibility of such decay induced effects is usually derived from comparing the recoil energy with the displacement energy of the host crystal.

3. Examples and comparison of radiotracer experiments in semiconductors

The spectroscopic techniques used up to now on radioactive tracers in semiconductors are well-established standard techniques of semiconductor physics: Deep Level Transient Spectroscopy (DLTS), Photoluminescence (PL) and the Hall effect. DLTS requires the preparation of Schottky or pn-diodes and detects 'deep' levels by applying electrical excitation pulses and observing the subsequent relaxation of the diode's capacitance. The sensitivity is high: the minimum concentration of deep levels to be detected is typically 10^{-3} N_S, where N_S is the shallow doping concentration. The Hall effect in an extrinsic semiconductor can be used to measure the free carrier concentration and thus to derive the dominant shallow doping concentration N_S. Thus, the technique is essentially complementary to DLTS with respect to the 'shallow' or 'deep' nature of bandgap states. The sensitivity is much worse than for DLTS since the defect of interest should be the dominating shallow level or be at least a considerable fraction of N_S. In Photoluminescence (PL), the light emission after optical excitation is detected spectroscopically and specific emission lines are detected that may be related to a certain impurity.

Both DLTS and PL detect bandgap states in the near-surface region and are therefore well suited to ion implanted samples. Both Hall effect and DLTS are quantitative regarding the concentration of bandgap states. Therefore, a direct relation between the observed concentration changes and the known nuclear half life can be established. In contrast, the PL signal is not necessarily proportional to the concentration, i.e. the signal height versus decay time may deviate from the exponential nuclear decay law [9,10].

Because of the advantages like sensitivity and quantitativeness, the first radiotracer experiments were performed by DLTS. After implantation of a precursor Hg at ISOLDE, the [195]Au→Pt transmutation was observed in Si by Peterson et al.[11] in order to contribute to the discussion about the Au lattice site in Si. After 'off-line' implantation of ^{111}In, its transmutation to Cd was observed by Lang et al. [12,2] to establish the deep levels of Cd in Si. In both experiments, the decay-induced recoil energy of the daughter atom is low (< 0.5 eV) and the main conclusions are based on the argument, that the transmuting atoms rest on their substitutional sites during transmutation. Further experiments in Si by Peterson et al. include the elements Ir and Os [13]. Further DLTS experiments using recoil-implanted isotopes will be described in chapter 4.

Hall effect measurements by Gwilliam et al.[14] demonstrated the acceptor property of Cd in GaAs: during the ^{111}In to Cd decay, an increasing hole concentration was detected due to the formation of substitutional Cd. Very similar results were found for the ^{67}Ga to Zn transmutation [15]. These implantations were 'off-line'. The sequential decay from [71]As via Ge to Ga was investigated [15] in GaAs as well. In this case, the interpretation involves lattice site changes of the radioisotopes.

Photoluminescence studies after

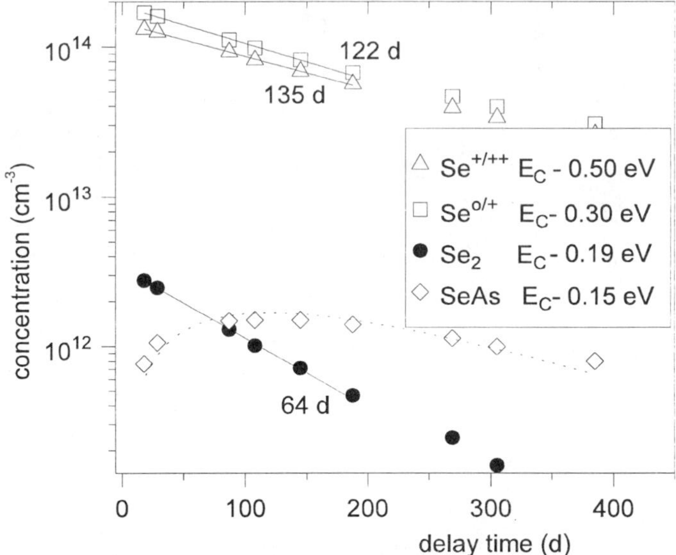

Fig. 2: Concentration of deep levels detected in ^{75}Se implanted n-type Si during the ^{75}Se to As transmutation. The solid lines are fits of exponential functions labeled by their half live. The dashed line is calculated and reflects the expected concentration of the intermediate state.

'off-line' implantation of ^{111}In are reported by Daly et al.[9] in silicon and by Magerle et al.[10] in GaAs. The latter work gives a detailed discussion about the correlation between PL peak height and the increasing Cd concentration during the transmutation.

4. Radiotracer-DLTS Experiments in Si and SiC

All measurements reported in this section are performed on Schottky contacts on recoil implanted and annealed samples. The DLTS spectra display the first Fourier coefficient of the capacitance transient in a predefined time window of 100ms. All level energies given reflect the thermal activation energy for the emission of majority carriers. They are obtained as usual from an Arrhenius plot (T^2-corrected) [16]. The height of a DLTS peak is directly proportional to the concentration of the deep level.

In addition to the chemical identification of bandgap states, the radiotracer technique also reveals how many radioisotopes are involved in a defect as it is demonstrated by the different curves in Fig.1. This was verified experimentally with recoil implanted ^{75}Se isotopes in n-type silicon [3]. After a temperature treatment (1300K, 20min plus 870K, 13d), subsequent DLTS measurements reveal 4 different deep levels. Their concentrations during the ^{75}Se→^{75}As transmutation are given in Fig.2. In addition to the two donor levels of the single Se atom, which expectedly decrease with a half life close to the nuclear half life of 120d, a level at E_C-0.19 eV was detected. Its concentration decreases twice as fast. This directly proves the involvement of two ^{75}Se atoms in this defect. A fourth level at E_C -0.15 eV was found to initially increase and to decrease later on. As its concentration closely follows the curve for the intermediate state (see Fig.1), it is identified to be a SeAs pair. The high isotopic purity (>95%) of the recoil implantation (^{72}Ge (α,n)^{75}Se, isotopically enriched ^{72}Ge foil) is the basic prerequisite for this pair identification. The unavoidable coimplantation of Ge (isoelectronic to Si) and As (shallow donor) does not disturb a DLTS measurement in n-type Si.

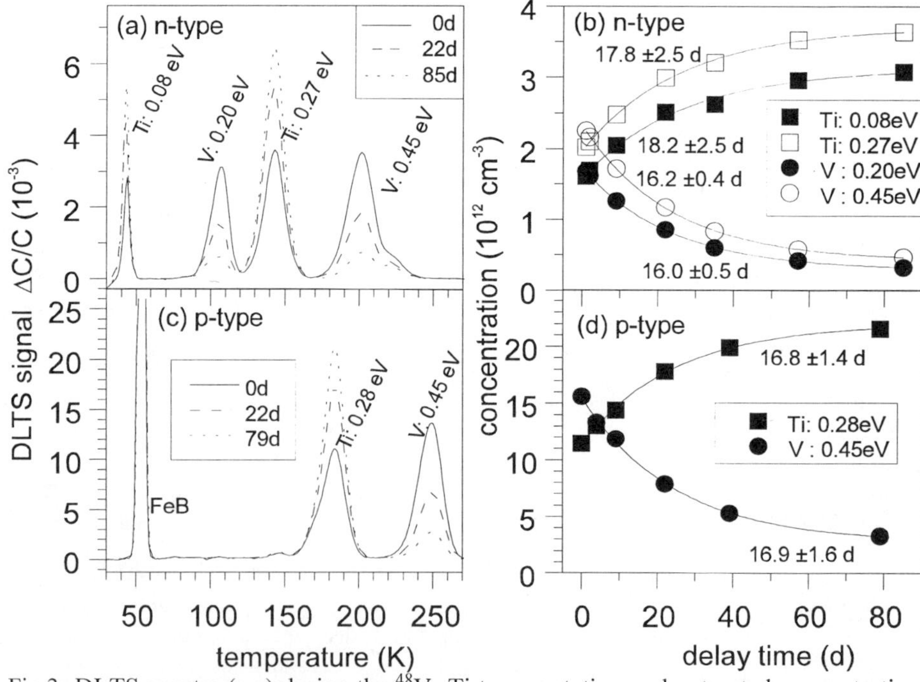

Fig.3: DLTS spectra (a,c) during the ^{48}V→Ti transmutation and extracted concentration (b,d) of deep levels versus delay-time in n-type (a,b) and p-type (c,d) silicon. The concentration data are fitted by exponential functions and labeled by the resulting half-life. The nuclear half-life is 16.0 d. The recoil implantation was done via the ^{48}Ti(p,n)^{48}V reaction.

Transition metals are frequent contaminations in semiconductors and generally create deep levels. Both the sequence of elements from Sc to Cr as well as the element Co have been investigated recently in silicon by radiotracer-DLTS on recoil implanted samples [17]. The results confirm the identification of Ti, V and Cr levels, but strongly contradict to the previous interpretation of Co related levels. The concentration versus time curves measured exactly reflect the nuclear decay, i.e. they are exponentially increasing or decreasing curves with the half life of the nuclear decay as it is demonstrated in Fig.3 for the case of the ^{48}V to Ti transmutation. The V concentration does not vanish completely because of stable, coimplanted V isotopes [6]. The initial Ti concentration originates from the above mentioned unavoidable co-implantation of the reaction foil's atoms. A quantitative transmutation from V to Ti levels takes place at though the recoil energy ($\approx$ 30eV) of the Ti daughter atom exceeds the displacement energy of Si. There are no hints to decay-induced defects.

The same transition metals are important impurities in silicon carbide as well [18]. Deep levels of V and Cr are identified by the following radiotracer DLTS experiment: the radioactive isotopes ^{48}V or ^{51}Cr are recoil implanted into epitaxial, n-type 4H-SiC and annealed at 1600 K for 4 h. Subsequently measured DLTS spectra during the Cr to V decay are shown in Fig. 4. Three peaks are decreasing exponentially with the proper half life and thus prove the involvement of one Cr atom in these defects. The increasing peak in Fig.4 (E_C-0.97 eV) has also been observed as a decreasing peak during the ^{48}V to Ti transmutation and is

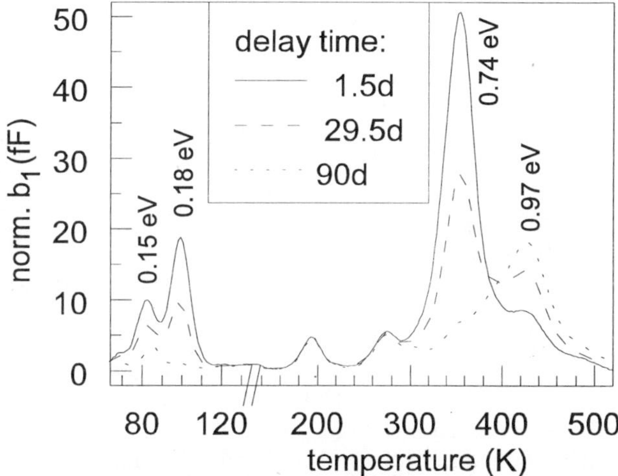

Fig. 4: DLTS spectra of n-type 4H SiC during the transmutation ^{51}Cr$\rightarrow$V. The nuclear half-life is 27.7d. The recoil implantation was done via the ^{51}V(p,n)^{51}Cr reaction.

therefore identified to be V related. By comparing the magnitude of the peak height changes (proportional to the concentrations) in Fig. 4, it is obvious that the transmutation is not complete. This is probably due to the high recoil energy of the ^{51}V daughter of 100 eV which is likely to create defects and thus to prevent most of the daughter atoms from forming the V level at E_C -0.97 eV. The resulting defects themselves, however, are obviously not detectable by DLTS.

5. Conclusions

The combination of radiotracer-doping with a classical, 'non-nuclear' technique of semiconductor spectroscopy (DLTS, PL, Hall) allows a definite chemical identification of bandgap states which is not available from the spectroscopic technique alone. The main problem of these experiments is the radioactive doping process and the interpretation of eventually arising decay-induced defects if daughter atom related levels are of interest. Since the choice of the radioisotope is restricted by half life and availability only, the technique is expected to be widely applicable for the identification of bandgap levels in semiconductors.

Acknowledgement

The assistance of J. Grillenberger, M. Kaltenhäuser, T. Licht, U. Reislöhner, M. Rüb and W. Witthuhn during the recoil implantations is greatly acknowledged.

The work was partially funded by the German BMBF (contract 03WI4JEN9) and by the *Deutsche Forschungsgemeinschaft* (SFB 196).

References

[1]   *Hyperfine Interaction of Defects in Semiconductors*; ed. G.Langouche, (Elsevier, Amsterdam, 1992).
[2]   M.Lang, G.Pensl, M.Gebhard, N.Achtziger, M.Uhrmacher, Mat.Sci.Forum **83-87**, 1097 (1992).
[3]   N.Achtziger, W.Witthuhn, Phys. Rev. Lett. **75**(24), 4484 (1995).
[4]   E.Kugler, D.Fiander, B.Jonson, H.Haas, A.Przwloka, H.L.Ravn, D.J.Simon, K.Zimmer, Nucl.Instr.Meth. **B70**, 41 (1992).
[5]   World Wide Web: http://www.cern.ch/ISOLDE/welcome.html.
[6]   N.Achtziger, H.Gottschalk, T.Licht, J.Meier and M.Rüb, U.Reislöhner and W.Witthuhn, Appl. Phys. Lett. 66, 2370 (1995).
[7]   N.Achtziger, Journal of Applied Physics, at press, scheduled 12/01/96.
[8]   R.Sielemann, L.Wende, G.Weyer, Phys. Rev. Lett. **75**(8), 1542 (1995).
[9]   S.E.Daly, M.O.Henry, K.Freitag, R.Vianden, J. Phys.: Condens. Matter **6** L643 (1994).
[10] R.Magerle, A.Burchard, M.Deicher, T.Kerle, W.Pfeiffer, E.Recknagel, Phys. Rev. Lett. **75**, 4484 (1995).
[11] J.W.Petersen, J.Nielsen, Appl.Phys.Lett. **56**(12) 1122 (1990).
[12] M.Lang, G.Pensl, M.Gebhard, N.Achtziger, M.Uhrmacher, Appl. Phys. **A53,** 94 (1991).
[13] J.W. Petersen, J.Nielsen, Nucl. Instr. and Methods B63, 186 (1992).
[14] R.Gwilliam, B.J.Sealy, R.Vianden, Nucl Instr. and Methods **B63**, 106 (1992).
[15] G.Rohrlack, K.Freitag, R.Vianden. R.Gwilliam, B.J.Sealy, J.G.Correia, D.Forkel-Wirth, Nucl. Instr. and Methods **B106**, 267 (1995).
[16] N.P.Blood and J.W.Orton, *The electrical characterization of semiconductors: majority carriers and electron states*, (Academic Press, London, 1992) p.426.
[17] N.Achtziger, Th.Licht, M.Rüb, U.Reislöhner, W.Witthuhn, in *Proceedings of the Int. Conf. on the Physics of Semiconductors, Berlin, July 1996,* (World Scientific, Singapore) at press.
[18] A.Uddin, H.Mitsuhashi, T.Uemoto, Jpn. J. Appl. Phys. **33**, L908 (1994).

Materials Science Forum Vols. 248-249 (1997) pp. 119-124
© 1997 Trans Tech Publications, Switzerland

Hall Effect Measurements on Transmutation Doped Semiconductors

G. Rohrlack[1], K. Freitag[1], Ch. von Nathusius[1], R. Vianden[1],
R. Gwilliam[2] and B.J. Sealy[2]

[1] Institut für Strahlen- und Kernphysik der Universität Bonn,
Nussallee 14-16, D-53115 Bonn, Germany

[2] University of Surrey, Guildford, GU25HX, UK

Keywords: Radioactive Isotopes, Hall Effect, Antisites, Transmutation Doping

Abstract

Transmutation doping of semiconductors is a well known technique, mainly applied to achieve an extremely uniform low level n-doping of large Si crystals via the $^{30}\text{Si}(n,\gamma)^{31}\text{Si} \Rightarrow {}^{31}\text{P}$ nuclear reaction. Recently, however, it has been shown that the doping of semiconductors by implanting radioactive isotopes can yield valuable information about the processes occurring during the incorporation of dopant atoms into the lattice as well as the defect - dopant interactions occurring after the decay of the instable isotope to a daughter isotope with usually different elemental properties. In this contribution, Hall effect measurements carried out so far on transmutation doped GaAs and InP will be presented. The specific problems and the potential of the method will be discussed.

Introduction

Regarding the intensity and time devoted to the study of intrinsic and extrinsic defects in elemental and compound semiconductors the number of open questions concerning their nature and behaviour is quite surprising. In this situation the application of nuclear methods to the study of these problems has recently attracted much attention [1]. Such methods like e.g. the perturbed angular correlation (PAC) or the Mößbauer technique yield information about the hyperfine interaction at the site of a probe atom and thus allow to draw conclusions about the lattice structure of its immediate environment. However, these techniques require the use of radioactive probe atoms which, due to the energy released locally by their decay, can cause lattice or electronic defects. In order to assess the validity of conclusions drawn from such hyperfine interaction measurements it is therefore desirable to study the fate of the radioactive atoms after the decay. This can be done by measuring the electrical activity of the new element resulting from the transmutation of the originally implanted radioactive atoms e.g. by Hall effect measurements.

Further the implantation of radioactive isotopes also allows the selective creation of defects, e.g. antisite defects, in compound semiconductors far from thermal equilibrium which can then be studied by a variety of classical methods [2, 3].

Finally this technique offers the possibility to implant e.g. InP with ^{111}In which then decays after the annealing process to Cd and thus p-dope such a material where still difficulties in the direct activation of a Cd implant are encountered.

Experiments

As substrate LEC semi-insulating GaAs and InP was used. In some cases the samples were implanted with a stable dopant prior to the radioactive implantation in order to improve the contacts and achieve a well defined position of the Fermi level.

The radioactive isotopes could either be obtained commercially (^{111}In, ^{67}Ga, ^{125}I) or were produced at the Bonn cyclotron (^{71}As). The implantation took place at the Bonn isotope separator or the ISOLDE/CERN facility (Geneva) which also allows the production of a wide range of isotopes (here ^{71}As, ^{77}Br) for such studies. Implantation energies ranged between 160 keV at the Bonn separator and 60 keV at ISOLDE/CERN.

The lattice damage caused by the implantation was removed by rapid thermal annealing (RTA) in a graphite strip heater. Prior to annealing, the GaAs and InP samples were capped with a Si_3N_4 layer of typically 800 Å thickness in order to avoid the loss of the group V component during the heating process. After the annealing the capping layer was removed in HF. Dots of Au:Zn (99:1) for GaAs, Au:Sb for Ge and Au:Zn for InP were evaporated onto the samples in order to fabricate Ohmic contacts after heating. The Hall effect and resistivity measurements were carried out in a *van der Pauw* geometry using a completely guarded system [4].

Results and Discussion

a. Isoelectronic implants

The first successful Hall effect measurement on radioactive dopants were reported by Gwilliam et al. [5]. There it could already be shown that ^{111}In, after implantation and annealing in GaAs, showed the electrical behaviour expected for a single acceptor once it had decayed to the daughter nucleus ^{111}Cd. This is only possible if In occupies a site on the Ga sublattice and the EC decay to Cd neither changes its lattice position nor leads to defects in the vicinity. Further the carrier concentration measured after the complete decay agreed well with the implanted dose, indicating that the radioactive In is incorporated quantitatively onto the Ga sublattice. This was the first proof that the conclusions drawn from PAC results about the microscopic environment of the probe atoms are valid for electrically active dopants.

To further support this point a similar experiment using ^{67}Ga ($t_{1/2}$ = 78h), not only isoelectronic but completely identical with one lattice constituent, which decays to ^{67}Zn was carried out. The results of the Hall effect and resistivity measurements of p-GaAs implanted with ^{67}Ga and annealed as described above are shown in Fig. 1. The sheet resistivity ρ_s and carrier concentration N_s as a function of time show the expected behaviour: N_s increases as a result of the decay of isoelectronic Ga atoms to the single acceptors Zn and D_s decreases consequently. The curves in the drawings are fitted to the data according to the known half-life of ^{67}Ga. The fitting function is given by:

$$N_S = N_{Ge} + N_{Ga}\,(1 - e^{-\lambda t}) \tag{1}$$

where N_{Ge} is the initial carrier concentration resulting mainly from the preimplanted Ge, N_{Ga} is the concentration of the implanted ^{67}Ga (decay constant $\lambda = \ln2/t_{1/2} = 0.0089$ h^{-1}). The fit yields $N_{Ge} = 7.59(2)\cdot10^{12}$ cm^{-2} and $N_{Ga} = 0.59(5)\cdot10^{12}$ cm^{-2}. This is in good agreement with the implantation data if we assume that nearly all of the Zn acceptors resulting from the radioactive decay of Ga are electrically active. The sheet resistivity ρ_s decreases according to the function:

$$\rho_S = \rho_{Ge} + \rho_{Ga}\,e^{-\lambda t} \tag{2}$$

where ρ_{Ge} is the contribution of the preimplanted Ge and ρ_{Ga} the contribution of the Ga atoms to the resulting resistivity. The parameters obtained from the fitted data are $\rho_{Ge} = 10287(21)$ $\Omega/\square$ and $\rho_{Zn} = 1056(47)$ $\Omega/\square$.

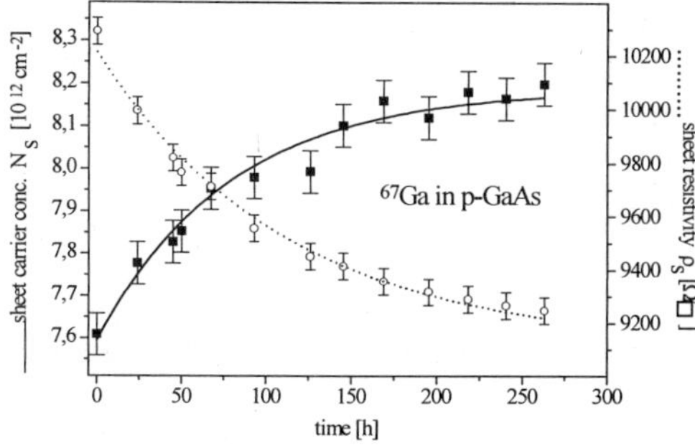

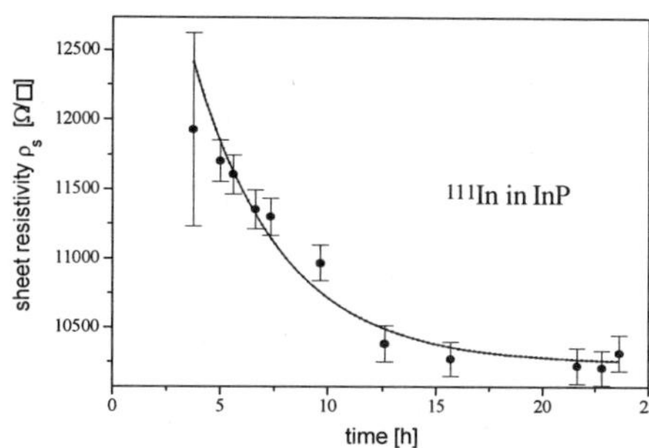

Fig. 1 Time dependence of the sheet carrier concentration and the sheet resistivity derived from Hall effect measurements for ^{67}Ga in GaAs. Thecurves are a fits to the data.

Fig. 2 Time dependence of the sheet resistivity measured for ^{111}In in InP. The full curve is a fit to the data.

The carrier mobility μ_s remains constant at about 80 cm^2/Vs as the decrease in ionized impurity scattering is small because of the small variation in the carrier concentration. Summarizing, the time dependent electrical properties of ^{67}Ga in GaAs are in good agreement with the results obtained earlier for ^{111}In implanted into GaAs [5]. Accordingly, in both cases the radioactive decay of the isoelectronic parent atoms does not disturb the transformation to stable single acceptors, so that these acceptors are electrically active on regular lattice sites. In fact, no passivating defects in the vicinity of the probe atoms are created by the decay, although in the case of ^{67}Ga the electron capture decay recoil energy with 8 eV is about one order of magnitude larger than in the case of ^{111}In (0.9 eV).

Finally, as mentioned above, the implantation of atoms identical to a lattice constituent, e.g. ^{111}In into InP, allows the doping of such a semiconductor by the transmutation of ^{111}In to Cd. In a first experiment s.i. InP was implanted with 5·10^{12} In/cm^2. After a recrystallisation step at 473 K the capped sample was annealed for 10 s at 1053 K. Subsequently Au:Zn contacts were fabricated and the resistivity measured over a period of 25 days. The result is shown in Fig. 2. Again the data could be well described by Eq. (2) with the given half-life of ^{111}In. Due to the relatively bad contacts, Hall effect measurements were not possible in this case. These difficulties are probably due to the fact that during the implantation the relatively low amorphization dose for InP [6]was surpassed and no perfect regrowth could be obtained.

b. The Ga - antisite in GaAs

By implanting ^{71}As ($t_{1/2}$ = 64h) which decays via ^{71}Ge ($t_{1/2}$ = 11.2d) to stable ^{71}Ga it is possible to produce selectively Ga-antisites in GaAs. In a first experiment an ^{71}As implantation was performed at the ISOLDE implanter at CERN, Geneva. Semiinsulating GaAs was implanted at 60 keV to a dose of ≈4·10^{12}cm^{-2}. After annealing at 900° C for 30 s Indium contacts were alloyed to the sample.

Fig. 3 shows the time dependent changes of the electrical properties observed subsequently. Due to the two-step radioactive decay the situation is more complicated than in the former cases. The development of the concentrations of ^{71}As, ^{71}Ge and ^{71}Ga can be described by:

$$N_{As}(t) = N_{0As}e^{-\lambda_1 t}$$

$$N_{Ge}(t) = N_{0As}\frac{\lambda_1}{\lambda_2 - \lambda_1}[e^{-\lambda_1 t} - e^{-\lambda_2 t}] + N_{0Ge}\,e^{-\lambda_2 t} \tag{3}$$

$$N_{Ga}(t) = N_{0As}\left[-\frac{\lambda_2}{\lambda_2 - \lambda_1}\,e^{-\lambda_1 t} + \frac{\lambda_1}{\lambda_2 - \lambda_1}\,e^{-\lambda_2 t}\right] - N_{0Ge}\,e^{-\lambda_2 t} + N_{0As} + N_{0Ge} + N_{0Ga}$$

where N_{0As}, N_{0Ge} and N_{0Ga} are the amounts of the corresponding elements in the sample at the beginning of the measurements (t=0).

The full curve in Fig.3 shows the best fit to the data. The fit values obtained, indicate that the sample has a slight initial p-type doping. The implanted As, incorporated onto the As sublattice, is assumed to be electrically neutral. Moreover, the Ge_{As} atoms resulting from the decay of As are treated as acceptors, which leads to an increase of the acceptor concentration. However, in order to describe the data, it has to be assumed that the finally resulting Ga_{As} atoms have compensating character which decreases the p-type doping. This behaviour is quite unexpected but in good agreement with the development of the sample resistivity (dotted line in Fig. 3). The ratio N_{0As}/N_{0Ge} ≈ 0.5 derived from the resistivity curve is reasonable, since the time elapsed between the generation/implantation of ^{71}As and the first measurement was more than one half-life, so that more than half of the As nuclei had already decayed to Ge. A value of $N_{0As} \approx 1 \cdot 10^{12}$ cm^{-2} is obtained from the fit.

In a second experiment the ^{71}As activity was produced at the Bonn cyclotron via the ^{72}Ge(d,3n)^{71}As reaction. After a destillation step to separate the activity from the bulk of the Ge target ^{71}As was implanted with 160 keV into GaAs predoped with 10^{13} Zn$^+$/cm^2. The predoping was carried out in order to improve the quality of the Ohmic contacts. After an annealing step identical to the sample describe above the time dependence of carrier concentration and sheet resistivity shown in Fig. 4 was found.

In contrast to the previous measurement here only one decay component is observed, indicating that only the decay from As to Ge influences the electrical parameters of the sample, whereas the final decay to Ga yields no further increase in the carrier concentration. Thus the final Ga antisite acts as a simple acceptor like Ge. At present we attribute the differences between the two samples to the fact that in the latter case the use of preimplanted material (10^{13} Zn$^+$/cm^2) led to a

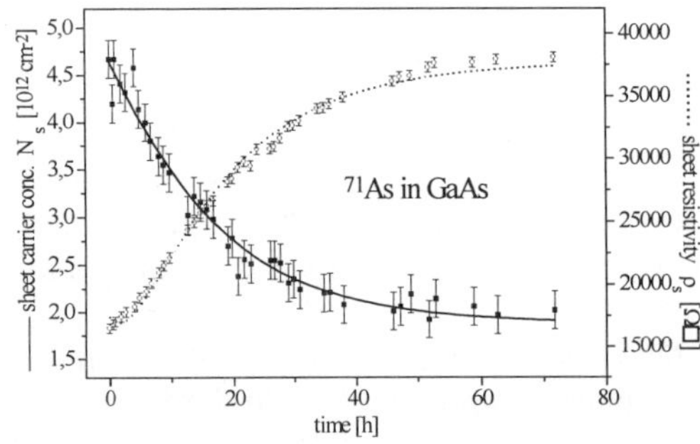

Fig. 3 Time dependence of the sheet carrier concentration and the sheet resistivity derived from Hall effect measurements for ^{71}As in GaAs. The curves are a fits to the data.

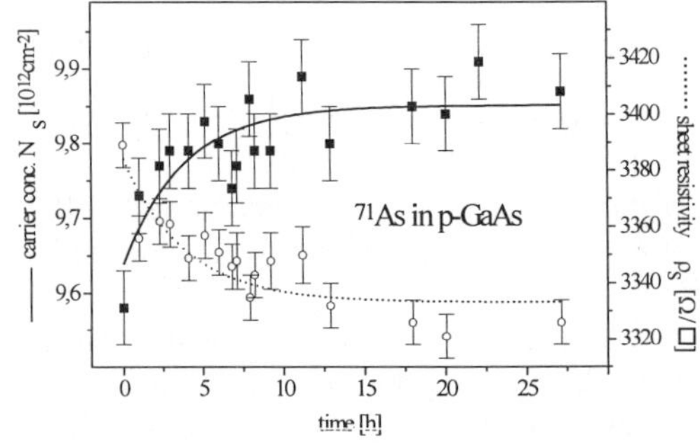

Fig. 4 Time dependence of the sheet carrier concentration and the sheet resistivity derived from Hall effect measurements for ^{71}As in GaAs. The curves are a fits to the data.

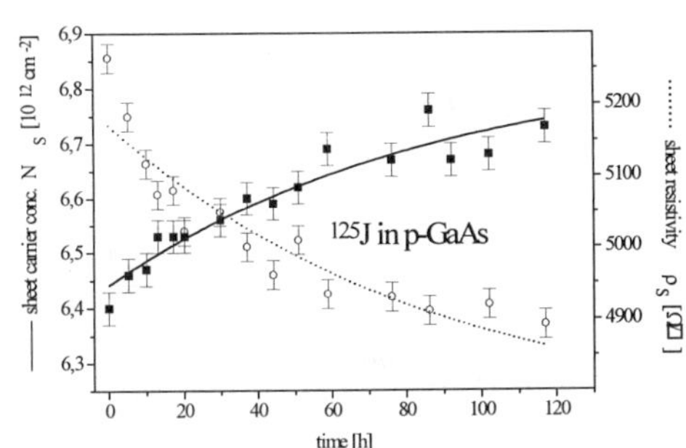

Fig. 5　Time dependence of the sheet carrier concentration and the sheet resistivity derived from Hall effect measurments for ^{125}I in GaAs. The curves are a fits to the data.

well defined Fermi level near the valence band edge, whereas in the first experiment semiinsulating material was used. Therefore the Fermi level was only determined by the implanted radioactivity, and thus might have varied during the experiment, which could cause a different behaviour of the Ga-antisite. The dominance of donors in the ISOLDE/CERN sample after the decay to Ga was independently confirmed by photoluminescence measurements carried out by Daly et al. [7]. However, for a final interpretation of these results additional experiments seem to be necessary.

c. Double donors in GaAs

Recently Bemelmans et al. [8] found that I in GaAs showed a DX-like behaviour after implantation into GaAs. In order to study the behaviour of a halogenic impurity, a double donor on an As site, in GaAs Hall effect measurements on [77]Br and ^{125}I implanted samples were carried out.

After annealing the ^{125}I implanted p-type GaAs samples the time dependent variations of carrier concentration and resistivity shown in Fig. 5 were measured. Again carrier concentration and resistivity could be fitted using eq. 1 and 2, respectively, and assuming the known half-life of 60.1 d for the ^{125}I dopants. The result is consistent with the assumption that I is incorporated on As sites where it acts primarily as a double donor compensating the stable acceptors. After the decay to Te only single donors are left, i.e. the compensation is reduced resulting in an increase of the p-type carrier concentration and a corresponding drop in the resistivity.

Similar experiments were carried out using [77]Br as a dopant. [77]Br is also expected to be incorporated onto the As lattice sites and is an interesting dopant since it has been used for PAC defect studies in several III-V semiconductors [9]. However, in contrast to I, up to now Br implanted GaAs samples showed no electrical activation of the Br dopant, a surprising behaviour which is subject to further investigation.

Summarizing, these first observations of elemental transmutations of dopants in semiconductors by means of Hall effect measurements have shown that such Radio-Hall measurements are feasible and can make valuable contributions to the understanding of the defect - dopant interaction. In the future especially the study of antisite defects produced in such a way or the transmutation doping of semiconductors which normally show compensating behaviour upon doping seems to be most promising.

Acknowledgement

This work has been partially supported by the Bundesminister für Forschung und Technologie (Grant No. 03BO-BON/02) and the EU (CHRX-CT93-0363). One of us (G.R.) acknowledges the award of a scholarship by the Graduiertenförderung Nordrhein-Westfalen.

References

[1]　Proceedings of the Workshop on Submicroscopic Investigations of Defects in Semiconductors (Elsevier (North Holland), Amsterdam) 1991

[2] J.W. Petersen, G. Weyer, S. Svane, E. Holzschuh W. Khndig, Nucl. Instr. and Methods B63 (1992) 179 and M. Lang, G. Pensl, M. Gebhard, N. Achtziger and M. Uhrmacher, Met. Sci. Forum 83-87 (1992) 1097

[3] S.E. Daly, M.O. Henry, K. Freitag and R. Vianden, J. Phys. Cond. Matter 6 (1994) L 643 and R. Magerle, A. Burchard, M. Deicher, T. Kerle, W. Pfeiffer and E. Recknagel, Phys. Rev. Lett 75 (1995) 1594

[4] P.M. Hemenger, Rev. Sci. Instr. 44 (1973) 698

[5] R. Gwilliam, B.J. Sealy and R. Vianden, Nucl. Instr. and Methods B63 (1992) 106

[6] E. Wendler, T. Opfermann, P. Müller and W. Wesch, Nucl. Instr. and Methods B106 (1995) 303

[7] S.E. Daly, M.O. Henry, C.A. Frehill, K. Freitag, R. Vianden, G. Rohrlack and D. Forkel, Mat. Sci. Forum 196-201 (1995) 1497

[8] H. Bemelmans, C. Borghs and G. Langouche, Phys. Rev. Lett. 72 (1994) 856

[9] M. Wehner, P. Friedsam, R. Vianden, S. Jahn amd D. Forkel-Wirth, Mat. Sci. Forum 196-201 (1995) 1419

Materials Science Forum Vols. 248-249 (1997) pp. 125-128
© *1997 Trans Tech Publications, Switzerland*

Characterization of Cu/Al$_2$O$_3$ Interfaces after Heavy Ion Irradiation

K. Neubeck[1], M. Rodewald[1], R. Riemenschneider[2], D.M. Rück[3], H. Baumann[4], H. Hahn[1] and A.G. Balogh[1]

[1] FB Materialwissenschaft, Technische Hochschule Darmstadt, D-64287 Darmstadt, Germany

[2] Inst. f. Hochfrequenztechnik, Technische Hochschule Darmstadt, D-64283 Darmstadt, Germany

[3] Gesellschaft f. Schwerionenforschung, (GSI), D-64291 Darmstadt, Germany

[4] Inst. f. Kernphysik, J.W. Goethe Universität, D-60486 Frankfurt, Germany

Keywords: Ion Beam Mixing, Metal/Ceramic Interfaces, Diffusion, Electron Microscopy, Rutherford Backscattering Spectroscopy, Sputtering, Interface Roughness

Abstract

Ion beam irradiation allows to modify the physical properties of surfaces and interfaces in materials. In this paper we focus on ion beam effects such as ion beam mixing, radiation enhanced diffusion (RED) and phase stability in Cu-Al$_2$O$_3$ marker interfaces. Specimens were prepared by vapour deposition: a copper-film of 70 nm thickness was evaporated onto a polished single crystal alumina substrate and additionally irradiated with 150 keV Ar$^+$ ions. Film thickness and ion energy were adjusted to obtain maximum nuclear stopping power at the interface. Two ion doses were used: $3.6*10^{16}$ and $7.2*10^{16}$ Ar$^+$/cm^2. The mixing behaviour was analysed by Rutherford Backscattering Spectroscopy (RBS) and X-ray Photoelectron Spectroscopy (XPS). Scanning Electron Microscopy (SEM), High Resolution Transmission Electron Microscopy (HRTEM) and Electron Energy Loss Spectroscopy (EELS) were used to obtain information about film quality and phase stability after irradiation.

1. Introduction

Ion beam mixing has been investigated systematically for a number of metal/metal bilayer and multilayer systems [1]. During the past decade, the research emphasis has shifted towards the modification of metal/ceramic interfaces by irradiation [2-6]. It has been found that the mixing efficiency is temperature-dependent for several material combinations. To describe this correlation, different transport processes such as cascade mixing and radiation enhanced diffusion have to be considered. Thermally enhanced mixing was found in Cu/Al$_2$O$_3$ interfaces of both marker and bilayer samples when irradiated in the temperature range of 77-673 K [7-9]. The mobility of irradiation-induced copper vacancies above room temperature is thought to be responsible for this effect. In this paper, we report more detailed studies on the Cu/Al$_2$O$_3$ interfaces irradiated at room temperature. ‚Different spectroscopic and microscopic methods were used to verify the data previously obtained.

2. Experiments

An electron beam evaporator operated at a pressure of $5*10^{-6}$ mbar was used to deposit Cu films on single crystal α-alumina. The film thickness was determined by a quartz thickness monitor and a profilometer. Prior to evaporation, the substrates were pre-sputtered with 500 eV Ar$^+$ ions to remove any contamination present on the surface. Beam heating was minimised using 150 keV Ar$^+$ ions with a maximum current of 200 nA and dose-ranges of $3.6*10^{16}$ and $7.2*10^{16}$ Ar$^+$/cm^2. Ion energy and film thickness were adjusted to obtain maximum nuclear stopping power ($F_{D(Cu)}$ = 125eV/Å, $F_{D(Al2O3)}$ = 110eV/Å) of the Ar$^+$ ions at the metal/ceramic interface as calculated by TRIM-89 [10]. Concentration depth profiles were measured by RBS using 2 MeV ^{4}He$^+$ ions and a backscattering angle of 171 degrees. The computer code RUMP [11] was used to fit the data. X-ray photoelectron spectroscopy (XPS) was performed using Al Kα radiation. The photo electron energy range of Cu ($L_{2,3}$ -2p), O (1s) and Al (2s) was detected.

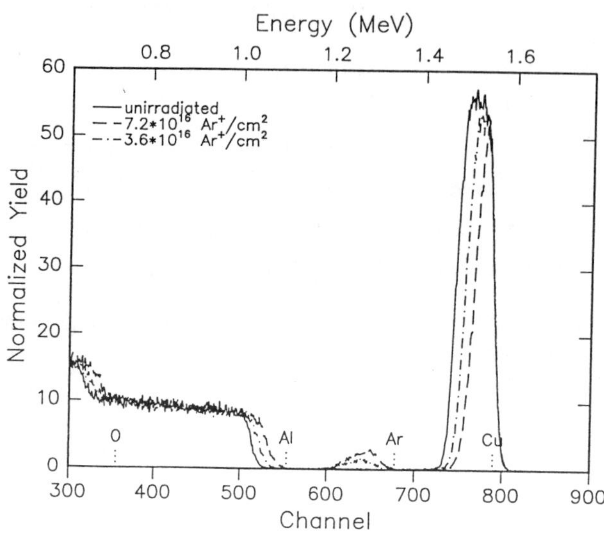

Fig 1: RBS spectra on irradiated Cu/Al_2O_3 specimens in comparison with unirradiated area.

3. Results and Discussion

In Fig. 1, RBS measurements of the irradiated area and the unirradiated area on the Cu/Al_2O_3 samples are compared. The mixing effect is revealed by a decrease in slope at the back edge of the Cu-peak as well as at the front edge of the Al related signals. The reduction in width of the Cu peaks is caused by sputtering during irradiation. A sputtering yield of 4.3 atoms/ion was calculated for the copper layers using the RUMP computer code. The following calculation is based on the assumption, that sputtering did not result in strong surface roughness, which can not be distinguished from

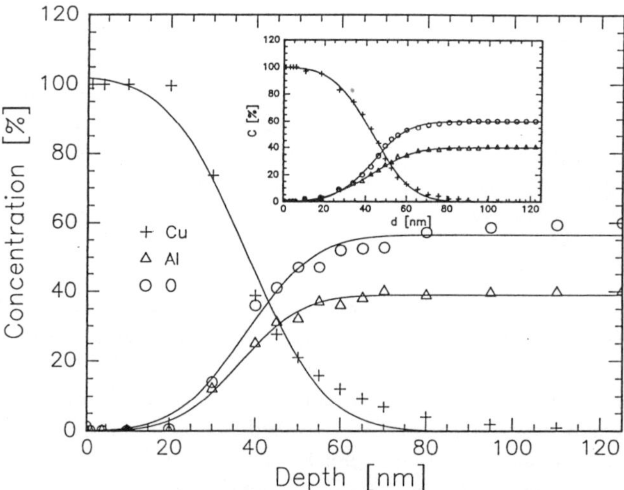

Fig 2: XPS depth profile of Cu/Al_2O_3 specimens irradiated with $7.2*10^{16}$ $Ar^+/cm2$ in comparison with the corresponding RBS evaluation (inserted graph).

For depth profiling, sputtering with 3 keV Ar^+ ions at an angle of 45° was used with intermittent data acquisition. The high resolution transmission electron microscopy study was carried out with a Philips CM 20 UT operating at 200 kV. Its point to point resolution was 0.19 nm. Specimens for the TEM were mounted in cross-section. With mechanical grinding and polishing the sample thickness was reduced to a thickness of ~ 50 µm. Electron transparency of the samples was achieved by ion milling with Ar^+ ions. In addition, elemental mapping by energy-filtered electron microscopy was performed. The quality of the film before and after irradiation was investigated with a Philips XL 30 FEG high resolution scanning electron microscope operated in the secondary electron mode.

interface mixing by RBS. The small reduction of Cu peak in the case of $3.6*10^{16}$ and $7.2*10^{16}$ Ar^+/cm^2 is due to argon incorporation and detector resolution. To quantify the mixing effect, concentration depth profiles for all the constituents were determined. Using diffusion theory, the depth profiles were fitted by an error function (Fig.2) with an argument $(x-c)/\sigma$, where x is the depth, c is the center of the interface and σ is equal to $(4Dt)^{1/2}$, where D is the diffusion coefficient and t is the irradiation time. In order to obtain values which are comparable to those given in literature [1], the mixing efficiency ε_{mix} was defined as $\varepsilon_{mix} = Dt/\phi F_D$, where F_D is the nuclear stopping power and ϕ is the ion dose. A mixing efficiency of

14±2 Å^5/eV was found for both specimens. This value is low in comparison with other material combinations, especially metal/metal interfaces [1]. Destructive depth profiling with XPS confirmed that the mixing efficiency lies within the experimental error bars. The effective broadening of the interface, defined as mixing length l_{mix}, was 22,8±4,6 nm for XPS analysis and 29,8±6,0 nm in the case of the RBS evaluation (Fig.3.). Chemical analysis by XPS of the interface did not show the formation of a spinell $CuAl_2O_4$ structure or CuO_2, since no shake-up satellite peak, which occurs in case of Cu^{2+} bonding, could be found between the Cu $2p_{3/2}$ and Cu $2p_{1/2}$ peaks [12]. However, the shape of the $Cu2p_{3/2}$ distribution suggests that $CuAlO_2$, a thermodynamically stable phase with negative formation enthalpy [13], is formed.

Fig 3: HRTEM micrograph of Cu/Al_2O_3 interface irradiated with $7.2*10^{16}$ Ar^+/cm^2 (a) in comparison with unirradiated interface (b).

The mixed area at the interface was investigated by TEM to establish whether it has a homogeneous structure or wether a new phase has been formed. Fig. 3(b) shows a high resolution micrograph of the unirradiated area. A sharp interface without any contamination or intermediate oxide layer is clearly visible between the single crystal alumina and the poly-crystalline copper film. In case of the irradiated ($7.2*10^{16}$ Ar^+/cm^2) interface, the mixing effect is obvious: the sharp interface has disappeared and dark areas at the alumina surface indicate the inhomogeneous mixing of Cu in Al_2O_3. Below the mixed area, the alumina is amorphous. The Cu-segregations are formed by diffusion during irradiation. The increased Cu concentration in the alumina has been confirmed by electron energy loss spectroscopy with a Gatan GIF. No Argon-

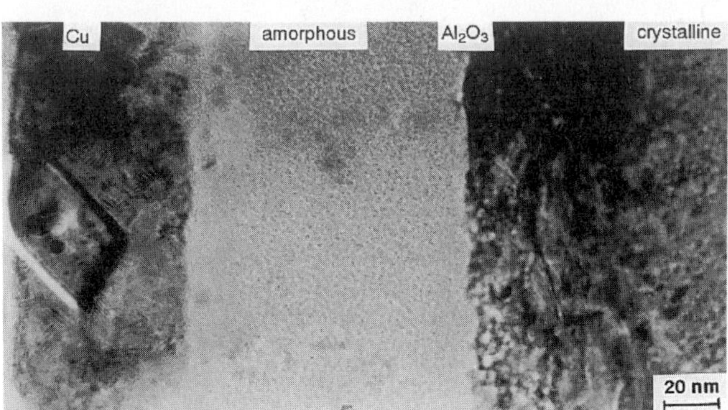

Fig 4: TEM micrograph of Cu/Al_2O_3 specimen irradiated with $3.6*10^{16}$ Ar^+/cm^2 . Cross-section of interface is shown.

bubbles were found at the interface by this method. The broadening of the interface calculated from the image is about 20-25 nm. This result confirms the mixing length calculated from the RBS and XPS analysis. The formation of a $CuAlO_2$-phase could not be observed in the HRTEM images and diffraction patterns. The complete cross-section of the Argon-affected zone after irradiation with an ion dose of $3.6*10^{16}$ Ar^+/cm^2 is shown in fig. 4. The remaining Cu-film is 61 nm thick and still

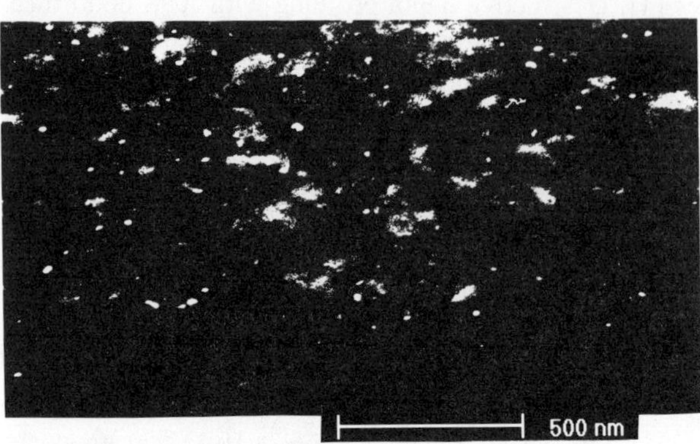

Fig 5: SEM micrograph of Cu/Al_2O_3 specimen irradiated with $7.2*10^{16}$ Ar^+/cm^2.

polycrystalline. The alumina region adjacent to the mixed area is amorphous up to a total depth of 153 nm. It is followed by a crystalline region of 70 nm which contains irradiation induced defects such as dislocations and point defects. The amorphous structure of the alumina is attributed to the incorporation of a high amount of Argon (up to 12%) which is thought to enhance the diffusion of Cu in this region.

Scanning electron microscopy revealed neither holes nor a strong increase in surface roughness due to sputter erosion in the irradiated Cu-film (Fig.5). Therefore, no strong influence of the surface conditions is expected on the results obtained by both the spectroscopic methods. This observation is in good agreement with irradiation experiments on Cu-films reported in [14].

Conclusions

Ion beam mixing in Cu/Al_2O_3 interface after heavy ion irradiation, observed by RBS and XPS techniques, are clearly confirmed by the HRTEM. The mixing was found to be inhomogeneous: partly segregation of Cu in the mixed area and amorphous structure of the Al_2O_3, which is thought to be due to an incorporation of Argon atoms during the irradiation process, was observed by HRTEM. SEM micrographs show no severe damage due to the irradiation.

References

[1] B.M. Paine, R.S. Averback, Nucl. Instr. and Meth. B 7/8, 666, (1985).
[2] A. Perez, E. Abonneau, G. Fuchs and M. Treilleux, C.J. McHargue and D.L. Joselin, Nucl. Instr. and Meth. B 65, 129, (1992).
[3] J.E.E. Baglin, Nucl. Instr. and Meth. B 65, 119, (1992).
[4] A.G. Balogh, M.-P Macht, V. Naundorf, J. Mat. Res. 9, 406, (1994).
[5] E.A. Cooper and M. Nastasi, Nucl. Instr. and Meth. B 91, 558, (1994).
[6] C.J. McHargue, D.L. Joselin and C.W. White, Nucl. Instr. and Meth. B 91, 549, (1994).
[7] K. Neubeck, C.E. Lefaucheur, H. Hahn, A.G. Balogh, H. Baumann, K. Bethge, D.M. Rück, Nucl. Instr. and Meth. B106, 589, (1995).
[8] K. Neubeck, H. Hahn, A.G. Balogh, H. Baumann, K. Bethge, D.M. Rück, N. Angert, J. Mater. Res. Vol. 11, No. 5, 1277, (1996).
[9] K. Neubeck, H. Hahn, A.G. Balogh, R. Hausner, H. Baumann, K. Bethge, D.M. Rück, Nucl. Instr. and Meth. B 113, 186, (1996).
[10] L.R. Doolittle, Nucl. Instr. and Meth. B 9, 344 (1985).
[11] J.F. Ziegler, J.P. Biersack, U. Littmark, The Stopping and Range of Ions in Solids ed. J.F. Ziegler, Pergamon Press, New York, (1985).
[12] A.W. Czanderna, S.P. Wolsky, methods of surface analysis, Vol.1, Elsevier Scientific Publishing Company, (1975).
[13] K. P. Trumble, acta metall.mater. 40, Suppl.,105, (1992).
[14] M.El. Bounanani, A. Chevarier, N. Chevarier, E. Gerlic, H. Jaffrezic and M. Stern, Nucl. Instr. and Meth. B50, 431, (1990)

Materials Science Forum Vols. 248-249 (1997) pp. 129-134
© *1997 Trans Tech Publications, Switzerland*

In-Depth Characterization of Damage Produced by Swift Heavy Ion Irradiation Using a Tapping Mode Atomic Force Microscope

L.P. Biró[1], J. Gyulai[1] and K. Havancsák[2]

[1] KFKI - Research Institute for Materials Science, P.O. Box 49, H-1525 Budapest, Hungary

[2] Eötvös University Budapest, Institute for Solid State Physics,
Múzeum krt. 6-8, H-1088 Budapest, Hungary

Keywords: In-Depth Damage Characterization, Swift Heavy Ions, Tapping Mode, AFM, Parallel Irradiation, Si, HOPG, Mica, Cleavage

Abstract

The damage produced by 209 MeV Kr ions in three different materials, muscovite mica (MM), an insulator; Si, a semiconductor; and Highly Oriented Pyrolithic Graphite (HOPG), a semimetal is investigated in the entire depth of penetration. The comparison of damage in the three materials shows similarities between MM and HOPG, while the damage found on Si is completely different from that found in the previous two. In all three materials damage produced around the trajectory of individual particles was identified. Production mechanisms are proposed for the observed features. The contrast mechanism of Tapping Mode AFM is discussed for some of the observed features.

Introduction

Ion beam techniques are tools that can induce and characterize atomic scale modifications in the properties of materials. The nature, amount, and location of the modification can be kept under a very precise control by selecting the type and the energy of the ions used, and by the beam parameters. Every ion beam technique produces defects in the material which is irradiated. These defects may be desired or not, in most of the practical applications the defects are not desired and have to be eliminated. In some of the applications the alteration in the physico-chemical properties produced by the presence of the defects is useful. Irrespective whether the defects are useful or not, it is very important to have as much information about their distribution and type as possible. Unfortunately, due to the dimensions of point defects - the most frequent primary defects produced during the slowing down of ions - most of the methods used in defect studies are not able to identify and characterize damage produced around an individual ion trajectory.

A relatively new family of methods, the Scanning Probe Methods (SPM) has proved its ability to identify and characterize damage produced by single energetic particles. Scanning Tunneling Microscopy (STM), has been applied to defect studies on Highly Oriented Pyrolithic Graphite (HOPG) [1 - 8] and dichalcogenides of transitional metals [9], Atomic Force Microscopy (AFM) has been applied to muscovite mica (MM) to investigate track formation around the path of swift heavy ions [10 -13]. The major drawback of SPM is that they are surface bound, i.e., due to their inability to probe deeper regions of the target, these methods give little insight in the processes occurring along the ion trajectory. To overcome this difficulty we used earlier a new irradiation geometry in combination with Tapping Mode AFM (TM-AFM) [14] to investigate damage produced in the entire penetration range in Si irradiated with 209 MeV Kr ions. In the present report we apply this method to compare damage produced by 209 MeV Kr ions in three different types of materials: an insulator: MM; a semiconductor: Si; and a semimetal: HOPG.

Experimental results and discussion

The two layered materials MM and HOPG were cleaved after being mounted on the sample holder, right before irradiation. Si samples were cleaved from 4" wafers in as-supplied state. During the same run, irradiations were carried out with a normal incidence on the surface intended for TM-AFM examination (N), and with the beam incoming in a direction parallel to this surface (P). The geometry used for parallel irradiation was described earlier [14]. A dose of 10^{12} cm^{-2} of 209 MeV Kr ions was used.

A Nanoscope III TM-AFM was used to image the samples. Etched Si tips with resonance frequency in the range 279 - 300 kHz were used. In the case of P samples the edge of the face on which the beam was falling was located first to have the "0" of the depth scale. On N samples several regions were selected for scanning at random.

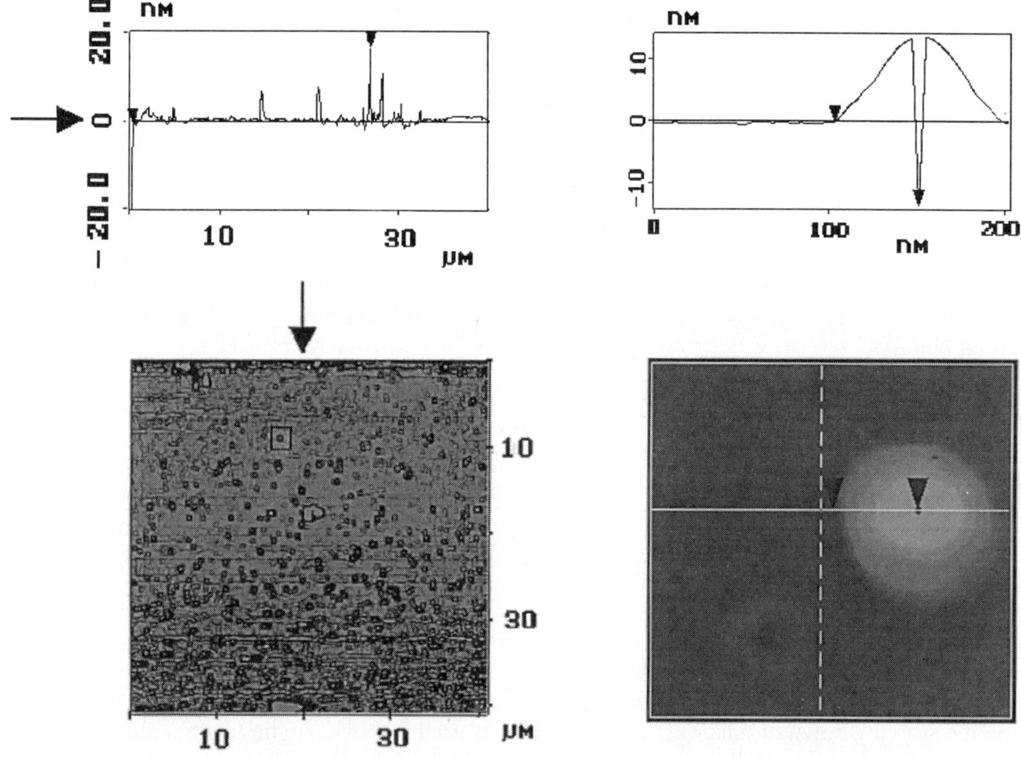

Fig. 1

Plane view AFM image of P-Si sample. Artificial illumination from top. The arrows indicate the direction of the ion beam. Black square marks a B4 type structure.

Fig. 2

Detailed AFM image showing a B4 type structure with a top channel.

In Fig. 1 a large scale image is shown for P-Si. An artificial "top illumination" type image was generated to evidence more clearly the features (B4 type) which give the most important contribution to the Root Mean Square (RMS) roughness measured in the TM-AFM images [15].One can notice that starting from the edge of the face on which the ions enter the target there is a region of 5 μm width in which there are only a few features of this type. The damage in this depth region consists mainly of elongated holes with a depth of the order of 1 nm which run parallel with the beam direction, and it is attributed to electronic stopping effects, in an earlier report these features were called A type features [14]. Deeper than 5 µm a different family of features is observed, called B type features, produced by nuclear cascades. The most significant contribution to the RMS value is given by the B4 type features, their number increases continuously till the region around the projected range, R_p = 28 µm (TRIM code [16]). The line cut in Fig. 1 shows that at R_p dramatic modifications of the surface topography take place. A more detailed analysis of B4 type features showed that they a have rounded conical shape, and some of them have a narrow and deep vertical channel in their topmost part, see Fig. 2. Although B4 features give the most important contribution to the RMS roughness value measured in areas of 1 $µm^2$, the most numerous features are shallow craters, called B2 type features, shown in Fig. 3. Their contribution to the RMS roughness value is reduced because their typical extension in the vertical direction is of the order of 1 nm. The B2 features appear for the first time in the transition region around 5 µm, where the value of electronic stopping power decreases below the damage production threshold. In the region from 20 to 40 µm a strong overlap of very shallow B2 features is found, here their typical vertical extension is 0.1 nm. The B1 and B3 features not treated here in detail because of their relatively small number are transitional features which appear around 5 µm, these have a similar size as the B2 features. The family of B type features is attributed to cascades which are produced around the trajectories of knocked on Si atoms which emerge, or come to rest close to the target surface. The TM-AFM is sensitive not only to the topography of the sample, but to the local modifications of the elastic properties too [17]. The "intensity" of the contact of the tip on the sample is measured not by the bending of the cantilever, but by the modification of the amplitude of its vibration. Due to this, in regions with great variations in density, narrow channels like that observed in the top region of some of the B4 type structures may appear. Fig. 4 shows schematically the influence of structural modifications around the trajectory of an emergent particle on the TM-AFM image.

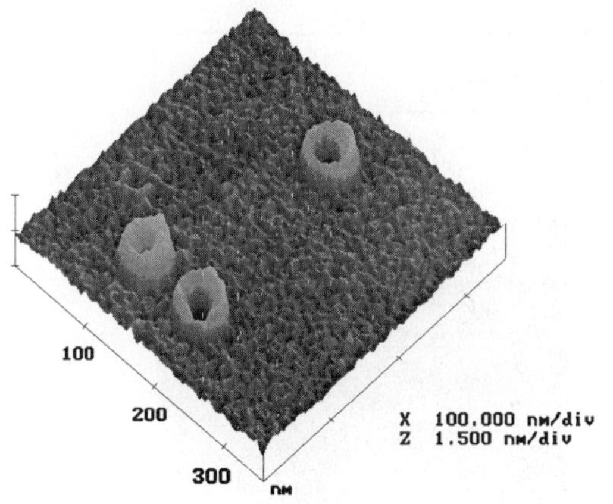

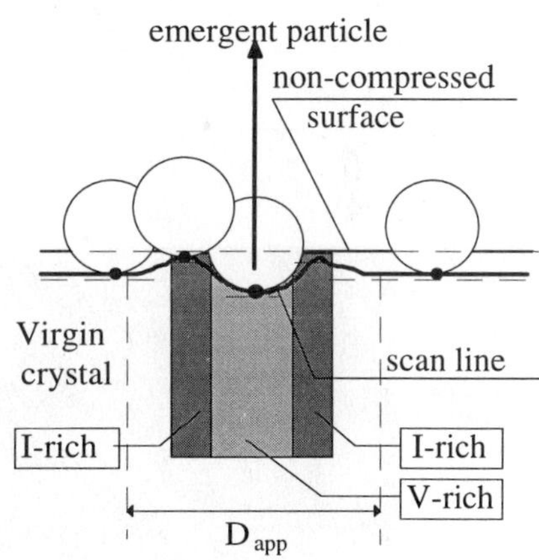

Fig. 3
B2 type features on P-Si sample. 3D presentation.

Fig. 4
Mechanism of image contrast for TM-AFM while scanning a B2 type feature. The AFM tip is modeled as a sphere.

In P-Si samples at depths greater than 40 µm, the damage is produced mainly by the well channeled fraction of the beam. It consist of a decreasing number of shallow craters that differ from B2 type features by not having a rim extended over the original surface. On the N-Si samples only B4 type features were found. Some effect of the native oxide on Si may not be excluded. To decide the magnitude of this influence experiments carried out under ultrahigh vacuum are needed.

The P-MM and P-HOPG samples showed cleavage in the region extending from the irradiated face to a depth of the order of 50 μm. Because of mechanical instability this made difficult the TM-AFM examination. The cleavage was much stronger in the case of MM, which is an electrical insulator with a particular structure: negatively charged layers built up of SiO_4 tetrahedra with relatively high atomic density, alternate with layers of K^+ ions. The cohesion of the material is given by electrical attraction between the oppositely charged layers. When a high energy particle moves in a direction parallel with the layers of this structure, the high energy electrons - produced by the inelastic interaction of the penetrating ion with the electrons of target atoms - will neutralize the K^+ ions at distances of the 1 μm order from the particle trajectory. If this process takes place close to the surface, in the region where the neutralization took place the material will explode due to the mutual repulsion of SiO_4 layers. As a result long channel-like structures are produced, like the one shown in Fig. 5. One can see that the depth and width of the channel decreases as the ion penetrates deeper. Most channels end before the depth of 50 μm is reached. A higher resolution investigation of regions free of channels showed that the surface is undulated with a periodicity of the order of 1.5 μm, and an amplitude of 10 - 20 nm. If the topmost layer of MM is removed by cleaving and the freshly prepared surface is examined, a relatively flat region which extends to a depth of 3 μm is followed by a very heavily damaged material extended to about 8 μm. Deeper, features similar to those reported in Ref. [13] are found.

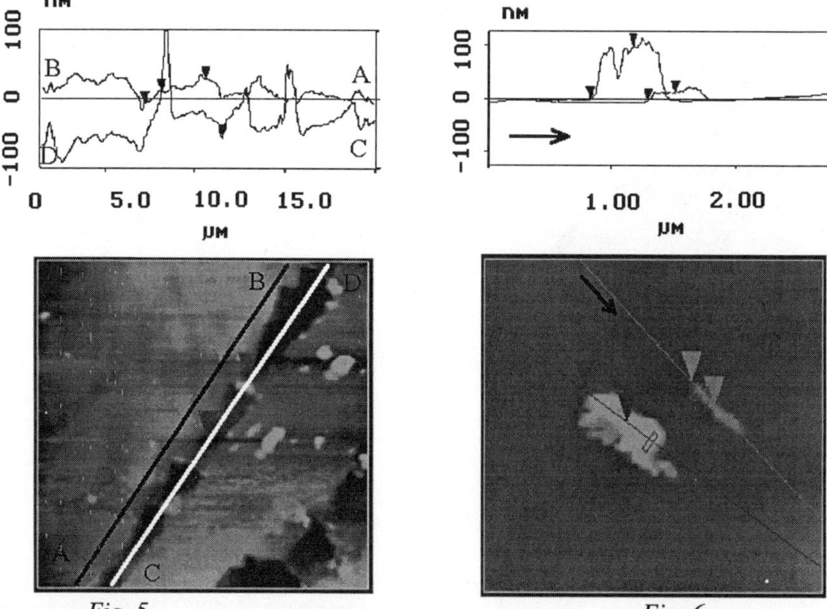

Fig. 5
Channel like feature in P-MM. The two line-
cuts show the surface profile in the channel,
(CD), and nearby, on the surface (AB).
The ion moves from D to C.

Fig. 6
Feature in the flat region of P-HOPG.
The line cuts show the significant dif-
ference in vertical extension. The
arrows show the ion beam direction.

In P-HOPG, the first 11 μm from the irradiated edge are flat as compared with the MM, however, undulation of the surface is also found. The periodicity is of the order of 1 μm, while the amplitude is of the order of 5 nm. This is an indication that although the interlayer cleavage is also present, like in MM, the amount of cleavage is significantly smaller. No undulations were found at depths exceeding 50 µm. Some features elongated in the beam direction are found in the first 10 µm. In Fig. 6, the two main types of features are shown. The irregularly shaped features have a vertical extension roughly one order of magnitude higher as compared with the smaller ones which have a regular "rod-like" shape. The lateral extension of the rod-like objects can only be estimated as their dimensions are comparable to those of the tip radius, 78 nm. Based on a simple geometrical model that uses a sphere for the tip and a longitudinally cut half cylinder for the object, where the measured height is taken as the cylinder radius, a lateral extension of the order of 20 - 30 nm can be estimated for the rod-like objects. This value is in good agreement with earlier STM results of track dimensions on HOPG irradiated with 215 MeV Ne ions [8]. The high resolution investigation of the flat region of HOPG shows that there are present objects with a rounded conical shape, with an apparent diameter of 200 nm and height in the range of 3 - 5 nm. These features are a result of the convolution of the tip shape with an object of comparable size on the sample. As order of magnitude the dimensions of these object coincide with those found in the case of Type I features, during STM examination of HOPG irradiated under a normal incidence with 215 MeV Ne ions [7]. The high features with irregular shape (Fig. 6) are attributed to branched collisional cascades initiated by the knocking on of target atoms by the incoming ions. The somewhat elongated shape of the feature is attributed to channeling in HOPG (the spacing of graphene layers is 0.335 nm). The shape of the rod-like object strongly suggests that it is produced around the ion trajectory where a certain threshold is exceeded in the deposited energy. The exact nature of the modification induced is a question. Recently, the possibility of diamond formation from graphite as a consequence of bombardment with fission fragments was reported [18]. If indeed a transition of this kind is responsible for the production of rod-like objects, the TM-AFM will react to the change of elastic properties and the object may not be a real topographical object. However, earlier STM work on 215 MeV Ne irradiated HOPG showed various kinds of surface distortions around the tracks of energetic particles. These may range from the apparition of a superstructure [7] to formation of tub-like features [8], the width of these features was found to be in the range of 20 -30 nm. The small features with rounded conical shape are a result of imaging the very end of the tip by Type I features that are produced on the surface where nuclear cascades generated in deeper regions reach the sample surface. Similar features were found on the N-HOPG sample, one may conclude that these structures are not strictly related with the particular irradiation direction used.

Conclusions

The value of electronic stopping power for all three materials used is close to 10^4 keV/µm, when the ions fall on the target. The damage structures found are rather different, but surprisingly the effect of the electrical conductivity seems to be overruled by differences in the nature of bonding. MM and HOPG are very different concerning electrical conductivity, but both are so-called layered materials that have a weak bonding in one direction, and can be easily cleaved. And indeed both samples showed undulations that indicate extensive interlayer cleavage in the bulk.

The very particular nature of bonding in MM produced the long and very broad channels that have dimensions over two orders of magnitude higher than the usual track dimensions in MM [19] irradiated under normal incidence. It is worth pointing out that while the extension of these channels in the cleavage plane is a hundred times higher than the track diameter in normal irradiation, the vertical extension, i.e., in a plane perpendicular to the cleavage plane - the plane of high atomic density - is only 10 times higher than the track diameters reported in [19].

The intensive cleavage of MM and of HOPG, too, is an indication that for better imaging conditions the dose for these materials should be lowered by one or two orders of magnitude.

Silicon, a semiconductor, which has a strong covalent bonding, exhibited very different damage as compared with HOPG and MM. Except the first 5 µm, every feature observed can be attributed to variations of the target density around particle trajectories. By the method presented it was possible to characterize the distribution of damage, and to achieve the high resolution investigation of the individual features, too.

Acknowledgments

The authors are indebted to Prof. H. Ryssel, and Dr. L. Frey for helpful discussions. L. P. Biró gratefully acknowledges the Eötvös Hungarian State Fellowship. The work in Hungary was partly supported by OTKA grant T017344, the sample irradiations in Dubna were supported by the Hungarian Academy of Sciences.

References

1   R. Coratger, A. Claverie, F. Ajustron and J. Beauvillain, Surf. Sci. **227,** 7 (1990)
2   L. Porte, M. Phaner, C. H. de Villeneuve, N. Moncoffre and J. Tousset, Nucl. Instr. Meth. B **44,** 116 (1989)
3   L. Porte, C. H. de Villeneuve and M. Phaner, J. Vac. Sci. Technol. B **9,** 1064 (1991)
4   S. Bouffard, J. Cousty, Y. Pennec and F. Thibaudau, Rad. Eff. and Defects **126,** 225 (1993)
5   R. Coratger, A, Claverie, A. Chahboun, V. Landry, F. Ajustron, and J. Beauvillain, Surf. Sci. **262,** 208 (1992)
6   T. Li, B. V. King, R. J. MacDonald, G. F. Cortelli, D. J. O'Connor, and Q. Yang, Surf. Sci. **312,** 399 (1994)
7   L. P. Biró, J. Gyulai, K. Havancsák, Phys. Rev. B. **52,** 2047 (1995)
8   L. P. Biró, J. Gyulai, K. Havancsák, Nucl. Instr. Meths. B **112,** 270 (1996)
9   M. -H. Whangbo, J. Ren. S. N. Magonov, H. Bengel, B. A. Parkinson, and A. Suna, Surf. Sci. **326,** 311 (1995)
10 F. Thibaudau, J. Cousty, E. Balanzat, and S. Bouffard, Phys. Rev. Lett **67,** 1582 (1991)
11 T. Hagen, S. Grafström, J. Ackermann, R. Neumann, C. Trautmann, J. Vetter, and N. Angert, J. Vac. Sci. Technol. B **12,** 1555 (1994)
12 D. D. N. Barlo Daya, A Hallén, J. Eriksson, J. Kopniczky, R. Papaleo, C. T. Reimann, P. Hakansson, B. U. R. Sundqvist, A. Brunelle, S. Della- Negra, Y. Le Beyec, Nucl. Instr. and Meth. B **106,** 38 (1995)
13 D. D. N. Barlo Daya, C. T. Reimann, A Hallén, B. U. R. Sundqvist, P. Hakasson, Nucl. Instr. and Meth. B **111,** 87 (1996)
14 L. P. Biró, J. Gyulai, K. Havancsák, A. Yu. Didyk, S. Bogen, L. Frey, in press Phys. Rev. B **54**(14) (scheduled for 1st October, 1996)
15 L. P. Biró, J. Gyulai, K. Havancsák, A. Yu. Didyk, L. Frey, and H. Ryssel, presented at the 10th International Conference on Ion Beam Modification of Materials, IBMM '96, 1 - 7 Sept. 1996, Albuquerque, New Mexico, USA, submitted to: Nucl. Instr. and Meth. B
16 J. Biersack and L. Haggmark, Nucl. Instr. and Meth. **174,** 257 (1980)
    J. F. Ziegler, J. P. Biersack, and U. Littmark, The Stopping and Ranges of Ions in Solids (Pergamon, New York, 1985), Vol. 1
17 J. P. Spatz, S, Sheiko, M. Möller, R. G. Winkler, P. Reineker, O. Marti, Nanotechnology **6,** 40 (1996)
18 T. L. Daulton, M. Ozima, Science **271,** 1260 (1996)
19 M. Toulemonde, S. Bouffard, S. Studer, Nucl. Instr. and Meth. B **91,** 108 (1994)

Materials Science Forum Vols. 248-249 (1997) pp. 135-146
© *1997 Trans Tech Publications, Switzerland*

Hardness Enhancement and Crosslinking Mechanisms in Polystyrene Irradiated with High Energy Ion-Beams

E.H. Lee, G.R. Rao and L.K. Mansur

Oak Ridge National Laboratory, P.O. Box 2008, Oak Ridge, TN 37831-6376, USA

Keywords: Ion-Beam, Surface Modification, Radiation Effects, Crosslinking, Scission, LET Effects, Hardness

Abstract

Recent work at ORNL has demonstrated that surface hardness values several times larger than that of steel can be produced by employing high energy ion-beams (HEIB) in the several hundred keV to a few MeV range. In the present study, detailed crosslinking mechanisms have been studied for polystyrene irradiated with 350 keV H^+ and 1 MeV Ar^+, by analyzing hardness variations in response to irradiation parameters such as ion species, energy, and fluence. In addition, ion track analysis has been used to interpret the experimental data. It has been demonstrated that high *linear energy transfer* (LET) is important for crosslinking. For low LET conditions, induced active free radicals along the track are so sparsely dispersed that little interaction occurs among radicals. Input energy tends to localize at an intra-molecular chain segment, leading to chain scission. On the other hand, for high LET conditions, massive ionization induces a high concentration of free radicals over many neighboring molecular chains thereby facilitating crosslinking. These mechanisms are explored in detail in the present study and effective crosslinking radii at hardness saturation are derived based on experimental data.

Introduction

Organic polymers are inherently soft due to the absence of chemical bonds between polymer chains. When polymeric materials are subjected to *high energy ion-beams* (HEIB) however, a high number density of crosslinks are introduced as well as numerous double and triple bonds, and free radicals. As a consequence, HEIB processed polymers become hard, electrically conductive, and optically dense, with improved resistance to chemicals and mechanical wear [1-5]. Hardness often improved over 50 times, exceeding hardness values of steels. The property variations were analogous to each other because the improvements originated from the same cause namely *crosslinks*. Hardness measurements have been used to investigate the materials' response to HEIB, since elastic properties and hardness are related to crosslink number density [6] and can be measured accurately and conveniently [7-9]. In this paper, therefore, ion beam-polymer interaction mechanisms are investigated by scrutinizing the hardness variations in response to HEIB processing parameters, namely ion species, energy, and fluence. Since other property changes mentioned above have been reported previously in references [1-5], they are not discussed in this paper.

Mechanisms

When an energetic particle impinges into a polymer medium with high velocity, its orbital electrons are stripped to varying degrees depending upon the velocity of the particle. The maximum achievable charge state of the particle depends upon the number of protons. The velocity required to achieve the highest ionized state also depends upon the atomic number. The higher the number of protons the particle has, the higher would be the coulombic force exerted when orbital electrons are stripped off, but the higher mass ion would also require higher energy to be accelerated to a given velocity. The stripped ion eventually re-acquires its orbital electrons as it slows down below the velocity of the Bohr electron [10]. When a charged particle passes through a medium, it sets nearby electrons in motion by its electric field and also creates a large number of secondary electrons by knock-on collisions. The particle thus loses its energy by transferring its energy to the medium until it slows down and stops. These energy loss processes are collectively known as *electronic stopping*. The particle also loses its energy by displacing atoms in the medium by nuclear collisions. These energy loss processes are known as *nuclear stopping*. While most electronic energy loss stems from primary ions, most nuclear energy loss comes from recoil atoms created by primary ions. This is because several recoil atoms can be created by a primary ion with lower energies and consequently larger nuclear stopping cross-section. Although recoil atoms also lose their energies by electronic processes, their contributions are small compared to those of primary ions because low energy recoil atoms have small cross section for electronic stopping.   For small atoms such as H or He, nuclear stopping is negligible because their nuclear collisional cross-sections are very small at most energies of interest. Nuclear stopping however, becomes important for ion species with a larger number of nucleons. The unit eV/nm/ion or simply eV/nm is used for the energy loss per unit path length or linear energy transfer (LET).

Finally, excited lattice atoms and electrons lose their energy through vibrational decay as phonons and plasmons, respectively. Phonons and plasmons decay mostly through thermal and radiative processes. Although energy loss by phonon decay becomes significant for large atoms, no obvious thermal effect was observed within the range of irradiation conditions employed in this work. Energy loss by plasmon decay is very small and its contribution would be negligible even if it affects material properties. Experimental evidence showed that modification of polymer properties depended mainly upon electronic LET_e and nuclear LET_n [2].

During irradiation, various physical and chemical processes take place in the polymer. Coulombic interactions between ions and electrons of host atoms, excessive bond stretching due to localized energy deposition, and atomic displacement by nuclear collision can release pendant atoms such as hydrogen, and cause bond breakage or chain scission. Thus, various gaseous molecular species are released during irradiation. The most prominent species are hydrogen, molecular scission products from the end groups as well as pendant groups of the polymer, and their reaction products. Radicals or dangling bonds are created by the release of pendant atoms such as hydrogen. Crosslinking occurs when two free dangling bonds on neighboring chains unite, whereas double or triple bonds are formed if two neighboring radicals in the same chain unite.

It has been well established that mechanical, physical, and chemical property changes in polymers are determined by the magnitude of crosslinking and scission, and that crosslinking enhances mechanical stability while scission degrades mechanical strength [11]. Although both electronic and nuclear energy transfer can induce crosslinking as well as scission as would be intuitively expected, experimental evidence suggests that electronic stopping causes more crosslinking while nuclear stopping causes more scissions [2, 12]. This provides much of the basis for understanding the materials' response to irradiation.

Experimental

Polystyrene (PS) sheets used in this experiment were purchased from Goodfellow Advanced Materials Company, Malvern, PA. The polymer was separately irradiated with 350 keV He^+ and 1 MeV Ar^+ ions at the Triple Ion-beam Irradiation Facility in the Metals and Ceramics Division at ORNL. Details about the facility can be found in reference [13]. Beam currents under 1 mA m^{-2} were used maintaining the temperature of the polymer surface below 100°C to avoid overheating and degradation of the polymer surface.

Surface hardness values were measured at 100 nm depth using the Nanoindenter®. In this technique, load is sensed continuously as a function of depth. Since the indenter diamond tip geometry is accurately known, hardness can be determined as load divided by the projected contact area as a function of depth, since the projected contact area can be determined for a specific depth. The entire process is computer controlled. Further details about the hardness measurement procedure can be found in reference [14].

Results

In Fig. 1(a), surface hardness values of polystyrene irradiated with 350 keV He[+] and 1 MeV Ar[+] ions are plotted as a function of fluence (number of ions injected per unit area of target material, ions m[-2]). Although comparing hardness values at an equivalent fluence is useful when one is concerned with the number of ions required to cover the subject target area, the effect of energy deposition or LET cannot be clearly delineated because the energies deposited by the 350 keV He[+] and 1 MeV Ar[+] are different at the same fluence. Therefore, hardness values are replotted as a function of dose (energy deposition per unit mass of target material) in Fig. 1(b). Here, the SI unit of Gray (Gy) is obtained by multiplying the fluence (ions m[-2]) with LET (eV/nm/ion) at the surface and dividing by the specific gravity of polystyrene (1.05×10^3 kg/m^3) and finally using a conversion factor of 1 Gy = 6.24×10^{18} eV/kg.

The results showed that surface hardness increased with fluence and reached a saturation value of ~20 GPa near a fluence of 2×10^{20} m^{-2} for 1 MeV Ar[+], and near 7×10^{20} m^{-2} for 350 keV He[+] as indicated by dotted vertical lines in Fig. 1(a). When plotted as a function of dose, however, the trends of hardness value appeared to saturate at the same dose of ~3×10^{10} Gy for both He[+] and Ar[+] ions as indicated by a dotted vertical line in Fig. 1(b). However, hardness values for 350 keV He[+] were lower at all doses than those for 1 MeV Ar[+] until hardness saturated. When the hardness data were extrapolated to the x-axis as indicated by the solid lines in Fig. 1(b), apparent incubation doses (here it is defined to be a dose required to improve hardness to 10 % above the pristine unirradiated value) were found to be ~2×10^8 Gy for 350 keV He[+], and ~5×10^7 Gy for 1 MeV Ar[+] indicating that He[+] required about 4 times higher incubation dose than Ar[+].

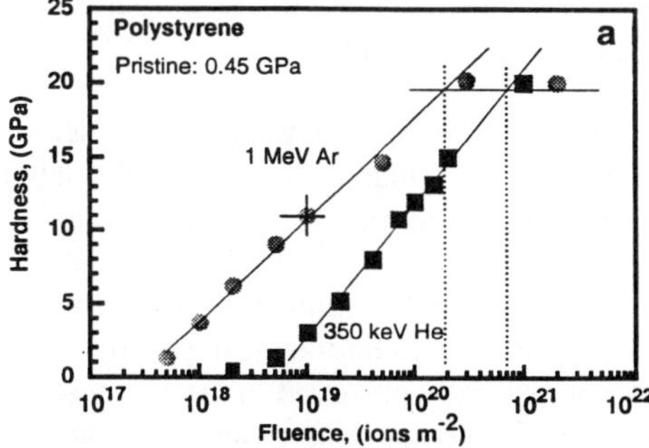

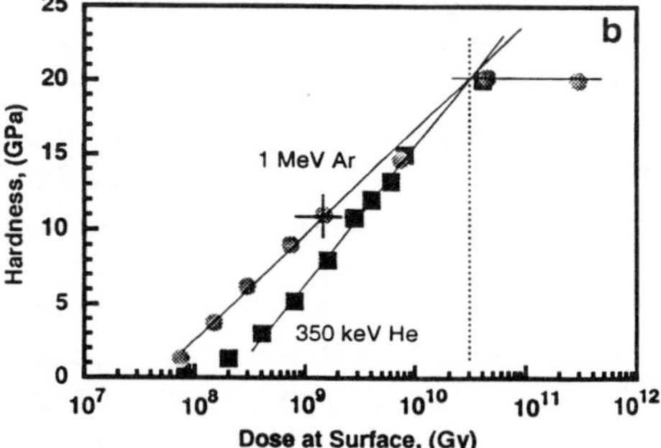

Figure 1. Surface hardness variation of polystyrene irradiated with 350 keV He[+] and 1 MeV Ar[+] ions as a function of (a) fluence and (b) dose.

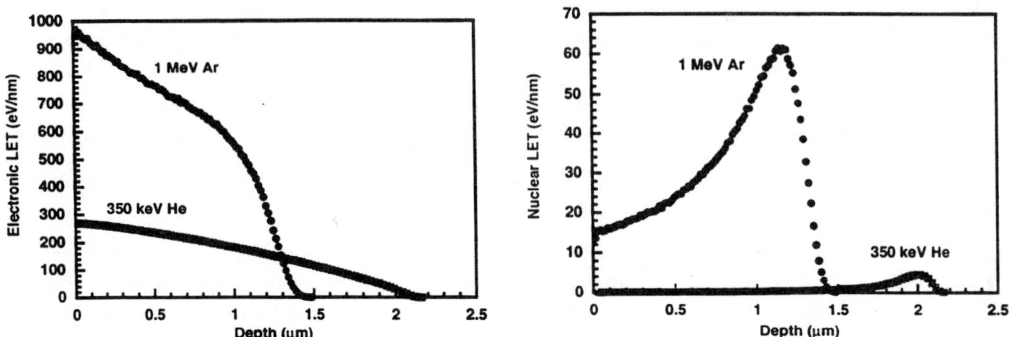

Figure 2. TRIM calculations of electronic and nuclear LET as a function of depth for 350 keV He and 1 MeV Ar for polystyrene.

Discussion

The hardness data presented above are examined in terms of three critical parameters, ion species, ion energy, and fluence. In Fig. 2, the electronic and nuclear energy loss profiles for 350 keV He^+ and 1 MeV Ar^+ are plotted to examine the effects of ion species and energy using the Monte Carlo simulation code *TRansport of Ions in Matter* (TRIM) [10]. All energy loss calculations in the following section are made using the TRIM version 95.06.

Electronic LET_e values for 350 keV He^+ and 1 MeV Ar^+ are about 260 and 960 eV/nm at the surface, respectively. Most nuclear energy loss occurs near the end of the ion track where recoil atoms are prevalent and have low energies. At high energies, nuclear cross-section is small and only few of the many encounters in the target cause significant deflection from the straight path of flight. Since the nuclear energy loss at the surface region is small compared to the electronic energy loss, less than a few percent for Ar and almost negligible for He, the hardness changes at the surface region can be considered mostly due to the electronic energy loss.

It is interesting to note that the surface LET_e ratio for He^+ to Ar^+ is about $260(He)/960(Ar) \approx 1/3.7$, which is approximately equal to the inverse of the saturation fluence ratio, $7 \times 10^{20}\,(He)/2 \times 10^{20}\,(Ar) \approx 3.5$. This result suggests that the higher the LET, the lower would be the fluence required to achieve a saturation hardness such that total deposition energy per unit mass of target would be about equal at saturation. This point is further verified in Fig. 1(b), which shows that the trends of hardness value appear to saturate at the same dose near $\sim 3 \times 10^{10}$ Gy for both He^+ and Ar^+.

An important question then is '*do we need high LET to achieve a desired hardness?*' It is seen from above that 350 keV He^+ would require 3.5 times higher fluence than 1 MeV Ar^+ at the same ion flux to attain the saturation hardness. Irradiation with 1 MeV electrons ($LET_e \approx 0.19$ eV/nm) would require 10^4 times higher fluence if the same assumption of linearity applies. For this reason, although radiation sources such as electron beams and γ-rays have been employed successfully in areas where moderate crosslinking is sufficient for applications as in radiation curing and photolithography, a large improvement in hardness was not generally observed in these cases. However, as will be discussed further below, for low LET, radical pairs or dangling bonds are produced so sparsely that the probability of two neighboring radicals to be close enough for crosslinking is very low. In such cases, a significant fraction of deposited energy remains within chains, leading frequently to chain scission and thus degradation of material.

Table 1. G(s) values for scission in irradiated PMMA [15]

Ion Energy	LET (eV/nm)	G(s)
^{60}Co γ-rays	0.2	1.5
45 MeV He^{2+}	20	0.8
240 MeV Ne^{7+}	476	0.3
90 MeV O^{4+}	577	0.3

Schnabel et al. [15] observed an increasing trend of scission with decreasing LET for PMMA irradiated by various high energy irradiation sources as summarized in Table 1. G-value is defined to be a quantitative measure of the specific chemical yield (here scission) per 100 eV of radiation energy absorption. In their experiment, G(s) values for random main-chain scission were determined from molecular weight distribution via gel permeation chromatography. Consistent with the reasons mentioned above, γ-rays produced the highest scission because of low LET despite the purely ionizing nature of the radiation (no nuclear displacements).

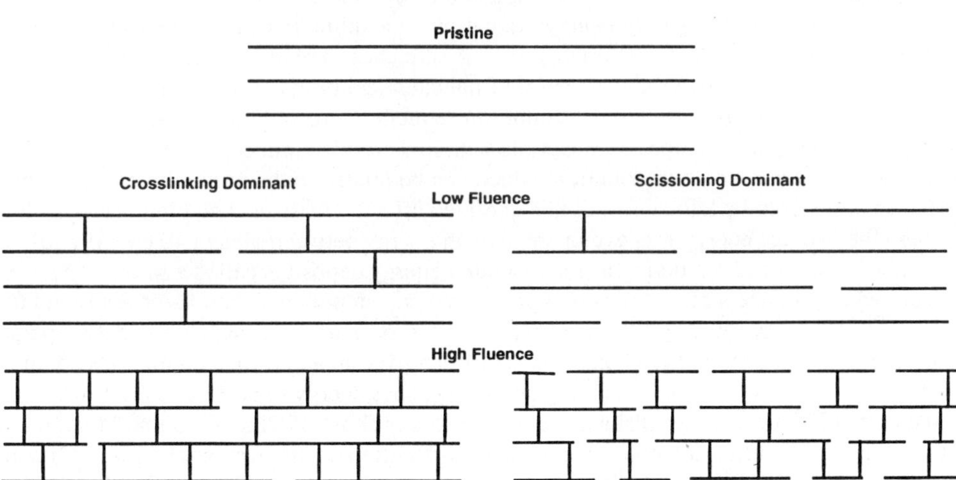

Figure 3.   Schematic illustration of microstructure evolution for crosslinking and scission dominant regimes. Crosslinking and scission occur simultaneously during irradiation. The ratio of crosslinking to scission depends upon polymer structure but can be altered by using different combinations of ion species, energy, and fluence in HEIB process. Under low LET condition, scissioning type polymers such as PMMA become more solvable at low fluence due to dominant scission but become insoluble at high fluence due to network formation by crosslinking. Under high LET condition, however, scissioning regime can be shifted to crosslinking regime by enhanced radical formation and crosslinking.

A second question then is '*why do 350 keV He⁺ and 1 MeV Ar⁺ give the same saturation hardness despite the difference in LET?*' To answer this question, the example of PMMA can be examined again. PMMA is a scissioning type polymer [16]. When PMMA was subjected to low LET beam sources such as 1.5 MeV H^+, 1 MeV He^+, and 20 keV electron beam, initially it became more soluble in a solvent at low fluences but eventually became insoluble at high fluences [17]. The reason for this peculiar solubility behavior of PMMA is illustrated schematically in Fig. 3. For scissioning type polymers, at low fluences, average molecular chain length decreases because of predominant scission. At high fluences however, scissioned segments are connected by random crosslinks making them difficult to dissolve. For this reason, at low LET, the crosslinked region may not have the same degree of mechanical integrity as for the high LET case, even if an apparent saturation hardness could be attained at high fluences. It should be noted that hardness of a polymer is affected by crosslink density as well as by chain mobility. Thus, a polymer can be hard even if numerous scissions occur in the chains when overall chains are locked up by crosslinks as shown in Fig. 3. Although PMMA degraded under low LET conditions, when PMMA was treated with high LET ions such as 2 MeV Ar^+ ($LET_e \approx 1130$ eV/nm), a smooth film with hardness of 10.8 GPa was obtained at a fluence of 1×10^{19} ions m^{-2} because of enhanced crosslinking [5]. There is evidence that, when further crosslinking is prevented by the formation of a rigid network, such as at saturation, additional irradiation may induce scissions [4,18]. In such a case, hardness reduction may not be noticeable until very severe degradation occurs because of restricted chain mobility. However, in an earlier study, electrical conductivity, which is much more sensitive to carrier mobility or chain connectivity, decreased after a maximum conductivity was reached indicating that additional irradiation indeed degraded chain connectivity [4,18].

Effects of ion irradiation in polymers can be examined from a radiation chemistry view point which can yield useful insights [19,20]. High energy radiations, including ions, form tracks in a material as their energy is transferred to the material atoms. A sequence of nonhomogeneous processes result in creation of intermediate products before final chemical products are formed [20]. The structure of these tracks can be used to determine subsequent microstructural effects in the material and corresponding property changes. In materials with low atomic numbers, such as organic polymers, the oscillator strength distribution, which can be related to the minimum energy required to dislodge an atom from its lattice position, is confined effectively to an excitation energy below 100 eV, and the average energy loss events involve an energy between 30 and 40 eV [19, 20]. An energy loss event in a track involving a small isolated energy deposit is called a *spur*. An average energy required to produce a pair of ions or a pair of radicals in polymers has also been found to be similar to the spur energy [19-21]. For low LET irradiations, in a track, spurs are widely separated like a string of beads. For high LET condition, spurs can overlap and form a continuous column, thereby facilitating cross-linking. Assuming an average spur energy to be 35 eV, for 350 keV He^+ with 260 eV/nm LET, the average distance between spurs is about $35/260 \approx 0.13$ nm, larger than C-H bond distance (~0.11 nm). It would, therefore, be difficult to create two dangling (C- -C) bonds nearby simultaneously and crosslink them together in a single ion track. On the other hand, for 1 MeV Ar^+, the average distance between spurs is $35/960 \approx 0.037$ nm, almost 3 times shorter than the C-H bond distance ($0.11/0.037 \approx 3$). In this case, spurs are so closely spaced that almost three spurs can overlap within a C-H bond distance (~0.11 nm).

It required many passes of ions, approximately 700 He^+ ions/nm^2 or 200 Ar^+ ions/nm^2 to attain the saturation hardness, see Fig. 1(a). Though it is unlikely, *if there were no overlaps of ion tracks,* effective radius of ion track at saturation can be calculated to be 0.02 nm and 0.04 nm for 350 keV He^+ and 1 MeV Ar^+, respectively (using $n_i \pi r_e^2 = 1$ nm^2 relationship where n_i is the number of ions injected to 1 nm^2 area at saturation). The reason for the small effective radius can be further elucidated if one considers the fact that polystyrene has 18 monomer units in a rhombohedral cell with lattice constants of a_o=2.2 nm, b_o=2.2 nm and c_o=0.663 nm [22, 23]. Although the polystyrene studied in this work is not a crystalline form, it is expected that the average distance between two polymer chains is larger than 1 nm. In this situation, it would be very difficult for two radicals to approach to a cross-linking distance, say the C-C bond distance (~0.15 nm). Experimental evidence, however, suggests that crosslinking occurs after substantial emission of hydrogen atoms followed by material compaction or an increase in specific gravity. In one of our unreported experiments, measurements of specific gravity using a density gradient column showed that specific gravity of pristine polycarbonate was 1.2×10^3 kg/m^3 and that of the bombarded layer (extracted by dissolving away the unbombarded substrate) was 1.54×10^3 kg/m^3 after irradiating to a fluence of 1×10^{19} ions/m^2 with 2 MeV O$^+$ ions. The result suggests that significant removal of hydrogen was required to move polymer chain segments to a cross-linking distance. When the saturation dose 3×10^{10} Gy was converted to 1.97×10^5 eV/nm^3 and a 35 eV per spur conversion factor was used, it turned out that almost 5600 spurs (1.97×10^5 /35 $\approx$ 5600) were deposited per nm^3 at saturation for both He and Ar. When one considers that there are about 6 monomer units per nm^3 in polystyrene, every monomer unit encountered 933 spurs (5600/6 $\approx$ 933) at saturation. This result suggests that most crosslinking occurred in the region where ion tracks overlap rather than along a single ion track.

A third question then would be *'why were hardness values for 350 keV He$^+$ lower at all doses than those for 1 MeV Ar$^+$ until hardness saturated?'* and *'why was the apparent incubation dose for 350 keV He$^+$ higher than that for 1 MeV Ar$^+$?'* as seen in Fig. 1(b). As discussed above, the experimental results suggest that most crosslinking occurred in the area where ion tracks overlap. An overlap of ion tracks should occur earlier for ions with larger effective crosslinking radii, or with higher spur density. In a stochastic process of ion track development, since an average distance between two ion track centers for the same fluence would be about equal regardless of track radii, the probability of track overlap increases with increasing track radii, as shown in Fig. 4. Actual crosslinking radius for 350 keV He$^+$ and 1 MeV Ar$^+$ could not be derived in this experiment. However, since the surface LET$_e$ ratio for He$^+$ to Ar$^+$ is about 260(He):960(Ar) $\approx$ 1:4 and the ratio of effective radius square *at saturation* (r_s^2) was found to be about $(0.02)^2$ (He):$(0.04)^2$ (Ar) =1:4, it is reasonable to assume that the ratio of effective crosslinking radius for He$^+$ and Ar$^+$ is approximately

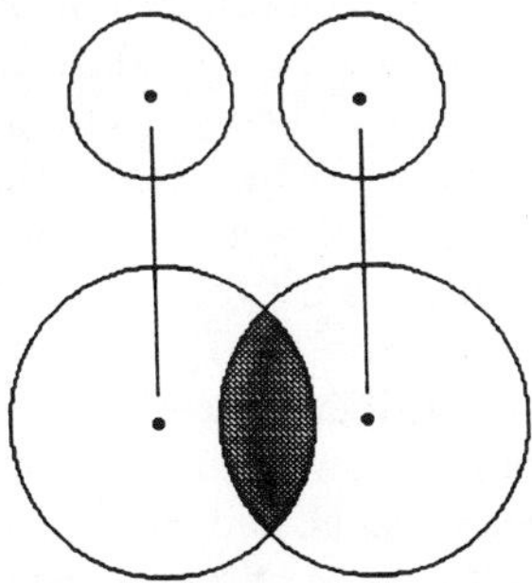

Figure 4. Overlap of two circles for two different radii. The probability of overlap increases with increasing radii for a given distance between the two circle centers.

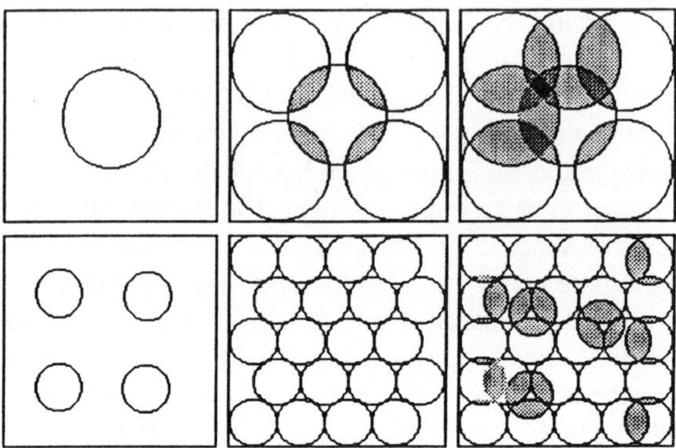

Figure 5. Track overlap for circles with radius ratio of 1:2. When circles are laid to cover the same area using an area ratio of 4:1 for small and large circles respectively, track overlap occurs earlier for the larger circles as indicated by increasing shade density for single, double, and triple overlap areas.

r_e(He):r_e(Ar) $\approx$ 1:2 and that the energy density within the radius is uniform. Then, track overlap progress can be illustrated at three different equivalent doses (e.g., four He ion tracks to one Ar track) as shown in Fig. 5. The tracks have been forced to fill the unbombarded area first to visualize size of ion track overlaps. In the figure, single, double, and triple overlap regions are indicated with increasing shade density. Indeed, multiple track overlap occurs earlier for the high LET case as expected.

It has been reported however, that the crosslinking G(x) value for polystyrene is invariant with LET for a wide range of LET values, contradicting the current observation. The final question then is how this disagreement arises. Before answering this question, it is appropriate to review the previously reported data first. The G(x) values were found to be within a range of 0.03 to 0.05 [15,24,25], when polystyrene was subjected to 1.25 MeV (average) ^{60}Co γ-rays (0.2 eV/nm), 275 MeV Ne^{7+} (370 eV/nm), and 180 MeV Ar^{8+} (2260 eV/nm) as summarized in Table 2. On the other hand, Calcagno et. al [26] reported a G(x) value of 0.3 for 100 keV He$^+$ (210 eV/nm), 200 keV Ne$^+$ (450 eV/nm), and 400 keV Ar$^+$ (850 eV/nm) irradiated polystyrene showing an order of magnitude larger value than those reported previously. In their work, 200-300 nm thin spin cast films were used to ensure uniform energy deposition and to minimize nuclear contribution. The discrepancy between the data of Calcagno et. al and the previous data appears to originate perhaps from computational difference rather than actual difference in G value because both data were derived from a similar range of LET. Calcagno et al. attributed their high G value to high LET by comparing with low LET γ-rays data without acknowledging the data of Schnabel et al. Nonetheless, if we consider the data of Calcagno et al. and previous data separately, all data can be interpreted consistently to indicate that the G(x) value of polystyrene is almost invariant with LET.

Table 2. G(x) values for polystyrene subjected to a wide range of LET irradiation sources

Ion Energy	LET (eV/nm)	G(x)	References
^{60}Co γ-rays 1.25 MeV avg.	0.2	0.05 0.03-0.05	Burlant [24] Parkinson [25]
275 MeV Ne^{7+} 180 MeV Ar^{8+}	370 2260	0.05 0.03	Schnabel [15]
100 keV He$^+$ 200 keV Ne$^+$ 400 keV Ar$^+$	210 450 850	0.3 0.3 0.3	Calcagno [26]

In most experiments, G(x) values were derived from the molecular weight distributions determined by gel permeation chromatography (GPC) [15,26] or viscosity measurements [24] after dissolving irradiated polymers in a solvent. This requires polymers to be still soluble after irradiation. Therefore, very low fluence of ions, generally much below 10^{16} ions/m^2 (or less than one ion per 1000 nm^2), was used to ensure solubility of the irradiated polymer. In this low fluence regime, ion tracks are so widely separated that most crosslinking may occur in a single ion track, probably in regions where two polymer chains are tangled or cross over each other such that they are within crosslinking distance. In such a case, higher LET ions would of course produce more crosslinks per ion but the number of crosslinks produced per unit deposited energy would be the same regardless of LET. Thus G(x) value is independent of LET for low fluence or in the single ion track regime.

From the data shown in Fig. 1(b) and discussions presented above, it can be considered that there are three regimes of crosslinking based on fluence: '*low fluence or single ion track regime*' where ion tracks are isolated and the number of crosslinks produced is proportional to the deposited energy only, but G(x) is invariant with LET; '*intermediate fluence or track overlap regime*' where enhanced crosslinking occurs in track overlap regions and G(x) varies with LET (track radius or spur density); and finally '*high fluence or saturation regime*' where most crosslinkable sites are exhausted and further irradiation may induce more scission than crosslinking and degrade the material. Idealized, these three regimes are illustrated in Figure 6. In the '*single ion track regime*' G(x) is low. In the '*track overlap regime*' crosslinking is enhanced and thus G(x) is high. In the '*saturation regime*' G(x) would decline again though this may not be detected by hardness measurement. If we interpret the incubation fluence to be the onset of track overlap, it can now be understood that the low LET He$^+$ showed a larger incubation fluence because track overlap was delayed due to the smaller track radii.

It is, however, interesting to note that although incubation period or onset of track overlap is delayed for the low LET case, the higher slope of the curve in the track overlap regime suggests that crosslinking rate is higher once track overlap is initiated. In the discussion above, in the demonstration of track overlap effect for low and high LET situation (Fig. 5), it was assumed that the effective radius ratio was r_e(He):r_e(Ar) $\approx$ 1:2 based on the effective radius values derived at *saturation* if there were no track overlap and that energy density within the track radii is constant. Within a track, depending on physical and chemical processes that occur, a chemical core and a physical core can be defined [20]. Approximately half the energy is deposited in the physical core and the other half is deposited in the tracks of knocked out electrons in the region referred to as the penumbra. The chemical core radius within which chemical products are formed, is intermediate

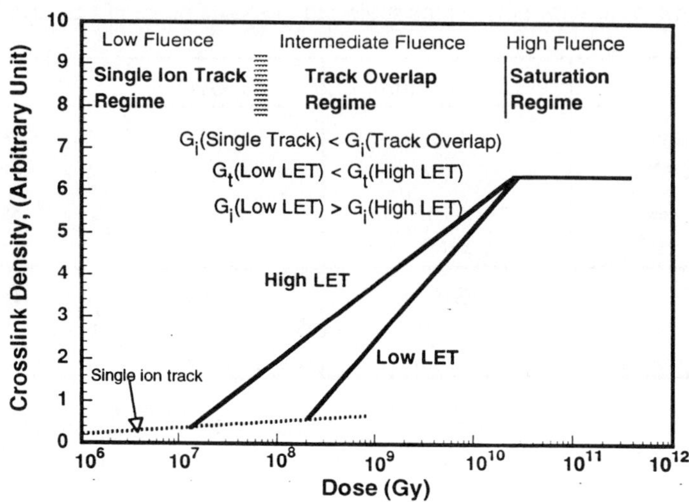

Figure 6. Idealized representation of crosslinking regimes as a function of dose. At low fluence, ion tracks are isolated and crosslinking occurs at a slower rate in single ion track regime. At intermediate fluence, enhanced crosslinking occurs in regions where tracks overlap. Therefore, the slope of the curve, instantaneous G_i(Single Track) is lower than G_i(Track Overlap). With increasing LET, effective crosslinking radius increases and track overlap occurs earlier resulting in a shorter incubation period, but crosslinking rate (slope of the curve) decreases because energy density within the track decreases. Therefore, the total G_t(High LET) is higher than G_t(Low LET) but the instaneous G_i(High LET) becomes lower than G_i(Low LET). At high fluence, scission becomes dominant as crosslinking is prevented due to rigid network formation. The transition from single ion track to track overlap and from track overlap to saturation regime would occur gradually.

 between physical core radius and the penumbra [20]. In reality the chemical core radius (r) increases almost linearly with LET within the energy range of interest (e.g., up to 0.4 MeV per nucleon for He and 1 MeV per nucleon for Ar) and decreases to a physical core radius with further increase of LET beyond these energies [see Fig. 6-1 in Ref. 19]. Thus the real ratio of r_e(He):r_e(Ar) might have been smaller than 1:2, probably 1:4 if track radii could be measured, and track overlap for low LET might have been delayed even further. If the increase in radius is linear with LET, energy density would decline more rapidly for high LET since the radial area increases as r^2. For the higher LET case (e.g., 1 MeV Ar^+), a total or an accumulated crosslinking yield, $G_t(x)$ value, would be higher because of earlier track overlap. On the other hand, an instantaneous crosslinking yield (slope of the curve) in the intermediate dose range, $G_i(x)$ value, would be lower compared to that of lower LET (e.g., 350 keV He^+) due to lower energy density in the track overlap regions, such that saturation hardness occurs for both cases at a similar dose.

High energy ion beam irradiation of polymers provides a unique opportunity for industrial applications in areas where light weight polymeric materials with high surface hardness, wear and scratch resistance, oxidation and chemical resistance, diffusion or permeation barrier, and improved electrical conductivity are required. Specific examples for the property improvements can be found in references [1-5]. The present study indicates that the degree of crosslinking and the depth of the ion-beam modified layer can be controlled based on physical principles, namely by controlling the three major processing parameters, ion species, energy, and fluence. Although apparent saturation hardness was the same for both low and high LET, experimental evidence suggested that materials may degrade in a very low LET environment because of dominant scissions, as in weathering by UV radiation. Besides, low LET irradiations require longer processing times, making the process uneconomical. It is therefore recommended that high energy ion-beams (HEIB) be used, when highly crosslinked and mechanically sound surfaces are desirable.

Conclusions

Crosslinking mechanisms in polymeric materials were investigated by examining hardness variation in polystyrene irradiated separately with 350 keV He^+ ions ($LET_e \approx$ 260 eV/nm) and 1 MeV Ar^+ ions ($LET_e \approx$ 960 eV/nm). The results showed that hardness improvement was greater for higher LET ion species at equivalent fluences as well as at equivalent doses, indicating that high LET is more effective in producing crosslinks. However, the saturation hardness was attained at the same dose of 3×10^{10} Gy regardless of LET. Ion track analysis indicated that spurs within tracks are separated by 0.13 nm for 350 keV He^+ and by 0.037 nm for 1 MeV Ar^+. When spurs are widely spaced as for the He case, a larger incubation fluence was required before the onset of hardness enhancement. This was explained in terms of effective crosslinking radii. Crosslinking rate is low in single ion track regime and high in track overlap regime. Lower LET has smaller effective track radii, and thus track overlap or hardness enhancement is delayed. In single ion track regime, G(x) is proportional to total accumulated energy and is invariant with LET. In track overlap regime, crosslinking yield, G(x), is enhanced because of track overlap. However, although the higher LET (1 MeV Ar^+) showed the higher total crosslinking yield, $G_t(x)$, because of earlier track overlap, the instantaneous $G_i(x)$ was lower compared to that of the lower LET (350 keV He^+) once in the track overlap regime, because energy density decreases as r^{-2} with increasing track radius. At high fluences, when crosslinkable sites are exhausted, scission becomes dominant and the polymer degrades. Based on these mechanisms, a detailed theoretical model should be developed in the future to better define and understand the track overlap mechanisms and the data.

Acknowledgment

This research was sponsored by the Division of Materials Sciences, U.S. Department of Energy, under contract No. DE-AC05-96OR22464 with Lockheed Martin Energy Research Corporation. The authors wish to thank Dr. M. B. Lewis and Dr. J. S. Lin for manuscript review and helpful discussions.

References

[1] E. H. Lee, M. B. Lewis, P. J. Blau, and L. K. Mansur, J. Mater. Res., **6,** 610 (1991).
[2] E. H. Lee, G. R. Rao, M. B. Lewis, and L. K. Mansur, J. Mater. Res., **9,** 1043 (1994).
[3] G. R. Rao, E. H. Lee, R. Bhattacharya, and A. W. McCormick, J. Mater. Res. **1,** 190 (1995).
[4] E. H. Lee, Chapter 17 in"Polyimide: Fundamental Aspects and Technological Applications," eds. K. Mittal and M. Ghosh, Marcel Dekker Inc., New York, 471 (1996).
[5] E. H. Lee, G. R. Rao, and L. K. Mansur, Trends in Polymer Science, **4,** 229 (1996).
[6] A. Charlesby, Radiat. Phys. Chem., **40,** 117 (1992).
[7] F. J. Baltá Calleja, Adv. Polym. Sci. **66,** 117 (1985).
[8] E. H. Lee, Y. Lee, W. C. Oliver, and L. K. Mansur, J. Mater. Res. **8, 377** (1993).
[9] G. R. Rao, L. Riester, and E. H. Lee, Mat. Res. Soc. Symp. Proc., **345,** 363 (1995).
[10] J. F. Ziegler, J. P Biersack, and U. Littmark, "The Stopping and Range of Ions in Solids," Vol. **1,** (Pergamon Press, 1985).
[11] J. M. Cowie, in "Polymers: Chemistry & Physics of Modern Materials," Intertext Books, Billings & Son Ltd. Worcester, Great Britian, 283 (1973).

[12] M. B. Lewis, E. H. Lee, and G. R. Rao, J. Nucl. Mater. **211**, 46 (1994).

[13] M. B. Lewis, W. R. Allen, R. A. Buhl, N. H. Packan, S. W. Cook, and L. K. Mansur, Nucl. Instrum. Methods B**43,** 243 (1989).

[14] E. H. Lee, Y. Lee, W. C. Oliver, and L. K. Mansur, J. Mater. Res. **8**, 377 (1993).

[15] Wolfram Schnabel, Siegfried Klaumünzer, Hideto Sotobayashi, Frithjof Asmussen, and Yoneho Tabata, Macromolecules **17**, 2108 (1984).

[16] A. Shapiro, in "Radiation Chemistry of Polymeric Systems," Interscience Publishers, John Wiley & Sons, New York, 353 (1962).

[17] T. M. Hall, A. Wagner, and L. F. Thompson, J. Appl. Phys. **53**, 3997 (1982).

[18] Y. Wang, S. S. Mohite, and L. B. Bridwell, J. Mater. Res. **8**, 388 (1993).

[19] J. L. Magee and A. Chatterjee, "Radiation Chemistry, Principles and Applications," Farhataziz and Michael A. J. Rodgers, Eds., VCH Publishers, New York (1987).

[20] J. L. Magee and A. Chatterjee, "Kinetics of Nonhomogeneous Processes," Gordon R. Freeman, Ed., John Wiley & Sons, New York, 184 (1987).

[21] ICRU Report 31, "Average Energy Required to Produce An Ion Pair," International Commission on Radiation Units and Measurements, Washington D.C. 1979.

[22] R. L. Miller and L. E. Nielsen, J. Polymer Science, **55**, 643 (1961).

[23] Polymer Handbook, 3rd edition, J. Brandrup and E. H. Immergut, eds., John Wiley & Sons, Inc., New York, (1989).

[24] W. Burlant and J. Neerman, J. Polymer Science, **58**, 491 (1962).

[25] W. W. Parkinson and R. M. Keyser in "The Radiation Chemistry of Macromolecules," ed. M. Dole, Vo. II, p 72 (Academic Press, New York, 1973).

[26] L. Calcagno, G. Foti, A. Licciardello, and O. Puglisi, Appl. Physs. Lett. **51**, 907 (1987).

Materials Science Forum Vols. 248-249 (1997) pp. 147-154
© *1997 Trans Tech Publications, Switzerland*

Diffusion Studies in Polymers Using MeV Ion Beams

A.S. Clough, D.W. Drew, M. van der Grinten, P.M. Jenneson and T.E. Shearmur

University of Surrey, Guildford, Surrey GU2 5XH, England

Keywords: Diffusion, Polymers, Ion Beams

Abstract: The principles of MeV ion beam energy loss and traversing techniques for studying diffusion in polymers are outlined. Attention is drawn to particular experimental precautions necessary for the study of polymer matrices. Reference is made to particular applications of the techniques of relevance to industry.

1.Energy loss techniques

1.1 Rutherford Backscattering

The principal ion beam technique used over many years for the study of distributions of elements in a matrix is Rutherford Backscattering (RBS) of α-particles, the nuclei of helium atoms. At Surrey a beam of ^{4}He+ ions from a 2 MeV Van de Graaff accelerator is transported under vacuum to a scattering chamber. Here it is incident on a target sample suspended on a liquid-nitrogen back-cooled goniometer rotatable in the horizontal plane. The α-particles scatter from elements in the sample and are detected in a silicon surface barrier detector located in the horizontal plane at an angle of 165° to the incident beam. At this near-backward angle their energy is proportional to both the mass of the element scattered from and the energy of the α-particles at interaction. Thus if an element is located at the surface of a sample the detected scattered α-particle will have more energy than if the element is located at depth - ions lose energy through interaction with electrons as they traverse the sample. If a particular contaminant element has a mass greater than those comprising the sample matrix then it is a relatively simple matter to measure its concentration profile in the matrix to a depth determined by the heaviest matrix element.

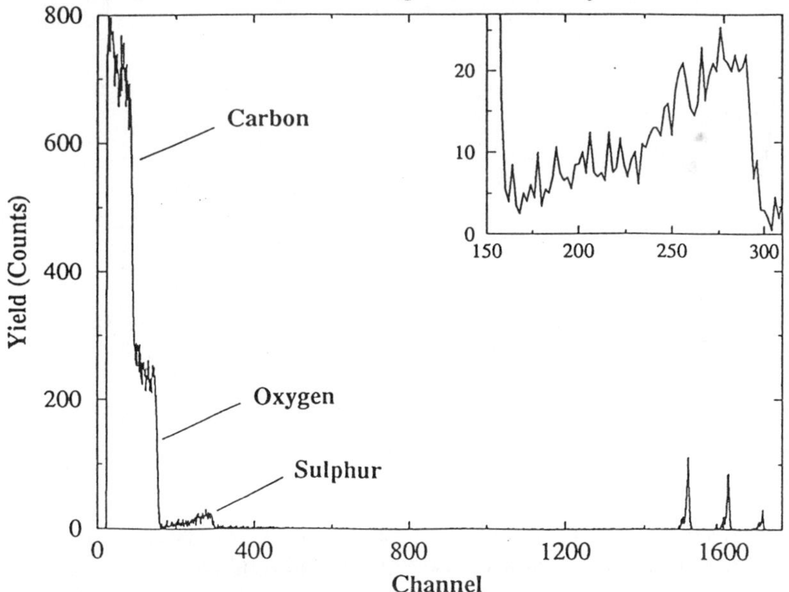

Fig. 1. RBS spectrum from carbon-coated sulphur-labelled dye diffused into polymer; inset is an expansion of the sulphur spectrum.

Thus for CHO polymers any element heavier than oxgen can be profiled in the vicinity of the matrix surface. A typical spectrum is shown in Fig. 1, where the backscatters of sulphur present in dye molecules introduced into the polymer surface are clearly observable at greater energies than backscatters from oxygen. Also present are backscatters from carbon (at 165° scatters from hydrogen are kinematically prohibited) and, in the highest channels, α-particles from a source permanently mounted near the detector containing three radioactive elements. These α-particles of known energy are used to calibrate the horizontal energy axis.

Simple precautions have to be taken when using polymeric matrices:

(1) As they are insulators they need to be carbon-coated by vacuum deposition to a depth of 200-300 Å to prevent charge build-up. The charge is led away from the surface by silver-dag at the edge of the sample to a (usually) silicon substrate and hence through the insulated target goniometer to an external earth. A spectrum from an uncoated sample is shown in Fig. 2. There is a general smearing of the spectrum.

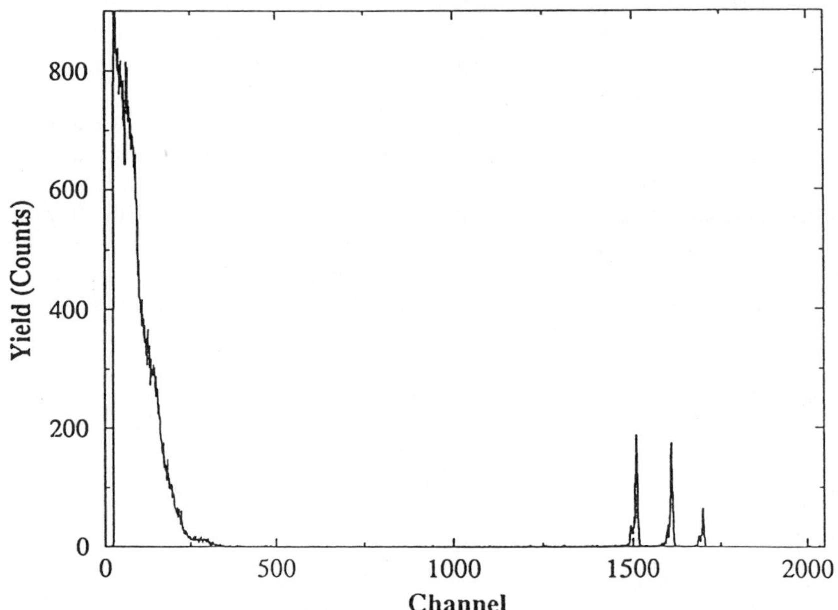

Fig. 2. RBS spectrum from uncoated sulphur-labelled dye diffused into polymer.

(2) Current from the sample is not monitored as the charge collection efficiency varies between polymers. Instead the current from a grid of vertical gold wires intercepting about 5% of the beam upstream of the sample is monitored for inter-sample normalisation purposes.

(3) To inhibit beam heating of the polymer low current densities are used (of order 0.1nA mm^{-2}) and the samples are routinely back-cooled with liquid-nitrogen.

We have successfully used this technique to profile sulphur-labelled dye diffusing into polymeric matrices [1] and fluorinated monomers segregating to the surface of resin[2].

1.2 Nuclear Reaction Analysis

There is a lot of interest in the self-diffusion and mutual diffusion of polymer pairs. Usually to study this one of the polymers has to be isotopically labelled to distinguish it from the other. A nuclear reaction technique can then be used to depth profile the isotope. The isotopic substitution that causes minimum disturbance to polymer properties is deuteration and fortunately the deuterated analogues of most polymers can be routinely purchased. To study the deuteron depth profile we use[3] a 0.7 MeV ^{3}He beam to induce a fusion reaction with the deuteron causing the production of protons and α-particles:

$$^3\mathrm{He} + d \rightarrow p + \alpha$$

As the final nuclear masses are less than the initial masses this reaction results in a substantial energy release - the protons are ejected with an energy ~ 12 MeV, the alphas of order ~ 4 MeV.

Just as in Rutherford Backscattering a ^{3}He ion loses energy with depth and the resultant proton and α energies change. A peculiar difference is that due to the light deuteron being knocked forward by the ^{3}He nucleus in the collision, as the ^{3}He ions lose energy the protons and the alphas emerge from the resultant compound nucleus at 165° with more energy! However just as in RBS the energy spectrum of the outgoing particles can be directly related to the depth distribution of the deuterons. A full spectrum from a uniformly deuterated sample is shown in Fig. 3. From the left are the backscattered ^{3}He ions, alphas from their interaction with deuterons, α peaks from the calibration source and the proton spectrum. We use the detected protons for depth profiling rather than the alphas because the depth resolution is better.

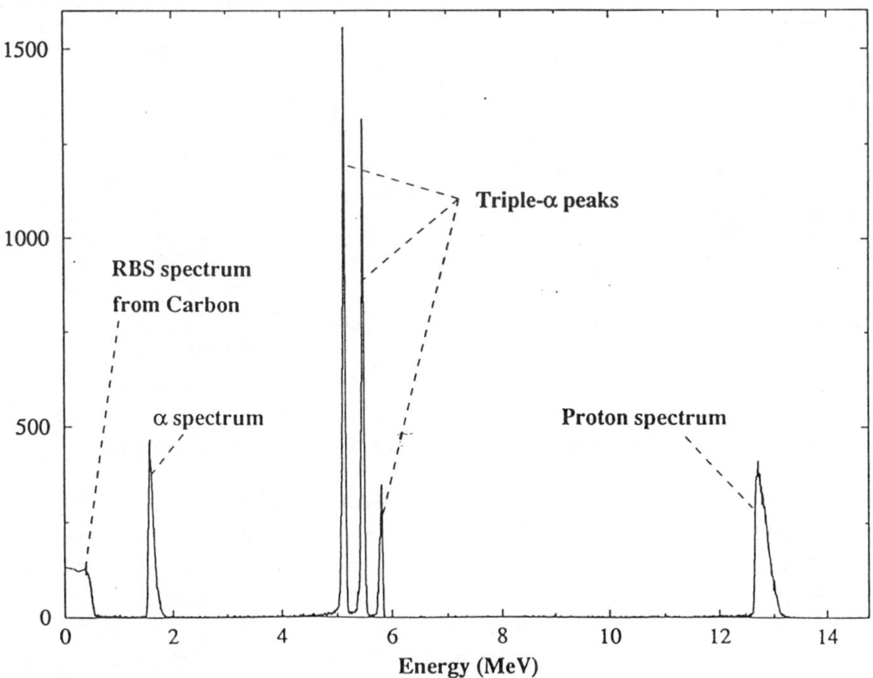

Fig. 3. Spectrum resulting from 0.7 MeV ^{3}He interactions with deuterated polystyrene.

Polymers such as polystyrene are resistant to beam heating effects but others - notably polymethylmethacrylate (PMMA) - degrade very rapidly at room temperature and have to be back-cooled to liquid nitrogen temperatures. To illustrate this we prepared a 'sandwich' of three equally thick layers of d-PMMA,PMMA and d-PMMA and exposed them to the ^{3}He beam for different times: (a) at room temperature, (b) back-cooled with liquid nitrogen. Results are illustrated in Fig. 4. As can be seen at room temperature (4(a) and 4(b)) the surface d-PMMA layer is destroyed rather rapidly. As it gets thinner the inner layers move towards the surface. Contrast this with the behaviour of the sample back-cooled with liquid nitrogen (4 (c) and 4(d)) - the surface layer maintains its integrity under the same current density for times well in excess of that needed to do a useful measurement (typically < 200 seconds). When using back-cooling a simple point is that it is important to change the sample holder each time it is extracted from the chamber - otherwise if the sample is put onto a cold holder condensation occurs and there is the possibility of an ice layer forming when it is reintroduced to the back-cooled goniometer.

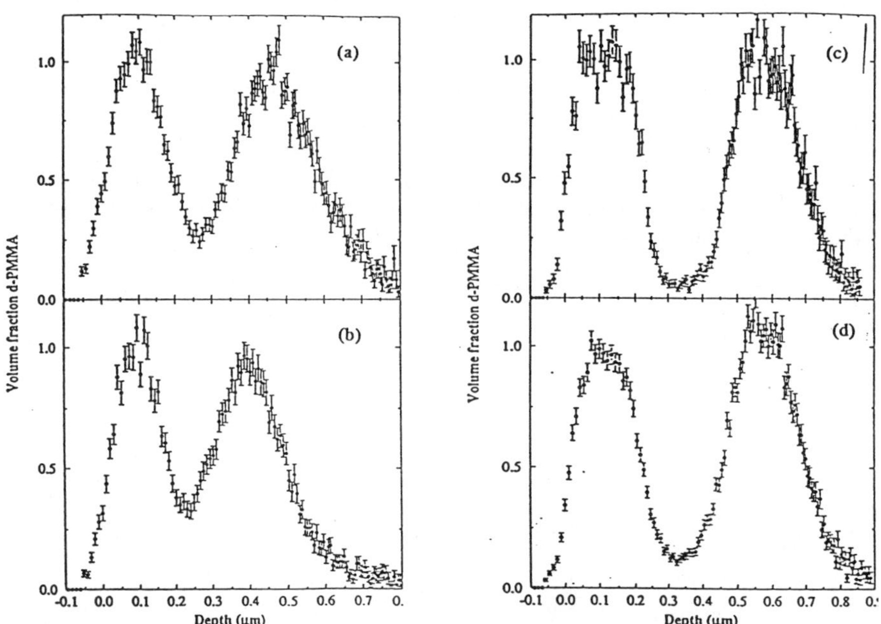

Fig. 4. Depth profiles of a d-PMMA, PMMA, d-PMMA trilayer. (a) and (b) were obtained after 250 seconds and 1000 seconds from an uncooled sample. (c) and (d) were obtained after 250 and 1000 seconds from a liquid nitrogen-cooled sample.

We have used this technique to investigate a variety of interdiffusion and segregation problems in polymer couples of industrial interest [4-11]. One example is the study of tracer diffusion of deuterated polystyrene (d-PS) or an admixtureof d-PS and PαMS into a blend of PS and PαMS (Poly alpha methyl styrene)[12]. Results obtained using the ^{3}He energy loss technique for these two situations are shown in Fig. 5. When a surface layer of pure d -PS is annealed to a receiver layer of PS/PαMS at 190C for various times diffusion is rather slow . If however the surface layer is an admixtureof d-PS and PαMS then diffusion is much more rapid.

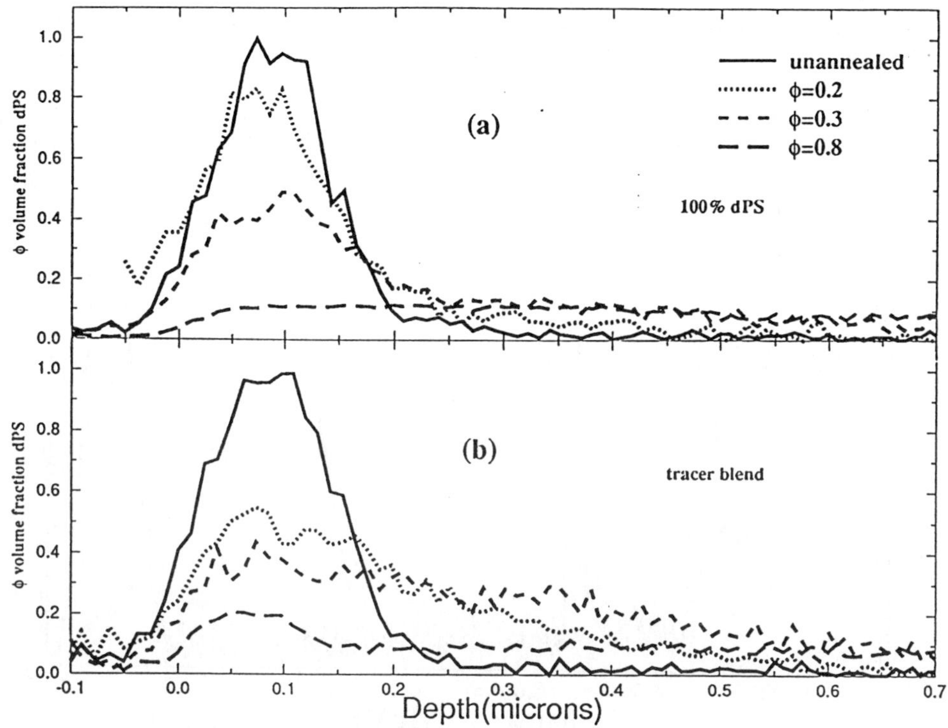

Fig. 5. Depth profiles of:
(a) d-PS diffused into a PS/PαMS blend
(b) d-PS/PαMS tracer blend diffusing into PS/PαMS blend.

2. Beam traversing techniques

For the study of small molecule diffusion in polymers the depth limitations of energy loss techniques (a few microns) can be a hindrance. Small molecules can often rapidly diffuse distances of mm from the surface . To measure their profiles we use a traversing beam technique on a cross-section of material cut perpendicular to the surface.

The ion beam is a focussed microbeam: the beam from the Van de Graaff is focussed on an object aperture of diameter 1000, 200,50 or 25 microns and passes through horizontal and vertical transputer-controlled raster scanned steering plates. It is then demagnified in diameter by a factor of five by a set of four quadrupole magnets prior to incidence on the sample. Scanning can be over a square of side length ranging from 100 microns to millimetres.

Any combination of ion beam (protons, ^{3}He or ^{4}He ions) and reaction product (backscattered particle, charged nuclear reaction product or characteristic X-ray) can be scanned. This allows the spatial profiling of practically every element except helium over distances from tens of microns to millimetres with spatial resolution from a few microns to 200 microns We can study systems as disparate as elemental inclusions in the insulator layer of power cables (of order 10 microns), deuterated surfactant diffused into hair (of order 100 microns diameter) and heavy water into polymer (millimeter distances).

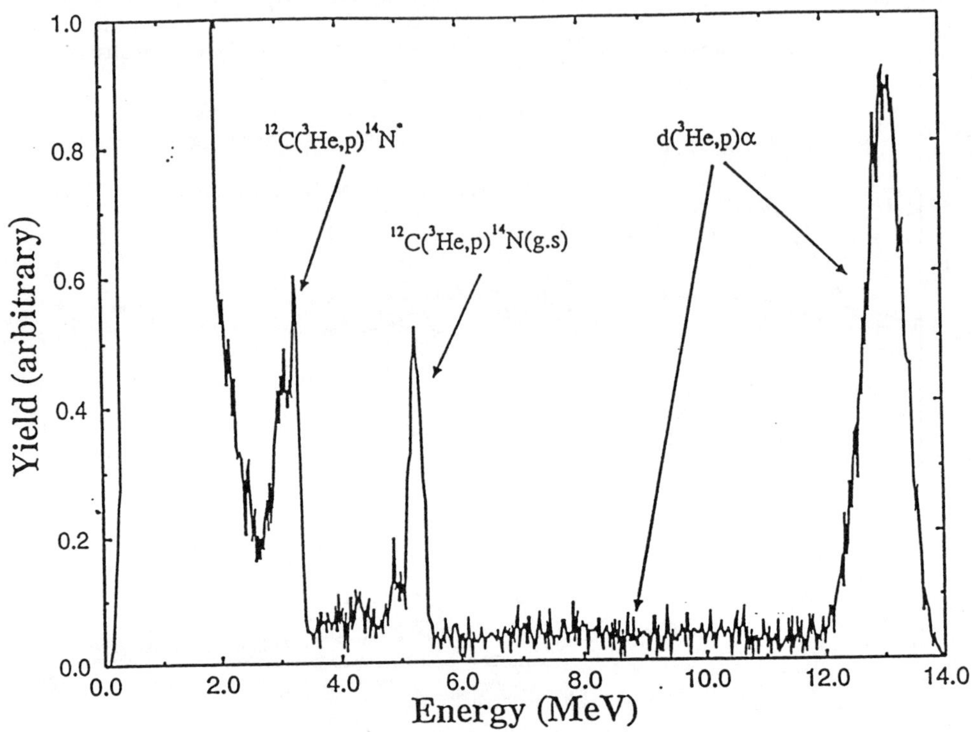

Fig. 6. Spectrum from deuterated water diffused into polymer. The 'window' of protons scanned is from 6-14 MeV.

As an example of the latter application[13] a sheet of hydrophilic polymer is immersed in heavy water (D_2O) for a certain time. The sample is then rapidly frozen by plunging it into liquid nitrogen and a clean break made perpendicular to the surface. The exposed edge is mounted on a plate in the scattering chamber back-cooled by liquid nitrogen. It is then scannned with a ^3He beam and a spectrum of charged reaction products (Fig 6) recorded in a pulse height analyser linked to software enabling a 'window' of the high energy proton spectrum to be correlated with the position of the ion beam scan (Fig 7a). A depth profile folded with the system resolution is easily obtained (Fig 7b) and hence a diffusion coefficient can be measured. In addition the proton yield from the beam reaction with the matrix carbon, $^{12}C(^3He,p)^{14}N$, can be used for inter-sample concentration comparison.

We are presently upgrading the system to allow the simultaneous profiling in three dimensions of heavy water by NRA and surfactants (elementally labelled) by RBS by scanning multiple 'windows' within the energy loss spectra to investigate the problems of drying of latex films.

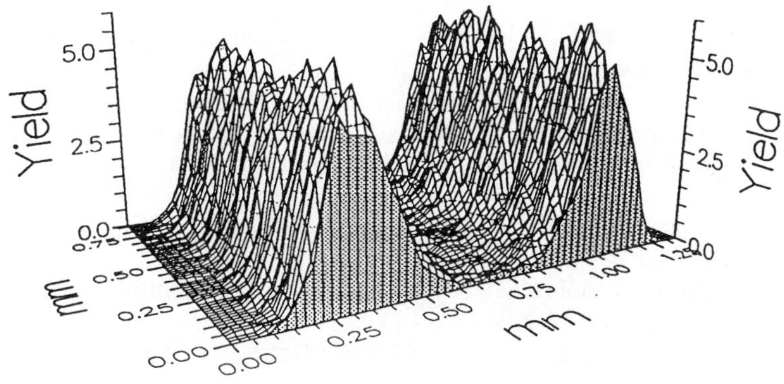

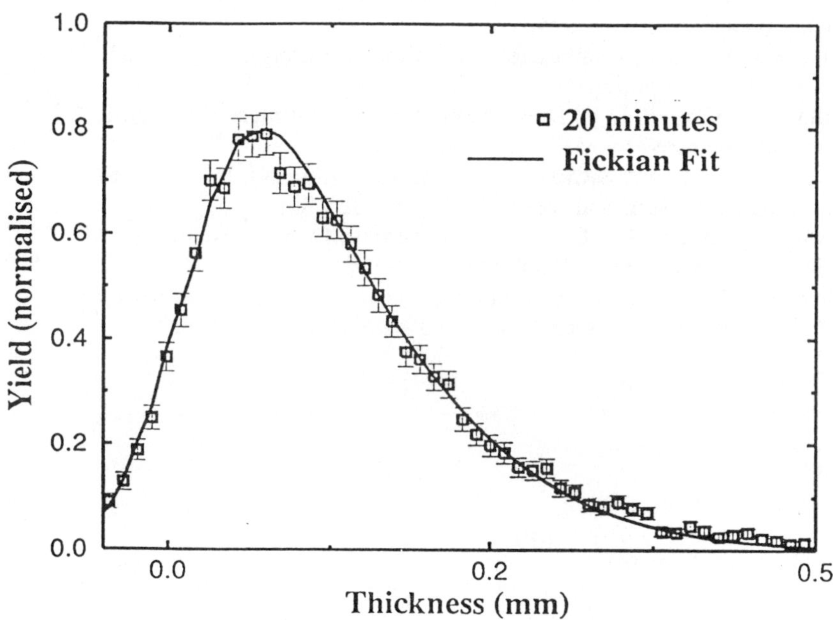

Fig. 7. (a) A 3-d spectrum of deuterated water diffused into the two sides of a polymer sheet. (b) A plot of proton yield versus depth extracted from one side of (a) and fitted with an appropriate diffusion profile.

Conclusion

MeV ion beam analysis can be used successfully and straightforwardly to investigate many problems of diffusion in both commercial and biological polymers. Diffusion depths from fractions of a micron to millimetres and diffusants from light molecules to polymers can be profiled.

References

[1] T E Shearmur, D W Drew, A S Clough, M G D van der Grinten (Surrey), A T Slark (ICIImagedata). Polymer 37, 2695 (1996)

[2] M G D van der Grinten, A S Clough, T E Shearmur (Surrey), R Bongiovanni and A Priola (Politecnico di Torino). Accepted by J Colloid and Interface Science

[3] R S Payne, A S Clough, P Murphy and P J Mills. Nucl Instrum Meths B42, 130 (1989)

[4] M Geoghegan, R A L Jones, R S Payne (Cambridge), P Sakellariou (ICI), A S Clough (Surrey), J Penfold (RAL). Polymer 35, 2019 (1994)

[5] M Geoghegan, R A L Jones (Cambridge), A S Clough (Surrey), J Penfold (RAL). J Polym Sci B: Polym. Phys. 33, 1307 (1995)

[6] M Geoghegan, R A L Jones (Cambridge), A S Clough (Surrey). J Chem Phys 103, 2719 (1995)

[7] M Geoghegan, R A L Jones (Cambridge), D S Sivia, J Penfold (RAL), A S Clough (Surrey). Phys Rev E 53, 825 (1996)

[8] C J Clarke, R A L Jones (Cambridge), J L Edwards (ICI), A S Clough (Surrey), J Penfold (RAL).Polymer 35, 4065 (1994)

[9] C J Clarke, R A L Jones (Cambridge), A S Clough (Surrey). Polymer 37, 3813 (1996)

[10] T Shearmur, A S Clough, D W Drew, M G D van der Grinten (Surrey), R A L Jones (Cambridge). Accepted by Macromolecules.

[11] M G D van der Grinten, A S Clough, T E Shearmur (Surrey), R A L Jones (Cambridge). Accepted by Surf. and Interf. Anal.(ECASIA 95).

[12] M G D van der Grinten, A S Clough, T E Shearmur, D W Drew(Surrey), M Geoghegan (Saclay), R A L Jones (Cambridge). In preparation.

[13] D W Drew, A S Clough, P M Jenneson, T E Shearmur, M G D van der Grinten (Surrey) and P Riggs (London Hospital). Accepted by Nucl Instrum Meths.

Materials Science Forum Vols. 248-249 (1997) pp. 155-158
© *1997 Trans Tech Publications, Switzerland*

Ion Induced Passivation of Metal Surfaces:
The Phenomenon and its Origins

J.E.E. Baglin[1], M.H. Tabacniks[2], A.J. Kellock[1],
N.S. Somcio[1,3] and T.T. Bardin[1,3]

[1] IBM Almaden Research Center, San Jose, CA 95120, USA

[2] Institute of Physics, University of Sao Paulo, Brazil

[3] Dept. of Mat. Eng., San Jose State University, San Jose, CA 95192, USA

Keywords: Copper, Passivation, Ion Beam, Oxidation, Corrosion

Abstract

The mechanism of ion beam passivation of a copper surface has been investigated. While inhibition of Cu oxidation is found to be correlated with irradiation in the presence of contaminant carbon, this experiment shows that films formed and irradiated under UHV conditions (no carbon) are not passivated. Ion beam passivation is attributed to radiation degradation of adventitious adsorbate hydrocarbons or solvent residues, leading to ~ 2 nm of graphitic carbon that may block chemically reactive sites on the Cu surface.

INTRODUCTION

It has long been recognized that irradiation of metal surfaces (Cu, Fe, Al, and others) with energetic beams of ions (e.g. H^+, He^+, B^+, Ar^+) significantly inhibits subsequent oxidation or aqueous corrosion of those surfaces [1-5]. However, the mechanism underlying this technologically interesting phenomenon has not been determined. Initially, it was widely assumed that ion bombardments conducted in poor vacuum must be producing a plaque-like barrier on the metal, consisting of beam-degraded pump oil. While it is indeed possible to build such thick protective coats by intentional dosing (e.g. ref [6]), it has also become evident that oxidation protection still occurs after irradiation in a reasonably clean vacuum, and some reports [4] have implied that a carbonaceous coating may be less relevant than structural or chemical changes induced within the metal.

This experiment was designed to test the nature and role of surface coatings of carbon in the passivation of copper surfaces.

EXPERIMENT

The passivation of copper is illustrated in Fig.1. In this test, a film of copper (400 nm thick) was e-beam deposited on a polished amorphous carbon substrate. The sample was exposed to room air while being transferred to the ion implanter, and one half was then irradiated with 2 MeV Ar^+ ions, to a dose of 2×10^{16} Ar/cm^2, in a turbo-pumped vacuum at approx. 10^{-6} Torr. The sample was then heated in air at 170°C for 18 h. The irradiated and non-irradiated areas were then analyzed by RBS. Fig. 1. shows the copper surface signal from those RBS spectra. The unimplanted area was oxidized to a depth of ~ 200 nm (apparently forming Cu_2O and CuO), while the implanted area showed oxidation to 5 nm or less. The use of a carbon substrate enabled RBS to show not only the oxygen profile for the non-irradiated portion, but also a surface carbon layer about 3 nm thick, exclusively on the implanted portion. In other similar runs [5] this thin carbon layer was found consistently to accompany passivation of the copper. But was the carbon the cause of the passivation, or an incidental accompaniment to some other effect of the ion beam? To answer that question, a new experiment was devised that would exclude carbon from the sample, both before and during irradiation.

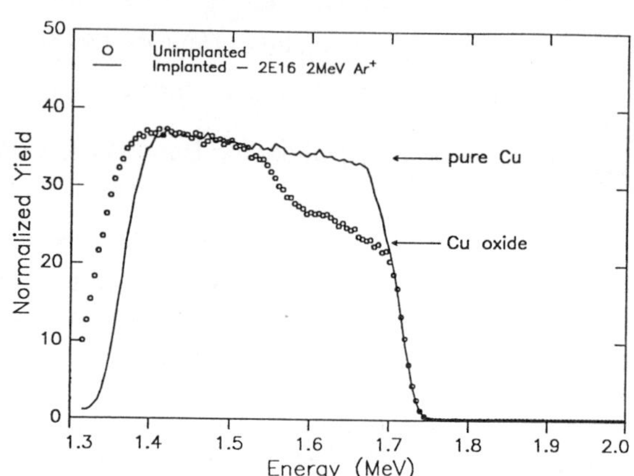

Fig.1. Ion beam passivation of Cu. The figure shows RBS spectra from air-exposed Cu films ~ 400 nm thick, after heating in air at 170°C for 18 h. The dotted data refer to an unimplanted sample, where a thick layer of oxide is seen. The solid line represents a sample implanted with 2×10^{16} Ar/cm² before heating, on which virtually no oxide has grown.

Carbon-free experiment

The layout of the UHV chamber is schematically shown in Fig. 2. The system was evacuated with sorption and turbo pumps, N_2 purged and baked, yielding a pressure $< 10^{-11}$ Torr, with all carbon-related RGA signals totalling less than 10^{-12} Torr. A Mo boat evaporator was used to deposit a 400 nm Cu film on a polished Cu substrate, which was cooled to maintain room temperature at all times.

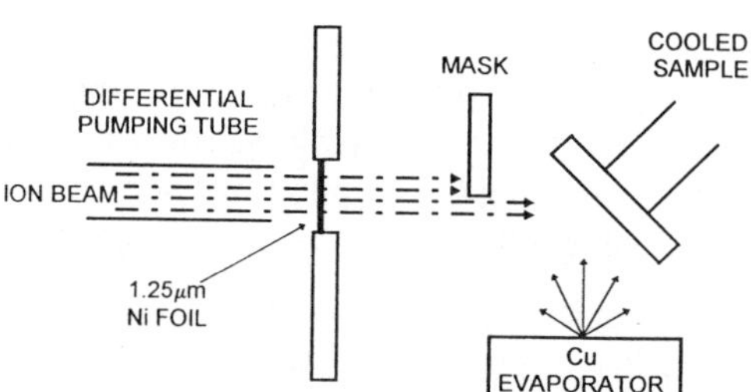

Fig.2. Experimental layout within the UHV chamber.

The ion beam from the accelerator (normal beam line pressure 10^{-8} Torr) entered through a differential pumping tube. In order to intercept line-of-sight contaminants from the beam line, the ion beam was passed through a 1.25 μm Ni foil before reaching the sample. (This foil degrades the energy of the incident He⁺ ions used for this experiment to 1.5 MeV at the sample.) The beam dose was calibrated with reference to an electron suppressed Faraday cup. The dose of 2×10^{16} He⁺/cm² was delivered in 90 minutes, immediately following deposition of Cu; thus, contaminant species landing on the surface could amount to no more than $\sim 10^{15}$ at/cm² during the implantation.

After implantation, the sample was transferred (with a brief exposure to room air) to an ESCA system, where line scans would enable comparison of surface carbon in the irradiated and non-irradiated areas, and full spectrum analyses at selected points could identify the bonding states for Cu in both surface areas. The sample was then oxidized in air at 170°C for 18 h, and the implanted and non-implanted regions were examined by RBS to quantify the oxidation.

Controlled contaminants

As a reference for comparison with sample surfaces exposed to contaminants before irradiation, the same procedure was repeated, differing only by a brief exposure to room air and a rinse with ethanol before irradiation.

Fig. 3. Oxidation of "carbon-free" Cu samples, deposited at UHV and implanted in situ (solid line) or not implanted (dots) prior to heating in air. Implantation without carbon has given little or no protection.

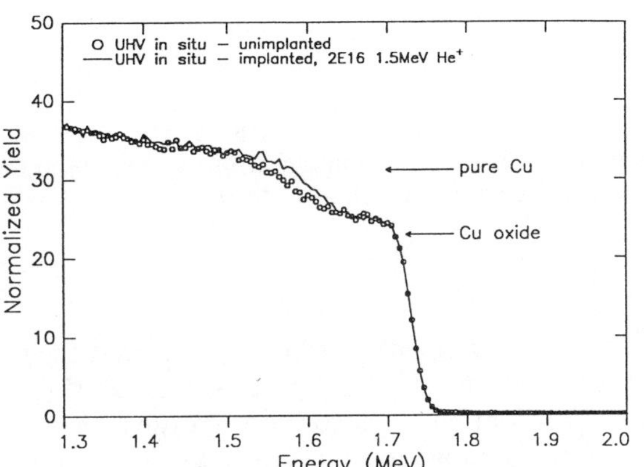

RESULTS

Fig. 3 displays the RBS spectra from the "carbon-free" sample following its oxidation. The extent of oxidation is large, and differs only marginally between irradiated and non-irradiated areas. Visually, no difference between the regions could be observed. Evidently, the passivation effect is absent when carbon is rigorously excluded.

Fig. 4(a) shows the line scan ESCA profiles across the two regions of the sample, based on total intensities of C 1s, O 1s and Cu 2p3 lines, and quoted as atomic percent within the electron escape depth (< 2 nm). C and O amounts like those seen on the unimplanted area are found for all samples exposed to room air, and are attributed to hydrocarbon complexes physisorbed on the surface. From the components of the Cu Auger LMM line, it was deduced that the large Cu signal represents approximately equal contributions of Cu^{1+} (bonded presumably to oxygen) and Cu^{0} (virgin metal). It thus appears that, in transit to analysis, pure Cu gains (i) a monolayer of surface oxide and (ii) about a monolayer of hydrocarbon adsorbate. The important feature of Fig. 4(a) is that the implanted surface is indistinguishable from that of the non-implanted area, apparently accumulating no extra carbon by virtue of the irradiation step.

By contrast, Fig. 4(b) shows line scan ESCA data from the sample that was exposed to air/ethanol prior to implantation. Here, the implant has evidently attached extra carbon at the surface, that had been acquired during room exposure. (The profile reflects the non-uniformity of the beam spot.) Presumably, the ions ballistically dissociate the hydrocarbon adsorbate, liberating hydrogen and leaving graphitic carbon. Furthermore, in the implanted area, the Cu^{1+} signal proved to be smaller than that in the virgin area, consistent with inhibition of Cu oxidation by the carbon during transport to analysis.

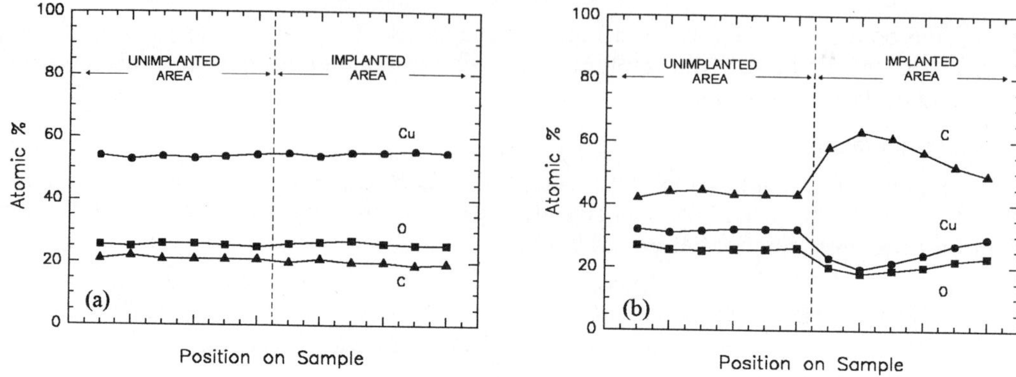

Fig. 4.   ESCA spatial scans across (a) the UHV sample of Fig. 3;   (b) a Cu sample exposed to air/ethanol rinse before implantation. The implanted area of sample (b) with increased carbon showed little oxide growth after heating.

CONCLUSIONS

The studies described are reproducible, and they show a clear correlation of Cu passivation with the existence of a carbon coat ~ 2 nm thick after irradiation. This layer arises by ion beam degradation of adsorbate hydrocarbons, which in their normal state offer no protection against oxidation. When carbon is excluded, the ion beam does not passivate the Cu.

Such a thin layer can offer macroscopic protection only by creating a surface on which reagent molecules or ions do not chemisorb. All that we have seen would be consistent with the process reported by van Kooten et al. [7], in which oxidation of the Cu(111) surface could be inhibited by a sub-monolayer of graphitic carbon physisorbed selectively at the threefold hollow sites, thereby blocking the sites required for oxygen to chemisorb and react. Such a mechanism appears likely to account for the observed ion beam passivation of copper. This understanding should provide a basis for the design of a robust process, with practical technological application.

REFERENCES

1.   B.L. Crowder and S.I. Tan, "Prevention of Corrosion of Copper by Ion Implantation", IBM Technical Disclosure Bulletin **14,** no.1 (1971)

2.   H.M. Naguib, R.J. Kriegler, J.A. Davies and J.B. Mitchell, J. Vac. Sci. Technol. **13,** 396 (1976)

3.   P.J. Ratcliffe and R.A. Collins, Phys. Stat. Solidi A **78,** 547 (1983)

4.   P.J. Ding, W.A. Lanford, S. Hymes and S.P. Murarka, J. Appl. Phys. **74,** 1331 (1993)

5.   N.S. Somcio, "The Effect of Ion Implantation on Copper Oxidation", M.S. Thesis, Dept. of Materials Eng., San Jose State University, 1997.

6.   Y. Itoh, S. Hibi, T. Hioki and J. Kawamoto, J. Mater. Res. **6,** 871 (1991)

7.   W.E.J. van Kooten, O.L.J. Gijzeman and J.W. Geus, Surface Science **303,** 16 (1994)

Materials Science Forum Vols. 248-249 (1997) pp. 159-162
© *1997 Trans Tech Publications, Switzerland*

Flux Pinning by Columnar Defects in $Bi_2Sr_2CaCu_2O_8$-Thin Films

F. Hillmer[1], G. Wirth[2], P. Haibach[1], U. Frey[1], Th. Kluge[1],
G. Jakob[1] and H. Adrian[1]

[1] Institut für Physik, Johannes Gutenberg Universität, Staudinger Weg 7, D-55099 Mainz, Germany

[2] Gesellschaft für Schwerionenforschung, Planckstr. 1, D-64291 Darmstadt, Germany

Keywords: Bi-2212 (High Temperature Superconductor) Thin Films, High Energy Heavy Ion Irradiation, Columnar Defects, Transport Measurements, Pinning

Abstract *In $Bi_2Sr_2CaCu_2O_8$-thin films continuous amorphous columnar defects were created by heavy ion irradiation. Three of four identical striplines on the same sample were irradiated in a different way which includes irradiations under different angles with respect to the film $\vec{c}$-axis. After irradiation the normal state resistivity is enlarged and the films exhibit a reduction of T_c proportional to the volume of the damaged material. Furthermore the activation energy for flux movement shows an enhancement which is best for magnetic fields close to the matching field. Measurements of the transport critical current density, the activation energy and the irreversibility field in dependence of the angle between magnetic field and the film $\vec{c}$-axis show no anisotropic pinning due to the induced columnar defects. This can be related to the intrinsic two-dimensionality of $Bi_2Sr_2CaCu_2O_8$.*

PACS numbers: 74.25 Fy, 74.60 Ge, 74.60 Jg, 74.72 Hs

Introduction

Among the high temperature superconductor cuprates $Bi_2Sr_2CaCu_2O_8$ is one of the most sensitive compound to external magnetic fields originating from its layered structure and high mass anisotropy. Due to this intrinsic two-dimensionality the pinning of magnetic flux is weak. Therefore already in the presence of small transport currents flux lines are set in motion which then has the consequence of dissipation. A possibility to improve this unfavourable property is the deposition of artificial defects into the crystal lattice, which according to the local reduction of superconductivity, represent pinning centers for flux lines. Very effective pinning centers are columnar defects produced by heavy ion irradiation [1, 2] here of approximately 8 nm in diameter in the $Bi_2Sr_2CaCu_2O_8$-thin films. Only by using ions of high atomic numbers with high, well defined kinetic energy the electronic energy loss in the material will be high enough to create a continuous amorphous track along the rectilinear path of each ion. Confirmed by transmission electron microscopy the columnar defects extend through the whole thickness of the film. Because of their linear geometry they are able to interact along their entire length with magnetic flux lines explaining their high pinning activity and strong influence on the vortex system [3]. In the present case irradiations were performed in different orientations with respect to the film $\vec{c}$-axis. With the exploitation of transport measurements in magnetic fields for different angles between the magnetic field and the film $\vec{c}$-axis we intend to draw conclusions on the vortex dynamics.

Irradiation of the films

In order to compare different irradiation effects it is necessary to perform irradiation experiments on samples showing similar results of transport measurements before irradiation. For this purpose four identical striplines have been patterned on the same film which can be measured independently of each other in the standard four point technique. The preparation of the high quality dc-sputtered $\vec{c}$-axis oriented $Bi_2Sr_2CaCu_2O_8$-thin films is described elsewhere [4]. For the lateral homogenous films with the size of 9×9 mm^2 and T_c values around 87 K transport measurements on the four striplines of a given sample yield identical results before irradiation. By individual irradiation of three of the four striplines on the same film it is possible to compare the different irradiation effects with each other and with respect to the unirradiated case. The analysis is thus unaffected by the deviations in sample quality.

The heavy ion irradiations with ^{197}Au with an energy of 11.4 MeV/u were carried out at the UNILAC of the *Gesellschaft für Schwerionenforschung (GSI)* in Darmstadt. At room temperature we performed two types of irradiation: Type I is the irradiation on three of four striplines of the same sample in the direction of the film $\vec{c}$-axis with different doses corresponding to the dose equivalent fields B_ϕ of 0.3 T, 1 T and 2 T. Type II is the irradiation on three of four striplines of the same sample under different angles of 45°, 30° and 15° with respect to the film $\vec{c}$-axis but with equal doses corresponding to the dose equivalent fields B_ϕ of 1 T. The dose equivalent or matching field B_ϕ is the magnetic field obtained from the condition that in average the film contains equal numbers of columnar defects and flux lines per unit area.

Resistive transitions

Before irradiation resistive measurements were performed in different magnetic fields $\vec{B}||\vec{c}$. In Arrhenius representation ($\Rightarrow \ln\rho$ in dependence of $1/T$) from the linear behaviour over four decades of resistivity, the resistivity can be described by $\rho = \rho_o \cdot \exp(-U(T,B)/k_BT)$ with an activation energy $U(T,B) \propto (1/\sqrt{B}) \cdot (1 - T/T_c)$ in accordance with the model of plastic thermally assisted flux flow with double vortex-kink formation [5].

After type I irradiation the resistive transitions in zero magnetic field show an enlargement of the normal state resistivity and a reduction of T_c which depends linearly on the ion dose. After type II irradiation the behaviour of resistive transitions in zero magnetic field is qualitatively the same but seems to be independent of the direction of irradiation. From these measurements we conclude that the irradiation causes a T_c reduction proportional to the volume of the damaged material. Ascertained from resistive transitions after type I irradiation in external magnetic fields the field dependence of the activation energy shows deviations from the $1/\sqrt{B}$ proportionality. Compared to the unirradiated situation the activation energy is increased. This increase is strongest for magnetic fields close to B_ϕ showing that only in the case of coinciding density of vortices and columnar defects the highest pinning is expected.

Transport measurements in dependence of the angle between magnetic field and the film $\vec{c}$-axis

For the rest of this paper all measurements are performed after type II irradiation. Magnetic field and columnar defects are always oriented perpendicular to the fixed direction of the transport current. Fig. 1 shows the behaviour of the critical current density during a complete rotation of the magnetic field with respect to the film $\vec{c}$-axis. Owing to the intrinsic pinning resulting from the layered structure of $Bi_2Sr_2CaCu_2O_8$ the values of J_c exhibit strong peaks at $\pm 90°$. These are slightly reduced after irradiation according to the degradation of the material. In difference to similar measurements on $YBa_2Cu_3O_7$-thin films [6] there is no indication to anisotropic pinning due to the induced columnar defects since the values of J_c display no extra peaks in the direction of irradiation. But what is remarkable is a J_c enhancement within a wide-angle range around 0° regardless of the orientation of the columnar defects. To understand this behaviour measurements of J_c in dependence on the magnetic field $\vec{B}||\vec{c}$ give further

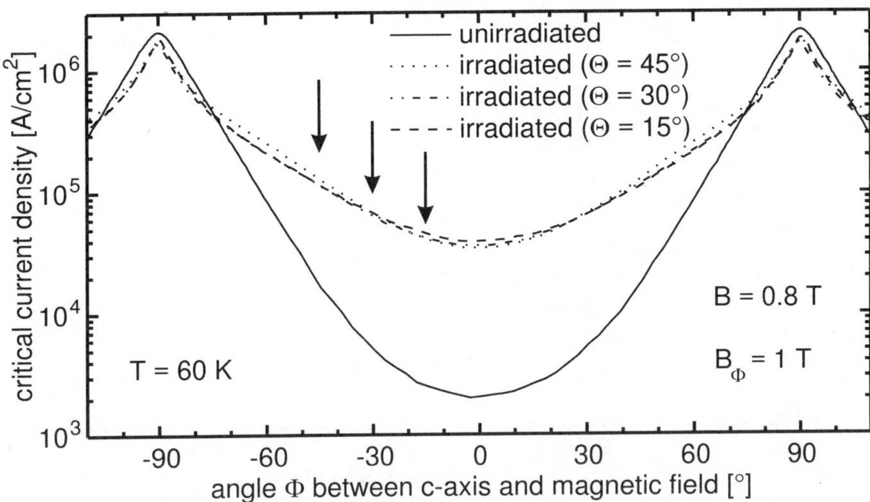

Figure 1: Critical current density in different magnetic field orientations after type II-irradiation. Θ is the angle of irradiation with respect to the film $\vec{c}$-axis which in addition is shown by the arrows for each stripline.

information. In Fig. 2 at low fields the measurements of the irradiated striplines show a reduction of J_c according to the degradation of the material. At higher fields the pinning of the columnar defects comes into play resulting in an enhancement of J_c. From these measurements with $J_c(\Phi) = J_c(B_{\|\vec{c}} = B_o \cdot \cos \Phi)$ the calculation of a complete turn of $J_c(\Phi)$ yield exactly the same result as measured in Fig. 1. That means the dependence of $J_c(\Phi)$ is determined only by the field component in $\vec{c}$-direction which is a consequence of the intrinsic two-dimensionality. Now the J_c enhancement in the wide-angle range can be explained by the fact that in this angle range the field component in $\vec{c}$-direction takes values at which from Fig. 2 the highest enhancement can be obtained. Supposing the existence of pancake vortices pinned by disklike defects created by the irradiation in each CuO_2-plane the highest pinning is expected if in the CuO_2-planes the number of pancake vortices is equal to the number of disklike defects. This can be verified from the measurements of Fig. 2 by calculating the enhancement factors of J_c in dependence on the magnetic field. The result is a typical maximum near B_ϕ (not shown). With increasing angle of irradiation this maximum is shifted toward smaller field values in consistence with the decreasing density of disklike defects.

As another possibility to look for anisotropic pinning resulting from the induced columnar defects the activation energy ascertained from resistive transitions was investigated in dependence on the field orientation. The measurements after type II irradiaton of course show an enhancement of the activation energy compared to that of the unirradiated stripline but no dependence on the direction of irradiation can be seen.

According to the relation [7] $B_{irr}(T, \Phi) = B_{irr}(T, \Phi = 0°)/\epsilon$ with $\epsilon^2 = \cos^2 \Phi + 1/\gamma^2 \cdot \sin^2 \Phi$ and $\gamma^2 = m_c/m_{ab} \simeq (150)^2$ for $Bi_2Sr_2CaCu_2O_8$ [8] the irreversibility field B_{irr} multiplied with the cosine of the angle Φ between $\vec{c}$-axis and magnetic field should be constant. The evaluation of these measurements after type II irradiation (not shown) reveal an enhancement of the irreversibility field but for the irradiated striplines the expected deviation from the constant behaviour was not found. Here the irreversibility field B_{irr} is obtained from a resistivity criterion like $\rho(T, \Phi, B_{irr}) = \rho_{crit}$.

The fact that in this work the measurements do not show evidence of anisotropic pinning

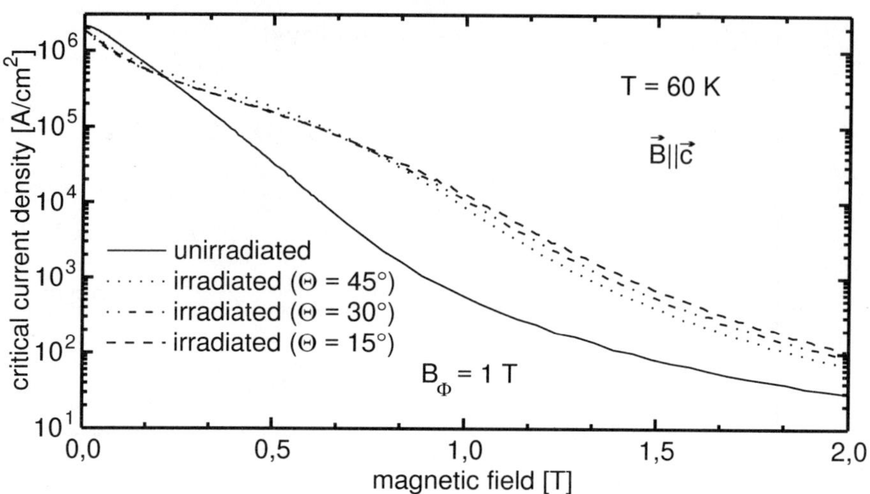

Figure 2: Behaviour of the critical current density after type II-irradiation in magnetic fields parallel to the film $\vec{c}$-axis. Θ is the angle of irradiation with respect to the film $\vec{c}$-axis.

due to the induced columnar defects is in contrast to measurements on $Bi_2Sr_2CaCu_2O_8$ single-crystals containing irradiation induced columnar defects [9]. This can possibly explained by the higher natural defect concentration of films compared to single crystals.

Acknowledgments

This work is financially supported by the Ministerium für Bildung, Wissenschaft, Forschung und Technologie der Bundesrepublik Deutschland and by means of the Gesellschaft für Schwerionenforschung.

References

[1] L. Civale, A. D. Marwick, T. K. Worthington, M. A. Kirk, J. R. Thompson, L. Krusin-Elbaum, Y. Sun, J. R. Clem and F. Holtzberg, *Phys. Rev. Lett.* **67**, 648 (1991).

[2] W. Gerhäuser, G. Ries, H. W. Neumüller, W. Schmidt, O. Eibl, G. Saemann-Ischenko and S. Klaumünzer, *Phys. Rev. Lett.* **68**, 879 (1992).

[3] S. Behler, S. H. Pan, P. Jess, A. Baratoff, H.-J. Güntherodth, F. Lévy, G. Wirth and J. Wiesner, *Phys. Rev. Lett.* **72**, 1750 (1994).

[4] P. Wagner, F. Hillmer, U. Frey, H. Adrian, T. Steinborn, L. Ranno, A. Elschner, I. Heyvaert and Y. Bruynseraede, *Physica C* **215**, 123 (1993).

[5] V. Geshkenbein, A. Larkin, M. Feigel'man and V. Vinokur, *Physica C* **162-164**, 239 (1989) and V. M. Vinokur, M. V. Feigel'man, V. B. Geshkenbein and A. I. Larkin, *Phys. Rev. Lett.* **65**, 259 (1990).

[6] B. Holzapfel, G. Kreiselmeyer, M. Kraus, G. Saemann-Ischenko, S. Bouffard, S. Klaumünzer and L. Schultz, *Phys. Rev. B* **48**, 600 (1993).

[7] G. Blatter, V. B. Geshkenbein and A. I. Larkin, *Phys. Rev. Lett.* **68**, 875 (1992).

[8] P. Wagner, F. Hillmer, U. Frey and H. Adrian, *Phys. Rev. B* **49**, 13184 (1994).

[9] D. Zech, S. L. Lee, H. Keller, G. Blatter, B. Janossy, P. H. Kes, T. W. Li and A. A. Menovsky, *Phys. Rev. B* **52**, 6913 (1995).

Materials Science Forum Vols. 248-249 (1997) pp. 163-166
© 1997 Trans Tech Publications, Switzerland

Depth Resolved Oxygen Analysis at the Interfaces of Au-Al Layers

A. Markwitz and G. Demortier

Laboratoire D'Analyses par Réactions Nucléaires, Facultés Universitaires Notre-Dame de la Paix, 22 rue Muzet, B-5000 Namur, Belgium

Keywords: Oxygen Analysis, Au-Al Layers, Backscattering Spectroscopy, Ion Beam Bombardment, Interface Behaviour

Abstract.
Oxygen at the interfaces of multilayered Au-Al systems was investigated using 7.6 MeV ^{4}He backscattering spectroscopy (evaporation on polished glassy carbon). Due to the high depth resolution of this analysis method, oxygen was successfully detected at the specimen surface, the Au-Al interface (two layer system) and at the substrate. After an ion beam bombardment with high-energy ^{4}He$^+$ ions for a few minutes, a complete Au-Al interdiffusion in the Au-Al layers was measured (in-situ RBS) and in addition oxygen at the Al-Al interface disappeared.

1. Introduction

During deposition of thin gold and aluminium layers on carbon substrates using a conventional evaporation set-up, oxygen contaminations especially at the interfaces between the substrate and the first deposited layer, and the layers itself, can appear. It is known that aluminium and oxygen are forming an ultra-thin Al_2O_3 layer immediately, e.g., when aluminium get in contact with air, whereas in the case of gold and oxygen no special compound is formed. However, the first gold atom layers are usually contaminated with the elements H, C and O.

Multilayered Au-Al systems [1,2] were used for investigating the Au-Al interdiffusion due to high-energy ion beam bombardment [3,4,5]. As a matter of fact, the presence of ultra-thin Al_2O_3 layers at the Au-Al interfaces can have a disturbing effect on the interdiffusion behaviour (diffusion barrier by Al_2O_3). Due to this fact, a depth resolved oxygen detection in the nanometer scale is strongly required. A possibility for this oxygen analysis, e.g., even in high-Z substrates, is to use the (α,α) elastic scattering resonances at 7.3 to 7.65 MeV [6,7], where the non-Rutherford cross section is more than 100 times larger than the Rutherford cross section (laboratory angle 170°). In this contribution, backscattering measurements (BS) concerning ^{16}O at the interfaces in as-deposited and ion beam bombarded Au-Al layers are presented.

2. Experimental

Thin Au-Al multilayered systems were deposited by evaporating gold (thermal heating) and aluminium (electron gun) in the same chamber under high vacuum conditions ($P_0 < 10^{-7}$ hPa, LN$_2$ trap) on polished glassy carbon substrates maintained at room temperature. Al was deposited first on carbon. Au layers were chosen as top-layers in order to avoid the formation of oxide compounds at the specimen surface. Inhomogeneities in the thickness of the gold and aluminium layers are avoided by the large distance (25 cm) between the evaporation sources and the substrates. In that geometry 200 nm thick gold and aluminium films are deposited in less than 10 min [8].

Ion beam bombardment (IBB) using 2.0 MeV ^{4}He$^+$ ions with the current of 300 nA (beam diameter 0.7 mm) was performed to investigate the interdiffusion behaviour of the Au-Al multilayered systems (incident ion range 6.5 μm [9]). In the experiments, the RBS spectra, indicating the interdiffusion behaviour, were acquired simultaneously [10,11]. Carbon substrates having a low Z were used to measure the aluminium signals in the nearly background-free region of the RBS spectra [12]. Furthermore, apertures with diameters of 0.5 mm were attached in front of the detector and at the distance of 1.5 cm to minimise multiple scattering which yields in a high background in the RBS spectrum, especially in the region of the aluminium signal.

The oxygen analysis was performed using the (α,α) elastic scattering resonances at 7.3 to 7.65 MeV [5,6], where the elastic cross section is more than 100 times larger than the Rutherford cross section (used laboratory angle 171°). Using the ^{4}He energy of 7.6 MeV, oxides at the surface and interfaces (e.g., in 700 nm) can be measured without changing the incident ion energy, as has to be performed when using the „classic" sharp $^{16}O(\alpha,\alpha)^{16}O$ resonance reaction at 3.045 MeV for the detection of oxygen in heavy matrices [13]. Furthermore, at 3.045 MeV the σ/σ_R enhancement factor is only 24 [5]. As backscattering spectra (BS) spectra may be fully calculated using the known physical parameters (e.g., ion sort, ion energy, backscattering angle, charge, beam current, etc.), the depth distributions of gold, aluminium, as well as oxygen can be extracted from the experimental spectra using the RUMP simulation program [14] considering non-Rutherford cross sections.

The phases described in this contribution are concluded by the ratios of elements measured by RBS. Au and Al layer thicknesses are calculated by using the standard densities of gold and aluminium, whereas the thickness of the interdiffused AuAl layer is calculated considering the specific atomic ratios.

3. Results and discussion

Fig. 1 shows a part of the BS spectrum (oxygen region) of a two layer Au-Al system in the as-deposited state. Oxygen impurities at the specimen surface (backscattering energy E = 2.734 MeV, areal density A = $2\;10^{15}$ at./cm^2), the Au-Al interface (E = 2.659 MeV, A = $21\;10^{15}$ at./cm^2) and the substrate (E = 2.611 MeV, A = $40\;10^{15}$ at./cm^2) were measured. Assuming that oxygen is bonded with aluminium, the thickness of the Al$_2$O$_3$ layer at the Au-Al interface is 5 ± 0.5 nm, and the thickness of the Al$_2$O$_3$ layer at the substrate is 10 ± 0.5 nm. In the Au-top having a thickness of $6.1\;10^{17}$ at./cm^2 an oxygen contamination of 1.5 ± 1 at.% was measured, whereas in the aluminium layer (thickness $1.1\;10^{18}$) the oxygen contamination amounts 6 ± 2 at.%. Carbon at the specimen surface is located in the BS spectrum at E = 1.905 MeV.

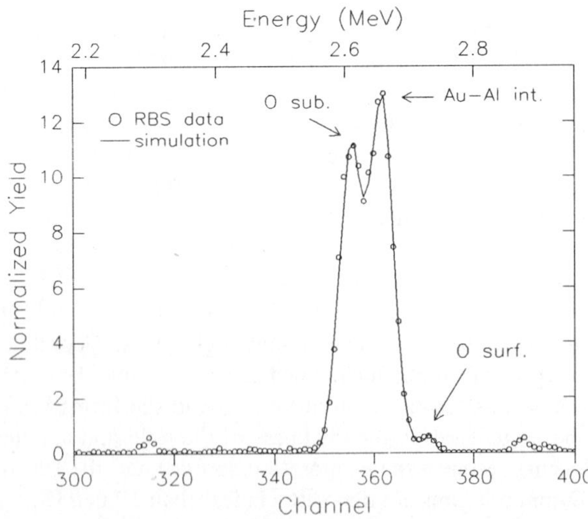

Fig 1 : A part of the BS spectrum in the oxygen region of an Au-Al specimen in the as-deposited state

A very fast AuAl interdiffusion is observed when multilayered Au-Al systems were irradiated with 300 nA 2.0 MeV ^{4}He$^+$. Figs. 2 shows the RBS spectra measured during IBB.

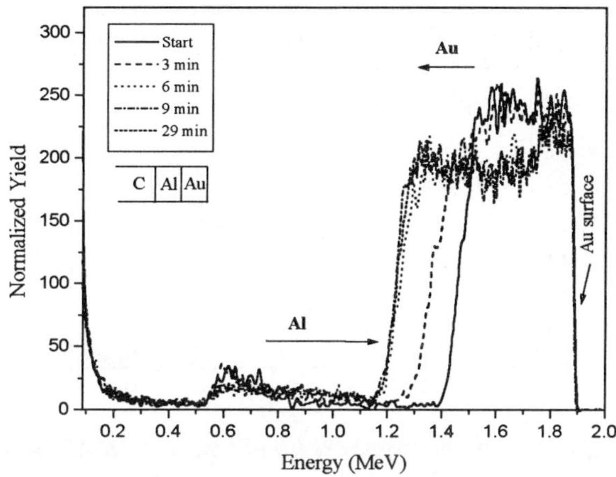

Fig. 2 : RBS spectra of the two layered Au-Al system irradiated with 300 nA in dependence on the irradiation. Arrows are indicating the AuAl interdiffusion

A two layer system containing of a 310 nm Au top-layer and a 290 nm Al layer was deposited on polished glassy carbon. The interdiffusion of gold and aluminium is indicated, e.g., by the shift of the gold edge in the RBS spectrum from the Al/Au interface to the Al/C interface (indicated by arrows in Fig. 2). After about 8 min, Au atoms reached the Al/C interface (diffusion speed 36 nm/min). Using the relation $x = 4(D't)^{1/2}$, the diffusion coefficient amounts to $D' = 1.2 \cdot 10^{-13}$ cm^2/s, which corresponds to the diffusion coefficient of the thermal annealing at about 450 °C (the solid solution forms gradually at temperatures over 300 °C). This diffusion is very fast compared to the interdiffusion with half of the ion beam intensity [10] where the diffusion speed of about 1 nm/min was measured. During IBB it was observed that the sample was red gleaming where the ion beam hit the target indicating a high temperature that was responsible for the thermal heating of the sample. Calculations using the Cookson relations [15] yielded a temperature difference between the irradiation spot and the sample holder of about 800 °C.

Fig. 3 shows a part of the BS spectrum in the oxygen region of the IBB Au-Al multilayered system (ref. to Fig. 1). After the irradiation the former separated Au/Al layers are interdiffused (ref. to Fig. 2). The remaining gold to aluminium ratio is 2:1, indicating the Au$_2$Al phase. In Fig. 3 the gold and aluminium signals can be clearly identified but with a low depth resolution due to the high ^{4}He energy of 7.6 MeV. Three peaks appear in the BS spectrum at the energies of 2.73, 2.34, and 1.91 MeV. The first peak is related to ^{16}O located at the specimens surface, whereas the second peak belongs to an oxygen contamination at the interface to the substrate. At 1.91 MeV, carbon deposited during the oxygen analysis by the analysis beam itself, located at the specimens surfaced can also be detected with a high precision due to the (σ/σ_R) factor of 51 (132 µb/sr at 170°) [7]. The oxygen signals were evaluated assuming oxygen bond with aluminium to Al$_2$O$_3$, due to the fact that only Al$_2$O$_3$ layers can resist the thermal heating above 500 °C for minutes. Oxygen bonded, e.g., in H$_2$O, O$_2$ or CO$_x$ phases could clearly not be maintained at the interfaces between the substrate and the aluminium layer. The thickness of the Al$_2$O$_3$ layer at the specimen surface is 4.9 ± 0.3 nm, and the thickness of the Al$_2$O$_3$ layer at the interface to the substrate is 9.0 ± 0.5 nm. The detection limit of oxygen located in the interdiffused AuAl layer is about 1.5 at.%. It has to be noted, that the background in this part of the BS spectrum originates from ^{27}Al(α,p)^{30}Si reactions and was subtracted.

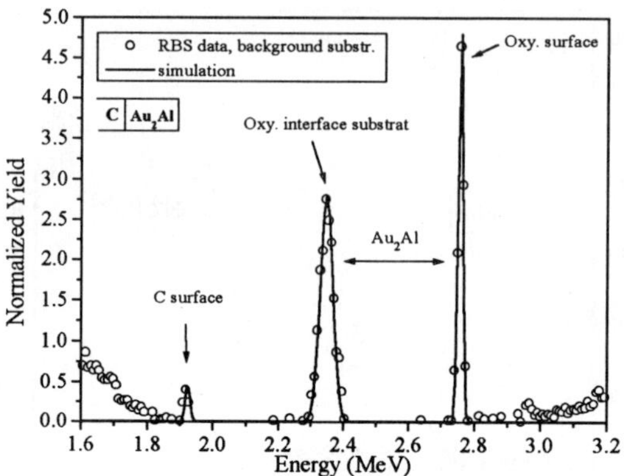

Fig. 3 : A part of the BS spectrum in the oxygen region of an IBB Au-Al specimen

An astonishing result of this depth resolved oxygen analysis is the absence of the ultra-thin Al_2O_3 layer between the deposited Au and Al layers in the IBB specimen (would be located at 2.55 MeV) in contrast to the Au-Al layer system in the as-deposited state (ref. to Fig. 1).

Acknowledgements

The BS measurements were performed at Institut für Kernphysik, J. W. Goethe-Universität Frankfurt am Main, Germany. This work is supported by the Human Capital and Mobility Program of the EC (9191).

References

[1]   J. M. Poate, K. N. Tu and J. W. Mayer (1978) *Thin Films-Interdiffusion and Reactions*, Whiley, New York

[2]   C. Hass and R. F. Thun (1977) *Physics of Thin Films*, Academic, New York

[3]   B. X. Liu, Phys. Stat. Sol. (A) **94**, 11 (1986)

[4]   L. S. Hung and J. W. Mayer, Nucl. Instr. And Meth. **B7/8**, 676 (1985)

[5]   S. U. Campisano, Chu-Te Chang, A. Logiudice and E. Rimini, Nucl. Instr. And Meth. **209/210**, 139 (1983)

[6]   H. Cheng, H. Shen, J. Tang and F. Yang, Nucl. Instr. and Meth. **B83**, 449 (1993)

[7]   J.A. Davies, F.J.D. Almeida, H.K. Haugen, R. Siegele, J.S. Forster and T.E. Jackman, Nucl. Instr. and Meth. **B85**, 28 (1994)

[8]   N. Vandesteene and G. Demortier, Societe Belge de Physique, General Scientific Meeting, May 1994, Université de Mons-Hainaut, Belgium

[9]   J. P. Biersack and W. Eckstein, Comput. Phys. Commun. **A34**, 73 (1984)

[10] A. Markwitz and G. Demortier, Trans. Tech Publications, accepted for publication, 1997

[11] A. Markwitz and G. Demortier, Surface and Interface Analysis, in print, 1997

[12] A. Markwitz, N. Vandesteene and G. Demortier, Fresenius J. of Anal. Chem., accepted for publication, 1997

[13] J. R. Cameron, Phys. Rev. **90**, 839 (1953)

[14] L. Doolittle, Nucl. Instr. and Meth. **B9**, 334 (1985)

[15] J. A. Cookson, Nucl. Instr. and Meth. **B30**, 324 (1988)

Materials Science Forum Vols. 248-249 (1997) pp. 167-170
© *1997 Trans Tech Publications, Switzerland*

Substrate Influence on Topography and Chemical Composition of Thin Metal Films Produced by Means of Self-Ion Assisted Deposition

I.S. Tashlykov[1], V. Vishnyakov[2] and A. Wirth[2]

[1] Belorussian State Technological University, 13-a Sverdlova St., 22063 Minsk, Belarus

[2] MRI, Sheffield Hallam University, City Campus, Pond Street, Sheffield S11WB

Keywords: Thin Films, Self-Ion Assisted Deposition, Silicon, Rubber

Abstract

SEM, RBS, EDX and EDX mapping were used to characterise rubber and silicon surface modified by means of self-ion assisted deposition (SIAD). It was found that SIAD of Cr, Zr and W are accompanied by incorporating of C, O, Al and Si into a coating, what can promote a creation of observed grainless layers. Modified rubber surfaces have quasi-periodical topography which may be due to build in stress as a result of self-ion bombardment. Evolution of RBS signals from metal and incorporated elements are associated with an intermixing effects in the layer/substrate region.

Introduction

Over the last two decades thin films deposition on a surface has become widely used in microelectronics to produce conductive or isolated layers [1,2], in electrochemistry to improve corrosion resistivity and promote catalytical properties [3,4]. In mechanical engineering they are used to increase hardness of working surfaces of tools and to lubricate moving assembles in precision mechanisms [5,6]. It was reported recently that thin film deposition leads to reduction in friction force and friction coefficient of rubber against steel [7].

Properties of systems with modified surfaces are in some aspects defined by properties of a coating-substrate interface. However, even the most recent review papers promoting ion assisting deposition [8,9] do not address deeply enough the nature of an interface region. Lack of the knowledge dramatically increases when the substrate is not a rigid matter but an elastomer or rubber. It is widely acknowledged that the adhesion of a deposited film in a PVD processes in this case is poor. In general an improvement in adhesion can be achieved by energetic condensation, for example using ion-beam assisted deposition (IBAD) [5,8,9]. In the last case evaporation and reactive or noble gas ion irradiation are used simultaneously providing atomic mixing in the interface region. One may expect even further improvement in coating quality using an self-ion assisted deposition (SIAD). Coatings produced by this last means have some unique properties and it may be therefore considered that analysis of composition and structure of modified surfaces by means of SIAD is actual.

Experimental techniques

Chromium,Zirconium and Tungsten layers were deposited on rubber plates (GOST 7-IRP-1068) and (100) Si using a resonance vacuum arc ion source which provided simultaneous deposition of a metal film and irradiation by the same metal ions. The ion source is build in an industrial system which is pumped by a diffusion pump and supply vacuum conditions better then 10^{-2} Pa. Self-ion irradiation energy can be varied from 5 to 20 keV. The ion-to-atom ratios and deposition rates were measured as 0.1-0.3 and 1-4 A/sec.

The Rutherford backscattering (RBS) technique was employed for an investigation of the modified layers composition and for depth profiling. The energy of He^+ ions was 2 MeV, scattering, entry and escape angles were 168^0, 0^0 and 12^0 respectively. The energy resolution of the analysing system was 25 keV. RBS data were compared with RUMP code computer simulation [10].

Scanning and transmission electron microscopy (SEM), (TEM) were used to investigate structural and topographical properties of the modified surfaces. Energy dispersive x-ray analysis (EDX) and EDX mapping were used to investigate phase and compositional properties. Analytical Scanning and Transmission Electron Microscopes JEOL-840 and JEOL-3010 were used.

Results and discussion

The surface of virgin silicon and rubber shows no periodical or aligned topography. However after surface modification by means of SIAD a complicated quasi-periodical structure can be seen in case of the rubber when examined in the SEM, as shown on Fig.1. White spots seen on the image indicate a nonuniform concentration of Zn under the coating, as confirmed by EDX mapping at different electron energies. All the coatings examined were homogeneous and did not reveal visible grain structure. The reasons for grainless structure will be discussed later.

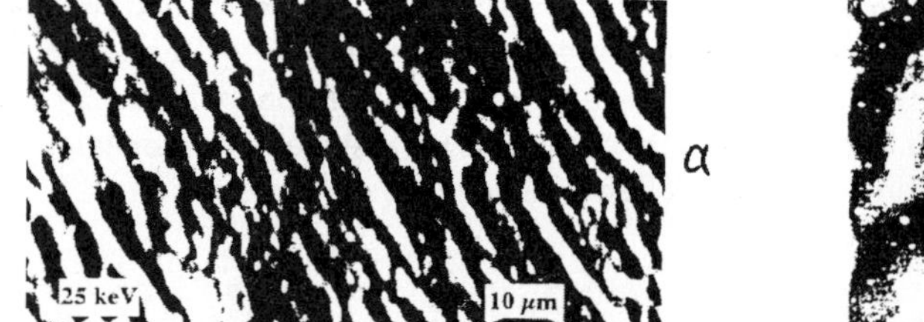

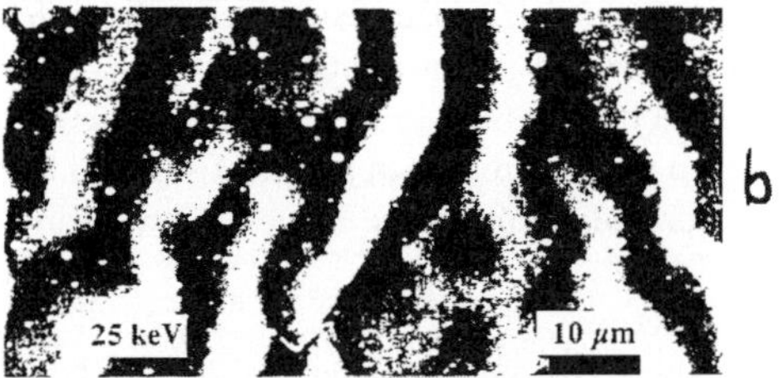

Fig.1. SEM emage of a rubber surface modified by: (a) W, (b) Cr. Energy of self ion assistance (SIA): (a) 20 keV, (b) 10 keV.

The presence of quasi-periodical topography on layers produced by ion-assisted deposition is known [11] and was explained on a basis of build up in stress [12]. In the case of a modification of rubber surfaces by SIAD 'wrinkling' of the deposited layers can be significantly promoted by the nonrigid nature of substrate.

EDX at two different electron energies from W coated rubber was undertaken. At lower electron energy EDX spectrum is mostly generated in coating, and shows high concentrations of Carbon, Oxygen and Tungsten. Deeper penetration of primary electron at 25 keV leads to enlargement of C, S and Zn signals in the spectrum. This reflects the basic chemical composition of the rubber.

Fig.2 shows an RBS spectrum of a sample with Cr modified surface imposed over a spectrum of a virgin rubber. The spectrum of virgin rubber contains three major steps which are associated with Carbon, Sulphur and Zinc. Hydrogen, which is presented in a rubber, can not be observed in a RBS spectra because of its low atomic mass, but it's content is estimated using the RUMP programm [10]. Small peaks in the regions of 125th and 144th channels were to be expected due presence of N and O on the surface of rubber. The following chemical composition $Zn_{0.18}S_{0.47}O_{1.5}N_{1.5}C_{50}H_{46.35}$ can be derived from the spectrum.   EDX spectra and EDX mapping show that O, S, Zn are distributed nonuniformly on the surface. There is a correlation between O and S maps suggesting the presence of SO_x. The examination of peaks related to S and Zn on the RBS spectrum indicate higher concentration (ten times for S and fiften times for Zn) of those dopants in a thin (100-120 A) layer on the rubber surface. These high concentrations can not be explained at the present time.

Spectrum of the rubber with a Cr modified surface on Fig.2 shows that SIAD of Chromium is accompanied by incorporating of C, O and Al/Si into the modified surface. Similar shape signals of these elements to the shape of the Cr signal suggests uniformity of the coating composition. EDX mapping at different electron energies supports this suggestion. Compositions of the investigated coatings according to the RUMP calculations are presented in Table 1.

Table 1. Composition of a rubber surface modified by means of SIAD. E - energy of self-ion assistance, FWHM - full width at half-maximum of Me peak. Data presented are calculated using RBS spectra and RUMP.

Metal	E	Ion/Atom	FWHM	Content of Chemical Elements, at%				
	keV	ratio	keV	H	C	O	Al+Si	Me
Cr	5	0.3	35	24-40	16-65	2-28	3-5.5	6-10
Cr	10	0.1	85	50-60	30-35	15-20	2-2.5	3-6
W	20	0.2	30	10-16	30-60	8-15	6-8	7-9
Zr	5	0.2	312	6-30	40-55	15-30	2-3	4-8.5

Presence of Carbon, Oxygen and Hydrogen in the modified layers can be explained by poor vacuum conditions in the target chamber. A probable source for the Al and Si are the ceramic isolators in the ion source. It is known that ion implantation of metalloids into metals leads to amorphisation of the metal structure [13,14]. The influence of Carbon and Oxygen on the formation of grainless layers and chemical composition are currently under investigation and will be reported elsewhere.

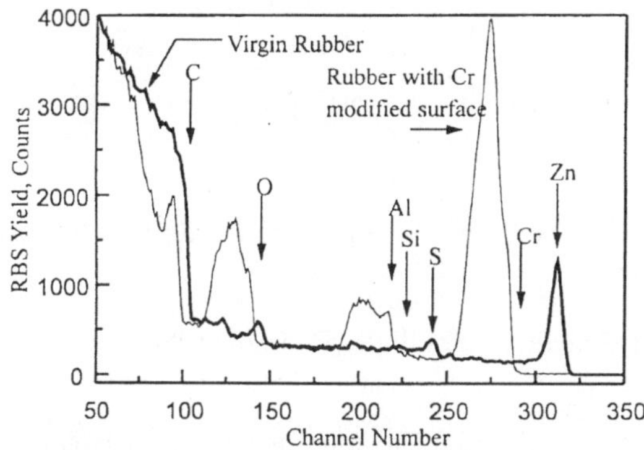

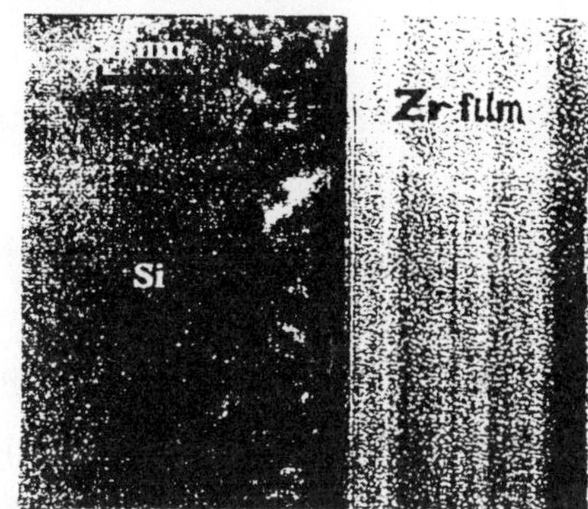

Fig.2. RBS spectra of rubber. SIA energy is 10 keV.

Fig.3. Cross-sectional TEM bright field image of Zr film SIAD on (001) Si.

Structural information on the coatings deposited on (100) Si under self ion bombardment by Zr^+ ions at 5 keV is presented in Fig. 3. Defects appear in the silicon near the Zr/Si interface as a result of irradiation by the Zr^+ ions during the initial stages of deposition. The coating on the sample is of uniform thickness but its structure changes from polycrystalline at deeper layers to amorphous in the interface region.

Increasing the ion-to-atom ratio produces a drop of metal content in a coating, as displayed in Tab.1. This behaviour can be explained on a basis of a sputtering conception by energetic ions.The high concentration of C is due to the fact that the RBS signal is taken from the modified rubber surface.Information about depth distribution of the elements can be extracted from the shape of the peaks in an RBS spectrum. The presence of a 'tail' in a low energy side of the Carbon (from the rubber) signal and change of slope in the region of channel number from 60 to 90 is firm evidence of deep penetration of the Chromium and accompanied elements into rubber. This penetration is the result of ion implantation and ion-assisting mixing in the interface region of a modified rubber surface. Deposition of W and different metals by means of SIAD leads to the similar behaviour of Carbon (from the rubber) signal. Therefore the concept of atomic intermixing can be also applied to nonrigid substrates. Deep penetration of metal atoms into rubber provides an improvement in adhesion of the films deposited by means of SIAD, as demonstrated in wear resistance experiments [7].

Conclusions

Deposition of Cr and W layers on rubber by means of SIAD produces the layers with a quasi-periodical surface topography which may be due to build in stress as a result of self-ion bombardment. Future promotion of topography development can be expected from the nonrigid substrate (rubber), which can not meditate tensions build up by ion irradiation in the deposited layer and the layer/substrate interface. Incorporating C, O, Al and Si can promote a creation of grainless layers deposited on rubber and silicon. Evolution of RBS signals from metal and incorporated elements are associated with intermixing effects in the layer/substrate region. The intermixing in turn provides enhanced adhesion of the deposited layer to a rubber, also to a silicon.

The structure of SIAD coating on Si comprises an amorphous layer near to the Si interface and layer containing polycrystalline regions of metals. The change in contrast near the boundary between silicon and coating results from intermixing of components causing elastic stress in the substrate. The thickness of this altered layer is some 30-50 nm which is several times higher than the projected range of 5-10 keV Zr^+ ions in silicon..

Acknowledgements

One of the authors (I.S.T.) is grateful for financial support from the Royal Society of Great Britain.

References

[1] J.S. Gailliard, in *Surface Engineering*, ed. by R.Kossowsky and S.Singhal, p.32. Martinus Nijhoff Publishers, Dortrecht (1984).

[2] R.S. Averback, L.J. Thompson, J. Moyle and M. Schalit, J.Appl.Phys. **53**, 1342 (1982).

[3] G.K. Wolf, J.Vac.Sci.Technol., **A10**, 1757 (1992).

[4] I.S. Tashlykov, O.A. Slesarenko, J.S. Colligon and H. Kheyrandish, Surf.J.Int. **1**, 106 (1986).

[5] G.K. Wolf, Surf.Coat.Technol. **43/44**, 920 (1990).

[6] M.N. Gardos, Tribology Lett.**1**, 67 (1995).

[7] V.S.Kulikauskas, V.I. Kasperovich, I.I. Tashlykova, N.V.Alov, M.G. Shadruchin and S.P.Shot, in Papers of XXI Intern. Conf. on Physics of ICPC, ed. A.F.Tulinov, p.78, MSU, Moscow (1995).

[8] J.S. Colligon, J.Vac.Sci.Technol. **A13**, 1649 (1995).

[9] G.K. Wolf, J.Vac.Sci.Technol. **A10**, 1757 (1994).

[10] L. Dolittle, RUMP simulation programme, Version 3.30 (1985).

[11] F.M. d'Heurle, *Metall.Trans.***1**, 725 (1970).

[12] Yu.V. Martynenko and G. Carter, Rad.Eff. and Def. in Solids **132**, 103 (1994).

[13] Z.Y.A. Al-Tamimi, W.A. Grant and G. Carter, Nucl.Instr. and Meth. **209/210**, 363 (1983).

[14] C. Cohen, A. Benyagoub, H. Bernas, J. Chaumont, L.Thome, M.Berti and A.V. Drigo, Phys.Rev. **B31**, 5 (1985).

Materials Science Forum Vols. 248-249 (1997) pp. 171-176
© *1997 Trans Tech Publications, Switzerland*

Radiation Damage and Amorphization Mechanisms in Xe⁺ Irradiated CuInSe₂ Single Crystals

M.V. Yakushev[1], I.S. Tashlykov[2], R.D. Tomlinson[1], A.E. Hill[1] and R.D. Pilkington[1]

[1] Dept. of Electronic and Electrical Engineering, University of Salford, Salford M5 4WT, UK

[2] Physics Department, Belorussian State Technological University, 13-a Sverdlova St., 220630 Minsk, Belarus

Keywords: CuInSe₂, Radiation Damage, RBS/Channelling, Xe Implantation

ABSTRACT

The damage evolution in ion bombarded $CuInSe_2$ single crystal has been studied using the RBS/channelling analysis with 2MeV He^+ ions. 40keV Xe^+ ions were implanted with fluences in the range from 10^{13} to 10^{16} cm^{-2} and an ion current density of $1.9\mu A/cm^2$ at room temperature. It was found that the radiation accumulation follows a linear function in a double logarithmic plot with slope m=1.5. The saturation level of the damage was achieved at a fluence of about 10^{15} cm^{-2}. A heterogeneous mechanism of the damage accumulation in the $CuInSe_2$ crystal irradiated with ions with mass equal to or greater than xenon mass is suggested.

INTRODUCTION

Ternary compound $CuInSe_2$ (CIS) is a semiconductor material with chalcopyrite structure and is used as an exceptionally efficient absorber layer in thin-film solar cells. The efficiency achieved to date for the CIS based solar cells has been reported to be in excess of 17% [1]. One of the most important properties of these cells (which could find applications in space) is very high resistance to radiation - the best among known materials used for absorber layers [2,3]. The classical approach to the consideration of the radiation hardness property involves studying of the fluence dependence of the accumulation of radiation damage and amorphization processes. One of the first characterisations of radiation damage following implantation with various ions, made with the Raman technique [4], showed that implantation fluences at which amorphization was observed usually exceeded the theoretical (TRIM [5] predicted) values by 2-3 orders of magnitude.

EXPERIMENTAL

The sample used for the study was p-type conducting CIS single crystal with dimensions $2x1cm^2$, and thickness 2mm. It was cut from the middle part of the ingot grown by the vertical Bridgman technique from a near stoichiometric charge [6]. The sample was mechanically polished with different grade diamond pastes and finished in a vibrating bath with $0.05\mu m$ alumina slurry. Polished sample was etched in 0.1% Br in methanol solution for 1min and annealed in vacuum for 30s at $300^{\circ}C$ to remove a selenium layer left at the surface due to the etch [7]. The orientation of the crystal surface was found to be within 5° from the (112) plane. The prepared surface was irradiated at room temperature with 40keV Xe^+ to fluences: $10^{13}, 3x10^{13}, 10^{14}, 3x10^{14}, 10^{15}$ and $3x10^{15}$. Different fluences were implanted into different stripes of 3mm width. One stripe was left undamaged for reference. The ion current density was $1.9\mu A/cm^2$. Rutherford

Backscattering/channelling (RBS/C) measurements were carried out using 2MeV He$^+$ ions in the normal incident beam geometry and with 168° backscattering angle. The energy resolution of the analysis was about 25keV. The homogeneity of the lattice quality was established by taking the aligned spectra at various points on the surface prior to implantation. The dechannelling parameter $\chi_{min}=Y_{amin}/Y_{rIn}$ (where Y_{amin} is the minimum backscattering yield in the aligned spectra and Y_{rIn} is backscattering yield in the random spectrum) was found to be 6.2±0.2%. The RBS aligned spectra were taken along the <221> axial channel straight after the implantation. RBS aligned and random spectra for fluences in the range 10^{14} - 3×10^{15} cm^{-2} along with one from virgin area are shown in Fig.1. The sample was kept at room temperature for 8 months. The measurements were repeated to establish whether any annealing process had taken place at room temperature.

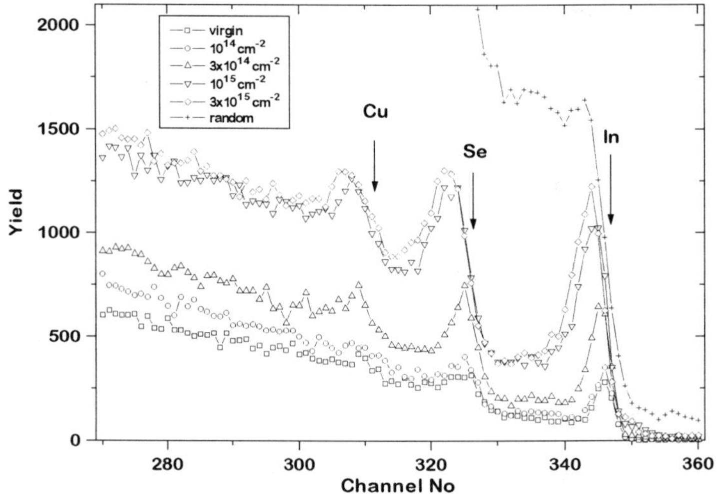

Fig.1. Effect of 40keV Xe$^+$ irradiation on RBS aligned spectra from CuInSe$_2$ crystal.

CALCULATION PROCEDURE

A model to explain dechannelling in multinary crystal lattices with a thin near-surface layer containing atoms displaced from lattice sites has been developed. This model gives a quantitative interpretation for the changes which occur in the RBS aligned spectra following damage in the near-surface layer. The model was used to calculate concentration depth profiles for indium atoms displaced from the lattice sites as a result of 10 keV hydrogen implantation into CuInSe$_2$ single crystals [8]. The aligned spectra of ternary compounds can be considered in terms of an overlay of the three separated aligned RBS spectra for each of the atomic species. The spectra are shifted in the energy scale according to the element mass. The high energy part of the spectra just before the surface peak was approximated to be linear. Parameters of this part of the spectrum were calculated from a mean square fit (regression line) of the experimental yield of the aligned spectrum. For other lines the parameters were calculated assuming that the slope coefficients of the lines and minimum values of the aligned yields (Y_{aCu}, Y_{aIn}, Y_{aSe}) are proportional to the Rutherford scattering cross sections of the corresponding atomic species. The random yields corresponding to different elements (Y_{rCu}, Y_{rIn}, Y_{rSe}) were derived using the same approach. As a result three normalised functions $\chi(x)= Y_a/Y_r$ (where x is the channel number) for the three elements can be calculated. After this procedure the energy scales were converted into depth using a standard technique [9]. Separated and normalised aligned spectra for the three elements from virgin and Xe^+ implanted areas of the CIS crystal have been obtained. The difference between the energy losses for channelling and none-channelling ions was not taken into account. The depth available for the analysis depends on the difference in masses, type of ion used for the RBS probe and its energy. In the case of a 2 MeV He^+ analysing beam in $CuInSe_2$ the mass separated depth analysis can be achieved up to approximately 100nm. The depth concentration profiles for each sublattice were derived using the iterative calculation approach [10]. The derived concentration depth profiles of scattering centres for the three sublattices are shown in Fig.2 (a).

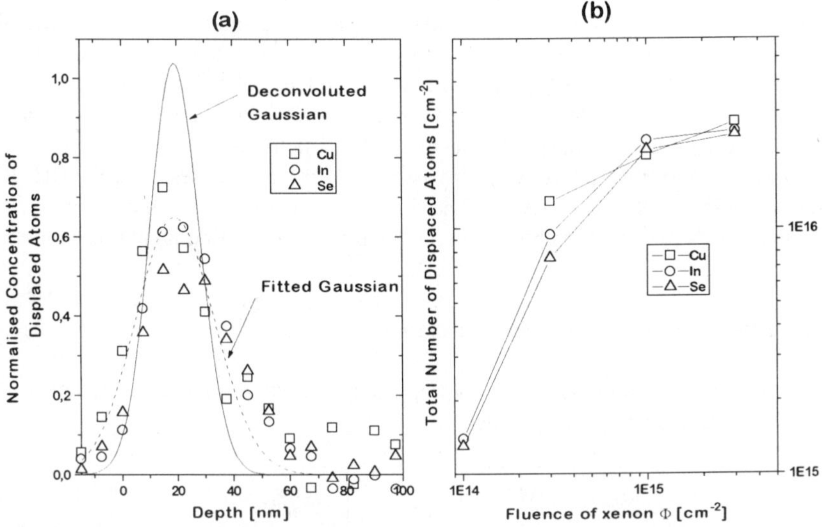

Fig.2. (a) - concentration of displaced atoms depth profiles with fitted and deconvoluted Gaussian curves for $CuInSe_2$ implanted with 10^{15} cm- Xe^+ ; (b) - the total number of displaced atoms accumulated in the near surface layer following Xe^+ implantation .

The resolution of the spectrometer is measured to be 5 channels or in terms of depth for indium near surface layer it is $2\sigma=38nm$. The shape of the concentration profiles is close to a Gaussian curve with $2\sigma=40nm$, very close to the resolution. This implies that the shape has been considerably modified. The deconvoluted concentration profile along with the fitted Gaussian curve are shown in Fig.2(a). We still can not evaluate the depth location of the maximum because of the poor depth resolution but we can judge the maximum volume concentration of the damage. For a more accurate calculation of the concentration of the displaced atoms it is necessary to know the dechannelling cross-section which depends on the nature of the defects and the exact locations of the interstitial atoms in the lattice. Without this additional information derived values can be used only as a first order estimation of the concentration depth profiles and for relative comparison of the amount of accumulated damage. All the calculation procedures were integrated into a computer programme written in TURBO PASCAL.

RESULTS AND DISCUSSION

Irradiation of the crystal with fluences up to 10^{14} cm^{-2} was found not to affect the aligned spectrum. For fluences larger than 10^{14} cm^{-2} three distinguishable peaks related to defects in the indium, selenium and copper sublattices appeared in the aligned spectra. Such peaks suggest that dechannelling in the damage layers mostly occurred due to direct scattering processes. The model described above has been employed to derive the number of displaced atoms and to estimate the maximum concentrations after the implantations. The distributions of displaced atoms for all the elements for the implantation fluences from 10^{14} to 3×10^{15} cm^{-2} were found to be similar within the accuracy of the experiment. It confirms a suggestion that during irradiation all the sublattices of compounds are damaged to the same degree [11]. The amount of the indium sublattice related damage has also been estimated with the RUMP programme. Indium related peaks of the simulated random spectra for CIS films were fitted to the peaks of the experimental aligned RBS spectra following the procedure of Mader [12]. From the RUMP calculation the increase in the number of displaced atoms as a function of implantation dose was obtained. This was achieved by calculating the increase in the thickness of a film which produced a similar indium related peak in the aligned spectrum. The results are shown in Table 1. The good agreement with the numbers derived using the model described above suggests that the model can be used for such evaluations. The amount of xenon incorporated into the near surface layer was also estimated for higher doses using the RUMP programme [13]. The results are shown in Table 1.The discrepancy between the amount of xenon detected and the fluence at 3×10^{15} cm^{-2} can be ascribed to significant sputtering taking place in the case of xenon irradiation. The presence of a considerable fraction of xenon atoms suggests the existence of clusters of xenon atoms or even of small gas bubbles. Such defects has been established [14] to cause direct scattering observed in the aligned spectra.

Table 1. Fluence of xenon implanted into CIS crystal, doses of xenon and total numbers of displaced atoms derived from RBS/c spectra versus.

Fluence of Xe$^+$ [cm^{-2}]	10^{14}	3×10^{14}	10^{15}	3×10^{15}
RUMP derived Dose Xe$^+$ [cm^{-2}]			10^{15}	1.5×10^{15}
RUMP derived number of In displaced atoms [cm^{-2}]	10^{15}	8×10^{15}	1.9×10^{16}	2.6×10^{16}
The programme calculated number of displaced atoms of In [cm^{-2}]	1.3×10^{15}	9.1×10^{15}	2.3×10^{16}	2.8×10^{16}

The total numbers of displaced atoms derived for all the sublattices as a function the implanted xenon fluence are shown in Fig.2 (b). The curves show a saturation behaviour at a fluence of 10^{15} cm^{-2}. The maximum concentration of the atoms displaced due to implantation at this dose is very close to the atomic densities of the elements. Therefore it may be concluded that at this dose the crystal/amorphous phase transition has occurred. CIS crystals irradiated with 10^{16} cm^{-2} of 40keV xenon have been characterised with TEM and SIMS analysis [15]. SIMS measurements of the xenon depth distribution showed a good agreement with the TRIM simulated data (the mean range is 16nm, the straggling is 7.2nm). The implanted layer was found to contain a high density of stacking faults and microtwins but not amorphous material. This fact was explained as a result of self healing during the implantation or, possibly, as solid-state recrystallization due to TEM specimen preparation. Therefore we can not confidently conclude that amorphization detected in terms of RBS/C actually takes place. Amorphous zones in CIS have been observed after a 10^{17} cm^{-2} dose of 40 keV oxygen, but unlike xenon, oxygen creates chemical bonds with CIS and with maximum volume concentration of 4% can already form secondary phases such as In_2O_3 which may prevent healing process. The number of displacements created by one ion of xenon according to a TRIM calculation, without taking in account annealing processes is 1111. The derived value of the remnant damage for a $3x10^{14}$ cm^{-2} xenon fluence is only 126 in the present work, almost 9 times smaller. This discrepancy may be explained as in terms of damage annealing taking place during implantation. In order to estimate the temperature in the area of the collision cascades during the thermal spike stage we calculated Θ - the average energy deposited per target atom [16] has been estimated to be about 1eV/at assuming 84 as the mean mass in the compound. It was found that the effective thermal spike temperature could reach 11000K over a period of $5x10^{-12}$ s.

The radiation damage accumulation within range from 10^{14} to $3x10^{14}$ cm^{-2} was observed to follow a linear function described by m=$\Delta log\Sigma_d/\Delta log\Phi$ = 1.5. Where Σ_d is the total number of displaced atoms following irradiation with fluence Φ. According to the approaches [17,18] on damage production in semiconductors the experimental data suggest a heterogeneous mechanism of the crystal/amorphous phase transition of the $CuInSe_2$ crystal irradiated with Xe^+.

We can compare CIS, silicon and GaAs radiation hardnesses. The amorphization of silicon implanted with 40keV Sb^+ ions, which is only 10% lighter than xenon, was observed after dose $2x10^{14}$ cm^{-2} [19]. In the case of GaAs implanted with 80keV ions of Sb^+ amorphization was observed [20] after dose $2x10^{13}$ cm^{-2} more than two order of magnitude lower than in the case of CIS. The origins of such extraordinary radiation tolerance of the CIS crystal lattice possibly are related to the high concentration of point intrinsic defects. These defects which are mostly vacancies and antisite defects, could reach concentrations of up 10% of the atomic density [21]. High quality CIS crystals are usually copper deficient therefore they should contain a substantial concentration of copper vacancies. The collision cascade simulation showed [22] that the presence of stoichiometric vacancies in semiconductor and insulators can hamper focusing processes decreasing the damaging effect of the collision. Also a role in the healing process must be related to the partly ionic nature of the bonds formed by copper atoms.

CONCLUSION

- Xenon ions implantation into CIS crystals produces severe damage to the lattice.
- All three sublattices were found to be damaged to the same degree.
- The measured dose of xenon atoms in the samples considerably differs from the irradiating fluence possibly due to sputtering process.
- The rate of damage for the linear part of the dose dependence curve is about one order of magnitude lower than the theoretical prediction. This is clear evidence of the radiation hardness of the material and the existence of effective annealing processes of the point defects during the irradiation due to high temperature in the post collision thermal spikes.
- The defect layer is located at a depth which is in good agreement with TRIM prediction. This agreement allows to draw the conclusion that remnant defects, left after the recombination during

implantation, are not mobile at room temperature. The statement is confirmed by the fact that after 8 months at room temperature there is no evidence of annealing of the damage.
- The dose dependence curve reaches a saturation at a fluence of 10^{15} cm^{-2} , the displaced atom concentration maximum achieves 100% at the same fluence. We can therefore conclude from the RBS/C data that the crystal/amorphous transition has occurred at this fluence.

ACKNOWLEDGEMENT

The authors gratefully acknowledge Dr V Vishniakov for the help with implantation of xenon and Professor G Carter for the discussion.

REFERENCES

[1] H.W. Schock, in Proc. of the 12th Europ. Photovoltaic Solar Energy Conf. ed. by R.Hill and H.S.Stephance, p.834, Amsterdam (1994).
[2] J.R .Woodyard and G.A. Landis, Solar Cells **31,** 297 (1991).
[3] R.W. Burges, W.E. Devaney and W.S. Chen, in Conf. Record of the 23 IEEEP photovoltaic Specialists Conf., p.1465, Louisville (1993).
[4] G. Lippold, M. V. Yakushev, R. D. Tomlinson, A. E. Hill and W. Grill, Cryst. Res. Technol. **95,** 385 (1995).
[5] J.P. Biersack and L. G. Haggmark, Nucl. Instr. and Meth. **174,** 257 (1980).
[6] R. D. Tomlinson, Solar Cells **16,** 17 (1986).
[7] M.V. Yakushev, A. E. Hill, R. D. Pilkington, R. D .Tomlinson and G.Lippold, J. Mater. Sc., Materials in Electronics **7,** 155 (1996).
[8] M.V. Yakushev, G. Lippold, A.E. Hill, R.D. Pilkington and R.D. Tomlinson, Cryst. Res. Technol. **31,** 357 (1996).
[9] Wei-Kan Chu, J.W. Mayer and M.A. Nicolet, Backscattering Spectrometry (N.Y. Academic Press, 1978).
[10] E. Bogh, Can. J.Phys. **46,** 653 (1968).
[11] C. Ascheron., J.P. Biersack, D. Fink., P. Goppelt, A. Manuaba., F.Paszti and N.Q. Khanh, Nucl. Instr. and Meth. **B64,** 203 (1992).
[12] A. Mader , J.D. Meyer and K.Bethge, Nucl. Instr. and Meth. **B68,** 266 (1992).
[13] L. R. Doolittle, Nucl. Instr. and Meth. **B15,** 227 (1986).
[14] H. Kudo and M. Mannami, J. Phys. Soc. Jap. **40,** 1654 (1976).
[15] C. A. Mullan, C. J. Kiely, M. V. Yakushev, M. Imanieh, R. D. Tomlinson and A. Rockett, Phil. Mag. 1131 (1996).
[16] P. Sigmund , Appl. Phys. Lett. **25,** 169 (1974).
[17] R.R. Dannis and E. B. Hale, Rad. Effects **30,** 219 (1976).
[18] G.Carter, Rad. Effects Lett. **86,** 25 (1983).
[19] L. Eriksson, J.A. Davies and J. Denhartog , Appl.Phys.Lett. **10,** 323 (1967).
[20] V. S. Vavilov and I. S. Tashlykov, Preprint 103, Moscow, FIAN (1989).
[21] H.Neumann, Cryst.Res.Technol. **29,** 985 (1994).
[22] V. M. Koshkin, L.P. Gal'chinetskii, V.N.Kulik and U.A.Ulmanis, Rad. Effects **29,** 1 (1976).

Materials Science Forum Vols. 248-249 (1997) pp. 177-180
© 1997 Trans Tech Publications, Switzerland

Ion Beam Mixing in Epitaxial Ag/Fe/Ag-(001)-Layers Investigated with Ferromagnetic Resonance and X-Ray-Diffraction

D. Kurowski[1], J. Pelzl[1], K. Brand[2], P. Sonntag[3] and P. Grünberg[4]

[1] AG Festkörperspektroskopie, Ruhr-Universität-Bochum, D-44780 Bochum, Germany

[2] Dynamitron-Tandem-Laboratorium, Ruhr-Universität-Bochum, D-44780 Bochum, Germany

[3] Expeimentalphysik IV, Ruhr-Universität-Bochum, D-44780 Bochum, Germany

[4] IFF4, Forschungszentrum (KFA) Jülich, D-52410 Jülich, Germany

Keywords: Ag/Fe-Multilayer, Ion Beam Mixing, Magnetic Anisotropies, Strain, Ferromagnetic Resonance, X-Ray-Diffraction

Abstract——Ion beam mixing in epitaxial Ag/Fe/Ag-(001)-layers by 95 keV-Fe-ions at room temperature has been investigated by means of ferromagnetic resonance (FMR) and out-of-plane x-ray-diffraction (XRD). The XRD-data show for fluences less than 10^{15}cm^{-2} a strain relaxation of the Fe-layer and interface roughness of 1-3 monolayers. The influence on the magnetic parameters can be well described by a magnetoelastic effect due to the strain relaxation indicating rearrangements and demixing at the interfaces. At a fluence of 10^{16} cm^{-2} we observe a strong reduction of the perpendicular-uniaxial-anisotropy and a change of the crystalline-anisotropy from cubic towards that of a tetragonal symmetry. Simultaneously a lattice widening above the Fe-bulk value and an increase of the interface roughness occur.

INTRODUCTION

Ion beam mixing in epitaxial Ag/Fe/Ag-(001)-layers by Fe-ions attracts attention for different reasons. First, recent ion beam mixing studies mainly deal with layers of miscible elements. The investigated system is thermally immiscible, where ballistic mixing and thermal spike diffusion operate in reverse directions. Second, only little experimental work has been done on specific modifications of the magnetic properties due to ion beam treatments. The magnetization, volume- and surface-anisotropies are sensitively dependent on the ion-induced composition change, crystal defects and structural changes of the interfaces. These modifications of the magnetic properties were studied by FMR, which allows one to determine the basic magnetic properties, in particular the anisotropy fields, with a high accuracy [1]. For the structural characterization out-of-plane x-ray-diffraction methods were performed.

The main aim of this work was to study the correlation between structural and magnetic changes, which are also of basic interest for the design of magnetic thin films. Additionally the possibilities for a directed ion beam modification of magnetic properties in thin magnetic films is explored.

EXPERIMENTAL TECHNIQUES

The MBE grown epitaxial (001)-multilayers consisted of Ag 20Å/Fe 200Å/Ag 1500Å/Fe 10Å/GaAs-substrat, where the last Fe-layer act as a seed-layer [2]. To protect the first Ag film from sputtering a polycrystalline 500Å thick ZnS-layer was added. During the growth process the substrates were bent, which caused strain fields as established by FMR and XRD data.

The ion irradiation were carried out in Bochum with the 100 keV-accelerator of the Dynamitron-Tandem-Laboratory equipped with a Cs-sputter-source and an electrostatic scanning device. The films were irradiated with negatively charged 95 keV Fe-ions at room temperature and the ion current was limited to 0.9 µA/cm^2 to avoid strong sample heating. The fluences were varied from 10^{13} cm^{-2} up to 10^{16} cm^{-2}. According to ballistic calculations carried out with the program TRIM a modification of the whole volume of the ferromagnetic layer by the irradiating Fe ions and an ion mixing mainly of the first 2 nm is expected at this ion-energy [3]. Channeling was neglected for these considerations because of the polycrystalline ZnS-layer.

XRD-measurements were performed with a thin film diffractometer using either $Mo_{K\alpha}$ radiation or $Cu_{K\alpha}$ radiation. With high angle Bragg scattering the out-of-plane lattice constant $a_\perp$ was determined. Because of the close (004)-GaAs-substrate peak and (002)-Fe peak fittings of Lorenz curves to the Bragg data were necessary. As criterion of the mosaicity of the crystallites the HWFM of rocking scans across the (002)-Fe fundamental peak were used. The interface roughness, defined as the rms-deviation from the ideal interface, was estimated by fitting the x-ray small angle reflectivity measurements according to the Paratt-formalism and using a Gauß-shaped height fluctuation [4, 5]. Because of the complex sample structure the fitting results of the roughness change can be regarded as a tendency only.

The FMR setup consists of a conventional X-band-ESR-spectrometer equipped with a magnet for fields up to 2.55 T and a sample rotating device.

The magnetic parameters were determined by analyzing the resonance fields B as function of the orientation of the external magnetic field. The following orientations were used: varying the external magnetic field 1. in the film plane, 2. out-of-plane from (100) to (001) and 3. out-of-plane from (110) to (001).

For a ferromagnetic single layer the theoretical description of the resonance field B can be expressed in terms of the partial derivative of the free energy density F [6]. The free energy density F can be described by (1a), which consists of the contributions: the Zeeman energy, the uniaxial-perpendicular-anisotropy ($B_u^\perp$) and the effective magnetocrystalline-anisotropy with tetragonal symmetry ($K_{1\,eff}^{\parallel}$ and $K_1^\perp$) [7]:

$$F = -\vec{M} \cdot \vec{B} + \frac{1}{2} M B_u^\perp \alpha_z^2 - \frac{1}{2} K_1^\perp \alpha_z^4 - \frac{1}{2} K_{1eff}^{\parallel}\left(\alpha_x^4 + \alpha_y^4\right) \tag{1a}$$

with $\quad B_u^\perp = \mu_0 M - \dfrac{4K_s}{Md} + \dfrac{2B_1}{M}\left(e_\perp - e_\parallel\right) \quad$ (1b) and $\quad K_{1eff}^{\parallel} = K_1^{\parallel}\left(1 + 4e_\parallel\right) \tag{1c}$

M is the saturation magnetization and α_x, α_y and α_z are the directional cosines of M with respect to the [100], [010] and [001]. The uniaxial-perpendicular-anisotropy $B_u^\perp$ (1b) includes the demagnetization-, the surface- and the perpendicular-magnetoelastic-anisotropy. The strain contributions in (1b) and (1c) with the $e_\parallel$ in-plane- and $e_\perp$ out-of-plane-strain-tensor components can be derived from the phenomenological description of the magnetic anisotropies as introduced by Néel [8] by assuming lateral strains equal in the film plain [7]. Note that in a FMR-experiment the deduced values of the anisotropies are effective magnetic fields as $K_1^\perp/M$ with the unit T [6].

RESULTS AND DISCUSSION

In Fig. 1 the x-ray small angle reflectivity data for the as-prepared and two irradiated samples ($5*10^{14}$ cm^{-2} and 10^{16} cm^{-2}) are presented. From the fitting results the roughness of both Ag/Fe-interfaces are estimated to 2-4 monolayers for the as-prepared sample. For fluences up to 10^{15}cm^{-2} the roughness is unchanged or may slightly decrease to 1-3 monolayers. At higher fluences the interface roughness increases to 4-6 monolayers.

Fig. 2 shows the high angle Bragg-scattering data for the same three samples. Due to the used $Cu_{K\alpha}$ radiation two GaAs-(400)-peaks occur corresponding to the separation of the radiation in $Cu_{K\alpha1}$ and $Cu_{K\alpha2}$, which reflects the high crystalline quality of the substrates. Because of a non-

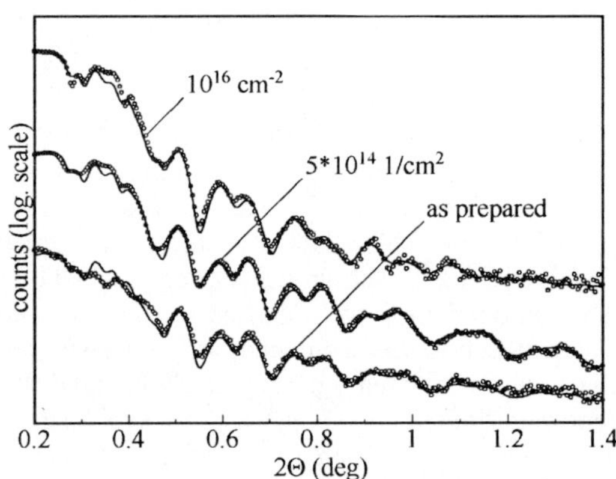

Fig.1: Small angle reflectivity of the as prepared and two irradiated samples measured with $Mo_{K\alpha}$-radiation. Lines are fits to the data.

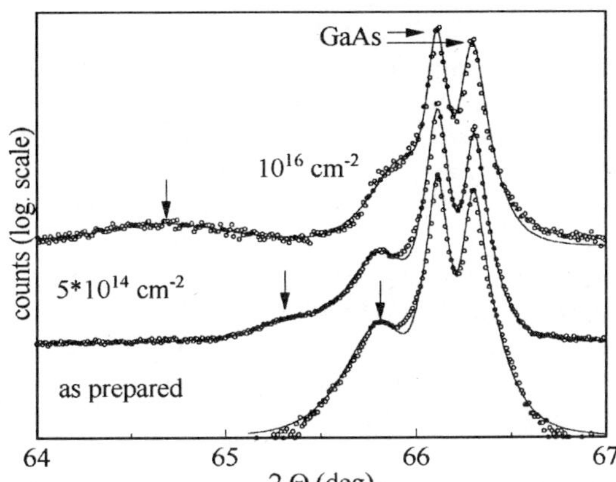

Fig. 2: Bragg-scattering data for the as-prepared and two irradiated samples measured with Cu$_{K\alpha}$-radiation. The Fe-(200)-peaks are marked with arrows. Because of a non-irradiated part of the samples the as-prepared Fe-(200)-peak is observed in each measurement.

irradiated part of the samples due to the shadow of the sample holder, the as-prepared Fe-(200)-peak is observed in each measurement. From this data we obtain for the as-prepared Fe layer an out-of-plane strain of $e_\perp = -1\%$ compared to the bulk lattice constant $a_0 = 2.866$ Å. With fluences up to 10^{15} cm^{-2} the out-of-plane strain is reduced, which is also indicated by the magnetic parameters (see below). At a fluence of 10^{16} cm^{-2} a lattice widening above the bulk-value is observed with $e_\perp = 0.6\%$.

The rocking scans across the Fe-(200)-peaks give an increase of the mosaicity from 0.2° for the as-prepared Fe layer up to 0.46° for the sample with the highest fluence. These values demonstrate the very good alignment of the crystallites even after irradiation.

As an example of the FMR-data, which were published and discussed in [9], Fig. 3 shows angular dependent FMR measurements of the as-prepared and the 10^{16} cm^{-2} implanted film. The orientation of the external magnetic field was varied from [100] in-plane to [001] out-of-plane (ϑ is the polar angle of the external magnetic field starting in the (001)-direction). For the as-prepared sample two resonance lines were observed, when the external magnetic field was applied along the (001)-direction or $\vartheta=0°$, respectively. Due to the crystalline anisotropy the as-prepared film shows FMR-absorptions only within an angle interval of 2 degrees around the (001)-direction. For the sample with a fluence of 10^{16} cm^{-2} a resonance line is observed for every orientation, indicating a reduction of the crystalline anisotropy.

The magnetic parameters determined from the FMR-measurements show for fluences up to 10^{15} cm^{-2} a decrease of the effective magnetocrystalline- and perpendicular-uniaxial-anisotropies, which are attributed to the observed strain relaxation in the Fe layers. This is demonstrated in Fig. 4, where the perpendicular-uniaxial-anisotropy and the effective magnetocrystalline-anisotropy values are shown as a function of the out-of-plane lattice constant.

According to (1b) and (1c) these effective anisotropy values depend linearly on the strain tensor and consequently linearly on the out-of-plane lattice constant. Here we used the assumption that $e_\parallel \sim e_\perp$, which follows from the linear stress theory of crystal defects as vacancies, interstitials or dislocations. The proportional factor depends on the nature of the defects (e.g. interstitials in bcc Fe form a (110)-dumbbell in the unit-cell) and the polarisation of the defects due to external strain [10].

In Fig. 4 the expected linear dependence is shown. For a fluence of $5*10^{14}$ cm^{-2} the evaluated magnetic properties approach values which correspond to those of a high quality Fe film with monolayer roughness at the interfaces, which confirm the XRD-results. Therefore we conclude that the ion irradiation up to 10^{15} cm^{-2} leads to a strain relaxation, rearrangement and

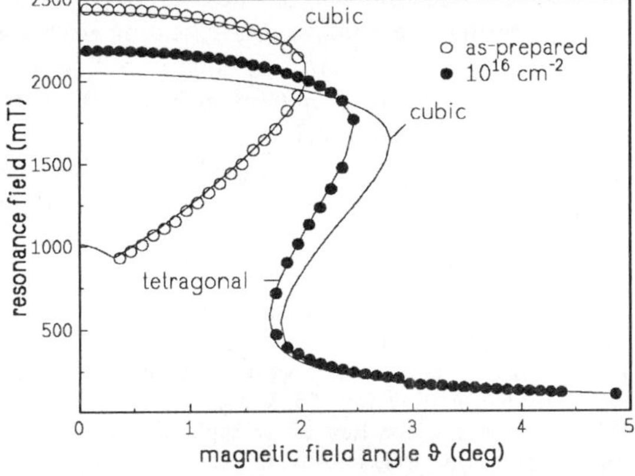

Fig. 3: Resonance fields of the as-prepared Ag/Fe/Ag-sample and the film with a fluence of 10^{16} cm^{-2} for varying the external magnetic field out-of-plane from (001) to (100). Also added are the theoretical graphs (—) for the magnetocrystalline anisotropy with cubic and tetragonal symmetry. Microwave frequency = 9.225 GHz

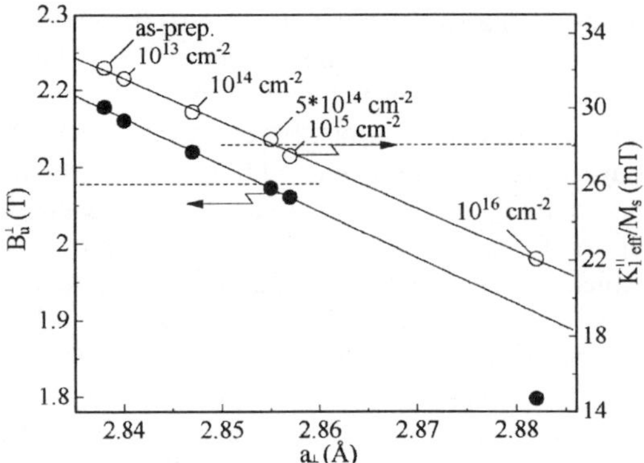

Fig. 4: The effective in-plane magnetocrystalline aniso-tropy $K_{1\,eff}^{\|}/M$ and the perpendicular-uniaxial-anisotropy $B_u^{\perp}$ as a function of the out-of-plane lattice constant $a_\perp$. The linear dependence of these parameters are expected for a magnetoelastic effect. For the sample with the highest fluence a strong deviation is observed, which indicates an additional effect of the irradiation. The dashed lines display the anisotropy values of a high quality Ag/Fe 200 Å/Ag film with monolayer roughness at the interfaces.

demixing at the interfaces. Corresponding results have been found by [11].

From the change of the magnetic parameters and the change of the lattice constant and using $K_1/M = 28$ mT and $M = 1,71*10^6$ A/m we can calculate the magnetoelastic value B_1 as $(3\pm0.5)*10^6$ J/m^3, which agrees with the bulk value [12].

At higher fluences of 10^{16} cm^{-2} we observe a strong reduction of the perpendicular-uniaxial-anisotropy, which may be a result of the increasing interlayer roughness. Further, a change of the magnetocrystalline-anisotropy towards that of a tetragonal symmetry occurs. The difference between cubic and tetragonal magnetic symmetry is demonstrated in Fig. 3. The graph for the cubic symmetry is calculated with the results of the in-plane measurements [9] with $K_{1\,eff}^{\|} = K_1^{\perp}$, which agree with the other samples. For the film with the highest fluence we found $K_{1eff}^{\|}/M = 22.8$ mT and $K_1^{\perp}/M = -29$ mT.

We interpret these results as a formation of FeAg, which has undergone a long range ordering along the film normal. With this process a small lattice widening of the Fe due to miscible Ag atoms should be connected, which has been observed by XRD analysis. Due to the observed small lattice widening we can rule out a tetragonal distortion to be responsible for the strong change of magnetocrystalline anisotropy. Further we can exclude a phase transition to γ-FeAg, which has been found for polycrystalline multilayers [11].

CONCLUSION

High fluence implantation of 95 keV Fe-ions in epitaxial Ag/Fe/Ag-(001)-multilayers was investigated by means of ferromagnetic resonance and X-ray diffraction-techniques. For fluences less than 10^{15} cm^{-2} the XRD-data show strain relaxation of the Fe-layer and interface roughness of 1-3 monolayers. The influence on the magnetic parameters can be well described as a magnetoelastic effect due to the strain relaxation indicating rearrangements and demixing at the Ag/Fe interfaces. At higher fluences of 10^{16} cm^{-2} a lattice widening above the Fe-bulk value and an increase of the interface roughness occur. Simultaneously we observe a change of the crystalline-anisotropy from cubic towards that of a tetragonal symmetry. We interpret these results as a formation of FeAg, which has undergone a long range ordering along the film normal.

Acknowledgment: This work is supported by the Deutsche Forschungsgemeinschaft, SFB 166.

REFERENCES

[1]   B. Heinrich, J.F. Cochran, Adv. Phys. Vol. 42, No. 5, 523 (1993)
[2]   J. A. Wolf, et al., J. M. M. M., 121, 253 (1993)
[3]   J.F. Ziegler, et al., „The stopping and range of ions in solids", Pergamon Press New York (1985)
[4]   L. G. Parrat, Phys. Rev. 95, 359 (1954)
[5]   L. Névot, P. Croce, Rev. Phys. Appl. 15, 761 (1980)
[6]   J. Smith, H.G. Beljers, Adv. in Electr. 6, 70, (1954)
[7]   B. Heinrich, et al., J. Appl. Phys. 70 (10), (1991)
[8]   M. L. Néel, J. de physique et radium 15, 225 (1954)
[9]   D. Kurowski, J. Pelzl, et. al, Proc. IEEE Ion Impl. Techn. (1996), accepted for publication
[10] P. Ehrhart, „Physics of Radiation Effects in Crystals", ed. by R. A. Johnson, A. N. Orlov (1986), (North Holland)
[11] H.U. Krebs, et al., Appl. Phys. A61, 591 (1995)
[12] Chikazumi, „Physics of Magnetism" (1986), (Malabar, Florida: Krieger)

Materials Science Forum Vols. 248-249 (1997) pp. 181-184
© *1997 Trans Tech Publications, Switzerland*

Thermal Depth Profiling of Nickel after He-Ion Irradiation

D. Kurowski[1], G. Kalus[1], K. Brand[2], B.K. Bein[1] and J. Pelzl[1]

[1] AG Festkörperspektroskopie / Solid State Spectroscopy, Ruhr-Universität-Bochum,
D-44780 Bochum, Germany

[2] Dynamitron-Tandem-Labor, Ruhr-Universität-Bochum, D-44780 Bochum, Germany

Keywords: He-Ion Irradiation, Nickel, Thermal Properties, Thermal Waves, Plasma Wall Interaction, Nuclear Fusion

Abstract - For Nickel, a candidate material for pump limiter plates in Tokamak reactors, the effects of He-ion implantation on the thermal properties have been measured by means of IR detection of thermal waves. For ion energies of 25 - 100 KeV considerably reduced thermal properties have been found just at the surface, for ion energies of 1 - 3 MeV the reduced thermal properties are mainly deeper below the surface.

1. He-ion contributions to plasma-wall-interaction in nuclear fusion

In nuclear fusion based on the D-T reaction in Tokamaks, the He ashes have to be removed from magnetic confinement and a low impurity content has to be maintained in the plasma core. Losses of MeV He ions due to the magnetic field ripple and the diffusion of thermalized He particles followed by selective He pumping at the plasma boundary represent two loss channels [1,2]. The active control of the He profile by Ion Cyclotron Resonance (ICR) heating at the plasma boundary would represent a third efficient loss channel with energetic He ions removed from magnetic confinement [3]. This technique to control the He-ion profile could contribute to an enlarged operation window of stable plasma burning, on the other hand however, it will have consequences for plasma wall interactions: He ions in the energy range of 10 to 100 keV would hit the first wall and contribute to stronger surface erosion of the plasma-facing materials and to an increased plasma impurity content.

Here, the radiation effects of He$^+$ ions on Nickel, a candidate material for pump limiters which serve to extract the He ashes from the fusion plasma, are analyzed by means of thermal depth profiling. This method gives depth-resolved information about the thermal properties and can be used to characterize the radiation effects. Thermal depth profiling of solids is here briefly described, then the results are presented and discussed. In the irradiation experiments, the ion energies have been varied from 25 to 100 keV, the range which is representative for He ions removed by ICR heating, and from 1 to 3 MeV, the energy range of He ions escaping from the magnetic confinement owing to the magnetic field ripple. The fluences have been varied between 10^{16} and $2.5 \cdot 10^{17}$ He$^+$ cm^{-2}.

2. Thermal depth profiling of soild surfaces

Thermal waves, which are excited in solids by intensity-modulated heating, e.g. by a modulated laser beam, are governed by the heat diffusion equation. Their amplitude is damped exponentially and the phase shift increases with the penetration depth x, as can be seen in Eq. (1) for a thermal wave

$$\delta T(x,t) = \frac{I_0}{e\sqrt{2\pi f}} \exp(-x/\mu) \cos(2\pi ft - x/\mu - \pi/4) \tag{1}$$

in an opaque semi-infinite solid. I_0 is the intensity of the absorbed laser beam and f is the heating modulation frequency. As shown by Eq. 1 the penetration depth of thermal waves, which is proportional to the thermal diffusion length, $x \propto \mu = \sqrt{\alpha/\pi f}$, is limited by exponential damping and can be controlled by varying the modulation frequency. Since the amplitude and phase shift depend on the thermal properties, frequency-dependent measurements of amplitude and phase give depth-resolved information about the thermal diffusivity α and effusivity e, which are related to the thermal conductivity λ, mass density ρ, and specific heat capacity c by $e = \sqrt{\lambda \rho c}$ and $\alpha = \lambda/\rho c$. For higher frequencies f, information is obtained from the region just at the surface, for low frequencies corres-

ponding to large penetration depths, information is obtained from deeper subsurface regions
[4]. The method is sensitive to the effective
thermal parameters: thus it is sensitive to the
lattice damage affecting the thermal transport
on the microscopic scale, as well as to meso-
or macroscopic voids, e.g. porosity or roughness due to He bubbles or blistering. The measured thermal parameters can be used to calculate the temperature response to the absorption
of heat pulses $F_s(t)$. As can be seen in Eq.2

$$\Delta T_s(t) = \frac{1}{e\sqrt{\pi}} \int\limits_0^t dt' \frac{F_s(t-t')}{\sqrt{t'}} \qquad (2)$$

the effusivity e is the only thermophysical parameter affecting the surface temperature in
transient surface heating of semi-infinite solids
and is thus the relevant thermophysical parameter in plasma surface interactions [5]. The
thermal diffusivity α is required to calculate the
propagation of heat or temperature within a
semi-infinite solid.

The schematic of the measuring system used
for thermal depth profiling is shown in Fig. 1.
To excite thermal waves, an Ar$^+$ laser beam is

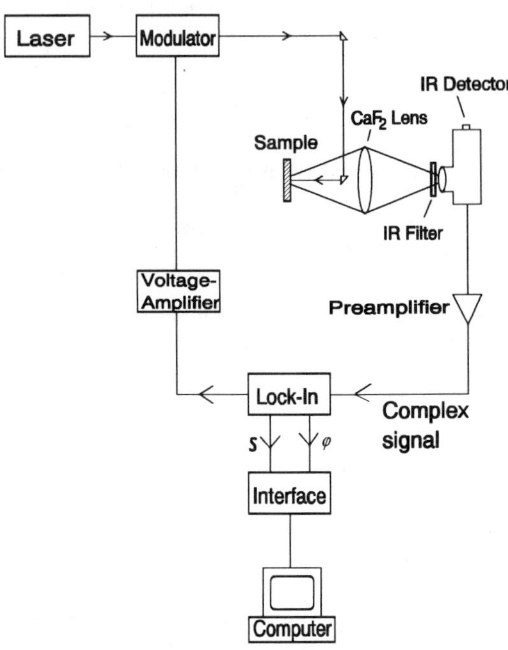

Fig. 1: Schematic of the thermal wave system

used and intensity-modulated by an electro-optical modulator. For the resolution of the thin surface
and subsurface layers affected by He$^+$ irradiation, the modulation frequencies of the thermal waves
are varied in the range of 1 Hz - 80 kHz. For the detection of the thermal response a photoconductive HgCdTe detector and an IR optical system consisting of a CaF$_2$ lens and a Si cut-on filter are
used, allowing to detect an IR interval of 1 - 9 µm. A two-phase lock-in amplifier serves to filter the
small thermal wave response from the relatively high radiation level related to the average sample
temperature and to analyze the thermal waves with respect to amplitude and phase lag relative to the
heating modulation. The measurements are computer-controlled, the data are registered as function
of the modulation frequency and interpreted in the frame of a two- or three-layer model [6]. For the
lower range of implantation energies, the thermal parameters of the first layer characterize the effects
of irradiation. For the MeV range of implantation energies, the properties of the second layer mainly
describe the effects of irradiation on the material. To calibrate the measured signals and to eliminate
the frequency response of the measuring system, the signals have been normalized with the help of
reference signals obtained for smooth homogeneous samples of glassy carbon. Details about the
measuring system and thermal depth profiling have already been published elsewhere [4,7].

3.   Irradiation effects on the thermal properties of nickel

The samples of polycrystalline nickel (Ni 99.2%) have been exposed to He$^+$ ions of 25, 50, 75 and
100 keV. At an ion current density of 2.7 µA/cm^2, the irradiation doses have been varied between
$5 \cdot 10^{16}$ and $2.5 \cdot 10^{17}$ He$^+$/cm^2 on homogeneous spots of 10 mm diameter. During the irradiation the
temperature rise at the rear surface of the samples remained nearly unchanged.

Figure 2 shows results obtained for samples, which have been exposed to the ion energies of 50 keV
and 100 keV at a constant dose of $5 \cdot 10^{16}$He$^+$/cm^2. The amplitudes are presented in the inverse normalized form $S_n^{-1}(f^{-1/2}) = S_{ref}(f^{-1/2})/S_s(f^{-1/2})$ and as functions of the inverse of the square root of the
modulation frequency. The signals S_s correspond to the irradiated Ni samples and S_{ref} to the reference
samples. As the   penetration   depth of the thermal wave is proportional to the thermal diffusion

length μ and inversely proportional to the square root of the modulation frequency, $x \propto \mu = \sqrt{\alpha/\pi f}$, and as the thermal wave (Eq. 1) is inversely proportional to the effusivity e, Fig. 2 gives information about the depth profile of the effusivity. The experimental results are approximated by a two-layer model. At small penetration depths corresponding to high modulation frequencies f, low effusivity values have been found, which increase with growing penetration depths $x \propto f^{-1/2}$. Based on the two-layer model, values of $e_1 = 11000$ Ws$^{1/2}$m^{-2}K^{-1} and 6500 Ws$^{1/2}$m^{-2}K^{-1} are obtained for the first layer after irradiation at ion energies of 50 and 100 keV, respectively. These reduced values, which are a measure for the effect of the radiation damage on the thermal transport, have to be compared to the value $e_2 = 19000$ Ws$^{1/2}$m^{-2}K^{-1} of the Ni bulk material. The thermal diffusion times $\tau_1 = (d^2/\alpha)_1$, which are a measure for the thermal thickness of the layer affected by the irradiation damage, are found to be $\tau_1 = 12.4$ µs and 75.5 µs for 50 and 100 keV, respectively. If we assume that it is mainly the thermal conductivity of the first layer which is affected by the irradiation and not the heat capacity, $k_1 \ll k_2$, $(\rho c)_1 \approx (\rho c)_2$, limit values for the thermal diffusivity α_1 and for the thickness d_1 of the first layer can be obtained. For the thermal diffusivity, values of $\alpha_1 = 5.8 \cdot 10^{-6}$ and $2.0 \cdot 10^{-6}$ m^{2}s^{-1} are found for 50 and 100 keV, respectively, which have to be compared to the value $\alpha_2 = 18.2 \cdot 10^{-6}$ m^{2}s^{-1} of the unirradiated material. For the thickness of the layer, where the thermal transport has been affected by irradiation, values of about $d_1 = 8.5$ and 12.3 µm are obtained for 50 and 100 keV, respectively. These lengths are much larger than the penetration depths of the He ions predicted by TRIM code simulation, which are of the order of 0.17 µm and 0.3 µm for 50 and 100 keV, respectively. The differences can be explained if we consider that the assumption of a rather undisturbed heat capacity, $(\rho c)_1 \approx (\rho c)_2$, can not be valid for higher fluences, bubble formation and blistering and that the effects of irradiation on the thermal transport properties extend deeper below the solid surface than the implantation depths estimated by a model of binary collisions.

Figure 3 shows the thermal depth profiles obtained for a constant ion energy (100 keV) at different fluences. Although the presented measurements seem to be very different, the quantitative interpretation based on the two-layer model reveals that the thermal depth profile, which is characterized quantitatively by the ratio of the effusivities of the first layer to that of the bulk material, e_1/e_2, and by the thermal diffusion time τ_1 of the irradiated surface layer remains nearly constant.

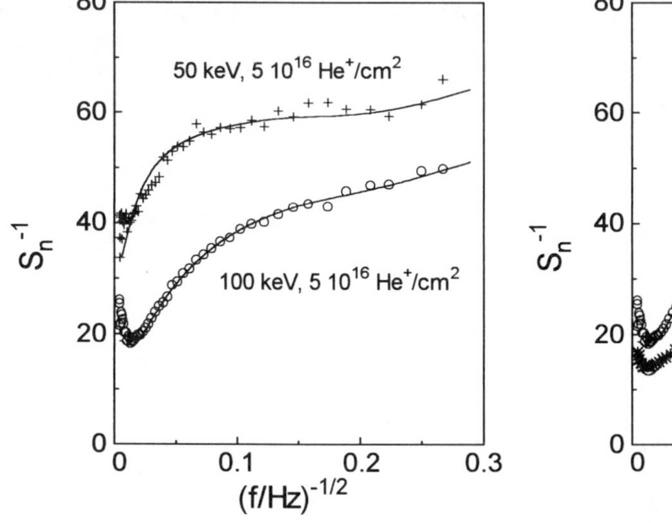

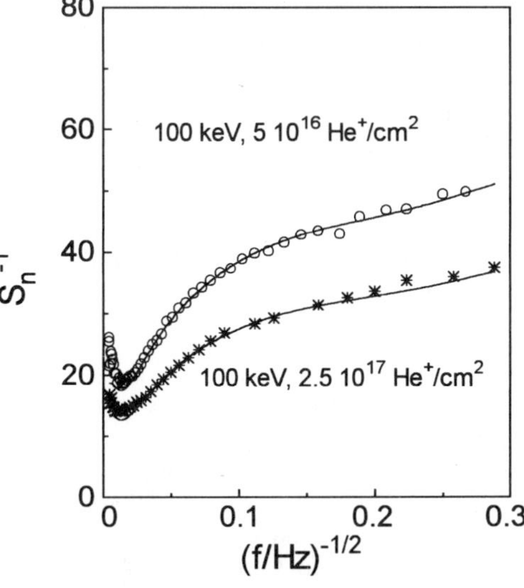

Fig.2: Thermal depth profile of Ni after irradiation - Ion energy variation at constant fluence.

Fig.3: Thermal depth profile of Ni - Variation of the fluence at constant ion energy.

The changes observed in Fig. 3 are thus mainly due to optical changes of the surface: the optical absorptivity at the used laser wavelength still increases with the higher radiation dose. A stable thermal depth profile has thus been reached for the He-irradiated Ni sample, while the optical properties continue to change owing to a continuing fragmentation of the surface.

For the MeV ion energies, the thermal properties are mainly reduced deeper below the surface, whereas the material just at the surface is less affected. Figure 4 shows the thermal depth profiles for energies of 1, 2 and 3 MeV at a fluence of 10^{16} He$^+$/cm^2. The measured signals have here been normalized with the help of reference signals obtained for a Ni sample without ion implantation. For 1 MeV a very slight decrease of the effusivity profile is observed at a value of $f^{-1/2} = 2 \cdot 10^{-2}$ s$^{1/2}$. For the ion energies of 2 and 3 MeV, the curve minima are more pronounced and shift to larger penetration depths $x \propto f^{-1/2}$. The signals measured for 3 MeV are compared to an approximation by a three-layer model, where unchanged thermal properties for a first layer and considerably reduced thermal transport for the second layer are assumed. For the first layer, a thickness of $d_1 \approx 4.8$ μm is obtained, a value of the order of magnitude expected for the penetration depth of the He ions. For the second layer, the values obtained for the effusivity and thermal diffusion time are $e_2 = 13000$ Ws$^{1/2}$m^{-2}K^{-1} and $\tau_2 = (d^2/\alpha)_2 = 160$ μs, respectively. From the latter value one can conclude, that the layer of reduced thermal transport properties has a thickness of about 40 μm if a thermal diffusivity of $\alpha_2 = 10 \cdot 10^{-6}$ m^{2}s^{-1} is assumed.

4. Conclusions

In the range of He$^+$ energies from 25 to 100 keV and for fluences of $2.5 \cdot 10^{16}$ - $2.5 \cdot 10^{17}$ He$^+$/cm^2, the thermal depth profiles of nickel exhibit a marked dependence on ion energy and fluence. The effusivity, the relevant thermal parameter for transient surface heating, decreases linearly with increasing ion energy and fluence. The measured thermal depth profiles have been interpreted in the frame of a two-layer model, where the reduced thermal transport properties of the first layer serve to characterize the radiation effects. The characteristic thickness of the layers of reduced thermal properties is much larger than the penetration depths deduced from TRIM code simulations. For the range of He$^+$ energies from 1 to 3 MeV, the measured thermal depth profiles can be approximated by assuming unchanged thermal properties in a first layer and considerably reduced thermal transport parameters in the second layer.

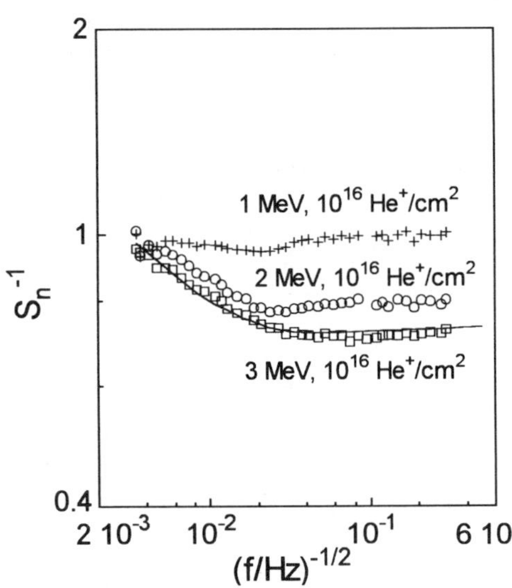

Fig.4: Thermal depth profile of He$^+$ irradiated Ni - Ion energy variation at constant fluence.

[1] L.M. Hively, J.A. Rome, Nucl. Fusion 30 (1990) 1129.

[2] J.N. Brooks, A. Krauss, R.E. Nygren, B.L.Doyle, K.H. Dippel, K.H. Finken, J. Nucl. Mater 196-198 (1992), 664.

[3] C.S. Chang, J.Y. Lee, H. Weitzner, Phys. Fluids B 3 (1991), 3429.

[4] B.K. Bein, J. Pelzl, in *Plasma Diagnostics*, V. 2, *Surface Analysis and Interactions*, (Eds. O. Auciello, D.L.Flamm), Academic Press (1989), 211.

[5] R. Behrisch, J. Nucl. Mater 93&94, (1980), 498.

[6] B.K. Bein, S. Krueger, and J. Pelzl, J. Nucl. Mater. 141-143 (1986), 119-123.

[7] B.K. Bein, J. Gibkes, J.H. Gu, R. Hüttner, J. Pelzl, D.L. Balageas, A.A. Deom, J. Nucl. Mater. 191-194, Part A, (1992), 315.

Materials Science Forum Vols. 248-249 (1997) pp. 185-188
© *1997 Trans Tech Publications, Switzerland*

The Role of Surface Carbon Contamination on the Tribological Properties of Ion-Implanted 100Cr6 Bearing Steel

A. Faussemagne[1], A. Benyagoub[1], Ph. Kapsa[2], H. Jaffrezic[1] and G. Marest[1]

[1] Institut de Physique Nucléaire de Lyon, IN2P3-CNRS, F-69622 Villeurbanne Cedex, France

[2] Laboratoire de Tribologie et Dynamique des Systèmes, Ecole Centrale de Lyon, F-69131 Ecully Cedex, France

Keywords: Ion Implantation, 100Cr6 Steel, Friction Coefficient, Boron Nitride Formation, Carbon Contamination

Samples of 100Cr6 bearing steel were implanted with nitrogen and boron in order to study the evolution of their tribological properties. Even though, the physico-chemical characterizations show an important modification of the nature of the sample surfaces after B and N ion implantation, the tribological tests do not exhibit any significant change of friction. However, on a set of samples accidentally contaminated with carbon during ion implantation, friction was highly improved. We then decided to continue this kind of surface treatment but by provoking deliberately a carbon pollution not only during B and N ion implantations but also during separate Ne ion irradiations. The tribological tests reveal a clear improvement of friction (by a factor of 5) whatever the implanted ion.

I-Introduction :

It is well-known that the weak point of any mechanical equipment is the bearing system because it is at this point where friction and wear phenomena are concentrated. As generally this system is made with 100Cr6 steel, it is very attractive to improve the tribological properties and thus the surface characteristics of such a material without altering the interesting mechanical peculiarities (*i.e.*, hardness) of the bulk. In order to achieve such a goal, ion implantation was used intensively during the last three decades. Hence, a lot of work was devoted to the study of physico-chemical, mechanical and tribological effects of nitrogen and boron ion implantations. The literature is abundant on the subject and it is sometimes suggested that the implantation of nitrogen increases the wear resistance by hardening the surface while boron implantation decreases friction as a consequence of the amorphization of the implanted layer. The aim of the present study is to combine these two effects by performing double implantation of nitrogen and boron with the additional possibility of synthesizing boron nitride which has also very interesting tribological properties. It is worth to mention that up to now very few work [1,2] dealt with such a specific subject.

II Experimental :

In order to be as close as possible to reality, the samples were directly cut from commercial bearing tracks made of 100Cr6 steel. They were all polished in the same manner to ensure that they have the same initial surface state. This is necessary to ensure good reliability of the subsequent surface treatments and tribological tests.

With the aim of favoring BN formation by ion implantation in 100Cr6 steel, it is worthwhile to superimpose the boron and nitrogen profiles. This was achieved by implanting boron at 65 keV and nitrogen at 110 keV both at a nominal fluence of 2.5×10^{17} ion cm^{-2} and with a current density of $3 \, \mu A$ cm^{-2}. The influence of the implantation sequence (B+N or N+B) on these profiles was also studied.

The elemental analysis of the samples was performed by Nuclear Backscattering Spectrometry (NBS) [3] using α particles of 6.04 MeV energy. This energy is well suited for profiling, all at once, boron, nitrogen and also carbon in a heavy matrix due to the presence of resonances with increased cross-sections relative to the Rutherford ones by more than an order of magnitude. A prior knowledge or measurement of these cross-sections is, of course, necessary to ensure reliable quantitative analyses. The chemical characterizations of the sample surfaces were realized by X-ray Photoelectron Spectroscopy (XPS), Grazing Incidence X-ray Diffraction (GIXD) and Conversion Electron Mössbauer Spectroscopy (CEMS).

The tribological tests were conducted on a tribometer employing a ball-plate geometry with a rectilinear alternate motion of the ball made of 100Cr6 steel and having a diameter of 10 mm. The oscillation frequency was 0.5 Hz and the track length was fixed at 4 mm, leading to a velocity of 4 mm/s. The tests were performed in air, without lubrication, under pure sliding conditions and with a normal load of 5 N. Furthermore, the ball was changed after each tribological test.

III Results and Discussion :

Fig. 1a presents the boron and nitrogen profiles obtained by NBS on the implanted samples following either the B plus N or the N plus B sequence. It is clear in both cases that the affected depth is about 230 nm and that the implanted ions have nearly the same ion projected range (~ 110 nm). The maximum atomic concentrations of the implanted species range between ~ 15 and ~ 25 %. These results were confirmed by the analysis of the XPS spectra recorded after different sputtering time. Moreover, a careful examination of the B 1s signal revealed the presence of a subpeak located at 190.3 eV which is the signature of boron nitride formation in these samples [4,5]. The ratio of the area of this subpeak to that of the total B 1s signal gives the proportion of boron linked to nitrogen (*i. e.*, boron nitride). The evolution of this proportion with the sputtering time is presented in Fig. 1b for the two implantation sequences. From this figure one can see that at maximum nearly one half of boron atoms participate to the boron nitride phase. The GIXD experiments show an amorphization of the surface layers independent from the implantation sequence. This result is confirmed by the CEMS study which shows additionally the formation of the following iron-boron nitride phases: $Fe_{1-x}(B,N)$, $Fe_{2-x}(B,N)$, $Fe_{3-x}(B,N)$ [6].

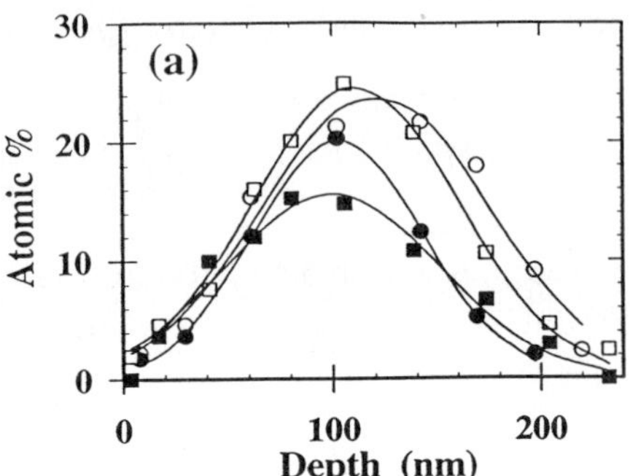

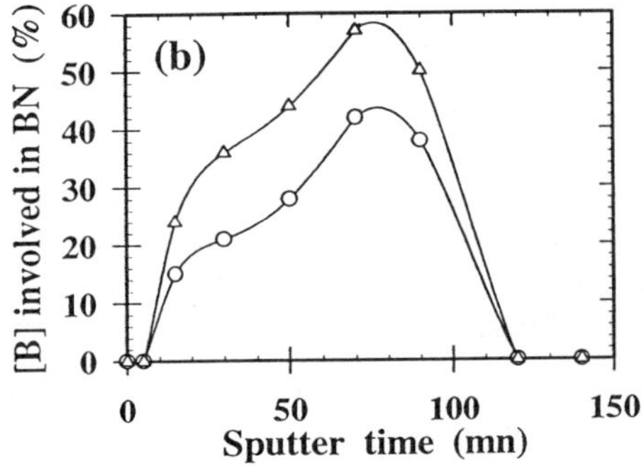

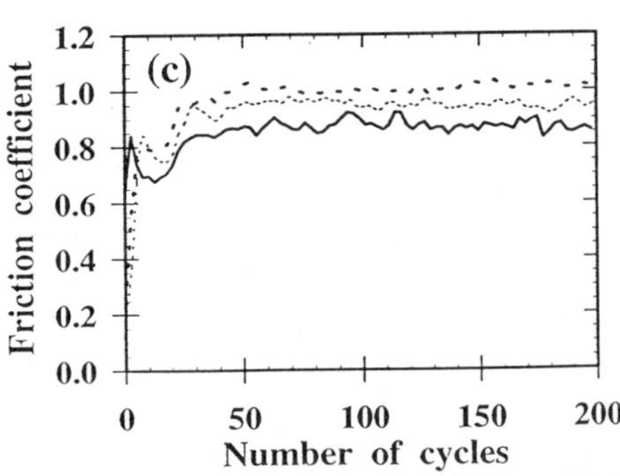

Fig 1 : (a) *Distribution profiles of boron (open symbols) and nitrogen (full symbols) obtained by NBS after implantation according to the B+N (circles) and N+B (squares) sequences.*

(b) *Evolution of the proportion of boron linked to nitrogen as function of the sputtering time deduced from XPS analysis of the samples prepared according to the B+N (triangle) and N+B (circle) implantation sequences.*

(c) *Evolution of the friction coefficient with the number of cycles for a virgin (solid line), a B+N implanted sample (dashed line) and a N+B implanted sample (dotted line).*

The evolution of the friction coefficient (μ) with the number of cycles obtained during tribological tests are presented in Fig. 1c. In the case of an unimplanted sample, μ reaches a value

~0.85 after 50 cycles. It is clear from the figure that the double implantation of boron and nitrogen is far away from inducing any significant change of the coefficient of friction of 100Cr6 steel which becomes nearly equal to 0.9 (*i. e.*, still within the error bars) after 50 cycles. These results are rather surprising since, according to the enlightenment provided by the elemental and chemical characterizations presented above, the double ion implantation severely modifies the structure and the composition of the surface layers and is then expected, at a first glance, to induce an important change of their tribological properties. In addition, these results seem to be in contrast with those obtained by other authors [1,2] who noticed a clear decrease of the friction coefficient after a similar surface treatment.

With the aim of understanding the origin of this discrepancy, we continued our investigations with other series of boron and nitrogen implantations. Thus, on a set of samples accidentally contaminated with carbon during implantation, we noticed a considerable decrease of the coefficient of friction in agreement with the results obtained by M. Braun [7] for nitrogen implantations. Fig. 2a presents a NBS spectrum recorded on such a sample. One can see the occurrence of an appreciable carbon peak besides the boron and nitrogen signals. As the NBS technique is also well adapted to the detection of carbon [3], it is then quite straightforward to quantify this surface pollution. Fig. 2b displays the evolution of the coefficient of friction versus the number of cycles in the case of two samples having different carbon pollutions. The coefficient of friction of the sample with the low carbon contamination (6×10^{16} C cm^{-2}) is around 0.2 till 300 cycles while that of the specimen with higher carbon pollution (4×10^{17} C cm^{-2}) remains stable and ~ 0.2 at least up to 2000 cycles. These results in comparison with those presented in Fig. 1c are clear evidence that the amount of carbon pollution plays an important role in the tribological properties of ion-implanted 100Cr6 steel. As this pollution is commonly encountered in surface treatment by ion beams but generally not considered due to the difficulty lying in its detection and especially in its evaluation, one can invoke this phenomenon to explain the origin of the discrepancy reported by the literature.

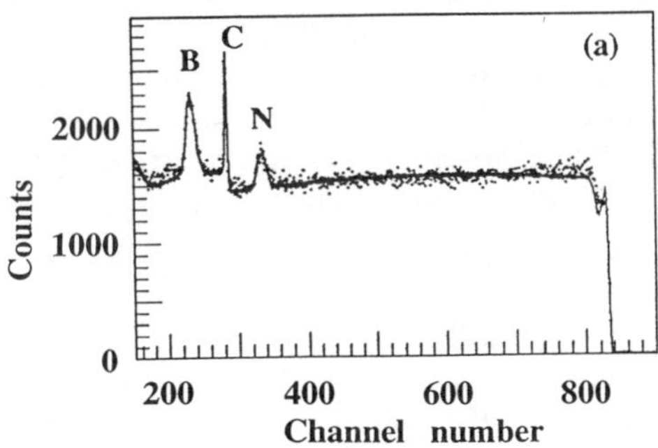

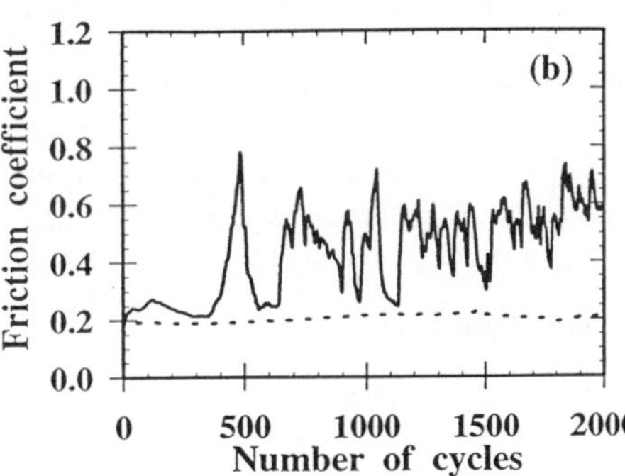

Fig 2 : (a) *NBS spectrum obtained on a sample implanted according to the N+B sequence and containing a surface pollution of 6×10^{16} C cm^{-2}; analysing particles: 6.04 MeV-^{4}He^{2+} ions.*
  (b) *Evolution of the friction coefficient with the number of cycles for the sample described in* (a) *(solid line) and a sample implanted according to the B+N sequence and containing a surface pollution of 4×10^{17} C cm^{-2} (dotted line).*

In order to go a step further, it is worthwhile to check whether the improvement of the coefficient of friction is due to a combination effect of the boron and nitrogen implantations with the carbon pollution or is caused solely by the latter phenomenon. These investigations were handled by implanting 100Cr6 samples with a rare gas, namely 220-keV Ne$^+$ at a fluence of 10^{17} at. cm^{-2} under good (~ 3×10^{-7} Torr) and bad (~ 8×10^{-6} Torr) vacuum conditions. This latter case was achieved by removing some liquid nitrogen traps from the ion implanter and the implantation chamber. Here again, the samples were characterized by NBS after Ne implantation in order to evaluate the amount of their surface carbone contamination. Fig. 3 presents the evolution of the friction coefficient with the number of cycles of a sample free from any carbon pollution (curve a) and another one contaminated with 1.2×10^{17} C cm^{-2} (curve b). The friction coefficient of the first sample is similar to that of the virgin one while that of the second specimen remains stable and around 0.2 at least up to 1000 cycles. This latter sample has thus a behavior close to the samples implanted with boron and nitrogen and contaminated with carbon. These results clearly demonstrate that the carbon contamination has a notable direct influence on the improvement of the coefficient of friction of 100Cr6 steel by ion implantation.

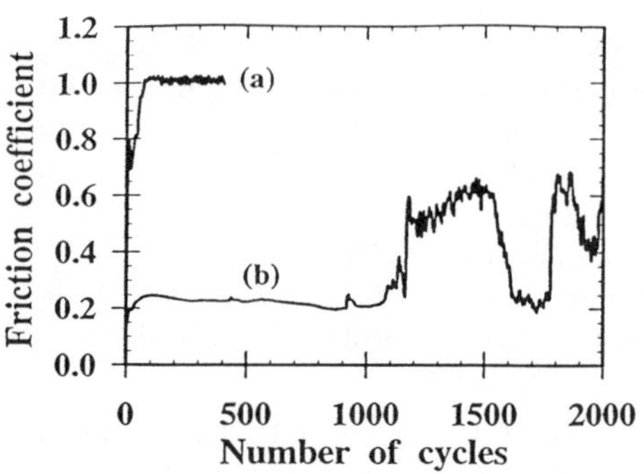

Fig 3 : *Evolution of the friction coefficient with the number of cycles for two samples implanted with 220-keV Ne$^+$ at a fluence of 10^{17} at.cm^{-2}. Curve (a) corresponds to a sample free from any carbon contamination and curve (b) to a sample containing 1.2×10^{17} C cm^{-2} at the surface.*

It is worth to mention that very recent Raman experiments [8] performed on the samples described in this paper indicate that the surface carbon contamination consists of a diamond-like carbon layer having a good adhesion to the substrate. This result could explain the important improvement of the friction coefficient of the samples contaminated with carbon during the implantation process.

IV-Conclusion :

The important results obtained from the experiments reported in this paper is the observation of a drastic reduction (by a factor 5) of the coefficient of friction of 100Cr6 steel when submitted to ion implantation. Such a phenomenon seems not to be related to the nature (boron, nitrogen or neon) of the implanted species but is caused directly by the carbon contamination arising during ion implantation. Therefore, in order to avoid any artefact, one has to pay attention to the carbon pollution in the studies of the modifications of the tribological properties of 100Cr6 steel by ion implantation. Additional experiments with different amounts of carbon contamination are now in progress in order to confirm definitely the present results.

Acknowledgements

The authors would like to thank A. Plantier for performing the ion implantations with the Ion Separator of the Institut de Physique Nucléaire de Lyon. One of the authors (A. F.) is indebted to the Région Rhône-Alpes for its financial assistance for this work which is a part of her PhD thesis.

References

[1]   S. Ohtani, Y. Mizutani and T. Takagi, Nucl. Instrum. Meth. B **80/81**, 336 (1993).

[2]   Y. Dehua, Z. Xushou, X. Qunji and W. Hanqing, Surf. Coat. Technol. **56**, 119 (1993).

[3]   See, for example, the *"Handbook of Modern Ion Beam Analysis"*, Eds. J. R. Tesmer and M. Nastasi (MRS, Pittsburgh, 1995).

[4]   G. M. Ingo and G. Padeletti, Thin Solid Films. **228**, 276 (1993).

[5]   J. P. Rivière, Y. Pacaud and M. Cahoreau, Thin Solid Films. **227**, 44 (1993).

[6]   A. Faussemagne, G. Marest, A. Benyagoub and N. Moncoffre, in *"ICAME-95"* Conference *Proceedings Vol. 50*, Ed. I. Ortalli (SIF, Bologna, 1996), p. 691.

[7]   M. Braun, Nucl. Instrum. Meth. B **39**, 544 (1989).

[8]   A. Faussemagne, PhD dissertation, Université Claude Bernard Lyon I (1996).

Materials Science Forum Vols. 248-249 (1997) pp. 189-192
© *1997 Trans Tech Publications, Switzerland*

Ion Implantation into Stainless Steel - Depth Selective Phase Analysis with an Improved Mössbauer Technique

G. Walter[1], B. Stahl[1], R. Gellert[1], R. Nagel[1], D.M Rück[2], G. Klingelhöfer[1] and E. Krankeleit[1]

[1] Institute of Nuclear Physics, Technical University Darmstadt, Germany

[2] GSI Darmstadt, Darmstadt, Germany

Keywords: Ion Implantation, Depth Selective Conversion Electron Mössbauer Spectroscopy (DCEMS), Irradiation Induced Phase Transformation

Abstract

The main point of interest is to study the dynamic evolution of the depth profile of implanted ions. Basically, this is determined by the mean projected range of the implanted ions, by sputtering of surface atoms and by radiation induced mixing effects. The quantities describing these processes change dynamically as function of the implanted ion dose with changing composition in the implanted depth region. To analyze the influence of phase composition **Depth Selective Conversion Electron Mössbauer Spectroscopy (DCEMS)** is applied. The system under investigation is ^{151}Eu implanted in high-austenitic stainless steel of composition Fe62Ni20Cr18.

Introduction

Mössbauer techniques play an important role in the analysis of ion implanted materials. Especially the Depth Selective Conversion Electron Mössbauer Spectroscopy (DCEMS) allows a non-destructive depth dependent analysis of the chemical and magnetic phase composition in the approximately topmost 200 nm of a sample, which is typical for ion implanted systems. The method correlates the phase information due to the hyperfine interactions of the ^{57}Fe probe nucleus with the depth information due to the energy loss of the conversion electrons (transport tensor) in the solid sample. To improve this method a new experimental setup, an ultra high vacuum orange-type magnetic electron spectrometer was developed and built up [1] during the last years, which is described in the next section. It is based on an instrument that worked successfully during the last three decades [2] and promises further applications of this method in the many fields of materials science, such as improvement of wear resistivity or corrosion of steels or temperature dependent studies in the field of nanostructured materials.

Experimental

Ion Implantation

In extension of our systematic studies of ion implantation into stainless steel [3], a 2 μm thick foil of composition Fe62Ni20Cr18 (^{57}Fe enrichment of 90%) was implanted with 400 keV ^{151}Eu with a dose of $3 \cdot 10^{16}$ ions/cm^2. The implantation was done at the 300 kV implantation facility at GSI, Darmstadt. The temperature during implantation was 77 K to avoid thermally driven diffusion processes, the residual pressure in the implantation recipient was about $8 \cdot 10^{-7}$ mbar. The analysis of the sample was done with the new DCEMS setup described below.

DCEMS Setup

Using an orange-type magnetic electron spectrometer, a set of Mössbauer spectra containing the phase information is measured at different electron energy settings corresponding to the depth information. The most important properties of the spectrometer are the following:

- High transmission of 20 % of 4π at an energy resolution of $\Delta E/E = 2$ % as a typical value for DCEMS applications.
- UHV conditions ($< 5\cdot10^{-9}$ mbar) using a differentially pumped Aluminum double chamber system reduce adsorbate covering of the samples drastically and allow depth selective Mössbauer studies at low temperatures.
- Optimized focal geometry allows simultaneous measurement of $\Delta E/E = 8$ % in 5 different detection channels.
- A miniaturized electromechanical Mössbauer drive, developed for a space mission [4], is placed along the central axis of the spectrometer.

The DCEMS experiment as described above was done using a 100 mCi ^{57}Co(Rh) source with an isomer shift of 0.102 mm/s relative to an Fe-absorber. Mössbauer spectra were measured at 17 different electron energy settings.

Results and Discussion

The result of a depth selective Mössbauer experiment is a two-dimensional array of data points $\dot{Z}(E_\gamma, E_{el})$ with E_γ and E_{el} being the γ and electron energy, respectively. The data analysis has to be done by a deconvolution procedure in both energy variables. With respect to γ energy this results in the **hyperfine interaction parameters** and thus in a **phase characterization** of the sample. With respect to the **electron energy** this procedure results in the **depth distribution** [5,6] of the phases as characterized above. This problem is visualized in figure 1. For the implanted sample a set of 17 ^{57}Fe Mössbauer spectra have been measured in an electron energy interval of [2.5 - 14 keV]. This corresponds to a depth region of about 0 to 300 nm. In the upper part of figure 1a.) one typical Mössbauer spectrum is plotted. The velocity axis that corresponds to the energy scale of the hyperfine interaction parameters characterizes the different observed phases as follows:

- a single line corresponding to the original high-austenitic stainless steel host matrix,
- a doublet identified as Fe in a low oxidation state (Fe_xO, $x\approx1$),
- a broad distributed magnetically split component corresponding to a martensitic phase.

Without discussing details of the least square analysis of this two dimensional $\dot{Z}(E_\gamma, E_{el})$ surface, in figure 1b.) a contour plot of the data is shown. The non-resonant background is subtracted and each Mössbauer spectrum is normalised to its total areal intensity. This contour plot shows the relative intensities of the three contributing Mössbauer components mentioned above (stainless steel, iron oxide and martensite) as function of the electron energy. On the right hand side, figure 1c.), the weight functions with respect to the depth scale of sensitivity are shown exemplary for some electron energies.

The DCEMS experiment shows the following results:

The doublet has a significant contribution only at electron energies of (7.2 ± 0.1) keV whereas the weight functions (figure 1c.), first graph) show the high surface sensitivity of this electron energy region. From this result one can conclude a surface oxidation of the implanted sample.

Figure 1: a.) Mössbauer spectrum at electron energy setting of 7.2 keV, b.) contour plot of normalised resonant count rates versus velocity and electron energy. c.) shows the depth weight functions, the actual depth resolution is highly improved by the deconvolution procedure. For details, see text.

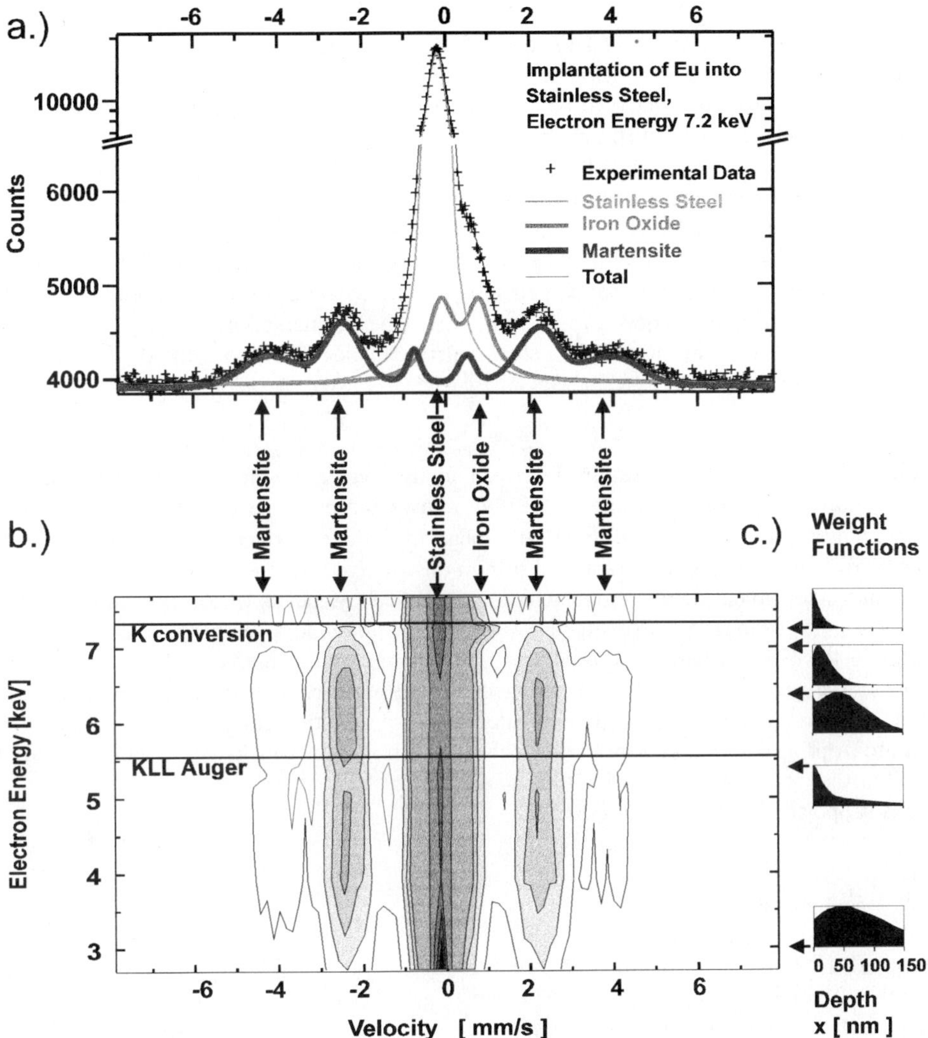

The single line corresponding to the initial stainless steel host matrix shows a significant contribution in this electron energy interval, in contrast to the the magnetically split component. To lower electron energies the signal of the martensitic component increases to a maximum in the electron energy interval of (6.3 ± 0.3) keV, corresponding to the depth range shown in figure 1c.), third graph. This behaviour is interpreted as a buried martensitic transformed layer.

At energy settings lower than 5.4 keV (KLL Auger electron energy of iron) one finds a mixture of surface (Auger) and bulk (internal conversion) sensitivity due to high energy loss K conversion electrons. In this energy region we can observe nearly equal contributions of the stainless steel and the martensitic component to the Mössbauer spectra. Going to very large energy losses, i.e. to an electron energy of 3 keV the stainless steel component dominates the Mössbauer spectra, coming from the deeper bulk of the sample lying beyond the mean projected range of the implanted Eu ions.

As the weight functions shown in figure 1c.) demonstrate, the quantitative analysis of DCEMS is a very complex problem, since also the Mössbauer excitation including secondary effects as well as the response of the spectrometer have to be taken into account. In principle we have chosen the following mathematical treatment of this problem:

- The parameters of the depth profile and those describing the hyperfine interactions are introduced simultaneously into a least square routine whereas
- the electron transport is considered using an extensive Monte-Carlo code.

The detailed analysis that leads to the depth profile of phases is still in progress. The potential of this method of analysis is already demonstrated in [3], were systematic implantations of Eu into stainless steel are discussed. They turned out, that the implantation induced martensitic transformed depth region [7] in the stainless steel matrix coincides well with the depth profile of Eu, analysed by Rutherford Backscattering Spectroscopy.

Perspectives

Further systematic studies of the system described in this paper are in progress, using a new implantation setup which was biult up at GSI. It allows temperature dependent implantation experiments at a residual pressure of $< 1{\cdot}10^{-7}$ mbar to reduce surface contamination during implantation. In general, DCEMS as a method of non-destructive phase analysis is sensitive to the topmost 200 nm, the typical depth region which is modified by means of ion implantation. The new developed UHV orange spectrometer allows experiments at low temperatures, which is of great interest in the field of nanostructured and thin film materials (Superparamagnetism). It is possible to use the great potential of this method also in research fields relevant for industrial applications, e.g. first experiments with implanted high speed steel samples with natural ^{57}Fe enrichment are in progress [8]. The multi channel detection system which will be built up during this year will further reduce the measuring time by a factor of 5.

(This work is supported by BMBF and GSI.)

References

[1] B. Stahl, E. Kankeleit, Nucl. Inst. Meth. B, in press

[2] E. Moll, E. Kankeleit, Nucleonic **7**, 180 (1965)

[3] G. Walter, R. Heitzmann, D.M. Rück, B. Stahl, R. Gellert, O. Geiß, G. Klingelhöfer, E. Kankeleit, Nucl. Inst. Meth. B **113**, 167 (1996)

[4] G. Klingelhöfer, P. Held, R. Teucher, F. Schlichting, J. Foh, E. Kankeleit, Hyp. Int. **95**, 305 (1995)

[5] D. Liljequist, M. Ismail, Phys. Rev. B **31/7**, 4131 (1985)

[6] R. Gellert, O. Geiß, G. Klingelhöfer, H. Ladstätter, B. Stahl, G. Walter, E. Kankeleit, Nucl. Inst. Meth. B **76** 381 (1993)

[7] E. Johnson, A. Johansen, L. Sarholt-Kristensen, H. Roy-Poulsen, A. Christiansen, Nucl. Inst. Meth. B **7/8**, 212 (1985)

[8] F. Jährling, D.M. Rück, G. Walter, H. Fueß, see this Volume

Materials Science Forum Vols. 248-249 (1997) pp. 193-196
© *1997 Trans Tech Publications, Switzerland*

Formation of Aluminum Gradient Films on Stainless Steel by Ion Implantation

R.M. Hausner[1], H. Baumann[1], K. Bethge[1], F. Noli[2] and P. Misaelides[2]

[1] Institut für Kernphysik, Johann Wolfgang Goethe-Universität, D-60486 Frankfurt, Germany

[2] Department of Chemistry, Aristotle University, GR-54006 Thessaloniki, Greece

Keywords: Ion Implantation, Gradient Films, Nuclear Resonance Analysis, High Energy Backscattering, Corrosion Behaviour

Abstract:

High dose implantation (fluence $\geq$ $2 \cdot 10^{18}$ ions/cm^2) of 40 keV aluminum ions into stainless steel (AISI 321, composition: Fe / Cr 18 / Ni 10 / Ti) under increased partial pressure of oxygen (P $\geq$ $1 \cdot 10^{-8}$ hPa/O$_2$) leads to the formation of pure and thick aluminum films on the sample surface with a transition region inside the stainless steel sample where the concentration of the implanted aluminum increases from 0 up to 100 at.%. The films produced by this implantation process have the properties of pure aluminum bulk material with regard to corrosion behaviour and microhardness.

The aluminum depth distributions and the areal density of the aluminum films were measured by means of the sharp resonance in the excitation function of the ^{27}Al(p,γ) ^{28}Si nuclear reaction at 992 keV. The oxygen content was determined by (α,α) high energy backscattering HEBS (171°) with a 7.6 MeV He beam. The strong and broad resonance in the energy range between 7.3 and 7.6 MeV enables a sensitive depth profiling of the oxygen.

SEM analyses, Vickers microhardness measurements and the potentiodynamic polarization technique in 1 normal H$_2$SO$_4$ with slow and fast scan rates were used to characterize the properties of the formed aluminum films.

Introduction:

The properties of near-surface layers can be modified by ion beam techniques, for example by ion beam mixing [1,2] or by ion implantation [3]. Especially metal ion implantation is known to be capable of improving the behaviour of metals in the fields involving surface properties as wear and friction [4] and corrosion [5,6]. In high temperature alloys substantial additions (amounting to 5 wt.%) of aluminum lead to the formation of highly protective films consisting predominantly of Al$_2$O$_3$ [7]. Owing the fabrication difficulties associated with a high aluminum content ion implantation is a promising alternative since it allows a surface concentration of the aluminum well above thermodynamic equilibrium levels in order to obtain oxide films with a greater protective power.

The aluminum concentration in the near surface region of stainless steel being attainable by implanting aluminum ions of some tens keV does not exceed 50 at.% due to the sputtering effect. However, if the aluminum ion implantation is carried out under increased partial pressure of O$_2$ an aluminum film is formed on the sample surface with a high durability [8]. The adsorbed oxygen molecules were dissociated by the energetic aluminum ions and as a consequence a low temperature oxidation process was stimulated which reduces the partial sputtering yield of aluminium.

The properties of these aluminum films at the sample surface were investigated by SEM and Vickers microhardness measurements and the corrosion behaviour by the potentiodynamic polarization technique in 1 normal H$_2$SO$_4$.

Results and discussion:

Mechanically polished stainless steel samples (AISI 321, composition: Fe / Cr 18 / Ni 10 / Ti) and thin evaporated aluminum films on glassy carbon substrates were implanted close to room temperature with 40 keV aluminum ions and ion fluences up to $9 \cdot 10^{18}$ ions/cm^2 using the 50 KV implanter of the Institut für Kernphysik. An ion flux of 20 µA/cm^2 and an increased O$_2$ partial pressure of about $1 \cdot 10^{-6}$ hPa/O$_2$ was used for all implantations. The base pressure at normal vacuum conditions using a liquid nitrogen trap was $1 \cdot 10^{-6}$ hPa which corresponds to a O$_2$ partial pressure of $5 \cdot 10^{-9}$ hPa/O$_2$.

The aluminum depth distributions were measured by means of the sharp resonance in the excitation function of the $^{27}Al(p,\gamma)^{28}Si$ nuclear reaction at 992 keV [9] and the oxygen depth distributions inside the produced aluminum films up to the depth of some microns with the extremely strong and broad $^{16}O(\alpha,\alpha)^{16}O$ elastic scattering resonance at 7.6 MeV (170 times enhanced over the Rutherford cross section, 300 keV FWHM) [10].

Fig. 1a shows aluminum depth distributions of stainless steel samples implanted under backfilling O_2 for aluminum ion fluences from $1 \cdot 10^{18}$ up to $9 \cdot 10^{18}$ ions/cm^2. Due to a reduced sputtering yield the implantation process leads to the formation of pure and thick aluminum films on the sample surface with a transition region where the concentration of the implanted aluminum increases from 0 up to 100 at.%. For comparison the the aluminum depth distribution of 40 keV aluminum ions implanted under normal vacuum conditions is shown. In this case only a maximum attainable concentration of 42 at.% is reached as a consequence of the sputtering effect.

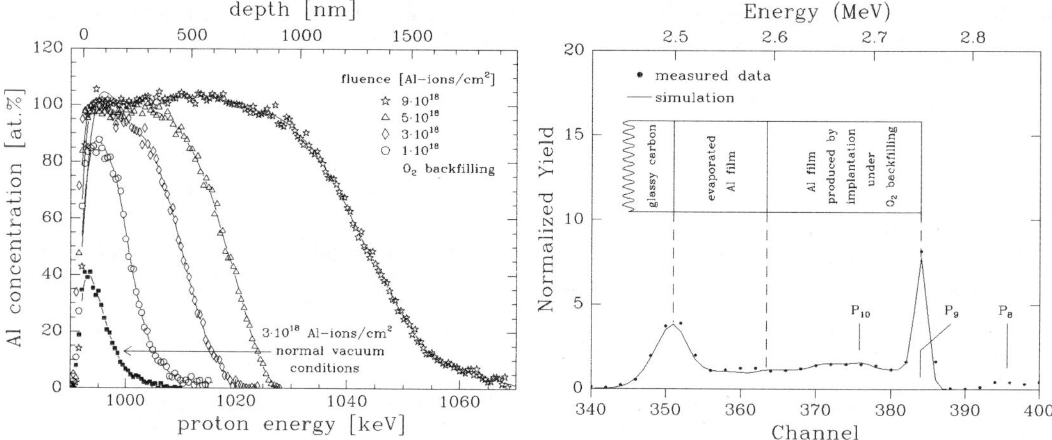

Fig. 1a: Depth distributions of 40 keV aluminum ions (fluence $1 \cdot 10^{18}$-$9 \cdot 10^{18}$ ions/cm^2) implanted in stainless steel under increased O_2 partial pressure ($P \approx 1 \cdot 10^{-6}$ hPa/O_2). For comparison the depth distribution obtained at normal vacuum conditions corresponding to the aluminum saturation dose is shown.

Fig. 1b: Oxygen depth distribution of an aluminum film produced by implantation of 40 keV aluminum ions into an evaporated aluminum film (fluence $3 \cdot 10^{18}$ ions/cm^2) under backfilling O_2 determined by HEBS. P_x indicates proton groups from the nuclear reaction $^{27}Al(\alpha,p)^{30}Si$.

The oxygen content inside the aluminum film is lower than 1 at.% as measured by the $^{16}O(\alpha,\alpha)^{16}O$ elastic scattering resonance at 7.6 MeV (scattering angle 171°). The aluminum film was produced under backfilling O_2 by implantation of 40 keV aluminum ions (fluence $3 \cdot 10^{18}$ ions/cm^2) into an aluminum film evaporated on glassy carbon. The analysing α-beam hits the sample under 60° to normal incidence and the backscattered α-particles were measured by a surface barrier detector (energy resolution 13 keV, detector solid angle 0.34 msr). The oxygen concentration was determined by fitting the measured backscattering spectrum with a simulated one (RUMP computer code [11]), see fig 1b.

Fig. 2a shows a SEM photography of an aluminum film (thickness 1 μm) on stainless steel produced by implantation of 40 keV aluminum ions (fluence $9 \cdot 10^{18}$ ions/cm^2) under increased O_2 partial pressure. During the formation of the aluminum film some pores were build with a typical diameter of 1 μm and a depth of about 300 nm. The pore density was determined to 1 per 30 μm^2 but nevertheless the aluminum film is closed as can also be seen from the corrosion measurements.

The Vickers microhardness measurements indicates that the hardness of the aluminum films produced by the implantation process under backfilling O_2 is equivalent to the pure aluminum bulk material. The microhardness of stainless steel, pure aluminum and two different aluminum films on steel (ion fluence $3 \cdot 10^{18}$ and $5 \cdot 10^{18}$ ions/cm^2) dependent on the stylus force are shown in fig. 2b.

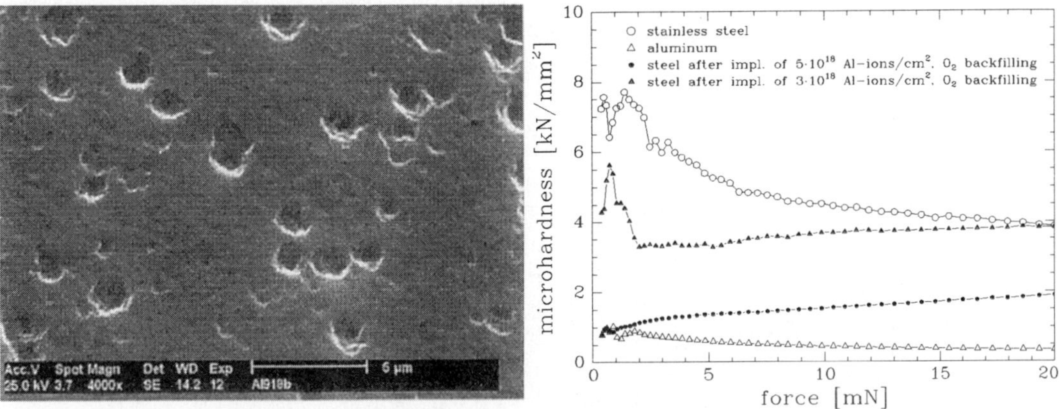

Fig. 2a: SEM photography of the aluminum film on stainless steel produced by implantation of 40 keV aluminum ions (fluence $9 \cdot 10^{18}$ ions/cm^2) under increased O$_2$ partial pressure.

Fig. 2b: Vickers microhardness of stainless steel, pure aluminum and aluminum films on stainless steel produced by implantation of 40 keV aluminum ions (fluence $3 \cdot 10^{18}$ and $5 \cdot 10^{18}$ ions/cm^2) under increased O$_2$ partial pressure.

The corrosion behaviour of the aluminum implanted steel samples was studied in aqueous solution of 1 normal H$_2$SO$_4$ by means of the potentiodynamic polarization technique [11] with slow (0.6 V/h) and fast (10 V/h) scan rates. Fig. 3 shows plots for the polarization of the stainless steel, pure aluminum and the stainless steel implanted with 40 keV aluminum ions under normal vacuum conditions up to the saturation dose (fluence $4.5 \cdot 10^{17}$ ions/cm^2, 42 at.% aluminum) as well as under increased O$_2$ partial pressure (fluence $2.5 \cdot 10^{18}$ ions/cm^2, 100 at.% aluminum).

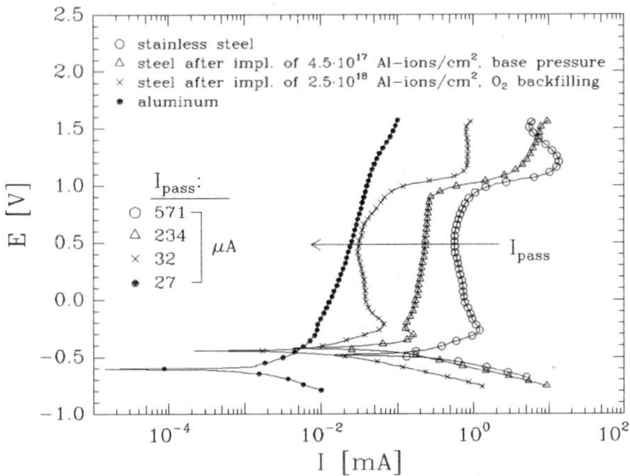

Fig. 3: Potentiodynamic polarization curves in 1 normal H$_2$SO$_4$ of stainless steel, pure aluminum bulk material and stainless steel implanted with 40 keV aluminum ions under normal vacuum conditions ($4.5 \cdot 10^{17}$ ions/cm^2) and under backfilling O$_2$ ($2.5 \cdot 10^{18}$ ions/cm^2).

Polarizing the samples to more electropositive potentials initially leads to a higher corrosion rate. Above the passivation potential a reduction in current density occurs due to the formation of a film which acts as an efficient barrier to the corrodent. This passivity is seen to markedly reduce corrosion to a low level indicated by the passivation current density I_{Pass}. At higher oxidizing potentials the passive film breaks down . The measurements indicate that I_{Pass} for stainless steel was reduced as a consequence of the aluminum implantation and depending on the aluminium concentration down to the value of pure aluminum, see tab. 1.

Sample:	slow scan rate 0.6 V/h				fast scan rate 10 V/h			
	E_{Corr} [mV]	I_{Pass} [μA]	I_{Crit} [μA]	I_{Max} [mA]	E_{Corr} [mV]	I_{Pass} [μA]	I_{Crit} [μA]	I_{Max} [mA]
stainless steel	-475	150	684	7.0	- 480	571	1.344	14.6
aluminum	-495	42	165	< 0.1	- 610	27	9	< 0.1
Al in steel base pressure	-408	95	121	8.4	- 430	234	158	7.5
Al in steel O$_2$ backfilling	-400	50	78	4.0	- 420	32	71	0.8

Tab. 1: Corrosion potential E_{Corr}, passivation current density I_{Pass}, critical current density I_{Krit} and maximum current density I_{Max} in 1 normal H_2SO_4 of the studied samples for slow and fast scan rates.

Conclusion:
The aluminum ion implantation into stainless steel with ion energy of 40 keV under increased partial pressure of oxygen leads to the formation of aluminum films on the sample surface. The films produced by this implantation process have the same properties of pure aluminum bulk material with regard to corrosion behaviour and Vickers microhardness. So it is possible to create a so called gradient material with totally different properties on the sample surface and in the bulk material and with a transition region where the aluminum concentration increases from 0 up to 100 at.%.

References:
[1]   C. Jaounen, J.P. Eymery, E.L. Mathe, J. Delafond, Mater. Sci. Eng. **69** (1985) 483
[2]   K. Neubeck, H. Hahn, D.M. Rück, R.M. Hausner, H. Baumann, K. Bethge, A.G. Balogh, Nucl. Instr. and Meth., **B113** (1996) 186
[3]   P. Goode, I.J.R. Baumvol, Nucl. Instr. and Meth., **189** (1981) 161
[4]   F. Alonso, J.L. Vivente, J.I. Onate, B. Torp, B.R. Nielsen, Nucl. Instr. and Meth., **B80/81** (1993)
[5]   F. Noli, P. Misaelides, P. Spathis, M. Pilakouta, H. Baumann, Nucl. Instr. and Meth, **B68** (1992) 398
[6]   F. Noli, P. Misaelides, P. Spathis, G. Giorginis, H. Baumann, R.M. Hausner, Nucl. Instr. and Meth, **B113** (1996) 171
[7]   U. Bernabi, M. Cavallini, G. Bombara, G. Dearnaley, M.A. Wilkins, Corr. Sci., **20** (1980) 19
[8]   R.M. Hausner, H. Baumann, K. Bethge, Nucl. Instr. and Meth., **B113** (1996) 176
[9]   L.C. Feldman, S.T. Picraux, Ion Beam Handbook for Material Analysis, Eds. J.W. Mayer, E. Rimini, Academic Press (1977) 211
[10] F.J.D. Almeida, J.A. Davies, T.E. Jackman, Nucl. Instr. and Meth., **B82** (1993) 393
[11] L.R. Doolittle, Proc. High Energy and Heavy Ion Beams in Mat. Anal., Eds. J.R. Tesmer et al., Mater. Res. Soc. (1990) 175
[12] C.R. Clayton, Nucl. Instr. and Meth., **182/183** (1981) 865

Materials Science Forum Vols. 248-249 (1997) pp. 197-200
© *1997 Trans Tech Publications, Switzerland*

Studies on Ion Beam Modification and Analysis Using the Bucharest Cyclotron

B. Constantinescu, C. Sarbu, E. Ivanov, D. Plostinaru and R. Bugoi

Institute of Atomic Physics, PO. Box MG-6, Bucharest, Romania

Keywords: Cyclotron, Blistering, Helium Bubbles, Amorphisation, Absorbed Halogens

Abstract

Studies on dose and energy dependence of blistering and flaking on stainless steels, Ni, Cu and Mo, produced by 3.0, 4.7 and 6.8 MeV He^+ ion irradiation are presented. Using SEM (Scanning Electron Microscopy) and TEM (Transmission Electron Microscopy) techniques, irradiation phenomena such as sponge- and wave-like structures, submicronic cracks, microcraters, helium bubbles on matrix, grain boundaries, loops and TiC precipitates, are illustrated. The appearance of an amorphous phase in the Ti-modified austenitic steel 12KH18N10T is discussed. Two cases of surface absorbed fluorine and chlorine determination by PIGE (Particle Induced Gamma-ray Emission) and RBS (Rutherford BackScattering) are reported.

Introduction

During the past few years, the subharmonic acceleration regime of the Bucharest classical U-120 cyclotron was used for various material studies: 3-7 MeV alpha-particles were used to study helium influence on the properties of fusion reactor materials, while 4.7 MeV alpha-particles were employed in elemental analysis methods (PIXE-Particle Induced X-ray Emission, PIGE, RBS) [1]. A renewed interest in radiation damage problems has arisen in the last decade with the advent of fast neutron reactors and design proposals for future fusion reactors. A considerable attention has focused on the behavior of helium in metals as helium is produced in reactor materials by direct ion injection, (n, α) transmutation reactions, or tritium decay. Since helium is known to have an extremely low solubility in metals, generally much less than 1 appm, it strongly tends to participate in helium bubbles. Thus, the formation of helium bubbles and subsequent bubble growth result in a swelling of metals containing helium and may lead to He-embrittlement. Surface damages such as blisters, exfoliation and sputtering are also produced. A detailed description of these damages is presented in [2].

An important problem which can be solved using nuclear analysis methods is the absorption of halogens (fluorine and chlorine, especially) on metal surfaces, which is very dangerous for the mechanical resistance of the installation, mainly through induced brittleness.

Experimental

To investigate the dose and energy dependence of blistering and flaking, different kinds of candidate fusion reactor first wall materials such as Soviet 12KH18N10T (17-19% Cr, 9-11% Ni, 2% Mn, 0.8% Ti) and Japanese W-4541 (17-19% Cr, 8.0-9.5% Ni, 0.6% Mn, 2% Ti) austenitic steels, Romanian W-4016 (16.04% Cr, 0.13% Ni, 0.86% Mn) ferritic steel, Romanian SS-304 (18-20% Cr, 8-12% Ni, 2% Mn) austenitic steel, Ni, Cu and Mo were irradiated at room temperature with 3.0, 4.7 and 6.8 MeV He$^+$ ions at the Bucharest classical Cyclotron. The effects were investigated by means of a JEOL TEMSCAN 200 CX electron microscope (SEM and TEM) and a metallographic Olympus microscope.

For fluorine determination we examined a stainless steel component from the Romanian Cernavoda Nuclear Power Plant, suspected of halogen infiltration at the surface, as a result of a fire. The presumably affected (contamined) surface area was filed 0.1-0.2 mm in depth. Pressure deposited pellets on lead backing (20 mm in diameter and 1 mm deep) were prepared as samples for analysis. The PIGE method (see [3]) was used through the reaction:

$$(100\%) \; {}^{19}F(\alpha, \alpha'){}^{19}F, \; E_\gamma = 110; \; 197 \; keV$$

induced by 4.7 MeV α-particles. The 583 and 1275-1280 keV doublet γ-rays were also observed. A 40-50 nA intensity beam was used for 25-30 min. direct spectra (beam charge 100 μC), detected and recorded by a 70 cc Ge(Li) detector coupled to a CANBERRA-80 MCA. CaF$_2$ mixed with filed iron standards (fluorine concentration from 100 ppm to 4%) were used.

For chlorine determination, we examined a Romanian oil chemistry plant duraluminium component, also suspected of halogen surface infiltration (from salty sea air). In this case, usual analytical methods (neutron activation included) are unsuitable because the potential chlorine is located 1000-2000 A near the surface. Therefore, a specific surface analysis method (RBS-see [4]) was used. The samples, cut into 10×10×2mm pieces taken from the mechanical fractured areas, have been analyzed using a 50 nA intensity beam of 4.7 MeV α-particles and an ORTEC 500 μm depleted zone silicon surface barrier detector.

Results and discussion

Beside blistering (see Figure 1), irradiation phenomena such as sponge- and wave-like structures (Figure 2), submicronic cracks, multilayer flaking, micro-conglomerates, secondary blisters, microcraters (Figure 3) were observed. Average blister diameter values between 150 and 1200 microns, depending on irradiation energy and material, were obtained. Critical doses for blistering appearance were in the range 0.45-6.2×10^{18} ions/cm^2. A potential explanation of the energy dependence of average blister diameter d and blister critical dose could be the stressed-induced model, described in [5], which predicts the relationship: $d \sim t^m$, where t is the blister skin thickness. Measurements by Kaminsky, Das and Fenske [6] for He in several models yielded for the relation $d \sim t^m$ values for m of 1.25 for Be, 0.85 for V, 1.15 for Ni, and 1.22 for Nb at low energy, but 1.50 at high energy. For our high energy data, a m=1.8 value is relatively convenient, except for Mo, where m=1.5 is more suitable. Similarly, concerning the dependence of our experimental values for critical fluence, we can assume a relationship of the form: $f \sim t^2$, therefore concluding that not only the resistance (proportional to t) of the blister skin itself is important, but also other internal resistance factors, such as those involving the inter-submicronic bubbles fracture mechanism, in the stage of damage preceding the gas accumulation at the end of the He ions range. Using the TEM technique, microstructural changes such as helium bubbles on matrix and grain boundaries (Figure 4), on loops (Figure 5) and on TiC precipitates (Figure 6), were observed. A very interesting effect is an amorphous phase in the Ti-modified austenitic steel 12KH18N10T, which is formatted during irradiation with 6.8 MeV He$^+$ ions up to a dose of crater occurrence (7×10^{18} ions/cm^2, beam density on target 0.3-0.8 μA). The temperature of the specimens (100 μm thick) was maintained during irradiation within the

30-60°C range. X-ray microanalysis showed the presence of the major alloy elements in the amorphous area, whereas in that embedding a microcrystallite Ni depletion and the presence of Si and Mo could be noticed. To confirm the tendency of the m values to shift from 1-1.5 for low energy to 1.5-2 for high energy (our data), we intend to enlarge the variety of samples (other metals and alloys resulting from various metallurgical preparation conditions) and of bombarding beams (other ions, with different energies).

As for F determination in stainless steel, a 40 ppm sensitivity was reached. A 15% measurement error was estimated. Unfortunately for the mechanical resistance, fluorine was detected in all of the analyzed samples. For Cl absorption in duraluminium, the presence of chlorine was detected and estimated at a level of ca. 0.5% atoms in the analyzed sample 2000 A near the surface layer. We can conclude that nuclear analysis methods such as PIGE and RBS could be successfully applied to investigate halogen absorption in various metal surfaces.

REFERENCES

[1] B. Constantinescu, S. Dima, V. Florescu, E. Ivanov, D. Plostinaru and C. Sarbu, Nucl. Instr. Meth. B **16**, 488 (1986).
[2] H. Ullmaier, Nuclear. Fusion **24**, 8, 1039, (1984).
[3] J. F. Ziegler, New uses of ion accelerators, Plenum Press, New York and London, (1975)
[4] I. S. Giles and M. Peisach, J. Radioanal. Chem. **50**, 1, 307 (1979)
[5] U. Scherzer, Sputtering by particle bombardment II. In Topics in Applied Physics, vol. 52, p. 480, Springer Verlag, Berlin, Heildelberg, New York, Tokyo(1983).
[6] S. K. Das, M. Kaminsky and G. Fenske, J. Appl. Phys., **50**,3304 (1979).

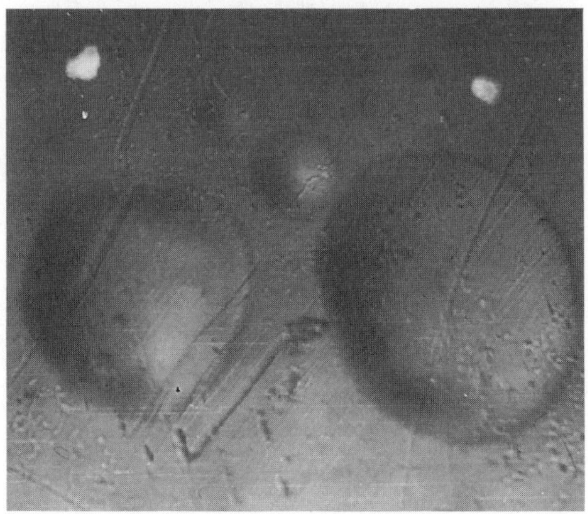

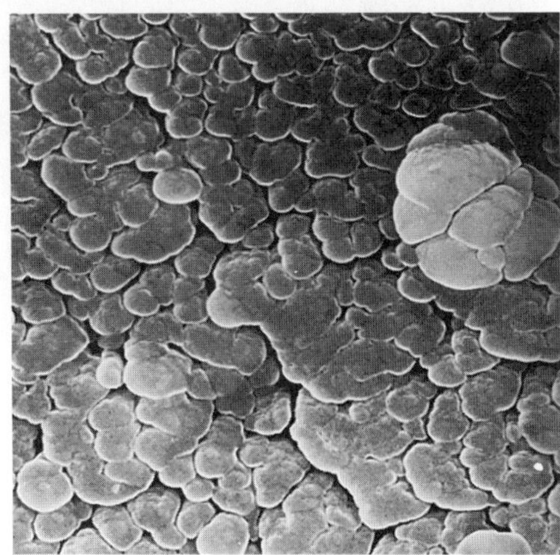

Figure 1 Circular blisters in nickel, E_{He}=3.0 MeV, 0.9×10^{18} ions/cm^2, ×240

Figure 2 Wave like structures on the bottom of an opened blister in molybdene, E_{He}=3.0 MeV, 1.5×10^{18} ions/cm^2, ×6000

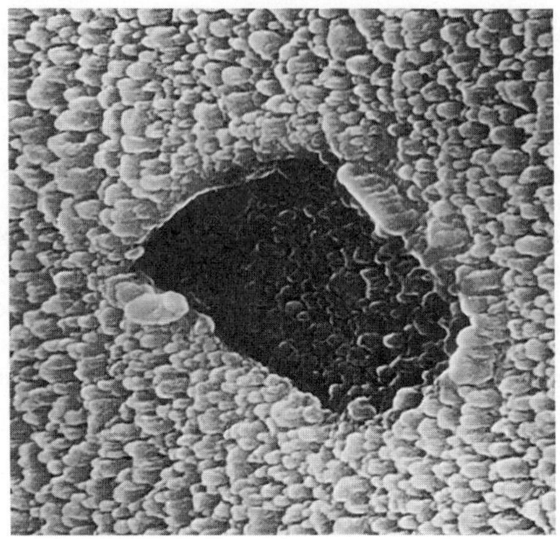

Figure 3 Microcraters in an open blister in 12KH18N10T, E_{He}=4.7 MeV, 3.6×10^{18} ions/cm^2, ×5000

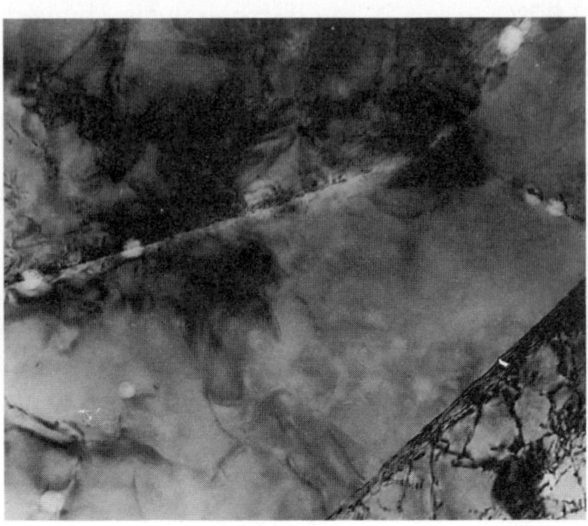

Figure 4 Helium bubbles on grain boundaries in 12KH18N10T, E_{He}=6.8 MeV, 6.2×10^{18} ions/cm^2, ×50000

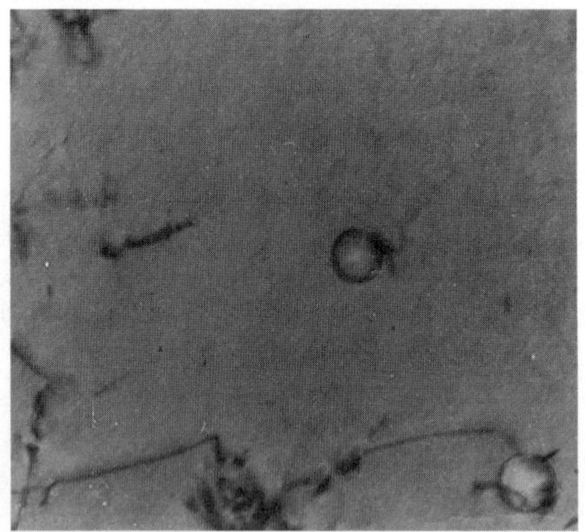

Figure 5 He bubbles on loops in 12KH18N10T, E_{He}=6.8 MeV, 6.2×10^{18} ions/cm^2, ×73000

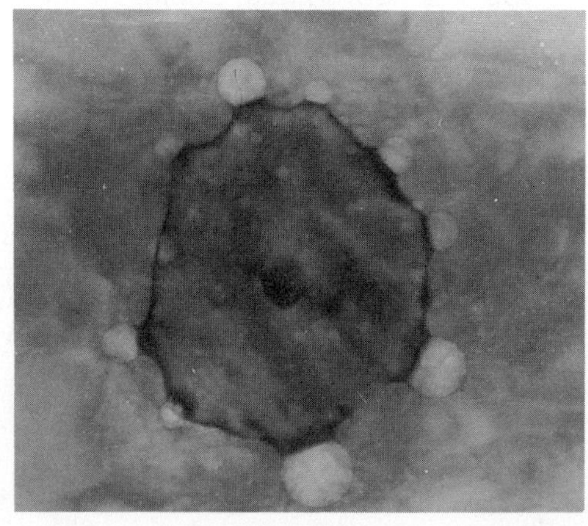

Figure 6 He bubbles on TiC precipitate in 12KH18N10T, E_{He}=6.8 MeV, 6.2×10^{18} ions/cm^2, ×100000

Materials Science Forum Vols. 248-249 (1997) pp. 201-204
© 1997 Trans Tech Publications, Switzerland

Microstructure and Tribology of Carbon Implanted High-Speed-Steel

F. Jährling[1], G. Walter[2], D.M. Rück[3] and H. Fuess[1]

[1] Fachbereich Materialwissenschaft, TH Darmstadt, Petersenstr. 23, D-64287 Darmstadt, Germany

[2] Institute of Nuclear Physics, TH Darmstadt, Schlossgartenstr. 9, D-64289 Darmstadt, Germany

[3] Gesellschaft für Schwerionenforschung mbH, Planckstr. 1, D-64291 Darmstadt, Germany

Keywords: Ion-Implantation, High-Speed-Steel, Tribology, Microhardness

Abstract

AISI-M2 high-speed-steel was implanted with different doses of carbon at energies of 60 kV and 100 kV. Tribological behaviour and microhardness were studied. The measurements reveal a dose dependence of the friction coefficient μ. For 60 kV the decrease of μ is more pronounced than for 100 kV. Microhardness does not show a dose dependence for both energies but a strong energy dependence. For both friction and hardness the carbon concentration seems to be the dominant factor, whereas carbide formation does not influence the hardness considerably.

Introduction

Ion implantation in steels may result in strong enhancements of wear-resistance and an increase of hardness. To understand these changes it is necessary to separate the influence of the single implantation-parameters dose and energy. Especially implantation with carbon-ions has been shown to cause considerable effects on tribology and hardness of martensitic steels whereas the widely investigated nitrogen-implantation does not lead to distinct changes [2,3].
The present work discusses results concerning carbon-implanted AISI M2 high-speed-steel.

Experimental

AISI M2 steel discs (approximate composition in wt% : C 0.8, Cr 4, W 6, V 2, Mo 5, Mn 0.4, Si 0.5 and balance Fe) were mechanically mirrorpolished and implanted with C^+-ions at the GSI 300 keV - implanter [1] using a Duopigatron ion source with CO_2 as feeding gas. For the sake of homogeneinity the ion beam was scanned over the implanted area with a diameter of 20 mm. Sample heating was limited by use of a watercooled targetholder and beam currents lower than 15 $\mu A/cm^2$.
Studies of tribology were carried out on a CSEM-pin-on-disc-tester. AISI 52100 steel balls with a radius of 5 mm were used as pin. The normal load F_N was chosen to be 2N, the relative velocity was 0.1 m/s. The tests took place at room temperature and air with approximately 60% humidity.
For determination of microhardness a FISCHERSCOPE microindentation device was used. The load was increased from 0.4 mN to 10 mN and then unloaded. Indentation was measured under load.
For each sample the hardness values were averaged over ten measurements.
For microstructural investigation grazing incidence x-ray diffraction was used. The measurements were carried out with Cu-Kα radiation (λ=1.5406 Å) on a thin-film diffractometer designed in the authors' institute.

Results and Discussion

Dose series at two energies (100 kV and 60 kV) were implanted. For 100 kV the doses range from 1×10^{17} to 3×10^{18} C^+/cm^2 and from 1×10^{17} to 1×10^{18} C^+/cm^2 for 60 kV, respectively.

Pin-on-disc measurements give informations about wear behaviour and friction. Here the friction coefficient μ, defined as ratio of frictional force (which is recorded by the pin-on-disc machine) and normal force F_N is discussed. Fig. 1 shows μ as a function of implanted dose for both series.

For the 100 kV series μ decreases with increasing dose with a slight change in slope around 5×10^{17} C^+/cm^2. The values at 60 kV do not show a pronounced dose dependence in the observed range. The initial decrease from unimplanted steel to a dose of 1×10^{17} C^+/cm^2 is higher by a factor 1.5 than in the 100 kV-series. So μ can be lowered by nearly a factor 2 by implantation of 1×10^{17} C^+/cm^2 at 60 kV while for 100 kV a dose of approximately 7×10^{17} C^+/cm^2 is needed to obtain the same result. To understand this it has to be taken into account that implantation energy mainly affects the range of ions in the solid and the width of the implanted distribution. For 60 kV the range is smaller and the profile narrower than for 100 kV. This implies that the implanted carbon concentration per volume plays a major role for friction.

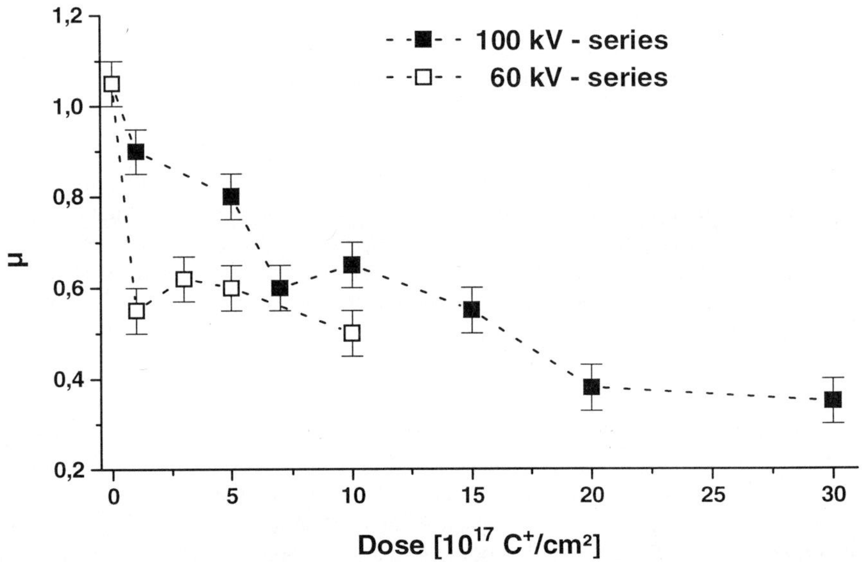

Fig. 1. Friction coefficient μ as a function of dose for 60 kV (□) and 100 kV (■). Dotted lines are guides to the eye.

Grazing incidence x-ray diffraction (GIXD) reveals the formation of ε-carbide at a dose of approximately 5×10^{17} C^+/cm^2 for an energy of 100 kV (Fig. [2]). This coincides with the slight change in slope of μ mentioned above. It is likely to assume that ε-carbide is at least partially responsible for the change in slope. For 60 kV it is only known up to now that no ε-carbide is formed at a dose of 1×10^{17} C^+/cm^2. Nevertheless μ shows a signficant decrease. This indicates that ε-carbide is not decisive for the numerical value of μ.

Visual observation of the wear tracks showed the appearance of red-brown powder on the samples indicating oxidation of the wear partners. This indicates that tribo enhanced oxidation takes place.

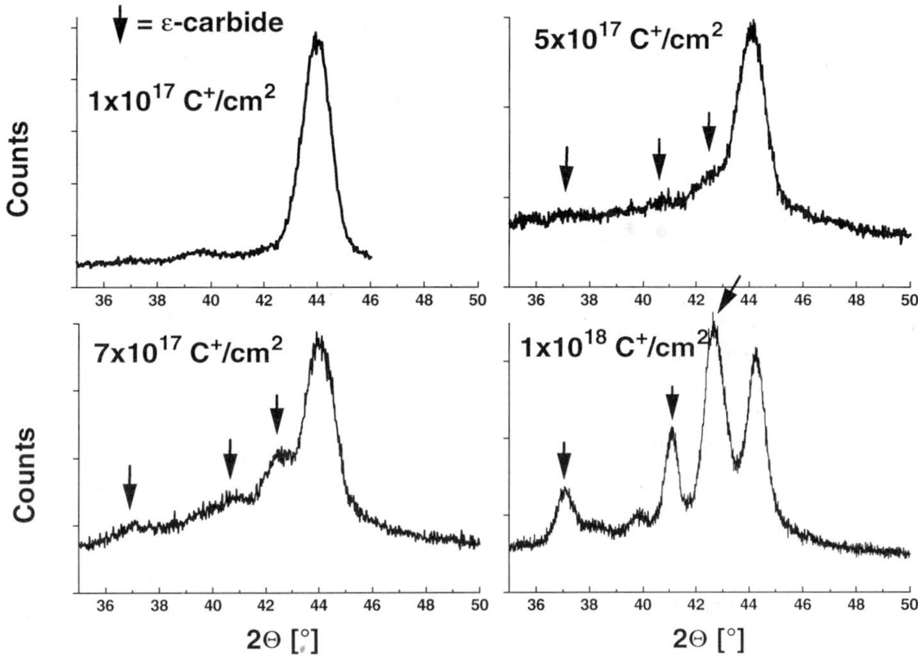

Fig. 2. GIXD patterns at an incidence angle of 1° revealing the formation of ε-carbide (⇨) in the 100 kV series. The mean scattering depth was approximately 70 nm.

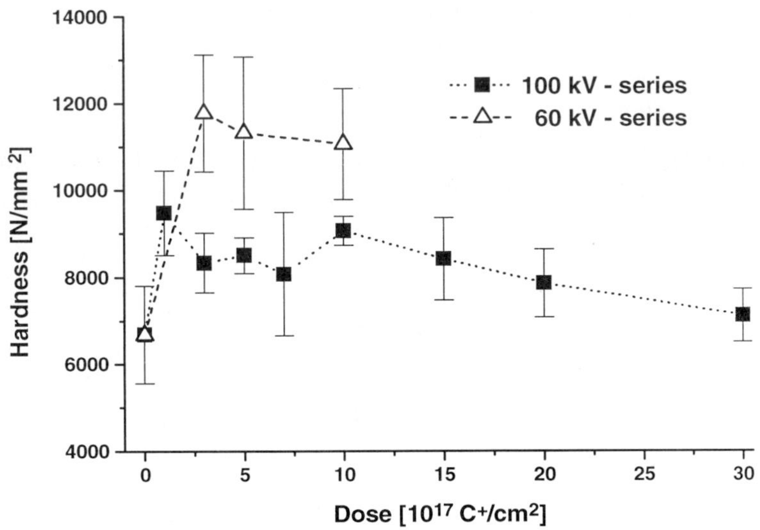

Fig. 3. Microhardness as a function of dose for 60 kV (△) and 100 kV (■). The dashed lines are guides to the eye.

Microhardness as function of dose for both energies is shown in Fig. 3. Except the initial increase there is no significant change in hardness over the discussed dose-range within the experimental errors. These results are different from those for carbon-implanted AISI 52100 steel studied by Kobs et al. [3]. They found a clear dose dependence of hardness with a maximum around 1×10^{18} C^+/cm^2 for 100 kV. This gives evidence that the influence of initial microstructure and composition is not to be neglected. Nevertheless an energy-dependence of microhardness can be observed. Hardness is reduced by a factor 1.6 for 60 kV and a factor 1.2 for 100 kV. This energy dependence can be explained in terms of ion range and implanted distribution in the same way as discussed above for friction. Near surface carbon concentration seems to be the dominating factor for microhardness. It is more difficult to interpret the lack of a dose dependence. Probably the precipitation of graphitic carbon [2,3] with low hardness compensates hardening by dislocations and interstitials. Second phase hardening does not seem to play a role, because there is no distinct change in hardness at the onset of ε-carbide formation around 5×10^{17} C^+/cm^2 for 100 kV. To understand this in detail more microstructural investigations will be performed.

Moreover it is obvious that there is no simple correlation between friction and hardness.

Summary and Outlook

It has been shown that carbon-implantation in AISI-M2-high-speed-steel has the capability of improving friction and hardness. There is a clear dose dependence in friction at least for an energy of 100 kV which can not be found in microhardness. Friction as well as microhardness is showing a clear energy dependence implying that the near surface carbon concentration could play a major-role for the mentioned changes. Formation of ε-carbide has no influence on hardness but possibly affects friction at an implantation energy of 100 kV.

The underlying mechanisms on the microscopic scale are not yet clear. For this purpose microstructural methods like X-ray-diffraction, Mössbauer-spectroscopy, Rutherford backscattering and XPS will be used to obtain complete information about the microscopic state of the implanted layers. Furthermore wear will be studied using profilometric measurements and scanning electron microscopy.

Acknowledgements

This work is financially supported by GSI. The authors would like to thank Dr. K. Thoma from Gesamthochschule Kassel for providing the raw AISI M2 samples.

References

[1] D.M. Rück, N. Angert, H. Emig, K.D. Leible, P. Spädtke, D. Vogt, B.H. Wolf, Nucl. Tracks Radiat. Meas. **19**, 951 (1991)

[2] J-P. Hirvonen, D. Rück, S. Yan, R. Lappalainen, P. Torre, Mat. Res. Soc. Symp. Proc. **316**, 507 (1994)

[3] K.Kobs, H. Dimigen, C.J.M. Denissen, E. Gerritsen, J. Politiek, R. Oechsner, A. Kluge, H. Ryssel, Nucl. Instr. Meth. Phys. Res. B **59/60**, 746 (1991)

Materials Science Forum Vols. 248-249 (1997) pp. 205-208
© *1997 Trans Tech Publications, Switzerland*

Investigation of High Fluence Carbon Ion Implanted Titanium

A. Königer, A. Wenzel, C. Hammerl and B. Rauschenbach

Institut für Physik, Universität Augsburg, D-86135 Augsburg, Germany

Keywords: Ion Implantation, In-Situ Electrical Resistivity, Transformation Kinetics, Titanium Carbide

Abstract
High-fluence carbon ion implantations into titanium were performed at temperatures of -40°C. Fluences of up to $1.2 \cdot 10^{18}$ C$^+$/cm² were used. The concentration distribution of carbon was measured by Rutherford backscattering and the structure and morphology were studied using transmission electron microscopy and X-ray diffraction. Formation of titanium carbide was found after implantations at low temperature. During implantation and sub-sequent annealing „in-situ" resistivity measurements were performed. Thus the kinetics of transformation process could be compared with time-dependent heterogeneous isothermal transformations. Similarities were found and hints for a dominant intra-cascade mechanism can be given.

Introduction
Ion implantation is a powerful technique to improve mechanical and chemical properties of the surface and near-surface regions of materials. A lot of papers have been published dealing with influence of non-metal ion implantation on properties of metals [1-4]. In recent time also new materials attracted attention because of qualities like high strength at low weight or good biological compatibility. Titanium carbide (TiC) became an important material for wear or corrosion protection. Formation of titanium carbide has been found for carbon ion implantation into titanium alloys[5,6]. Only a few papers on carbon ion implantation into pure titanium have been presented [7,8].
In this paper the TiC phase formation is studied by high fluence carbon ion implantation into titanium at low temperatures. A method was developed to understand the elementary processes during ion implantation by „in-situ" measurement of resistivity. Measurement of film resistivity has already been applied in the field of ion beam mixing [9-11] and implantation induced defect generation [12]. It proved to be a precise tool to study transformation kinetics. In this work the formal theory of phase transformation kinetics is applied to processes occurring during ion implantation. The scope is to investigate fundamental processes by means of resistivity measurements.

Experimental
Titanium films were electron-beam evaporated onto sapphire (102) substrates. A special geometry well-suited for four-point resistance measurements was patterned by means of masking technique. Films were deposited at 4-6 nm/s to a final thickness of 340 nm at a base pressure of $2 \cdot 10^{-5}$ Pa and a substrate temperature of 350°C. The oxygen impurity concentration in these films was lower than 0.3 at.-% oxygen. Implantation was performed at a target temperature of -40°C with fluences of up to $1.2 \cdot 10^{18}$ C$^+$/cm². Ion energy was 100 keV and ion beam current density was approximately 12 µA/cm².
Elemental composition and ion range profiles were determined using Rutherford backscattering (RBS). Structure of implanted specimens was investigated using both X-ray diffraction in a Seeman-Bohlin geometry (XRD) and transmission electron microscopy (TEM).
Measurement of resistivity was performed in a 4-point geometry. A current of 10 mA was supplied by a constant current source. Voltages across the current bridge were read out by a digital multimeter with an accuracy of ±0.01 mV. The whole equipment was controlled via personal computer. In order to eliminate contact and thermal voltages the current direction was

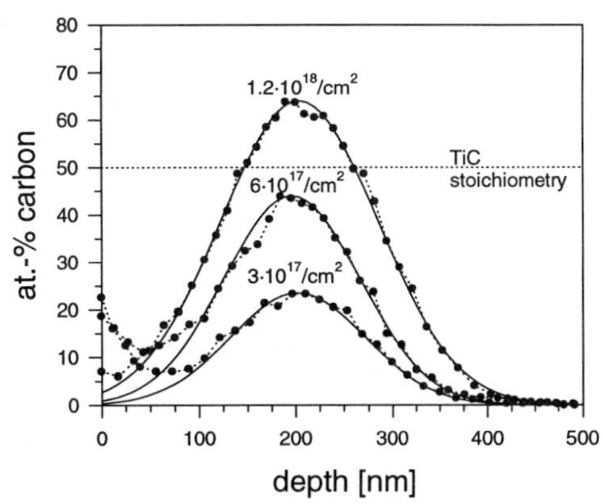

Fig. 1: Carbon concentration profiles after implantation of $3 \cdot 10^{17}$ C$^+$/cm² - $12 \cdot 10^{17}$ C$^+$/cm² in titanium with an energy of 100keV at a temperature of -40°C.

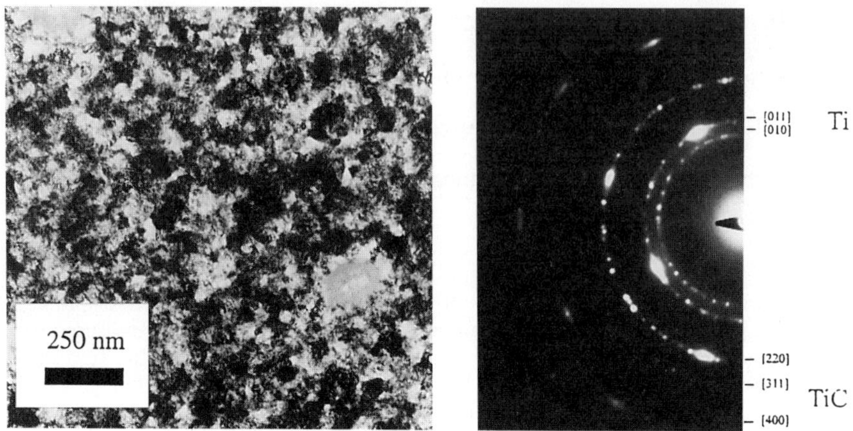

Fig. 2:   TEM micrograph of titanium implanted with $6.5 \cdot 10^{17}$ C$^+$/cm². The polycrystalline diffraction pattern exhibits both reflections from titanium and titanium carbide (strongest reflections are indicated on the right).

alternatingly reversed and corresponding voltages were averaged.

Results

Implantation of carbon ions into titanium films leads to Gaussian-like concentration distributions of carbon. Fig. 1 shows carbon concentration profiles obtained for samples implanted with fluences from $3.0 \cdot 10^{17}$ C$^+$/cm² to $1.2 \cdot 10^{18}$ C$^+$/cm². Up to the highest fluence a very good fit to a Gaussian curve is possible (see solid lines in Fig. 1). This is remarkable since in most systems at even lower doses a strong deviation from such an ideal behaviour is found. It has to be emphasised that very high amounts of carbon can be deposited in titanium. For a fluence of $1.2 \cdot 10^{18}$ C$^+$/cm² 65 at.-% carbon are reached. Experiments at even higher fluences (up to $3.6 \cdot 10^{18}$ C$^+$/cm²) showed maximum concentrations of 88 at.-% carbon (for details see ref. [13]).

Both X-ray diffraction and transmission electron diffraction confirmed the implantation induced phase formation of titanium carbide (TiC) for fluences larger than $3 \cdot 10^{17}$ C$^+$/cm² and at temperatures of -40°C. Fig. 2 shows a TEM micrograph of a titanium film implanted with $6.5 \cdot 10^{17}$ C$^+$/cm², leading to a broad profile with about 50 at.-% carbon. A polycrystalline structure can be observed. The average grain diameter is about 90 nm compared to a value of 20 nm before implantation. The diffraction pattern shows clear titanium peaks and additionally reflections arising from TiC.

During ion implantation the voltage over the current bridge was measured every 30 s. After conversion into resistivity this can be plotted as function of implanted fluence. Fig. 3 shows a typical plot. First a steep rise in resistivity, which decreases more and more and finally pronounced saturation behaviour is always found. Resistivity increases from initially 54 μΩcm to about 73 μΩcm. The initial value of agrees well with data from literature [14]. Only deviations of less than 10 μΩcm are observed, which can be assigned to the preparation dependent defect density. As the implanted volume is smaller than the whole current bridge the resistivity increase in the modified volume is even higher. The resistivity in this region can be calculated to be about 120 μΩcm. After implantation of $5 \cdot 10^{17}$ C$^+$/cm² the resistivity remains nearly constant. Fluctuations within this curve are due to changes in the ion beam current and/or the target temperature. As the specimen temperature raises about 100°C within the first minutes of implantation a special correction was done to eliminate this temperature variations.

Fig. 4 exhibits a special presentation of these values referring to the theory of phase transformation kinetics [15]. As the saturation behaviour was observed it was possible to estimate the fraction of transformed volume c due to ion bombardment. The untransformed volume fraction is therefore (1-c). Resistivity was converted into transformed volume fraction taking into account a model of randomly distributed spherical inclusions in a 3-dimensional matrix [16,17]

$$c = 1 - \frac{(\rho/\rho_2 - 1)(2 + \rho/\rho_1)}{3(\rho/\rho_2 - \rho/\rho_1)} \qquad (1)$$

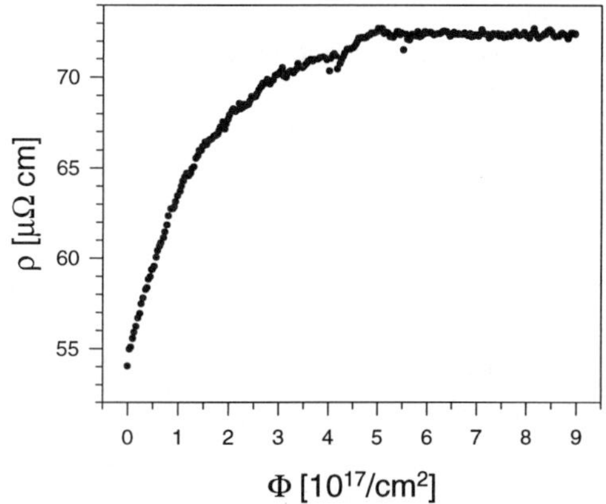
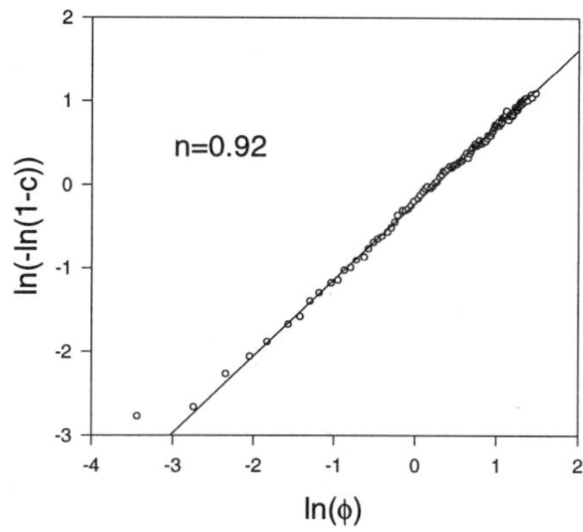

Fig. 3: Plot of resistivity during 100keV carbon ion implantation at -40°C versus fluence.

Fig. 4: Avrami plot versus $\ln(\phi)$ corresponding to data from Fig.3. Results from linear regressions are indicated in the picture

where the resistivity of the high resistance phase is denoted ρ_2, and ρ_1 corresponds to the low resistance phase. Traditionally phase transformation under isothermal conditions has been described by the Johnson-Mehl-Avrami equation:

$$c(t) = 1 - \exp[-(kt)^n] \qquad (2)$$

where the transformed volume fraction c is given as a function of time t. k is an effective rate constant and n the so-called Avrami exponent which takes values from 1 to 4 depending on characteristics of transformation. The transfer to the problem of ion implantation is realised by replacing the time t by the implanted fluence ϕ. Plotting $\ln(-[\ln(1-c)])$ versus logarithm of fluence allows to check whether kinetics behave like described in this formal theory. If a linear behaviour is found it is possible to determine the Avrami exponent from the slope of the curve. As can be seen from Fig. 4 experimental data fits to a straight line. The solid line shows the result of a linear regression. A value of n=0.92 is extracted from the slope of this line. Additional experiments were done after implantation during subsequent annealing. Here resistivity is plotted as function of temperature (Fig. 5). In order to distinguish between reversible and non-reversible processes during annealing several heating/cooling cycles were performed. First heating with 10 K/min was carried out up to 150°C. This temperature was held constant before cooling at equal rate. Second followed a similar heat process up to 300°C and finally an heating process up to 450°C. The first cycle up to 150°C shows a linear increase during heating and linear decrease during cooling (not presented in Fig. 5). First changes in resistivity occur at higher temperatures. The second cycle leads to a slight decrease of resistivity. The value at room temperature decreases from 90 $\mu\Omega$cm to 88 $\mu\Omega$cm. An even stronger effect is seen within the last cycle. Above 400°C a decrease of resistivity is observed. After cooling down to room temperature 76 $\mu\Omega$cm is reached. This resistivity is 13 $\mu\Omega$cm higher than that before implantation.

Discussion

Implantation of carbon ions at fluences larger than $3 \cdot 10^{17}$ C^+/cm^2 into pure titanium at temperatures as low as -40°C led to the formation of TiC. Phase formation was proved by XRD and TEM analysis. This phase formation after implantation at low temperatures can be understood both due to the simple structure of TiC and due to the large heat of formation for TiC (-180 kJ/mol). The carbon concentration distribution is Gaussian-like up to fluences of $1.2 \cdot 10^{18}$ C^+/cm^2. This phenomenon can be explained since calculations showed that effects due to sputtering and swelling directly compensate up to doses of $1.2 \cdot 10^{18}$ C^+/cm^2 [13]. Concentration profile measurements after implantation and annealing at higher temperatures showed that effects of carbon diffusion can nearly be neglected up to 600 °C [13]. Thus the film thickness stayed constant and no correction has to be done for the resistivity analysis.

In situ resistivity measurements during ion implantation always led to a typical curve. These could be well described in terms of transformation kinetics for heterogeneous phase transformation (see Fig. 4). Nearly the

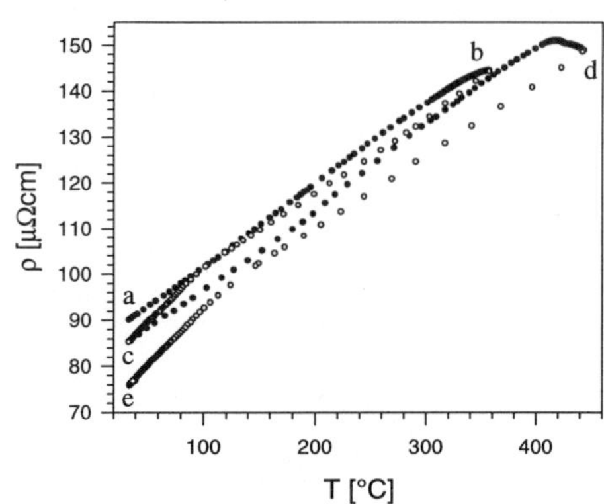

Fig. 5:   Change of resistivity during subsequent annealing. First cycle (a-b-c) heating up to 300°C and second (c-d-e) up to 450°C.

whole range covers a straight line in plot of ln(-[ln(1-c)]) vs. ln(ϕ) according to the equation (2) of Johnson-Mehl-Avrami [15]. An Avrami exponent of n=0.92 was determined. In this case the situation of n=1 seems to be valid. Karpe [9] showed that inhomogeneous damage profiles, like in the case of ion implantation, lead to a slight decrease of theoretically expected values for n. Results have to be interpreted in the way that intra-cascade mechanisms dominate kinetics of observed phase transformation. Resistivity measurements during several anneal cycles led to the following conclusions. Up to 150 °C and also up to 300 °C no strong changes are visible. In this temperature regime no further volume changes occur and no radiation induced defects vanish. At heat up to 400 °C distinct changes in resistivity are observed. This could be caused by global rearrangement and phase growth of TiC. Another possibility is the annealing of radiation defects. One hint for this explanation are the parallelity of the lines a-b, b-c and d-e in Fig. 5. This must be interpreted in terms that the temperature coefficient seems not to be influenced by all annealing cycles. Therefore only the residual resistance ρ_0 was decreased which can be caused by thermally stimulated annihilation of defects.

Conclusion

Carbon ion implantation into pure titanium led to formation of titanium carbide. The concentration distribution of carbon was found to be Gaussian-like up to fluences of $1.2 \cdot 10^{18}$ C^+/cm^2. In-situ resistivity measurements during ion bombardment were performed. The same kinetic was found which was predicted for heterogeneous phase transformations. From the results it can be concluded that direct intra-cascade mechanisms play a major role at transformation processes during carbon ion implantation into titanium.

References:

[1]   I.L. Singer, Vacuum **34**, 853 (1984)

[2]   D.M. Follstaedt, Nucl. Instr. Meth. B **10/11**, 549 (1985)

[3]   T. Weber, L. de Wit, F.W. Saris, A. Königer, B. Rauschenbach, G.K. Wolf, S. Krauss, Mater. Sci. Eng. A **199**, 205 (1995)

[4]   K. Baba, R. Hatada, Surf. Coat. Technol. **66**, 368 (1994)

[5]   R.G. Vardiman, Defect and Diffusion Forum **57/58**, 135 (1988)

[6]   T. Fujihana, Y. Okabe, M. Iwaki, Surf. Coat. Technol. **66**, 419 (1994)

[7]   B.X. Liu, J. Wang, X.Y. Cheng, Phys. Stat. Sol. A **128**, K71 (1991)

[8]   M. Guemmaz, A. Mosser, L. Boudoukha, J.J. Grob, D. Raiser, J.C. Sens, Nucl. Instr. Meth. B **111**, 263 (1996)

[9]   N. Karpe, Nucl. Instr. Meth. B **95**, 485 (1995)

[10]   J.P. Rivière, J. Delafond, C. Jaouen, A. Bellara, J.F. Dinhut, Appl. Phys. A **33**, 77 (1984)

[11]   W.Miehle, A.Plewnia and P.Ziemann, Nucl. Instr. Meth. B **80/81**, 424 (1993)

[12]   S.Coffa, F.Priolo and A.Battaglia, Phys. Rev. Lett. **70**, 3756 (1993)

[13]   A. Wenzel, A. Königer, C. Hammerl, B. Rauschenbach, submitted to Nucl. Instr. Meth.

[14]   Landolt-Börnstein, *Numerical Data and Functional Relationship in Sciene and Technology*, Springer, Berlin (1993)

[15]   J.W. Christian, *Theory of Transformations in Metals and Alloys*, Pergamon, Oxford (1975)

[16]   R. Landauer, in: *Electrical Transport and Optical Properties of Inhomogeneous Media*, eds. J.C. Garland and D.B. Tanner, AIP Conf. Proc. **40**, 2 (1978)

[17]   R. Landauer, J. Appl. Phys. **23**, 779 (1952)

Materials Science Forum Vols. 248-249 (1997) pp. 209-212
© *1997 Trans Tech Publications, Switzerland*

Depth Distributions and Wear Behaviour of Ion Implanted Ti6Al4V

H. Schmidt[1], H. Baumann[2] and D.M. Rück[3]

[1] Technical University of Darmstadt, Department of Physical Metallurgy,
Petersenstr. 23, D-64287 Darmstadt, Germany

[2] University of Frankfurt, Institute for Nuclear Physics,
August-Euler-Str. 6, D-60486 Frankfurt, Germany

[3] GSI Center for Heavy Ion Research, Planckstr. 1, D-64291 Darmstadt, Germany

Keywords: Ion Implantation, Titanium, Wear, Backscattering Spectrometry, Joint Prostheses

Abstract

The effect of ion implantation of carbon, nitrogen, oxygen and noble metals (Pd, Ir, Pt, Au) with different energy-dose combinations on the tribological behaviour of the alloy Ti6Al4V was investigated. Depth distributions were determined using Rutherford and high energy backscattering spectrometry. Wear tests were performed against UHMWPE. Implantation of noble metals leads to reduction of wear provided the enrichment of implanted atoms starts at the alloy surface. After nitrogen or carbon implantation the most pronounced wear reduction occurs when the enrichment is high directly below the oxide film on the surface, followed by a continuous decrease of the concentration of implanted atoms with increasing depth.

1. Introduction

The titanium alloy Ti6Al4V is widely used for hip and knee joint prostheses. When the alloy is combined with ultrahigh molecular weight polyethylene (UHMWPE) in a sliding couple, wear can lead to black staining of the surrounding tissue and failure of the joint [1]. The tribological behaviour can be improved when the surface of the alloy is modified by ion implantation of nitrogen, as demonstrated by several groups [2-5]. However, only few energy-dose combinations were investigated in these studies, and it is not yet clear which implantation parameters lead to an optimized tribological behaviour.

In the present study, various biocompatible elements (C, N, O, Pd, Ir, Pt, Au) are implanted into Ti6Al4V under systematically varied conditions. The aim is to correlate the wear behaviour with the depth distributions and the implantation parameters.

2. Experimental Procedure

Flat Ti6Al4V annuli (inner diameter 14 mm, outer diameter 20 mm) and discs (diameter 38 mm) were polished to an average surface roughness less than 0.03 μm. During ion implantation the specimens were mounted on watercooled target holders, and the residual gas pressure was kept below $3 \cdot 10^{-6}$ mbar to avoid sample heating and surface contamination.

The discs were implanted with nitrogen, carbon, carbon monoxide, palladium, iridium, platinum and gold, as described in an earlier paper [6]. C^+-, N_2^+- and CO^+-implantation was performed at energies E of approximately 80 keV per atom and with integral doses of $6 \cdot 10^{17}$ cm^{-2}. The noble metals were implanted at E=200 keV and with doses of 0.05 and $0.1 \cdot 10^{17}$ cm^{-2}.

The implantation parameters used for the annuli are summarized in Table 1. In addition to single implantations, nitrogen and carbon were also implanted at two and three different energies (double and triple implantations), in order to achieve a broad distribution of implanted atoms. During nitrogen triple implantation energies and doses as used by Sioshansi [2] were applied.

Rutherford backscattering (RBS) and high energy backscattering (HEBS) spectrometry were carried out using a scattering angle of 171° and helium ions for RBS (specimen tilt angle δ=0°; incident projectile energy E_0=2 MeV) as well as for HEBS (δ=70°; E_0=7.6 MeV). Depth distributions were calculated with the RUMP program [7]. To simulate the actual depth profiles, the implanted layers were divided into a finite number of sublayers, each with uniform composition. To evaluate the HEBS spectra non-Rutherford elastic scattering cross sections [8-10] were used for C, N and O.

Tribological tests were performed with the alloy discs against UHMWPE pins (constant sliding velocity 5 cm/s) as well as with the annuli against UHMWPE discs (oscillatory movement, maximum sliding velocity 4.2 cm/s), using contact pressures of approximately 10 MPa and physiological Ringer solution as the lubricant. Each Ti6Al4V annulus was tested with several discs successively. The UHMWPE wear volume was determined using surface profilometry.

TABLE 1. Ion implantation parameters of the Ti6Al4V annuli.

Ion	Energy/ atom E (keV)	Atomic dose D (10^{17}cm^{-2})	Specimen designation
Pt^+	200	0.05	Pt 0.05 single
Pt^+	200	0.1	Pt 0.1 single
CO^+	C: 68.6 O: 91.4	C: 3 O: 3	CO single
N_2^+	80	6	N single
N_2^+ C^+	80 80	3 3	N+C single
N^+ N_2^+	130 60	3 5	N double
C^+ C^+	120 60	3 4	C double
N^+ N_2^+ N_2^+	160 80 35	7 3 2.5	N triple
C^+ C^+ C^+	200 100 50	4.2 1.6 2.2	C triple
Non-implanted			Non-impl.

3. Depth Distributions

After noble metal implantation with doses of $1\cdot10^{16}$ cm^{-2} (E=200 keV) the thickness of the implanted layers is approximately 100 nm, and the maximum concentrations approach 5 at.% (Fig. 1). The enrichment of gold starts at the titanium surface and that of palladium at a depth of 15 nm. Gold, platinum and iridium implanted with the same energy and dose show similar depth distributions owing to the fact that these elements have approximately the same atomic mass.

The HEBS spectrum of carbon monoxide implanted Ti6Al4V is shown in Fig. 2. Owing to the large scattering cross sections the yields of carbon and oxygen are strongly enhanced with respect to the substrate. The formation of an oxide film in air after implantation leads to a splitting of the oxygen peak. The thickness of the implanted layer is approximately 350 nm. The maximum concentration of implanted atoms is at a depth of 170 nm. Nitrogen or carbon implanted at E=80 keV and with the same integral dose as used for carbon monoxide implantation ($6\cdot10^{17}$ cm^{-2}) show similar depth distributions (Fig. 2). After double and triple implantations the implanted layers are thicker (400-650 nm), and the maximum concentrations of implanted atoms are closer to the surface.

4. Wear Behaviour

As discussed in detail in an earlier paper [6], non-implanted alloy discs show severe wear with scratching of the titanium surface and abrasion of black particles when tested in the pin-on-disc device. After ion implantation with the appropriate parameters a strong reduction of friction and metallic wear as well as an inhibition of scratching are observed. Wear is reduced after implantation of C, N, CO, Ir, Pt or Au while the Pd implanted discs show scratching. Thus noble implantation reduces wear when the enrichment of implanted starts at the surface. This condition is not fullfilled after 200 keV Pd implantation (Fig. 1).

The wear behaviour in the annulus-on-disc device is illustrated in Fig. 3. As described to more detail in a forthcoming paper, UHMWPE wear is high (Fig. 4) when non-implanted Ti6Al4V annuli are tested. A polyethylene film adhering to the alloy surface is formed. When the film is removed narrow scratches are observed (Fig. 5). The number of scratches and the UHMWPE wear are reduced after implantation of Pt, C or N (Fig. 3). Carbon or nitrogen implanted annuli show a strong wear reduction when a large fraction of the dose is implanted at energies below 70 keV, i.e. when the enrichment of implanted atoms is high directly below the oxide film. After the single implantations the wear rates are higher than after the multiple implantations. The most pronounced reduction of scratches and wear is achieved by double implantation, i.e. when a continuous decrease of the concentration of implanted atoms with increasing depth occurs.

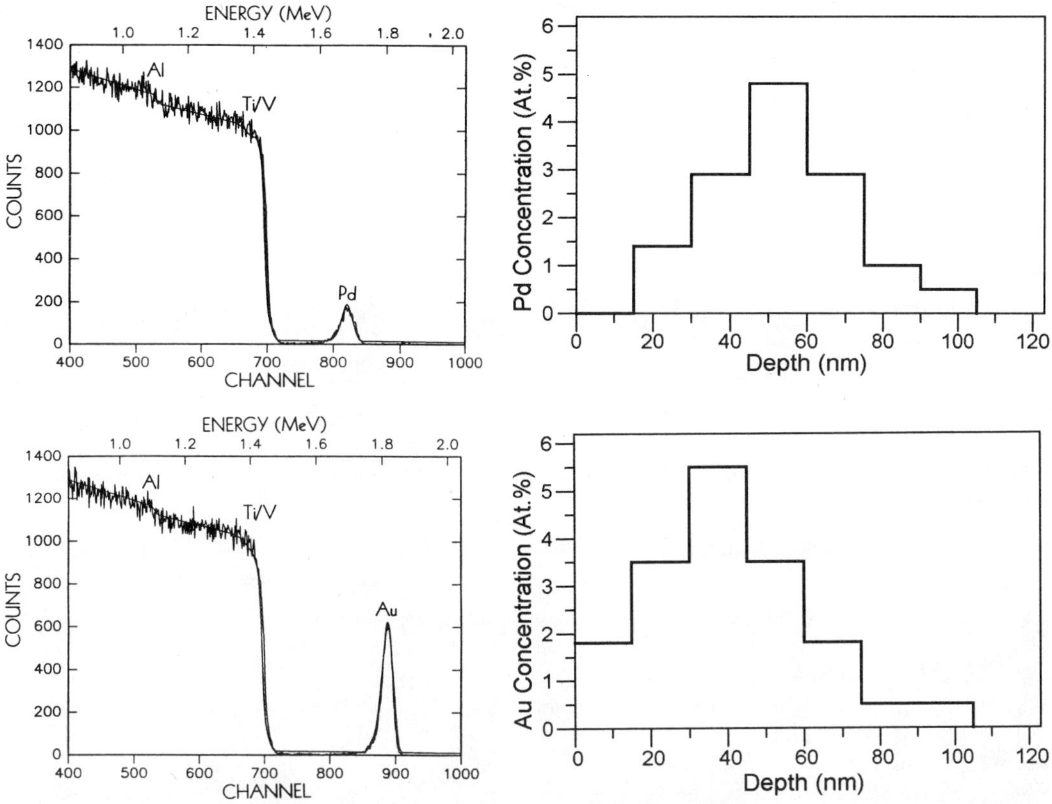

FIGURE 1. RBS spectra of Pd implanted (upper left) and Au implanted (lower left) Ti6Al4V and RUMP calculations (smooth curves) as well as depth distributions of palladium (upper right) and gold (lower right).

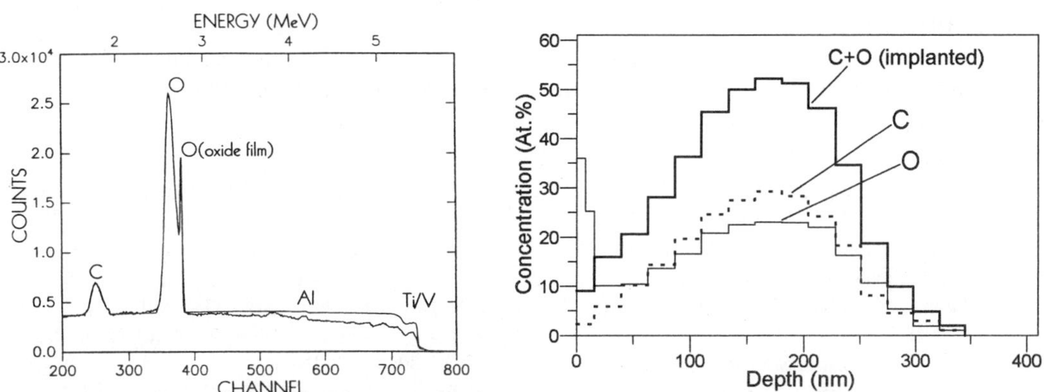

FIGURE 2. HEBS spectrum (left) of CO implanted Ti6Al4V (CO^+; integral dose $6 \cdot 10^{17}$ cm^{-2}) and RUMP calculation (smooth curve) as well as depth distributions (right) of carbon, oxygen and of implanted atoms which was estimated at depths of 0 to 15 nm.

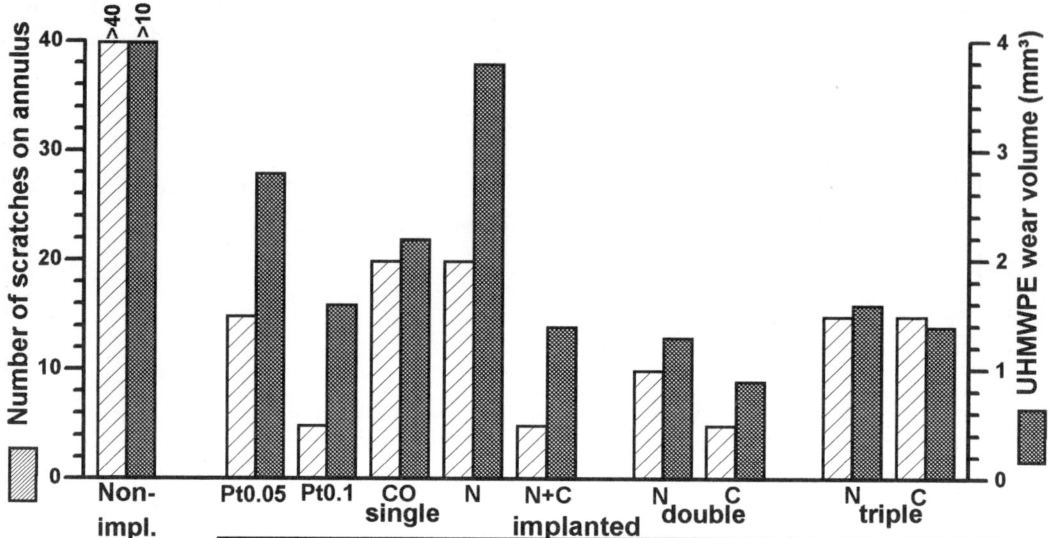

FIGURE 3. Wear behaviour of the Ti6Al4V annuli. The implantation parameters corresponding to the specimen designations are listed in Table 1.

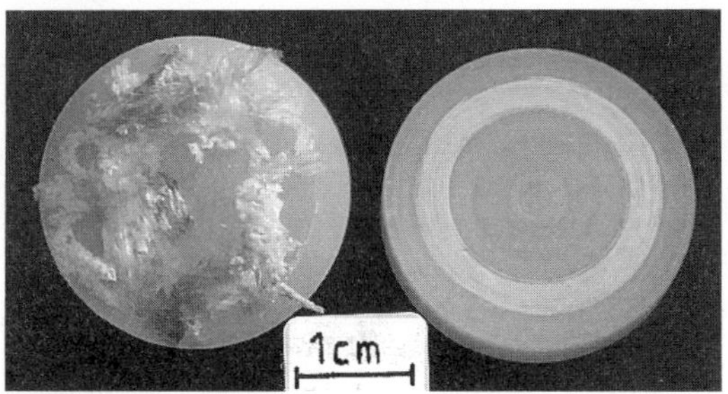

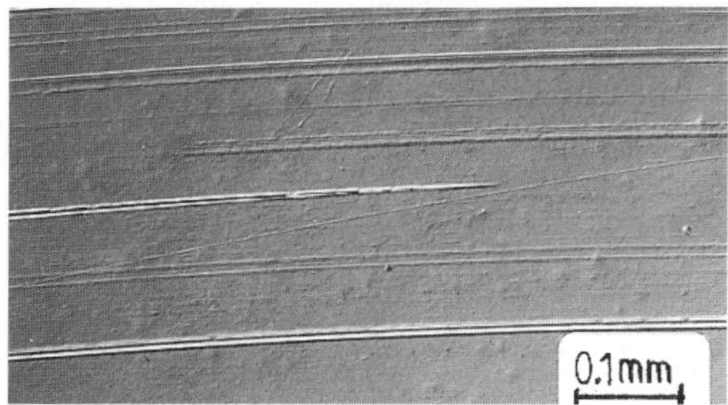

FIGURE 4. UHMWPE discs after the wear tests: high polyethylene wear with metallic black particles (left); low polyethylene wear (right).

FIGURE 5. Sliding surface of the CO implanted Ti6Al4V annulus after removal of the polyethylene film with a polishing paste.

Acknowledgements

The authors thank Dr. U. Fink (Aesculap AG, Tuttlingen) for providing the specimens. This project is financially supported by GSI (Contract No. DAEXM) and by Deutsche Forschungsgemeinschaft (Contract No. Ex 8/3-1).

References

[1] S. Nasser , P. Campbell, D. Kilgus, N. Kossovsky, H. Amstutz, Clin. Orthop. **261**, 171 (1990)

[2] P. Sioshansi, Mater. Sci. Eng. **90**, 373 (1987)

[3] J. Rieu, A. Pichat, L.-M. Rabbe, A. Rambert, C. Chabrol, M. Robelet, Mater. Sci. Technol. **8**, 589 (1992)

[4] F. Alonso, A. Arizaga, S. Quainton, J.J. Ugarte, J.L. Vivente, J.I. Onate, Surf. Coat. Technol. **74-75**, 986 (1995)

[5] R. Martinella, S. Giovanardi, G. Palombarini, Wear **133**, 267 (1989)

[6] H. Schmidt, H.E. Exner, D.M. Rück, N. Angert, U. Fink, in R. Kossowsky and N. Kossovsky (eds.), *Advances in Materials Science and Implant Orthopedic Surgery*, Kluwer, Dordrecht, 207 (1995)

[7] L.R. Doolittle, Nucl. Instr. Meth. **B9**, 344 (1985)

[8] Y. Feng, Z. Zhou, Y. Zhou, G. Zhao, Nucl. Instr. Meth. **B86**, 225 (1994)

[9] Y. Feng, Z. Zhou, G. Zhao, F. Yang, Nucl. Instr. Meth. **B94**, 11 (1994)

[10] F.J.D. Almeida, J.A. Davies, T.E. Jackman, Nucl. Instr. Meth. **B82**, 393 (1993)

Materials Science Forum Vols. 248-249 (1997) pp. 213-216
© *1997 Trans Tech Publications, Switzerland*

Defect Formation in Irradiated Nanostructured Materials

M. Rose, A.G. Balogh and H. Hahn

Technical University Darmstadt, Department of Materials Science, Darmstadt, Germany

Keywords: Nanostructured Materials, Electron Irradiation, Defect Density, Defect Cluster, Grain Size, Transmission Electron Microscopy

Abstract

Defect evolution during electron irradiation of nanostructured Pd and ZrO_2 using a high voltage transmission electron microscope is reviewed. Nanocrystalline powders, synthesized by the inert gas condensation technique, were compacted and subsequently sintered. Samples with grain sizes from 10 to 300 nm, were irradiated in situ with an 1 MeV electron beam. A correlation between grain size and defect density was observed. Small palladium grains (20 - 50 nm) show a reduced defect density compared with the larger grains (> 80 nm). At higher temperatures (400 °C) the defect density decreases by one order of magnitude. Nanocrystalline ZrO_2 samples with an average grain size of 10 - 50 nm show no defects over the entire temperature range from room temperature to 400 °C.

Introduction

First results of irradiation effects in nanophase materials were published recently in the literature [1,2]. In situ radiation damage experiments using a high voltage transmission electron microscope (HVTEM) have the advantage that the evolution of defects can be followed dynamically in time. This paper presents results of in situ electron irradiation experiments of n-Pd and $n\text{-}ZrO_2$ with the aim to point out the enhanced radiation resistance of these materials. MeV electron irradiation creates Frenkel pairs (interstitials and vacancies) in a high concentration. At elevated temperatures, when point defects become mobile, radiation enhanced diffusion and formation of clusters and dislocation loops can deteriorate the material properties.

Nanocrystalline materials consist of small grains with diameters in the range from 5 to 100 nm. The volume fraction of the interfacial regions (grain boundaries) can reach up to 40 % if the grain size is below 10 nm. Grain boundaries act as sinks for point defects. Because of the large volume fraction, point defects, such as vacancies and interstitials, can diffuse to the sinks and annihilate there. Consequently, a substantial reduction of radiation damage is expected in comparison to conventional solids.

Experiments

Nanocrystalline Pd and ZrO_2 powders were prepared by the inert gas condensation technique (IGC) [3]. Pd and ZrO were evaporated by resistivity heating from crucibles under a He gas pressure in the range of 5 - 10 mbar. Due to the supersaturation of the evaporating species, clusters form in the vicinity of the source and are transported by the convective gas flow to a liquid nitrogen cooled rotating cylinder. The zirconia nano-powders were subsequently oxidized to obtain the stoichiometric composition ZrO_2. The $n\text{-}ZrO_2$ powders (grain size < 10 nm), compacted at room temperature (1GPa) were pressureless sintered in air at temperatures below 1000 °C with sintering times ranging from 2 to 68 hours, to obtain bulk samples with different initial grain sizes (10 to 300 nm) and densities (80 % - 95 %). The n-Pd powders were compacted in situ without exposure to atmosphere at room temperature under high pressure (1 GPa), resulting in a density of 80 % and grain sizes in the range from 10 to 80 nm. Electron microscopy samples were prepared by mechanical grinding to a thickness of 30 μm and subsequent ion milling with Ar ions at room temperature (preparing conditions: 3-5 kV, 1-2 mA, angle 10°). The irradiation experiments were performed using a HVTEM with 1 MeV electron beam and a flux in the range from 0.8 - $5*10^{21}$ e cm^{-2} s^{-1} at the Max-Planck-Institute, Stuttgart.

Results and Discussion

Fig. 1 shows a TEM micrograph of a coarse grained ZrO_2 sample (grain size about 270 nm) in the early phase of   the e-irradiation at 400 °C. The contrast inside the grain shows no indication for defect clusters. Increasing the irradiation dose, strain contrasts from small defect clusters appear as "black dots" with a diameter of about 2 nm. Fig. 2 shows this sample after an irradiation time of 15 min., which corresponds to a total damage dose of 35 dpa (displacement per atom). The arrows indicate some of the defects. The defect density increased from room temperature to 200 °C by one order of magnitude. In the temperature range from 200 to 400 °C no further changes were observed (Fig.7). In contrast to the coarse grained ZrO_2 sample, no defects were found over the whole temperature range from room temperature up to 400 °C in the n-ZrO_2 sample with grain sizes from 10 to 50 nm. In the nanocrystalline sample the sink concentration (volume fraction of grain boundaries) is sufficiently high to cause the annihilation of all point defects created during electron irradiation. The micrograph of Fig. 3 shows the n-ZrO_2 sample, irradiated at 400 °C with a total damage dose of 115 dpa.

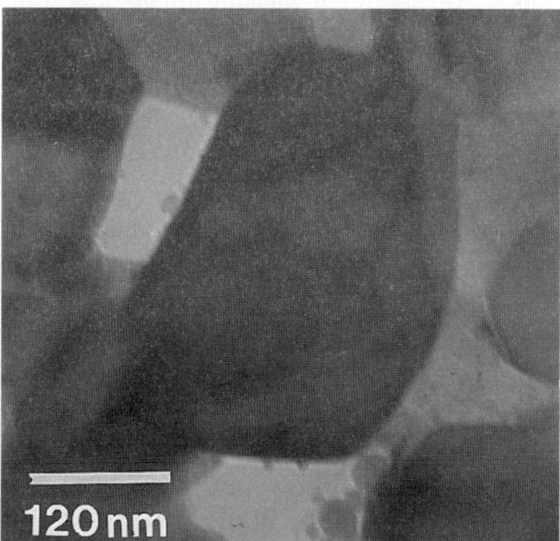

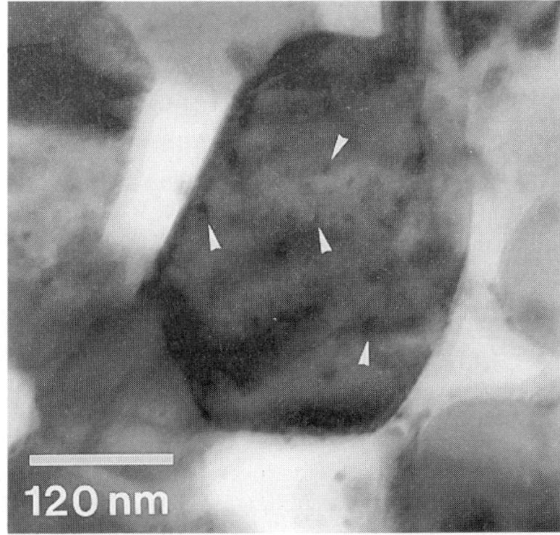

Fig 1: TEM micrograph of coarse grained zirconia at the beginning of the irradiation

Fig 2: Formation of defect clusters in coarse grained ZrO_2 at 400 °C (15 min / 35 dpa).

The measured defect concentration in palladium shows a strong temperature dependence for grains with a grain size in the range from 20 to 50 nm. The defect density measured in n-Pd samples irradiated at 400°C (Fig. 7) is reduced by a factor of 7 compared to a sample irradiated at room temperature. The smallest grains (< 20 nm) show no defects at all. At higher temperature (400 °C) all grains with grain sizes smaller than 30 nm were defect free.

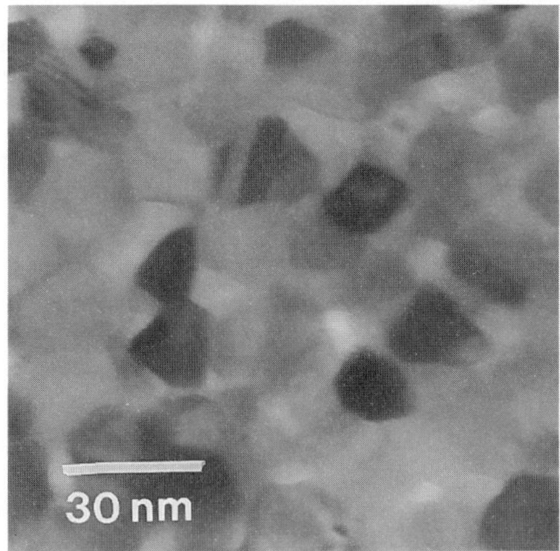

Fig 3: Nanocrystalline zirconia , irradiated
15 min at 400 °C (115 dpa).

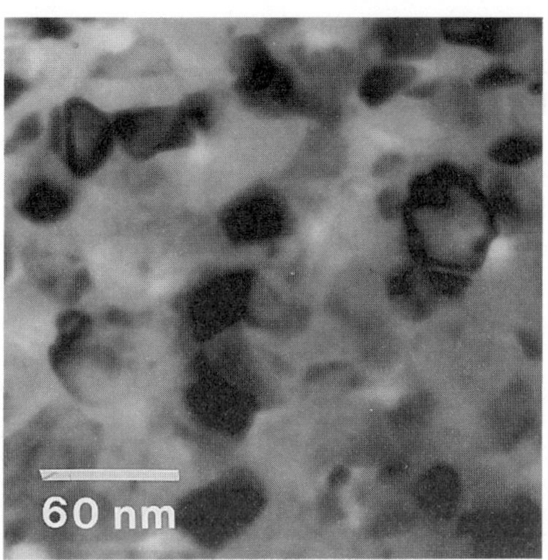

Fig 4: Nanocrystalline palladium irradiated
30 min at 400 °C (145 dpa).

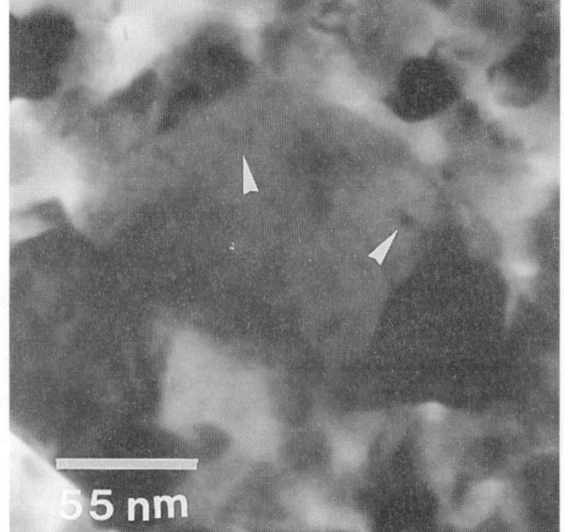

Fig 5: Defect clusters in palladium, irradiated
2 min. at 300 °C (10 dpa)

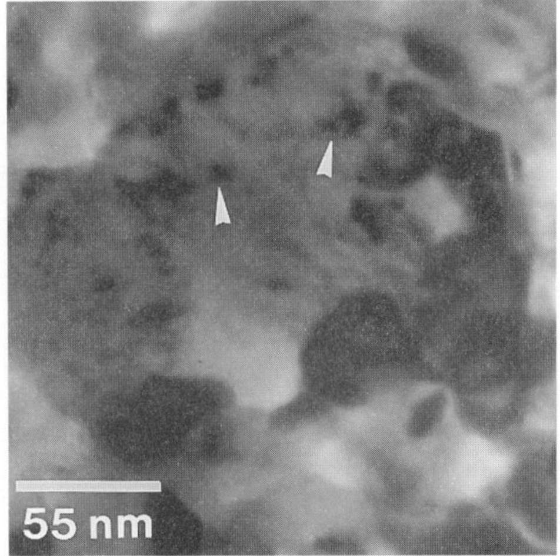

Fig 6: Defect formation in palladium after
25 min. at 300 °C (120 dpa).

Fig. 4 shows a TEM micrograph of a nanocrystalline Pd sample irradiated for 30 min. (145 dpa) at 400 °C. Defect clusters were only found in a few grains with sizes larger than 30 nm. Fig. 7 shows the drastical reduction ot the defect density after irradiaton at 300 °C by more than one order of magnitude comparing the small grains (< 50 nm) with the large ones (> 100 nm).
In the large Pd grains (> 100 nm) only a weak temperature dependence was observed (Fig. 7). In the very early stage of the e-irradiation, defect clusters occur and a steady state defect density is reached. Fig. 5 shows small defect clusters (~2nm) after 2 min irradiation, corresponding to a damage dose of 10 dpa. As a function of irradiation time the diameter of the defect agglomerates increases and decreases continously, while the number of defects remains unchanged. The Pd grain with a size of 100 nm in Fig. 6 is imaged after an irradiation time of 21 min. (120 dpa). Inside the grain strain contrasts from defect clusters with different defect sizes are shown.

The effect of the strong temperature dependence for the small Pd grains, is due to the correlation between diffusion length and grain size. In the case of palladium the point defects (interstitials and vacancies) are mobile at room temperature [4]. If the diffusion length is comparable to the grain size, the point defects can reach the sinks (grain boundaries) before agglomeration to observable defect clusters occurs. A critical grain size for the agglomeration, which is correlated to the diffusion length can be estimated to 20 nm. At higher temperatures the diffusion length increases resulting in a reduced defect density and in a higher critical grain size of 30 nm.

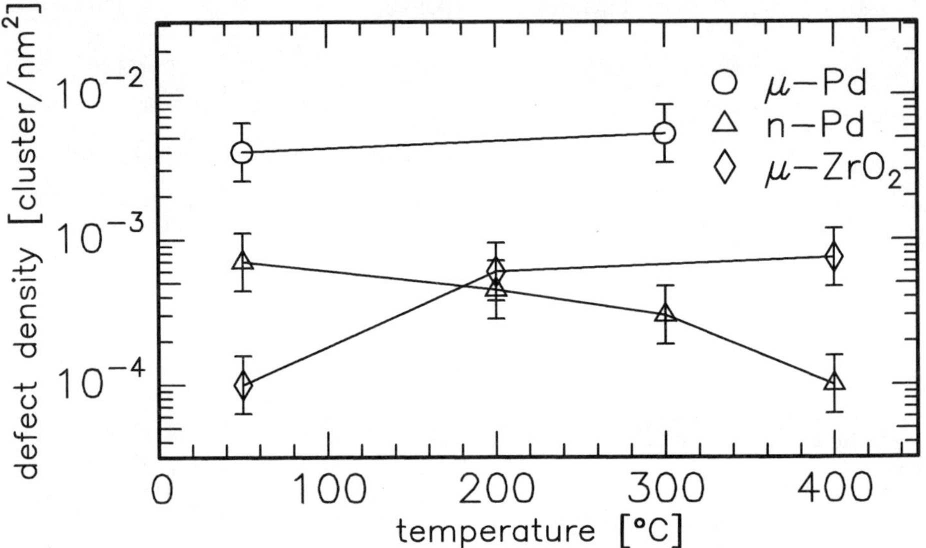

Fig 7: Defect density vs. temperature for zirconia and palladium samples with different grain sizes.

Conclusions

Nanocrystalline zirconia samples (grain size 10-50 nm) show improved stability against 1 MeV e-irradiation. In the coarse grained zirconia sample (grain size 100 - 300 nm) small defect clusters (cluster size $\approx$ 2 nm) could be observed during irradiation.

Defect formation in palladium samples with grain sizes in the range of 20 - 50 nm shows a strong temperature dependence. At higher temperatures the defect density decreases drastically. The critical grain size for defect free grains shifts from 20 nm at room temperature to 30 nm at 400 °C.

Acknowlegements

This research is supported by the Gesellschaft für Schwerionenforschung Darmstadt. The authors thank W. Sigle, MPI Stuttgart, for his support during the e-irradiation experiments and the opportunity to use the 1 MeV HVTEM.

References

[1] M. Rose, G. Gorzawski, G. Miehe, A. G. Balogh, H. Hahn, Nanostructured Materials
       Vol. 6 (1995) 731 - 734
[2] M. Rose, A. G. Balogh, H. Hahn, Nucl. Instr. Meth. (1997) in press
[3] H. Hahn, R. S. Averback, J. App. Phys. 67 (1990), 1113
[4] Ehrhart, Landolt-Börnstein, New Series III/25 (19) 254 - 256

Materials Science Forum Vols. 248-249 (1997) pp. 217-220
© *1997 Trans Tech Publications, Switzerland*

TiN Coatings Formed by Dual Beam IBAD Technique

B. Rajchel[1], J. Stanek[2], E. Wantuch[3], T. Burakowski[4],
J. Fedotova[1,5] and A. Sellmann[1]

[1] Institute of Nuclear Physics, ul. Radzikowskiego 152, PL-31-342 Cracow, Poland

[2] Institute of Physics, Jagiellonian University, ul. Reymonta 4, PL-30-059 Cracow, Poland

[3] University of Technology, al. Jana Pawła II, Cracow, Poland

[4] Institute of Precision Machines, ul. Duchnicka 3, Warsaw, Poland

[5] on leave from: Institute of Physics, Belarussian State University

__Keywords:__ TiN, RBS, Conversion Electron Mössbauer Spectroscopy, Ion Beam Assisted Deposition

Abstract: Thin TiN layers were created by the Dual Beam IBAD technique on pure iron enriched at the surface in ^{57}Fe. The depth distribution of composition of the obtained layers was determined by the detection of He$^+$ ions or protons backscattered at angle 170^0 after the ^{14}N(p,p)^{14}N resonance reaction. The Conversion Electron Mössbauer Spectroscopy (CEMS) was used for determination of the local electronic states of Fe atoms at interference.

Introduction

The physical, chemical and mechanical parameters of technological materials can be changed by formation of thin coating layer. Hard coatings, for example of TiN and DLC type, are frequently created in technological processes based on the PVD and CVD methods. However, the adhesion of the obtained in that way covers is sometimes not sufficient being influenced by the poorly controlled physical conditions during their formation. In contrary, the IBAD technique groups, still not widespread, offer the possibility of formation of the thin coatings with well determined and reproducible parameters. Such specimens are also ideal objects for basic research.

Our group is involved in formation and studies of TiN$_x$ layers on different substrates such as such the PROTASUL steel for endoprothesis or high speed tool steels. Recently we initiated the study of DLC type coatings on the Si <111> single crystal. However, it appeared certain, that the success in search for the optimal technology depends on the better understanding of chemistry and thermodynamics of the processes taking place at the interfaces. For this reason we have chosen rather simple case, i.e. TiN on metallic iron. In this paper the preliminary results of the RBS and ^{57}Fe Mössbauer spectroscopy study of that system are reported.

Samples and analysis

Coating layers of TiN on metallic iron were created by the dual beam IBAD technique. As IBAD setup, the 75 kV ion implanter of the Institute of Nuclear Physic, Cracow, working in the dual beam geometry [1] has been used, see Fig.1. Before formation of the coating layer the substrate has been implanted by ^{57}Fe$^+$ ions (dose: 10^{16}/cm^2, energy: 30 keV) to enrich the interface in the Mössbauer probe. Deposited titanium ions were sputtered from auxiliary Ti target by 35 keV Ar$^+$ ions and simultaneously bombarded by nitrogen ions beam with energy of 25 keV. The flux of N$^+$ ions was varied and a set of TiN$_x$ ($0.3 \leq x \leq 1$) samples with different nitrogen concentration was prepared. The thickness of the TiN covers was about 80 nm. The final depth distributions of Ti, N and Fe in sample were determined by detection of backscattered at angle 170^0 He$^+$ ions or protons. For profiling of N concentration the ^{14}N(p,p)^{14}N resonance reaction was used. The proton initial energy was varied from 1700 to 2000 keV. By adjusting the flux of N^+ ions we succeeded in the formation of the cover with 1:1 Ti to N atomic concentration. This sample, of the blue colour, (in contrary to the "yellow" Ti reach samples) was used for the further Mössbauer study.

The typical RBS spectrum recorded for determination of the depth distribution of elements in the TiN/Fe layer is shown in Fig. 3. It is worth to note, that the nitrogen penetrated the iron layer up to 300 nm in spite of its low energy (25 keV).

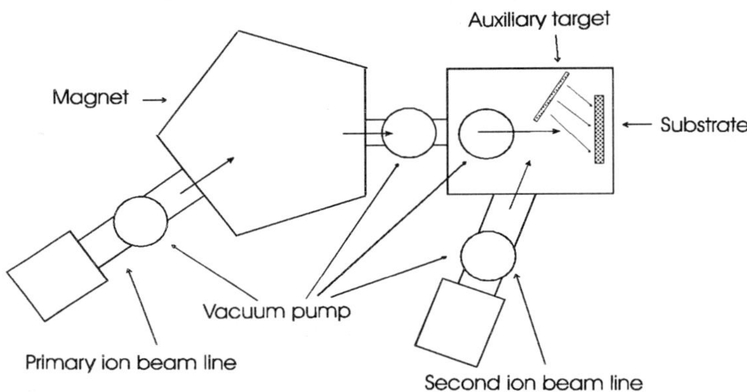

Fig. 1. The 75 kV INP ion implanter in dual beam geometry used as the IBAD setup for formation of coatings.

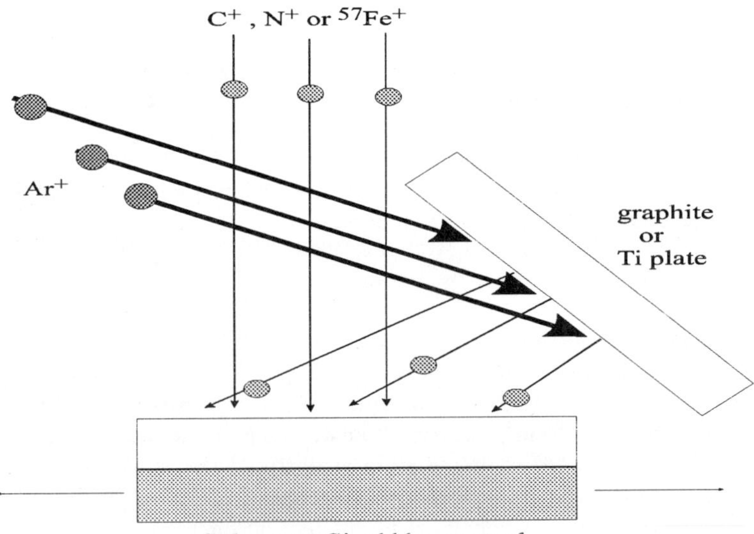

Fig.2 The main idea of formation of the TiN/Fe and DLC coatings by dual beam IBAD technique.

For the analysis of the local states of iron the Conversion Electron Mössbauer spectroscopy (CEMS) has been applied. This technique, based on the detection of the remitted after resonant nuclear absorption of gamma radiation low energy (7.3 keV) electrons is restricted to the surface layers (100 nm) of the investigated samples, being especially suitable for the studies of surface treatments. However, in the case of coatings, the protecting layer must be thin enough to make the interface accessible for the study. Moreover, the interface has normally the thickness of few atomic layers and its contribution to total CEMS spectrum does not exceed a few percent. To avoid that problem the surface of the substrate has been enriched in the ^{57}Fe isotope by low energy implantation of 10^{16} ions/cm^2. The CEMS spectrum of such substrate was identical to that of the initial material that proves that the enrichment of the iron surface in ^{57}Fe isotope by this rather low dose implantation does not influence the crystal structure. The experiments have been performed at room temperature with the use of the proportional gas flow counter with the sample placed inside of the counter. The details of the CEMS technique are given, for example, in [2].

From the parameters of ^{57}Fe Mössbauer spectra (isomer shift δ, quadrupole splitting Δ, and nuclear Zeeman splitting H) the chemical bonding of the probe atoms, here ^{57}Fe, could be identified.

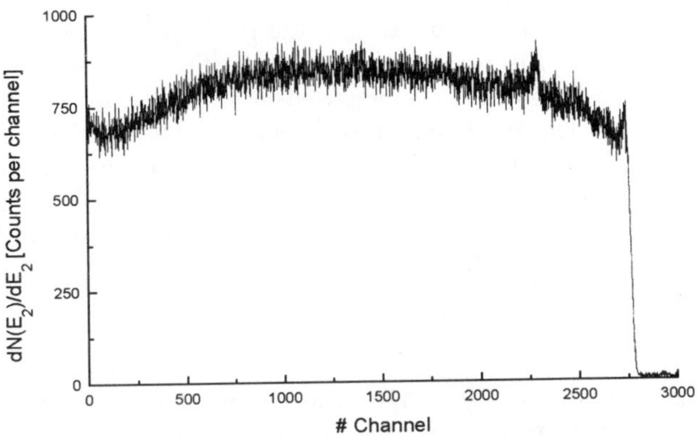

Fig. 3. The RBS spectrum of protons backscattered at angle 170⁰, recorded for determination of the depth distribution of elements in the TiN/Fe layer. The $^{14}N(p,p)^{14}N$ resonance is observed

The CEMS spectra of the freshly prepared TiN/Fe samples shown, in general, the typical six line pattern of metallic iron, see Fig. 3a, but with significantly (by 40%) broaden line width, Γ. This points out some distribution of the hyperfine magnetic fields acting on ^{57}Fe nuclei. In addition, some tiny paramagnetic fractions with the very broad resonance lines near zero velocity were detected. After annealing at 500⁰C the metallic iron lines became narrow again and a clear quadrupole doublet was observed, see Fig. 3b. The hyperfine parameters of this component are as follows: δ= 0.95(1) mm/s, Δ= 1.02(1) mm/s, Γ= 0.69 mm/s.

Discussion

The observed broadening of the ^{57}Fe Mössbauer lines of metallic iron shown that the coated surface layer of the substrate of at least 100 nm thickness is modified during coating process. This is in agreement with RBS data that showed that nitrogen penetrates up to 300nm, i.e. about 10 times deeper than the implantation range. This state of the substrate is unstable, and recovers at 500 ⁰C.

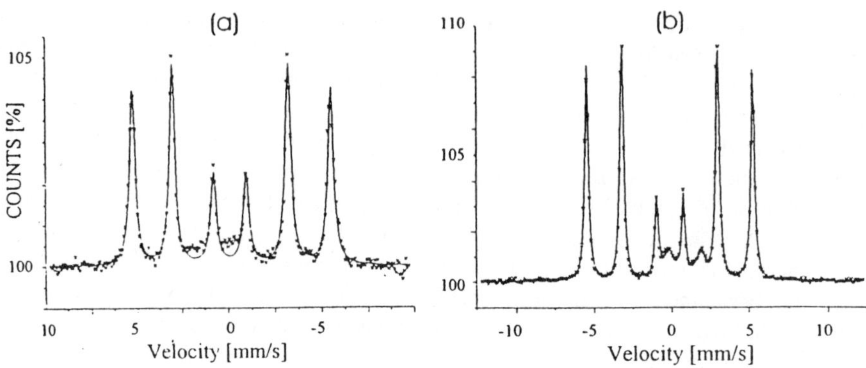

Fig.3 The CEMS spectra of the TiN(^{57}Fe) sample. (a) before annealing, (b) after annealing at 500^{0}C for 2 hours in vacuum.

The interface effects are visible in CEMS spectrum by the paramagnetic fractions. The local states of Fe impurities in TiN have been studied in the past [3] for TiN samples implanted with ^{57}Fe ions. It turned out that one third of implanted ions replaces N atoms being surrounded by Ti neighbours. The rest of iron implants replaces Ti forming Ti-N complexes. These two types of iron surroundings in TiN are easily distinguished in Mössbauer spectra by negative and positive δ, respectively. The phase formed after annealing at the TiN-Fe interface is characterised by highly positive isomer shift and should be related to Fe-N compounds. The exact identification of this fraction is not possible yet but significant broadening of the corresponding Mössbauer lines, which comes from some distribution of the hyperfine parameters δ, and Δ, point out nonstoichiometry of this compounds and/or its defect-type structure. This supposition should be verified by careful determination of the variation of the composition depth distribution using RBS experiments.

Conclusion

The composition of the TiN$_x$ layers formed by IBAD technique may be exactly regulated by the flux of N$^+$ ions. The N$^+$ ions penetrate deep layers of Fe substrate forming metastable structures that recover at 500 °C. At the TiN/Fe interface layer some Fe-N compounds are formed at that temperature. The further annealing experiments are necessary to observe the formation of the stable interface region in relation to the atomic migration.

Acknowledgements

This work was partially supported by State Committee of Scientific Research - Grant no. 2P03B 11009 and Grant no. 7 T08C 031 08. Authors wish to thank Dr Lucyna Jaworska from the Institute of Metal Cutting for her advice.

Literature

[1] B. Rajchel, M. Drwięga, E. Lipińska, M. Wierba, NIM **89** (1994) 342

[2] B.D. Sawicka, J.A. Sawicki, in. "Topics in Current Physics", Mössbauer Spectroscopy II, Ed. U.Gonser, Springer - Verlag, Berlin, Heidelberg, New York, 1981

[3] K. Burda, Ch. Dauwe, P. De Bakker, E. De Grave, P. Fornal, J. Stanek, J. App. Phys. **78**, (1995) 5067.

Materials Science Forum Vols. 248-249 (1997) pp. 221-224
© *1997 Trans Tech Publications, Switzerland*

Creation of the DLC Layer by the Dual Beam IBAD Technique

B. Rajchel[1], J. Stanek[2], E. Wantuch[3], T. Burakowski[4], J. Fedotova[1,5],
M. Drwięga[1], E. Lipińska[1] and A. Sellmann[1]

[1] Institute of Nuclear Physics, ul. Radzikowskiego 152, PL-31-342 Cracow, Poland

[2] Institute of Physics, Jagiellonian University, ul. Reymonta 4, 30-059 Cracow, Poland

[3] University of Technology, al. Jana Pawła II, Cracow, Poland

[4] Institute of Precision Machines, ul. Duchnicka 3, Warsaw, Poland

[5] on leave from Institute of Physics, Belarussian State University

Keywords: DLC, RBS, Conversion Electron Mössbauer Spectroscopy, Ion Beam Assisted Deposition

Thin carbon films created as coating layers exhibit interesting physical, chemical and mechanical properties. Especially, the DLC (Diamond Like Coatings) can be used as extremely hard coatings for any kind of mechanical tools or, for example, in endoprothesis. Final parameters of the carbon coatings are determined by method used for their creation. In particular, the dual beam IBAD technique (Ion Beam Assisted Deposition) can be used for building of carbon based coatings. In this method, any impurity, like Fe, Ni, H, etc., may be dynamically added into created layer. Here, the transfer of ion's mass, charge and energy, is fully controlled which makes this technique very perspective for the new material technology.

In presented work the structure of carbon layer and their modifications by Ni and Fe impurities were studied. Complex carbon coatings were created. As substrates pure Si <111> single crystal and also Si<111> crystals covered by thin ($\approx$10nm) Ni layer were used. Deposited carbon ions were sputtered from graphite target by 35 keV Ar^+ ions and simultaneously bombarded by carbon ions beam with energy of 25 keV. Finally, the 30 keV ^{57}Fe ions (instead carbon ions) were added into the IBAD coatings. The depth distribution of composition of the obtained layers was determined by the detection of He^+ ions or protons backscattered at angle 170^0. The proton initial energy was varied from 1700 keV to 2000 keV. For determination of the carbon distribution the resonance reaction of $^{12}C(p,p)^{12}C$ was used. The Conversion Electron Mössbauer Spectroscopy was used for determination of the local electronic states of Fe atoms.

Introduction

Thin carbon films created as coating layers exhibit interesting physical, chemical and mechanical properties [1] and can be used as extremely hard coatings for any kind of mechanical tools or, for example, in endoprothesis. The DLC (Diamond Like Coatings) type coatings on technological materials are frequently created by low energy ionic methods as the PVD and CVD[2]. These methods are nowadays commercially used. However, the adhesion of the obtained in that way covers is sometimes not sufficient being influenced by the poorly controlled physical conditions during their formation. Final parameters of the carbon coatings are determined by method used for their creation. In particular, the dual beam IBAD technique (Ion Beam Assisted Deposition) can be used for building of carbon based coatings with very good adhesion. In this method, any impurity, like Fe, Ni, H, etc., may be dynamically added into created layer. Here, the transfer of ion's mass, charge and energy, is fully controlled which makes this technique very perspective for the new material technology. The IBAD technique groups, in contrary for the PVD or for the CVD methods, offer the possibility of formation of the thin coatings with well determined and reproducible parameters.

In presented work the structure of thin carbon layer and their modifications by Ni and Fe impurities were studied. Both mentioned elements could change the growth microcrystals in the created coatings layer. Complex carbon coatings were created by dual beam IBAD methods. As substrates pure Si <111> single crystal and also Si<111> crystals covered by thin Ni layer were used. The 30 keV ^{57}Fe ions were added into the IBAD coatings. The final depth distribution of all elements (C, Ni, Fe) of the obtained coating layers was determined by the detection of He$^+$ ions or protons backscattered at angle 170^0. The local electronic states of Fe atoms were studied by the Conversion Electron Mössbauer Spectroscopy. After initial investigation of the local electronic states and the depth distribution of elements all samples were annealed in vacuum by 2 hours in 500^{0}C. After annealing also the depth distribution of elements and the local electronic states of Fe atoms were determined.

Samples and analysis

The complex carbon coatings doped by Ni and by ^{57}Fe ions were created by dual beam IBAD and/or by IBSD (Ion Beam Sputter Deposition) techniques. The 75 kV INP ion implanter working in the dual beam geometry[3] has been used for creation complex coatings on the Si <111> single crystal. Before creation of the DLC - IBAD layer the Si single crystal has been covered by thin Ni layer. The Ni layer was formed by the IBSD technique. For sputtering of the Ni atoms the Ar$^+$ ion beam with energy of 25 keV was used. The impact angle of the Ar$^+$ ions beam was 67.5^0. Next the carbon coating was build. Deposited carbon ions were sputtered from graphite target by 35 keV Ar$^+$ ions and simultaneously bombarded by carbon ions beam with energy of 25 keV. Finally the 30 keV ^{57}Fe ions (instead carbon ions) were added into the IBAD coatings.

The final depth distribution of all elements (C, Ni, Fe) of the obtained coating layers was determined by the detection of He$^+$ ions or protons backscattered at angle 170^0. For better determination of the carbon distribution the resonance reaction of ^{12}C(p,p)^{12}C was used and by this reason the proton initial energy was varied from 1700 keV to 2000 keV. Initial energy of He$^+$ ions was 1750 keV.

The local electronic states of Fe atoms were studied by the Conversion Electron Mössbauer Spectroscopy (CEMS). From the parameters of ^{57}Fe Mössbauer spectra (isomer shift δ, quadrupole splitting Δ, and nuclear Zeeman splitting H) which are related to the s-electron density at nuclei, electric field gradient and hyperfine magnetic field, respectively, the chemical bonding of the probe atoms, here ^{57}Fe, may be concluded and, the in most cases the formed iron compounds could be identified. The applied technique, CEMS, based on the detection of the remitted after resonant nuclear absorption of gamma radiation low energy (7.3 keV) electrons is restricted to the surface layers (100 nm) of the studied samples, being especially suitable for the studies of surface treatments. The experiments have been performed at room temperature with the use of the proportional gas-flow counter with the samples placed inside of the counter. the details of the CEMS technique is given, for example, in [4].

The CEMS spectrum of the as-prepared sample is shown in fig. 2a. The amplitude of resonant scattering effect is only 0.6%. The spectrum consist of two party overlapping quadrupole doublets with the following parameters (Γ is the line width):

Doublet I: δ= 0.22 mm/s, Δ=1.40 mm/s, Γ= 0.28 mm/s, fraction 33%
Doublet II: δ= 0.20 mm/s, Δ=0.74 mm/s, Γ= 0.44 mm/s, fraction 66%

Next the IBAD sample was heated by 2 hours in 500^{0}C in vacuum.
After annealing the local electronic states of Fe atoms were studied by the Conversion Electron Mössbauer Spectroscopy. After annealing the effect of resonant scattering increased by one

order of magnitude. The spectrum consist of magnetically split pattern and quadrupole doublet. The hyperfine parameters of these components are as follows:

Sextet: H= 26.8 T, δ= 0.0 mm/s, Δ=0.0 mm/s, Γ= 46 mm/s, fraction 70%
Doublet: δ= 0.19 mm/s, Δ=0.41 mm/s, Γ= 0.36 mm/s, fraction 30%

Corresponding the CMS spectrum of the heated sample is shown in fig. 2b.

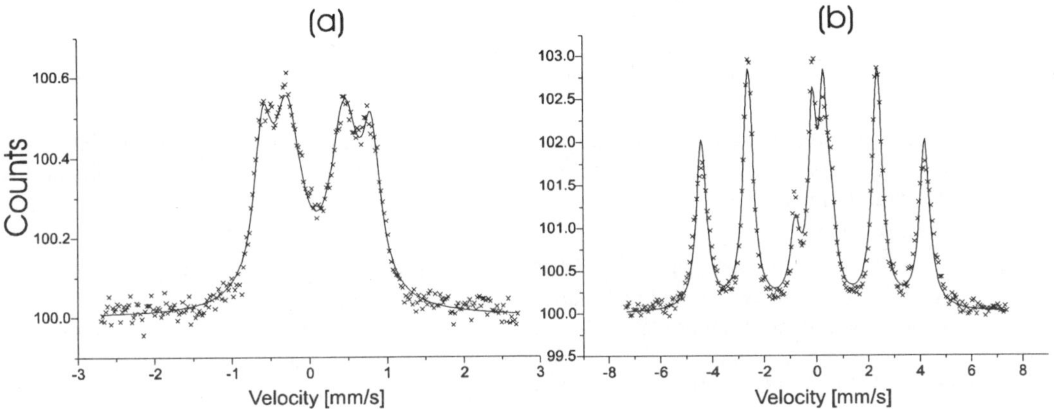

Fig.1 The CEMS spectrum before annealing (a) and after annealing (b).

After annealing the new depth distribution of C, Ni and Fe atoms were investigated by detection of He$^+$ ions (initial energy of 1750 keV) backscattered at angle 170^0. In fig. 3 two RBS spectra are shown: one before annealing and second after annealing in 500^{0}C.

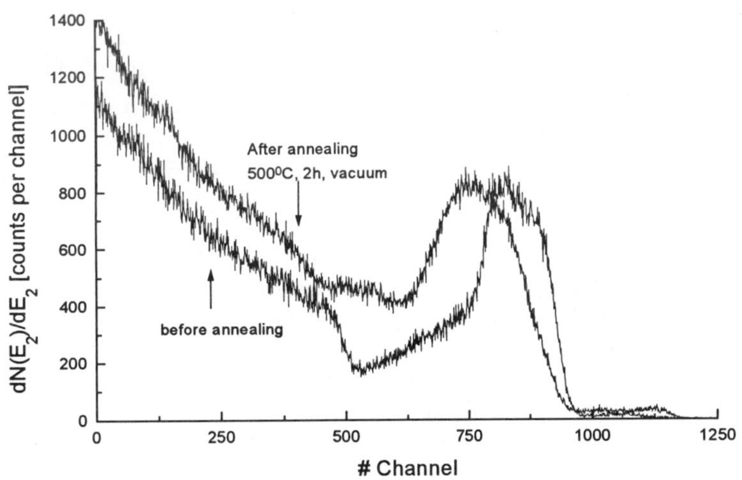

Fig.3 Comparison of the RBS spectrum before annealing and the RBS spectrum after annealing in 500^{0}C by 2 hours.

Discussion

The very weak resonance scattering effects points out that the iron atoms are extremely weakly bounded to the host (carbon). Iron occupies two types of states with similar δ but distinct Δ. The states with bigger Δ (1.4 mm/s) can be described as iron located in amorphous carbon. The hyperfine parameters are in that case quite similar to these obtained for ^{57}Fe implanted into graphite or diamond [5]. However, the narrow line width ($\Gamma = 0.28$ mm/s) is an evidence that the local atomic arrangement around Fe atoms is rather uniform in a contrary to the Fe implantation case where the line width is reached $\Gamma = 1$ mm/s. The states of Fe in carbon with smaller Δ (0.74 mm/s) are not known and these fraction could not be identified. It should be noted, that although in that case the local symmetry of the Fe surroundings is less perturbed, this state may not be assigned to Fe in diamond lattice which shows a single Mössbauer line, e.i. vanishing quadrupole interaction and big negative $\delta = -1$ mm/s[6].

At 500^0C the weakly bounded Fe atoms may easily migrate and finally the more stable phases are formed. The dramatic increase of resonant effect after annealing of the sample illustrates convincingly. The dominating magnetic fraction may be easily identified as iron impurity in Ni. The value of H and δ are identical to these known for Fe diffused in Ni. A part of Fe remains in C layer. The unchanged δ value suggests that the chemical bonds are basically the same as in initial sample but the decrease of Δ points out some local rearrangement of the Fe surroundings.

This supposition corresponding with data obtained from the RBS/NRA experiments. We observed that the depth distribution of Ni atoms after annealing was dramatically changed and thin surface layer, initially doped only by ^{57}Fe ions, was enriched in Ni atoms.

Conclusion

Annealing of Ni/Fe/DLC systems can help us for creation of the stable phases in coating layers formed by IBAD technique. Investigation of the structure of the DLC layers doped by metals will be continued by using also another experimental methods (for example: the Raman Spectroscopy) and by computer simulation of processes of formation of the coating layers.

Acknowledgements

This work was partially supported by State Committee of Scientific Research - Grant no. 2P03B 11009 and Grant no. 7 T08C 031 08. Authors wish to thank dr Lucyna Jaworska from the Institute of Metal Cutting for her advice and help.

Literature

[1] G. Dearnaley, NIM. B 40/41 (1989) 571

[2] S. Okoti, R. Haubner, B. Lux, Diamond Relat. Mater. **1** (1992) 955

[3] B. Rajchel, M. Drwięga, E. Lipińska, M. Wierba, NIM **B 89** (1994) 342

[4] B.D. Sawicka, J.A. Sawicki, in. "Topics in Current Physics", Mössbauer Spectroscopy II, Ed. U.Gonser, Springer - Verlag, Berlin, Heidelberg, New York, 1981

[5] J.A. Sawicki, B.D. Sawicka, NIM **194**, (1982) 465

[6] B.D. Sawicka, J.A. Sawicki, H. de Waard, Phys. Lett. **85A**, (1981) 303

Materials Science Forum Vols. 248-249 (1997) pp. 225-228
© *1997 Trans Tech Publications, Switzerland*

Application of High Energy Ion Beams for Local Lifetime Control in Silicon

P. Hazdra and J. Vobecký

Department of Microelectronics, Czech Technical University in Prague,
Technická 2, CZ-166 27 Prague 6, Czech Republic

Keywords: Helium Irradiation, Deep Levels, Lifetime Control

Abstract   Low-dose energy-dispersed irradiation of silicon power diodes by $^4He^{2+}$ ions in the energy range of 4 - 15 MeV was used to create a deep damage region which locally changes the carrier lifetime. Properties and profiles of vacancy-related defects were studied by DLTS and C-V measurement. Substantial improvements of the diode dynamic parameters were achieved and the advantages of the novel technique, multiple energy-dispersed irradiation, are also shown.

INTRODUCTION

At present, the high energy ion beams are widely utilized for lattice modifications in various devices. One of the most emerging applications is the irradiation by means of fast light ions for local control of carrier lifetimes in silicon power devices. Whilst the proton irradiation in the energy range of several MeV became standard technological procedure [1], less attention was paid to the characterization and application of lattice damage produced by heavier projectiles - alpha particles [2, 3]. The aim of this paper is to present the results of detailed characterization of radiation damage produced in silicon by high energy helium ions and to show its beneficial influence on resulting device characteristics. On the top of it, advantages of both the energy-dispersed and multiple energy-dispersed irradiations are presented as a novel technique.

EXPERIMENTAL

The device under test was a conventional p^+pnn^+ power diode (100A/2.5kV) fabricated on <111> FZ NTD 110Ωcm n-type silicon. The device was irradiated by a beam of $^4He^{2+}$ ions from cyclotron. The primary 34MeV single energy beam was dispersed on 170μm thick set of Al and Fe foils thus giving the average energy of 15 ± 0.5 MeV at the output from vacuum into air. The energy distribution of particles entering the d.u.t.(see Fig.1), which was placed in air, was varied in the range 4 - 15 MeV by changing the distance between the sample and the outlet from vacuum. In order to reach both the optimal damage profile in respect of the resulting device parameters and the possibility of damage inspection by means of electrical methods the final energies 10, 12 and 14 MeV were chosen for irradiation [6,7]. To show the benefits of multiple energy-dispersed irradiation the devices were also irradiated by a mixture of two (11 and 12 MeV, dose ratio 1:1 - M2) and three (7, 8.5, and 10 MeV, dose ratio 3:2:1 - M3) energies. The integral irradiation dose range selected for each experiment varied within 1×10^9- 1.6×10^{10} cm^{-2}. After irradiation the properties of resulting damage were investigated by Deep-Level Transient Spectroscopy (DLTS) and capacitance-voltage (C-V) measurements and compared with simulation of primary damage provided by Monte-Carlo code TRIM [4]. The impact of irradiation on real devices is presented by means of the relevant static and dynamic electrical parameters that were measured prior to and after irradiation.The most relevant static parameter was the forward voltage drop V_F measured at I_F=100A. The dynamic parameters - the reverse recovery charge Q_{rr}, and the maximum reverse voltage V_{RM} - were derived from the reverse recovery process which was performed by use of the dc reverse voltage source of 100V connected in series with inductance assuring the current decrease dI/dt=100A/μs.

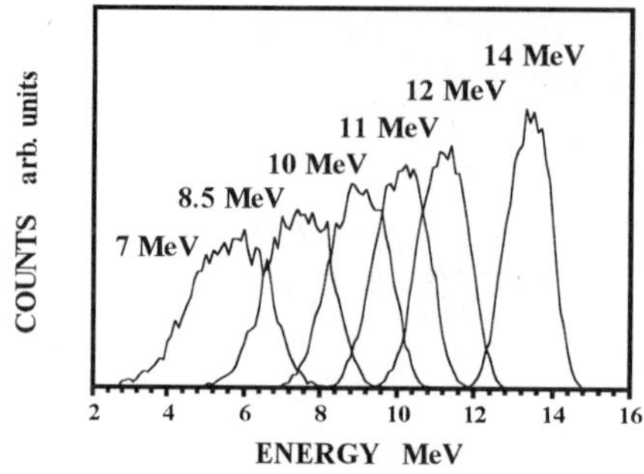

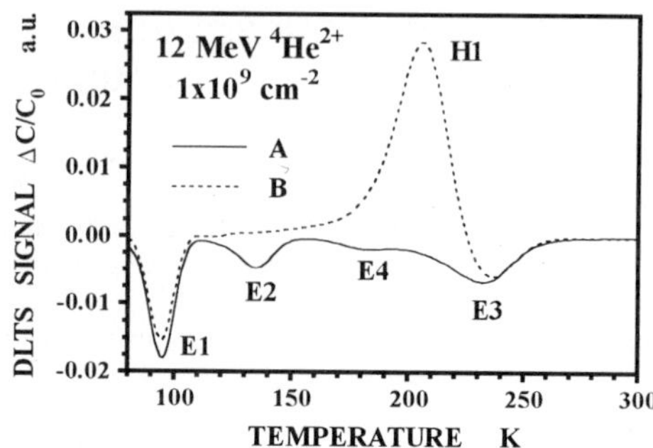

Fig. 1　The alpha particle energy distribution within the beam - irradiation with different final energies (simulation TRIM90).

Fig. 2　DLTS majority (A) and minority (B) carrier spectrum of 12 MeV ^{4}He^{2+} irradiated diode (rate window 260 s^{-1}).

RESULTS AND DISCUSSION

The majority and minority carrier DLTS spectra (Fig.2), typical for all ^{4}He^{2+} irradiated diodes, give the information about the defects produced. Five distinctive peaks labelled E1, E2, E3, E4, and H1 corresponding to the different deep defect levels can be resolved. Table 1 gives an overview over the band-gap positions of the levels, their capture cross sections, and annealing temperatures. The levels E1, E2, E3 and H1 were already observed in electron, gamma and ion irradiated silicon. They are connected with the pure radiation damage, since only these levels were found after silicon self-implantation [5]. The most pronounced electron trap E1 is traditionally assigned to the acceptor level of vacancy-oxygen pair VO$^{(0/-)}$. Levels E2 and E3 are attributed to the divacancy, to the double-acceptor V$_2$$^{(-/2-)}$ and the single-acceptor V$_2$$^{(0/-)}$ level, respectively. The energy level of E3 also coincides with the acceptor level of vacancy-phosphorus pair, but this defect was omitted, since we did not observe any decrease of E3 level signal after annealing at temperatures higher than 420 K, where VP pair dissociates. The profiles of VO$^{(0/-)}$, V$_2$$^{(-/2-)}$, and V$_2$$^{(0/-)}$, resulting from DLTS measurement on the diode irradiated by 12 MeV ^{4}He^{2+} 1x10^9 cm^{-2}, are shown in Fig. 3 and compared with the simulated distribution of primary vacancies. The distributions of all three levels follow fairly well the profile of primary vacancies giving thus a further evidence that all the levels are vacancy-related. Moreover, the coincidence between the maxima of the simulated and measured profiles implies a very good prediction capability of the TRIM code. We come to the same conclusion, if we compare the profile of the deep acceptors (VO centers) received from C-V measurement on the diode irradiated by 14 MeV ^{4}He^{2+} 4x10^9 cm^{-2} (Fig.4) with the corresponding distribution of primary vacancies. Figs.3 and 4 also indicate a slight broadening of all the measured secondary defect profiles. If we exclude the influence of limited resolution of both profiling methods, the diffusion of primary defects due to their concentration gradient appearing nearby the end of the ion range may be responsible for this effect, as reported in [8] for silicon irradiated by MeV protons.

Table 1　Deep levels in FZ *n*-type silicon irradiated by high energy ^{4}He^{2+} ions

Trap	Energy level	Capture cross section	Annealing temperature	Generation rate (defect/vacancy)	Identity
E1	E$_c$-0.163 eV	σ_n= 3x10^{-19} m^2	620 K	0.032	VO$^{(0/-)}$
E2	E$_c$-0.229 eV	σ_n= 8x10^{-20} m^2	570 K	0.0076	V$_2$$^{(-/2-)}$
E3	E$_c$-0.421 eV	σ_n= 1x10^{-19} m^2	570 K	0.0097	V$_2$$^{(0/-)}$
E4	E$_c$-0.338 eV	σ_n= 1x10^{-19} m^2	~570 K	-	?
H1	E$_v$+0.361 eV	σ_p= 3x10^{-19} m^2	620 K	-	COV$^{(0/+)}$

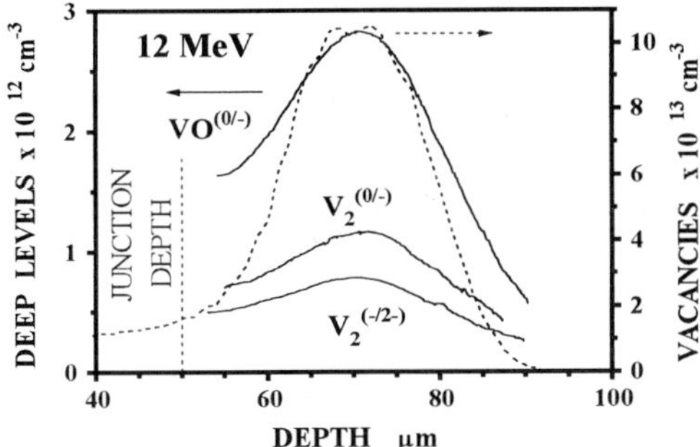

Fig. 3 Profiles of secondary defects (thick) resulting from DLTS measurement compared with simulation of primary vacancies distribution (dotted): 12 MeV ^{4}He^{2+} 1x10^9 cm^{-2} irradiation.

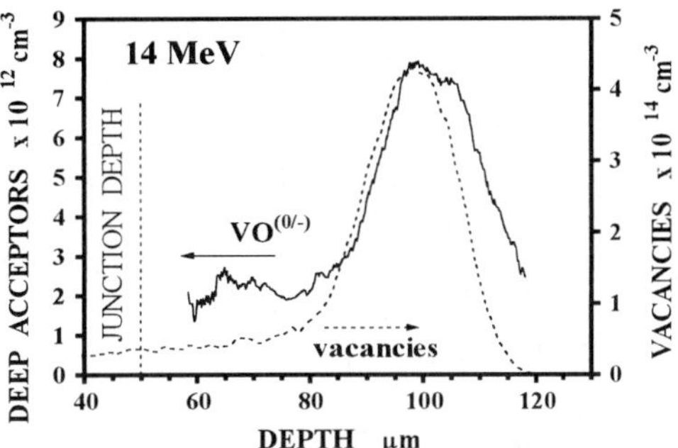

Fig. 4 Profile of deep acceptors (VO) from C-V measurement at 78K (thick) compared with simulation of primary vacancies distribution (dotted): 14 MeV ^{4}He^{2+} 4x10^9 cm^{-2} irradiation.

Both the VO pair and the divacancy show the linear production dependence on the ion dose within the interval under study. The generation rates shown in Tab.1 were calculated by comparing the integral defect concentration of the primary defects and the secondary ones received from simulated (vacancies) and measured (V$_2$,VO) profiles. The hole trap H1 was attributed to the donor level of the carbon-vacancy-oxygen (CVO) complex [5]. Our experimental methods could not provide the exact profiling of this level, but one may expect that the defect behaves as VO pair and V$_2$, since it is also vacancy-related. The weak level E4, which was previously observed in medium-energy He$^+$ irradiated silicon [3], has a different spatial distribution compared to the vacancy-related defects - the peak is shifted deeper towards the projected range of helium ions. It was considered that E4 is also connected with the pure damage [3], but further investigations have to prove this claim.

The section above showed that ^{4}He^{2+} irradiation produces in particular a pure vacancy-related damage thermally stable up to 570 K. The distribution of the relevant defects follows the profile of primary vacancies and their concentration depends linearly on the dose. This facilitates the design of irradiation and enables one to project more complex and advanced irradiation schemes, e.g. energy-dispersed and multiple energy/dose irradiation. In this paragraph we show how the device operation can be improved using a customer-specific defect profile attainable only by these techniques. Fig.5 shows the simulated spatial distribution of VO pairs (solid line) after irradiations described in the experimental section together with the doping profile (p+n) of the anode region. The profiles were created in the anode p-n junction area by irradiation from the anode side (on the left in Fig.5) that assures the optimal device electrical parameters [7]. It is clear that the multiple energy/dose irradiation (M2 and M3) enables one to bring the defect profiles with nearly arbitrary shape. Fig.6 shows the measured dependence between the forward voltage drop and the reverse recovery charge which represents the typical trade-off relation between static and dynamic parameters. In general, the best trade-off curves are those ones closely passing the lower left corner of the graph. In this respect the unirradiated device (marked as N.I.) always appears in the lower right corner. In other words any ion irradiation means a remarkable improvement. The remaining task is a choice of the best irradiation parameters. In this respect the Fig.6 clearly implies the irradiation M3 as superior. However, there are other parameters playing an important role. One of them is the maximal reverse voltage that appears on diode during the reverse recovery. If the maximal value is too high the diode may damage itself or undesirable oscillations may appear. Fig.7 shows that except for the 14 MeV irradiation all the irradiation treatments are beneficial. The M3 irradiation is very close to the best case, so it provides superior characteristics.

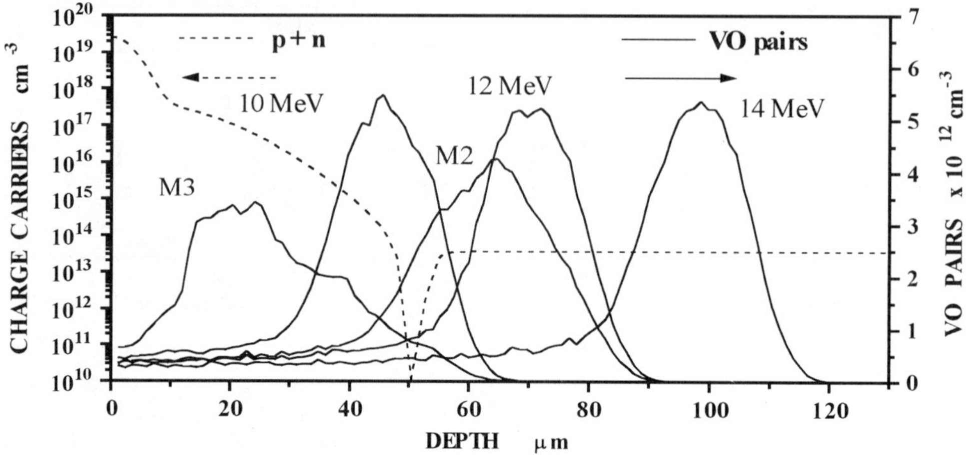

Fig.5 Simulated profiles of free carriers (doping) and VO defects after 2×10^9 cm^{-2} ^{4}He^{2+} irradiations with different final energies (10, 12, 14 MeV, M2: 11 and 12 MeV, M3: 7, 8.5, and 10 MeV).

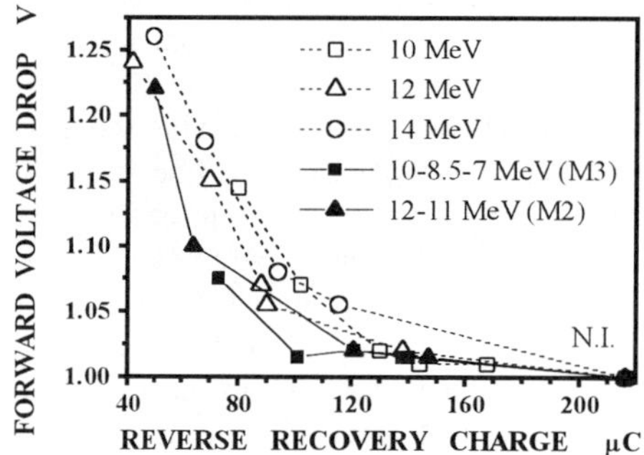

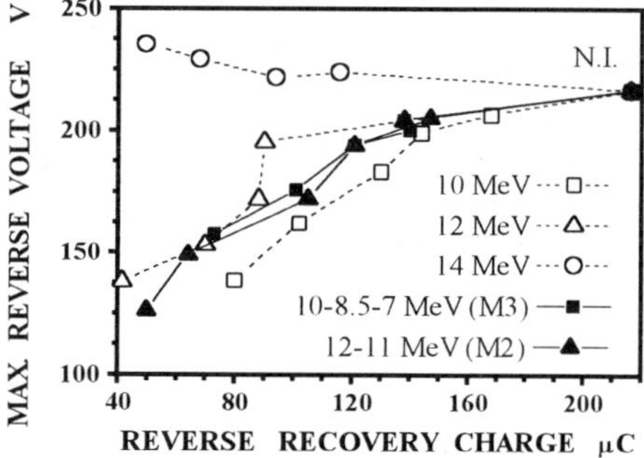

Fig. 6 Comparison of measured V_F vs. Q_{rr} trade-off curves for different energies and types of ^{4}He^{2+} irradiation.

Fig. 7 Comparison of measured V_{rrm} vs. Q_{rr} trade-off curves for different energies and types of ^{4}He^{2+} irradiation..

CONCLUSIONS

The application of ^{4}He^{2+} irradiation in the MeV range is advantageous for the local lifetime control in silicon devices. Both the damage and its distribution can easily be predicted what enables one to design and use more advanced irradiation techniques (e.g. multiple energy/dose irradiation) the benefits of which were demonstrated for optimization of the power diode characteristics.

REFERENCES

[1] D. C. Sawko, J. Bartko, IEEE Trans. Nucl. Sci. **NS-30**, 1756 (1983).

[2] W. Wondrak, A. Boos, Proc. of the ESSDERC'87, Bologna, 649 (1987).

[3] L. Palmetshofer, J. Reisinger, J. Appl. Phys., **72** , 2167 (1992).

[4] J.P. Biersack, L.G.Haggmark, Nucl.Instrum. Methods, **174**, 257(1980).

[5] P.Hazdra, V. Hašlar, M. Bartoš, Nucl.Instrum. Methods B **55**, 637 (1991).

[6] P. Hazdra, J. Vobecký, Solid-State Electron., **37**, 127 (1994).

[7] J. Vobecký, P. Hazdra, J. Homola, IEEE Trans. Electron Devices, **43**, (1996), to be published.

[8] N. Kestikalo, A. Hallén, Solid-State Electron., **37**, 55 (1994).

Materials Science Forum Vols. 248-249 (1997) pp. 229-232
© *1997 Trans Tech Publications, Switzerland*

Ion Implantation Induced Damage Accumulation Studied by Rutherford Backscattering Spectrometry and Spectroscopic Ellipsometry

T. Lohner[1], N.Q. Khánh[1], P. Petrik[1], M Fried[1], E. Kótai[2] and J. Gyulai[1]

[1] KFKI-Research Institute for Materials Science, P.O.B. 49, H-1525 Budapest, Hungary

[2] KFKI-Research Institute for Particle and Nucler Physics, P.O.B. 49, H-1525 Budapest, Hungary

Keywords: Ion Implantation, Damage Characterization, Rutherford Backscattering Spectrometry, Spectroscopic Ellipsometry

ABSTRACT

In this study, the damage created by ion implantation of N_2^+ ions into single crystalline silicon is characterised using Rutherford Backscattering Spectrometry (RBS) and Spectroscopic Ellipsometry (SE). Samples were implanted with ion energy of 400 keV to create surface-damaged layer and buried disorder. High-depth-resolution RBS combined with channeling has been used to examine the surface damage. For the analysis of ellipsometry data we applied the method of assuming appropriate optical model and fitting the model parameters (thickness of surface oxide and damaged silicon layer and the volume fraction of amorphous silicon component in the partially damaged layer). The thickness of the disordered surface layer was extracted from the high photon energy range of the ellipsometric spectra, the buried disorder was studied by evaluating the interference oscillations of the low photon energy range of the spectra. The optical model construction was independently checked by RBS experiments.

INTRODUCTION

In the past years spectroscopic ellipsometry was applied to materials science problems as an optical technique for nondestructive depth profiling and characterization of multilayer structures and interfaces with considerable success [1-3]. The measured optical response of a multilayer structure under investigation can only be related to actual material properties by a model calculation. The successful application of ellipsometry is not only determined by the quality of the measurements, but more importantly by the quality of the optical model. The object of this paper is to compare the damage profiles obtained by RBS and SE on ion implanted silicon samples.

EXPERIMENTAL

To investigate damage profiles 400 keV N_2^+ were implanted at room temperature in $<100>$ silicon. The implantation dose was varied from 2.5×10^{15} atoms/cm^2 to 4×10^{16} atoms/cm^2. RBS and channeling techniques with 1.5 MeV He^+ ions were used in the experiments. Two detectors were placed to detect ions scattered through 165° and 97° (i.e. with a glancing exit angle of 7° to the surface). In the latter geometry, the depth resolution was better than 5 nm. To evaluate the spectra we used the RBX code written by Kótai [4], which can also handle channeled spectra. The SE spectra were obtained at Fraunhofer Institut für Integrierte Schaltungen in Erlangen in the range of 270-850 nm with a rotating polarizer ES4G SOPRA ellipsometer.

RESULTS AND DISCUSSION

Figure 1 presents the high-depth resolution RBS spectra recorded with a detector placed at scattering angle of 97°. The thicknesses of the surface oxide and disordered layers were deduced from the oxygen peak and the surface damage peak; the thicknesses extracted in units of atoms/cm² were converted into nm assuming the surface disorder being totally amorphous.

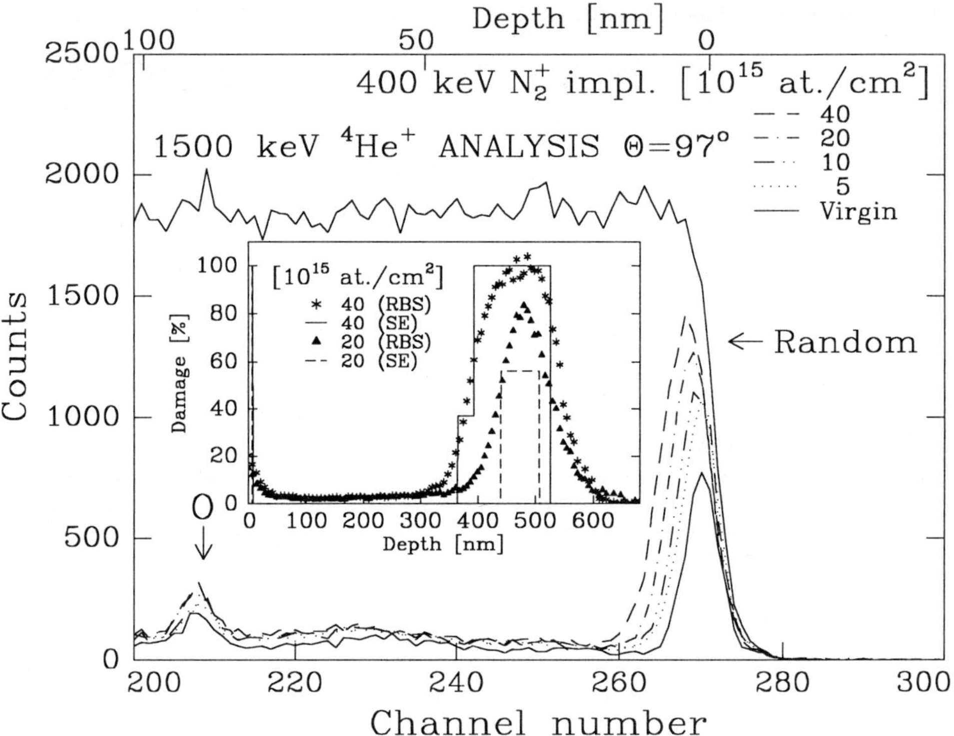

Figure 1. Random and <100>-aligned high-depth-resolution RBS spectra recorded with the detector placed at 97° scattering angle. The insert shows the comparison between the damage depth dependencies extracted from RBS measurements taken at 165° scattering angle, and spectroellipsometry, respectively. The spectra are shown for different implantation doses according to the lists within the figure.

For the evaluation of SE data concerning the surface damage, the 420 nm wavelength was chosen as an upper limit because in this case the optical penetration depth (even in crystalline Si) is not larger than 200 nm. The optical model consists of a native oxide layer, a thin amorphous silicon layer and a slightly damaged semiinfinite layer. For the analysis of SE data we used the conventional method of assuming an appropriate optical model and fitting the model parameters (layer thicknesses and volume fraction of the constituents in the slightly damaged semiinfinite layer) by linear regression. The details of the evaluation is given in [5]. It is important to note that we used the complex dielectric function of implanted amorphous silicon [6]. The slightly damaged semiinfinite layer was modeled as a mixture of crystalline silicon and fine-grain polycrystalline silicon [7], i.e. the complex refractive index of the slightly disordered layer was calculated by Bruggeman effective medium approximation using crystalline and fine-grain polycrystalline silicon as end-points. Measured ellipsometric spectra together with the results of multiparameter fitting are shown in Fig. 2. For comparison, a reference spectrum of virgin (unimplanted) silicon was also represented.

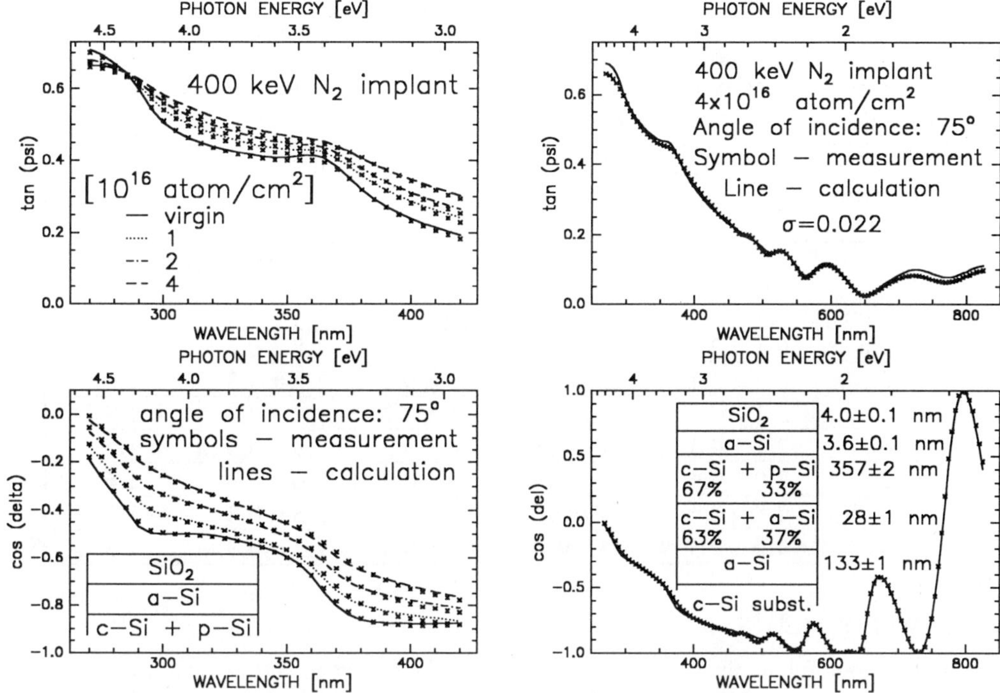

Figure 2. Measured and fitted SE spectra. The insert shows the optical model.

Figure 3. Result of SE fitting together with experimental data for the highest dose implant. The insert shows the optical model.

Table 1 summarizes the layer thickness values resulting from the evaluation of RBS and SE measurements. For RBS analysis we used the density of silicon (5×10^{22} atoms/cm^3) to calculate the thickness of the surface disordered layer. Evaluation of ellipsometric data yields thickness values for surface damage that are in good agreement with those obtained by RBS.

To investigate the buried disorder, the SE data in the wavelength range of 270 - 825 nm were evaluated. In silicon the optical penetration depth of light depends on its wavelength, the longer the wavelength, the deeper the penetration is, and so the measurements at longer wavelength contain more information on the buried layer. Fig. 3. displays the measured and the best fitted spectra for the highest dose irradiation together with the optical model. The buried disorder was described by two layers: a transition layer (mixture of c-Si and a-Si) and a completely amorphised layer. In the evaluation of SE data the main problem is the choice of the initial parameters. It is essential to define good starting values for the thickness parameters; otherwise the program can find a false minimum.

The insert in Fig. 1 displays the damage profiles deduced from RBS measurements taken at 165° scattering angle and from SE data.

The results demonstrate the applicability of spectroscopic ellipsometry together with a proper optical model construction for probing both surface damage and buried disorder. RBS, as an independent cross-checking method, basically supported the optical models.

Table 1. Implantation conditions together with the results of RBS/channeling analysis and SE fitting procedure. $D_{a\text{-}Si}$ is the thickness of the surface amorphous silicon, D_{oxide} is the thickness of native oxide, σ is the unbiased estimator for SE.

400 keV N$_2$ Implant	$D_{a\text{-}Si}$ [nm]		D_{SiO_2} [nm]		σ [10^{-2}]
Dose [10^{15} at./cm^2]	SE	RBS	SE	RBS	
Virgin	0.1 ± 0.2	0	2.2 ± 0.3	1.8 ± 0.2	1.3
2.5	0.5 ± 0.1	0.9 ± 0.2	2.3 ± 0.1	1.5 ± 0.3	0.8
5	2.1 ± 0.3	2.1 ± 0.3	2.1 ± 0.2	1.4 ± 0.3	1.3
10	2.1 ± 0.3	2.4 ± 0.3	2.4 ± 0.3	1.9 ± 0.3	1.5
15	3.0 ± 0.4	3.6 ± 0.3	2.9 ± 0.3	2.0 ± 0.3	1.6
20	2.8 ± 0.4	3.4 ± 0.3	3.2 ± 0.3	2.5 ± 0.3	1.7
40	4.5 ± 0.5	5.3 ± 0.4	3.4 ± 0.3	2.3 ± 0.3	1.5

ACKNOWLEDGMENTS

P. Petrik is grateful to the Soros Foundation for a scholarship during his research work. Partial support from OTKA grants (No. TO16506, No. TO16821 and No. TO17344) is greatly appreciated. For his assistance with implantation Ing. J. Waizinger is acknowledged. The authors wish to thank the team operating the accelerator for their help in experimental procedures.

REFERENCES

[1] P.J. McMarr, K. Vedam and J. Narayan, J. Appl. Phys. **59** (1986) 694.

[2] S. Lynch, M. Murtagh, G. M. Crean, P.V. Kelly, M. O'Connor and C. Jeynes, Thin Solid Films **233** 199 (1993)

[3] M. Fried, T. Lohner, J.M.M. de Nijs, A. van Silfhout, L. J. Hanekamp Z. Laczik, N.Q. Khánh and J. Gyulai, J. Appl. Phys.**66** 5052 (1989).

[4] E. Kótai, Nucl. Instr. Meth. B **85** 588 (1994)

[5] T. Lohner, E. Kótai, N. Q. Khánh, Z. Tóth, M. Fried, K. Vedam, N.V. Nguyen, L.J. Hanekamp and A. Van Silfhout, Nucl. Instr. and Meth. B**85** 335 (1994).

[6] M. Fried, T. Lohner, W. A. M Aarnink, L. J. Hanekamp and A. van Silfhout, J. of Appl. Physics **71** 5260 (1992)

[7] G.E. Jellison, Jr., M.F. Chisholm, and S.M. Gorbatkin, Appl. Phys. Lett. **62** 3348 (1993).

Materials Science Forum Vols. 248-249 (1997) pp. 233-236
© *1997 Trans Tech Publications, Switzerland*

Plasma Immersion Ion Implantation of Nitrogen into Porous Silicon Layers

A. Manuaba[1], I. Pintér[2], E. Szilágyi[1], G. Battistig[2], C. Ortega[3], A. Grosman[3] and G. Amsel[3]

[1] KFKI-Research Institute for Nuclear and Particle Physics, P.O.Box 49, H-1525 Budapest, Hungary

[2] KFKI-Research Institute for Materials Science, P.O.Box 49, H-1525 Budapest, Hungary

[3] Groupe de Physique des Solides, Universités Paris 7 et Paris 6, URA17 du CNRS, Tour 23, 2 place Jussieu, 75005 Paris, France

Keywords: Ion Beam Analysis, Porous Silicon, Plasma Immersion, Ion Implantation

Abstract

Nitrogen plasma immersion ion implantation (PIII) (with bias voltage, pressure and duration up to 1000 V, 1 mbar and 240 seconds, respectively) was applied to modify the composition of porous silicon layers prepared on p and p^+ type single crystal Si substrates. The amounts of nitrogen and other impurities like O and C were determined using the $^{14}N(d,\alpha)^{12}C$, $^{14}N(\alpha,\alpha)^{14}N$, $^{16}O(d,p_1)^{17}O^*$ and $^{12}C(d,p)^{13}C$ nuclear reactions. For a 30 sec immersion, the obtained nitrogen amount in the 2 μm thick columnar-type porous layers increased with plasma pressure and bias voltage. To introduce nitrogen into sponge-like porous layer, nitrogen PIII with bias voltage of 1000 V and 1 mbar pressure was applied. As measured by Rutherford Backscattering Spectrometry (RBS), the introduced nitrogen is always in the near surface region and its amount reaches a saturation value after ~30 sec treatment of direct PIII. An example is shown for remote PIII (1000 V, 1 mbar and 180 sec) that is able to introduce nitrogen into the whole porous layer.

Introduction

Nowadays porous silicon (PS), a very promising material for optoelectronic devices and sensor applications, is widely studied by RBS [1] and other ion beam analytical techniques, like nuclear reaction resonance experiments combined with ^{15}N isotopic tracing. Methods based on energy loss fluctuation measurements provide information on the morphology of PS layers [2]. Using electrochemical etching two basically different morphologies of PS layers with thicknesses in the μm range can be prepared. The columnar-type structure formed on heavily doped p^+ silicon contains long pores of typically 15 nm diameter, surrounded by 5-10 nm thick silicon walls and run parallel to the <100> direction of the silicon lattice. In the case of sponge like morphology obtained on p type silicon substrates, small pores are randomly distributed in the film, their diameter being about 2 nm and the wall thickness around 2-5 nm depending on the porosity of the layer. Porous layers have very high specific internal surface areas of typically up to 600 m^2/cm^3 for sponge-like and 200 m^2/cm^3 for columnar-type layers. In the case of columnar films the internal specific surface area is independent of the porosity in the range of 35-70% [3].

A stabilised structure and passivated internal surface are crucial requirements for technological applications of PS layers [4,5,6]. The classical methods of passivation based on high temperature oxidation or nitridation might involve the danger of partial or complete collapse of the delicate structure of the pores [2,7].

Plasma immersion ion implantation (PIII), where the ions are accelerated by a bias voltage across the ion sheath of a plasma, is one of the most promising techniques in IC technology. One of its advantages is homogeneous doping on large samples independently of the surface morphology. This gives the chance to perform modifications of the composition of the whole internal surface of PS layers at moderate temperature. In this paper we present experimental results on the feasibility of nitrogen PIII into porous Si.

Experimental

Porous silicon layers were produced with two main morphologies by electrochemical etching. The columnar type PS films with thickness of 2 μm and porosity of 66 % were grown in a 1:1:2 solution of 50% HF:water:ethanol. The current density and etching time was 38 mA/cm^2 and 96 sec, respectively. The 0.5 μm thick sponge-like PS samples of 68 % porosity were anodised in a 1:1 solution of 50% HF:ethanol using 10 mA/cm^2 current density for 50 sec. The porosities of the samples were determined by gravimetry method [8]. The samples were cut into pieces of 6 $\times$ 25 mm^2, each containing a 6 mm wide compact Si part as well.

The PIII treatments were carried out at KFKI on the water cooled sample holder of a specially built parallel plate RF plasma chamber [9] that was evacuated below 10^{-5} mbar. The nitrogen plasma generated in the pressure range of 0.1-1 mbar was excited with a RF frequency of 13.56 MHz with a maximum RF power density of 3 W/cm^2. The bias voltage was varied between 250 and 1000 V and the duration of immersion varied from 30 to 240 sec. The temperature of the whole sample during PIII treatment was estimated by mounting metal pieces of different melting points on the top of a silicon sample. This experiment showed that the average temperature of the silicon wafer does not exceed 330 °C during PIII treatment, even at highest bias voltage and immersion times.

The first set of experiments with samples exposed to plasma immersion aimed at answering the question whether this treatment can be applied to modify the composition of porous silicon. For this purpose 2 μm thick columnar layers were used. After 30 sec PIII of bias voltage between 250 and 1000 V the retained amount of nitrogen was measured in Paris with the ^{14}N(d,α)^{12}C nuclear reaction at a deuteron energy of 1500 keV and a detection angle of 150°. Using liquid N$_2$ cold traps during the measurements the vacuum in the scattering chamber was better than 10^{-6} mbar. The sample tilt was 0° and the high intensity scattered deuterons were filtered out by covering the detector with a 23 μm thick Mylar foil. The beam dose was 20 μC. The absolute quantity of nitrogen was determined using reference Si$_3$N$_4$ samples containing 486 and 847 $\times$ 10^{15} at/cm^2 nitrogen. ^{16}O and ^{12}C contents were measured with the ^{16}O(d,p$_1$)^{17}O* and ^{12}C(d,p)^{13}C nuclear reactions at 850 and 960 keV, respectively, using a Ta$_2$O$_5$ reference sample with 707 $\times$ 10^{15} ^{16}O/cm^2 and the cross section ratio of the above reactions as described in ref. [10].

The second set of experiments aimed at determining the immersion duration needed for a non-destructive treatment. For this purpose sponge-like layers of 0.5 μm thickness and 68% porosity were chosen, since the very fine pores of this structure is expected to be closed easily during the PIII treatment. In this case a nitrogen plasma of 1000 V bias voltage and 1 mbar pressure was applied and the duration of immersion was varied from 30 to 240 sec. In the case of 180 sec immersion the sample was partially shadowed by a Si wafer mounted over the sample at a distance of 2 mm, i.e., the plasma did not reach directly the sample: this is referred to as *remote PIII*. The amount and depth profile of impurities were measured at KFKI by RBS. The chosen He beam energy was 3.555 MeV where the cross sections for nitrogen and carbon are about 2 and 7 times higher than the Rutherford values [11]. The spectra were taken with a charge of 80 μC at tilt and scattering angles of 30° and 165°, under similar vacuum conditions as in Paris.

Results and discussion

The total amounts of nitrogen and other main impurities (O and C) found in the 2 μm thick columnar porous layer after immersion at different plasma energies and pressures for 30 seconds, are shown in Table 1. In PIII treated compact Si the total amount of nitrogen was undetectable, while in porous silicon layers it was of the order of 10^{17} N/cm^2. One may reason that this high quantity of nitrogen is distributed over a much larger area, i.e., the nitrogen covers the pore walls. Furthermore, in porous Si the higher was the plasma bias voltage and pressure, the higher was the nitrogen content in the sample. Due to the very poor depth resolution of the applied analytical method we cannot deduce from these measurements how deep the nitrogen penetrates. The internal surface area of the 2 µm thick porous layer before the treatments is about 400 cm^2 for a sample area of 1 cm^2. Assuming homogeneous distribution over the whole internal surface the highest nitrogen content corresponds to ~1 × 10^{15} atoms/cm^2 that is less than one monolayer.

Table 1. Impurity content expressed in 10^{15} at/cm^2 in 2 µm columnar porous Si of 66% porosity after 30 sec immersion for different bias voltages and pressures.

Bias [V]	non-treated			250			500			1000		
Element	N	O	C	N	O	C	N	O	C	N	O	C
Pressure 0.1 mbar	-	100	160	50	280	160	60	440	150	110	540	160
1.0 mbar	-	100	160	70	510	150	220	950	160	430	1140	140

After the PIII treatment, a high amount of oxygen was also found in the porous samples, which might come from the residual gas of the plasma process chamber. The amount of carbon in the samples was independent of any PIII treatment. Presumably, it originated from contamination by organic species both from atmosphere and residual gas of the scattering chamber [12].

The amount of main impurities as a function of the immersion time in sponge-like PS layers of the second set of experiments, studied by RBS, is shown in Fig 1. The nitrogen was always found in the near surface region of the films (as shown in Fig. 2 for 180 sec immersion time) and its content reaches a saturation value after a PIII treatment of 30 sec duration or shorter. The oxygen content gradually increases, while the carbon content of the samples decreases with increasing PIII time. The silicon content of the porous layers is 7.9×10^{17} Si/cm^2 in good agreement with the calculated value for a 0.5 µm thick layer of 68 % porosity and it remains constant within experimental errors of ±2 %, showing that in this case the sputtering process is negligible.

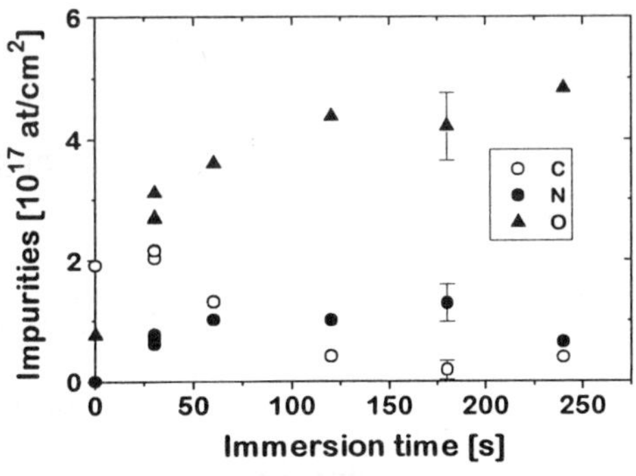

Fig. 1 Impurity contents in sponge-like porous Si of 0.5 µm thickness and 68 % porosity vs. PIII time at 1000 V bias voltage and 1 mbar pressure. (0 sec immersion time stands for non-treated sample.)

To explain the decreasing carbon contents with the increase of PIII durations assuming that the carbon content originated from the contamination by atmosphere and from the scattering chamber [12] we have to assume that i) the pores gradually close or ii) the internal surface of the porous layers decreases with increasing PIII time. Such processes might be caused by oxidation during the PIII treatment that always starts at the surface of the sample and leads to swelling or by the densification process of porous materials that is induced by higher energy ions and discussed in ref. [13]. Further investigation is necessary to understand better the role of the above processes and the effect of plasma heating that increases the local temperature of the PS layer.

Fig 2. shows two RBS spectra of the sponge-like PS layer after 180 sec PIII treatment. The spectrum recorded on the region of the sample exposed directly to the plasma, reveals that the nitrogen is located near the surface. The other spectrum, however, taken on the shadowed part (remote PIII), shows that the nitrogen is more or less evenly distributed along the whole depth of the PS layer. These results encourage us to continue our study on surface coverage of the PS layers by remote PIII.

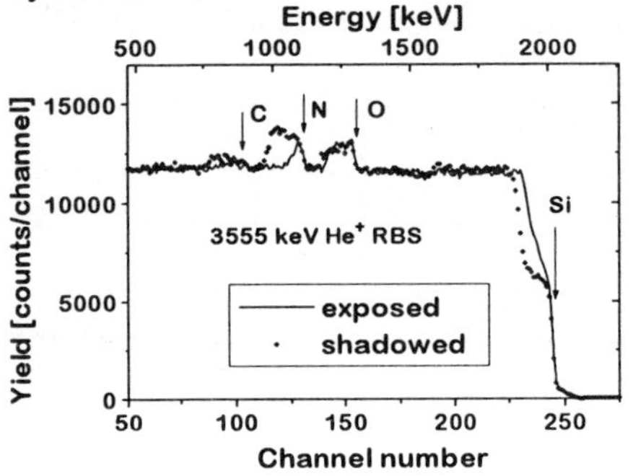

Fig. 2. RBS spectra recorded at 30° tilt angle and with 80 μC beam dose on a sponge-like PS sample after 180 sec PIII treatment at 1000 V bias voltage and 1 mbar N pressure (solid line - unshadowed, dots - shadowed region of the sample).

Conclusion

It was shown that plasma immersion ion implantation is applicable to introduce nitrogen into PS layers. With increasing plasma bias voltage and pressure, increasing amounts of nitrogen were found in the 2 μm thick columnar PS layer. Nitrogen plasma immersion at 1000 V bias voltage and 1 mbar pressure introduces nearly the same amount of nitrogen into the near surface region of sponge-like PS layers independently of the time step chosen for the PIII treatment. Remote PIII results in nearly homogeneous nitrogen profile in sponge-like porous silicon.

References

[1] G. Battistig, V. Schiller, E. Szilágyi and É. Vázsonyi, Nucl. Instr. and Meth. B 118, 654 (1996).

[2] G. Amsel, E. d'Artemare, G. Battistig, V. Morazzani and C. Ortega, in print in Nucl. Instr. and Meth. B.

[3] G. Bomchil, R. Herino, K. Barla and JJ.C. Pfister, J. Electrochem. Soc. **130**, 1611 (1983).

[4] A. Halimaoui, Porous Silicon Science and Technology, eds.: J.-C. Vial and J. Derrien, Springer-Verlag, Berlin, 33, (1995).

[5] R. Herino, Porous Silicon Science and Technology, eds.: J.-C. Vial and J. Derrien, Springer-Verlag, Berlin, 54, (1995).

[6] W. Lang, P. Steiner and F. Kozlowski, Porous Silicon Science and Technology, eds.: J.-C. Vial and J. Derrien, Springer-Verlag, Berlin, 293, (1995).

[7] V. Morazzani, M. Chamarro, A Grosman, C. Ortega, S. Rigo, J. Siejka and H.J. von Bardeleben, Journal of Luminescence **57**, 45 (1993).

[8] Porous Silicon Science and Technology, eds.: J.-C. Vial and J. Derrien, Springer-Verlag, Berlin, (1995).

[9] I. Pintér, M. Ádám, Cs. Dücső, N.Q. Khanh and I. Bársony (CIP '95, 10th International Colloquium on Plasma Process, Antibes, France, 1995) Supplement a la Revue Le Vide, No 275, 410, (1995).

[10] G. Amsel, J.P. Nadai, C.Ortega, J. Siejka, Nucl. Instr. and Meth. 149, 705 (1978).

[11] Ion Beam Handbook for Material Analysis, eds.: J.M. Mayer and E. Rimini, Academic Press, New York 1977.

[12] A. Grosman, Thesis, University of Paris 7, 1995

[13] F. Pászti, A. Manuaba, E. Szilágyi, É. Vázsonyi and Z. Vértesy, Nucl. Instr. and Meth. B 117, 253 (1996)

Materials Science Forum Vols. 248-249 (1997) pp. 237-240
© *1997 Trans Tech Publications, Switzerland*

Formation of Buried SiC Layers in Silicon by Ion Beam Synthesis

K. Volz, J.K.N. Lindner and B. Stritzker

Universität Augsburg, Institut für Physik, D-86135 Augsburg, Germany

Keywords: Ion Implantation, Silicon, Silicon Carbide, Ion Beam Synthesis

Abstract.
The structural and compositional evolution of Si(111) and Si(100) wafers which takes place during keV high-dose carbon ion implantation for the formation of buried SiC layers in silicon is studied using XTEM and RBS. The formation of epitaxially grown 3C-SiC is already observed for low doses. The depth distribution of crystalline SiC precipitates is monitored up to overstoichiometric doses, where the surplus carbon is found to cause amorphization of previously formed 3C-SiC. Annealing of stoichiometrically or near-stoichiometrically implanted silicon at 1250° C leads to the formation of well-defined buried SiC layers in silicon, while excess carbon in overstoichiometrically implanted wafers seems to impede the formation of homogeneous SiC crystals.

1. Introduction
Silicon carbide is a wide band gap semiconductor with excellent thermal conductivity, high saturation drift velocity, and outstanding chemical and mechanical stability. SiC can be n-type and p-type doped with high dopand activation and it forms a thermal oxide, another prerequisite for the application of SiC in high-temperature and high-frequency semiconductor devices [1,2]. Therefore, different techniques are being investigated at present to produce thin silicon carbide films, which might be applied in electronics and micromechanics. Ion beam synthesis (IBS), i.e. high-dose carbon ion implantation and subsequent annealing has been investigated for more than 20 years [3,4] as a possible route to obtain buried SiC layers in silicon, and at present is matter of intensive investigations [5-8]. Different to the IBS of buried SiO_2 layers in silicon it has been found that the pile-up of the impurity concentration profile into a box-like depth distribution is hampered by the low diffusivity of carbon in silicon [5]. However, recently it has been demonstrated [9-11] that buried epitaxial 3C-SiC films with homogeneous composition and sharp interfaces to the Si top layer and the bulk can be formed using unexpectedly low annealing temperatures, if a limited dose interval is maintained. It is well known that in the IBS of buried layers the structural state after implantation has a crucial influence on the redistribution behaviour of impurities during the annealing step. However, little is known about the structure of high dose carbon implanted silicon at conditions suitable for the succesful IBS of buried SiC Layers. The aim of this paper is to elucidate the structural development of high dose carbon ion implanted silicon and to compare with it the structure after annealing.

2. Experimental
Float zone (100) and (111) Si wafers were implanted with 180 keV C^+ ions using an EATON medium current implanter. Implantations were performed in dose range of 2 to 12 x 10^{17} C^+/cm^2, which includes the dose interval of 7.6 to 9 x 10^{17} C^+/cm^2 which has been found [10] to be suitable for the IBS of homogeneous carbide layers. Beam heating of the ~500 µA ion beam scanned over an 3˝ diameter area was used to obtain target temperatures of 270-440 °C, which were measured by a thermocouple clamped to the target holder. Some samples were annealed at 1250°C for up to 10 h in the flowing Ar atmosphere of a quartz tube furnace. Characterization of wafers in the as-implanted and in the annealed state was done using Rutherford Backscattering (RBS) and channeling analysis with 1.5 and 1.8 MeV α-particles as well as plane-view and cross-sectional transmission electron microscopy (XTEM) at 300 keV.

3. Results and Discussion

3.1 The as-implanted state

According to TRIM [12] simulations, 180 keV C$^+$ ion implantation in silicon results in the generation of a Gaussian-like carbon depth distribution with a peak at a depth of approximately 500 nm beneath the surface. A stoichiometric composition (50 at.%) is reached at a dose of 9 x 10^{17} C$^+$/cm^2.

For the smallest dose of 2 x 10^{17} C$^+$/cm^2, electron diffraction using XTEM samples reveals the presence of epitaxial 3C-SiC embedded in crystalline silicon (c-Si) in a depth interval, which is 250 nm thick and covered with 300 nm c-Si. Due to the very small size of precipitates it has not been possible to obtain micrographs of these particles using conventional imaging techniques. The silicon matrix in the SiC rich zone is strained and contains a high density of defects, but due to the high implantation temperatures no indications of amorphization are observed.

A dose increase by a factor of two leads to significant changes in the structure of as-implanted samples: the c-Si top layer is followed by a 10 nm thin amorphous zone on top of a 90nm thick interval with a high SiC precipitate density and a 30 nm interval, where individual SiC precipitates can be observed within a crystalline silicon surrounding.

In a dose interval from 6.5 to 9 x 10^{17} C$^+$/cm^2 this layer sequence is supplemented by an additional amorphous zone underneath the SiC precipitate-rich depth interval. Both amorphous zones strongly increase in thickness with increasing dose, while the thickness of the SiC-rich interval increases only by approximately 20 % to 125 nm at the higher dose. The resulting depth structure is displayed in the XTEM micrographs of Fig.1 for the dose of 9 x 10^{17} C$^+$/cm^2 and depths greater than 200 nm. It is interesting to note that the Si top layer is subdivided into a scarcely damaged upper part (not visible) and a heavily damaged lower part. Considerable damage is also found in a 90 nm zone underneath the lower a-Si layer. The most striking feature is that the depth distribution of epitaxial 3C-SiC precipitates, which have a typical diameter of about 4 nm, seems to be nearly uniform within the depth interval which is surrounded by the amorphous layers. No crystalline silicon inclusions have been detected in this SiC rich zone using silicon dark field images. The SiC precipitates exhibit an epitaxial orientation relationship with the silicon substrate, as can be seen from the presence of two single crystalline diffraction patterns of Si and SiC superimposed in Fig.1(c). The <111> streaks of some Si and SiC spots in this diffraction pattern are due to Si stacking faults in the Si top layer and due to platelet shaped SiC microtwins on {111} lattice planes in the SiC rich zone, respectively. The limited depth interval in which one finds SiC to be homogeneously distributed seems to be in contrast to the Gaussian-like depth distribution of carbon which is observed by RBS also for this dose. The most likely explanation is that outside the SiC-rich zone, i.e. in the Si top layer, carbon is present in precipitates smaller than the detection limit and within the SiC-rich layer, a portion of implanted carbon is either on SiC interstitial sites, or in SiC precipitates smaller than the detection limit, or in the interspace between SiC precipitates.

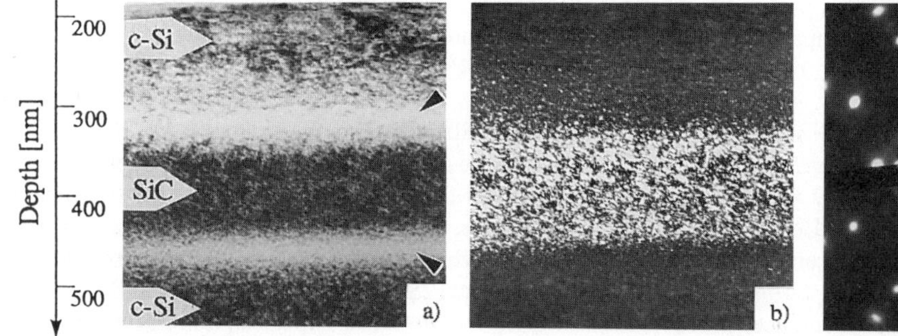
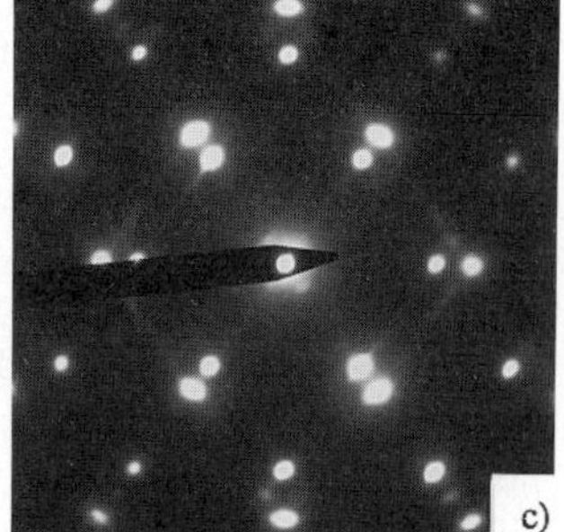

Fig.1: *XTEM bright field (a) and (111)SiC dark field micrograph (b), diffraction pattern (c) of a Si(111) wafer as-implanted with 9 x 10^{17} C/cm^2. Amorphous zones are marked by arrows.*

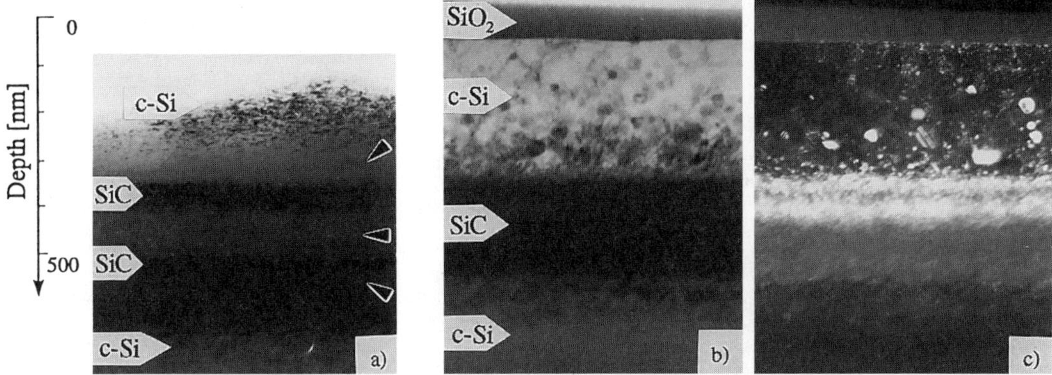

Fig.2: *Layer structure for a dose of 12 x 10¹⁷ C/cm²: (a) and (b) bright field images for the as-implanted and the annealed state, respectively, (c) (220)SiC dark field image corresponding to (b). Amorphous zones are marked by arrows.*

A dose increase to 12 x 10^{17} C^+/cm^2 leads to the formation of an additional 70 nm thick amorphous zone, located whithin the SiC rich depth interval (Fig.2(a)). A 50 nm amorphous zone and 55 nm of crystalline SiC precipitates are found on top of the central amorphous region and 25 nm of crystalline SiC precipitates plus 35 nm of amorphous material are observed underneath. According to RBS studies of overstoichiometrically carbon implanted silicon wafers the peak carbon concentration amounts to 67 at.%. Therefore it is concluded that the formation of the central amorphous region is due to the necessity of incorporating overstoichiometrically implanted carbon.

The origin of the outer amorphous zones is more difficult to explain. Due to the small mass of carbon ions, amorphization of pure silicon is not expected for the high implantation temperatures used. However, carbon impurities may stabilize defects in silicon and therefore promote amorphization. A few permill of carbon have been observed to reduce the crystallization velocity of amorphous silicon measurably [13]. Once crystalline SiC precipitates have gained a sufficiently large size they might be more stable against amorphization than crystalline silicon and serve as a sink for diffusing carbon atoms. It seems to be important that the formation of crystalline SiC is observed at doses below the amorphization threshold, since such SiC precipitates may provide the orientation information necessary to obtain a completely aligned SiC precipitate layer at the stoichiometry dose. SiC precipitate formation in the early stages of carbon implantation, however, is very likely due to the negligibly low solid solubility [14] of carbon in silicon at the temperatures of interest.

3.2 The annealed state

Annealing for 5 h at 1250° C leads to a redistribution of implanted carbon atoms for implantation doses up to stoichiometry dose, as demonstrated earlier [9,11] by RBS. For a (111)Si wafer implanted with 4 x 10^{17} C^+/cm^2 a distribution of SiC is obtained which is significantly narrower than in the as-implanted state. According to RBS, the maximum carbon concentration is 42 at.%.

Homogeneous and RBS-stoichiometric SiC layers have been obtained after 5 h annealing at 1250° C for doses of 7.6 x 10^{17} C^+/cm^2 using Si(111) wafers and for doses of 9 x 10^{17} C^+/cm^2 in Si(111) and Si(100) wafers. Longer annealing times lead to more abrupt interfaces of the SiC layer, as observed using Si(100) wafers implanted with a dose of 9 x 10^{17} C^+/cm^2 (Fig.3). Some isolated large SiC precipitates exist in the Si top layer and in the silicon bulk underneath this 170 nm thick crystalline SiC layer. As in all annealed samples, a thin surface oxide layer is observed.

5 h/1250° C annealing of overstoichiometrically implanted silicon wafers with a dose of 12 x 10^{17} C^+/cm^2 is observed to result only in a minor redistribution of carbon atoms. The maximum concentration of 67 at.% is unchanged and only a small step at 50 at.% carbon occurs in the upper

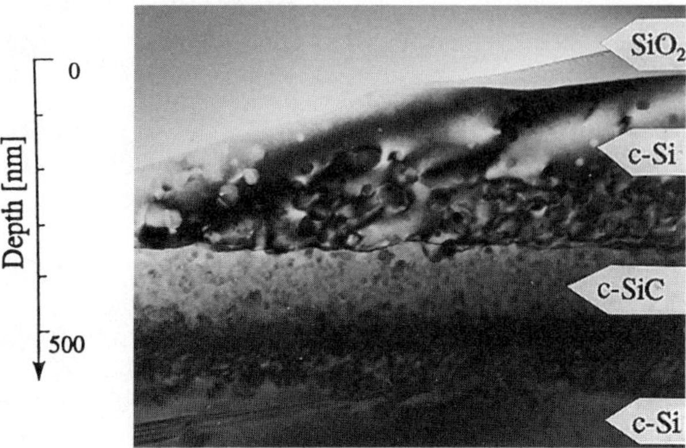

Fig.3: Bright field micrograph of a Si/SiC/Si layer system synthesized by implantation of 9 x 10¹⁷ C/cm² in Si(100) and 10 h annealing at 1250° C.

wing of RBS measured concentration profiles [9,11]. XTEM investigations show that a buried layer is formed in between a c-Si top layer and the bulk, both containing numerous SiC precipitates (Fig.2(b,c)). The layer is subdivided into an upper part, which mainly consists of epitaxially oriented 3C-SiC (bright in Fig.2(c)) and a lower part, which cannot be strongly excited using an reflection of epitaxial 3C-SiC. The diffraction pattern of this layer does not show any amorphous portions. Therefore it is assumed that the lower part of the layer consists of highly disordered crystalline SiC with a high density of carbon interstitials. Since recent Raman investigations [15] on 1000° C annealed overstoichiometrically carbon implanted silicon crystals have revealed the presence of carbon clusters with an estimated diameter of 2.5 nm, further investigations are necessary to clarify this point.

4. Summary and Conclusions

The structure and composition of high-dose 180 keV carbon ion implanted silicon wafers have been studied by XTEM and RBS in the as-implanted state and after thermal annealing at 1250° C. In the as-implanted state the evolution of a layer structure with increasing dose is observed, which results in the formation of a well-defined precipitate layer of epitaxially oriented 3C-SiC crystallites in between two thin layers of amorphous material at the stoichiometry dose. Annealing of this layer leads to the formation of a homogeneous layer of 3C-SiC with sharp interfaces. Overstoichiometrically implanted carbon is observed to cause amorphization of previously formed 3C-SiC and is stable against thermal redistribution during the anneal.

References

[1]   H. Morkoç, S. Strite, G.B.Gao , M.E.Lin, B.Sverdlov, and M.Burns, *J. Appl.Phys.* **76** (1994) 1363-1398.

[2]   P.L.Dreike, D.M.Fleetwood, D.B.King, D.C.Sprauer, and T.E.Zipperian, *IEEE Trans Comp. Pack. Manuf. Techn.* **17** (1994) 594-609.

[3]   J.A.Borders, S.T.Picraux, and W.Beezhold, *Appl. Phys. Lett.* **18** (1971) 509.

[4]   W.Rothemund and C.R.Fritzsche, J. Electrochem. Soc. 121 (1974) 587.

[5]   K.J.Reeson, P.L.F.Hemment, J.Stoemenos, J.R.Davis, and G.K.Celler, *Appl. Phys. Lett.* **51** (1987) 2242.

[6]   P.Martin, B.Daudin, M. Dupuy, A.Ermolieff, M.Olivier, A.M.Papon, and G.Rolland, *J. Appl. Phys.* **67** (1990) 2908.

[7]   A.Nejim, P.L.F.Hemment, and J.Stoemenos, *Appl. Phys. Lett.* **66** (1995) 2646-2648.

[8]   C.Serre, A.Pérez-Rodríguez, A.Romano-Rodríguez, J.R.Morante, R.Kögler, W.Skorupa, J.Appl. Phys. 77 (1995) 2978.

[9]   J.K.N.Lindner, A.Frohnwieser, B.Rauschenbach, and B.Stritzker, Mater. Res. Soc. Symp. Proc. **354** (1995).

[10] J.K.N.Lindner, K.Volz, and B.Stritzker, Proc. 6th. Int. Conf. on Silicon Carbide and Related Materials, Kyoto (1995), Inst. Phys. Conf. Ser. No **142** (1996) 145.

[11] J.K.N.Lindner, K.Volz, U.Preckwinkel, B.Götz, A.Frohnwieser, B.Rauschenbach, and B.Stritzker, Mat. Chem. and Phys. (1996), in the press.

[12] J.F.Ziegler, J.P.Biersack, and U.Littmark, in: *The Stopping and Range of Ions in Matter*, Vol.1,ed. by J.F.Ziegler (Pergamon Press, New York, 1985)

[13] E.F.Kennedy, L.Csepregi, J.W.Mayer, and T.W.Sigmon, J.Appl.Phys. **48** (1971) 4241.

[14] A.R.Bean and R.C.Newman, J.Phys. Chem. Solids **21** (1971) 1211.

[15] L.Calcagno, G.Compagnini, M.G.Grimaldi, G.Foti, and P.Musumeci, E-MRS Spring Meeting Strasbourg (1996).

Materials Science Forum Vols. 248-249 (1997) pp. 241-244
© *1997 Trans Tech Publications, Switzerland*

Lattice Damage of Relaxed $Si_{1-x}Ge_x$ Alloys of Various Composition Implanted with 2 MeV Si Ions

J.K.N. Lindner

Universität Augsburg, Institut für Physik, D-86135 Augsburg, Germany

Keywords: Ion Implantation, MeV, Radiation Damage, SiGe Alloys, Semiconductors

ABSTRACT

The accumulation of damage in relaxed $Si_{1-x}Ge_x$ alloys ($0.04 \leq x \leq 0.36$) after 2 MeV Si ion implantation at 300 K has been studied by optical reflectivity depth profiling (ORDP). Damage is observed to increase with increasing Ge content of the alloy. While for sufficiently low doses the shape of damage profiles is similar to that of the nuclear stopping power for all Ge concentrations, composition dependent changes in profile shape are observed at doses where defects produced in different collision cascades may interact. For the depth of maximum damage the dose dependence of damage is described by a model. From the model parameters it is concluded that the stronger damage in the Ge rich alloys is mainly due to a stronger production of amorphous portions in individual collision cascades.

1. INTRODUCTION

The radiation damage of solids is well known to be dependent on implantation parameters such as ion mass, energy and dose, dose rate, target temperature and target material. The use of relaxed $Si_{1-x}Ge_x$ alloy layers as target material allows to vary target properties continuously along with the Ge content x. However, only few articles have appeared on the radiation damage of SiGe alloys so far [1-5]. In the present paper the radiation damage produced by 2 MeV Si ions in MBE grown relaxed $Si_{1-x}Ge_x$ alloys with various compositions is studied by a sophisticated optical technique called Optical Reflectivity Depth Profiling (ORDP). The samples stem from an European Community founded IBOS (Ion Beam Processing of Semiconductors) network project, where the same samples have been investigated by a variety of different methods [6].

2. EXPERIMENTAL DETAILS

Relaxed, homogenous $Si_{1-x}Ge_x$ films of 3 μm thickness and with different Ge contents were grown in Aarhus (DK) on Si(100) by molecular beam epitaxy using compositionally graded $Si_{1-x}Ge_x$ buffer layers. It has been shown by different techniques that layers produced under equivalent conditions are of high structural quality. The exact compositions were determined by Rutherford Backscattering Spectroscopy (RBS) in Bologna (I) to be x=0.04, 0.13, 0.24, and 0.36, respectively. The alloys were implanted at Catania (I) with 2 MeV Si ions at 300 K in a dose range from 10^{12} to 4 x 10^{15} Si/cm^2 with dose rates of 10^{10} and 10^{11} $cm^{-2}s^{-1}$. Details of the sample preparation are given in Ref. [6] and references within.

For depth resolved investigations of the radiation damage with the ORDP technique [7], a cut through the damage profile was produced by small angle bevelling at about 1°. The bevel angle of each sample was measured exactly (± 0.5 %) using a DECTAC surface profiler to define the depth scale. For the ORDP analysis, measurements of the reflected light intensity at 650 nm were performed across the beveled surface at normal incidence. For this a CCD camera attached to a microscope is used, each CCD channel corresponding to a depth interval of ~11 nm. For a constant Ge concentration, the optical reflectivity change $\Delta R/R_c$ of implanted $Si_{1-x}Ge_x$ with respect to the reflectivity R_c of undamaged crystalline silicon increases with increasing damage, until saturation is reached at $\Delta R_{a\text{-}SiGe}/R_c$ due to amorphization [9]. The difference of optical signals from damaged and undamaged $Si_{1-x}Ge_x$ alloys ($\Delta R_{c\text{-}SiGe}/R_c$), related to the difference of amorphized and undamaged material, is used as a measure of the depth dependent lattice damage s, which is $s=0$ for undamaged and $s=1$ for amorphous alloys:

$$s(z) = \frac{\Delta R(z)/R_c - \Delta R_{c\text{-}SiGe}/R_c}{\Delta R_{a\text{-}SiGe}/R_c - \Delta R_{c\text{-}SiGe}/R_c} \qquad . \tag{1}$$

3. RESULTS AND DISCUSSION

Fig. 1 shows on a logarithmic scale selected depth profiles of the optically determined lattice damage for Ge contents of 0.04 (a) and 0.36 (b) and for doses between 3×10^{12} and 4×10^{15} Si/cm^2. The detected damage extends from the surface down to a depth of more than 3 μm with a maximum at a depth of about 1.8 μm. As already shown for a Ge content of 0.24 [8], the profile shape observed for all x for the lower doses ($< 3 \times 10^{13}$ Si/cm^2) corresponds well to the shape of the nuclear stopping power profile calculated using TRIM90 [9], disregarding a slight shift (0.1 μm) in depth scales. Therefore it is concluded that damage production in Si$_{1-x}$Ge$_x$ alloys is governed by the energy deposition in nuclear collisions. For the higher Ge concentrations (x=0.36 and x=0.24), the profile shapes stay approximately the same even for doses up to 1×10^{14} Si/cm^2. Amorphization saturation is observed to set in at the depth of profile maximum at doses of 3×10^{14} Si/cm^2 (x=0.36) and 5×10^{14} Si/cm^2 (x=0.24, not shown). For the lower Ge concentrations some deviations of the profile shape exist at higher doses, most prominent for x=0.04. A slight damage saturation is observed in profiles for a Ge content of 0.04 when a dose of about 0.3×10^{14} Si/cm^2 is exceeded, connected with comparatively flat profile shapes. At doses above 3×10^{14} Si/cm^2 a strong and pointed peak evolves at the profile maximum, indicating a sharp transition into amorphization at this depth. The upper interfaces of amorphous layers in low Ge content material are generally sharper than those of comparatively thick amorphous layers in higher Ge content alloys.

The dose dependence of damage measured at the peak of the depth distribution is displayed in Fig. 2 for the different Si$_{1-x}$Ge$_x$ alloys, along with calculated damage curves described later. For a fixed dose beneath the amorphisation threshold the damage is the higher the larger the Ge content of the alloy is. The damage curves exhibit a sigmoidal shape which gets more pronounced with decreasing Ge concentration, similar to Ref. [4] where Si$_{1-x}$Ge$_x$ alloys where implanted with 100 keV Si ions and similar to MeV implantations of pure silicon [10]. In the latter case damage curves get more s-shaped with increasing target temperature.

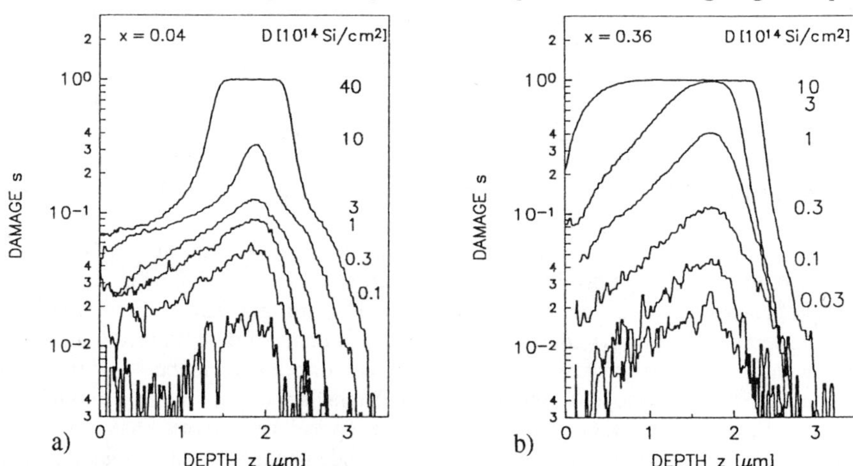

Fig. 1 Depth profiles of the optically detected lattice damage in Si$_{1-x}$Ge$_x$ alloys implanted with 2 MeV Si$^+$ ions for x=0.04 (a) and x=0.36 (b).

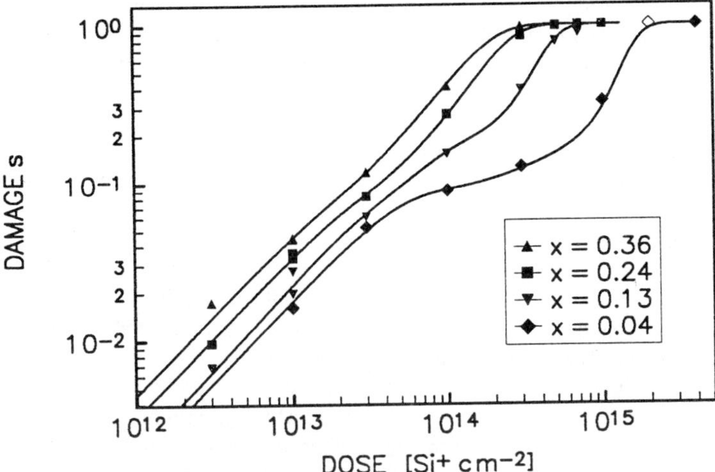

Fig. 2 Dose dependence of the lattice damage at the profile peak for $Si_{1-x}Ge_x$ alloys of different Ge content x implanted at 300 K with 2 MeV Si^+ ions. Point $\Diamond$ from Ref. [6].

For doses up to 3 x 10^{13} Si/cm^2 damage increases linearly with dose in all $Si_{1-x}Ge_x$ alloys considered, the damage efficiency ds/dD increasing with increasing Ge content. A linear dose dependence is expected in a low dose region as long as defects from individual collision cascades do not interact with defects from other collision cascades. Therefore the higher damage efficiency in higher Ge content material has to be attributed to a stronger defect generation per collision cascade. Since according to TRIM simulations the peak of the nuclear stopping power of 2 MeV Si ions is nearly the same in all the alloys (25 eV/Å), the different damage efficiencies in the materials considered reflect a different susceptibility for damage rather than a different amount of deposited energy.

At doses above 3 x 10^{13} Si/cm^2, defects generated in different collision cascades may interact. A dominating intercascade point defect recombination may lead to a sublinear dose dependence of damage, while a dominating clustering of point defects may lead to a superlinear dose dependence, since clustered point defects are stable against intercascade recombination. Intercascade defect recombination is suggested to be responsible for the sublinear damage increase with dose observed mainly for x=0.04 and x=0.13 alloys. The superlinear increase of damage with dose observed for all alloys at the transition into the amorphous state is ascribed to intercascade point defect clustering and to the stimulated growth [11] of isolated amorphous regions. For all alloys, damage saturation due to the formation of a continuous amorphous layer occurs at sufficiently high doses.

A damage model developed by Hecking et al. [11] to describe the damage of Si implanted with MeV Si ions has been used to calculate the damage curves shown in Fig. 2. In this damage model the total damage s is assumed to be the sum $s = s_p + s_a$ of volume fractions s_p and s_a damaged by point defects or clusters and by amorphization respectively. The dose dependence of s_p and s_a is treated by a set of coupled differential equations containing six model parameters, which describe the production of point defect type damage in individual collision cascades, the generation of amorphous volume fractions in individual cascades, the recombination and the clustering of point defects originating from different collision cascades, the saturation of point defects and the stimulating effect of existing amorphous volume portions on the generation of additional amorphous regions during the deexcitation of a spike, respectively. The dependence of these parameters on the Ge content has been discussed elsewhere in more detail [12]. It should be stated that by far the most prominent influence of the Ge concentration is observed to be on the production of amorphous material in individual cascades, which increases by nearly two orders of magnitude when the Ge content is increased from 0.04 to 0.13, indicating that the strong increase of damage and damage efficiency with increasing Ge content has to be attributed mainly to an enhancement of amorphous volume production. This interpretation is supported by the fact that in EPR studies only an amorphous line has been detected in high Ge content material [6], different to pure silicon, where the lines of different types of point defect clusters have been identified after MeV ion implantation [13].

4. SUMMARY AND CONCLUSIONS

The radiation damage of relaxed $Si_{1-x}Ge_x$ layers of different composition after 2 MeV Si ion implantation has been studied as a function of dose by optical reflectivity depth profiling (ORDP). Significant differences in the shape of damage profiles are detected for different Ge contents. For a fixed dose the damage increases with increasing Ge content at any depth. For a fixed composition the dose dependence of damage at the depth of damage maximum can be described by the damage model proposed by Hecking et al. using fit parameters to account for the strength of different defect production and defect interaction mechanisms. From a consideration of model parameters it is concluded that the main reason for the stronger damage in $Si_{1-x}Ge_x$ alloys of higher Ge concentration is a stronger formation of amorphous volumina.

Acknowledgements

The author is grateful to all colleagues from the IBOS network, especially A. Nylandsted Larsen (Aarhus, DK) for providing the $Si_{1-x}Ge_x$ samples, F. Priolo (Catania, I) for performing the implantations, G. Lulli (Bologna, I) for RBS measurements and to P.L.F. Hemment (Guildford, UK) for organizing the network. The support of B. Stritzker (Augsburg, D) with the participation in the project is gratefully acknowledged.

References

[1]   O.W.Holland and T.E.Haynes, Appl. Phys. Lett. 61 (1992) 3148.

[2]   T.E.Haynes and O.W.Holland, Appl. Pyhs. Lett. 61 (1992) 61.

[3]   M.Vos, C.Wu, I.V.Mitchell, T.E.Jackman, J.-M. Baribeau, and J.P.McCaffrey, Nucl. Instr. and Meth. B 66 (1992) 361.

[4]   D.C.Lie, A.Vantomme, F.Eisen, T.Vreeland, M.-A.Nicolet, T.K.Carns, V.Arbet-Engels, and K.L.Wang, J. Appl. Phys. 74 (1993) 6039.

[5]   Z.Atzmon, M.Eizenberg, Y.Shacham-Diamand, J.W.Mayer, and F.Schäffler, J.Appl.Phys. 75 (1994) 377.

[6]   A.Nylandsted Larsen, C.O`Raifeartaigh, R.C.Barklie, B.Holm, F.Priolo, G.Franzo, G.Lulli, M.Bianconi, R.Nipoti, J.K.N.Lindner, A.Mesli, J.J.Grob, F.Cristiano, and P.L.F.Hemment, submitted to J.Appl. Phys. (1996).

[7]   K.F. Heidemann, Philos. Mag. B**44** (1981) 465.

[8]   J.K.N.Lindner, E-MRS Spring Meeting Strasbourg 1995, accepted for publ. in Nucl. Instr. and Meth B (1996).

[9]   J.F. Ziegler, J.P. Biersack, U. Littmark, *The Stopping and Range of Ions in Matter*, vol. 1, ed. J.F. Ziegler (Pergamon, New York, 1985).

[10]   J.K.N.Lindner, R.Zuschlag, and E.H.te Kaat, Nucl. Instr. and Meth. B 62 (1992) 314 and references within.

[11]   N.Hecking, K.F.Heidemann and E.teKaat, Nucl. Instr. and Meth. B 15 (1986) 760.

[12]   J.K.N. Lindner, presented at the IBMM96, submitted to Nucl. Instr. and Meth. B

[13]   V.S.Varichenko, A.M.Zaitsev, J.K.N.Lindner, R.Domres, N.M.Penina, D.P.Erchak, A.R.Chelyadinskii, V.A.Martinovitsh, Nucl. Instr. and Meth. B 94 (1994) 240.

Materials Science Forum Vols. 248-249 (1997) pp. 245-248
© *1997 Trans Tech Publications, Switzerland*

Auger Depth Profiling with Good Depth Resolution of Low Energy Implantation Induced Ion Mixing

A. Sulyok[1], A. Galisova[2] and M. Menyhard[1]

[1] Research Institute for Technical Physics of the Hungarian Academy of Sciences,
P.O. Box 76, Budapest, H-1325, Hungary

[2] Department of Microelectronic Faculty of Electrical Engineering and Information Technology
Slovak Technical University, Slovakia

Keywords: Depth Profiling, Ion Mixing, Ion Implantation

Abstract

Auger depth profiling using rotating specimen, grazing angle of incidence and low ion energy was carried out on multilayered Ge/Si specimen ion implanted by various fluences. Even the raw depth profiles showed qualitatively the increase of the mixing with the increasing fluence. A trial-and-error type of deconvolution was developed and checked by T-DYN simulation, to restore the concentration distribution of ion mixed region from the depth profile. The variance versus deposited energy curve was calculated and found to be quasi linear.

Introduction

Ion implantation is frequently used for engineering the surface near regions of material. One of the beneficial or unwanted effect is the ion mixing, which alters the distribution of the elements close to the interfaces.

At high (> 100 keV) ion energies and fluences the broadening of an originally abrupt interface is in the range of some tens of nanometers. Such a compositional profile can easily be measured by RBS having similar depth resolution [1]. With decreasing ion energies or/and fluences the broadening decreases. The analysis of the broadened concentration distribution needs better depth resolution technique. We will show that one of the other depth profiling techniques (SIMS, AES, XPS, that are generally cheaper and easier on the other hand) can easily be applied for this case.

The analytical sensitivity of the mentioned methods varies from some ppm (SIMS) to about percentage (AES, XPS). Their depth resolution is practically the same and depends on the quality of the sputter removal of the material. Using low ion energy and rotating specimen the depth resolution is in the nm range being better than that of the RBS. Cirlin et al used SIMS to study the mixing of delta layers imbedded in GaAs [2]. Based on the well resolved depth profile they could even get conclusion concerning the type of mixing.

In this work we will discuss the possibility of using Auger depth profiling for the study of the ion mixed region. It is obvious that considering the rather poor sensitivity of the AES (it is in the range of percentage) we can not study the profile of e.g. doping elements in semiconductors, or the extended tail of the mixing profile, but rather it is restricted to the study of the distribution of the major components close to the interfaces. The experimental study of this region is also necessary for several reasons. First of all one should realize that the mixing causes major concentration change in the interface near region. Next, the usual dilute alloy approximation is obviously non valid in this region and thus the experimental determination of the mixing has great importance.

A Ge/Si multilayer system mixed by Ar ions of 30 keV was Auger depth profiled using rotating specimen and low (0.7 keV) ion energy of various doses. Both the mixing process and depth profiling was simulated by means of the dynamic TRIM code [3]. We will show that at least at higher doses the depth profiling does not alter significantly the elemental distribution produced by the ion mixing, that is, without additional evaluation process of the experimental data the ion mixing can be studied.

Experimental procedure

The multilayer structure was grown in a dual-target dc magnetron sputtering system (Thin Film Physics Division, Linköping University) with a base pressure of about 10^{-7} mbar [4]. The films were deposited on unheated silicon substrate with its native oxide, at a rate of 0.2 nm s^{-1} and by Ar sputtering gas of 5 mbar. To prevent a columnar structure growth a negative substrate bias of approximately 140 V was applied [4]. The nominal structure of the specimen was as follows:

10 nm Si - 10 nm Ge - 14 x (1 nm Si -10 nm Ge) - substrate

The first interface will be referred later as thick layer transition, while the others as first, second δ layer transition.

Cross sectional high resolution transmission electron microscopy (XTEM) studies preceded the depth profiling, and only specimens having sharp and flat interfaces were ion implanted. Layer thicknesses were determined on the XTEM images and these values were used for obtaining the sputtering rate.

The ion implantation took place in KFKI-Research Institute for Materials Science, Budapest by using a Varian implanter and Ar ions. The energy of the ions were 30 keV, the angle of incidence 7°. Three fluences were applied: $2*10^{14}$, $6*10^{14}$, $2*10^{15}$ ion/cm^2 with ion currents 16, 16, 63 μA., respectively.

The depth profiling were carried out in our dedicated depth profiling device [5]. A special Teletwin type ion gun was used (described in [5]). The incidence angle of Ar$^+$ ions was 86° relative to the normal of the surface, the incidence energy was 0.7 keV. The diameter of the ion beam was about 0.3 mm. The ion current was in the range of some μA. The specimen was rotated at a speed of 15 rpm during sputtering. The slope of the sputtered crater was calculated (by measuring the distances of the subsequent layers on the wall of the crater) to be less than 10^{-5} radian.

Dynamic TRIM code [3], T-DYN (1991) version 4.0, was used to simulate depth profiling using the following parameters: lattice binding energies for both elements were 1 eV, the ions and recoils moved while their energy was larger than 3 eV. The surface binding energies were 3.88 and 4.70 eV for Ge and Si, respectively. The results are rather insensitive to the above parameters, thus their actual values are not too crucial. The channel thickness in the simulation was chosen to be 5Å, and it was shown that the results are independent of the changes of this value also. Si concentration was calculated from the number of atoms being in the topmost channel.

Results and discussion

Figs 1a and 1b show the measured and simulated depth profiles (Si concentration versus removed depth) for fluences of 0, $2*10^{14}$, and $2*10^{15}$ ion/cm^2. It is clear considering the depth profile of the non-implanted specimen in Fig 1a , that the depth profiling process causes serious broadening. On the other hand the depth profiles of the implanted specimen are strongly different from each other and from that of the non-implanted one. Thus we can hope that the elemental distributions induced by implantation can be restored from the measured depth profiles.

To carry out this restoration one must know how the various broadening processes corrupt the original distribution. We have discussed in our previous papers that the presently available analytical models (Gaussian and Liau broadening) can not explain the depth profiles measured on various Ge/Si structures [6,7]. On the other hand we have found that the T-DYN simulation predicted various features of the depth profiles, but it still failed to give the absolute values. This problem was solved by developing a new deconvolution method for the restoration. This assumes that the mixing is independent on the local concentration (dilute approximation), consequently the Gauss convolution (the mathematical operation for modelling the ion implantation effect) can take place after the ion mixing caused by the ion depth profiling. We must note, that this order change can be done only in the model. Although the dilute approximation used in the model is not perfectly fulfilled, it is reasonable assumption regarding the concentration values. On the other hand, we will

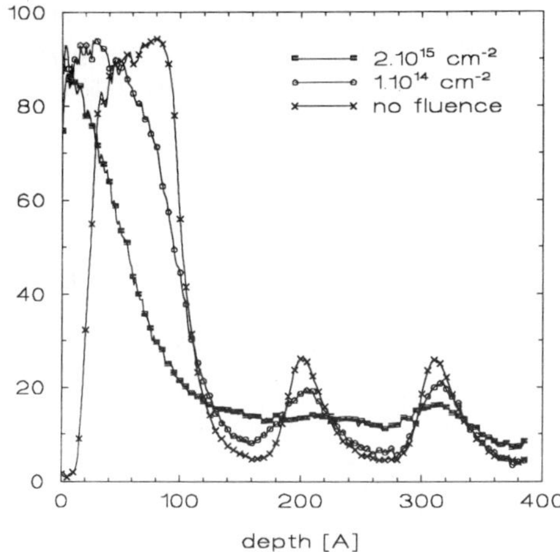

Fig. 1a. Measured depth profiles of variously implanted specimens.

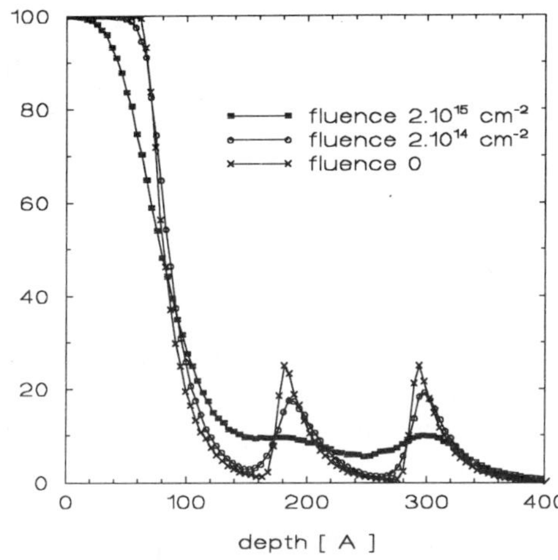

Fig. 1b. Simulated depth profiles of variously implanted specimens.

use a T-DYN simulation to prove the validity of this procedure for this given case of a Ge/Si layer system.

Comparing Figs 1a and 1b we conclude that the T-DYN simulation predicts well the tendencies of the depth profiling. It is also clear, however, that the absolute values are different. As it was found previously [7] the simulation predicts less mixing. In the simulation we can separate the elemental processes and thus we can check how the restoration of the original distribution can be made. In this case, we can use a trial-and-error type of "deconvolution". The procedure is as follows. According to the model described above we first simulate the depth profiles of the non-implanted specimen and that of the ion implanted one. The next step is to apply a Gaussian convolution on the depth profile of the non-implanted specimen that simulates the broadening effect of ion implantation. The width of the Gaussian (characterized by σ) was varied while the resulting

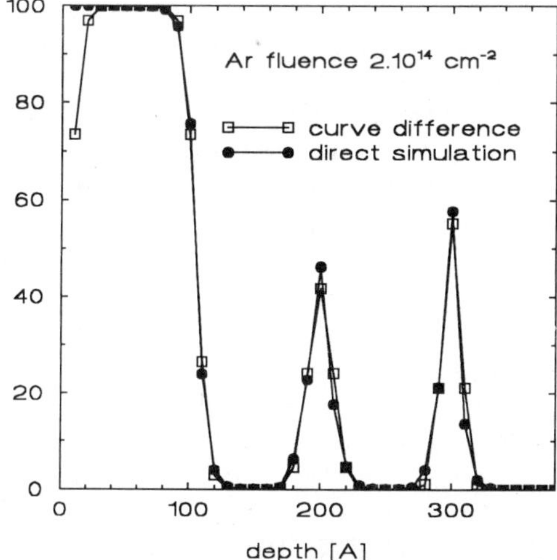

Fig. 2 The Si distribution after implantation; simulated and deconvoluted curves

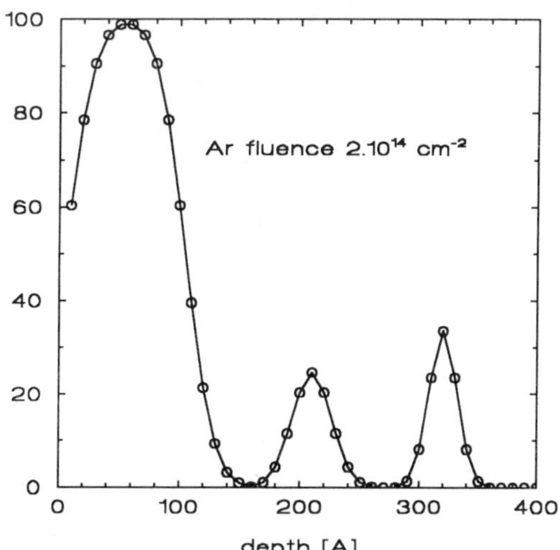

Fig. 3 The Si distribution after implantation; calculated from the measured depth profiles

curve was similar to the depth profile of ion implanted specimen. The concentration distribution produced by ion implantation was calculated by convoluting the original distribution using the above σ value. We can also calculate the ion implantation induced ion mixing directly by T-DYN.

This curve can be compared with the previous one as it is shown on Fig. 2. The amazingly good agreement supports the applicability of our approach.

The same calculation was applied for the measured depth profiles. Fig 3 shows the calculated silicon distribution produced by an implantation of a fluence of $2*10^{14}$ ion/cm^2. The result is similar to the measured profiles but shows a higher degree of mixing than that calculated by TRIM.

We calculated the σ^2 values for the thick layer transitions and for the second δ layer (the first overlap with the tail of the thick layer transition causing calculation problems) for all three fluences. Fig. 4 shows the ion bombardment induced variance, σ^2, versus deposited energy (given by the T-DYN simulation). The reasonable linear dependence also supports the applicability of our method.

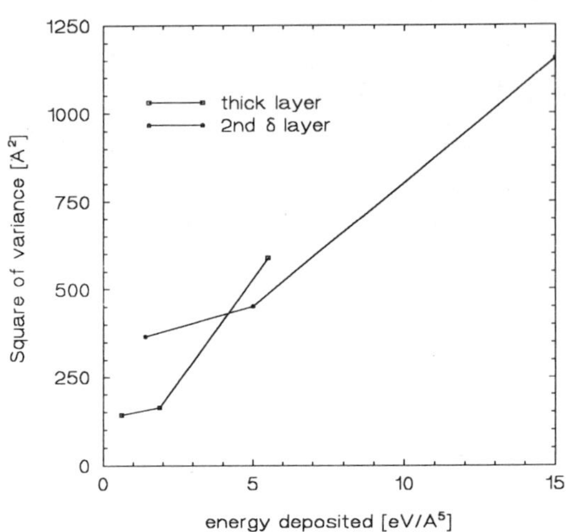

Fig. 4 Variance values calculated from measured profiles versus deposited energy

Conclusion

1. Auger depth profiling was successfully applied for the study of ion implantation induced mixing, including the determination of the broadened profile.

2. A procedure was developed to determine the variance of the broadened concentration distribution and to restore the implantation induced mixing from the depth profiles.

3. The implantation induced variance versus deposited energy curve was calculated and was found to be quasi linear.

Acknowledgements

This work was supported by the Hungarian Fund for Scientific Research through grants OTKA T14252 (A.S) and T15880 (M.M). Part of the work was completed by A. Galisova while she worked in Budapest. Her stay was possible by the help of the CEEPUS Mobility Grant Network A-2 . We are thankful for A. Debreceni and K. Pajer to perform the implantation.

References

1. W. Bolse, Mat. Sci. Eng, R12, 53 (1994)
2. J.J. Vajo, Eun-Hee Cirlin, R. G. Wilson, and T.C. Hasenberg JAP 72 90 (1992)
3. J.P. Biersack, Nucl. Instr. Methods **B27**, 21 (1987)
4. K. Jarrendahl, J. Birch, L. Hultman, L.R. Wallenberg, G. Radnoczi, H. Arwin and J-E. Sundgren, Proc. Mat. Res. Symp. **258**, 571 (1992)
5. A. Barna and M. Menyhard, phys.stat sol. (a) **145**, 263 (1994)
6. M. Menyhard, A. Barna, and J.P. Biersack, J. Vac. Sci.Techol. A **12** 2368 (1994)
7. M. Menyhard, A. Barna, J.P. Biersack, K. Järrendahl, and J-E Sundgren, J. Vac. Sci.Technol. A **13** 1999 (1995)

Materials Science Forum Vols. 248-249 (1997) pp. 249-252
© *1997 Trans Tech Publications, Switzerland*

Defect Formation by Low Energy Ions during Sputter Deposition of TiW and Au on Epitaxially Grown n-Si at Different Plasma Pressures

P.N.K. Deenapanray, F.D. Auret, G. Myburg, M. Hayes,
W.E. Meyer and C. Schutte

Department of Physics, University of Pretoria, Pretoria 0002, South Africa

Keywords: Sputter Deposition, Magnetron System, Defects, DLTS

Abstract: The electronic and annealing properties of the main defects introduced during RF sputter deposition of TiW and Au contacts on n-Si were investigated using current-voltage (IV) and deep level transient spectroscopy (DLTS) techniques, combined with isochronal and isothermal annealing, respectively. It was found that the barrier heights of the sputter deposited diodes were lower than that of a control diode fabricated using resistive evaporation. Furthermore, the barrier heights of the sputter deposited diodes were found to decrease with a decrease in the Ar-plasma pressure used. Sputter deposition of Au and TiW introduced five and four prominent levels below the conduction band, respectively. No direct evidence could be obtained for the introduction of V_2- and VP-centres during sputtering. The sputter-deposition-induced (SDI) defects may, amongst others, be attributed to impurities in the Si, and to energetic metal and Ar atoms from the plasma impinging the Si.

I. Introduction

Sputter deposition is a versatile technique used during metallization for the stoichiometric deposition of compounds and refractory metals on semiconductors. Compared to other deposition techniques such as resistive evaporation, sputtering also offers the possibility of control over the deposition rate and better film adhesion to the substrate [1]. During sputter deposition, both the metal contact and the semiconductor substrate material are subject to bombardment by energetic particles. These particles include low energy Ar ions from the plasma, neutral Ar atoms which are backscattered from the target, sputtered metal atoms, energetic electrons, X-rays and photons [2]. The interaction between these particles and the substrate results in the introduction of donor-type defects [3] at and close to the metal-semiconductor interface, which modify its barrier height. In addition, the electrical, optical and structural properties of the semiconductor, and hence the characteristics and performance of the devices fabricated on them, can be adversely affected. Hence, a reproducible and controlled introduction of these defects can be used to obtain barrier heights of specific values. Other Schottky barrier diode (SBD) parameters such as the reverse leakage current and ideality factor, are also affected by sputter-deposition-induced damage. The extent of this damage depends on the deposition parameters such as deposition time and rate, sputter mode and bias conditions, as well as gas pressure and species.

In order to either predict the influence of defects on device characteristics or be able to successfully use them in any form of defect engineering [4,5], it is essential that their electrical and structural properties be established. Several electron and hole traps introduced during sputter deposition on bulk grown Si have been studied using DLTS [6,7]. In this study, SDI defects were characterised and barrier height changes monitored after the sputter deposition of Au and TiW on epitaxially grown Si at different plasma pressures in a magnetron-type system.

II. Experimental

Epitaxially grown (111) oriented n-type Si layers with a free carrier concentration of 6×10^{15} /cm^3, grown on n$^+$-substrates, were used in this study. The samples were chemically cleaned after which circular TiW and Au Schottky contacts of 0.75 mm diameter, and 200 nm and 500 nm thick, respectively, were sputter-deposited through a metal contact mask in an RF magnetron-type sputtering system at plasma pressures of 2×10^{-3}, 7×10^{-3}, and 2×10^{-2} mbar, respectively. For control purposes, 300 nm thick circular Au contacts were resistively deposited through the same contact mask. Ohmic contacts were formed on the n^{+}-substrates by using a liquidus In-Ga alloy prior to diode and defect characterization.

The quality of the SBDs were evaluated by I-V and C-V measurements. The barrier heights (ϕ_b^{IV}) for TiW diodes were extracted from the log I-V curves by using the Beguwala and Crowell three-point method [8]. The C-V barrier height (ϕ_b^{CV}) was calculated from a graph of $1/C^2$ vs V_R by using a reverse bias, V_R, of between 0 and 1 V, while the section $1 < V_R < 3$ V of the same graph was used to determine the free carrier density. The defects induced during the sputter deposition process were characterised by deep level transient spectroscopy (DLTS) [9]. The DLTS "signatures" (energy level, E_t, and apparent capture cross-section, σ_a) of the defects were determined from DLTS Arrhenius plots of log (e/T^2) vs $1/T$, where e is the emission rate at a temperature T.

Isochronal annealing of the Au sputter-deposited SBD (7×10^{-3} mbar) was performed between 50 and 200°C (in steps of 50°C) in an argon atmosphere for 15 min periods. I-V and C-V measurements were used to monitor the changes in the non-ideality factor (n) of the diodes and their barrier heights, and the Si free carrier density. To study the removal rates of two defects, with overlapping DLTS peaks at approximately 63 K (see section on **"Results and discussion"**), isothermal annealing experiments were conducted at 75 °C.

III. Results and discussion
A. I-V and C-V results

The I-V barrier heights of the SBDs analyzed were found to be lower than those by the C-V method. The greater fluctuations observed for C-V than for I-V measurements are explained by the deviation from linearity of the $1/C^2$ vs V_R curve in the region below 1.0 V reverse bias.[10] Since this deviation leads to inaccuracies in the C-V measurements, the discussion that follows will be based on the I-V results.

Fig. 1 shows the variation of barrier heights (ϕ_b^{IV} and ϕ_b^{CV}) as a function of the argon plasma pressure for Au SBDs. It is observed that ϕ_b^{IV} increases from (0.731 ± 0.007) eV at a pressure of 2×10^{-3} mbar to approximately (0.74 ± 0.01) eV at a pressure of 2×10^{-2} mbar. In the case of TiW diodes, the barrier height was (0.54 ± 0.01) eV and (0.52 ± 0.01) eV at 2×10^{-3} and 2×10^{-2} mbar, respectively. Mullins and Brunnschweiler [3] have proposed that sputter damage creates donor-type traps close to the semiconductor surface which result in the lowering of the I-V barrier height in n-type material. Applying the above model to our results reveals that less sputter-deposition induced damage is created when the sputter pressure is increased. We explain this by noting that for a given target and plasma, the mean free path is inversely proportional to the pressure during sputtering. Hence, the mean free path between collision of the particles in the plasma and the target atoms will decrease when the sputter pressure is increased. The particles (mostly sputtered metal and argon atoms) which bombard the semiconductor surface in a denser plasma, are less energetic and therefore create less damage [11].

The changes in barrier heights (ϕ_b^{IV} and ϕ_b^{CV}) and non-ideality factor as a function of isochronal annealing temperature are shown in Fig. 2 for a diode deposited at a pressure of 7×10^{-3} mbar. It is observed that the ϕ_b^{IV} and ϕ_b^{CV} values follow opposite trends and converge for

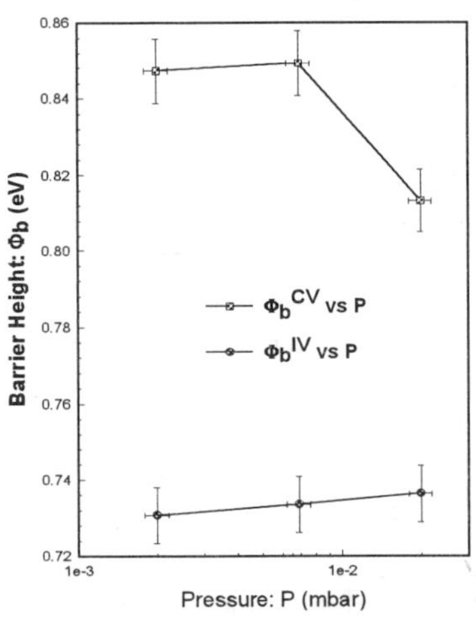

Fig. 1 Au/n-Si SBD barrier heights (I-V and C-V) as a function of argon plasma pressure.

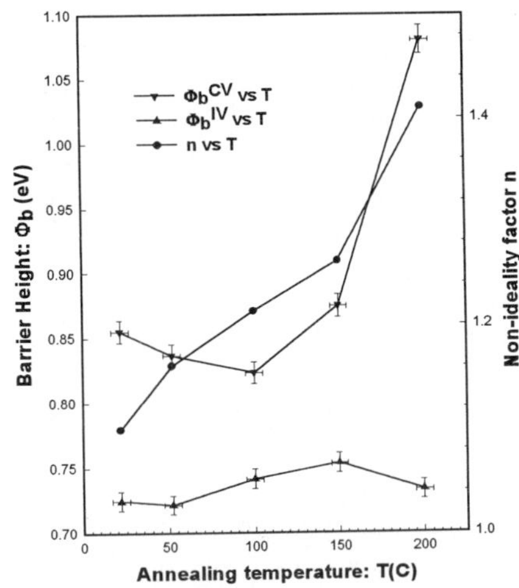

Fig. 2 IV and CV barrier heights and non-ideality factor as a function of annealing temperature for a Au sputter deposited diode at a plasma pressure of 7×10^{-3} mbar.

annealing temperatures up to 150 °C. The I-V barrier height is found to increase from (0.73 ± 0.01) eV before annealing to (0.76 ± 0.01) eV after annealing at 150 °C, which is close to the value of $\phi_b^{IV} = (0.76 \pm 0.02)$ eV for our resistively deposited control sample. As expected, this increase in ϕ_b^{IV} with temperature is characteristic of damage removal. Furthermore, the free carrier density was found to increase from its partially compensated value of 5.1×10^{15} /cm^3 after sputter deposition to approximately 5.9×10^{15} /cm^3 after annealing at 150 °C, which is also indicative of damage recovery during annealing. However, Fig. 2 also shows that the values of ϕ_b^{IV} and ϕ_b^{CV} start to diverge from each other for annealing temperatures higher than 200 °C. The sharp increase in the non-ideality factor shows substantial degradation of the diode with temperature such that the thermionic emission theory used to extract ϕ_b^{IV} is no longer valid.

B. DLTS results

Sputter deposition of Au and TiW introduced five and four prominent traps, respectively, below the conduction band. The electronic properties of these defects are summarised in Table1. Although we have determined the electronic properties of the main SDI defects, their structures remain unknown. To learn more about the structures of SDI defects we have compared them to the primary defects (most of which have been structurally identified) introduced during high energy alpha-particle irradiation [12] and those created during low energy Ar-ion bombardment. Fig. 3 shows the DLTS spectra of RF sputter-induced defects together with the spectra of 1 keV Ar ion bombarded and 5.4 MeV alpha-particle irradiated n-Si. It is observed that EAu2 and ETiW2 are identical to the VO-complex (A-centre). However, no direct evidence was obtained regarding the possible introduction of V_2- ($V_2^{-/0}$ and $V_2^{=/-}$) and VP-centres during sputter deposition of Au and TiW. It has been observed that EAu4 is due to the superposition of 2 or more defect peaks which may include both the $V_2^{-/0}$ and the VP-centre. These overlapping peaks could not be totally separated using different biases and frequencies, or narrow (20 ns) filling pulses. We are currently conducting further isochronal annealing studies (above 200 °C) to investigate the presence of the $V_2^{-/0}$ and VP-centres after sputter deposition. It can be argued that $V_2^{=/-}$ could not be detected because of the presence of stress fields due to extended defects in the region probed by DLTS [13]. EAu1 was observed to anneal out at room temperature. Fig. 4 shows the DLTS spectra of the isothermal annealing of EAu1 and EAu2 at 75 °C for three successive 10 min periods. In this figure it is demostrated that EAu1 is the result of two overlapping defects having almost similar activation energies. We are also conducting further isothermal annealing studies to determine the removal rates of the main SDI defects.

Table 1   Electronic properties of the main SDI defects.

Defect (Au)	E_t (eV)	σ_n (cm^2)	Defect (TiW)	E_t (eV)	σ_n (cm^2)	Defect structure
EAu1	0.112	5.0e-16	ETiW1	0.111	5.5e-16	Ar-related?
EAu2	0.172	1.1e-15	ETiW2	0.174	1.7e-15	VO-centre
EAu3	0.313	2.4e-16				
EAu4	0.379	1.2e-17	ETiW3	0.374	2.8e-17	
EAu5	0.392	3.5e-19	ETiW4	0.449	2.9e-18	

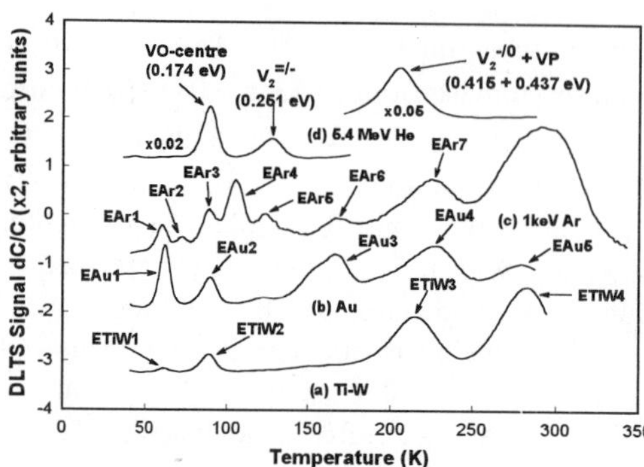

Fig. 3 DLTS spectra of (a) sputter deposited TiW, (b) sputter deposited Au, (c) 1 keV Ar ion bombardment, and (d) 5.4 MeV alpha-particle irradiation of n-Si.

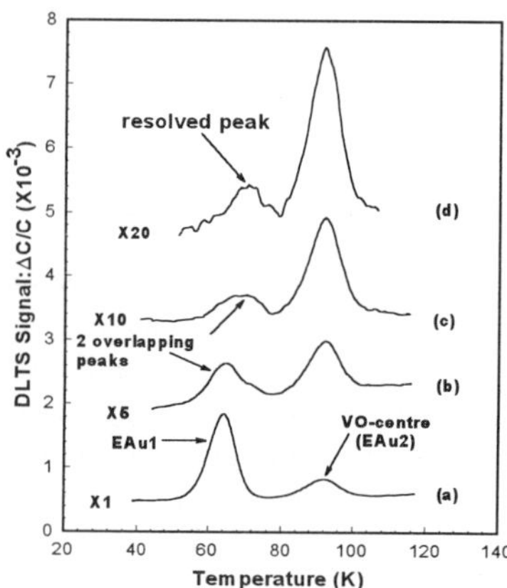

Fig. 4 DLTS spectra showing successive isothermal annealing periods of EAu1 and EAu2 at 75 °C for a Au sputter-deposited diode:(a) unannealed sample, (b) 10 minutes, (c) 20 minutes, and (d) thirty minutes.

III. Conclusions

We have demonstrated that RF sputter deposition of Au and TiW introduces defects near the semiconductor interface. I-V measurements confirm that sputter-induced damage results in a lowering of the barrier height on n-Si. The barrier height is also observed to decrease with a decrease in plasma pressure and this is attributed to an increase in the mean free path of the plasma particles at lower pressures. No direct evidence could be obtained to show the presence of V_2- and VP-centres in our samples. The absence of $V_2=/-$ could be due to the presence of stress fields caused by the presence of extended defects. EAu2 and ETiW2 have been identified to be the VO-complex. Using our Arrhenius diagram, EAu1 is due to the overlapping of two defects. An increase in the barrier height as well as in the free carrier density from its compensated value after sputter deposition during isochronal annealing are characteristic of damage removal. We are currently conducting further annealing experiments to investigate the presence of V_2-and VP-centres together with the removal rates of the main SDI defects.

Acknowledgments

We are grateful for the financial assistance of the South African Foundation for Research and Development and the Carl and Emily Fuchs Institute for Microelectronics. The technical assistance of C. du Toit and J. P. Le Roux is kindly acknowledged. We are thankful to DETEK for making available their sputtering system for this study.

References

[1] L. I. Maissel: Handbook of Thin Film Technology, ed. L. I. Maissel and R. Glan, McGraw Hill, New York, 1 (1970).

[2] E. Ohta, K. Kakishita, H. Y. Lee, T. Sato and M. Sakata, J. Appl. Phys., **65**, 3938 (1989).

[3] F. H. Mullins and A. B. Brunnschweiler, Solid-State Electron., **19**, 47 (1976).

[4] S. Brotherton and P. Bradley, J. Appl. Phys., **53**, 5720 (1982).

[5] V. Rianeri, G. Fallica and S. Libertino, J. Appl. Phys., **79**, 9012 (1996).

[6] O. Paz, F. D. Auret and J. F. White, J. Electrochem. Soc., **127**, 1573 (1980).

[7] F. D. Auret, O. Paz and N. A. Bojarczuk, J. Appl. Phys., **55**, 1581 (1984).

[8] M. Beguwala and C. R. Crowell, J. Appl. Phys., **45**, 2792 (1974).

[9] G. L. Miller, D. V. Lang and L. C. Kimerling, Ann. Rev. Mater. Sci.,**7**, 203 (1961).

[10] F. D. Auret, M. Nel and N. A. Bojarczuk, J. Vac. Sci. Techno., **B 4**, 1168 (1986).

[11] P. N. K. Deenapanray, F. D. Auret, W. E. Meyer and G. Myburg, unpublished.

[12] J. R. Troxell, Solid-State Electronics, **26**, 539 (1983).

[13] B. G. Svensson, B. Mohadjeri, A. Allen, J. H. Svensson and J. W. Corbett, Phys. Rev. B **43**, 2292 (1991).

Materials Science Forum Vols. 248-249 (1997) pp. 253-256

Pulsed Electron Beam Annealing of GaAs after High Dose Implantation of Hydrogen

T. Hauser[1], L. Bredell[1], H. Gaigher[1], H. Alberts[1], A. Botha[2], M. Hayes[1] and E. Friedland[1]

[1] Department of Physics, University of Pretoria, Pretoria 0002, South Africa

[2] Electron Microscopy Unit, University of Pretoria, Pretoria 0002, South Africa

Keywords: GaAs, PEBA, Hydrogen Implantation

Abstract

Undoped (100) GaAs crystals were implanted with 0.5, 1.0 and 2.0 MeV protons at high fluences of about 5×10^{17} ions cm^{-2}. After implantation the samples were subjected to pulse electron beam annealing (PEBA) at variable energy densities.

The samples were analysed with alpha particle channeling as well as with electron channeling patterns (ECP), in order to evaluate the damage introduced during implantation. Surface and compositional studies were performed by means of scanning electron microscopy (SEM) and energy dispersive X-rays (EDX) respectively.

Unimplanted samples exhibit a peculiar surface structure after pulsed electron beam annealing at energy densities above 0.9 J cm^{-2}, consisting of regular rectangles with variable side lengths in the range of 10 to 200 µm. After high dose proton implantation the PEBA threshold energy density for observing these structures could be reduced to 0.4 J cm^{-2}.

Cross sectional SEM samples reveal a cracked subsurface region at the position where these rectangular blocks separate from the bulk crystal. Deeper into the crystal the ion implanted damage distribution, which correlated well with TRIM calculations, was also observed.

1. Introduction

Semiconductors based on $A_{III}B_V$ materials like GaAs are frequently used in device and circuit fabrication. Through the implantation of different elements its electrical and mechanical behaviour can be modified. One element that is used to alter optoelectronic devices based on $A_{III}B_V$ semiconductors is hydrogen [1]. It is also used to create electrically isolating and optically wave-guiding layers [2].

During the implantation of impurities different kinds of defects are introduced into the lattice. In ref. [1] it is suspected that ionisation processes in the target lead to a reduction of the threshold energy E_d for atomic displacements of the target atoms.

There are different methods for defect annealing. Besides furnace annealing, short duration annealing processes such as pulsed electron beam annealing (PEBA), laser annealing or flash radiant heat sources can be used for the annealing of ion implanted GaAs. These methods have the advantage that the samples do not have to be covered with a suitable capping material or necessitate an arsenic overpressure in order to minimize the loss of arsenic from the surface.

Using PEBA after Cr^+ and Pb^+ implantation Alberts *et al.* [3,4] found that only in a relatively narrow range of pulse energy densities (0.38 J cm^{-2} - 0.75 J cm^{-2}) an amorphous layer, introduced during the implantation process, was no longer found. Using higher energy densities (> 0.9 J cm^{-2}) resulted in an arsenic loss at the surface region due to its high vapour pressure.

In this study pulsed electron beam annealing of uncapped GaAs (100) before and after the implantation of hydrogen was investigated by means of α-particle channeling, electron channeling, scanning electron microscopy (SEM) and energy dispersive X-rays (EDX).

2. Experimental

The investigations were performed on (100) oriented undoped semi-insulating GaAs which were implanted at 300 K with 0.5, 1.0 and 2.0 MeV protons incident at 5° to the surface normal. The implanted area on the sample was circular with a diameter of about 3.5 mm. In the channeling analysis a 2.0 MeV He$^+$ beam was used. The hydrogen implantation at different energies and fluences up to 10^{18} ions cm^{-2}, as well as the channeling analysis, were carried out by using a 2 MV Van de Graaff accelerator.

Before and after implantation the samples were subjected to an electron pulse of 300 ns duration as obtained from a pulsed electron beam generator which is similar to the one described by Geerk and Meyer [5].

A JEOL 840 SEM was used in the secondary electron mode at 5 kV in order to reveal fine surface detail. EDX spectra were obtained from a virgin GaAs single crystal, a 1 MeV proton bombarded sample as well as on a 0.9 J cm^{-2} PEBA annealed 1 MeV proton irradiated crystal.

Electron channeling patterns (ECP) were obtained from several rectangular blocks (front and back side) as well as on the region of substrate where the blocks were removed.

The defect profiles and implantation depths as measured on micrographs of SEM cross sections were compared with theoretical calculated profiles and projected ranges obtained with the TRIM code [6].

3. Results and discussions

High dose ($\phi > 10^{15}$ ions cm^{-2}) hydrogen implantation causes an expansion of the lattice due to the creation of lattice disorder and the additional volume required by the implanted atoms [7,8]. This leads to stresses within and around the bombarded region. For high fluences ($\phi \sim 4\times10^{17}$ ions cm^{-2}) the stresses in GaAs are mainly due to the implanted hydrogen[8]. Neethling and Snyman [9,10] observed the formation of hydrogen platelets after furnace annealing at temperatures higher than 500 °C. The strain in the area that is penetrated by energetic ions is positive and perpendicular to the sample surfaces [11].

After the implantation of Cr$^+$ and Si$^+$ into GaAs [3] an amorphous surface layer was observed with Rutherford backscattering, which could partly be annealed with PEBA energy densities of about 0.7 J cm^{-2}. In the present work even high fluences of hydrogen could not create an amorphous layer on the surface of the crystal, as was also seen in [12].

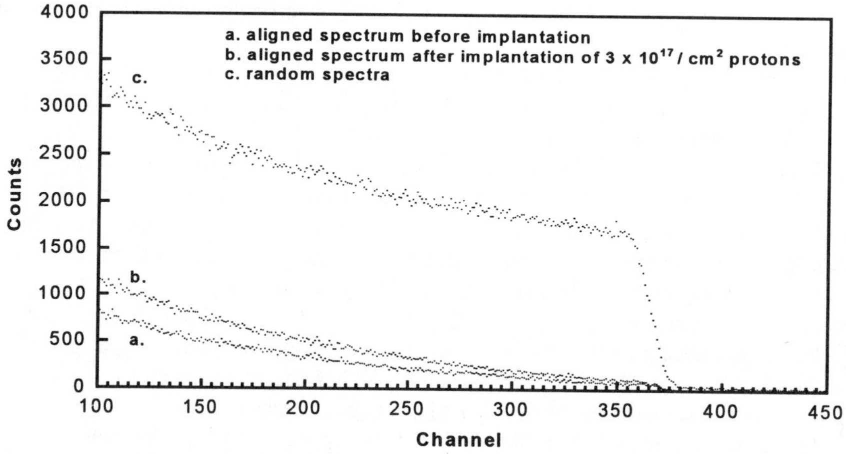

Fig. 1: Channeling spectra before and after implantation of 3 x 10^{17} hydrogen ions cm^{-2}.

 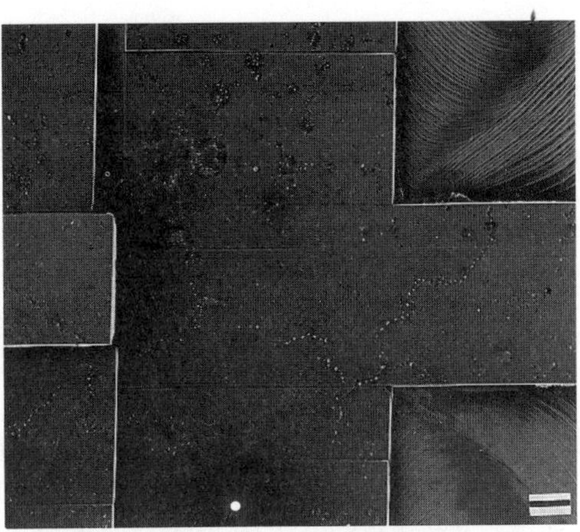

Fig. 2: SEM picture of rectangular blocks formed after proton bombardment and PEBA (Scale bar = 100 μm).

Fig. 3: An uneven surface due to the lifting of blocks. Stress patterns are visible on the subsurface (Scale bar = 10 μm).

In Fig. 1 the channeling and backscattering spectra obtained before and after implantation of 3 x 10^{17} hydrogen ions cm^{-2} are shown. The surface structure after the implantation is basically unchanged. The aligned spectra reveal a slightly higher dechanneling yield for implantation doses higher than 3 x 10^{16} hydrogen ions cm^{-2} (not shown). This might be due to strain and defects in the implanted region. In an attempt to restore the damage introduced in the surface region by the implanted hydrogen, the samples were annealed with a pulsed electron beam of different energy densities. Pulses of about 0.4 J cm^{-2} formed rectangular blocks with variable side lengths in the range of 10 to 200 μm in the implanted area (Fig.2). These blocks tend to split off the surface, producing an uneven surface which precluded further reliable channeling measurements. Characteristic stress patterns are discernible on the subsurface (Fig. 3). ECP's on the front (surface side) as well as the back of blocks revealed loss in crystallinity, while the subsurface (beneath the removed blocks) still remained crystalline. Thus, separation of blocks was confined to the zone between this shallower damage region (which is less than the projected range R_P) and the underlying crystalline region.

The thickness of the blocks does not correspond with R_P values (see Table 1). Clearly this ruled out a flaking mechanism found in implanted and furnace annealed samples [7], where the depth of craters produced on the surface corresponded with R_P.

It was also observed that the thickness of the blocks remained fairly constant (~3 μm) over the PEBA energy range (0.4 - 1.2 J cm^{-2}).

Different implantation energies with corresponding block thicknesses, as well as the experimental and theoretical calculated (TRIM) implantation depths are listed in Table 1 for an implantation dose of ~ 3 x 10^{17} ions cm^{-2}.

Implantation energy (MeV)	PEBA energy density (J cm^{-2})	Thickness of blocks (μm)	Measured depth of implantation (μm)	Calculated depth of implantation (TRIM) R_P (μm)
0.5	1.2	2.5	4.7	4.6
1.0	0.9	3.0	11.3	12
2.0	0.74	3.5	30.6	33.1

Table 1: Different implantation energies with the corresponding block thicknesses resulting from arbitrary PEBA energy densities (> 0.4 J cm^{-2}), as well as experimental and theoretically calculated (TRIM) implantation depths.

The experimental values for the ion range (R_P), obtained from SEM cross sections, corresponded fairly well to those obtained from TRIM predictions.

In order to establish whether ion implantation was a prerequisite for block formation, we treated an unimplanted sample with different pulse energy densities. The threshold energy density to create the blocks was significantly higher (~ 0.9 J cm^{-2}) compared to 0.4 J cm^{-2} for ion implanted crystals.

Energy densities higher than 0.9 J cm^{-2} on implanted samples created blocks in both the implanted and the unimplanted area. The sizes of the rectangles is, however, smaller in the implanted area.

The formation of blocks and their lifting and splitting off the surface is probably due to the pre-introduction of stress in the lattice through implantation. By increasing the stress in the surface region of the sample by the implantation process, the threshold energy density for PEBA to produce the observed blocks on the GaAs surface is reduced from ~ 0.9 J cm^{-2} to ~ 0.4 J cm^{-2}.

The most widely spaced planes in the GaAs crystal are the {110} cleavage planes. The block formation on the surface could be produced by cleavage due to stress components perpendicular to cleavage planes produced by surface heating and cooling during PEBA.

High thermal stresses were produced by rapid temperature changes during pulsed annealing [13]. In our study stresses introduced during PEBA are thought to interact with the stress profile due to high dose proton implantation, in order to produce blocks of a characteristic thickness.

This suggests a possible dose dependence for block thickness, which is still to be investigated.

About 2% arsenic loss was detected by EDX on the surface only of PEBA samples, which is consistent with previous observations [3].

4. Summary and Conclusions

GaAs (100) crystals were implanted with high fluences of hydrogen. After PEBA rectangular blocks formed at the surface. To our knowledge this block formation was not previously observed by other authors. The blocks tend to lift off from the surface due to strain in the lattice. The thickness of the blocks does not correspond with the projected range of the implanted hydrogen ions.

The PEBA energy density for block formation was found to be significantly higher on unimplanted samples.

References

[1] C. Asheron, J.P. Biersack, D. Fink, P. Goppelt, A. Manuaba, F. Paszti and N.Q. Khanh, Nucl. Instr. and Meth. B **68**, 443-449 (1992).

[2] J. Dyment, A.J. Spring, J. Thorpe, D.F. King and J. Straus, J. Electron. Mater. **6**, 173 (1977).

[3] H.W. Alberts, H.L. Gaigher and E. Friedland, Mat. Sc.and Eng. B **9**, 331-335 (1991).

[4] H.W. Alberts, H.L. Gaigher and L.J. Bredell, Nucl. Inst. and Meth. B **80/81**, 519-522 (1993).

[5] J. Geerk and O. Meyer, Radiat. Eff. **63**, 133 (1982).

[6] J.F. Ziegler, J.P. Biersack and U. Littmark, The Stopping and Range of Ions in Solids (Pergamon, 1985).

[7] C. Asheron, H. Bartsch, A. Setzer, A. Schindler and B. Paufler, Nucl. Inst and Meth. B **28**, 350-359 (1987).

[8] C. Asheron, A. Schindler, R. Flaggmeyer and G. Otto, Nucl. Inst. and Meth. B **36**, 163 (1989).

[9] J.H. Neethling and H.C. Snyman, J.Appl. Phys. **60**, 941 (1986).

[10] H.C. Snyman, J.H. Neethling and J.S. Vermaak, Appl. Phys. Lett. **39**, 243 (1981).

[11] B.M. Paine, N.N. Hurvitz and V.S. Speriosu, J. Appl. Phys. **61**, 1335-1339 (1987).

[12] H.C. Snyman and J.H. Neethling, Rad. Eff. **69**, 199 (1983).

[13] J.S. Williams, Laser Annealing of Semiconductors (Academic Press, 1982).

Materials Science Forum Vols. 248-249 (1997) pp. 257-260
© *1997 Trans Tech Publications, Switzerland*

Dose Rate and Temperature Dependence of the Crystallization Rate during IBIEC in InAs

R. Schulz[1], T. Fehlhaber, E. Glaser and F. Schrempel

Institut für Festkörperphysik, Friedrich-Schiller-Universität Jena,
Max-Wien-Platz 1, D-07743 Jena, Germany

Keywords: InAs, Ion Implantation, IBIEC, Ion Beam Induced Nucleation, Dose Rate Dependence

Abstract

(100) InAs was irradiated with 100 keV Se^+ ions to prepare a 100 nm thick amorphous surface layer. These as-implanted samples were subsequently irradiated with 1.0 MeV Ar^+ ions at temperatures between 0°C and 70°C to investigate ion beam induced epitaxial crystallization (IBIEC) at different ion dose rates but constant implantation dose. It was found that it becomes possible to reduce the crystallization rate by increasing the dose rate until interfacial amorphization of the layer (IBIIA) occurs. At higher temperatures the dose rate dependence vanishes due to enhanced ion beam induced nucleation and growth within the amorpous layer.

I. Introduction

Recrystallization of damaged and amorphous semiconductors is a major topic in the research field of ion implantation. Besides solid phase epitaxy (SPE) and liquid phase epitaxy (LPE) ion beam induced epitaxial crystallization (IBIEC) of amorphous layers has attracted a lot of interest in the last years. By this technique the generation of athermal growth sites at an amorphous/crystalline (a/c) - interface by high energy ions enhances crystallization. Therefore IBIEC needs lower temperatures than thermal annealing to reach the same crystallization rates. The IBIEC effect has been investigated extensively in silicon [1]. It was found that the crystallization rate follows an Arrhenius dependence on the implantation temperature. Furthermore, at temperatures lower than a reverse temperature T_R ion beam induced interfacial amorphization (IBIIA) can be observed, i.e. the thickness of the amorphous layer increases. T_R depends on the ion dose rate and the nuclear energy deposition by the impinging ions [2, 3].

Thermal annealing of amorphous layers in III-V-compounds is more complicated than in elemental semiconductors. In these materials a high amount of residual damage like stacking faults and microtwins remains. To anneal this damage temperatures are necessary which have to be higher than the disintegration temperature of the compound. It was found that IBIEC suppresses the twin formation [4]. Therefore IBIEC became an interesting alternative to recrystallize amorphous layers of III-V-compound semiconductors. Although a lot of experimental effort has been made the processes responsible for IBIEC and IBIIA in III-V-compounds are not yet clear. In competition to epitaxial regrowth, ion beam induced nucleation and growth within the amorphous layer arises [5]. First experiments have shown that IBIEC and IBIIA in InAs are very sensitive to the ion dose rate [4] .

II. Experimental

In order to produce a nearly 100 nm thick amorphous surface layer single crystalline InAs was irradiated with 100 keV Se^+ ions at liquid nitrogen temperature (LNT) using an ion

---

[1]corresponding author, Tel: +49 3641 635818, Fax: +49 3641 635854, e-mail: schulz@pinet.uni-jena.de

dose of $5 \times 10^{14} \text{cm}^{-2}$. These samples were subsequently irradiated with 1.0 MeV Ar$^+$ ions at temperatures between 0°C and 70°C. The samples were mounted on a copper target holder who's temperature was controlled by a Ni-CrNi thermocouple placed in the center of the holder. For optimum thermal contact the samples were additionally sticked with liquid silver paint. The holder was connected by copper cables to an electrical isolated metal pot of the implantation chamber. This pot could be heated by an 650 W halogen lamp or cooled by the steam of liquid nitrogen. The accuracy of the implantation temperature measurement is approximately 1K.

The temperature dependence of the interface movement during high energy Ar$^+$ implantation was measured at dose rates between $3 \times 10^{11} \text{cm}^{-2}\text{s}^{-1}$ and $2.5 \times 10^{12} \text{cm}^{-2}\text{s}^{-1}$. The dose of $2.5 \times 10^{15} \text{cm}^{-2}$ was the same for all implantations.

The interface position after high energy implantation was determined by Rutherford Backscattering Spectrometry under channeling conditions (RBS/C). The backscattering spectra were measured using 1.4 MeV He ions and a backscattering angle of 170°.

III. Results and Discussion

The results of the high energy Ar$^+$ implantations at different temperatures in comparison to the as-implanted sample are shown in Fig. 1. The dose rate was $5 \times 10^{11} \text{cm}^{-2}\text{s}^{-1}$. The initial thickness of the amorphous surface layer amounts to 95 nm. At 15°C, the thickness of the amorphous layer increases. This means that due to IBIIA mechanism the interface moves into the bulk material.

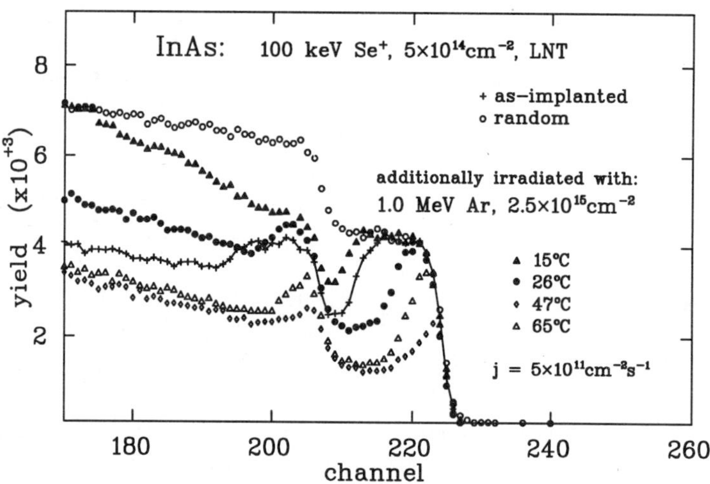

FIG. 1: RBS/C spectra of preamorphized InAs after irradiation with 1.0 MeV Ar using $2.5 \times 10^{15} \text{cm}^{-2}$ at 15°C, 26°C, 47°C and 65°C in comparison to the as-implanted sample. The dose rate is $5 \times 10^{11} \text{cm}^{-2}\text{s}^{-1}$.

At 26°C, the thickness of the amorphous layer decreases which indicates that IBIEC occurs. This means that the reverse temperature should be found in the temperature range between 15°C and 26°C. At $T = T_R$ the processes which are responsible for IBIEC and IBIIA are in balance and no interface movement can be observed. A further increase of the implantation temperature up to 47°C increases the crystallization rate which is in principal agreement with the behaviour of Si. In comparison to 47°C the thickness of the recrystallized layer is reduced at 65°C. This is in contrast to the results obtained for the elemental semiconductors.

This behavior becomes obvious if the thickness of the recrystallized layer is plotted as a function of the temperature as shown in Fig. 2 for three different ion dose rates. Note, that negative values of the "recrystallized" thickness are defined as an increase of the layer thickness by IBIIA.

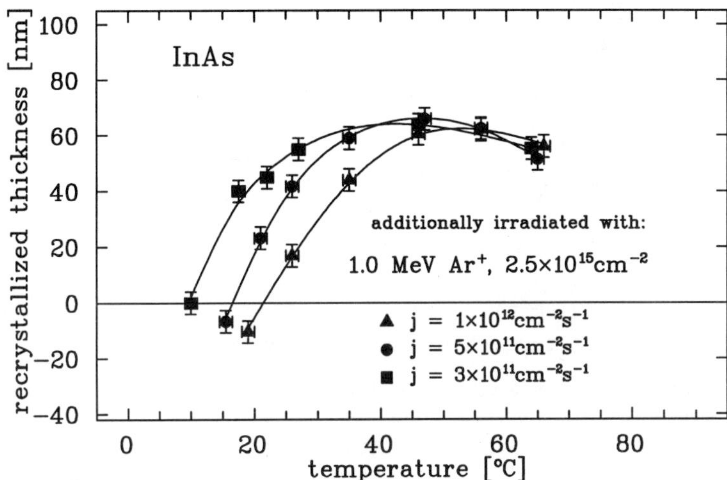

FIG. 2: Thickness of the recrystallized layer as a function of implantation temperature for three different ion dose rates.

Considering the curve for 1×10^{12}cm^{-2}s^{-1} one can see that at 19°C IBIIA is observed. The interface moves 10 nm into the bulk material. At 26°C a thickness of 17 nm is recrystallized by IBIEC. For higher temperatures the crystallization rate increases until at nearly 55°C the crystallization rate (thickness of the recrystallized layer per ion dose) reaches a maximum value. Further increase of the temperature leads to a slight decrease of the crystallization rate. The shape of the temperature dependence is the same for each dose rate and qualitatively the same temperature dependence was found for GaAs, too [6]. It is obvious that the temperature dependence does not follow an exponential function. This deviation from the Arrhenius dependence can be explained by ion beam induced nucleation and growth of crystallites within the amorphous layer which at last blocks the epitaxy of the interface. Because the nucleation rate also depends on the temperature the amount of generated crystallites increases with temperature. These crystallites grow by the IBIEC processes, until the initial amorphous layer becomes completely polycrystalline. If the temperature is high enough nucleation and growth proceed blocked earlier. Therefore a temperature T_C is considered where nucleation and growth become dominant in the amorphous layer and drastically disturb the IBIEC process. In InAs T_C amounts to 55°C for 1.0 MeV Ar$^+$ implantations at a dose rate of 1×10^{12}cm^{-2}s^{-1}.

Furthermore Fig. 2 shows that decreasing the ion dose rate to 5×10^{11}cm^{-2}s^{-1} and 3×10^{11}cm^{-2}s^{-1} the reverse temperature T_R reduces to 17°C and 10°C, respectively. Also the temperature T_C is shifted to lower values. A reduction of the reverse temperature due to a decrease of the ion dose rate also has been found in Si [2]. However, because of the ion induced nucleation the situation in InAs is more complicated.

In Fig. 3 the interface movement is depicted as a function of the ion dose rate for different implantation temperatures. As can bee seen, the crystallization rate decreases with increasing dose rate but the slope varies extremely with temperature. At 15°C it is possible to change

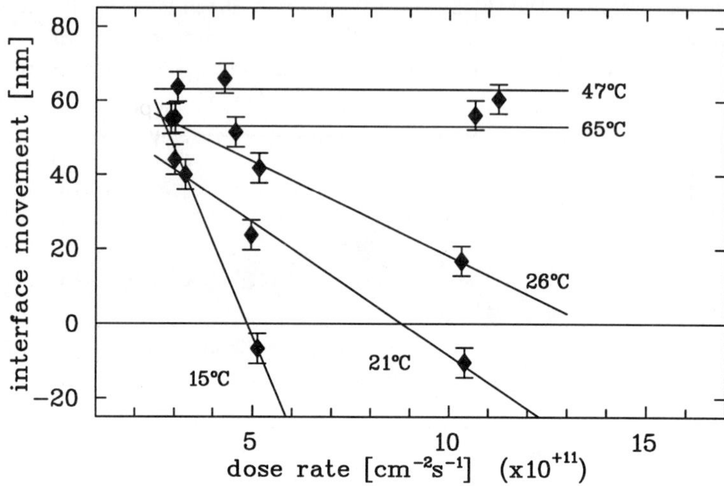

FIG. 3: Interface movement as a function of the ion dose rate for different implantation temperatures.

the direction of the interface movement, i.e. a transition from IBIEC to IBIIA occurs just increasing the dose rate by a factor of 1.66. If the implantation temperature increases the slope decreases. Therefore the critical dose rate where IBIEC and IBIIA processes are in equilibrium can be determined. Assuming a linear dose rate dependence the values amount to 9×10^{11}cm^{-2}s^{-1} and 1.3×10^{12}cm^{-2}s^{-1} for 21°C and 26°C, respectively. In the considered range of dose rates and for temperatures higher than 47°C, the thickness of the recrystallized layer is nearly independend on the dose rate. This is the temperature range behind the maximum in Fig. 2 where the remaining layer becomes polycrystalline. As shown in Fig. 3, at lower temperatures the crystallization rate can be increased by decreasing the dose rate. At higher temperatures, not only the rate of epitaxial regrowth but also the growth rate of the crystallites increases with decreasing dose rate. This leads to a stronger decrease of the rate of epitaxial crystallization at low dose rate as can be seen in Fig. 3. Thus, by competition of epitaxial regrowth and random nucleation and growth, a net growth rate independend on ion dose rate is obtained.

References

[1] see F. Priolo, E. Rimini, Mat. Sci. Rept **5**, 319 (1990) and references therein

[2] J. Linnros, R.G. Elliman, W.L. Brown, J. Mater. Res. **3**, 1208 (1988)

[3] R. Kögler, V. Heera, W. Skorupa, E. Glaser, T. Bachmann, D. Rück, Nucl. Instr. and Meth. **B80/81**, 556 (1993)

[4] E. Glaser, T. Bachmann, R. Schulz, S. Schippel, U. Richter, Nucl. Instr. and Meth. **B 106**, 281 (1995)

[5] T. Bachmann, E. Glaser, R. Schulz, U. Kaiser, G. Sáfrán, Nucl. Instr. and Meth. **B 113**, 214 (1996)

[6] R. Schulz, T. Bachmann, E. Glaser, P. Gaiduk, Nucl. Instr. and Meth., in press

Materials Science Forum Vols. 248-249 (1997) pp. 261-265
© *1997 Trans Tech Publications, Switzerland*

Damage Production in Ion Implanted III-V Compounds:
A Comparative Study

E. Wendler

Friedrich-Schiller-Universität Jena, Institut für Festkörperphysik,
Max-Wien-Platz 1, D-07743 Jena, Germany

Keywords: Ion Implantation, Damage Production, Critical Temperatures, InP, GaAs, GaP, InAs, Rutherford Backscattering Analysis

ABSTRACT
The damage production during ion implantation into InP, GaAs, GaP and InAs is investigated. The influence of the ion mass, the ion energy, the substrate temperature and of the dose rate is discussed. The results for the various materials can be understood within a uniform picture of critical temperatures which are in the order of 350-420K in the phosphides and 250-350K in the arsenides.

INTRODUCTION
In III-V compounds ion implantation can be used to introduce dopands, to produce insulating regions by defect or chemically induced compensation of free carriers, and it can also be used for intermixing of multi-quantum-well structures in order to get lateral confinements. In most of these cases, annealing processes are necessary after implantation to remove the damage and to incorporate the foreign atoms on lattice sites. Systematic investigations of the electrical properties of implanted and annealed layers (especially in GaAs [1] and InP [2]) as a function of the implantation conditions showed that the quality of the electrical parameters increases the lower the damage after implantation is. That means, amorphization should be avoided and implantation conditions have to be used, which keep the damage as low as possible. Therefore, the damage production in GaAs, InP, InAs and GaP has been systematically investigated for various ion energies E_I and target temperatures T_I.

DOSE DEPENDENCES
When comparing the damage evolution in different III-V compounds, which were implanted under identical conditions, significant differences occur [3,4]. To illustrate this, Fig. 1 shows the concentration of displaced lattice atoms, N_{pd}, as determined by RBS [5] versus the ion fluence N_I for 100keV B^+ implantation into InP (Fig 1a) and 280/325keV N^+ implantation into GaAs (Fig. 1b) at room temperature. In the case of the B^+ implantation, the number of displacements per ion and unit length, N^*_{displ}, in the maximum of the distribution amounts to $0.5 \times 10^8 \text{cm}^{-1}$ and in the case of the N^+ implantation to $0.8 \times 10^8 \text{cm}^{-1}$ (This quantity represents the energy deposition into nuclear processes and it is calculated with TRIM-87 [6].). Hence, the two implants are almost comparable. From Fig. 1 it can be seen that in InP N_{pd} increases continuously with the dose and reaches the maximum value $N_{pd}=N_0$ (N_0: atomic density) at $N_I \approx 6 \times 10^{14} \text{cm}^{-2}$. In difference to that, in GaAs the damage saturates for $N_I > 2 \times 10^{14} \text{cm}^{-2}$ at a level of $N_{pd}=3.5 \times 10^{21} \text{cm}^{-3}$ corresponding to $N_{pd}/N_0 \approx 0.1$. Up to an ion dose of $N_I \approx 6 \times 10^{15} \text{cm}^{-2}$ the defect concentration does not change, thus indicating that an equilibrium between defect production and defect recombination exists. It could be shown that for ion fluences below $6 \times 10^{15} \text{cm}^{-2}$ the existence of amorphous zones can be excluded and the defects formed are point defects and point defect complexes [8]. Only after implantation of $6 \times 10^{16} \text{cm}^{-2}$ a buried amorphous layer is formed [8]. Consequently, during implantation of InP and GaAs under similar conditions, the formation of amorphous layers requires an energy deposition into nuclear processes corresponding to 0.77dpa (**d**isplacements **p**er lattice **a**tom given by $N^*_{displ}N_I/N_0$) and 109dpa, respectively. In order to understand this difference, systematic investigations as a function of the implantation temperature were performed on InP [9]. They showed that a damage formation as shown in Fig. 1b for GaAs, also occurs in InP but at temperatures around 400K.

EXISTENCE OF CRITICAL TEMPERATURES
Summarizing all our experimental data we conclude that the damage evolution in InP, GaAs, GaP and InAs can be described within a uniform picture in terms of a <u>critical temperature T_∞</u> which depends on the respective ion-energy-target combination. Values of T_∞ are given in Table 1 for different implantation conditions. It is worth to note that in the phosphides T_∞ lies between ≈ 350K and ≈ 420K, whereas in the arsenides the values are always lower (between ≈ 250K and ≈ 350K). This correlates well with threshold temperatures of defect mobility of ≈ 250K in GaAs, ≈ 350K in InP and ≈ 420K in GaP determined by in situ emission channeling by Hofsäss et al. [16].

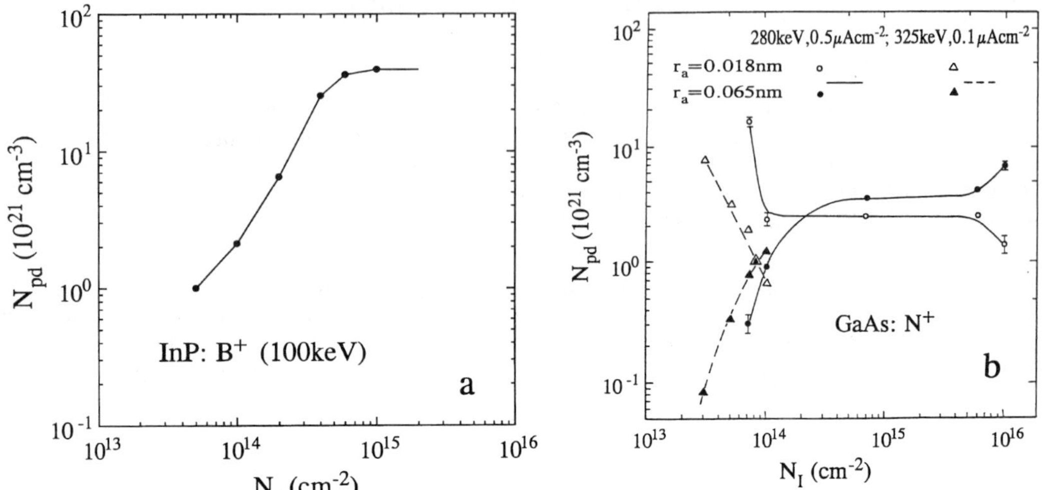

Fig. 1　*Concentration of displaced lattice atoms, N_{pd}, versus the ion fluence N_I.*

(a)　*InP implanted with 100keV B^+ ions at room temperature. N_{pd} is calculated assuming a random distribution of the displacement distance of the displaced lattice atoms.*

(b)　*GaAs implanted with 280/325keV N^+ ions at room temperature. The calculation is made assuming two preferred displacement distances r_a perpendicular to the axis of incidence. N_{pd} for $r_a=0.065nm$ represents the concentration of defects (and the corresponding curve is to be compared with that given in Fig. 1a) and N_{pd} for $r_a=0.018nm$ represents the lattice distortions in the surrounding of the defects (data are taken from Wesch et al. [7], see there for a more detailed discussion).*

	ion	E_I (keV)	T_∞ (K)			ion	E_I (keV)	T_∞ (K)	
InP	B	100	342	[11]	GaAs	N	280	278	[8]
	Si	100	373	[12]		Si	80	317	
	Si	300	396	[9]		Si	100	316	[14]
	O	330	373	[13]		Si	1000	300	
	Ar	200	413	[11]		Ar	200	330	
	Ar	2000	403			Se	600	358	
	Se	600	418	[9]		Se	3000	333	
GaP	B	150	334		InAs	Se/As	2000	300	[15]
	Ar	200	403						
	Se	300	418						

Table 1　*Critical temperatures T_∞ which were estimated on the basis of the defect-outdiffusion model introduced by Morehead and Crowder [10]. This model analyses the ion fluence, which is necessary to reach amorphization, N_I^{am}, as a function of the implantation temperature. T_∞ is that temperature for which N_I^{am} approaches infinity. The uncertainty of the values is different and varies between 2% and 10%.*

If the implantation is performed at $T_I \leq T_\infty$, the primarily produced damage clusters are almost stable and amorphization is achieved by a successive accumulation and overlap of these clusters (see e.g. Fig. 1a). Within this temperature range, an increase of T_I results in an increase of the ion dose necessary for amorphization, which is caused by a reduction of the region damaged by one ion due to relaxation and annealing of defects in the outer part of the damage cascades [10]. For a fixed temperature, the remaining damage is mainly determined by the energy density deposited into nuclear processes, but still depends on the ion mass, especially in the region of low ion masses. This is illustrated in Fig. 2 for ion implantation into GaP at room temperature. In this figure the relative defect concentration $n_{pd}=N_{pd}/N_0$

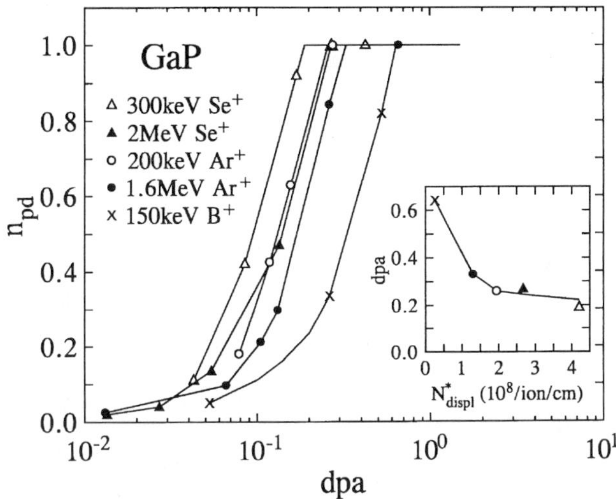

Fig. 2 Relative concentration of displaced lattice atoms, n_{pd}, in the maximum of the distribution versus the ion dose which was recalculated into displacements per lattice atom (dpa; see text) for various implantations into GaP at room temperature. The inset gives the number of displacements per lattice atom which is necessary to obtain maximum damage (i.e. $n_{pd}=1$) as a function of the number of displaced lattice atoms per incident ion and unit depth, N^*_{displ}.

is plotted versus the ion dose which is recalculated into displacements per atom in order to compare the results for different ion species. As can be seen, the maximum concentration $n_{pd}=1$ is reached at different values of dpa which tend to increase with decreasing ion mass and increasing ion energy. These effects are summarized in the inset of Fig. 2 which depicts the number of displacements per atom necessary to reach maximum damage as a function of N^*_{displ}, the number of displacements per ion and unit length. N^*_{displ} is a measure of the effect within a single collision cascade. The results in Fig. 2 show that in the case of B$^+$ ions a larger energy density has to be deposited into nuclear processes in order to obtain maximum damage. This indicates that in more dilute collision cascades, which are to be expected for such a light ion, recombination processes are promoted during the irradiation.

At $T_I \approx T_\infty$ a pronounced defect rearrangement and annealing during implantation takes place, indicating that point defects become mobile at these temperatures. The dose dependence of the relative defect density shows a broad plateau at 0.10-0.15 (see e.g. Fig. 1b, [9,15]), i.e. an equilibrium between defect production and defect recombination occurs. Amorphization can be reached only at very large ion doses, but within a small dose interval. By TEM it could be shown that at the end of the plateau mentioned above small dislocation loops are formed [11]. They may act as centres for the nucleation of amorphous seeds which grow rapidly during the following ion impacts resulting in a collapse-like transition to amorphousness [9]. For the formation of loops and the nucleation of amorphous seeds also the high number of implanted ions has to be taken into account. For example, the transition to amorphousness occurs in GaAs at $N_I \approx 5 \times 10^{16}N^+$cm^{-2} (see Fig. 1b; [8]), which corresponds to a nitrogen concentration of 4at%.

DOSE RATE EFFECTS

An important parameter during ion implantation is the dose rate j which may have a considerable effect on the damage production. In Fig. 3 the integral defect density $N_d = \int N_{pd}(z)dz$ is plotted as a function of j for InP (Fig. 3a) and GaAs (Fig. 3b). In the case of InP, the increase of damage is similar for B$^+$ and Si$^+$ if the implantation is performed at room temperature. For implantation at an elevated temperature (see data for Si$^+$ and $T_I=363$K in Fig. 3a) the slope of the curve $N_d(j)$ is much steeper than for $T_I=295$K. In the case of GaAs, already at room temperature remarkable differences in the dose rate dependence occur for the two different ion species shown in Fig. 3b. The implantation of the light element N$^+$ with a relatively high ion energy results in a very strong dependence of N_d on the dose rate around j=1-2μAcm^{-2}. In fact, the relative defect density changes from ≈ 0.1 (see Fig. 1b) to 1 [18]. On the contrary, for implantation with 100keV Si$^+$ ions the effect is much smaller (Fig. 3b). When comparing the implantation temperatures given in Fig. 3 with the critical temperatures for the corresponding ion/target combinations (see Table 1), it becomes obvious that the influence of the dose rate on the resulting defect concentration is stronger the closer the implantation temperature comes to T_∞. A similar effect was found in ion implanted Ge [14].

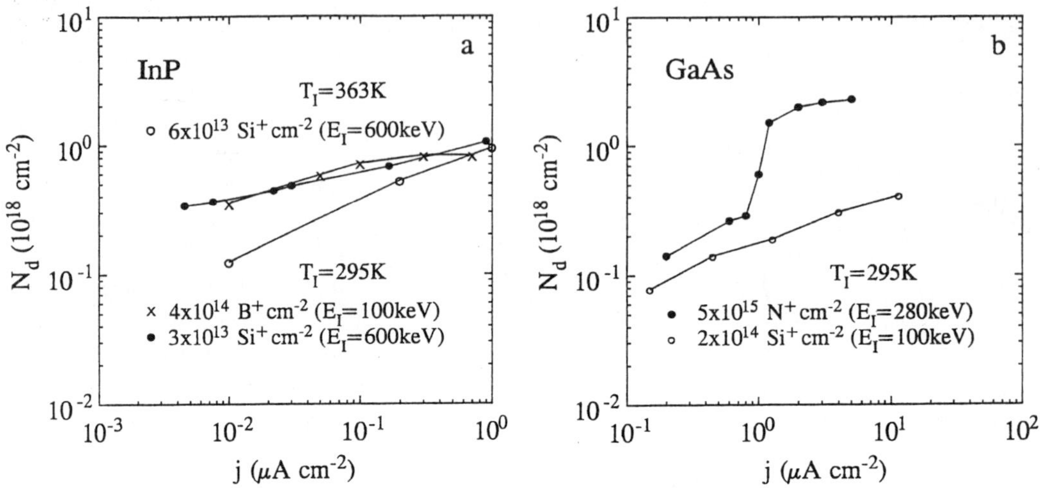

Fig. 3 *Integral defect density N_d versus the dose rate j.*
 (a) 100keV B^+ implantation at room temperature and 600keV Si^+ implantation at room temperature and at $T_I=363K$ into InP (data for the 600keV Si^+ implants are taken from Akano et al. [17]).
 (b) 280keV N^+ and 100keV Si^+ implantation into GaAs at room temperature (data for the 100keV Si^+ implantation are taken from Haynes and Holland [14]).

INFLUENCE OF IMPLANTATION ENERGY

When increasing the implantation energy up to a few MeV, two effects can be observed. First, in the near-surface region a strong defect annealing occurs, which is assumed to be connected with the high electronic energy loss in the corresponding depth region [15,19,20]. During room temperature implantation this self-annealing is found especially in the arsenides, but not in the phosphides [19,20]. And, secondly, the defect concentration in the maximum of the nuclear energy deposition decreases with increasing ion energy even if the same energy density is deposited into nuclear processes [19,20]. This can be seen in Fig. 2 for GaP (cp. the results for Ar^+ and Se^+ ions implanted with a high and a low energy, respectively). Similar results are plotted in Fig. 4 for Se^+ implantation into GaAs. At the low implantation temperature the difference for the two ion energies is low (similar as in the case of GaP, see Fig. 2), but at room temperature the curve for $E_I=3MeV$ is shifted to much larger numbers of dpa. Our results show that these effects of the ion energy become more pronounced, if the implantation temperature comes closer to T_∞.

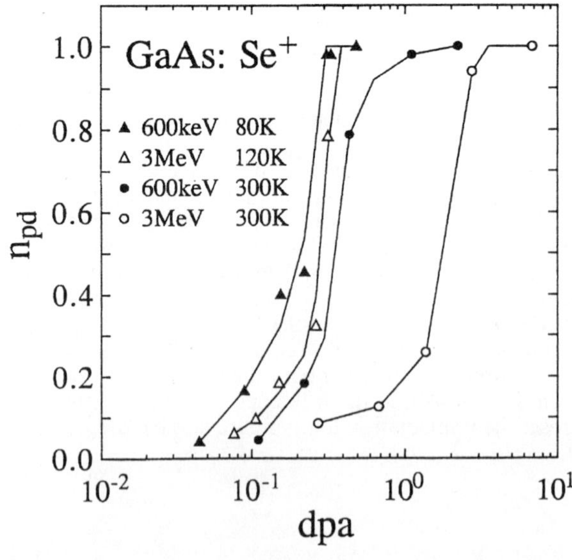

Fig. 4 *Relative concentration of displaced lattice atoms, n_{pd}, versus the ion fluence which was recalculated into displacements per lattice atom (dpa; see text) for Se^+ implantation into GaAs at 80/120K and 300K with 600 keV and 3MeV, respectively.*

CONCLUSIONS

Our systematic investigations of the damage production in various III-V compounds showed that in these materials similar damaging processes occur but at different temperatures. This can be described in the framework of critical temperatures which depend on the respective ion-energy-target combination. Starting from our results it becomes possible to predict the damage evolution which is to be expected in these materials for a certain implantation condition.

REFERENCES
[1]   see e.g. D.V. Morgan, F.H. Eisen, in: <u>Gallium Arsenide</u>, eds. H.J. Howes and D.V. Morgan (Wiley, New York 1985) chap. 5.
[2]   see e.g. P. Müller, T. Bachmann, E. Wendler and W. Wesch, J. Appl. Phys. **75**, 3814 (1994).
[3]   E. Wendler, W. Wesch and G. Götz, Nucl. Instr. and Methods **B55**, 789 (1991).
[4]   E. Wendler, W. Wesch and G. Götz, Nucl. Instr. and Methods **B63**, 47 (1992).
[5]   see e.g. G. Götz and K. Gärtner (Eds.), High Energy Ion Beam Analysis of Solids (Akademie, Berlin, 1988).
[6]   J.P.Biersack and J.F. Ziegler, in: The Stopping and Ranges of Ions in Matter, vol. 1 (Pergamon, 1995).
[7]   W. Wesch, K. Gärtner, A. Jordanov and G. Götz, Nucl. Instr. and Methods **B45**, 446 (1990).
[8]   E. Wendler, W. Wesch and G. Götz, phys. stat. sol. (a) **112**, 289 (1989).
[9]   E. Wendler, T. Opfermann, P. Müller and W. Wesch, Nucl. Instr. and Methods **B106**, 303 (1995).
[10]   F.F. Morehead, Jr. and B.L. Crowder, Rad. Eff. **6**, 27 (1970).
[11]   E. Wendler, T. Opfermann, P. Müller, W. Wesch, P.I. Gaiduk and F.F. Komarov, to be published.
[12]   T.E. Haynes, R. Morton and S.S. Lau, Mat. Res. Soc. Symp. Proc. Vol. **300**, 311 (1993).
[13]   U.G. Akano, I.V. Mitchell, F.R. Shepherd and C.J. Miner, Nucl. Instr. and Methods **B106**, 308 (1995).
[14]   T.E. Haynes and O.W. Holland, Appl. Phys. Lett. **59**, 452 (1991).
[15]   E. Wendler, R.J. Wilson, C. Jeynes, W. Wesch, K. Gärtner, R.M. Gwilliam and B.J. Sealy, Nucl. Instr. and Methods **B96**, 298 (1995).
[16]   H. Hofsäss, S. Winter, S. Jahn, U. Wahl, E. Recknagel, Nucl. Instr. and Methods **B63**, 83 (1992).
[17]   U.G. Akano, I.V. Mitchell, F.R. Shepherd, C.J. Miner and R. Rousina, Can. J. Phys. **70**, 789 (1992).
[18]   E. Wendler, W. Wesch and G. Götz, Nucl. Instr. and Methods **B52**, 57 (1990).
[19]   E. Wendler, T. Bachmann and W. Wesch, Mat. Res. Soc. Symp. Proc. Vol **300**, 331 (1993).
[20]   W. Wesch, E. Wendler, T. Bachmann and O. Herre, Nucl. Instr. and Methods **B96**, 290 (1995).

Materials Science Forum Vols. 248-249 (1997) pp. 267-270
© *1997 Trans Tech Publications, Switzerland*

The Behaviour of ^{12}B in CdTe Studied with β-NMR Techniques

E. Oldekop[1], V. Eyert[1], F. Niedermeyer[1], M. Wienecke[2] and W.-D. Zeitz[1]

[1] Hahn-Meitner-Institut, Glienicker Str. 100, D-14109 Berlin, Germany

[2] Humboldt Universität zu Berlin, Fachbereich Physik, Invalidenstr. 110, D-10115 Berlin, Germany

Keywords: β-NMR, Boron, CdTe, Defects in Semiconductors, DFT, ASW

Abstract

The ß-NMR technique is used to study the implantation and annealing behaviour of boron in CdTe at very low boron concentration. The measurements reveal that a fraction of about 20% of boron is incorporated on a substitutional lattice site at room temperature. With increasing implantation temperature this fraction grows and reaches about 45% at 600K. First principle electronic structure calculations indicate that boron occupies the Te substitutional lattice site, where it acts as an acceptor. The data also suggests that the population of the acceptor like B_{Te} configuration increases with temperature on the expense of readily depolarized boron fraction.

Method

The experimental results that are presented here are obtained by use of the ß-NMR technique. Since this technique is based on ß-active impurities, concentrations as low as 10^8 cm^{-3} are sufficient for these measurements and therefore, the method is suitable to study the behaviour of defects in semiconductors when measurements in an extremely diluted impurity concentration regime are required. Defects play a decisive role when doping semiconductors. Defects are impurities as well as vacancies and interstitials in the crystal lattice.

The method utilizes nuclear reactions to produce ß-active nuclei and takes advantage of the recoil to implant the probe nuclei into the semiconductor. By selecting an appropriate recoil angle a nuclear polarization is achieved which leads to an asymmetric distribution of ß-particles in the subsequent decay. The ß-asymmetry gives rise to different count rates in two detector telescopes which are positioned on each side of the plane defined by the reaction geometry. By applying radio frequency (rf) fields, the polarization may be changed and nuclear magnetic resonances are obtained which carry the information about the lattice site and the dynamics of the implanted probe nuclei.

An rf field with the Larmor frequency $\nu_L = \mu$ **B**/h depolarizes the probe nuclei that are located in undisturbed environments of cubic symmetry. In the equation, μ is the nuclear magnetic moment and **B** is the applied external magnetic field. The CdTe crystals are of zincblende structure and hence possess two substitutional and two tetragonal interstitial sites, all of them showing cubic symmetry. Nuclei implanted on sites of non-cubic symmetry are subjected to quadrupolar interactions and their respective resonance frequencies can be found off the Larmor frequency.

In the following sections, the capability of the method is demonstrated by investigations which have been carried out on the implantation and annealing behaviour of ^{12}B in CdTe single crystals. Here, the ß-active ^{12}B ($I^\pi=1^+$, $\tau=29.4$ ms) is produced by the nuclear reaction ^{11}B(d,p)^{12}B using 1.5 MeV deuterons [1].

Boron in CdTe

Electrically active boron in CdTe has not been detected up to now [2,3]. High dose boron implantations into CdTe were reported to be unsuccessful because expected doping effects were masked by crystal damage effects [2]. Unlike other group III elements like Ga or In which act as donors in CdTe [4], the implanted boron resisted a variety of subsequent annealing procedures, with the result that electrically active boron donor or acceptor levels have not been found [3]. The physical properties of boron in CdTe and even its lattice site upon implantation are not known. The goal of these ß-NMR investigations is to improve the understanding of the incorporation of B into CdTe.

Specially prepared CdTe single crystals are used in the investigation: Prior to the experiment, the CdTe crystals are annealed in an atmosphere with either Te or Cd excess to produce Cd rich CdTe (CdTe:Cd) or CdTe:Te. These annealing processes result in the incorporation of vacancies on the chalcogen or metal sublattices with a maximum concentration of several $10^{16}cm^{-3}$. It is known that chalcogen vacancies act as acceptors thus producing p-type CdTe, whereas metal vacancies act as donors [5].

Figure 1 shows an example of a resonance line of ^{12}B in the CdTe:Cd crystal recorded with low rf power ($B_{1rot}=10,1(3)\mu T$) at the Larmor frequency. The temperature is 614K and the orientation of the external magnetic field **B** is set parallel to one of the crystal's <100> axes. Similar resonance lines are taken at various temperatures between 300K and 614K. Following the approach intoduced by van Vleck [6], an analysis of the linewidths yields that these ^{12}B nuclei are located on substitutional, either Cd or Te, lattice sites [7].

First Principles Electronic Structure Calculations for the System CdTe:B

Based on the density functional theory (DFT) and the local density approximation (LDA), first principles electronic structure calculations are performed to further the knowledge on the substitutional incorporation site and the electrical properties of the boron. As a calculational scheme the augmented spherical waves (ASW) method is used [8]. Calculations are performed for the four possible cubic sites of the boron. The analysis of the density of states (DOS) and the crystal orbital overlap population (COOP) reveals that, due to strong antibonding Te 5sp - B 2sp orbitals, boron is most likely incorporated on the Te lattice site [7].

The most conspicuous result is derived from the calculated density of states (DOS) which are depicted in Fig. 2 and Fig. 3. As a reference for the subsequent boron substitution of a Te atom we first show the DOS of pure CdTe in Fig. 2. The 4.5eV wide valence band consists of twelve bonding Cd 5sp - Te 5sp bands. States above the gap result from the corresponding antibonding states. Obviously, the energy gap separating occu-

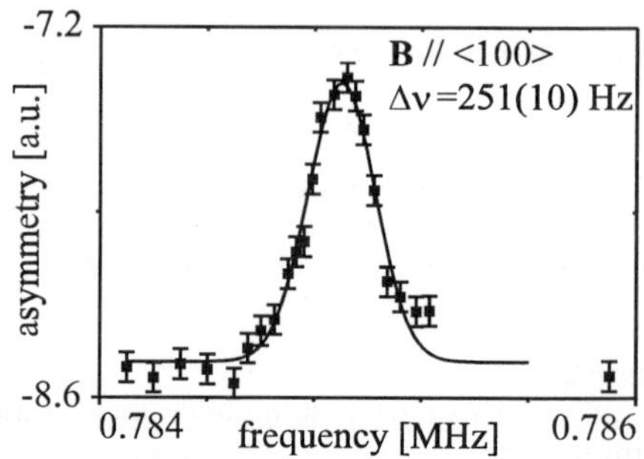

Figure 1: Resonance line recorded at the Larmor frequency in CdTe:Cd at 614K and low rf power.

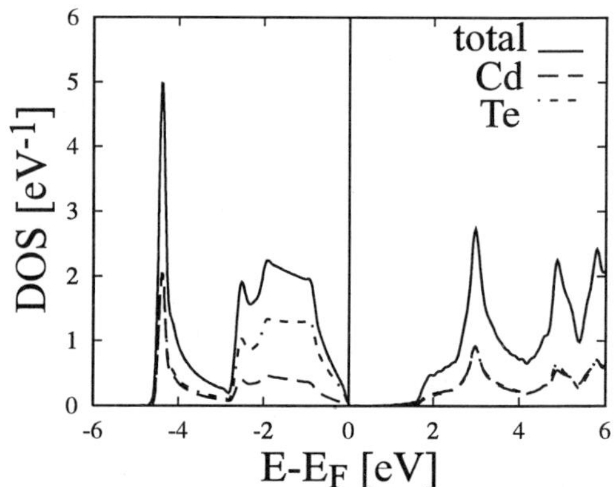

Figure 2: Total and partial densities of states (DOS) of pure CdTe.

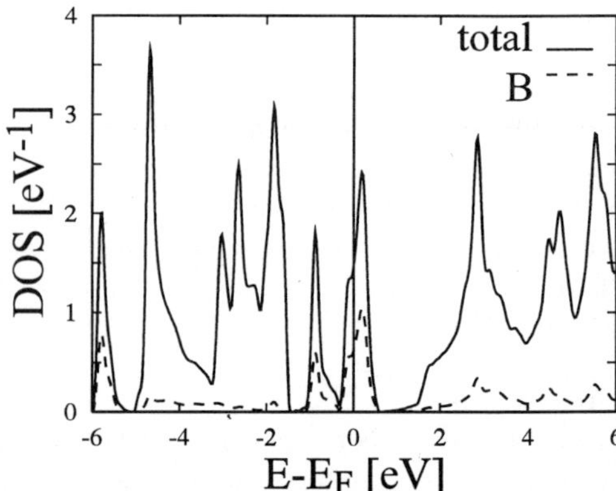

Figure 3: Total and partial densities of states (DOS) of CdTe$_{0.75}$B$_{0.25}$ calculated within a simple cubic super cell. Here, we assumed that boron is incorporated on the substitutional Te lattice site.

pied and empty states is somewhat smaller than the experimental value of 1.45eV. This deviation is known to appear with all DFT/LDA calculations and we point out that DFT, by construction, holds for the ground state only and is not intended to describe excitation processes as those involved in measurements of the insulating gap [9]. The DOS of boron substituted CdTe is displayed in Fig. 3. The previous valence and conduction bands are immediately identified. Due to reduction of symmetry the bands are now composed of several peaks, in contrast to the more plateau like upper half of the valence band of pure CdTe. Moreover, the twelve valence bands separate

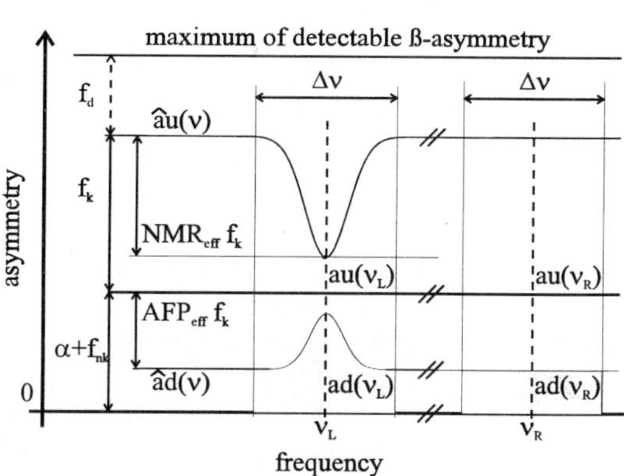

Figure 4: Illustration of the frequency modulated pulse technique.

in two groups: The first group ranges from -6eV to ~ -1.4eV and comprises nine bonding Cd 5sp - Te 5sp bands. The second group straddles the Fermi energy and is dominated by three Cd 5sp - B 2sp bonding bands. The corresponding antibonding bands start at +0.6eV and reflect the former conduction bands. The metallic character of $CdTe_{0.75}B_{0.25}$ is a consequence of the rather high boron content used in the calculation. To summarize the results of the calculation, substitution of Te for B leads to a splitting of the valence band into Te and B derived states which lie below and at the Fermi energy. The empty states related to the boron sites induce an acceptor like behaviour of these atoms.

Annealing Behaviour of the Implanted Boron

In order to study the annealing behaviour of substitutional boron, the CdTe crystals are subjected to temperatures in the range of 300K up to 600K whilst the ß-NMR measurements are performed. An increase of the fraction of boron on substitutional Te lattice sites is expected to manifest itself in deeper Larmor resonances. In the following measurements, frequency modulated rf signals are applied instead of recording frequency resolved functions $\hat{a}u(\nu)$ which can be obtained by use of unmodulated rf frequencies. The technique is illustrated in Fig. 4. There, a set of four different rf signal combinations with identical modulation width $\Delta\nu$ makes up one measurement cycle. The first signal is centered around the Larmor frequency ν_L and the second data point is taken at a reference frequency ν_R, where no depolarisation of the boron's nuclear spin occurs, thus no resonance signal is expected. The signal obtained

is $\int_{\nu'-\Delta\nu/2}^{\nu'+\Delta\nu/2} \hat{a}u(\nu)\,d\nu$ and we will conveniently label this

experimentally observed integral ß-asymmetry value $au(\nu_L)$ or $au(\nu_R)$, according to the central NMR frequency. The difference $au(\nu_L) - au(\nu_R)$ correlates directly with the fraction of boron resonant at the Larmor frequency. The measurements of the integral ß-asymmetry at ν_L and ν_R are repeated in conjunction with the application of an *adiabatic fast passage*

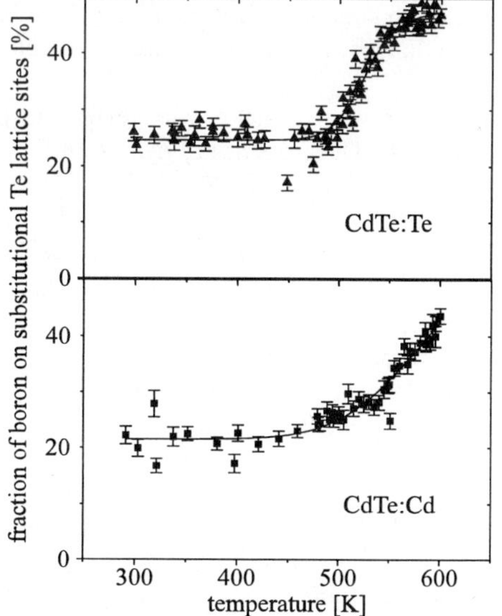

Figure 5: Increase with temperature of the fraction of B_{Te} in CdTe:Te and CdTe:Cd single crystals.

(AFP) pulse. The integral ß-asymmetry values are labeled ad(ν_L) and ad(ν_R). The AFP pulse is centered at the Larmor frequency and flips the orientation of the boron's nuclear spins. Figure 4 depicts how the values of the experimentally obtained integral ß-asymmetry are related to the efficencies NMR_{eff} and AFP_{eff} of the rf signals. In a further step the measurement serves to discriminate between the fractions of boron being in different environments in the CdTe lattice and to indicate how the increase of f_k with temperature is being fed. Here, f_k is the integral ß-asymmetry of the boron that is resonant at the Larmor frequency. The source of any gain in f_k might be the integral ß-asymmetry f_d of readily depolarized boron, i.e. boron that is depolarized during or after the incorporation into CdTe, or the integral ß-asymmetry f_{nk} that stems from the boron which is located at lattice sites of non-cubic symmetry. The method as presented here allows for a determination of the absolute value of f_k, whereas due to the emergence of an experimental ß-asymmetry value α, only relative changes in f_d and f_{nk} can be observed. However, the herein presented method is novel in the respect that it allows to draw conclusions about the source that governs the temperature dependent changes of f_k.

ß-NMR measurements as described above are performed in CdTe:Cd and CdTe:Te crystals in the temperature range between 300K to 600K. There is an annealing behaviour found in the CdTe:Cd and in the CdTe:Te crystal. Fig. 5 depicts the increase with temperature of the fraction of boron that resides on substitutional Te sites in either crystal. Knowing the maximum value of the experimental integral ß-asymmetry that corresponds to signals from the complete ^{12}B ensemble to be 7% [10], the fraction of boron population in undisturbed cubic environment is estimated to be approximately 20% at room temperature. In both CdTe crystals the following holds: After passing a threshold temperature of approx. 500K, a higher temperature leads to an increase of the fraction of B_{Te} and at 600K a fraction of approx. 45% of the implanted boron can be found on substitutional Te lattice sites. Furthermore, the shape of the curves suggests a first saturation of the annealing process at 550K in the CdTe:Te crystal. In both crystals it is also found that the fraction f_d decreases while f_k increases. We interpret this as a temperature dependent conversion of readily depolarized boron into the diamagnetic acceptor B_{Te}.

We want to conclude with the remark that for the first time a measurable quantity of boron located in an undisturbed environment has been detected in CdTe crystals.

The authors want to acknowledge fruitful discussions with H. Ackermann and H.-J. Stöckmann, both with the Department of Physics, Philipps-Universität Marburg and S. Hermann, HMI.

References

[1] H. Ackermann, P. Heitjans and H.-J. Stöckmann in: *Hyperfine Interactions of Radioactive Nuclei*, Topics in Current Physics Vol.31, J. Christiansen (ed.), (Springer) Berlin 1982

[2] R. C. Bowman, jr., R. L. Alt, P. M. Adams and J. F. Knudsen, J. Cryst. Growth **86**, 768 (1988)

[3] K. M. James and J. L. Merz, J. Appl. Phys. **60**, 3699 (1986)

[4] J. M. Francou, K. Saminadayar and J. L. Pautrat, Phys. Rev. **B 41**, 12035 (1990)

[5] M. Wienecke, H. Berger and M. Schenk, Mater. Sci. Eng. **B 16**, 219 (1993)

[6] J. H. van Vleck, Phys. Rev. **74**, 1168 (1948)

[7] E. Oldekop, W.-D. Zeitz, V. Eyert and M. Wienecke, to be published

[8] A. R. Williams, J. Kübler and C. D. Gelatt, jr., Phys. Rev. **B19**, 6094 (1979)

[9] V. Eyert, *Electronic structure calculations for crystalline materials*, in: *Density Functional Methods: Applications in Chemistry and Material Science*, M. Springborg (ed.), (Wiley) Sussex 1997

[10] E. Oldekop, *ß-NMR-Messungen an ^{12}B in CdTe*, (Shaker) Aachen, 1997

Materials Science Forum Vols. 248-249 (1997) pp. 271-277
© *1997 Trans Tech Publications, Switzerland*

Ion Beam and Electrical Conductivity Analysis of Nanocrystalline α-BN and α-Al$_2$O$_3$ Ceramics Implanted with Ti^{+n} Ions and Annealed

S.M. Duvanov[1,3], A.V. Kabyshev[2] and A.P. Kobzev[3]

[1] Institute of Applied Physics, National Academy of Sciences of the Ukraine,
58 Petropavlovskaja Str., 244030 Sumy, Ukraine

[2] High Voltage Institute, 2^a Lenin Ave., 634050 Tomsk, Russia

[3] Frank Laboratory of Neutron Physics, Joint Institute for Nuclear Research,
141980 Dubna, Moscow Region, Russia

Keywords: Ceramics, Pulsed Ion Implantation, Thermal Annealing, Ion Beam Analysis, Electrical Conductivity Analysis

Abstract   RBS, ERD, resonant elastic Backscattering Spectrometry of ^{1}H$^+$ and ^{4}He$^+$ ions, electrical resistivity analysis were used to study properties of near-surface layers of modified nanocrystalline α-BN and α-Al$_2$O$_3$ ceramics. High-fluence and high-flux pulse implantation of Ti^{+n} ions and post-implantation thermal annealing in vacuum were utilised for modification of the ceramics. As a result of implantation and subsequent annealing, thermally stable (up to 1600 K) low electrical conductive surface coatings were formed. Observed changes in elements depth profiles and in surface electrical resistivity of these coatings as a result of the post-implantation thermal treatment are discussed in connection with temperature evolution of the damage microstructure.

Introduction   Irradiation of α-BN and α-Al$_2$O$_3$ ceramics by high current ion beams with fluence of about 10^{17} cm^{-2} creates in the near-surface layers quasi-equilibrium structure from the amorphous and crystallite grinding state of the initial phase, new compounds and high concentration of defects [1,2]. This metastable state may be stabilised by post-irradiation treatment under vacuum. As a result of implantation and subsequent thermal annealing, high temperature (up to 1600 K) stable conductive surface layers were formed. Most significant changes in dielectric properties (up to dielectric-semiconductor transition) of the samples occur under frequency-pulse regime of ion implantation at current density of about 10^{-2} A/cm^2 [1]. Observed effect of dielectric-semiconductor transition in the ceramics consist particularly in the dramatic increase of conductivity.
Post-implantation annealing not only stabilises surface conductivity but also increases it.
What is main process that produces such considerable changes? It is not clear from the previous investigations.
The goal of this work is to investigate evolution of elements depth profiles and conductivity in the near-surface layers of the ceramics as a result of post-implantation thermal annealing under vacuum.

Experimental   Polished nanocrystalline α-Al$_2$O$_3$ and α-BN samples were implanted with Ti^{+n} ions in plasma-arc DIANA source [3] in the frequency-pulse regime. Optimal regime of implantation is following: implantation energy 50-150 keV; pulse ion current density 10^{-2} to 10^{-3} A/cm^2 ; pulse duration 250 µs; frequency 10 to 50 Hz; charge state of ions Ti$^+$ = 3 %, Ti^{++}=80 %, Ti^{+++} = 17 %; residual gas pressure 5x10^{-3}

Pa; fluence 10^{17} cm^{-2} [1]. For the considerable modification of chemical and phase composition of the ceramics, above mentioned regime is assisted by single or alternated deposition of Ti atoms.   Such modification regime reduces surface resistivity ρ_s by a factor of 2—8. Depth of mutually mixed coating-substrate layers is comparable with projected range of implanted ions. But this depth increases as a result of post-implantation annealing. However the property of the created coating is not stable and is unreversibly changed under the thermal treatment.

Implantation of ions increases the surface conductivity σ. Subsequent thermal treatment favours its further increase. Stabilisation of the coating properties occurs after annealing at 950—1500 K. As a result of ion beam and thermal treatment, thermostable electroconductive surface coating has been formed. Resistivity of this coating is controlled in the range of 10^4—10^{16} Ohm/□ and its temperature coefficient does not exceed the value of 10^{-3}—10^{-4} K^{-1} up to 1500 K [1].

Optimal annealing temperature for the oxide ceramics is between 950 K and 1200 K, and that for the nitride ceramics is between 1100 K and 1500 K. Annealing was carried out in vacuum or in inert gas. Such conditions exclude electronic exchange of the modified surface with active reactants of gas phase, and recovery of surface properties due to electron-ion reactions between defects [4].   More significant transformation of surface properties is observed for compounds with dominant covalent bonding fraction (nitride ceramics), and also under the ion beam mixing regime. Coatings on polycrystalline materials are more stable to a thermal annealing than those on monocrystals.

Depth profiling and electrical resistivity analysis   Depth   profiles of elements in subsurface layers of modified BN and Al$_2$O$_3$ ceramics have been investigated with combination of techniques: RBS of ^{4}He$^+$ ions, resonant elastic BS of ^{4}He$^+$ and ^{1}H$^+$ ions, and ERD of ^{1}H$^+$ ions [5]. Concentration profiles of heavy elements were analysed by Rutherford backscattering of ^{4}He$^+$ ions having 2 MeV initial energy. Well known elastic resonance of ^{16}O(^{4}He,^{4}He)^{16}O reaction near 3.045 MeV was used to obtain the profiles of oxygen atoms in surface layers of the samples. Elastic resonances of ^{12}C(p,p)^{12}C reaction near 1.726 MeV [6] and ^{14}N(p,p)^{14}N reaction near 1.740 MeV [7] were utilized for depth profiling of carbon and nitrogen atoms, respectively. Elastic recoil detection of ^{1}H$^+$ ions with 2.4 MeV initial ^{4}He$^+$ ions analyzed beam was used for hydrogen atoms depth profiling. The analytical techniques were described in more detail in ref. [5].

Annealing of the samples in the temperature range up to 2000 K and measurement of resistive properties of coatings at 300 K have been carried out using apparatus described in more detail in ref. [8].

Results and discussion   Three independent series of the alumina (Al$_2$O$_3$) samples were analyzed in this work. Each series was prepared as follows. One large plate of alumina implanted with Ti^{+n} ions (fluence of 10^{17} cm^{-2}) and then was divided in some pieces which were in turn annealed at various temperatures in vacuum.

Elastic resonance of ^{16}O(^{4}He,^{4}He)^{16}O reaction (BS) near 3.045 MeV was used for oxygen atoms depth profiling in surface layers of the samples. Concentration profiles of titanium atoms were mainly analyzed by Rutherford backscattering (RBS) of ^{4}He$^+$ ions having 2 MeV initial energy.

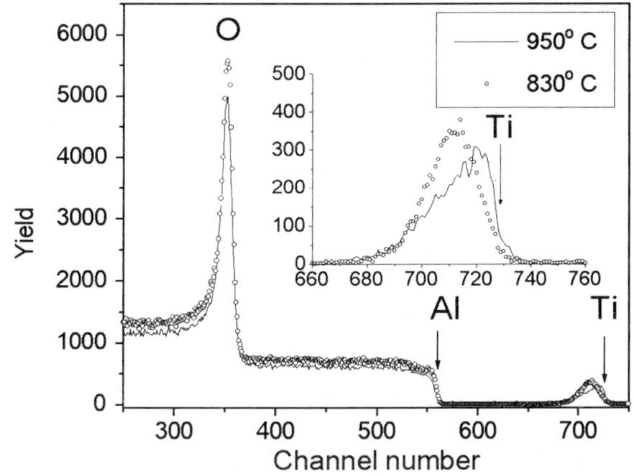

Fig. 1. $^4He^+$ ions BS spectra from Al_2O_3 implanted with Ti and subsequently annealed at different temperatures. The insets show values of annealing temperature and high energy titanium parts of the spectra.

Fig. 1 demonstrates typical BS spectra from the alumina samples implanted with Ti^{+n} ions (fluence of 10^{17} cm^{-2}) and subsequently annealed at different temperatures. Peak at 340 channel is due to the resonance yield (the elastic resonance in $^{16}O(^4He,^4He)^{16}O$ reaction at 3.045 MeV) of helium ions scattered by oxygen atoms present in the layers implanted with Ti atoms. The spectra show that titanium atoms are redistributed to surface as a result of post-implantation annealing at 950° C in comparison with those at 830° C.

Fig.2 shows typical annealing behaviour of titanium and oxygen depth profiles in alumina implanted with Ti^{+n} ions.

Originally as-implanted titanium depth profile are redistributed to surface as a result of post-implantation treatment at different annealing times at 830°C. The same redistribution of the titanium depth profile was obtained for the alumina samples annealed at 900, 950, 1050° C in comparison with those of the samples annealed at 400, 830° C. Time of annealing was about 30 min. After post-implantation annealing at 830° C (12 hours), the titanium depth profile is characterized by a two-step form (fig.2a). This form is typical for the profiles obtained at different annealing regimes at which redistribution of doped titanium atoms was observed.

At the same time the oxygen depth profiles is not changed after post-implantation thermal treatment (fig. 2b). Fig. 2b also illustrates enriched layer in oxygen atoms in the depth range between about 3000 Å and 5000 Å. The similar result on oxygen atoms enrichment had been observed in ref. [9] where glass samples are modified by ion beam assisted deposition of Ti ions.

Fig. 3 demonstrates isochronal annealing behaviour of surface resistivity measured at 300 K. Curves 1 and 2 correspond to the series of the samples with various concentration of doped titanium atoms, respectively. Differences between these

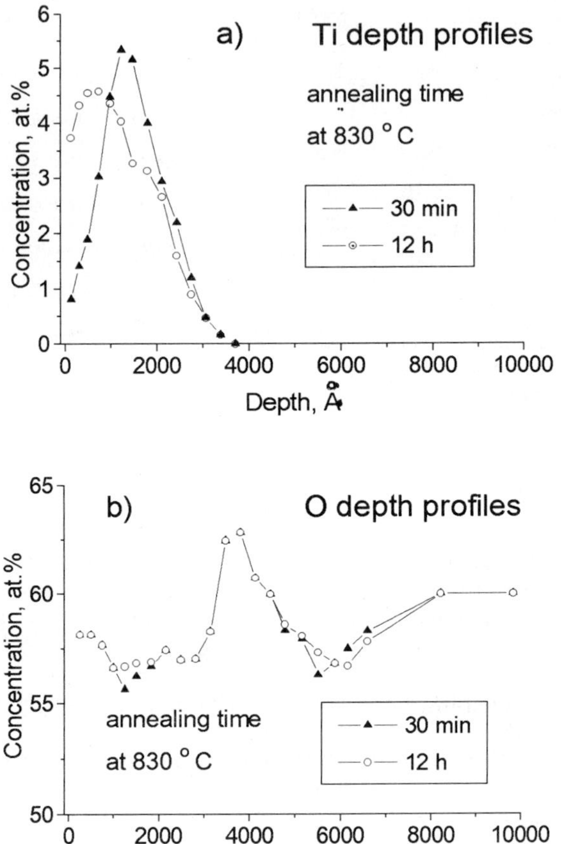

Fig. 2. RBS and BS depth profiles of titanium (a), and oxygen (b) in Al_2O_3 initially as-implanted with Ti ions and subsequently annealed at different annealing times at 830° C under vacuum. Time of annealing is shown in the inset.

curves at initial annealing temperature are caused by different concentrations of implanted atoms.

Complex behaviour of surface resistivity (fig.3, curve2) can be explained   as follows. Two concurrent dominant processes occur during post-implantation thermal annealing: implantation-induced defect evolution and new conductive phase formation.

According to the experimental results on the annealing behaviour of defects in neutron-irradiated sapphire [10], oxygen vacancies (F-centers) starts to anneal at ~ 470 K and proceeds to anneal out in several stages until annealing is completed at ~ 1000 K. The aluminium vacancies (V-centers) are much more stable and do not start to anneal until temperatures > 700 K and do not finally annealed at 1400 K.

The increase of resistivity in the temperature range between about 500 and 1000 K (fig. 3, curve 2) is likely related to the transformation of dislocation agglomerates and vacancies annealing.

At the same time subsequent decreasing of resistivity may be associated with the new conductive phase formation.

Transmission electron microscopy (TEM) of the implanted alumina shows that new synthesized phases start to form at T>750—780°C. Shape and concentration of these new phase agglomerates depends on a temperature value and an annealing time.

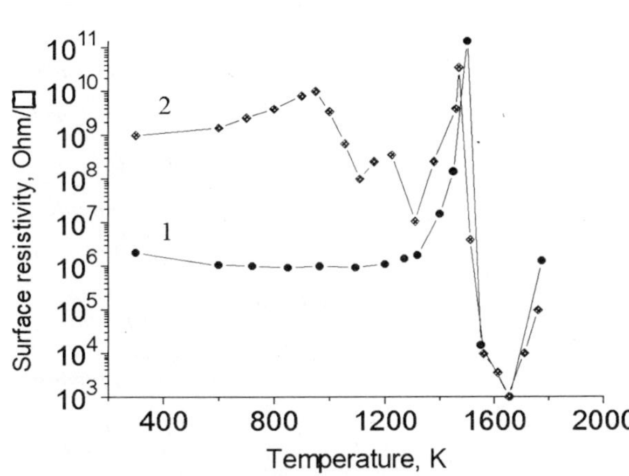

Fig. 3. Measured at 300 K surface resistivity of Al_2O_3 after post-implantation isochronal annealing at different temperatures under vacuum. Curves 1 and 2 corresponds to the series of the Al_2O_3 samples with different concentration of doped titanium atoms, respectively.

Observed redistribution of titanium atoms to a surface at 830 and 950° C is in correlation with V-centers annealing and voids formation. In polycrystalline alumina it has been shown in ref. [11] that voids show a tendency to line up along grain boundaries for high-dose and high-temperature irradiation. There was a 5 nm-wide zone denuded of voids next to the grain boundary. Void formation is enhanced adjacent to the grain boundary only if the grain boundary lies nearly parallel to the basal plane of the hexagonal crystal structure. This effect may result from anisotropic diffusion of vacancies and interstitials.

It is assumed that the fraction of implanted ions occupies substitutional lattice sites for the Al sublattice as a result of post-implantation annealing. This assumption is in agreement with RBS-channeling experiment [12] in which the degree of "substitutionality" in the sublattices of Al_2O_3 has been obtained for a given crystallographic direction by examining the backscattered yield from the impurity in an aligned and "randomly" oriented crystal. The fraction of implanted ions that have substituted for Al (FS) indicates value of 0.9 for Ti.

This assumption is also in accord with the obtained results on the unchanged oxygen depth profile via thermal treatment (fig. 2b).

Thus, the peak on the resistivity function (fig. 3, curve 2) near 1200 K may be associated with the above mentioned process.

Finally, observed dramatic decrease of conductivity near 1530 K is caused by phase transition of created conductive coating to a low-resistive state.

It is important to note that the annealing of implanted alumina samples initially considerably distinguished on the surface conductivity values finally forms coatings having identical conductivity properties.

Fig. 4 shows annealing behaviour of titanium and oxygen depth profiles in BN ceramics implanted with Ti^{+n} ions (fluence of 10^{17} cm^{-2}) and subsequently annealed at different temperature under vacuum.

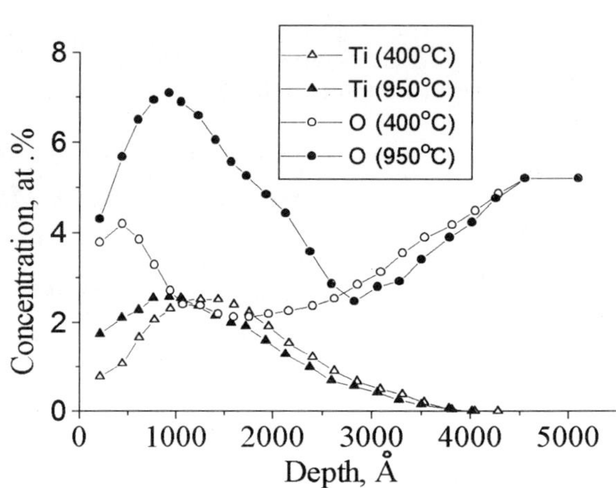

Fig. 4. RBS depth profiles of titanium (a), and oxygen (b) in BN implanted with Ti ions and subsequently annealed at different temperatures under vacuum. Curve symbol of atoms and corresponding annealing temperature are shown on the inset.

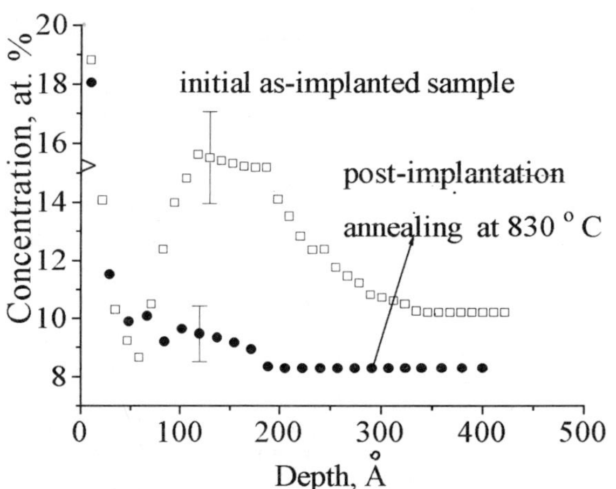

Fig. 5. ERD hydrogen depth profiles in BN implanted with Ti ions and subsequently annealed at temperature of at 830° C. One can be seen the considerable decrease of hydrogen atoms concentration as a result of annealing at 830° C with respect to the as-implanted sample.

In contrast to the unchanged oxygen depth profiles in alumina near-surface layers (fig. 2), those in BN is enriched in oxygen atoms as a result of post-implantation thermal annealing at 950° C. Moreover, depth profiling indicates a correlation between titanium and oxygen depth profiles. Such correlation was also observed in ref. [5,9] in which glass samples were modified by high-dose implantation and simultaneous deposition of Ti ions. This correlation of depth profiles may result from a formation of titanium oxides.

Surface 500 Å layer of BN sample annealed at 400° C is enriched in oxygen atoms. This result is in correlation with ERD analysis of BN implanted with Ti^{+n} ions and subsequently annealed (fig. 5).

Fig. 5 demonstrates that ~350 Å surface layer of as-implanted BN is enriched in hydrogen atoms. Subsequent thermal annealing significantly reduces both a concentration of hydrogen atoms and a depth of hydrogenous layer. Observed H and O concentration profiles behaviour may result from water absorption on the irradiated specimens and from water desorption during subsequent post-implantation annealing.

A decrease of hydrogen concentration via an increase of a temperature of the German nuclear waste glass GP 98/12 implanted with 10^{15} Kr-ions/cm^2 was also observed in ref. [13].

Conclusion   IBA analysis of nanocrystalline α-Al$_2$O$_3$ ceramics modified by high-dose and high-flux pulse Ti^{+n} ions implantation demonstrates redistribution of doped atoms to a surface as a result of post-implantation thermal annealing. Surface electrical conductivity analysis shows a complex annealing behaviour of surface resistivity of the samples and transformation of formed conductive coating to a low-resistive state.   Observed effects of implanted atoms redistribution and annealing behaviour of surface resistivity up to a temperature of 1500 K are in agreement with experimental literature results on the degree of "substitutionality" in the sublattices of Ti ions implanted crystalline Al$_2$O$_3$ and on annealing behaviour of defects in an irradiated sapphire.

In dominant covalent-bonded implanted nanocrystalline α-BN ceramics, observed redistribution of oxygen and hydrogen atoms in depth as a result of post-implantation annealing is significantly distinguished from that in nanocrystalline α-Al$_2$O$_3$ ceramics.

References

[1]   V.V.Lopatin, A.V.Kabyshev and L.S.Bushnev, Phys. Stat. Sol. A, **116**, K69 (1989).

[2]   L.S.Bushnev, A.V.Kabyshev, V.V.Lopatin, Phizika i khimija obrab. mat. **2**, 5 (1990)(in Russian).

[3]   A.I.Aksenov, et al., Prib. i Tekh. Exper. **3**, 139 (1987) (in Russian).

[4]   A.V.Kabyshev, V.V.Lopatin, Poverkhnost'. Phizika, khimija, mekhanika, **7**, 86 (1994).

[5]   S.M. Duvanov, A.P. Kobzev, A.M. Tolopa, D.M. Shirokov, Proc. 11th Int. Conf. on Ion Beam Analysis, Nucl. Instr. and Meth. B, **85**, 264 (1994).

[6]   S.M.Duvanov, A.P.Kobzev, pp.1—10. Preprint of JINR P15-96-69, JINR, Dubna, Russia, 1996 (in Russian).

[7]   E. Rauhala, Nucl. Instr. and Meth. B, **12**, 447 (1985).

[8]   O.I. Buzhinskii, V.A. Butenko, V.V. Lopatin, Prib. i Tekh. Exper. **3**, 236 (1981) (in Russian).

[9]   S.M. Duvanov, A.P. Kobzev and A.M. Tolopa, pp. 741—744 in Ion Implantation Technology—94, Edited by S.Coffa, G.Ferla, F.Priolo, E.Rimini. Elsevier Science B.V., Amsterdam, Netherlands, 1995.

[10] G.P. Pells, J. Am. Ceram. Soc. **77[2]**, 368 (1994).

[11] R.Yamada, S.J.Zinkle, and G.P.Pells, J. Nucl. Mater., **191—194**, 640 (1992).

[12] C.J. McHargue, P.S. Sklad, C.W. White, Nucl. Instr. and Meth. B, **46**, 79 (1990).

[13] Hj.Matzke, G.Della Mea, J.C.Dran, G.Linker and B.Tiveron, Nucl. Instr. and Meth. B, **46**, 256 (1990).

Materials Science Forum Vols. 248-249 (1997) pp. 279-283
© 1997 Trans Tech Publications, Switzerland

Nucleation and Growth of the Amorphous Phase in Na-Irradiated Quartz

F. Harbsmeier and W. Bolse

II. Physikalisches Institut and Sonderforschungsbereich 345, Universität Göttingen,
Bunsenstr. 7-9, D-37073 Göttingen, Germany

Keywords: SiO_2, Radiation Damage, Amorphisation, Ion Beam Analysis, Surface Profiling

Abstract α-quartz was irradiated with 50 keV Na^+ ions at fluences between $5\cdot10^{12}/cm^2$ and $5\cdot10^{16}/cm^2$ and a temperature of 77 K. The induced damage and the related density alterations were monitored by means of Rutherford backscattering spectrometry in channelling geometry and by mechanical surface profiling, respectively. The depth distribution of the implanted Na was determined employing resonant nuclear reactions analysis. Since the ion range distribution agrees well with TRIM95-simulations, the latter code was also used to estimate the energy density F_D deposited in primary and secondary elastic collisions. The damage accumulation is subdivided into two fluence regimes. In the "nucleation regime", a Gaussian-like damage profile was found, which increases in height with increasing fluence ϕ. The relationship between the apparent damage and the deposited energy density was found to be highly non-linear, $\chi(\phi,z) \propto (\phi\cdot F_D)^{3.2}$. At the critical fluence $\phi_c = 4.9(3)\cdot10^{13}/cm^2$ a coherent buried amorphous layer has formed at the depth of maximum energy deposition. This amorphous layer then grows towards the surface and into larger depths. The damage accumulation in this "growth regime" ($\phi > \phi_c$) can still be described by a power law, the exponent, however has changed to $m = 0.1(1)$.

Introduction In the course of a systematic study on the damage accumulation, amorphisation and short range order in covalently bound Si-based ceramics [1-4] we have investigated the structural modification of α-quartz by Na -ion irradiation. Within the context of this study SiO_2 is of special interest, since it is the most ionic compound and is of insulating character. Therefore any influence of the electronic structure on the ion beam amorphisation process should be best visible in this compound. In fact, quartz is amorphised already at very low displacement rates of about 0.06 dpa [5], while 0.2 dpa are necessary for the amorphisation of SiC [1,2]. To study in detail the amorphisation kinetics, we have investigated the microstructural evolution of α-quartz as a function of the applied fluence of light, almost insoluble Na-ions. The choice of the latter was to allow for measuring the distribution of the implanted foreign atoms via RNRA. The experimental results were compared to the theoretical predictions of the TRIM95-code [6].

Experimental details (0001) α-SiO_2 samples (1mm $\times$ 10mm $\times$ 10mm) were irradiated with 50 keV Na-ions at the Göttingen heavy ion implanter IONAS [7]. The applied fluences ϕ ranged between $5\cdot10^{12}/cm^2$ and $5\cdot10^{16}/cm^{-2}$. During implantation the samples were in thermal contact to a liquid nitrogen reservoir and the beam current was kept below 1 μA to avoid beam heating. Thus the target temperature was fixed between 80 K and 90 K. Homogeneous implantation over an area of 1cm $\times$ 1cm was achieved by means of an electrostatic beam sweep system. To avoid channelling during implantation the samples were misaligned by 4° with respect to the beam direction. Since part of the surface was masked with a square shaped Fe rod, the irradiation induced volume changes could be monitored by measuring the steps between the implanted and the covered non-implanted area using a DEKTAK ^{3}ST surface profiler. The depth distribution of the implanted Na ions was determined using resonant nuclear reaction analysis (RNRA), utilizing the $^{23}Na(p,\gamma)$ resonance at E_R

= 309 keV (Γ < 20 eV) [7]. A comprehensive description of the setup and the technique can be found in [8]. RBS-C was used to analyse the disorder, which was introduced by the irradiation, and to measure the thickness of the amorphous layer formed during implantation. The analyses were carried out with the 900 keV He^{2+}-beam of IONAS. For details of the technique and the setup refer to [9] and [10], respectively. The apparent damage profiles were calculated with the computer code DAMAGE, which is based upon a procedure given by Walker and Thompson [11] to eliminate the dechannelling effect.

Results and discussion α-quartz is very sensitive to bombardment with 50 keV Na-ions and becomes amorphous already at very low fluences of about $5 \cdot 10^{13}/cm^2$. Fig. 1 shows a series of apparent damage profiles $\chi(z)$ for samples implanted with fluences between $2 \cdot 10^{13}/cm^2$ and $6 \cdot 10^{15}/cm^2$. In fig. 2 the maximum of the damage profile, $\chi^{max}(\phi)$, is plotted versus the fluence ϕ, both on a linear (a) and a logarithmic scale (b). Up to the critical fluence $\phi_c = 4.9(3) \cdot 10^{13}/cm^2$ a Gaussian-like damage distribution appears with its maximum at a depth of about $500 \cdot 10^{15}$ at/cm^2 ($\approx$65nm). At ϕ_c a coherent buried amorphous layer has formed, which during further irradiation grows towards the surface and into larger depths. Obviously defect agglomerates and/or small incoherent amorphous zones are formed in the low fluence regime which therefore is referred to as the "nucleation regime". The high fluence regime on the other hand is dominated by planar growth of the amorphous layer and is labelled the "growth regime".

The S-like shape of $\chi^{max}(\phi)$ in fig. 2a clearly demonstrates, that direct impact amorphisation (each single ion directly generates an amorphous volume within its collision cascade) does not play an important role in the amorphisation of quartz under the given conditions. For such mechanism an initially linear increase of $\chi^{max}(\phi)$ is expected, which then saturates and approximates $\chi^{max}=1$ at higher fluences. However, only slightly better agreement was found with the damage overlap models [12], which accounts for the accumulation of point defects in the crystalline phase until amorphisation occurs after several individual collision cascades have spatially overlapped. A much better description is achieved by a power law, $\chi^{max} \propto \phi^n$, with the exponent n = 3.2 (1), as can be seen in the logarithmic plot of fig. 2b (closed symbols).

Such power law description is not only restricted to $\chi^{max}(\phi)$, but can be extended to the dependence of the entire damage profile $\chi(z,\phi)$ on the spatial distribution of the atomic relocation rate. The latter can be best represented by the deposited damage energy density $F_D(z)$, which was estimated according to $F_D = E_d \cdot (2n_d + n_r)$ [8] using TRIM95 simulations of the number densities of displacement, n_d, and replacement collisions, n_r. E_d =25 eV is the average displacement energy necessary to generate a stable Frenkel pair in α-quartz [5]. Although the simulated ion range distribution agrees quite well with the measured Na-profile, as can be seen in fig. 3, the depth distribution of the deposited energy density $F_D(z)$ does not resemble the apparent damage profile. Good agreement, however, is achieved when plotting $F_D^{3.23}$. In the nucleation regime the apparent damage profile as a function of ion fluence and depth can thus be described by $\chi_{max} = (\phi \cdot F_D(z)/E_c)^n$, where the total deposited damage energy per unit volume, ($\phi \cdot F_D$ (z)), is a measure of the number of displacements per

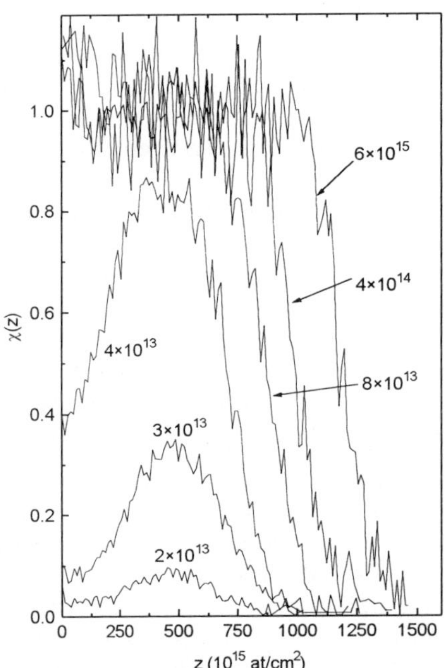

Fig. 1 Damage profiles as deduced from RBS-C

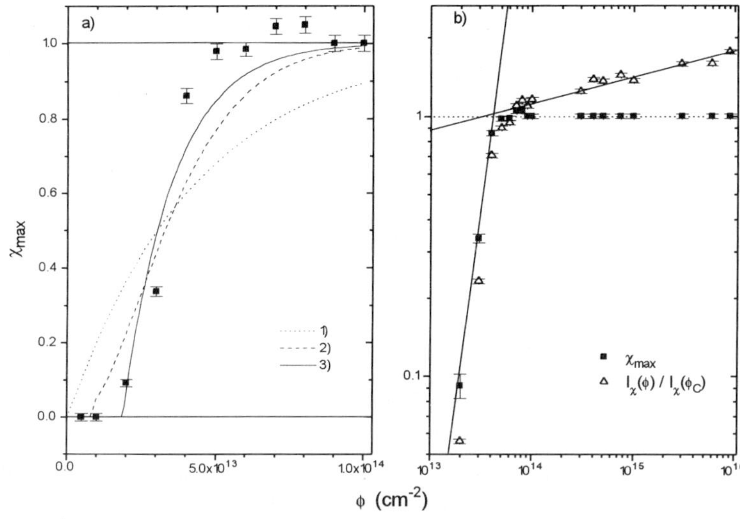

Fig. 2 *Maximum apparent damage χ^{max} and integrated damage I_χ as a function of the applied fluence ϕ. In the linear plot (a) the predictions of the direct impact model (1), the Gibbons damage overlap model (2) and the improved damage overlap model (3) are compared with the experimental data.*

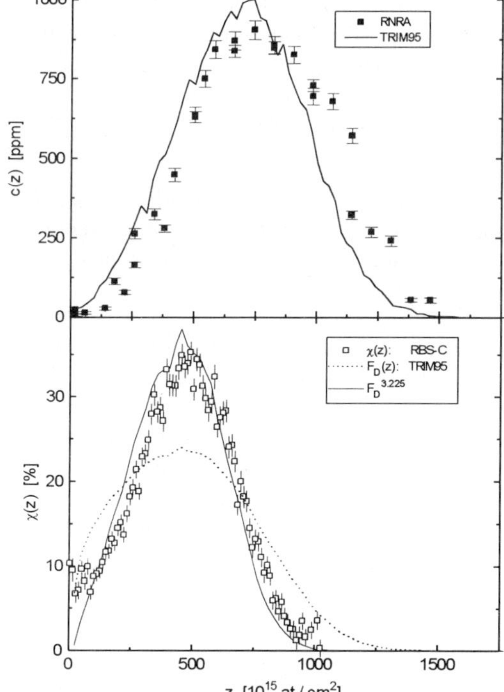

Fig. 3 *Na-distribution and damage profile of the sample irradiated with $3 \cdot 10^{13}$ Na/cm^2. In addition the ion range and deposited energy distribution as predicted by TRIM95 are given.*

target atom at depth z and $E_c = 1.7(2)$ eV/atom is the average critical energy necessary to be deposited per target atom to induce amorphisation. Almost the same power law behaviour with n=3.2(4) was found when evaluating the integral damage $I_\chi = \int \chi(z,\phi)dz$ (open symbols in fig. 2b).

The reason for the continuously increasing apparent damage rate in the nucleation regime is not yet clear. By using single and double alignment channelling techniques, Holland and coworkers [13] found that during self-ion irradiation of Si first vacancy type defects were generated, while a sudden increase of the amount of amorphous-like defect agglomerates was observed shortly before the critical fluence ϕ_c was reached. Because of the flux peaking on the channel axis, vacancy-type defects can only interact with the channeled ion via the induced lattice strain. Therefore, the sensitivity of RBS-C to such type of damage might be reduced as compared to interstitial or amorphous-like defect agglomerates. This would result in an increasing detection probability with increasing disorder.

Another reason for the accelerated damage accumulation below ϕ_c, which was also mentioned by Holland et al, could be the interaction of the vacancy type defects via their strain fields, which may cause a continuously increasing instability of the crystalline structure and a corresponding enhancement of the nucleation of amorphous-like defect agglomerates until a coherent amorphous layer has formed. In fact, we found some hints that defect induced mechanical stresses might play a significant role during damage accumulation and amorphisation in α-quartz. Fig. 4 shows the height of the steps observed at the boundary between irradiated and non-irradiated areas of the sample surface as a function of the number of disordered atoms (i.e. the thickness of the amorphous layer). As can be clearly seen the volume

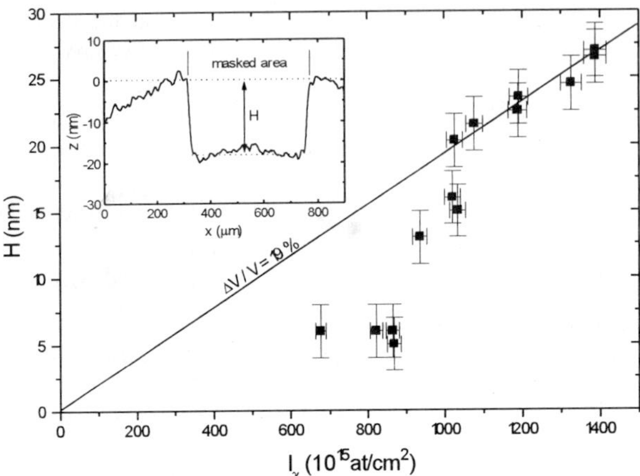

Fig. 4 Step height H between irradiated and non-irradiated areas as a function of the integrated damage I_χ. The insert shows the surface profile taken across the masked area of the sample implanted with $1 \cdot 10^{15}$ Na/cm^2.

swelling ΔV at high fluences, where an extended amorphous layer has formed ($I_\chi < 1000 \cdot 10^{15}$ at/cm^2), is proportional to the thickness of the amorphised zone. This can be explained by a homogeneous density reduction in the amorphous material by about 19%. At lower fluences on the other hand the step height is much less than expected for an amorphous phase of the above density and does not seem to depend on the amount of disordered atoms. This indicates the presence of significant compressive stresses in the irradiated zone even above the critical fluence, at which a buried amorphous layer has formed. This finding is strongly corroborated by the preliminary results of our study on 50 keV Ne-irradiated α-quartz [14], where we have measured the change of the bending radius of thin samples as a function of the ion fluence. These measurements clearly illustrate that irradiation below ϕ_c results in the generation of compressive stress, reaching its maximum of $\sigma \approx 0.7$ GPa at the critical fluence and then starts to be slowly released over a fluence range of about 10 ϕ_c.

In the growth regime ($\phi > \phi_c$) the total induced damage I_χ can still be described by a power law $I_\chi \propto \phi^m$. Its exponent, however, has significantly changed to a value far below unity, m = 0.10(1), which means that the amorphisation rate becomes strongly reduced with increasing ion fluence. This behaviour can be easily understood when taking into account that the deposited damage energy density F_D at the crystalline-amorphous interface is continuously diminished with increasing thickness of the amorphous layer.

Conclusion   In conclusion we have shown that the amorphisation of α-quartz during Na-ion irradiation can be subdivided into a "nucleation regime" at low fluences and a "growth regime", above the critical fluence, where a coherent buried amorphous layer has formed. In the nucleation regime disordering occurs most probably by the generation of point defects and subsequent nucleation of defect agglomerates in the still crystalline matrix. The apparent damage exhibits a highly non-linear power law dependence on the number of displaced atoms, which can only be understood when taking into account further driving forces such as mechanical stresses. The growth regime is characterized by planar growth of the amorphous layer towards the surface and into larger depths at decreasing growth rate.

References
[1] W. Bolse, J. Conrad, F.Harbsmeier, M. Borowski, T. Rödle, this conference.
[2] W. Bolse J. Conrad, T. Rödle, T. Weber, Surf. Coat. Techn. **74/75** , 927 (1995).
[3] M.Borowski, W. Bolse, J. Conrad, A.M. Flank, Proc. XAFS IX (Grenoble 1996).
[4] W. Bolse, S.D. Peteves, F.W. Seris, Appl. Phys. **A58** , 493 (1994).

[5] G. Götz, in P. Mazzoldi,G.W. Arnold (eds.), Ion beam modification of insulators, Elsevier, Amsterdam, Netherlands, (1987).

[6] J.P. Biersack, J.M. Manoyan, Nucl. Instr. Methods B **35** , 215 (1988).

[7] M. Uhrmacher, K. Pampus, F.J. Bergmeister, D. Purschke, K.P. Lieb, Nucl. Instr. Methods B **9** , 234 (1985).

[8] W. Bolse, Mat. Sci. Eng. R. **12** , 53 (1994).

[9] W.K. Chu, J.W. Mayer, M.-A. Nicolet, Backscattering Spectrometry (Academic Press, Orlando, FL. 1978).

[10]J. Conrad, doctoral Thesis, Universität Göttingen (1996).

[11] R.S. Walker and D.A. Thompson, Nucl. Instr. Methods **135** , 489 (1976).

[12] J.R. Dennis, E.B. Hale, J. Appl. Phys. **49** , 1119 (1978).

[13] O.W. Holland, S.J. Pennycook, G.L. Albert, Appl. Phys. Letters **55 (24)** , 2503 (1989)

[14] F. Harbsmeier , W. Bolse, to be published

Materials Science Forum Vols. 248-249 (1997) pp. 285-288
© 1997 Trans Tech Publications, Switzerland

Vacancy Type Defects in Proton Irradiated SiC

W. Puff[1], A.G. Balogh[2], P. Mascher[3] and H. Baumann[4]

[1] Institut für Kernphysik, Technische Universität Graz, Petersgasse 16, A-8010 Graz, Austria

[2] Department of Material Science, Technical University Darmstadt, Darmstadt, Germany

[3] Department of Engineering Physics, McMaster University, Hamilton, Ontario, Canada

[4] Institut für Kernphysik, J.W. Goethe Universität Frankfurt, Frankfurt/Main, Germany

Keywords: SiC, Irradiation, Defects, Positron Annihilation

Abstract Annealing of defects introduced by proton irradiation of n- and p-type 6H-SiC has been investigated by positron Doppler-broadening measurements. It is shown that in both materials positrons are sensitive to radiation induced defects which anneal out at about 150, 500, 900, and 1300 °C, respectively. During annealing, the formation of larger defect complexes can be observed.

Introduction

Investigations of structural and particularly of radiation induced defects in semiconductor materials are of fundamental importance for the physics of semiconductors. Nevertheless, the behaviour of impurities and especially native and radiation induced point defects in SiC has been studied insufficiently. Only in the recent years some progress has been made in this field, resulting in a few comprehensive reviews [1-3].

The nature of many radiation induced centres, however, has not yet been identified. In this context, it is important to not only apply conventional techniques, such as photo-luminescence spectroscopy (PL), electron spin resonance (ESR), or optically detected magnetic resonance (ODMR), but also newer methods that allow selective investigation of different types of structural defects. Recent years have seen extensive utilisation of the positron annihilation method, which is very sensitive and can be applied selectively to vacancy-type defects [4,5]. The data obtained by this method and their correlation with the results of other investigations have demonstrated the great promise of the positron annihilation technique in the diagnostics of structural defects in semiconductor materials.

Several studies concerning defects in 3C-SiC as well as 6H-SiC and 4H-SiC have been carried out, including a few positron investigations [6-15], but the microscopic identification of defects and the understanding of their electronic properties are far from complete. ESR data on 3C- and 6H-SiC suggest the formation of silicon (V_{Si}) and carbon (V_C) vacancies [16, 17] upon electron irradiation and V_C [18] upon proton irradiation. The charge state of the carbon vacancy and therefore the question if V_C is an effective trap for positrons is still under discussion and there are speculations why the carbon vacancy should not be detectable in positron experiments [6, 7]. In the literature it is suggested that after proton irradiation, V_C^0 exists in n-type material and V_C^+ in p-type material [18], while after electron irradiation, V_C^+ was found in n-type samples [11]. Theoretical considerations suggest, at least for native carbon vacancies, a double donor nature [19]. On the other hand, very recent positron annihilation experiments on electron irradiated 6H-SiC show in n-type material both carbon and silicon vacancies, whereas in p-type material no vacancies were detected [13], while other lifetime experiments could not identify carbon vacancies [3 - 6].

First experimental results of the annealing of non-equilibrium defects show that the stability of the defects depends on a number of factors such as the mode of introduction, radiation dose, growth condition, and type of impurity [6-10]. Because of the high activation energy for interstitial and vacancy diffusion in this binary compound, primary radiation defects, in particular isolated vacancies, are thermally stable at room temperature, and far above. It has been recently established, that the silicon vacancy in 3C-SiC is thermally stable up to 750 °C [17, 20]. Comparable data for the carbon vacancy are not known. Under annealing at higher temperatures the mono-vacancies become mobile and can form associated intrinsic or extrinsic point defect pairs, as divacancies, or vacancy-impurity complexes.

At temperatures above 750 °C, when the silicon vacancy in SiC becomes mobile, complexing of V_{Si} with other intrinsic point defects is likely to occur. As in the case of secondary radiation defect centres in silicon [21], complexes of the type V_{Si}-oxygen, V_{Si}-donor, V_{Si}-acceptor, as well as divacancy centres, may be formed. Similar complexes with V_C are likely to occur. However, a microscopic understanding of such defect complexes in SiC so far practically does not exist.

In this short contribution we report first Doppler-broadening results of our positron annihilation study (lifetime and Doppler-broadening measurements) to investigate the annealing behaviour of point defects in n- and p-type proton irradiated SiC in the temperature range between room temperature and 1600 °C.

Experimental details

The starting material used in this work were two sets of research grade 6H-SiC single crystalline wafers obtained from Cree Res. Inc., one with n-type conductivity due to 1.7×10^{18} cm^{-3} nitrogen doping, the other one with p-type conductivity due to 6.2×10^{18} cm^{-3} boron doping. The specimens were irradiated with 5 MeV protons to a fluence of 1×10^{16} cm^{-2}. The temperature of the specimens was kept below 220 K during irradiation. The 20 min isochronal annealing was done in air from room temperature to 1000 °C and from there to the highest temperatures in an Ar atmosphere.

The Doppler-broadening of the annihilation line was measured using an intrinsic Ge detector with a resolution (FWHM) of 1.18 keV at 497 keV (close to the 511 keV annihilation γ energy). Digital stabilisation on the annihilation peak and on a reference peak (^{207}Bi) was employed. Each measurement lasted 2000 seconds and was repeated at least 10 times. A total of about 4×10^7 counts were recorded for each spectrum. The numerical analysis of the Doppler spectra was performed by determining the shape parameter, S, defined as the ratio of the amount of counts in a fixed central portion of the spectrum and the total amount of counts.

Results and discussion

Fig. 1 shows the S-parameter values as a function of isochronal annealing temperature for both samples. Contrary to earlier positron studies on electron irradiated SiC [13] we observe in n-type as well as in p-type samples a substantial increase of the S-parameter upon proton irradiation. The annealing behaviour, however, is obviously quite different. In the p-type sample we see a distinct annealing stage at about 125 °C. This annealing stage is ascribed to the annealing of silicon vacancies [19]. In the temperature range 300 to 450 °C, we observe some agglomeration of the remaining defects, and thereafter, two annealing stages, one at about 500 °C and the second at about 1000 to 1200 °C. In this sample, all the defects are annealed out at 1300 °C. This is also documented by the lifetime measurements where after this annealing step, the mean lifetime value is the same as for the sample before irradiation ($\tau_m = 148$ ps).

The situation in the n-type sample is quite different. As can be seen from fig.1 the low temperature annealing stage is not observable in this sample. Clearly this is not due to the absence of the silicon vacancies in these samples, but to the predominance of the capture of positrons by the secondary irradiation induced defects (divacancies forming the cascades). In this sample, we observe an annealing stage at about 800 °C, followed by a region where vacancy clusters are formed (documented by the negative annealing stage). Since there is no clustering of defects in the p-doped sample one might arrive at the conclusion that the nitrogen impurity plays a fundamental role in the formation of the defect clusters, as was suggested in [8]. The main annealing stage is at about 1100 °C. At 1500 °C a constant value for the S-parameter is reached, however, the defects are not all annealed out. This is again manifested by the lifetime results. The mean lifetime at this stage is about 20 ps higher than for the as-received sample. This finding is in agreement with earlier results on neutron irradiated SiC [8], where the annealing occurs in two stages in the temperature range 1200 to 1900 °C.

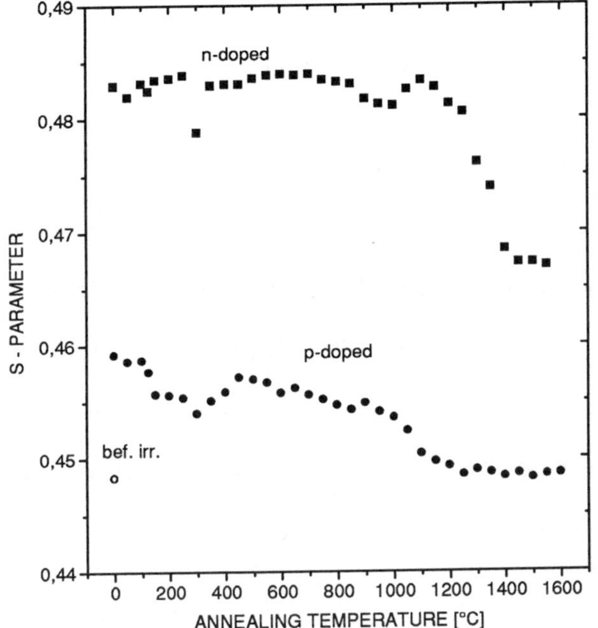

Fig.1: S parameter as a function of isochronal annealing for n (■)- and p-type (●) samples. Also shown is the S parameter value for the as received p-doped sample (O).

Acknowledgements

This work has been supported by the Natural Sciences and Engineering Research Council of Canada.

References

[1]   G. Pensl and R. Helbig, Festkörperprobleme **30** (1990) 133

[2]   J. Schneider and K. Maier, Physica B **185** (1993) 199

[3]   G. Pensl and W.J. Choyke, Physica B **185** (1993) 264

[4]   W. Brandt and A. Dupasquier, *Positron Solid State Physiscs*, North Holland, Amsterdam, (1983)

[5]   Y.-J. He, B.-S. Cao, and Y.C. Jean, *Positron Annihilation*, Trans. Tech. Publ. Aedermanns-dorf, (1995)

[6]   A.I. Girka, V.A. Kuleshin, A.D. Mokrushin, E.N. Mokhov, S.V. Svirida, and A.V. Shishkin, Sov. Phys. Semicond. **23** (1989) 790

[7] A.I. Girka, V.A. Kuleshin, A.D. Mokrushin, E.N. Mokhov, S.V. Svirida, and A.V. Shishkin, Sov. Phys. Semicond. **23** (1989) 1337

[8] A.I. Girka, A.D. Mokrushin, E.N. Mokhov, V.M. Osadchiev, S.V. Svirida, and A.V. Shishkin, Sov. Phys. JETP **70** (1990) 322

[9] A.D. Mokrushin, A.I. Girka, and A.V. Shishkin, phys. stat. sol. (a) **128** (1991) 31

[10] A.I. Girka, A.D. Mokrushin, E.N. Mokhov, S.V. Svirida, and A.V. Shishkin, Mater. Sci Forum **105-110** (1992) 1021

[11] H. Itoh, M. Yoshikawa, I. Nashiyama, L. Wei, S. Tanigawa, S. Misawa, H. Okumura, and S. Yoshida, Hyperfine Inter. **79** (1993) 725

[12] A.A. Rempel and H.-E. Schaefer, Appl. Phys. A **61** (1995) 51

[13] S. Dannefaer, D. Craigen, and D. Kerr, Phys. Rev. B **51** (1995) 1928

[14] W. Puff, M. Boumerzoug, J. Brown, P. Mascher, D. Macdonald, P.J. Simpson, A.G. Balogh, H. Hahn, W. Chang, and M. Rose, Appl. Phys. A **61** (1995) 55

[15] N. Hayashi, H. Watanabe, K. Sakai, K. Kuriyama, Y. Ikeda, H. Maekawa, and T. Miura,

[16] H. Itoh, N. Hayakawa, I. Nashiyama, and E. Sakuma, J. Appl. Phys. **66** (1989) 4529

[17] O. Chauvet, L. Zuppiroli, J. Ardonceau, I. Solomon, Y.C. Wang, and R.F. Davis, Mater. Sci. Forum **83-87** (1992) 1201

[18] H. Itoh, M. Yoshikawa, I. Nashiyama, S. Misawa, H. Okumura, and S. Yoshida, J. Electron. Mater. **21** (1992) 707

[19] J. Bernholc, S.A. Kajihara, C. Wang, A. Antonelli, and R.F. Davis, Mater. Sci. Eng. B **11** (1992) 265

[20] H. Itoh, M. Yoshikawa, I. Nashiyama, H. Okumura, and S. Yoshida, IEEE Trans. Nucl. Sci. **37** (1990) 1732

[21] G.D. Watkins, in: *Deep Levels in Semiconductors,* ed. S.T. Pantelides, Gordon and Breach, Yverdon, (1992) p. 177

Materials Science Forum Vols. 248-249 (1997) pp. 289-292
© *1997 Trans Tech Publications, Switzerland*

On the Amorphization of the Si/Ge Superlattices upon Ion Bombardment

N.A. Sobolev[1,2], K. Gartner[2], U. Kaiser[2], U. Konig[3], H. Presting[3], B. Weber[2], E. Wendler[2] and W. Wesch[2]

[1] Institute of Solid State and Semiconductor Physics, 220072 Minsk, Belarus

[2] Institut für Festkörperphysik der Friedrich-Schiller-Universität, D-07743 Jena, Germany

[3] Daimler-Benz Research Center, D-89013 Ulm, Germany

Keywords: Silicon, Germanium, Superlattice, Ion Implantation, Amorphization

Abstract The growth of damage induced by 150 keV Ar^+ ion implantation in a thin layer Si_9Ge_6 superlattice (SL) was examined by Rutherford backscattering/channeling (RBS) technique using 1.5 MeV He^+ ions and by cross-sectional transmission electron microscopy (XTEM) at voltages up to 200 keV. Supporting TRIM calculations of the damage profiles and layer intermixing were carried out. It was found that the Si and Ge layers in the SL are amorphized simultaneously at a dose very close to that of the amorphization of pure Ge contrary to previous results by other authors (*Appl. Phys. Lett.* **58**, *523 (1991);* **58**, *951 (1991)*) where a selective amorphization of the layers in the SiGe/Si SLs was observed. It is concluded that the main reason for the coherent amorphization behaviour of the Si and Ge layers in the Si_9Ge_6 SL is the fact that the typical dimensions of the individual damage clusters originating from the collision cascades induced by the primary recoil atoms are comparable to or exceed the SL period of only 22 A. A mutual influence of the very thin Si and Ge layers cannot be certainly excluded.

Introduction Recently there has been a great research activity around the possibility of a selective amorphization of the SiGe/Si and AlAs/GaAs SLs [1, 2]. In both SL systems, preferential damage of one layer type has been observed. Selective amorphization of the faster-damaging layer was demonstrated at higher doses, allowing formation of amorphous-crystalline SLs. However, multilayer structures with high individual layer thicknesses rather than true thin layer SLs were used in the studies cited. We have checked the situation in the case of the strain symmetrized Si_9Ge_6 SL. Contrary to the above mentioned results, the amorphization is found to occur simultaneously in the Si and Ge layers of the SL.

Experimental The Si_9Ge_6 SL samples that we use here are pieces of the wafer B2512 grown by MBE on (001) oriented nearly intrinsic Si substrates. They were characterized in detail elsewhere [3-5]. The buffer layer consists of a step-graded SiGe alloy with a thickness of 650 nm followed by a 500 nm thick $Si_{0.6}Ge_{0.4}$ alloy. The SL having 100 periods with a total thickness of 220 nm was deposited on top of the alloy buffer at 520°C. Prior to the growth of the SL, a monolayer of Sb was deposited to act as a surfactant.

The Ar^+ ion energy of 150 keV was chosen to ensure that the whole damage profile (calculated by TRIM-95) is located within the SL and the buffer layer is not damaged significantly. The implantation in the dose range 2.10^{12} to 5.10^{14} ions/cm^2 was carried out at room temperature in a non-channeling direction (7° off <001>) with the ion current density lower than 1 µA/cm^2 to avoid sample heating.

The damage analysis of the as-implanted samples was performed with Rutherford backscattering spectrometry (RBS) in combination with channeling techniques using 1.5 MeV He^+ ions and a

Fig. 1. RBS spectra taken with the 1.5 MeV He$^+$ ions along the [100] direction of the Si$_9$Ge$_6$ SL irradiated with 150 keV Ar$^+$ ions. Implantation doses are indicated in units of 10^{12} ions/cm^2.

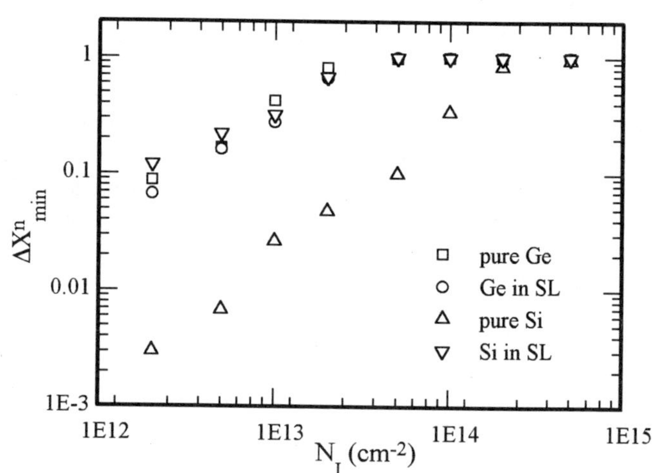

Fig. 2. Normalised minimum back-scattering yield (see text) for the pure crystalline Si and Ge as well as that for the Si and Ge layers in the Si$_9$Ge$_6$ SL plotted vs. the 150 keV Ar$^+$ ion dose.

Fig. 3. XTEM image of the Si$_9$Ge$_6$ SL bombarded with 5.10^{13} Ar$^+$/cm^2.

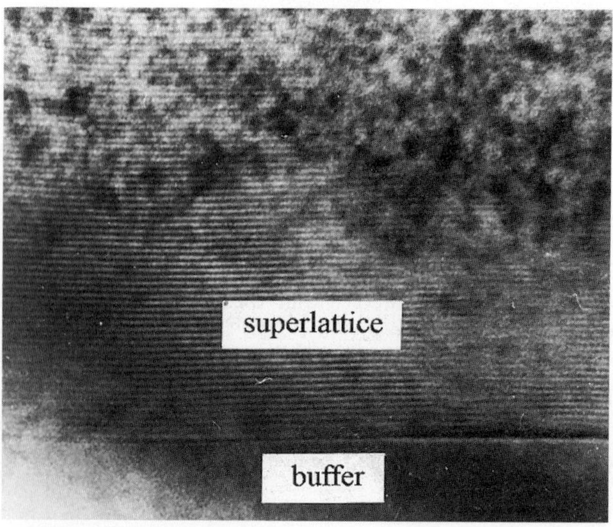

backscattering angle of 170°. For the analysis of the dose dependence of the damage production the normalised minimum yield was used (for further details see below). The cross-sectional transmission electron microscopy (XTEM) investigations were made at acceleration voltages up to 200 keV.

Results The RBS spectra of the as-grown and implanted SL are shown in Fig. 1. The spectrum of the virgin sample indicates a very high crystalline quality of the SL exhibiting a minimum backscattering yield lower than that of the $Si_{0.6}Ge_{0.4}$ alloy buffer layer (not shown in the figure). With increasing implantation fluence, two peaks are seen to grow in the spectrum. They are caused by the backscattering from displaced Ge and Si atoms in the SL (channels 330 - 395 and 230 - 270, respectively). The instrumental resolution is not sufficient to separate the individual Si and Ge layers within the SL having thicknesses as low as 12 A (Si) and 10.5 A (Ge). The backscattering yields in both peaks saturate at the same ion fluence of ca. 5.10^{13} Ar^+/cm^2 which indicates a simultaneous amorphization of the Si and Ge layers in the SL.

For the analysis of the fluence dependence of the damage production the minimum yield $\Delta X_{min} = (Y^{impl}_{al} - Y^{cryst}_{al})/Y_{ra}$ is used (Y^{impl}_{al}, Y^{cryst}_{al}: backscattering yield in channeling (aligned, subscript "al") direction of the implanted and virgin sample respectively, and Y_{ra}: backscattering yield in random (subscript "ra") direction). In Fig. 2 the normalised minimum yield ΔX^{n}_{min} (normalised to its maximum value at $Y^{impl}_{al} = Y_{ra}$) is depicted vs. the ion fluence. As is evident from Fig. 2, the pure Ge crystal as well as the Si and Ge layers in the SL exhibit nearly the same kinetics of the damage growth. Unlike those, the damage level in pure Si at a given fluence is much lower leading to a critical dose being an order of magnitude higher. (The fluence at which the backscattering yield saturates is usually interpreted as the critical dose of amorphization.) The difference in the critical doses for Si and Ge is indeed well known [1, 6].

The SL samples implanted with 2.10^{12}, 5.10^{13}, and 5.10^{14} Ar^+/cm^2 were investigated by the XTEM and electron diffraction. No amorphous regions could be observed in the first sample. In the XTEM image of the second sample, one sees heavy damage (Fig. 3). However, the SL is still crystalline even in the most damaged region though the RBS indicates a saturation of the backscattering yield (Fig. 1). Hence, the true critical dose of amorphization should be slightly higher. Thin layers immediately at the surface and near the interface SL/buffer are less damaged in a good agreement with the RBS profiling and the TRIM calculations. In the third sample having received the highest implantation fluence the SL is amorphous over the whole depth. Only near the interface in a layer as thin as 20 nm one can recognize some rests of the crystalline SL.

Discussion The most striking feature of the results obtained is the simultaneous amorphization of the Si and Ge layers within the SL. This experimental result contradicts the published observations [1, 2] of the selective amorphization of Si and SiGe layers in SLs upon ion bombardment. Hence, first of all one have to seek for the major differences between objects investigated by us and by others. The first difference is that our SL contains pure Ge layers instead of the $Si_{1-x}Ge_x$ ones in [1, 2]. However, the effect of selectivity does only rise with growing x [7, 8], and for x $\gtrsim 0.5$ the amorphization behaviour of the alloys approaches that of pure Ge [7].

The second difference could be the deviating pseudomorphic strain. However, no influence of the strain on the defect accumulation kinetics was found in [8].

And finally, the third difference is that in the thickness of the individual layers in the SL. The XTEM clearly shows that the amorphization in our Si9Ge6 SL proceeds through the accumulation and overlap of individual heavily damaged or amorphous clusters (Fig. 3). These clusters stem from the collision cascades created by the incident ions and the recoil host atoms, and their overlap [9]. It is important to note that in our case the characteristic dimensions of the clusters exceed the thickness of the Si and Ge layers in the SL. This is clearly seen in Fig. 3 and is caused by the fact that the mean projected range of the primary recoil atoms is comparable to or greater than the SL period of 22 A (see e.g. [9]). Thus, the simultaneous amorphization of the Si and Ge layers is caused by the development of the individual collision cascades through several alternating layers. Besides, the mutual influence of the very thin Si and Ge layers cannot certainly be excluded. Hence, with respect to the amorphization process a thin layer Si/Ge SL seems to behave like a $Si_{1-x}Ge_x$ alloy with the corresponding x value. In our case, x = 0.4. According to [7], the damage growth kinetics of a $Si_{0.6}Ge_{0.4}$ alloy should be very close to that of pure Ge (cf. Fig. 2). As to the previously published results [1, 2], SLs with much thicker layers were used which resulted in an independent behaviour of the Si and SiGe layers with respect to the amorphization.

The possible role of the compositional disorder as a result of the ion induced layer intermixing has been checked by means of a TRIM-95 simulation. Two model structures, each one containing a 2 nm thick Si or Ge layer embedded between two 100 nm thick $Si_{0.6}Ge_{0.4}$ layers have been considered using the displacement energy of 15 eV for both Si and Ge and the lattice binding energy of 1 eV. The distributions of Si and Ge recoil atoms upon 150 keV Ar^+ implantation have been simulated. As a result, the admixture of Si atoms to the Ge layer and vice versa does not exceed 1% for the fluence $\Phi = 5.10^{13}$ Ar^+/cm^2 and only 10% for $\Phi = 5.10^{14}$ Ar^+/cm^2, the highest fluence used in this work. Hence, the ion induced amorphization but not the Si/Ge intermixing causes the loss of the layer contrast in the XTEM picture for $\Phi = 5.10^{14}$ Ar^+/cm^2.

Conclusions In a thin layer Si_9Ge_6 superlattice, a simultaneous amorphization of the Si and Ge layers upon 150 keV Ar^+ ion implantation has been observed contrary to the data obtained by others on Si/SiGe SLs [1, 2]. As the main reason of the observed coherent amorphization behaviour of the whole SL we suggest the fact that the mean projected range of the primary Si and Ge recoils producing dense collision cascades is comparable to or exceeds the SL period of only 22 A. A mutual influence of the very thin Si and Ge layers cannot also be excluded. Hence, the reason of the selective amorphization phenomenon published in [1, 2] should be a high thickness of the individual Si and SiGe layers in the SLs resulting in an independent behaviour of the layers upon ion bombardment.

Acknowledgments The authors want to thank G. Lenk for operation at the implanter. One of us (N. S.) acknowledges the financial support by the Sonderforschungsbereich 196 of the Deutsche Forschungsgemeinschaft and by the Basic Research Foundation of Belarus.

References

[1] D. J. Eaglesham, J. M. Poate, D. C. Jacobson, M. Cerullo, L. N. Pfeiffer, and K. West, Appl. Phys. Lett. **68**, 523 (1991)

[2] M. Vos, C. Wu, I. V. Mitchell, T. E. Jackman, J.-M. Baribeau, and J. McCaffrey, Appl. Phys. Lett. **58**, 951 (1991)

[3] U. Menczigar, G. Abstreiter, J. Olajos, H. Grimmeis, H. Kibbel, H. Presting, and E. Kasper, Phys. Rev. B **47**, 4099 (1993)

[4] H. Presting, U. Menczigar, and H. Kibbel, J. Vac. Sci. Technol. B **11**, 1110 (1993)

[5] W. Jager, D. Stenkamp, P. Ehrhart, K. Leifer, W. Sybertz, H. Kibbel, H. Presting, and E. Kasper, Thin Solid Films **222**, 221 (1992)

[6] T. P. Sjoreen, O. W. Holland, M. K. El-Chor, and C. W. White, Mater. Res. Soc. **128**, 593 (1989)

[7] T. E. Haynes and O. W. Holland, Appl. Phys. Lett. **61**, 61 (1992)

[8] D. Y. C. Lie, A. Vantomme, F. Eisen, T. Vreeland, M.-A. Nicolet, T. K. Carns, V. Arbet-Engels, and K.-L. Wang, J. Appl. Phys. **74**, 6039 (1993)

[9] J. F. Gibbons, Proc. IEEE **56**, 295 (1968); **60**, 1062 (1972)

Ion Beam Analysis

Materials Science Forum Vols. 248-249 (1997) pp. 295-300
© *1997 Trans Tech Publications, Switzerland*

Application of HIERDA to Strontium Bismuth Tantalate Ferroelectric Films

P.N. Johnston[1], W.B. Stannard[1], J.F. Scott[2], I.F. Bubb[1], M. El Bouanani[1], D.D. Cohen[3] and N. Dytlewski[3]

[1] Department of Applied Physics, Royal Melbourne Institute of Technology, GPO Box 2476V, Melbourne 3001, Australia

[2] Faculty of Science, University of New South Wales, Sydney 2033, Australia

[3] Australian Nuclear Science and Technology Organisation, PMB 1, Menai 2234, Australia

Keywords: Heavy Ion Elastic Recoil Detection Analysis, Ferroelectric Thin Films, Strontium Bismuth Tantalate

Abstract The new ferroelectric material 'Y1', $SrBi_2Ta_2O_9$ or SBT, has been developed for applications in non-volatile random access memories (NVRAMs) because of its excellent fatigue resistance. In the production of other ferroelectric films developed for similar applications, e.g. BST [1], there has been evidence of interdiffusion during the annealing process. This study uses Mass and Energy Dispersive Heavy Ion Elastic Recoil Detection Analysis (HIERDA), otherwise known as Recoil Spectrometry (RS), with 75-100 MeV ^{127}I ions to determine the amount of interdiffusion in a standard treatment of SBT deposited on Pt electrodes on Si. A new analytical procedure, involving simulation of the Time of Flight and Energy (TOF-E) data from the HIERDA has been implemented. The analysis can be considered in two parts. Firstly, the simulation of the physical processes which take place in the sample and timing foils which yield the distribution of exiting ions arriving at the detector and secondly, the response behaviour of the detectors which determines the pulse height spectrum. The model takes into account straggling and stopping of the incident ions, scattering of the incident ions which gives rise to recoiling ions and scattered ions and determines both energy and intensity values, and stopping and straggling of the exiting ions in the target and timing foils. This approach leads to intermediate data which is stored in linked lists of exiting ions. This data is then combined with detector response functions and energy and time calibrations. The simulated spectra can readily be compared with the raw experimental data.

Introduction

Various new layered perovskite materials for use in NVRAMs, including strontium bismuth tantalate (SBT) and strontium bismuth niobium tantalate (SBNT) need high temperature annealing to ensure high quality ferroelectric properties, in particular high remnant polarisation (P_r).

Typically these layered perovskite materials are deposited on Pt, IrO_2 or RuO_2 electrodes or multi-layered electrodes containing one or more of these materials. Pt is preferred because its properties are well known and it is an element familiar to the semi-conductor industry. However it needs a Ti adhesion layer to avoid peeling from Si substrates. Transmission Electron Microscopy [2] of electrode structures shows major structural changes on annealing at 800°C. The studies show that Ti forms TiO_2 nodules throughout the Pt electrode and there is some evidence of $PtSi_2$ formation near the Si substrate. These problems may be detrimental to NVRAM fabrication if such high temperature annealing is used in the processing. However high temperature annealing may be essential.

In the current work HIERDA is used with 75.6 and 97.5 MeV ^{127}I ions to study the amount of interdiffusion in a standard treatment of SBT and SBNT deposited on Pt electrodes on Si. The emphasis of this paper involves demonstrating how 2-d simulation offers new ways of examining the data for analytical purposes. The analysis of the well separated elements is readily undertaken using the current method or earlier methodologies. The difficulties that we wish to overcome with the new technique relate to separation of overlapping data.

Experimental

Ferroelectric films
The samples analysed in this work are SBT ferroelectric films on Si wafer substrates, produced by Symetrix Corp., Colorado Springs USA., using the "mist" deposition technique. The samples consist of a 200 nm layer of SBT deposited on a 200 nm Pt layer above a 50 nm Ti adhesion layer and a silicon oxide/silicon substrate. They were annealed at approximately 800°C in an oxygen rich atmosphere.

Recoil spectrometry
HIERDA was carried out on the ferroelectric films using the heavy ion recoil facility on the ANTARES 8 MV Tandem Accelerator at the Lucas Heights Research Laboratories of ANSTO. Beams of 75.6 MeV ^{127}I ions were incident on the sample at an angle of 67.5° to the surface normal. Recoil nuclei were detected by a Time of Flight and Energy (ToF-E) detector at an angle of 45° to the incident beam direction. The ToF-E detector consists of a silicon surface barrier (SiSB) energy detector and two time pick-off detectors separated by a flight length of 495 mm as described in previous papers [3, 4].

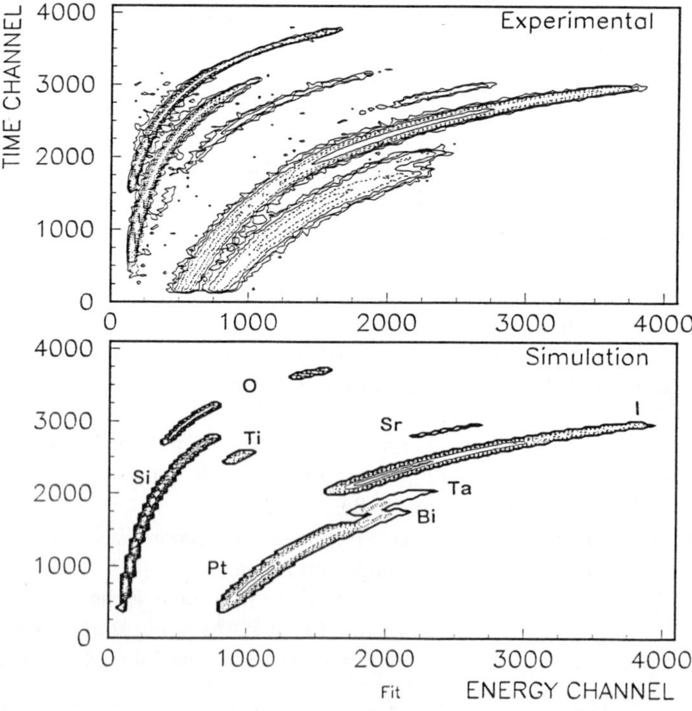

Figure 1. Raw experimental (ToF-E) data and 2-d simulation of the nominal SBT structure.

Analysis Procedure

Analysis using 2-d simulation of the spectral data consists of (i) calibration of the data for time and energy, (ii) simulation of the ToF-E data from physical principals and a model structure based on an estimated sample structure and (iii) refinement of the sample structure until features of the simulation match the experimental data. This allows a quantitative analysis of the depth profile of the sample. The first stages use the CERN analysis package PAW [5] and the macro package TASS [6]. The second and third stages use RMIT software for analysis of ToF-E data in 2 dimensions - 2DTOFE. A contour plot of raw data is compared with a simulation in figure 1

Calibration
Time calibration is performed by considering the ToF of recoiling and scattered ions from the sample surface as described by Stannard et al [1]

The energy calibration is based on pulse height distributions from narrow time slices and for known masses (i.e known energy) using the method of El Bouanani et al. [7] as described by Stannard et al [1]. The centroids and widths of the pulse height distributions are fitted to a model X(E,A) to provide the energy calibration using a weighted non-linear least squares fit (figure 2). This improves on an earlier analytical methodology [7] which relied on unweighted linear regression of the data which gave unwarranted weight to data from heavier elements. The fit of the calibration function to the data is greatly improved. Also, in this case we simulate the ToF-E data so it is necessary to convert energy, for a given nuclear mass, into SiSB detector pulse height, i.e. X(E,A), whereas previously E(X,A) was determined as well as a mass calibration.

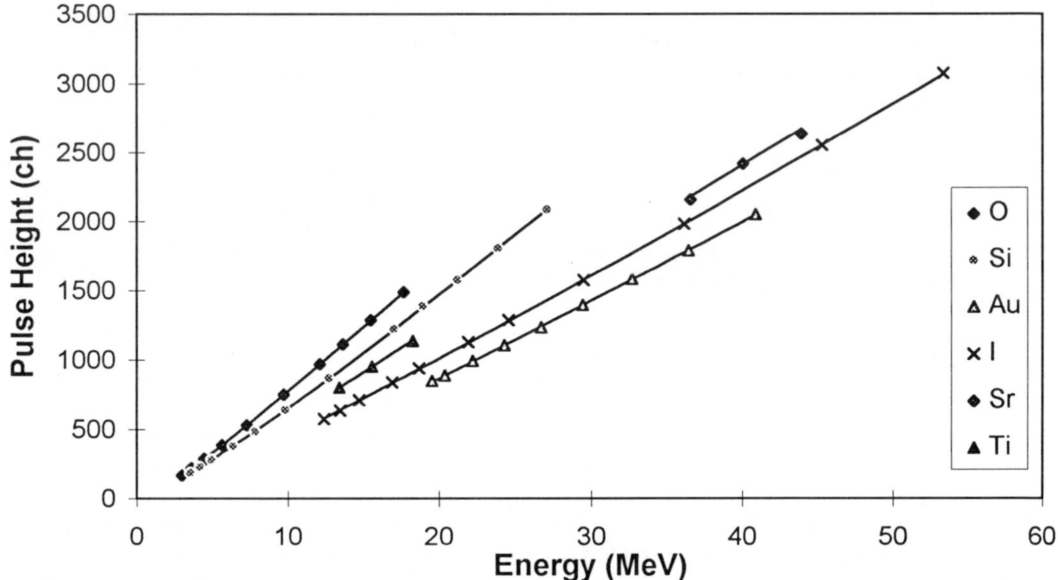

Figure 2. Energy Calibration. Data Points shown as indicated in the legend and fitted calibration shown as trendlines.

Simulation

The modelling of physical processes takes into account
a) stopping and straggling of the incident ion,
b) scattering of the incident ion which gives rise to recoiling ions and scattered ions, i.e. scattering determines the energy and intensity of exiting ions, and
c) stopping and straggling of the exiting ions in the target and timing foils.

Stopping powers are taken from the ZBL theory of Ziegler et al. [8]. Straggling of heavy ions is poorly described by the simple Bohr formula [9], so the straggling is estimated using the semi-empirical model of Yang et al. [10] This semi-empirical model divides straggling into two components: (i) straggling based on an effective charge model analogous to that used in ZBL stopping and (ii) a correlation component due to high ionisation density of heavy ions. The first component (i) comes from a model by Chu et al. [11, 12]. We have calculated effective charge (γ) using eqn. 3-16 from Ziegler et al. [9]. The second component (ii) uses Yang, O'Connor and Wang's semi-empirical formula [10]:

$$\frac{\Omega^2}{\Omega_B^2} = \gamma^2(Z_1, Z_2, v)\frac{\Omega_{CHU}^2}{\Omega_B^2} + \left[\frac{\Delta\Omega^2}{\Omega_B^2}\right]_{ion} \tag{1}$$

In generating the spectrum, each piece of exiting ion data is distributed into 10 parts which are handled separately to model the straggling. The simulation neglects multiple scattering which explains differences in the low energy tails, especially of scattered ^{127}I ions.

This leads to intermediate data which is stored in linked lists of exiting ions. This data is then combined with detector responses which include the:-
a) energy detector resolution of the Si detector extrapolated from Amsel's model [13, 14]:

$$\delta E = a + bE^{1/3} \tag{2}$$

The parameters *a* and *b* are functions of atomic mass *A*,
b) experimentally determined energy calibration,
c) experimentally determined time calibration and
d) time resolution of the detector.

The simulated spectrum is stored in an array which can be compared with the raw experimental data.

Results and Discussion

While many features in the raw data are replicated in the simulated spectrum in Figure 1, two features are clearly different. Firstly, the Ti signal in the raw data describes a long arc as the Ti has moved within the material structure and has probably formed TiO_2 nodules within the Pt electrode. This phenomenon has been observed in TEM studies of electrode structures without electroceramic covering [2]. Secondly, there are long tails on the arcs describing the I and Pt signals. These are due to multiple scattering, which is not included in the simulation.

In the case of SBT, the overlapping data is primarily due to recoiling atoms of Ta, Bi and Pt. We can however obtain information on these elements by consideration of scattering ^{127}I ions as well, as only these three elements have sufficiently heavy nuclei to scatter ^{127}I into the ToF-E detector.

The scattered ^{127}I spectrum is not very well defined in the energy projection, because of the poor resolution of the SiSB detector for these very heavy elements. However, the time projection (Figure 3) clearly differentiates the surface region of Bi from other elements, allowing extraction of Bi content. A similar result is available from conventional RBS.

The overlapping recoiled ions from Ta, Pt and Bi provide good information on the surface contribution of Ta when projected onto the time axis (figure 1) and clearly the three elements considered together could be analysed using the simulation technique. Combining the information from the scattered I signal with various time and energy projections allows unambiguous determination of the sample surface. However, but away from the surface, analysis is more difficult but provided the interactions are all understood, it should be possible analyse the data in a similar manner to RBS.

The other method which can be adapted is fitting the time contribution for the various elements to time projections corresponding to thin energy slices. This method appears to be promising for the extraction of information of diffusion of Bi into the electrode and substrate. However, in the present the lower limit for the detection of Bi is quite high, because of multiply scattered ions.

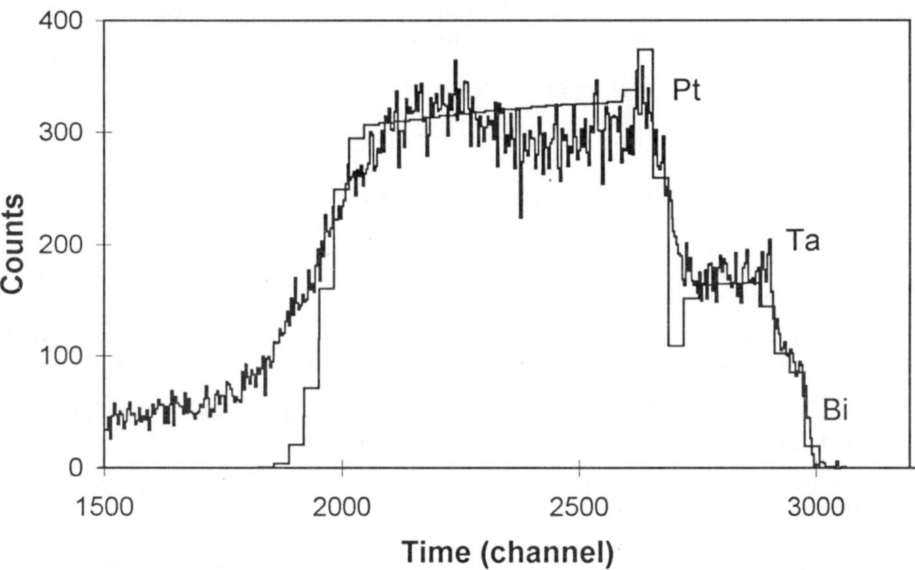

Figure 3. Time projection of scattered ^{127}I ions from the SBT sample compared with the projected part of the simulated spectrum assuming the nominal, as-deposited structure.

Conclusion

The analysis of SBT shows the presence of Ti within the Pt electrode which is totally consistent with TEM studies [2]. Further work is required to investigate Bi diffusion through the Pt electrode. This work demonstrates how 2-d simulation of ToF-E data can be applied to analysis of ToF-E recoil data. Further it shows that a good knowledge of multiple scattering is needed to examine the depth distributions of heavy elements in these materials. There is little evidence of large scale diffusion of elements from the SBT layer into the electrode.

pnj Thursday, 24 October 1996

References

[1]　'Experimental Studies of Interfacial Phenomena in Barium Strontium Titanate (BST) Devices', W. B. Stannard, P.N. Johnston, S. R. Walker, I.F. Bubb, J.F. Scott, D.D. Cohen, N. Dytlewski and J.W. Martin, Integrated Ferroelectrics (in press).

[2]　Kyu Ho Park, Cha Yeon Kim, Young Woo Jeong, Hyun Ja Kwon, Kwang Young Kim, Jeong Soo Lee and Sung Tae Kim, J.Mater.Res., 10, 1790 (1995).

[3]　W.B. Stannard, P.N. Johnston, S.R. Walker, I.F. Bubb, J.F. Scott, D.D. Cohen, N. Dytlewski and J.W. Martin, Integ. Ferroelectrics, 9, 243 (1995)

[4]　H.J. Whitlow, G. Possnert and C.S. Petersson, Nucl. Instr. Meth. B, 27, 448 (1987).

[5]　Application Software Group, Computing and Networks Division at CERN, the PAW Manual Version 1.14, (CERN, Geneva, Switzerland, 1992).

[6]　H.J. Whitlow, TASS, Internal Report, (Dept. Nuclear Physics, Lund Institute of Technology, Sölvegatan 14 S-223 62 Lund, Sweden, 1993).

[7]　M. El Bouanani, M. Hult, L. Persson, E. Swietlicki, M. Andersson, M. Östling, N. Lundberg, C. Zaring, D.D. Cohen, N. Dytlewski, P.N. Johnston, S.R. Walker, I.F. Bubb, and H.J. Whitlow, Nucl. Instr. and Meth. B **94,** 530 (1994).

[8]　P.N. Johnston, Nucl. Instr. and Meth. B **79**, 506 (1993).

[9]　J.F. Zeigler, J.P. Biersack, and U. Littmark, The Stopping and Ranges of Ions in Matter Volume 1 (Pergamon, New York, 1985).

[10]　N. Bohr, Mat. Fys. Medd. Dan. Vid. Selsk., **18**, No.8 (1948).

[11]　Q. Yang, D.J. O'Connor, and Z. Wang, Nucl. Instr. Meth. B **61**, 149 (1991)

[12]　W.K. Chu, Phys.Rev.A **13**, 2057 (1976).

[13]　W.K. Chu, In: Ion Beam Handbook for Materials Analysis, eds. J.W. Mayer and E. Rimini, Academic Press, New York, 1977.

[14]　G. Amsel, C. Cohen and A. L'Hoir, Ion Beam Surface Layer Analysis Volume 2, O. Meyer, G. Linker and F. Kappeler, eds. (Plenum, New York, 1976).

[15]　P.F. Hinrichsen, D.W. Hetherington, S.C. Gujrathi and L. Cliche, Nucl. Instr. and Meth. B **45**, 275 (1990).

Acknowledgments

This work is supported by the Australian Research Council (ARC) and the Australian Institute of Nuclear Science and Engineering (AINSE).

Materials Science Forum Vols. 248-249 (1997) pp. 301-312
© *1997 Trans Tech Publications, Switzerland*

Void Formation and Surface Rippling in Ge Induced by High Energetic Au Irradiation

H. Huber[1], W. Assmann[1], R. Grötzschel[2], H.D. Mieskes[1], A. Mücklich[2], H. Nolte[1] and W. Prusseit[3]

[1] Sektion Physik, Ludwig-Maximilians-Universität München, D-85748 Garching, Germany

[2] Forschungszentrum Rossendorf, Institut für Ionenstrahlphysik und Materialforschung, P.O.Box 51 01 19, D-01314 Dresden, Germany

[3] Technische Universität München, Physik-Department E10, D-85747 Garching, Germany

Keywords: High Energy Irradiation, Void Formation, Surface Roughening, Amorphization, Ge

Abstract

Ge wafer have been irradiated at room temperature with 100 - 266 MeV Au ions at different incidence angles. For an incidence angle of 15°, the Au bombardment leads to a surface roughening when fluences higher than 10^{14} ions/cm^2 are applied, while at normal incidence no surface modification could be seen up to a fluence of 1.2×10^{15} ions/cm^2. In the case of flat entrance angles the surfaces of the bombarded samples start to ripple with wave-lengths between 30 nm and 60 nm after a dose of 3.0×10^{14} ions/cm^2. With increasing fluence these ripples are transformed into a chaotic surface structure. Cross section TEM as well as channeling and in situ blocking measurements performed on the samples have shown that the modified surface layer has been amorphized during the irradiation, while the material underneath is damaged but still crystalline. These effects are completely absent after irradiation of Ge with 165 MeV Ni and 185 MeV I ions. Furthermore the heavy ion bombardment leads to the formation of a buried sponge-like layer. The formation depth of this layer depends on the ion energy as well as on the entrance angle θ of the ions. We discuss the amorphization of a surface layer with the underlying material being still crystalline and the dependence of the formation depth of the porous layer on the electronic and nuclear energy loss. Moreover we present a simple model which is capable to describe the wave-length of the surface-rippling observed.

1 Introduction

Ion beam induced material modification of semiconductors has been of great interest over the last two decades. One example of the great variety of ion beam assisted materials processings is the Separation by IMplanted OXygen (SIMOX), which is today routinely used in industry for Silicon-On-Insulator (SOI) technology. Another example of material modification is the formation of cavities in Si and Ge during high fluence He or H implantation,which can be used as very effective getter for transition metal impurities and interstitials [1]. Moreover it has been shown by Bruel [2] that bubble formation after hydrogen implantation can also be applied in SOI technology for in-depth micro-slicing of bonded wafers.

As early as 1957 Kleitman et al. reported on the radiation-induced expansion of GaSb and InSb crystals bombarded with 9 MeV deuterons [3]. More than 20 years later it was shown by Destefanis et al. [4] that amorphization is not the final state of heavy ion induced lattice damage of InSb and a sponge-like structure consisting of large voids forms within the amorphous phase of InSb. In the last decade many other authors reported also on a morphological instability of Ge which leads to the formation of a very porous damage structure during low energy heavy ion irradiation [5, 6, 7, 8, 9, 10]. It was shown by Holland et al. [7] that this structure does not form at temperatures below 70 K or above 620 K. Furthermore Wan et al. demonstrated that this sponge once formed cannot be annealed by thermal treatment up to 940 K [11]. In a previous paper we reported that irradiation of Ge even with high energetic heavy ions (e.g. 185 MeV I or 266 MeV Au) at room temperature leads to a buried sponge-like layer, consisting of large voids up to 350 nm in diameter [12]. The volume expansion versus fluence of the irradiated parts have been described using a one step swelling model. By comparing the damaging power of different heavy ions in Ge, ranging from 15 MeV C to 266 MeV Au, we found that a critical defect production rate is necessary to initiate the formation of the sponge-like layer.

In this paper we show new experiments in order to understand the formation depths of the porous structure as a function of the electronic and nuclear energy loss ratio of the implanted ions. Additionally we found that also the surface of Ge samples has been modified during irradiation with Au ions in the 100 - 300 MeV range, when fluences higher than 10^{14} ions/cm^2 are applied, but only for non normal incidence angles. This surface modification seems to be restricted to irradiations with very heavy ions. Furthermore we present a simple theoretical model which is capable to describe the periodicity of the surface modifications observed.

2　Experiment

The samples were 10×10 mm^2 rectangles cut from <100> 0.33 mm-thick polished wafer of n-type (5 Ωcm) germanium. The specimens have been irradiated at the ERDA setup at the Munich tandem accelerator laboratory [13] with 100 MeV, 252 MeV and 266 MeV Au ions and fluences in the range between 2×10^{14} ions/cm^2 and 1.4×10^{15} ions/cm^2 in the central part of the irradiated area. The target temperature never exceeded 40°C, because of the low beam current of max. 120 particle pA in a beam spot of 1×1 mm^2, leading to a flux of 7.5×10^{10} ions/(cm^{2}s). The entrance angle θ of the beam to the surface normal varied between 0° and 75°. A large area ionization chamber with particle and position resolution has been used for the detection of the elastically scattered recoils. Furthermore the calibrated count-rate of the ionization chamber was used for charge collection. The beam spot on the samples was not confined by apertures and it's size has been determined, using a CsI crystal for beam monitoring. The accuracy of the fluence measurements is therefore restricted to about 50% due to systematic errors like small movements of the beam and deviations of the flux profile from a gaussian which has been assumed for the dose calculations. After irradiation the samples were fractured in the beam spot for SEM and TEM measurements. A Hitachi S4000 microscope was used for the scanning electron microscopy and a Philips CM300 SuperTWIN-α microscope for the transmission electron microscopy.

3　Results and discussion

3.1　Void formation

The irradiated Ge samples show a step-like elevation of the beam spot which is due to the formation of a buried sponge-like structure [12]. Fig. 1 shows cross section TEM pictures of a Ge specimen irradiated with 266 MeV Au at an incidence angle θ of 75° to the surface normal and a total fluence of 1.4×10^{15} ions/cm^2. In fig. 1a one can clearly see the whole buried porous layer, having a width of 2.7 μm. The upper border of the sponge-like structure corresponds to a depth of 2.7 μm. It should be noted that the largest voids up to 350 nm in diameter can be detected in the central region of this structure, whereas the diameter of the cavities is continuously decreasing towards both ends. Fig. 1b shows a XTEM picture taken from the central region of the sponge-like structure shown in fig. 1a. XTEM as well as channeling and blocking measurements performed on the sample confirmed that a thin surface layer has been completely amorphized, followed by a 2.6 μm thick damaged but still crystalline layer ($\chi_{min} \approx 0.2$). The sponges in the upper part of the porous layer consist of polycrystalline Ge with amorphous parts, whereas the material is completely amorphized from a depth of 4.3 μm to 8.5 μm. Calculation of the mean ion range R_p using TRIM95 [14], gives 4.4 μm for the virgin and 6.0 μm for the damaged sample. In the later case the density of the porous layer was determined to be approx. 50% of the value for crystalline Ge. However radial scattering of the beam causes part of the ions to be implanted in much larger depths, explaining the amorphization of the sample up to a depth of 8.5 μm (see fig. 2a).

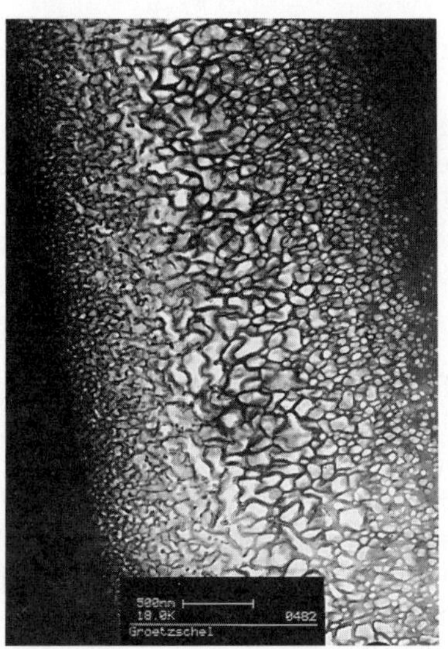
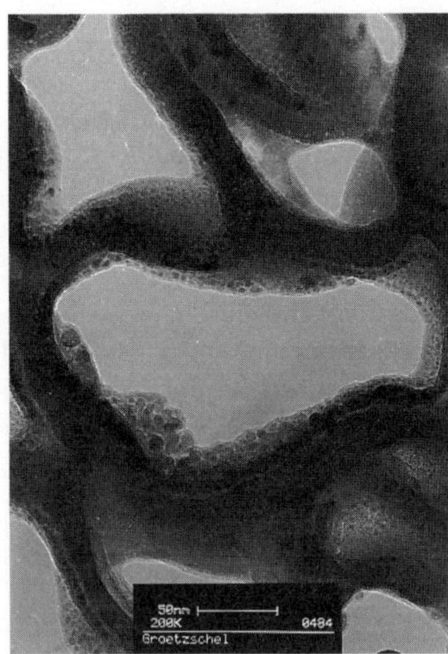

Fig. 1: Cross-section TEM (XTEM) pictures of Ge sample which has been irradiated with 266 MeV Au ions up to a total fluence of 1.4×10^{15} ions/cm^2.
a, In fig. 1a one can clearly see a very porous layer, having a with of 2.7 μm. The left border of the sponge-like layer corresponds to a depth of 2.7μm.
b, XTEM picture taken from the central region of the sponge-like structure of fig. 1a. The maximum diameter of the voids can reach values as high as 0.3 μm.

Fig. 2a shows the target vacancies/atom/ion versus depth as calculated by TRIM95 [14] for the case of 266 MeV Au ions impinging on Ge at an entrance angle θ of 75°. The measured starting depth of the porous layer (dashed vertical line) nicely fits with the calculated depth of the nuclear damage maximum. However in the case of nearly normal incidence angles (7°) the measured depths of the sponge-like structures differ by a factor of 1.6 from the depths of the calculated nuclear damage maxima (see fig. 2b). Table 1 shows the formation depths and widths of the sponge-like layers for different energies and entrance angles of the Au ions, the nuclear S_n^* and the electronic energy losses S_e^* of the ions entering the buried structure and TRIM95 calculations of the mean ion ranges R_p, assuming virgin Ge. The ratio S_e/S_n lies

Table 1

ion	energy [MeV]	fluence [10^{15} cm^{-2}]	θ	beginning of porous layer [μm]	end	R_p [μm]	S_e^*	S_n^* [keV/nm]
^{197}Au	266	1.4	75°	2.7	5.4	4.5	10.6	1.0
^{197}Au	252	1.4	75°	2.5	5.0	4.3	11.1	0.9
^{197}Au	252	1.2	7°	10.6	13.1	16.7	9.3	0.8
^{197}Au	100	1.4	75°	1.0	3.5	2.7	10.4	1.0
^{197}Au	100	1.0	7°	5.4	7.9	10.3	7.1	0.9
^{127}I	185	4.0	75°	2.7	4.9	4.3	7.0	0.6

Tab. 1 shows the formation depths and widths of the sponge-like layer for different energies and entrance angles of the ions, the nuclear S_n^* and the electronic energy losses S_e^* of the ions entering the buried structure and TRIM95 calculations of the mean ion ranges R_p, assuming virgin Ge.

around 10 in all cases. This behavior even does not change when going from Au ions to I ions. To study the behavior of a prior formed porous layer during a second irradiation with ions having a more than twofold electronic energy loss S_e in this depth region, we post-irradiated the beam spot of a 252 MeV Au irradiation at 75° (fluence 1.4×10^{15} ions/cm^2) with 252 MeV Au ions at normal incidence up to a total fluence of 2.8×10^{14} ions/cm^2. Fig. 3 shows SEM pictures of the twofold irradiated Ge which was fractured in the beam spot. In the upper part of the porous region one can see the voids created by the first irradiation. Surprisingly the voids formed by the post-irradiation cannot be found in about 11 μm depth but just at and beneath the depth of the prior damaged layer. Also the appearance of the sponge-like structure formed by irradiation at normal incidence completely differs from the appearance of the porous layer formed at flat incidence angles (see fig. 1a and fig. 3). In the case of normal incidence the voids are ordered in horizontal bands, whereas only a weak planar ordering at 50° to the surface normal can be detected in the case of 75° entrance angles. With increasing fluence the thickness of the bands is reduced, while their seperations and their lateral dimensions is increased (see fig. 3a and 3b), showing the Klaumünzer Effect [15] on a microscopic scale. The border between the top layer and damage structure is rather sharp, enabling an easy removal of the almost crystalline overlying material as shown in fig. 3c.

XTEM measurements performed on this sample showed that the amorphous sponges formed by the prior irradiation have been recrystallized from 2.5 μm to 3.4 μm depth. Then within 50 nm the structure of the sample changes from polycrystalline to completely amorphous. Thus the material is completely amorphous from a depth of 3.5 μm (first horizontal band from the top in fig. 3b) to a depth of 20 μm. This means that the amorphous layer formed by the prior irradiation has been recrystallized due to the electronic energy loss of the ions down to a S_e value of about 25 keV/nm. The recrystallization of amorphous Ge during high energetic heavy ion bombardment has also be seen in previous experiments made by Izui et al. [16] who measured the recrystallization of thin amorphous Ge and Si films along heavy ion tracks.

The observed formation depth of the cavities in the case of normal ion incidence angles stands in contrast with the void creation mechanism proposed by Destefanis et al. [4], assuming that the nuclear energy loss of the ions in a collision cascade induces a thermal spike which may lead to a local overpressure. If this overpressure is higher than the elastic limit, the material yields plastically, leaving a void. Therefore this model demands maximum void concentration at the depth of maximum ion induced nuclear damage. However we have shown that the porous layer starts in depth regions where the electronic energy loss S_e of the ions is about 10 keV/nm and exceeds the nuclear energy loss S_n by one order of magnitude. The formation of cavities is possibly enhanced by electronic excitations in this region. Therefore we emphasize on the void creation mechanism given by Wan et al. [11], who assumed that the driving force for void creation is the elimination of dangling bonds on void surfaces.

However the exact mechanism of void formation in amorphous materials remains still unclear, but S_n as well as S_e appear to be important parameters. Furthermore it seems that the porous structure once formed is a very effective getter for all kinds of defects, forcing vacancies created in the depth to diffuse up to this highly damaged layer.

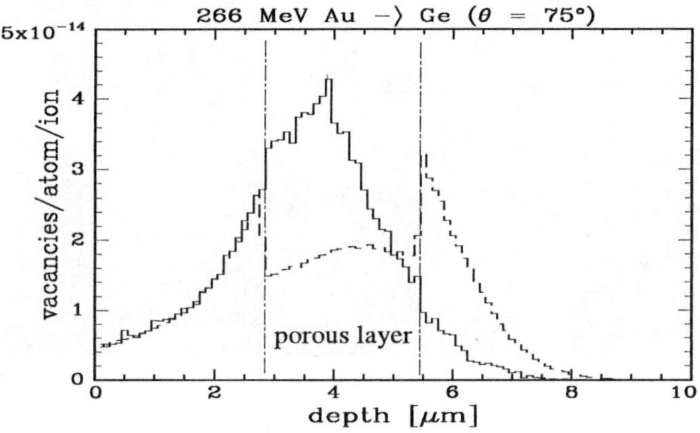

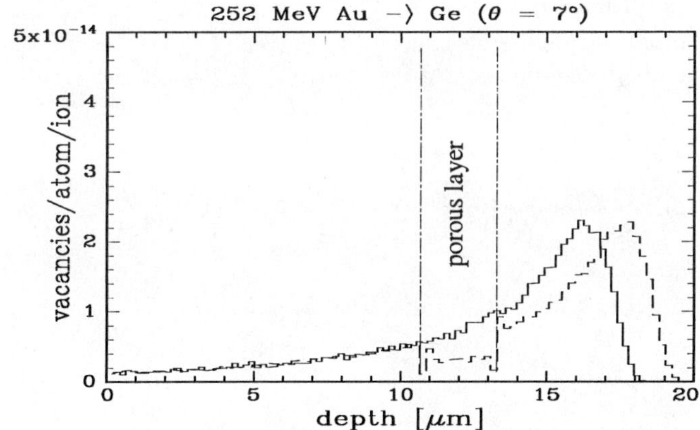

Fig. 2: TRIM95 calculations, showing the produced vacancies/atom/ion versus depth for high energetic Au implantation in virgin Ge (solid lines) and irradiated Ge (dashed lines) at different energies and incidence angles θ. To obtain an upper threshold of ion induced damage produced in the near surface region, a displacement energy of 6 eV has been used in the simulations (see text). The vertical dashed doted lines show the depth and width of the porous layer as measured by cross-section SEM.

a, Damage calculations for Ge implanted with 266 MeV Au ions at 75°.

b, Damage calculations for Ge implanted with 252 MeV Au ions 7°

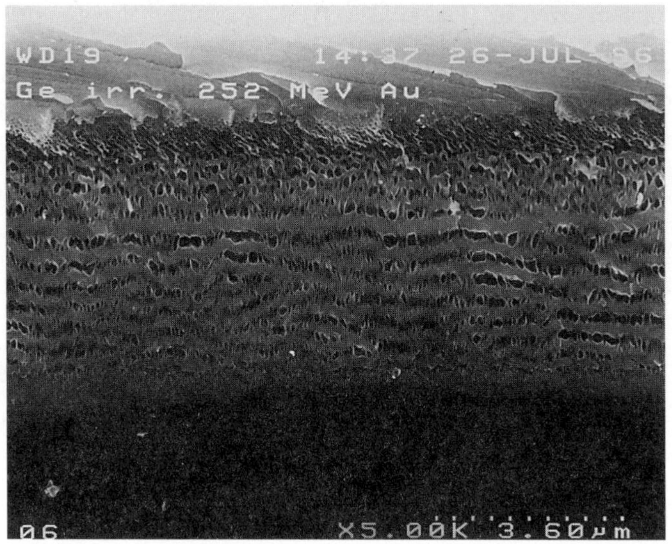

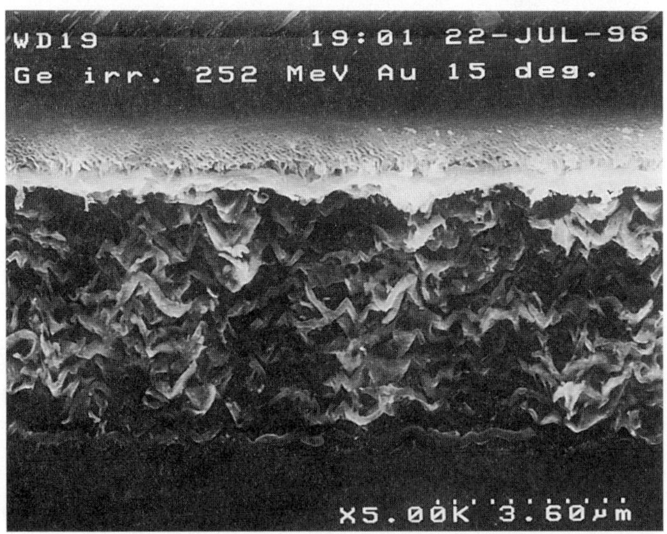

Fig. 3: Cross-section SEM pictures of a Ge sample irradiated with 252 MeV Au-ions at 75° and 0°.

a, b, The total fluence increases from fig. 3a to fig. 3b. Fig. 3b corresponds to a total fluence of 1.7×10^{15} ions/cm^2.

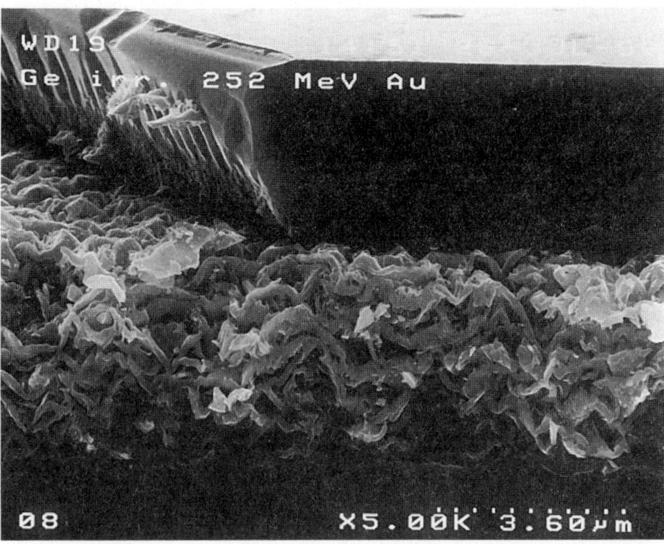

Fig. 3c: XSEM picture showing a twofold irradiated Ge where the crystalline overlying layer has been removed on the left side during the fractioning procedure.

3.2   Surface rippling

The surface of Ge irradiated with 252 MeV Au ions at an incidence angle of 75° up to a fluence of 3.0 $\times 10^{14}$ ions/cm^2 starts to ripple with a wave-lengths between 30 nm and 60 nm (see fig. 4a). Channeling and XTEM measurements performed on that sample showed that a approx. 10 nm thick surface layer has been amorphized. By further increase of the fluence these ripples transform into the chaotic structure shown in fig. 4b. No surface roughening could be measured for Ge irradiated with 252 MeV Au ions at an incidence angle of 7° up to a maximum fluence of 1.2×10^{15} ions/cm^2. Surface amorphization and surface roughening is also completely absent when Ge is irradiated with lighter ions like I or Ni.

It is now well established, that all amorphous solids exhibit dimensional changes under high energetic heavy ion irradiation [15]. Very recently the irradiation induced viscous flow of an amorphous surface layer has been explained in terms of a viscoelastic model [17]. Using this model, the off-diagonal elements of the tensor of strain ϵ and the shear flow δx of the surface layer can be calculated to be,

$$\frac{d\epsilon_{xz}}{dt} = \frac{1}{2}\left(\frac{\delta v_x}{\delta z} + \frac{\delta v_z}{\delta x}\right) = 3A_0\Phi sin\theta cos\theta \tag{1}$$

$$\delta x \approx 3\Phi t\tau_a A_0 sin\theta cos\theta \tag{2}$$

where Φt is the fluence in ions/cm^2, τ_a is the thickness of the amorphous layer, θ is the entrance angle of the ion beam to the surface normal, A_0 is the deformation yield of amorphous Ge and x is the direction parallel to the non normal velocity component of the beam to the surface. If a restoring force acts along x eq. (1) demands $v_z \neq 0$. However the formation of a periodic structure on the Ge surface cannot be explained with this viscoelastic model till yet. Therefore we apply a simple analytical model which was first developed by Pászti et al. [18] to explain wave pattern formation on the bottom of flaked Si surfaces after high fluence He bombardment. Using the equilibrium equations for a thin bent sheet [19] with the assumption that the variable y normal to the beam direction is constant the following condition has to be satisfied for a periodic solution:

$$\Gamma = -\Lambda k^2 - \Sigma/k^2 \tag{3}$$

where

$$\Gamma = -\tau_a \sigma \geq 0, \quad \Lambda = \frac{\tau_a^3 E}{12(1 - \nu^2)} > 0, \quad \Sigma \cong \frac{E}{3\tau_a}$$

k is the wave-number, σ is the lateral stress induced by lateral atomic displacements due to electronic excitations, ν is the Young's number and E is the Young's modulus. Eq. (3) has a solution only, if

$$\Gamma \geq \Gamma_{crit} = \sqrt{4\Lambda\Sigma} = -\tau_a \sigma_{crit} \tag{4}$$

Let us assume that $\sigma > \sigma_{crit}$ and $\Gamma = \Gamma_{crit}$. From eq. (3) we then obtain the wave-number k_{crit} of the surface pattern formation:

$$k_{crit} = \left[\frac{\Sigma}{\Lambda}\right]^{1/4} = \frac{(3(1 - \nu^2))^{1/4}}{\tau_a} \tag{5}$$

For $\tau_a = 10$ nm, corresponding to a fluence of 3.0×10^{14} ions/cm^2, we get 50 nm for the wave-length λ_{crit}, which is in good agreement with the wave-lengths observed.

The model was developed under the assumption that the amorphous surface layer starts to wrinkle when the irradiation induced lateral stress σ exceeds a critical value. However we should better speak in terms of stress induced viscous material flow. This means that the wave-structure observed is caused by material flow due to the radiation induced lateral stress. Using eq. (4) one can calculate the critical shear stress for wave-pattern formation,

$$\sigma_{crit} = \frac{E}{3\sqrt{(1 - \nu^2)}} \tag{6}$$

which is close to the known value of the shear module of Ge. Snoeks et al. [20] measured the lateral stress of a thin SiO$_2$ film on bulk Si during 4.0 MeV Xe irradiation. They also found a critical stress for plastic shear flow of 30 GPa which is comparable to the value of the shear module of Si. Therefore it seems that the simple model of a thin bent sheet can predict the critical lateral stress as well as the wave-length of the surface rippling to a highly satisfying degree.

It has been shown by Gutzmann et al. that this phenomenon occurs also during high energetic heavy ion irradiation of glassy Fe$_{40}$Ni$_{40}$B$_{20}$ and it was accounted to be a consequence of the ion-induced shear flow in amorphous solids [17]. Radiation-induced mass transport was also observed in a 8 μm thick amorphous InP layer on bulk InP by Cliche et al. [21]. Our measurements concerning wave-pattern formation differ from those made by Gutzmann and Cliche, because we started irradiateing crystalline Ge, whose surface and end-of-range region of the ions was then amorphized due to the ion bombardment. We have not observed a hitch and dike structure on the border of the irradiated and the non-irradiated part of the surface, since our beam spot was not confined by apertures. Therefore the edges of the irradiated part of the samples are not so well defined, resulting in a washing-out of the border structures.

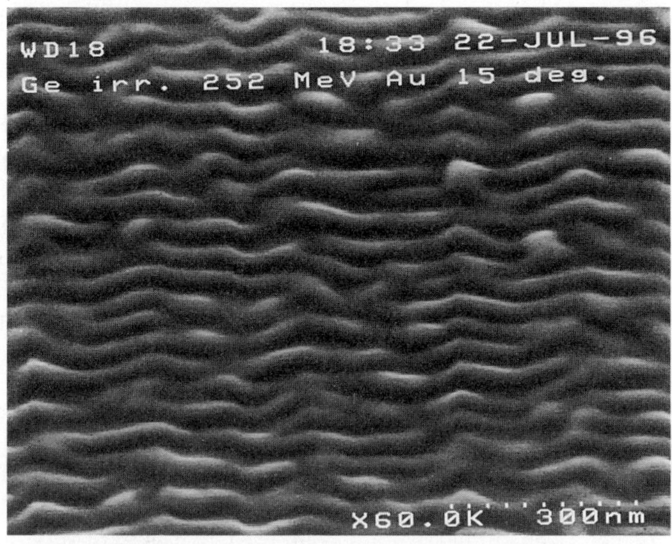

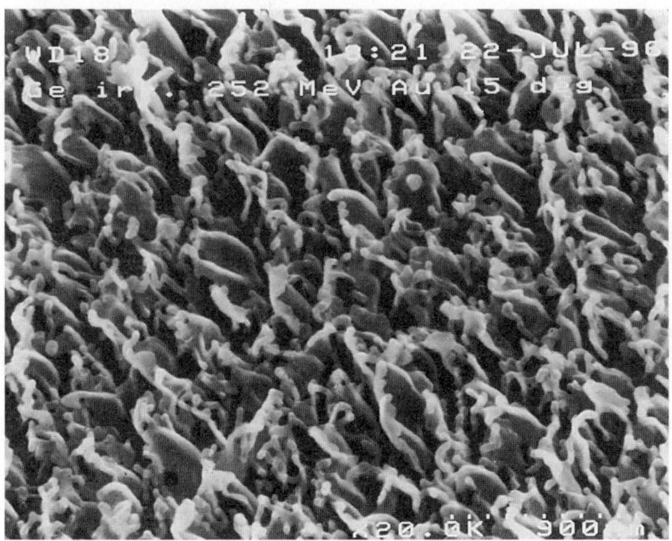

Fig. 4: SEM pictures of the surface of Ge samples irradiated with 252 MeV Au ions at an angle of 75° to the surface normal.
a, Surface ripples formed during 252 MeV Au irradiation up to a total fluence of 3.0×10^{14} ions/cm^2.
b, Irradiated surface after a total fluence of 1.4×10^{15} ions/cm^2.

3.3   Ion beam induced surface amorphization/recrystallization

From our experiments we obtained that the damaging power of very high energetic Au ions (E > 130 MeV) at the surface of Ge samples strongly depends on their incidence angles. For flat incidence the surface of the specimens was completely amorphized whereas for nearly normal incidence no amorphization could be measured. Taking into account that the amorphization of Ge starts in a very narrow 10 nm thick surface layer, it seems that the free surface has an important influence on the damage accumulation, leading to a reduced amount of damage required to turn the lattice amorphous. This was recently confirmed by Wan et al. [11] who measured that for 1.5 MeV Kr ions the amorphization fluence of 90 nm thick Ge foils is one order of magnitude larger compared to a 50 nm thick Ge specimen. Calculation of the upper threshold value of displaced atoms per atom (dpa), using TRIM95 [14] and a displacement energy E_d of 6 eV [22], yields 0.4 dpa produced near the surface for a total fluence of 3.0×10^{14} Au ions/cm^2. However 185 MeV iodine-irradiation of the samples up to 2.4 dpa ($E_d = 6$ eV) has not turned the surface layer amorphous [12], indicating a critical value in the nuclear energy deposition of the ions necessary for the production of heavily damaged and amorphous zones. Crystal damage due to electronic energy deposition can be ruled out, because 100 MeV Au bombardment of Ge gives similar results compared to the 266 MeV and 252 MeV Au irradiation and the electronic energy loss S_e of 100 MeV Au ions of 17 keV/nm lies below the S_e-value of 185 MeV I ions which is 21 keV/nm. For all measurements listed in table 1 we obtained surface amorphization, except for the case of 252 MeV Au ions at 7° incidence angle and for the case of I ions. Thus the value of the total critical nuclear energy deposition T_{n_c} in the first 10 nm from the surface lies between 3.5 keV (185 MeV I) and 4.5 keV (100 MeV Au). 252 MeV Au post-irradiation at normal incidence angle, performed on a beam spot of a previous Au bombardment at 75°, led to a recrystallization of the ion beam induced damage up to a depth of 3.4 μm as confirmed by XTEM measurements. However in this case the total nuclear energy deposition in a 10 nm thick surface layer is 2.1 keV, which is below the critical value for amorphization, whereas the value of the electronic energy loss of 28 keV/nm is well above the threshold value S_{e_c} for damage annealing due to electronic energy deposition of approx. 10 keV/nm [23, 24]. Therefore the damaging power of high energetic ions in the near surface region of Ge seems to be determined by two effects. For $T_n < T_{n_c}$ and $S_e > S_{e_c}$ most of the defects produced are annealed in situ due to the high electronic energy deposition. However for $T_n \geq T_{n_c}$ and $S_e > S_{e_c}$ the Au bombardment leads to a strong damage of the surface region. Thus above the critical nuclear energy deposition T_{n_c} the damage annealing due to the electronic energy loss of the ions seems to be not high enough to prevent surface amorphization.

4   Conclusion

We have shown that irradiation of crystalline Ge by high energetic Au ions leads to the creation of a buried sponge-like layer. The formation depth of this structure depends on the electronic as well as on the nuclear energy loss of the ions. The mechanism of void formation in amorphous materials is not understood till yet, but reduction of the free energy via elimination of dangling bonds on void surfaces may be the driving force for vacancy clustering in the amorphous phase of semiconductors. Also the amorphization of a thin surface layer during Au bombardment remains to be explained. Possibly the defect interaction with the specimen surface, leading to a reduced stability against radiation damage, is responsible for the surface amorphization. By further irradiation the amorphous layer shows anisotropic plastic deformation due to the electronic energy deposition of the ions. For flat ion incidence angles the release of stress induced by the dimensional growth of the sample perpendicular to the beam direction causes the formation of wave-patterns and chaotic surface structures on the target. This phenomenon cannot be observed for normal incidence and lighter ions. Moreover high energetic Au irradiation at normal incidence leads to a recrystallization of a prior damaged surface layer, indicating a critical value in the nuclear energy deposition for surface amorphization.

The ion beam induced modifications of Ge are unique in the way that a porous structure can be created simultaneously at the surface and in a depth of several μm. Thus the sponge-like structures of amorphous Ge may have some applications in device technology, e.g. as getter for transition metals or hydrogen.

5   Acknowledgment

This work was supported by BMBF (No. 06LM363 TP5). We like to thank Mrs. Morawitz for the preparation of the samples for XTEM measurements and Dr. R. Müller for her help in performing the SEM measurements.

References

[1] S. M. Myers, D. M. Follstaedt, G. A. Petersen, C. H. Seager, H. J. Stein, W. R. Wampler, Nucl. Inst. and Meth. B 106, 379 (1995).

[2] M. Bruel, Nucl. Inst. and Meth. B 108, 313 (1996).

[3] D. Kleitman and H. J. Yearian, Phys. Rev. 108, 901 (1957).

[4] G. L. Destefanis and J. P. Gailliard, Appl. Phys. Lett. 36 (1), 40 (1980).

[5] I. H. Wilson, J. Appl. Phys. 53(3), 1698 (1982)

[6] B. R. Appleton, O. W. Holland, J. Narayan, and O. E. Schow III, J. S. Williams and K. T. Short, E. M. Lawson, Appl. Phys. Lett. 41(8), 711 (1982).

[7] O. W. Holland, B. R. Appleton, and J. Narayan, J. Appl. Phys. 54(5), 2295 (1983).

[8] E. M. Lawson, K. T. Short, and J. S. Williams, B. R. Appleton, O. W. Holland and O. E. Show III, Nucl. Inst. and Meth. 209/210, 303 (1983).

[9] B. R. Appleton, O. W. Holland, D. B. Poker, J. Narayan, and D. Fathy, Nucl. Inst. and Meth. B7/8, 639 (1985).

[10] I. B. Khaibullin, G. G. Zakirov, M. M. Zaripov, T. Lohner, L. Pogány, G. Mezey, M. Fried, E. Kótai, F. Pászti, A. Manuaba, and J. Gyulai, Phys. Stat. Sol. (a)94, 371 (1986).

[11] L. M. Wang and R. C. Birtcher, Phil. Mag. A, Vol. 64, No. 6, 1209-1223 (1991).

[12] H. Huber, W. Assmann, S. A. Karamian, A. Mücklich, W. Prusseit, R. Grötzschel, M. Kokkoris, H. D. Mieskes and R. Vlastou, Nucl. Inst. and Meth. B, accepted for publication.

[13] W. Assmann, P. Hartung, H. Huber, P. Staat, H. Steffens and Ch. Steinhausen, Nucl. Inst. and Meth. B85, 726 (1994).

[14] J. F. Ziegler, J. P. Biersack, and U. Littmark, *The Stopping and Range of Ions in Matter*, Vol. 2-6 (Pergamon Press, New York, 1977-1985).

[15] Ming-dong Hou, S. Klaumünzer, and G. Schumacher, Phys. Rev. B41, 1144 (1990).

[16] K. Izui and S. Furuno, Proc. XIth. Int. Cong. on Electron Microscopy, Kyoto, 1299 (1986).

[17] A. Gutzmann, S. Klaumünzer, and P. Meier, Phys. Rev. Lett. 74 (12), 2256 (1995).

[18] F. Pászti, Cs. Hajdu, A. Manuba, N. T. My, E. Kótai, G. Mezey, M. Fried, Gy. Vizkelethy and J. Gyulai, Nucl. Inst. and Meth. B 7/8, 371 (1985).

[19] L. D. Landau and E. M. Lifshitz, Elastizitätstheorie, (Akademie Verlag, Berlin, 1965) Kap. II(14).

[20] E. Snoeks and A. Polman, C. A. Volkert, Appl. Phys. Lett. 65(19), 2487 (1994).

[21] L. Cliche, S. Roorda, M. Chicoine, ND R. A. Masut, Phys. Rev. Lett. 75(12), 2348 (1995).

[22] L. H. Crawford and L. M. Slifkin, *POINT DEFECTS IN SOLIDS,* Vol. 2 (Plenum Press, New York and London, 1975).

[23] M. Levalois, P. Bogdanski and M. Toulemonde, Nucl. Inst. and Meth. B63, 14 (1992).

[24] H. Huber, W. Assmann and S. A. Karamian, to be published.

Materials Science Forum Vols. 248-249 (1997) pp. 313-318
© 1997 Trans Tech Publications, Switzerland

Micro-IBA and Micro-AMS for Geological Materials

S.H. Sie, T.R. Niklaus and G.F. Suter

HIAF Laboratory, CSIRO Exploration and Mining, P.O.Box 136, North Ryde, NSW 2113, Australia

Keywords: IBA, Microbeam, AMS, Geological Applications, PIXE, Microprobe, Trace Elements

Abstract

Accelerator based microbeam methods, micro-IBA and micro-AMS are powerful and versatile and often unique tools for geological applications ranging from the basic science to economic geology. By far, micro-PIXE's ppm sensitivity for in-situ trace element analysis has been the most productive, contributing to the understanding of magma genesis and evolution, and leading to powerful practical applications in diamond exploration. Micro-AMS extends the detection limits for trace element analysis to sub-ppb level, and together with the in-situ isotopic microanalytical capability will open up new windows of opportunity.

1.Introduction

Geological samples are complex and inhomogeneous mixtures of intergrown phases, and can be admixtures of materials of multiple provenances. Their basic constituents are commonly microscopic, consisting of monomineralic grains or inclusions, which are glassy samples of parent magma typically a few microns in size. The monomineralic grains can also contain microscopic features such as zoning, due to episodic growth or to alterations by other melts, fluids or deformation processes, and fluid inclusions.

Microbeam techniques are essential for geological samples, to enable spatially resolved analysis of microscopic features of the complex samples and to test possible inhomogeneities. While mineral separation methods for sample preparation may be less expensive, they are at best very laborious and contamination problems are ever present.

IBA (ion beam analysis) techniques offer the sensitivity for chemical and structural analysis of geological samples, but the complexity of the samples demands a microbeam approach. Furthermore, even when the samples can be analysed by broad beam, IBA is not viable against other competing methods such as inductively coupled plasma mass spectrometry (ICP-MS) or neutron activation analysis (NAA). Micro-IBA, encompassing PIXE (particle induced X-ray emission), RBS (Rutherford backscattering spectrometry), ERDA (elastic recoil detection analysis), NRA (nuclear reaction analysis), PIGE (particle induced gamma-ray emission) delivered typically with spatial resolution $> \sim 0.5$ μm, typically between 1-30 μm, are standardless and quantitative, and a combination of the different methods can be applied simultaneously. The most useful data obtained are usually trace element abundances and lattice sites of minor or trace elements.

AMS (accelerator mass spectrometry) can potentially provide ultra trace (abundances $<10^{-9}$) and isotopic data, but again as a bulk method it is not viable against ICP-MS, NAA and against conventional mass spectrometry respectively. However, for in-situ analysis of complex samples, micro-AMS offers a distinct advantage in isotope geology, and in ultra trace measurements by obviating the problem of mass interference encountered in ion-microprobes.

Many reviews have been published covering mainly micro-PIXE applications [1-7]. The present paper is intended as a brief overview and review of micro-IBA applications in geological materials analysis, and a preview of micro-AMS in the same.

2. Instrument and Method

To be effective, micro-IBA and micro-AMS must be applied with considerations of the nature of geological samples. Essential features of the instrument include a high magnification (>x100) viewing system to locate the sample and identify microscopic features [8,9], and a microstage for positioning of the samples with μm precision. Considering the statistical nature of geological samples, the system must facilitate analysis of large numbers of samples, leading to features such as a quick sample loading facility, automated analysis and special features such as a facility for changing X-ray filters quickly in PIXE measurements [8,9].

Rock samples are prepared as thin or thick polished sections, the same method as that for mineralogical study by classical microscopy. Samples may also be fine detrital grains, or have been pre-processed as powders, either as part of a pre-concentration process or mineral separation. For such samples, setting in epoxy followed by polishing is the recommended method of sample preparation. The samples are usually coated with a thin layer of vacuum deposited carbon for conductivity.

3. Micro-PIXE

Micro-PIXE is the most readily applicable among all IBA methods, used primarily for trace element analysis. Beams of 2-4 MeV protons or alpha particles are typically used, with currents in the 5-50 nA range. The X-rays are measured in an energy dispersive spectrometer (EDS), usually Si(Li) detectors. The quest for best spatial resolution in the micrometer regime is constrained by the available beam intensity, defining the feasibility of the measurements, but it must also be balanced by other factors. The range in most minerals is about 40-60 μm, but the effective depth of analysis is less at typically 15-30 μm due to the rapid drop of ionisation cross-section and self-absorption effects for the X-rays. Depth averaging will occur unless samples are prepared as very thin sections. For thick samples the grains selected for analysis should preferably be greater than 50 μm in lateral dimension. This and the need to carry out the measurements in practical times (5-15 minutes), lead to an optimum beam diameter in the 10-30 μm range, allowing some latitude of beam currents to achieve the desired minimum detection limits (MDL). The logarithmic MDL values for the K lines and L lines vs. Z show an approximately parabolic behaviour , with minima at $Z \sim$ 30-40 and $Z \sim$ 80 respectively. The corresponding values for the minimum MDL, defined at 99% confidence limit, are 2-10 ppm and 5-10 ppm respectively for most minerals, for 10 μC beam charge and a 50 msr detector. The MDL value increases for low Z due to the use of absorbers. Similarly, for the light rare-earth elements (LREE), i.e. La to ~Sm, the MDL values are also poor at ~200 ppm caused by the rapid decrease of both the K shell ionisation and detection efficiency for the high X-ray energy (>~30 keV) with increasing Z, while their L line detection is handicapped by the usually ubiquitous Fe K-lines.

The minimum absorber required to protect the Si(Li) detector from the scattered proton beam limits the detection to elements with Z>~11. Even thicker filters are usually employed to optimise the detection of trace elements of interest and to reduce pile-up effects (e.g. Rb K line and U L beta lines are masked by Fe K pile-up peaks). Hence most major elements of most samples are not detectable by PIXE with protons, and must be supplied by electron microprobe (EMP) analysis. The EMP data also provide a convenient method to normalise the data when ion beam charge integration is not reliable.

While PIXE quantification is essentially standardless, the accuracy depends critically on reliable background estimation, and treatment of pile-up peaks and escape peaks in the spectral deconvolution procedure. An algorithm based on a peak 'filing' method for background [10,11] estimation appears to be reliable. The ability to treat layered targets is essential for analysis of thin samples [12], and special samples such as fluid or melt inclusions. Secondary fluorescence effects are significant in some minerals (e.g. Fe fluoresced by Zn in sphalerite), and must be taken into account in the quantification [12]. Ultimately the reliability of the data analysis system must be tested against known geological standards [11].

Analysis of igneous minerals, mainly from the earth's mantle sources, has proven to be a fertile area of study. Minerals studied include Cr-pyrope garnets $((Mg,Fe)_3(Al,Cr)_2(SiO_4)_3)$, chromian-spinels $((Fe,Mg)(Cr,Al)_2O_4)$, ilmenites $((Fe,Mg)TiO_3)$ and pyroxenes (e.g. diopside $CaMgSi_2O_6$). These minerals are usually collected in diamond exploration programs as detritus of weathered kimberlites, lamproites and other igneous rocks including their xenoliths, and as inclusions in diamonds [13]. Study of trace elements in mineral inclusions in West Australian diamonds reveals that they grow in an open system [14]. In these mafic minerals, geochemical marker trace elements such as large ion lithophile elements (LILE, e.g. Rb,Sr,K) and high field strength elements (HFSE, e.g. Zr,Nb,Ta) can be detected with ppm sensitivity with good efficiency. In ilmenites the distribution of these trace elements can be used to deduce effects of fractional crystallisation, and identify multiple sources [15]. In chromites classification schemes based on the trace elements have been developed to distinguish kimberlites, and lamproites which can be diamondiferous, from other usually barren igneous rocks. Specific trace elements such as Ni, Mn, Zn can be used in geothermobarometry. In garnets, thermometry based on the Ni [16] content is now used extensively in diamond exploration to assess prospectivity of kimberlites. Other trace elements are used to deduce mantle stratigraphy. Study of zoning features in garnets and pyroxenes from xenoliths reveal details of metasomatism, and are used to deduce timescales of geological events [17-19].

Micro-PIXE enables systematic quantitative study of partitioning of trace elements under controlled laboratory conditions simulating the mantle [20-23]. Studies of the effects of different agents of metasomatism (carbonate vs silicate, melt vs liquid) and immiscibility can now be carried out in detail [21-23].

For volcanogenic massive sulfide (VMS) mineral deposits, trace element distribution reflects conditions of deposition and subsequent deformation and metamorphism. Systematic studies of trace elements in pyrite and other sulfides, reveal characteristics of different styles of mineralisation [24-26]. A study of pyrites from a epithermal Au-Ag deposit reveal details of redox conditions throughout episodic depositions [27]. Precious metal (Au, Ag, PGE-platinum group elements) distribution in sulfides carries genetic information, and has direct applications in industry for reserve estimation and devising beneficiation strategy [28-35].

Fluid inclusions are relics of mineralising fluids, and thereby contain important genetic information. Destructive analytical methods for example by crushing and leaching will scramble any possible evidence of various episodes of deposition. Micro-PIXE enables analysis of individual inclusions without decrepitating the inclusion, which can introduce fractionation. Inclusions to be analysed are selected within 5-10 μm below the surface to avoid rupture during analysis [36-40]. Application in fluid inclusion studies is progressing rapidly, with precision and accuracy improving steadily with better techniques to determine the shape parameters. SXRF (synchrotron radiation X-ray fluorescence) is the main competing method at comparable levels of sensitivity [41,42]. Combined with PIGE [40], micro-PIXE might have a decisive advantage.

Study of trace elements in bedrock and detrital Au grains shows promise of enabling discrimination of different styles of mineralisation [3,4,43]. With the EDS system however, many elements of interest (Pb, Bi, Hg, PGE) are interfered with by the Au L lines themselves resulting in poor MDL values in the 100 ppm level. A crystal spectrometer system will help to avoid this problem.

4. Other Micro-IBA

Light elements are not amenable to detection by PIXE because of high self-absorption of the soft K-X-rays or interference by the major elements. They are however geochemically important: H and C in mantle minerals are of important consequence in mantle rheology, light elements such as B,Li,F affect magma generation and evolution, and control partitioning of other trace elements. IBA has an advantage over surface analytical techniques (e.g. SIMS-secondary ion mass spectrometry) for detection of these elements, by allowing analysis well below the surface area avoiding contamination problems, and eliminating possible ambiguities.

Both resonant and non-resonant NRA are in general suitable for detection of light elements. The latter include PIGE, which can be applied simultaneously with PIXE and yield data for C,N,O,F,Na,Mg,Al. Resonant scattering of an alpha beam on O can be used to determine low levels of O in pyrrhotite that affect the monoclinic phase. Hydrogen can be detected either through resonant reactions using ^{19}F or ^{15}N beams, or through ERDA. Advantages of the heavy ion beam resonant reactions include good depth resolution, and range up to 2-3 micrometers. Disadvantages include the high beam current requirement for reasonable reaction yield which can affect the measurements themselves due to the volatility of H compounds in the mineral itself. The 6.4 MeV ^{19}F resonance has been used to measure the hydration rate of quartz, and H content in melt inclusions in high Mg olivine (($Mg,Fe)SiO_4$) from ultramafic lava, believed to be samples of primitive mantle. Concentrations of several hundred ppm were obtained with detection limit of 50 ppm. Beam currents up to 25 nA (charge 3+) focussed to 100 μm were used. However, deterioration of the sample was observed which rendered the measurements suspect. ERDA promises to be more efficient in detection of H as has been demonstrated in a number of laboratories both in the transmission [44-46] and in the reflected geometries [47]. In the latter, the depth range is less but the technique can be applied more generally because it does not require special sample preparation beyond that required for PIXE or RBS analysis.

The reflected geometry was applied to samples from an experimental petrology, to determine water solubility in anorthite-albite system at pressures between 5-20 kb and temperatures between 500-1200 deg. The target was inclined at 16 degrees from the beam axis and the recoiling H detected at 32 degrees. With this technique, 2 nA of 3 MeV alpha beam, focussed to 20 microns produced an integrated count (corresponding to a depth of ~ 0.6 um) of ~ 9000 cts for 1% H content in ~ 15 min. At this level of beam current there was no target deterioration. Bulk detection limit of 5 ppm was obtained [47].

5. Micro-AMS

AMS has been applied mainly for the detection of rare cosmogenic isotopes (at abundances $<\sim 10^{-12}$) as a 'bulk' technique. It has also been applied successfully to detect low levels of trace of Au and platinum group elements (PGE) at the sub-ppb level. The special features that give AMS the high sensitivity can be applied more generally to other isotopes and geochronological systems, and in principle to all trace elements. AMS can be used to eliminate the problem of mass interferences in in-situ measurements carried out with ion-microprobes. At CSIRO, a microbeam AMS system designed to achieve a capability of in-situ microanalysis for geochronology and more sensitive trace element analysis, is under construction [48,49].

The sub-ppm sensitivity of micro-AMS will be applied to problems currently inaccessible by the proton microprobe, as discussed previously. Micro-AMS will facilitate measurements of PGE and rare-earth-elements (REE) fractionation patterns in petrogenetic studies, and the more practical applications of precious metal distribution in coexisting ore minerals for ore processing. Analysis of heavy minerals (e.g. tetrahedrite, monazite, zircon) will also benefit from the absence of background due to the high Z matrix. The high sensitivity is crucial in analysis of mantle minerals (olivine, garnets, spinels, pyroxenes), which are generally depleted with the majority of trace elements occurring at the sub-ppm level. Micro-AMS will facilitate studies of microscopic relics of primitive magma, trapped as inclusions in minerals, with greater sensitivity.

For isotopic studies, several interesting applications such as the Re-Os system which will became accessible more conveniently, U-Pb system becomes accessible for hydrous minerals, and Rb-Sr systems for Rb rich samples.

The ^{187}Re-^{187}Os system is potentially a powerful tracer for the mantle differentiation process as well as a sensitive geochronometer. From a ratio of ~1, Re/Os is highly fractionated during magma generation, with Re partitioning mainly into the melt, resulting in its enrichment in the crust (Re/Os ~ 3000). The effect on the ^{187}Os/^{186}Os ratio is large, varying from ~1 for the mantle to ~14 in the crust and providing a sensitive handle on the crustal-mantle mixing. AMS offers a specific advantage for the Re/Os system: the negative elemental ions of Re are highly suppressed reducing if not eliminating the isobaric interference effect. Micro-AMS will facilitate direct measurements on minerals (other than platinum group minerals) which are known to concentrate PGE (e.g. molybdenite).

With the facility to suppress interfering ^{87}Rb by selecting negative Sr hydride ions, micro-AMS enables analysis of Sr isotopes in high Rb phases, such as micas. Even in carbonates, where the Rb interference is low, micro-AMS offers an advantage over the ion microprobe measurement by eliminating interference from Ca_2 and $CaMgO$ molecules.

6. Conclusion

Micro-IBA and micro-AMS are powerful analytical tools for geological materials analysis. They complement other microanalytical tools such as the electron, laser ablation, synchrotron X-ray fluorescence and ion microprobes in delineating structural and compositional properties of material relevant to its function, its formation and its behaviour in processes. Their usefulness spans the whole spectrum of the geosciences, from basic to economic geology. It is important to understand the power and limitations of each of the IBA techniques and AMS arising from the use of the microbeam and to avoid using them in isolation. A number of applications uniquely exploit the micro-IBA, e.g. micro-PIXE analysis of fluid inclusions. With its ultra-high sensitivity for trace element and isotope in-situ analytical capabilities, micro-AMS opens up new windows of opportunity in geology. But to be effective in applying micro-IBA and micro-AMS in the geosciences, one must be selective in choosing the topics to address and ensure these methods have clear advantages over any other possible competing techniques.

7. Acknowledgment

Most of the micro-PIXE works referred to in this paper are the results of collaborations at the HIAF laboratory between the authors and their colleagues Ray Binns, Dave Cousens, Bill Griffin, John McAndrew, Chris Ryan, and with many others from various institutions. These include Trevor Green, John Adam, Ed Vicenzi, Anatoliy Chekmir, Alex Sobolev, Louis Cabri, David Green, Russell Sweeney, Ross Large, David Huston, Sam Garrett and many others from the Australian Minerals Exploration community too numerous to mention.

References

[1] S.H. Sie, Nucl. Instr. Meth. **B75**,403 (1993)

[2] S.H. Sie, C.G. Ryan, G.F. Suter, Scanning Microscopy **5**, 977 (1991).

[3] S.H. Sie, W.L. Griffin, C.G. Ryan, G.F. Suter and D.R. Cousens, Nucl.Inst. Meth. **B54**, 284 (1991).

[4] S.H. Sie, D.R. Cousens, C.G. Ryan and W.L. Griffin , Nucl. Instr. Meth. Phys. Res. **B45**, 604 (1990).

[5] S.H. Sie, C.G. Ryan, D.R. Cousens and W.L. Griffin , Nucl. Instr. Meth. **B40/41**, 690 (1989).

[6] C.G. Ryan and W.L. Griffin, Nucl. Instr. Meth. **B77**, 381 (1993).

[7] C.G. Ryan, Nucl. Instr. Meth. **B104**, 377 (1995).

[8] S.H. Sie and C.G. Ryan, Nucl. Inst. Meth. **B15**, 664 (1986).

[9] S.H. Sie, C.G. Ryan, D.R. Cousens and G.F. Suter, Nucl. Instr. Meth. **B45**, 543 (1990).

[10]C.G. Ryan, E. Clayton, W.L. Griffin, S.H. Sie and D.R. Cousens, Nucl.Instr. Meth. **B34**, 396 (1988).

[11]C.G. Ryan, D.R. Cousens, S.H. Sie and W.L. Griffin, Nucl. Instr. Meth. **B49**, 271 (1990).

[12]D.R. Cousens, C.G. Ryan, S.H. Sie and W.L. Griffin, Self Absorption and Secondary Fluorescence Corrections of PIXE Yields from Multilayered Targets, in Proc. of the 5th Australian Conference on Nuclear Techniques of Analysis, Lucas Heights,NSW., ISSN 0811-9422, 58-60 (1987).

[13]W.L. Griffin, C.G. Ryan, D.R. Cousens, S.H. Sie and G.F. Suter, Nucl. Instr. Meth. **B49**, 318 (1990).

[14]W.L. Griffin, L. Jaques, S.H. Sie, C.G. Ryan, D.R. Cousens and G.F.Suter, Contrib. Miner. Petr. **99**, 143 (1988).

[15]R.O. Moore, W.L. Griffin, J.J. Gurney, C.G. Ryan, D.R. Cousens, S.H. Sie, G.F. Suter, Lithos **29**, 1 (1992).

[16]W.L. Griffin, D.R. Cousens, C.G. Ryan, S.H. Sie and G.F. Suter, Contr.Min. Petr. **103**, 199 (1989).

[17]W.L. Griffin, D. Smith, F.R. Boyd, D.R. Cousens, C.G. Ryan, S.H. Sie and G.F. Suter, Geochim. Cosmochim. Acta **53**, 561 (1988).

[18]D. Smith, W.L. Griffin, C.G. Ryan and S.H. Sie, Contr. Min. Petr. **107**, 60 (1991).

[19]A. Greig, S.H. Sie, I.A. Nicholls, Nucl. Instr. Meth. Phys. Res. **B75**, 411 (1993).

[20]T.H. Green, S.H. Sie, C.G. Ryan and D.R. Cousens, Chemical Geology, **74**, 201 (1989).

[21]T.H. Green, J. Adam, S.H. Sie, Miner. Pet. **46**, 179 (1992).

[22]R.J. Sweeney, D.H. Green, S.H. Sie, Earth Plan. Sci. Lett. **113**, 1 (1992).

[23]E. Vicenzi, T.H. Green, S.H. Sie, Chemical Geology **117**, 355 (1994). Nucl. Instr. Meth . **B104** (1995) 470-476

[24]D.L. Huston, S.H. Sie, G.F. Suter, C.G. Ryan, Nucl. Inst. Meth. Phys. Res. **B75**, 531 (1993).

[25]D.L. Huston, S.H. Sie, G.F. Suter, D.R. Cooke, R.A. Both and J.L. Walshe, Economic Geology **90**, 1167 (1995).

[26]D.L. Huston, S.H. Sie, G.F. Suter, C.G. Ryan, Nucl. Inst. Meth. Phys. Res. **B104**, 476 (1995).

[27]W.L. Griffin, P.M. Ashley, C.G. Ryan, S.H. Sie and G.F. Suter, Can. Mineral. **29**, 185 (1991).

[28]D.L. Huston, R.S. Botrill, R.A. Creelman, K. Zaw, A.R. Ramsden, S.W. Rand, J. Gemmell, W. Jablonski, S.H. Sie and R.R. Large, Econ. Geol. **87**, 542 (1992).

[29]L.J. Cabri, H. Blank, A. El Goresy, J.H.G. Laflamme, R. Nobiling, M.B. Sizgoric, K. Traxel, Can. Miner. **22**, 521 (1984).

[30]L.J. Cabri, D.C.Harris and R. Nobiling, Trace Silver Analysis by Proton Microprobe in Ore Evaluation, in Precious Metals: Mining, Extraction and Processing, Kudryk V., Corrigan D.A., Liang W.W. eds, Conf. Proc. Metallurgical Soc. of AIME, pp.93-100 (1984).

[31]L.J. Cabri, J.L. Campbell, J.H.G. Laflamme, R.G. Leith, J.A. Maxwell and J.D. Scott, Can. Mineral. **23**, 133 (1985).

[32]L.J. Cabri, L.J. Hulbert, J.H. Gilles Laflamme, R. Lastra, S.H. Sie, C.G. Ryan and I.J. Campbell., Exploration & Mining Geology **2**, 105 (1993).

[33]K.J. Reeson, C.J. Stanley, C. Jeynes, G. Grime and F. Watt, Nucl. Instr.Meth. Phys.Res. **B45**, 327 (1990).

[34]L.J. Cabri, S.L. Chryssoulis, J.P.R. De Villiers, J.H.G. Laflamme, P.R.Buseck, Can. Mineral. **27**, 353 (1989).

[35]L.J. Cabri, S.L. Chryssoulis, J.L.Campbell, W.J. Teesdale, Appl. Geochem. **6**, 225 (1991).

[36]C.G. Ryan, D.R. Cousens, C.A. Heinrich, W.L. Griffin, S.H. Sie and T.P.Mernagh, Nucl. Instr. Meth. **B54**, 292 (1991).

[37]C.A. Heinrich, C.G. Ryan, T.P. Mernagh and P.J. Eadington, Economic Geology.87,1566 (1993).

[38]E.E. Horn and K. Traxel, Chemical Geology **61**, 29 (1987).

[39]C.G. Ryan, C.A. Heinrich, E. van Achterbergh, C. Ballhaus and T.P. Mernagh, Nucl. Inst. Meth. **B104**, 182 (1995).

[40]J.D. MacArthur, X.P. Ma, G.R. Palmer, A.J. Anderson and A.H. Clark, Nucl. Instr.Meth. **B45**, 322 (1990).

[41]J.D. Frantz, H.K. Mao, Y.G. Zhang, Y. Wu, A.C. Thompson, J.H. Underwood and R.D. Giauque, Chem. Geol. **69**, 235 (1988).

[42]J.B. Lowenstern, G.A. Mahood, M.L. Rivers, S.R. Sutton, Science **252**, 1405 (1991).

[43]S.H. Sie, S. Murao and G.F. Suter, Nucl. Instr. Meth. **B109/110**, 633(1996).

[44]M. Mosbah, J. Tirira, R. Cocchiatti, J. Gosset and P. Massiot, Nucl. Instr. Meth. **B49**, 340 (1990).

[45]M. Mosbah, R. Cocchiatti, J. Gosset, P. Massiot and D P. Trocellier, Nucl. Instr.Meth. **B54**, 298 (1991).

[46]J. Tirira, P. Trocellier, M. Mosbah and N. Metrich, Nucl. Instr. Meth. **B56**, 839 (1991).

[47]S.H. Sie, G.F. Suter, A. Chekhmir and T.H. Green, Nucl. Instr. Meth. **B104,** 261 (1995).

[48]S.H. Sie and G.F. Suter, Nucl. Inst. Meth. **B92** (1994) 221-226

[49]S.H. Sie, T.R. Niklaus and G.F. Suter, Microbeam AMS: Prospects of New Geological Applications, Nucl. Instr. Meth., in press (1996)

Materials Science Forum Vols. 248-249 (1997) pp. 319-325
© *1997 Trans Tech Publications, Switzerland*

Long and Short Range Order in Ion Irradiated Ceramics Studied by IBA, EXAFS and Raman

W. Bolse[1], J. Conrad[1], F. Harbsmeier[1], M. Borowski[1,3] and T. Rödle[2]

[1] II. Physikalisches Institut, [2] IV. Physikalisches Institut and [1,2] Sonderforschungsbereich 345, Universität Göttingen, Bunsenstr. 7-9, D-37073 Göttingen, Germany

[3] LURE, Université Paris-Sud, Bât. 209D, F-91405 Orsay Cedex, France

Keywords: Si, SiC, Si_3N_4, SiO_2, Ceramics, Radiation Damage, Amorphisation, Short Range Order, Ion Beam Analysis, Extended X-Ray Absorption Spectroscopy, Raman-Spectroscopy

Abstract   In the present paper we summarise the results of a recent study on the damage accumulation and the resulting microstructure in heavy ion irradiated single crystalline Si, SiC and SiO_2 and in polycrystalline Si_3N_4. By combining the ion beam analysis techniques Rutherford Backscattering Spectroscopy in Channelling geometry and Resonant Nuclear Reactions Analysis with X-ray absorption, Raman-scattering and mechanical surface profiling new insights into the development of both, long and short range order during ion bombardment of these covalent materials were achieved.

Introduction   Both, the crystalline as well as the amorphous phases of covalently bound Si-based compounds like SiC, Si_3N_4 and SiO_2 exhibit a unique combination of materials properties like high mechanical strength, small mass density, high chemical and thermal stability and wide band gap [1]. They have therefore gained great technological interest as the basic materials for structural and functional ceramics and for optical and electronical high temperature, high speed devices to be employed in or close to engines, turbines and reactors. Ion beams, on the other hand, were found as a valuable and efficient tool for surface analysis and materials modification and are almost indispensable for doping and processing of semiconducting devices [2]. The strong technological interest into the above materials as well as the ability of ion beams to process „new materials" with novel properties therefore suggest to perform detailed studies on the ion beam modification of the structure and the related properties of such compounds. In addition to the technological relevance of such studies there is also a strong fundamental interest into the alterations of the chemical short range order and bond types in the so-called amorphous semiconducting alloys with respect to their crystalline counterparts [3]. We have investigated the ion beam induced alterations of both, long <u>and</u> short range order in the above mentioned covalent materials by combining ion-beam analysis techniques (IBA) like Rutherford backscattering spectroscopy in channeling geometry (RBS-C) and resonant nuclear reaction analysis (RNRA) with mechanical surface profiling (SP), extended and near-edge X-ray absorption fine structure spectroscopy (EXAFS, XANES) and Raman-scattering (RAMAN) [4-8].

Structural properties of the crystalline materials   Silicon-Carbide,   Silicon-Nitride   and Silicon-Oxide can crystallise in a large number of different lattice structures [9]. SiC, for instance, exhibits a variety of cubic and hexagonal polytypes, the most prominent being 4C- and 6H-SIC. SiO_2 is best known in its α- and β-quartz modifications, but also exists as tridymit or cristobalit. As a common feature all these structures consist of a network of Si-centered tetrahedrons, where each Si is surrounded by either four C, four N or four O-atoms. According to its valence C is coordinated by

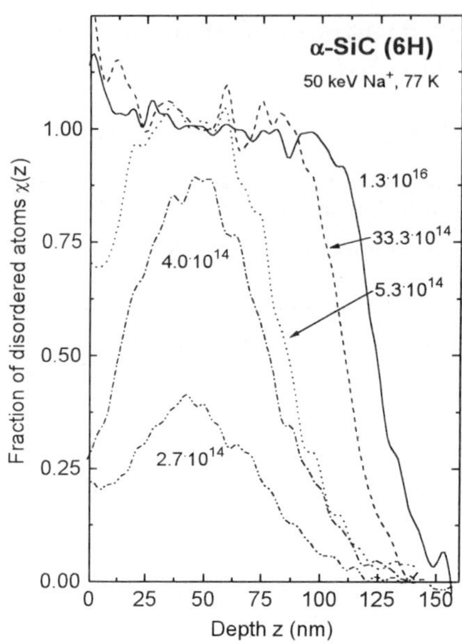

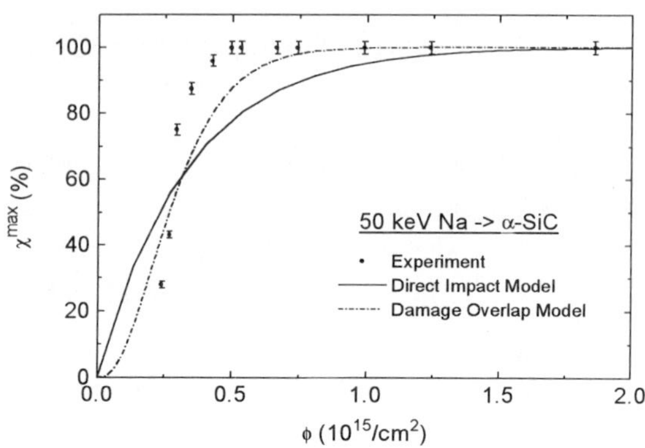

Figure 1a: (left) Apparent damage profiles $\chi(z)$ deduced from RBS-C data of α-SiC after bombardment with 50 keV Na⁺-ions at 77 K

Figure 1b: (top) Maximum χ^{max} of the apparent damage observed after 50 keV Na⁺-ion bombardment of α-quartz versus the ion fluence

four, N by three and O by two Si-atoms. The chemical short range order in the crystalline compounds is thus characterised by the exclusive presence of heteronuclear Si-C-, Si-N- and Si-O-bonds, and the complete absence of any homonuclear Si-Si-, C-C-, N-N- or O-O-bond.

Ion beam analysis of the long range order   The most common way to detect irradiation induced alterations of the long range order in ion implanted single crystalline ceramics and to monitor the respective defect density distribution is to employ the ion channelling technique [10]. Combined with Rutherford Backscattering Spectroscopy and (Resonant) Nuclear Reaction Analysis [11,12] channelling offers a fast and almost non-destructive technique to depth profile both, the implanted impurity concentration and the irradiation induced damage on a nm-scale. Fig. 1a shows the apparent damage profiles $\chi(z)$ deduced from RBS-C data of α-SiC after bombardment with increasing fluences of 50 keV Na⁺-ions. It clearly turns out, that first a Gaussian-like damage profile forms, which increases in height with increasing ion fluence. This fluence range is often referred to as the "nucleation regime", since it is characterised by the generation of point defects and the nucleation of defect agglomerates in a still crystalline matrix. At a critical fluence ϕ_c the maximum of the Gaussian reaches the random level, $\chi^{max}=1$, which is assigned to the formation of a coherent buried amorphous layer. Further irradiation with higher fluences results in planar growth of this disordered layer towards the surface and, with decreasing growth rate, towards larger depths. This fluence range is therefore labelled as the "growth regime". Similar results were obtained with other combinations of ion (H-Au) and target (Si, SiC, SiO₂). Because of the lack of single crystalline material, the radiation damage in Si_3N_4 can not be investigated with the channelling techniques. However, investigation of the mechanical properties and cross-section electron microscopy of Si bombarded polycrystalline Si_3N_4 ceramics suggest a similar behaviour [13-15].

In Fig. 1b the maximum χ^{max} of the apparent damage observed after 50 keV Na⁺-ion bombardment of α-quartz is plotted versus the ion fluence. The S-shaped development, which was also observed in the other cases mentioned above, clearly rules out direct impact amorphisation [16] (i.e. each single ion impact induces an amorphous volume) as an important disordering mechanism in these materials. However, also the damage overlap model [17] does not fit to the experimental data. This model accounts for the accumulation of point defects until a critical damage density is build up by several spatially overlapping recoil cascades and the crystalline structure becomes unstable in favour of the amorphous phase. In order to explain the continuously accelerated damage build-up in Si and Si-based compounds it has therefore been suggested to take into consideration further driving forces such as mechanical stresses and defect interaction via strain fields [8, 18]. In fact, the preliminary results of our recent study on 50 keV Ne bombarded α-quartz [19], where we have measured the bending radius of a 0.5 mm thin sample as a function of the

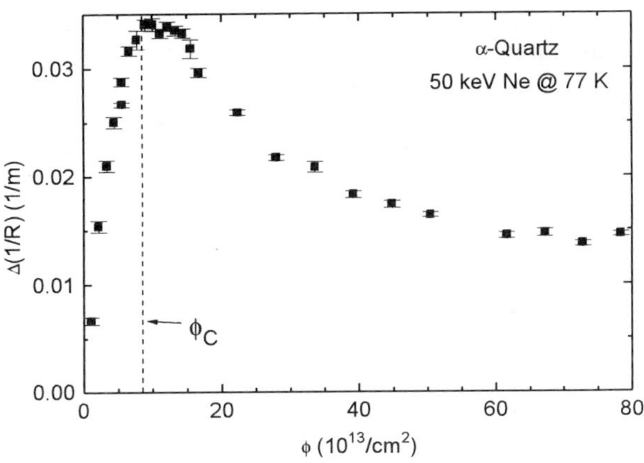

Figure 2: *Curvature change $\Delta(1/R)$ of a 0.5 mm thin quartz sample after irradiation with 50 keV Ne-ions*

applied ion fluence, clearly revealed the generation of strong compressive stresses at fluences below ϕ_c, which were slowly released after a buried amorphous layer has formed and grows towards larger depths. The results are illustrated in Fig. 2, which shows, as a measure of the induced stress, the curvature change $\Delta(1/R)$ as a function of the ion fluence (R = bending radius of the sample).

In SiC and SiO_2, comparison of the experimental results with TRIM95 cascade simulations [20] revealed that the dependence of ϕ_c on the energy and atomic numbers of the ion species can be explained by the corresponding density F_D of the energy deposited by primary and secondary elastic displacement collisions. The critical energy density model predicts the crystalline structure to break down as soon as the deposited energy ϕF_D reaches a critical level E_c, i.e. as soon as a critical defect density has been accumulated. In Fig. 3 ϕ_c is plotted versus $1/F_D$ for α-SiC bombarded with a variety of ions ranging from H to Au. In fact, it turns out that, independent on the nature and the energy of the incident ion, the deposition of $E_c = \phi_c\,F_D^{max} = 19(2)$ eV/atom is necessary to achieve amorphous SiC. In SiO_2 amorphisation is much faster and occurs after the deposition of about 2 eV/atom. This refers to damage levels of about 0.2 and 0.05 dpa (displacements/atom), respectively.

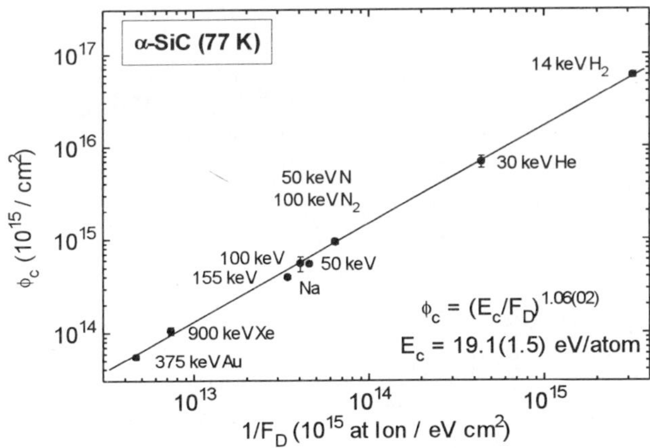

Figure 3: *Critical fluence ϕ_c for the formation of a buried amorphous layer in α-SiC during irradiation with ions ranging from H to Au versus $1/F_D$. F_D is the energy deposited in displacement collisions*

Investigation of the short range order

amorphisation in the covalent Si-based compounds occurs already after about 5 - 20 % of the target atoms have been displaced from their lattice sites. In view of such a small number of relocated atoms the very interesting question arises whether the chemical short range order, which consists of only heteronuclear bonds, is conserved during the crystalline-to-amorphous transition and if the amorphous structure can still be modified by further irradiation at higher fluences. We have therefore performed EXAFS- and RAMAN-studies in order to obtain information on the chemical short range order and the type of bonds formed during ion bombardment of Si, SiC, Si_3N_4 and SiO_2.

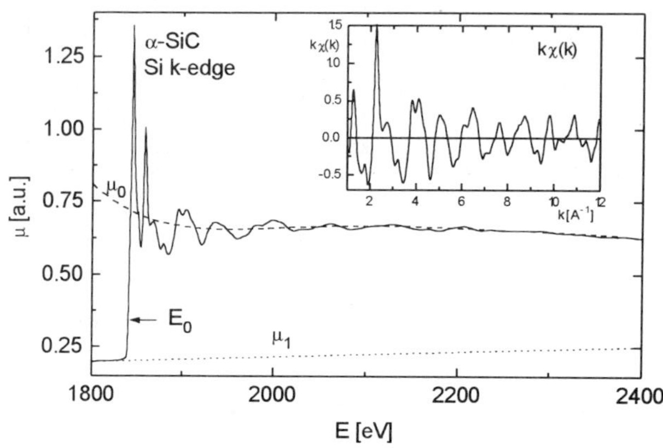

Figure 4: X-ray absorption coefficient μ in α-SiC as a function of the energy E at the Si K-edge. The clearly visible EXAFS oscillations are shown in the insert as a function of the wave number of the photo-electron.

X-ray absorption: In Fig. 4 the X-ray absorption spectrum $\mu(E)$ measured at the Si K-edge of a non-irradiated α-SiC sample is shown. All absorption spectra were taken by measuring the secondary electron yield instead of the transmitted X-ray intensity and therefore the origin of the EXAFS signal was restricted to a surface near layer of about 50 nm thickness. Note the clearly visible oscillations, which originate from scattering of the photo-electron at the atomic neighbours of the Si atom and the resulting interference of the emitted and reflected electron waves. For details of the technique and data evaluation refer to [21]. After proper background correction and

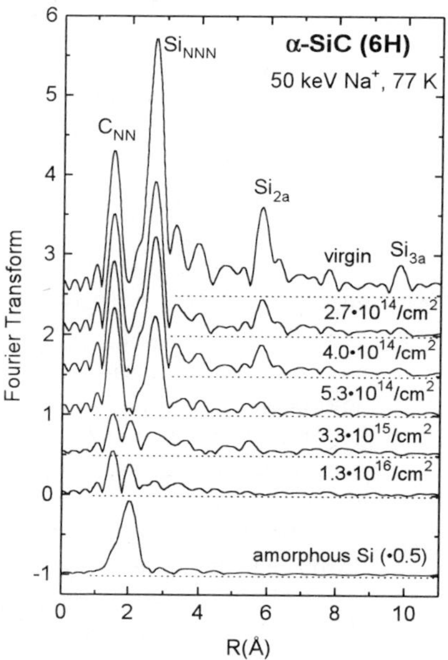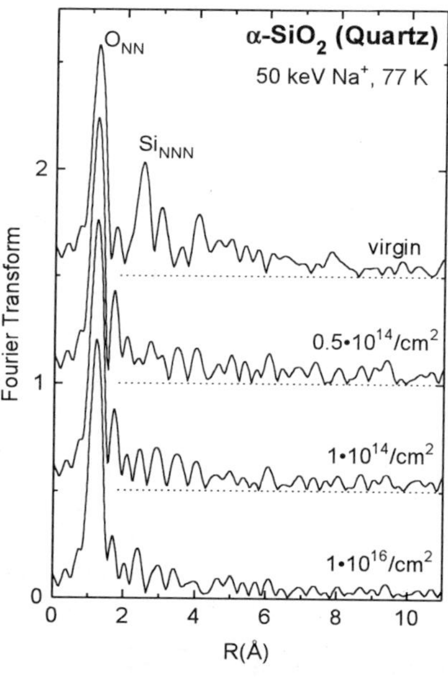

Figure 5: Fouriertransform of the EXAFS oscillations $k\chi(k)$ for a) α-SiC (6H) and b) α-Quartz after irradiation with 50 keV Na-ions at 77 K. For comparison the signal of amorphous Si is shown in a).

normalisation, the EXAFS-oscillations were extracted from the spectrum via $\chi(k)=(\mu(k)-\mu_0)/\mu_0$, where k denotes the wave number of the emitted photo-electron and μ_0 the absorption cross section of a free atom. In order to enhance the sensitivity at higher wave-numbers further data evaluation was performed using $k\,\chi(k)$ instead of $\chi(k)$. Fouriertransformation of this product provides a pseudo-radial distribution function (PRDF) with each peak corresponding to a specific shell of atoms surrounding the probe atom. The uppermost curve in Fig. 5a refers to the EXAFS oscillations shown in Fig. 4. The most intense peaks correspond to the C-tetrahedron around the Si atom (C_{NN}) and the nearest Si shell (Si_{NNN}). Further peaks refer to Si atoms at distances 2 a resp. 3 a (a = lattice parameter). Because of charging problems the EXAFS data taken with the insulating quartz samples exhibit a high noise level. However, as is shown in the uppermost curve of Fig. 5b, the peaks referring to the O-tetrahedron (O_{NN}) as well as the signal from the nearest Si-shell (Si_{NNN}) are clearly visible for the non-irradiated sample.

In spite of the similar amorphisation kinetics observed with RBS-C, EXAFS reveals a completely different microstructural evolution of α-SiC and α-SiO$_2$ during ion bombardment. In addition to the non-irradiated crystals the PRDF of 50 keV Na-irradiated samples are shown in Fig. 5. According to RBS-C a coherent buried amorphous layer has formed in the irradiated quartz sample at $\phi = 5 \cdot 10^{13}/cm^2$, which extends to the surface at $\phi = 1 \cdot 10^{14}/cm^2$ and has grown to a thickness of about 160 nm at $\phi = 1 \cdot 10^{16}/cm^2$. In all cases the peak corresponding to Si_{NNN} has disappeared, while the O_{NN}-peak has not been altered significantly. The latter observation clearly indicates that the chemical short range order of crystalline α-quartz (i.e. Si centred O-tetrahedrons) is completely conserved during ion bombardment. From the disappearance of the Si_{NNN}-peak, on the other hand, it can be concluded that the initially ordered network of Si-O$_4$-tetrahedrons has been transformed into a disordered network with highly distroted Si-O-Si bond angles and thus broadly distributed Si-Si distances. The results obtained for Si and Si$_3$N$_4$ also suggest the formation of a tetragonal random network, but here the reduction of the peaks corresponding to the nearest neighbour atoms indicates the presence of a significant number of broken Si-Si and Si-N-bonds, respectively.

For SiC (Fig. 5a) the situation is completely different. The distortion of the long range order in the "nucleation regime" is indicated by the attenuation of the peaks corresponding to the Si-shells, which is the stronger the larger is their distance. Yet, although according to RBS-C an amorphous layer has formed after implantation of $5.3 \cdot 10^{14}\,Na^+/cm^2$, the Si_{NNN} peak is still visible. Since almost no changes of the C_{NN} peak can be seen up to this fluence, we conclude that a network of Si-C$_4$ tetrahedrons has formed, which, in contrast to SiO$_2$, is not completely disordered and exhibits a much smaller distribution of the Si-C-Si bond angles as compared to Si-O-Si in amorphous SiO$_2$. Further irradiation with higher fluences, however, results in a drastic change of the microstructure. At $3.3 \cdot 10^{15}/cm^2$ the Si_{NNN} peak has almost completely disappeared and also the intensity of the C_{NN} peak is now strongly reduced. On the other hand a new peak appears at $R \approx 2\text{Å}$. Comparison with the EXAFS data of amorphous Si suggests to attribute this peak to homonuclear Si-Si bonds, which are not present in the crystalline structure. This observation is in good agreement with a molecular dynamics study on SiC quenched from the liquid [22]. According to these simulations the resulting structure is of rather complex nature, where more than 40% of the bonds formed by C-atoms are of homonuclear character. These C-atoms are arranged in chain-like structures, which are surrounded by tetragonal Si-Si- and Si-C-networks. The computed nearest neighbour distance ($d_{Si-C} = 1.89$ Å) and partial co-ordination number ($n_{Si-C} = 2.1$) of the Si-C configurations are in good agreement with the results of the quantitative analysis of our EXAFS data ($d_{Si-C} = 1.93(3)$ Å, $n_{Si-C} = 1.7(3)$). The simulated Si-Si distances were found to be broadly distributed with a first maximum close to the nearest neighbour distance $d_{NN} = 2.32$ Å in crystalline Si. Because of the broad distribution the partial co-ordination number can only be estimated from Fig. 1 in [22] to be about $n_{Si-Si} \approx n_{Si-C}/2$, which again is in good agreement with our experimental findings ($d_{Si-Si} = 2.35(2)$ Å, $n_{Si-Si} = 0.9(2)$).

Raman-Spectroscopy   The formation of new bonds in SiC due ion bombardment can also be monitored by means of Raman-scattering [23], as is illustrated in Fig. 6, which shows the Raman-spectra of α-SiC after irradiation with different fluences of 155 keV Na$^+$-ions. The uppermost spectrum was taken with a non-irradiated sample and exhibits the first order (768 cm^{-1} - 959 cm^{-1}) and a second order peaks (505 cm^{-1}) characteristic for crystalline α-SiC. The spectra were taken in backscattering geometry with an Ar$^+$ laser of λ = 514.5 nm. At this wave length α-SiC is transparent because of its wide band gap of about 3 eV. The Raman-scattered light therefore originates from the whole wafer thickness of 0.3 mm, which is about 2000 times the maximum ion beam modified depth. After irradiation the intensity of the "crystalline" peaks decreases with increasing fluence and has vanished after irradiation with $1.36 \cdot 10^{16}$/cm^2. This reduction can be assigned to the increasing absorption of the laser light in the damaged layer. In fact, correlation of the normalised "crystalline" Raman intensity I/I_0 with the thickness d_a of the amorphous layer reveals an exponential relationship, $I/I_0 = \exp(-\mu\, d_a)$ and the resulting absorption coefficient $\mu = 2.13(21) \cdot 10^5$/cm is in good agreement with the results of transmission and reflection measurements by Tsvetkova et al [24].

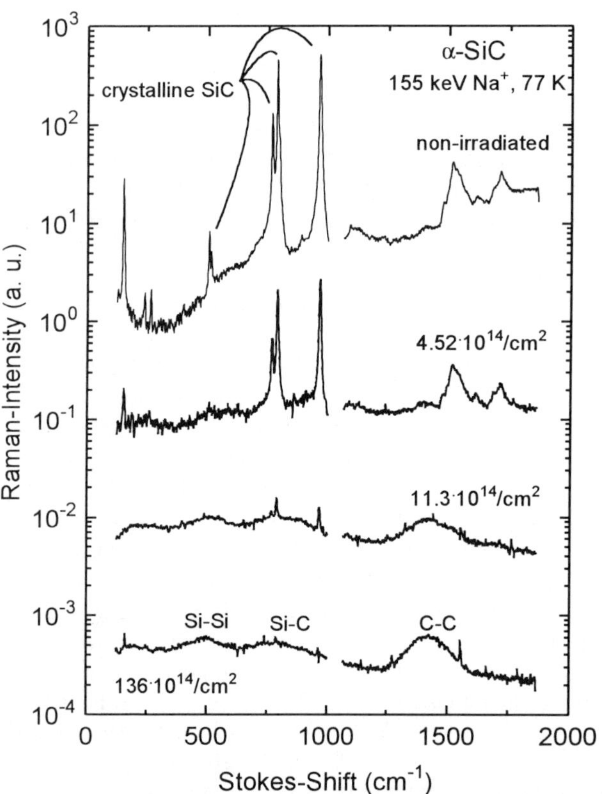

Figure 6: Raman-spectra of α-SiC irradiated with 155 keV Na ions at various ion fluences and a temperature of 77 K.

After irradiation with very high fluences, where the amorphous layer, according to RBS-C, has almost reached its saturation thickness and EXAFS suggests the formation of homonuclear Si-Si bonds, the Raman spectrum has completely changed. Instead of the sharp peaks, which are characteristic for the crystalline phase, now broad bands are visible, which indicate the formation of an amorphous structure. Two of these bands are similar to what is observed in a-Si (band located at 480 cm^{-1}) [25] and amorphous C (band located at 1450 cm^{-1}) [26] and can therefore be attributed to homonuclear Si-Si and C-C bonds, respectively. The weak band centred at about 800 cm^{-1} is assigned to distorted Si-C bonds. This is in nice agreement with the above mentioned MD simulations and with our EXAFS results.

Unfortunately no such measurements could be performed with the quartz samples because also the amorphous phase is transparent and the light scattered in the crystalline bulk of 1 mm hides the Raman signal from the modified layer of 200 nm maximum thickness.

Conclusions   In conclusion we have shown that an improved insight into both the ion beam induced amorphisation processes and the related microstructural changes can be achieved by combining IBA, which allows to investigate the alterations of the long range order, with EXAFS and RAMAN, which are sensitive to modifications of the short range order. In particular, we have demonstrated that amorphisation in the Si-based ceramics occurs by defect accumulation in the crystalline matrix until a critical damage density is achieved and the crystalline structure breaks down spontaneously. In all cases a structurally disordered, but chemically ordered tetragonal network forms at the critical fluence. In SiO_2, Si_3N_4 and Si this network seems be of random nature, while the Si-C_4 tetrahedrons in SiC are still correlated with respect to their orientation. In SiO_2, the short range order is conserved even after high fluence bombardment, while further irradiation of amorphous SiC at higher fluences results in a chemically disordered structure. The results are in good agreement with MD simulations on a-SiC quenched from the liquid state.

References

[1] G.L. Harris, C.Y. Yang (eds), Amorphous and crystalline Silicon Carbide and related materials, Springer-Verlag (New York, 1989)

[2] J. Singh, S.M. Copley, J. Mazumder (eds), Beam processing of advanced materials, ASM International (Materials Park OH, 1996)

[3] D. Emin, T.L. Aselage, C. Wood (eds), Novel refractory semiconductors, Materials Research Society (Pittsburgh, 1987)

[4] W. Bolse, J. Conrad, T. Rödle, T. Weber, Surf. Coat. Technol. **74/75**, 927 (1995)

[5] J. Conrad, T. Rödle, T. Weber, W. Bolse, Nucl. Instr. Methods B **118**, 748 (1996)

[6] M. Borowski, W. Bolse, J. Conrad, A.M. Flank, Proceedings of XAFS IX (Grenoble, 1996)

[7] J. Conrad, W. Bolse, M. Borowski, T. Rödle, S. Pietzonka, T. Weber, Phys. Rev. B (1996) submitted

[8] F. Harbsmeier, W. Bolse, this conference

[9] B.G. Hyde, S. Anderson, Inorganic crystal structures, John Wiley Sons (New York, 1989)

[10] L.C. Feldman, J.W. Mayer, Fundamentals of surface and thin film analysis, Elsevier Science Publishers (Amsterdam, 1986)

[11] W.K. Chu, J.W. Mayer, M.-A. Nicolet, Backscattering Spectrometry, Academic Press (Orlando, 1978)

[12] W. Bolse, Mat. Sci. Eng. R **12, 53** (1994)

[13] W. Bolse, S.D. Peteves, Nucl. Instr. Methods B **68**, 331 (1992)

[14] W. Bolse, S.D. Peteves, F.W. Saris, Euro-Ceramics **2**, 1091 (1993)

[15] W. Bolse, S.D. Peteves, F.W. Saris, Appl. Phys. A **58**, 493 (1994)

[16] J.R. Dennis, E.B. Hale, J. Appl. Phys. **49**, 1119 (1978)

[17] J.F. Gibbons, Proc. IEEE **60**, 1062 (1972)

[18] O.W. Holland, S.J. Pennycook, G.L. Albert, Appl. Phys. Lett. **55**, 2503 (1989)

[19] F. Harbsmeier, W. Bolse, to be published

[20] J.F. Ziegler, J.P. Biersack, W. Littmark, The stopping and range of ions in solids, vol. **1**, Pergamon Press (New York, 1985)

[21] D.C. Koningsberger, R. Prins (eds), X-Ray Absorption: Principles, Applications and Techniques of EXAFS, SEXAFS and XANES (New York, 1988)

[22] F. Finocchi, G. Galli, M. Parrinello, C.M. Bertini, Phys. Rev. Lett. **68**, 3044 (1992)

[23] R. Alben, D. Weaire, J.E. Smith Jr, M.H. Brodsky, Phys. Rev. B **11**, 2271 (1975)

[24] T. Tsvetkova, in: Beam Processing of Advanced Materials, J. Singh, S.M. Copley, J. Mazumder (eds), ASM International (Materials Park OH ,1996) p. 207

[25] J.E. Smith Jr, M.H. Brodsky, B.L. Crowder, M.I. Nathan, A. Pinczuk, Phys. Rev. Lett. **26**, 642 (1971)

[26] J.E. Smith Jr, M.H. Brodsky, B.L. Crowder, M.I. Nathan, J.Non-Cryst. Sol. **8-10**, 179 (1972)

Materials Science Forum Vols. 248-249 (1997) pp. 327-332
© *1997 Trans Tech Publications, Switzerland*

Progress in Non-Standard Defect Analysis by RBS

K. Gärtner

Friedrich-Schiller-Universität Jena, Institut für Festkörperphysik,
Max-Wien-Platz 1, D-07743 Jena, Germany

Keywords: Dechanneling, Point Defects, Defect Analysis, Rutherford Backscattering

ABSTRACT

Rutherford backscattering (RBS) combined with channeling techniques is widely used for defect analysis. The evaluation of the RBS data can be performed using standard methods or a non-standard method based on a master equation approach of dechanneling. This paper reports on the progress of the non-standard method of the defect analysis obtained by extending the master equation approach of dechanneling from elementary to compound crystals and by considering more complex point defect configurations. The procedure of the defect analysis is demonstrated for two examples of point defects in Se^+ irradiated GaAs.

1. INTRODUCTION

For about 30 years channeling has been used for studies of defects in crystals. The relative Rutherford backscattering yield for ions incident parallel to a crystal axis (RBS minimum yield χ_{min}) is increased if the crystal contains defects (displaced lattice atoms). The enhancement of the minimum yield is caused by direct backscattering and dechanneling of the ions due to the interaction with the displaced lattice atoms. The dechanneling mechanism depends on the correlation of the displaced lattice atoms which is different for different kinds of defects. With respect to dechanneling point defects can be characterized by uncorrelatedly displaced lattice atoms while dislocations and stacking faults cause a strong correlation of the lattice displacements. If the dechanneling is known information about the defects in the crystal can be obtained from the measured RBS minimum yield (defect analysis).

Standard methods of defect analysis by RBS are based on the description of dechanneling within the two beam approximation [1-6]. They can be applied in the cases of special point defects (positions of the displaced lattice atoms are uniformly distributed within the lattice cell, case of heavily damaged crystals) or special extended defects (edge dislocations, screw dislocations, stacking faults) with a pronounced (narrow) depth profile. If only one of the defects mentioned above is present its density as a function of the depth z can be calculated from the measured $\chi_{min}(z)$.

Based on a master equation approach of dechanneling [7-10] following the general dechanneling concept of Lindhard [11] and Bonderup et al. [12] a more elaborated method of defect analysis has been developed for elementary crystals. The corresponding computer code DICADA 1 (**D**echanneling **I**n **C**rystals **A**nd **D**efect **A**nalysis 1) allows to evaluate RBS data for crystals with point defects [7], dislocations [8,10] and stacking faults [9]. Especially, also point defects in weakly damaged crystals with preferred positions of the displaced lattice atoms can be analyzed [7,13] if the RBS data are obtained for different temperatures.

The aim of this paper is to improve the method of defect analysis mentioned above by extending it from elementary to compound crystals and by taking into account a second kind of point defects in the crystal. Dislocations and stacking faults are not considered.

2. MODEL

The master equation approach of dechanneling in elementary crystals with point defects [7] has been extended to the case of compound crystals. The procedure is given elsewhere [14]. With respect to dechanneling point defects are described by uncorrelatedly displaced lattice atoms. This paper deals with compound crystals containing two different kinds of point defects (displaced lattice atoms) characterized by different distributions $P_1(r_a)$ and $P_2(r_a)$ of the displacement distances r_a perpendicular to the channeling direction and by the relative densities of displaced atoms $n_{dat1}(z)$ and $n_{dat2}(z)$ as functions of the depth z. Two options for the distribution of the displacement distances are considered.

The first option for $P_1(r_a)$ and $P_2(r_a)$ is given by displaced lattice atoms, thermally vibrating around an average displacement distance of r_{a1} and r_{a2}, respectively. This option corresponds to preferred positions of the displaced lattice atoms which may be present in weakly damaged crystals. For simplicity it is assumed that the average displacement distances r_{a1} and r_{a2} do not differ for the different compounds of the crystal. The thermal vibration amplitude ϱ_{dat} of the displaced lattice atoms depends on the temperature and is different for the different components of the crystal. Usually, it is not known. In this paper the thermal vibration amplitude ϱ_{dat} of the displaced lattice atoms of a given component is assumed to be proportional to the thermal vibration amplitude ϱ_{perf} of the perfect lattice atoms of the same component and the same temperature ($\varrho_{dat} = \alpha\,\varrho_{perf}$) with the factor α being the same for different temperatures and components. If there is no information about ϱ_{dat} the analysis of the point defects starts with $\alpha = 1$.

The second option for the distributions of the displacement distances $P_1(r_a)$ and $P_2(r_a)$ is given by a random (uniform) distribution of the displaced lattice atoms within a unit lattice cell. This option corresponds to heavily damaged crystals or amorphous layers where no preferred positions of the displaced lattice atoms exist.

Based on the description of dechanneling in compound crystals [14] for the situation mentioned above, the computer code DICADA 2 for analyzing RBS data has been developed. The progress in defect analysis obtained by this code in comparison with DICADA 1 is that compound crystals can be analyzed and the crystal may contain a second distribution of displaced atoms. Using this code it is possible to calculate one of the densities of displaced atoms $(n_{dat1}(z))$ from the measured minimum yield $\chi_{min}(z)$ if the other density $(n_{dat2}(z))$ and the distributions $P_1(r_a)$ and $P_2(r_a)$ are known. As has been shown previously [7,13] the temperature dependence of the minimum yield is different for different displacement distances. Therefore, information about the distribution $P_1(r_a)$ can be obtained from the minimum yield $\chi_{min}(z,T)$ measured at different temperatures. The procedure of getting this information and the corresponding results are given in sections 3.1 and 3.2 for the case of only one kind of point defects and for a special case of the existence of a second kind of point defects, respectively.

3. PROCEDURE AND RESULTS

3.1 Example 1

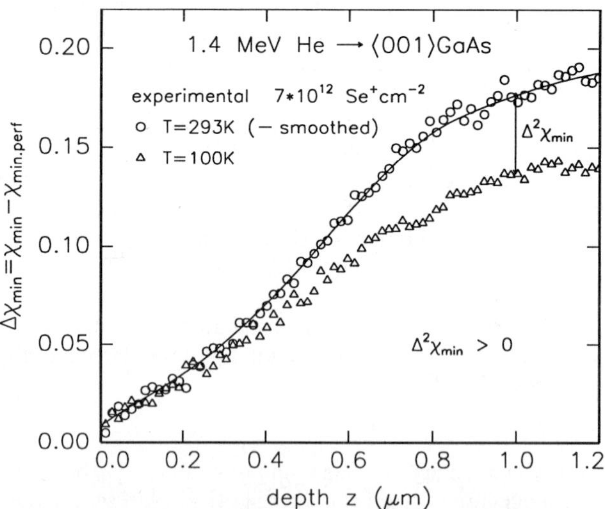

Fig.1. Difference in the RBS minimum yield $\Delta\chi_{min}$ as a function of the depth z for $\langle 001 \rangle$ GaAs irradiated with 2 MeV Se ions at a dose of $7{\times}10^{12}$ cm^{-2}, measured by Schrempel et al. [15] with 1.4 MeV He ions at two temperatures of 293K and 100K. For more explanation see text.

As an example for one kind of point defects in a compound crystal the experimental RBS data (Fig. 1) obtained by Schrempel et al. [15] for $\langle 001 \rangle$ GaAs irradiated with 2 MeV Se$^+$ ions at a dose of $7{\times}10^{12}$ cm^{-2} using 1.4 MeV He$^+$ ions, are analyzed. The quantity $\Delta\chi_{min}(z)$ depicted in Fig. 1 is the difference of the minimum yields after and before the irradiation. It has been measured for two different temperatures (293K, 100K). The temperature dependence of $\Delta\chi_{min}$ at $z=1\mu m$ is characterized by the second difference in the minimum yield $\Delta^2\chi_{min} = \Delta\chi_{min}\,(1\mu m,\ 293K) - \Delta\chi_{min}\,(1\mu m,\ 100K)$ (see Fig. 1). It is positive in this case.

The procedure of the defect analysis is the following. First, it can be stated that the GaAs is only weakly damaged ($\Delta\chi_{min} \lesssim 15\%$), i.e. preferred positions of the displaced lattice atoms are possible. Second, there is no hint for the

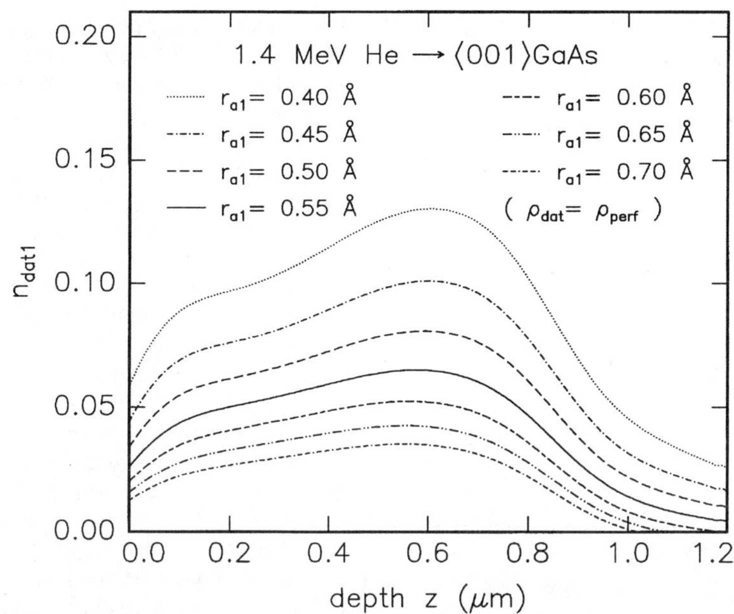

Fig.2. Relative densities of displaced atoms n_{dat1} as functions of the depth z calculated from $\Delta\chi_{min}(z)$ at 293K (smoothed curve in Fig. 1) for different values of the average displacement distance r_{a1}.

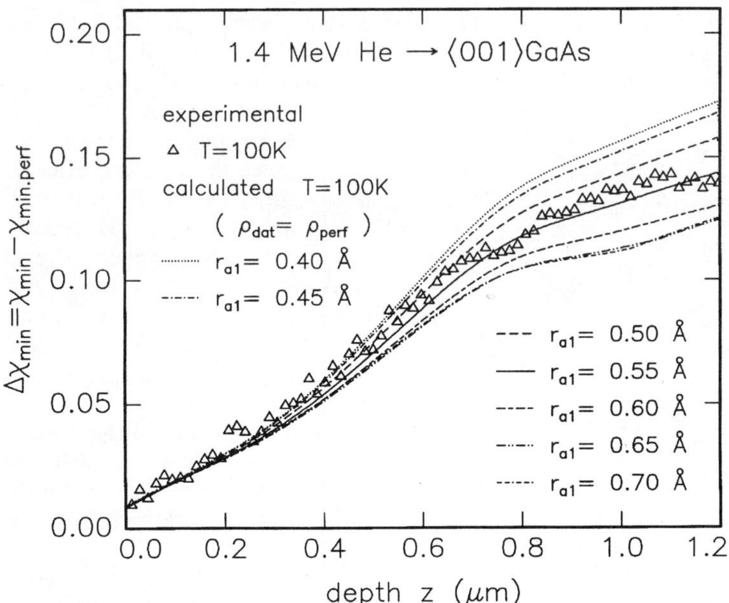

Fig. 3. Differences in the minimum yields $\Delta\chi_{min}$ at 100K as functions of the depth z calculated for the different values of r_{a1} and $n_{dat1}(z)$ given in Fig.2.

necessity to use two different distributions of displaced lattice atoms, Therefore, the analysis is performed using only one distribution of displaced lattice atoms with the relative density $n_{dat1}(z)$ and the option for the distribution $P_1(r_a)$ given by thermal vibration arround an average displacement distance r_{a1}. Because there is no other information, the thermal vibration amplitude of the displaced lattice atoms ϱ_{dat} is assumed to be equal to the thermal vibration amplitude of the atoms in the perfect lattice ϱ_{perf}. Under these assumptions the remaining unknown quantities are $n_{dat1}(z)$ and r_{a1}. They are determined in the following way. First of all a set of values of r_{a1} is chosen (see Fig. 2) and for each value of r_{a1} the corresponding relative density $n_{dat1}(z)$ is calculated from the measured $\Delta\chi_{min}(z, 293K)$ (smoothed curve in Fig. 1) using DICADA 2. Then for each value of r_{a1} and the corresponding calculated density $n_{dat1}(z)$ the difference in the minimum yield $\Delta\chi_{min}(z, 100K)$ at the lower temperature is calculated. As can be seen in Fig. 3 the calculated curves $\Delta\chi_{min}(z, 100K)$ differ remarkably for different values of r_{a1}. A good agreement with the experimental data for 100K is obtained only for r_{a1} between 0.50 Å and 0.55 Å. Another presentation of the results given in Fig. 4 shows the calculated second difference in the minimum

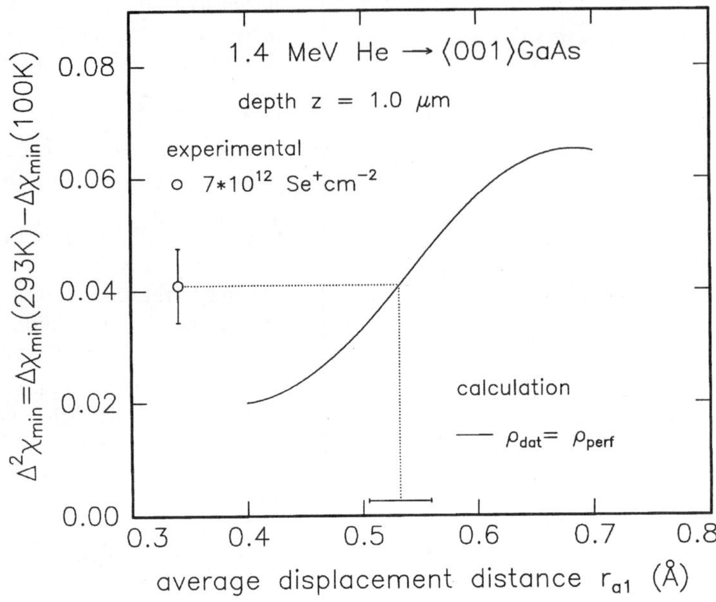

Fig.4. Second difference in the minimum yield $\Delta^2\chi_{min}$ (explained in text) as a function of the average displacement distance r_{a1} corresponding to the results in Fig.3.

yield $\Delta^2\chi_{min}(r_{a1}) = \Delta\chi_{min}$ (1μm, 293K) - $\Delta\chi_{min}$ (1μm, 100K, r_{a1}) as a function of the average displacement distance r_{a1}. The comparison of the calculated $\Delta^2\chi_{min}(r_{a1})$ with the measured value $\Delta^2\chi_{min}$ = 0.041 ± 0.007 (see Fig. 1) provides r_{a1} = (0.53 ± 0.03) Å. For this value $n_{dat1}(z)$ and $\Delta\chi_{min}(z, 100K)$ are calculated and depicted in Fig. 5 together with all experimental data. The calculated $\Delta\chi_{min}(z, 100K)$ agrees well with the measured curve and therefore the average displacement distance r_{a1} = 0.53 Å and $n_{dat1}(z)$ given in Fig. 5 are the final results of the defect analysis for this example.

3.2 Example 2

As an example for two kinds of point defects in a compound crystal the experimental RBS data obtained by Schrempel et al. [15] for ⟨001⟩ GaAs irradiated with 2 MeV Se⁺ ions at a dose of 10^{12} cm⁻² are analyzed. The differences in the minimum yield $\Delta\chi_{min}(z)$ for the temperatures 293K and 100K are depicted in Fig. 6. Compared with the results given in Fig. 5 the effect is smaller because of the smaller Se⁺ dose used and the temperature dependence of $\Delta\chi_{min}$ changed completely from a positive to a negative one ($\Delta^2\chi_{min}$ < 0). Moreover, the values of $\Delta\chi_{min}$ at depth z = 0 for the two temperatures differ clearly which indicates different states of the surfaces for the two temperatures. This can be understood by deposition of a thin impurity surface layer during the low temperature RBS investigation. With respect to the analysis of the defects in the bulk the impurity layer can be taken into

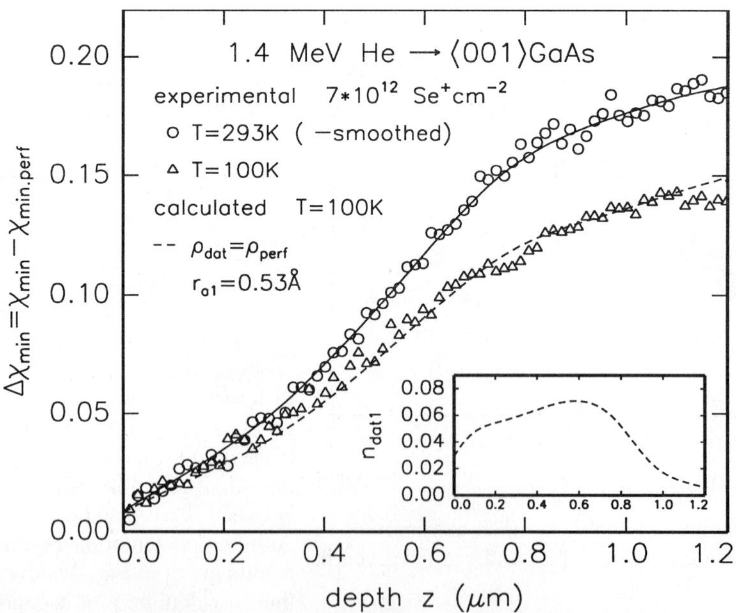

Fig. 5. Experimental data as given in Fig. 1 added by the results of the defect analysis (see text).

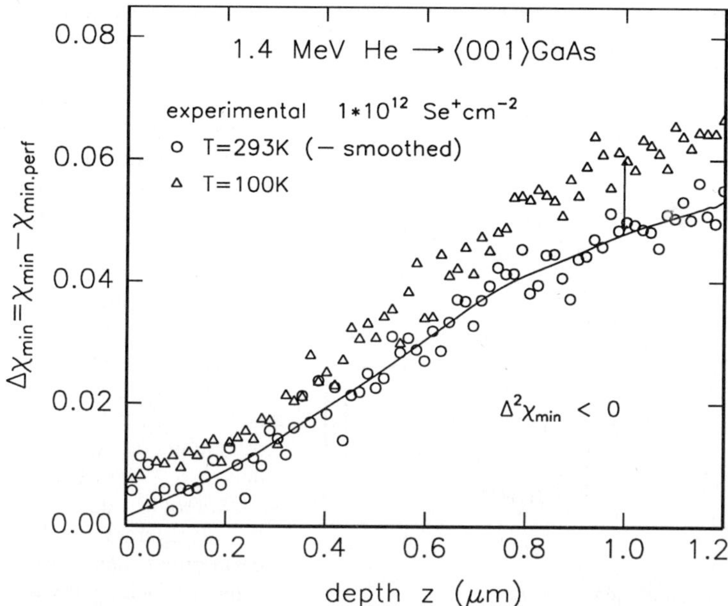

Fig.6. Difference in the RBS minimum yield $\Delta\chi_{min}$ as a function of the depth z for $\langle 001 \rangle$ GaAs irradiated with 2MeV Se ions at a dose of 10^{12} cm^{-2}, measured by Schrempel et al. [15] with 1.4MeV He ions at two temperatures of 293K and 100K.

The resulting $\Delta^2\chi_{min}(r_{a1})$ is shown in Fig. 7 (full line) together with the measured $\Delta^2\chi_{min}$ (see Fig. 6). As can be seen, for all r_{a1} the calculated $\Delta^2\chi_{min}(r_{a1})$ is larger than the measured one, i.e. there is no solution.

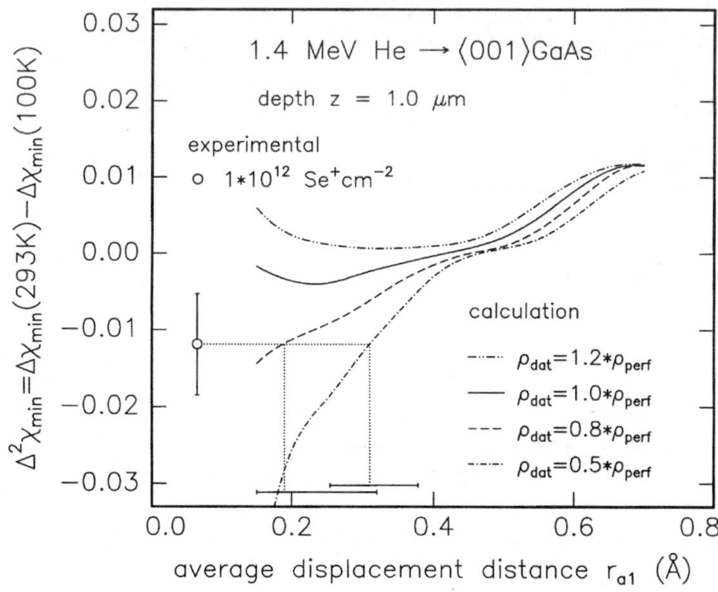

Fig.7. Second differences in the minimum yield $\Delta^2\chi_{min}$ as functions of the average displacement distance r_{a1} obtained from the evaluation of the data in Fig.6.

account as an equivalent amorphous surface layer with a thickness d_{am}, i.e. there as an additional distribution of displaced lattice atoms with $n_{dat2}(z) = 1$ for $z \leq d_{am}$ and $n_{dat2}(z) = 0$ for $z > d_{am}$ and with the option of $P_2(r_a)$ given by randomly distributed displaced lattice atoms. The defect analysis is performed step by step. First, neclecting the amorphous layer $n_{dat1}(z)$ and r_{a1} are calculated as in section 3.1 (not shown). Second, these zero order results are used to determine the thickness of the amorphous layer which results in $d_{am} = 15$ Å. Then, taking into account the amorphous layer, $n_{dat1}(z)$ and r_{a1} are again determined according to the procedure described in section 3.1. Therefore, the thermal vibration amplitude ϱ_{dat} of the displaced atoms is varied from $\varrho_{dat} = 0.5$ ϱ_{perf} to $\varrho_{dat} = 1.2$ ϱ_{perf}. The corresponding curves $\Delta^2\chi_{min}(r_{a1})$ are added in Fig.7. As can be seen the increase of the value of ϱ_{dat} does not provide a solution. However, for the two smaller values $\varrho_{dat} = 0.8$ ϱ_{perf} and $\varrho_{dat} = 0.5$ ϱ_{perf} there exist solutions given by $r_{a1} = 0.19$ Å and $r_{a1} = 0.31$ Å. For both values the calculated $\Delta\chi_{min}(z)$ for 100K agree well with measured ones (see Fig. 8). The corresponding relative densities $n_{dat1}(z)$ are shown in the insert of Fig. 8. Because the value of ϱ_{dat} is not known the results of the

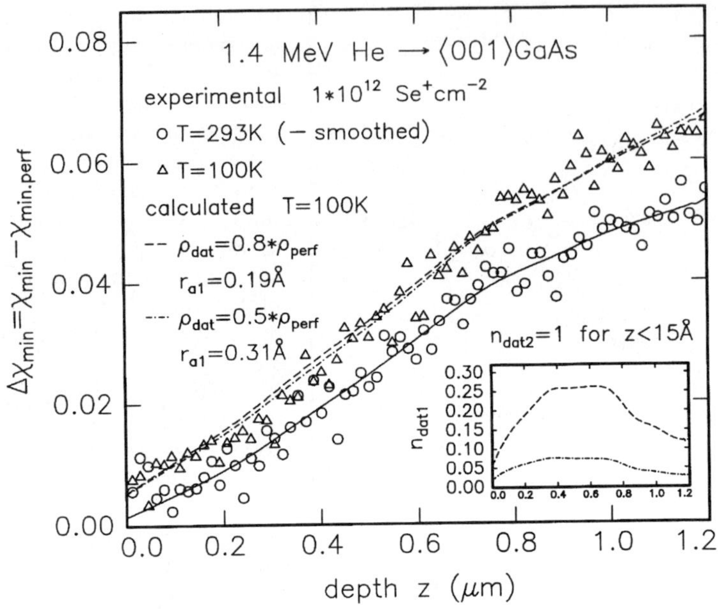

Fig.8. Experimental data as given in Fig.6 added by the results of the defect analysis (see text).

defect analysis are not definite. The information obtained is that the GaAs investigated contains point defects with preferred positions of the displaced lattice atoms with a small average displacement distance r_{a1} between 0.2 Å and 0.3 Å and a thermal vibration amplitude ϱ_{dat} at least 20% lower than that of the perfect lattice atoms (ϱ_{perf}). The resulting relative density $n_{dat1}(z)$ depends remarkably on the exact value of ϱ_{dat} (insert of Fig. 8). Moreover, during the low temperature RBS measurement there appears a defect surface layer which can be described by an equivalent amorphous GaAs layer of a thickness of 15 Å.

Summarizing, it can be stated that for average displacement distances $r_{a1} \gtrsim 0.4$ Å the non-standard defect analysis provides definite results $n_{dat1}(z)$ and r_{a1} from the measured $\Delta\chi_{min}(z,T)$. In the case of $r_{a1} < 0.4$ Å the results depend on the value of the thermal vibration amplitude of the displaced atoms ϱ_{dat} (see Fig. 7) which is usually unknown. However, unexpectedly the defect analysis provides also information about this value. If there exists a highly defect surface layer the analysis additionally provides its thickness. It is shown that for a correct defect analysis, defect surface layers as thin as 15Å must be taken into account.

REFERENCES
[1] E. Bøgh, Can. J. Phys. **46**, 653 (1968).
[2] J.E. Westmoreland, J.W. Mayer, F.H. Eisen and B. Welch, Radiat. Eff. **6**, 161 (1970).
[3] S.T. Picraux, J. Appl. Phys. **44**, 587 (1973).
[4] B.R. Appleton, Defects in Semiconductors, Proc. Mater. Res. Soc. Annual Meeting, Boston, 1980 (North. Holland, Amsterdam, 1981) p. 97.
[5] W.K. Chu, ibid., p. 117.
[6] S.T. Picraux, ibid., p. 135.
[7] K. Gärtner, K. Hehl and G. Schlotzhauer, Nucl. Instr. and Meth. **216**, 275 (1983) and **B4**, 55 (1984).
[8] K. Gärtner, K. Hehl and G. Schlotzhauer, Nucl. Instr. and Meth. **B4**, 63 (1984).
[9] K. Gärtner and K. Hehl, Nucl. Instr. and Meth. **B12**, 205 (1985).
[10] K. Gärtner and A. Uguzzoni, Nucl. Instr. and Meth, **B33**, 607 (1988) and **B67**, 189 (1992).
[11] J. Lindhard, K. Dan. Vid. Selsk. Mat. Fys. Medd. **34**, no. 14 (1965).
[12] E. Bonderup, H. Esbensen, J.U. Andersen and H.E. Schiøtt, Rad. Effects **12**, 261 (1972).
[13] K. Gärtner, W. Wesch and G. Götz, Nucl. Instr. and Meth. **B48**, 192 (1990).
[14] K. Gärtner, to be published.
[15] F. Schrempel, diploma thesis, Jena 1995; E. Wendler, F. Schrempel, P. Müller, K. Gärtner and W. Wesch, to be published.

Materials Science Forum Vols. 248-249 (1997) pp. 333-338
© *1997 Trans Tech Publications, Switzerland*

Heavy Ion Recoil Time of Flight Method for Materials Analysis

D. Cohen, N. Dytlewski, J. Martin and R. Siegele

Australian Nuclear Science and Technology Organisation, PMB1, Menai, NSW,2234 Australia

Keywords: Ion Beam Analysis, Materials Analysis, Time of Flight, Recoil Spectrometry, Heavy Ions

Abstract

The application of high energy, heavy ion recoil time of flight (RToF) techniques to the study of a range of materials including oxide films and catalytic converters in cars is discussed. In particular some results are presented on the modelling of RToF experiments including effects on depth profiling of roughness, straggling and multiple scattering of the incoming beam and recoiled ion.

Introduction

Materials analysis by IBA methods is a broad field and is covered by many well attended international conference series including particle induced X-ray emission (PIXE), Microprobe Applications and IBA series [1]. Ion beam interactions with atomic electrons produce characteristic X-rays while interactions with the nucleus may produce gamma rays and scattered and recoiled particles. The cross sections for these interactions vary with the atomic number, the mass and the energy of the ion as well as the atomic number and mass of target atom. Fig. 1 compares the cross sections for Rutherford backscattering (RBS) of 2 MeV helium ions (170° scattering angle), for 2.5 MeV PIXE and for recoiled spectrometry (RS) using a 65 MeV iodine beam (45° recoil angle) for a range of targets with atomic numbers spanning the periodic table. The figure demonstrates the need to tailor the ion beam to the analysis required. K-shell PIXE provides little depth information, has a larger cross section for the lighter elements below calcium but poorer sensitivity than RBS and RS for heavy elements above zirconium. Recoil spectrometry with 65 MeV iodine beams at 45° recoil angle has at least an order of magnitude larger cross section and hence better sensitivity than 2 MeV helium RBS at 170° scattering angle for most elements across the period table. For the lighter elements such as carbon and oxygen this difference reaches two orders of magnitude. RBS is most useful in elemental profiling when the element being profiled is heavier than the matrix elements themselves and decreases drastically when profiling light elements in a heavy matrix.

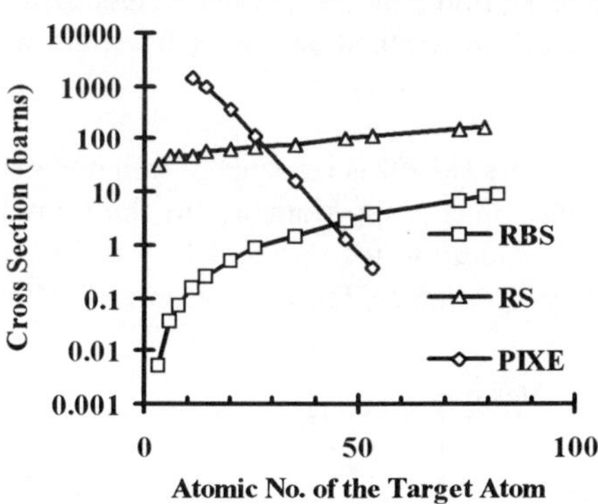

Fig. 1. Comparison of Rutherford backscattering (RBS), recoil spectrometry (RS), and K-shell PIXE.

In this paper we describe the high energy, heavy ion Recoil Time of Flight (RToF) method which takes advantage of the higher recoiled cross sections and uses fast time of flight techniques as well as a standard surface barrier detector to measure both the mass and the energy of the target ion recoiled by the incoming ion beam. This method produces two dimensional time-energy plots which are very useful in profiling and characterising surfaces, thin films and multilayers in materials analysis.

Experimental Setup

The series of experiments described here were performed on the 10 MV FN Tandem accelerator at the Australian Nuclear Science and Technology Organisation (ANSTO). Ion beams from carbon to iodine with energies in the range of 0.3 to 5 MeV/amu are typically used although the examples given here are restricted to 60 to 80 MeV iodine beams. The experimental setup has been described in detail elsewhere [2] and is based on principles described by the Group at the University of Lund, Sweden [3]. A schematic drawing of the forward recoil time of flight system is shown in Fig. 2.

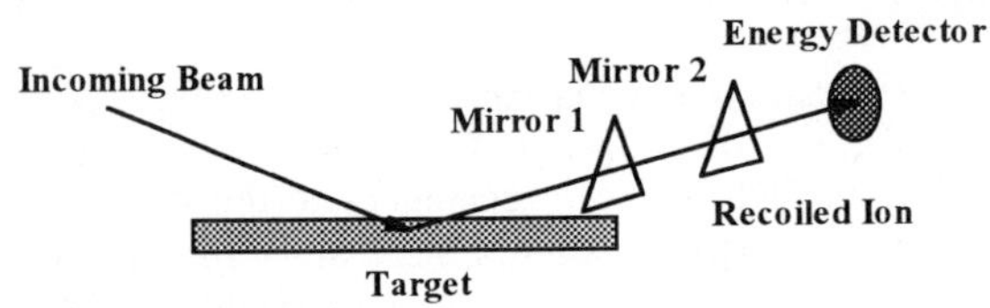

Fig. 2. Schematic drawing of the recoil time of flight target and detection system.

Heavy ions from the Tandem impinge on the target at an angle of 22.5° and recoiled particles are detected at a reaction angle of 45° to the incoming beam by two electrostatic timing mirrors (30-80 cm apart) and an ion implanted surface barrier energy detector subtending a solid angle of around 0.1 msr.

The start and stop timing mirrors are based on a design by Busch et al. [4] and use carbon foils (5-30 $\mu g/cm^2$) to generate a secondary electron timing signal which is detected by a set of 25 mm diameter microchannel plates (MCP's). The overall timing resolution of the mirror system was measured to be around 340 ps. Typical recoiled ion flight times in this experiment were tens of nanoseconds over 30 cm. Martin et al. [2] show that for timing foil thicknesses less than 30 $\mu g/cm^2$ there is little effect on measured mass and depth resolution for recoiled ions lighter than gallium obtained from incoming ion beams heavier than carbon and more energetic than 20 MeV. Fig. 3 shows the detection efficiency for the two ToF mirrors as measured for 77 MeV iodine beams on a series of thick standard targets from lithium to uranium. The efficiency for the heaviest recoiled ions was assumed to be 100%. Fig. 3. shows that the detection efficiency falls rapidly below oxygen to less than 20% for recoiled hydrogen. Standard techniques for extracting the recoiled ion mass from the timing signal were used [2,3,5] and will not be expanded on here.

The energy detection of heavy ions by surface barrier detectors (SBD) is reasonably well understood [3,6]. According to Hinrichsen et al. [6] the detector resolution is proportional to the third power of the ion energy. Fig. 4 shows the calculated SBD energy resolution for different ions and energies using Hinrichsen's tabulated values for the proportionality constants. The energy spread ΔE rises

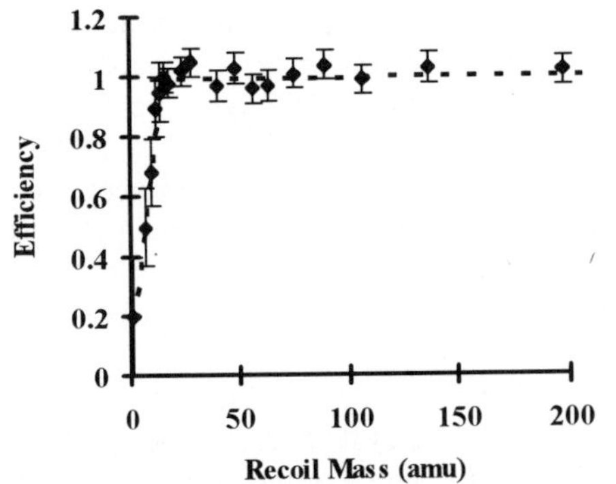

Fig. 3. The detection efficiency of the two time of flight mirrors for various recoiled ion masses.

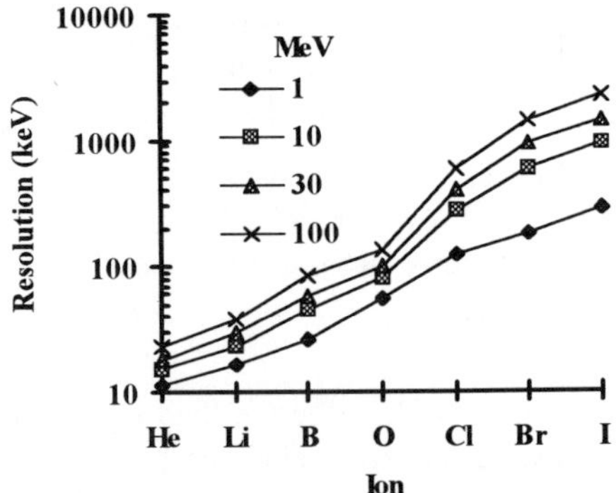

Fig. 4. SBD resolution for various ions and ion energies.

from 10 keV for 1 MeV helium to hundreds of keV for the higher energy heavy ions. However, the resolution is typically only 1-3% of the total recoiled ion energy. Work by Martin et al. [2] shows that for most practical situations the SBD resolution dominates the ultimate depth resolution obtained even when, straggling, roughness and multiple scattering in the foils and the target materials are considered. Hence a resolution of 1-3% of the recoiled ion energy is a major disadvantage of the heavy ion RToF method for depth profiling into target materials if nanometre resolution is required.

Applications of Recoil Time of Flight Analysis

The RToF analysis system on the FN Tandem accelerator has been used to characterise a wide variety of surfaces and materials. These include thin metal alloy contacts on gallium arsenide, metal ion implants into high temperature superconducting films, and indium-tin-oxide coatings on glass windows. To demonstrate the power of the techniques and its advantages over more conventional RBS, PIXE and nuclear reaction analysis method its application to characterisation of thin tantalum oxide films on tantalum and to the study of contaminants in catalytic converters in motor vehicles are discussed below.

Tantalum Oxide Films

For the lighter atoms in the sample bombarded with the high energy, heavy ion beams RToF has sufficient mass resolution to separate individual isotopes. This is clearly demonstrated in the two dimensional plot of Fig. 5, where time of flight against recoiled energy is shown for 77 MeV iodine beams recoiling tantalum and ^{16}O and ^{18}O ions from a 145 nm Ta_2O_5 film on pure tantalum substrate. The oxide film was made with 73% enriched ^{18}O. The third dimension, out of the page, contains the number of counts per channel and is reflected in the intensity of each curve mapped onto the page. The surface of the target is represented by the right hand extremity of the curve for each element. If the curve extends all the way down to zero energy then the target is thick with respect to that element as is the case for the Ta curve or 'boomerang'. However, for thin films these 'boomerangs' are shortened and for a given element the length is proportional to the elemental film thickness. The kinematics of this reaction are such that at the reaction angle of 45° the incoming iodine beam is scattered off the heavier tantalum and enters the detection system. If all elements in the target were lighter than zirconium then this scattered iodine 'boomerang' would not be present in Fig. 5. The multiparameter data acquisition and analysis system allows us to place software windows around each of the 'boomerangs' in Fig. 5 and extract individual elemental spectra for detailed depth information.

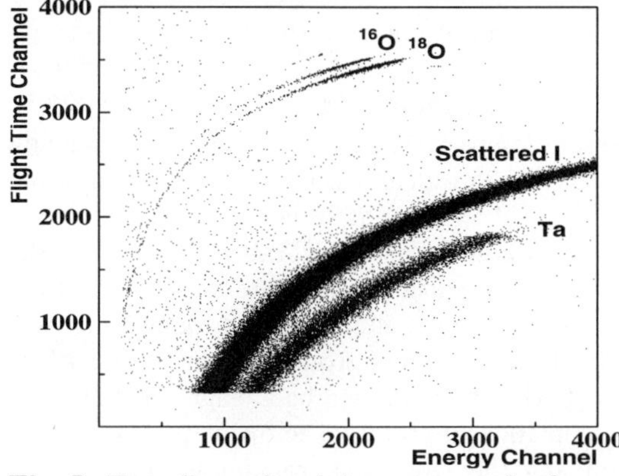

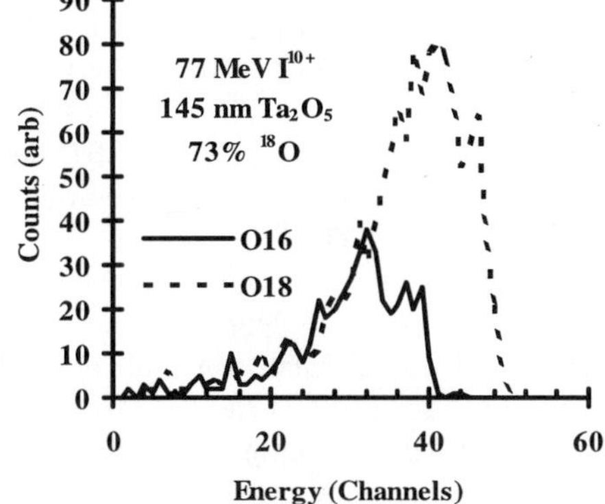

Fig. 5. Two dimensional time-energy plot for a 77 MeV I beam on an ^{18}O enriched Ta_2O_5 film.

Fig. 6. ^{16}O and ^{18}O recoil energy spectra extracted from Fig. 5.

Fig. 6 shows the ^{16}O and ^{18}O spectra extracted from Fig. 5. The front edges of the two spectra are well resolved reflecting the mass difference between the two isotopes. The total area under each curve relates to the isotopic enrichment. Note that one energy channel of Fig. 6 corresponds to 13 energy channels in Fig. 5. The full width half maximum of each spectra are similar and are a measure of the oxide thickness. The extraction of experimental data for individual elemental allows comparisons with full theoretical simulations of the recoil time of flight process.

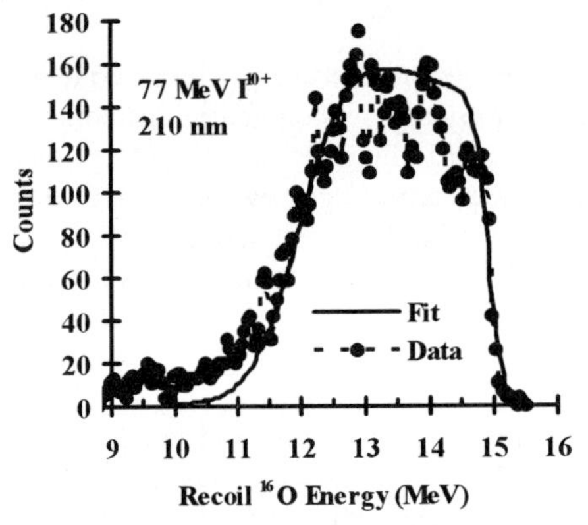

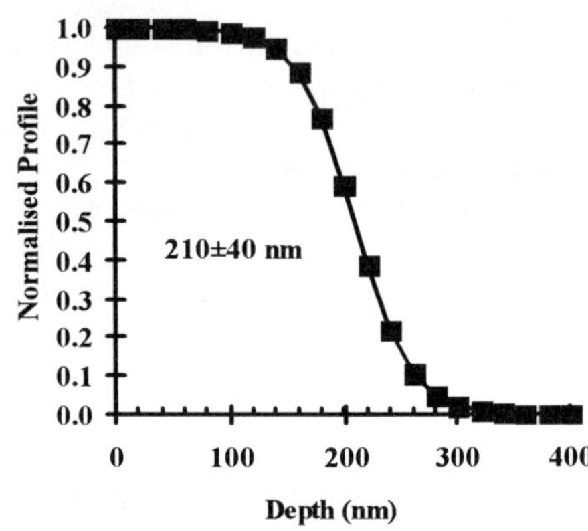

Fig. 7. A theoretical fit to the experimental data for a 210 nm Ta$_2$O$_5$ film bombarded with 77 MeV I^{10+} beams.

Fig. 8. The theoretical ^{16}O depth profile derived by simulating the experimental data of Fig. 7.

A simulation code which accounts for the full kinematics of the reaction, assumes Rutherford recoil cross sections, accounts for non-Bohr straggling [7,8] in the target and the timing foils and includes roughness and multiple scattering effects [9] in the target and in the timing foils has been written. The code calculates the elemental depth profile that simulates the experimental data taking into accounts all of the above mentioned effects in the energy detection and the timing detection. The results of such a simulation for a 210 nm Ta$_2$O$_5$ film bombarded with a 77 MeV iodine beam of 10 nA are shown in Fig. 7. An error function centred on 210 nm with a 10 to 90% rise at 250 nm and 170 nm respectively was used to represent the ^{16}O depth profile (see Fig. 8). Ellipsometric measurements gave the tantalum oxide thickness as (209±35) nm. The simulation calculates the effective energy and timing resolution at the surface (known as the front edge) and at depth in the sample (known as the back edge). In Table 1 we show typical contributions to the total energy spread of recoiled oxygen ions for the data of Fig. 7 from a depth of 210 nm the Ta$_2$O$_5$ film of density 8.2 g/cm^3. A 5 mm diameter beam of 77 MeV iodine ions was incident on the target at 22.5° and the reaction angle was 45°. The SBD detector resolution was 250 keV, the geometric straggling was 65 keV and the time of flight resolution was 340 ps. A 25 µg/cm^2 carbon foil was used in each of the ToF mirrors. The roughness on the target and the foils was set at 20 nm, this was equivalent to a 10% thickness variation in the target and the carbon foils. The Table shows that the contributions from straggling, roughness and multiple scattering in the foils are small. The total energy spread was calculated by adding all the components in quadrature and was found to be 410 keV at the surface and 671 keV at 210 nm giving a depth resolution of around 28 nm at the surface and 50 nm at 210 nm depth. This 60% difference between the front and back edge resolutions is clearly seen in the slope of the front and back edge simulation of Fig. 7. The major contributors to the total energy spread are the detector resolution and the straggling, roughness and multiple scattering of the incoming iodine beam in the target. Contributions from straggling and roughness in the timing foils were not significant.

The simulation shown in Fig. 7 has no floating parameters. The detector solid angle and beam current values used were experimentally measured. The only parameters varied to produce this simulation were the oxygen profile and the surface roughness of the target and the timing foils.

	Straggling (keV)	Roughness (keV)	Multiple Scattering (keV)
Front edge			
Foils (recoiled O)	37	60	<1
Target (incoming I)	0	287	0
Target (recoiled O)	0	136	0
Back edge			
Foils (recoiled O)	37	60	<1
Target (incoming I)	330	287	319
Target (recoiled O)	201	136	165

Table. 1. Typical contributions to the total energy spread of recoiled oxygen ions from a depth of 210 nm in Ta_2O_5 film of density 8.2 g/cm^3.

The simulation overestimates the average height of the experimental data by about 20%. This was probably due to a systematic overestimate of the target current or the detector solid angle. The oxide thickness was well reproduced by the simulation. The simulation underestimated the experimental data in the lower recoil energy portions of the backedge. This is probably due to the extra energy loss of the incoming ion or the recoiled atom by double scattering in the target. Incoming ions scattered just below and parallel to the surface before recoiling target atoms form the surface or recoiled target atoms travelling parallel to the surface and then scattering out of the target into the detector will produce an excess of recoiled atoms at the low energies shown below the backedge of the simulation shown in Fig. 7. These double scattering events are not accounted for in this model.

Catalytic Converters

Most automotive catalytic undergo some loss of activity as a result of thermal damage and/or poisoning by contaminants in the exhaust stream. The deposition of contaminants like P, Ca, and Zn from engine oils and Pb and Si from fuels occur within a porous alumina washcoat, usually 50 μm to 100 μm thick which is impregnated with active platinum group metals, rare earth and transition metal oxides. This washcoat is bonded to an underlying magnesium-aluminium silicate or cordierite substrate. These converters reduce carbon monoxide, oxides of nitrogen and hydrocarbon emissions and to be efficient need to operate at temperatures between 250°C and 450°C.

Ion beam methods have been used to study the deposition of these contaminants along the length of the converters and cross sectionally through them as a function of their known efficiency. The PIXE technique has been used to study the washcoat composition and the concentrations of associated contaminants. However, the range of 2.5 MeV protons, used in the PIXE analysis, was sufficient to pass through the variable contaminant layer and into the washcoat. This made for tedious iterative analyses for each sample analysed.

For high energy, heavy ion RToF analysis the ion range is considerably reduced and the energy can be selected so as not to penetrate the contaminant film, producing little or no signal from the underlying washcoat. This is well demonstrated by the time-energy plots of Figs. 9 and 10 for a clean and a used contaminated converter respectively. The spectrum from the clean converter shows 'boomerangs' for elements from carbon to cerium. Including a boomerang for the scattered iodine beam. Whereas the spectrum for the used contaminated converter does not contain curves for the Ba/La/Ce elements present in the underlying washcoat, but does show extra boomerangs for the contaminants P, K, Ca, Ni, Cu, Zn, and Pb. Also the background from the iodine beam is reduced because the kinematics only allows for scattered iodine from the trace quantities of Pb present whereas in the clean sample iodine scatters off all the major elemental washcoat components with masses heavier than zirconium.

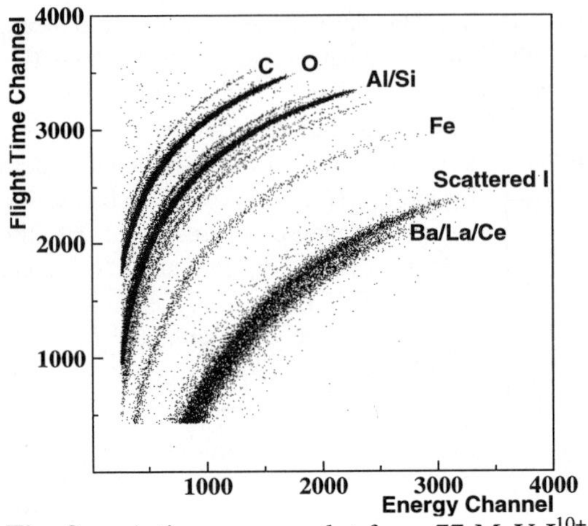

Fig. 9.   A time-energy plot for a 77 MeV I^{10+} beam on a clean converter.

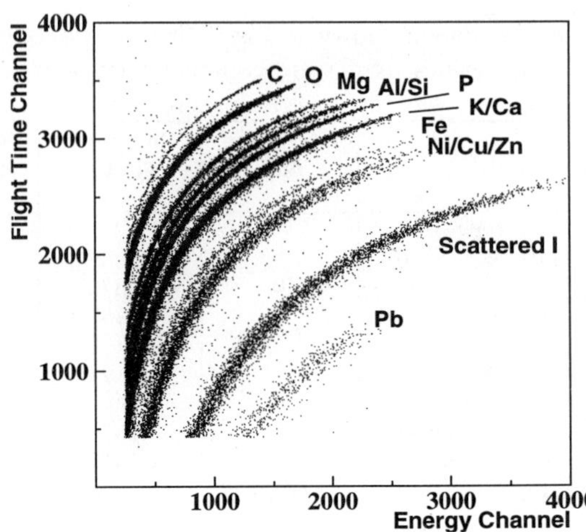

Fig. 10. As for Fig. 9 for a used converter.

Conclusion

Several aspects of the high energy heavy ion recoil time of flight techniques have been discussed in relation to materials analysis.   The technique has distinct advantages over conventional RBS methods for the analysis of light elements in a heavier matrix.   Depth profiles for elements from carbon to uranium as well as for some light element isotopes can be readily obtained by this method. High energy, heavy ion beams are required for the best mass resolution and reliable depth profiles over hundreds of nanometres are achievable with depth resolution of the order of tens of nanometres.   For many practical purposes the depth resolution is limited mainly by the energy measurement with a surface barrier detector and its response to energetic heavy ions.

Acknowledgements

The authors wish to acknowledge the collaborative assistance of Dr. Harry Whitlow and coworkers from the University of Lund, Sweden and Dr. Peter Johnston and coworkers from the RMIT, Melbourne throughout many phases of this work.   Also work is reported on here that was performed in collaboration with Professor Noel Cant and colleagues from the University of Macquarie.

References

[1]. Recent Proceedings of the: Int. PIXE Conf., Nucl.. Instr., and Methods, B109(1996)1-718, Int. Microprobe Conf., Nucl. Instr., and Methods B104(1995)1-662, Int. IBA Conf., Nucl. Instr., and Methods B85(1994)1-958.

[2]. J.W. Martin, D.D. Cohen, N. Dytlewski, D.B. Garton, H.J. Whitlow and G.J. Russell, Nucl. Instr., and Methods, B94(1994)277-290.

[3]. H.J. Whitlow, B. Jakobsson and L. Westerberg, Nucl. Instr., and Methods, A310(1991)636-648.

[4]. F. Busch, W. Pfeffer, B. Kohlmeyer and D. Schull, Nucl. Instr., and Methods, 171(1980)71.

[5]. R. Siegele, H.K. Haugen, J.A. Davies, J.S. Forster and H.R. Andrews, J. Appl. Phys. 76(1994)4524-4532.

[6]. P.F. Hinrichsen, D.W. Hetherington and S.C. Gujrathi, Nucl. Instr., and Methods, B45(1990)275-280.

[7]. D.D. Cohen and E.K. Rose, Nucl. Instr., and Methods, B64(1992)672-677.

[8]. C. Tschalar, Nucl. Instr., and Methods, 61(1968)141-156.

[9]. P. Sigmund and K.B. Winterbon, Nucl. Instr., and Methods, 119(1974)541.

Materials Science Forum Vols. 248-249 (1997) pp. 339-344
© *1997 Trans Tech Publications, Switzerland*

Transition-Metal Silicide Layer Formation, the Phases of Mixed Silicides

I. Dézsi[1,2], Cs. Fetzer[1], I.S. Szűcs[1], J. Dekoster[3] and G. Langouche[3]

[1] KFKI Research Institute for Particle and Nuclear Physics, H-1525 Budapest 114 Pf. 49, Hungary

[2] KFKI Research Institute for Materials Science, H-1525 Budapest 114 Pf. 49, Hungary

[3] IKS, Katholieke Universiteit Leuven, B-3001 Leuven, Belgium

Keywords: Transition-Metal Silicides, Thin Layers, Mössbauer Spectroscopy

Abstract The transition-metal silicides were investigated in order to further knowledge on the formation of epi- and mesotaxial (buried) layers on/in Si and on Si_xGe_{1-x} layers. The work was performed on synthesized $Co_xFe_{1-x}Si_2$ with the extremes of pure $CoSi_2$ and $FeSi_2$. The results showed that the layers of silicides can be formed by ion implantation in stoichiometric composition. No epitaxial $FeSi_2$ layer formed after implantation and annealing. The solubility of Fe in $CoSi_2$ is very low and no single phase is formed in the broad concentration range of Co and Fe. Co dissolves in β-$FeSi_2$, populating two lattice sites.

1. Introduction

Metal silicides have become important for fabricating interconnections in electronic devices. The demands for semiconducting silicides as sensors, detectors and for other electronic appliances are growing at a fast pace. Binary phased silicides, especially refractory and transition metal silicides on Si have proved their reasonably good compatibility for stable, reliable, fast interconnections in circuit fabrication. Ternary phases are far less studied than binary ones even though they are of greater technological significance. The practical motivation is an important factor in the studies but there is a great deal of pure scientific interest for a better understanding of the formation conditions of the heteroepitaxy on various surfaces and the mixed metal silicide phases.

The early Mössbauer studies of the Co implanted in silicon revealed the formation of the buried $CoSi_2$ precipitates in the host crystal after thermal annealing [1]. The condition of epitaxy of the transition metal silicide layer on Si was explained by the lattice symmetry and the close lattice constant between Si and the metal silicides [2]. Mössbauer spectroscopy proved higly suitable for studying the Co- and Fe-silicide structures because it could give direct information on the phases formed on and the structure around the metal atoms in the lattice. The site location of Co diffused in Si was already studied soon after the discovery of the effect [3], but the spectra observed could be explained much later by the metal silicide formation in Si [4].

The epitaxial Co-silicides were studied on samples of Co deposited on Si, and the formations of the different phases depending on the temperature were followed by Mössbauer effect [5]. The phases were identified by comparing the measured hyperfine interaction parameters with the parameters of the various stoichiometric Co-silicides. In many Mössbauer experiments on $CoSi_2$, in addition to a single resonance line, another line appeared whose identification motivated later studies.

Recently, numerous experiments were done in order to form epitaxial $FeSi_2$ layers on Si. The crystal structure of the semiconducting β-phase is orthorhombic and it proved difficult to orient its layers properly on the Si surface. Mössbauer spectroscopy, specially conversion electron Mössbauer spectroscopy (CEMS), was shown to be a useful method for studying silicide layers because of its high sensitivity (one iron monolayer is sufficient for the studies) [6-8].

Ternary silicides are now attracting more and more attention, because they may be beneficial to the epitaxy. Applications of ternary systems for growing epitaxial films are now under evaluation by device manufacturers [9]. The properly doped semiconducting metal silicides are ternary, homogeneous silicide phases.

We present results on the formation of the $FeSi_2$ on/in Si, $CoSi_2$ and on the Co-Fe ternary metal silicides. The formations of $FeSi_2$ structures on Si_xGe_{1-x} layers were also studied.

2. Experimental

The layer samples were prepared by ion implantation and by MBE deposition. ^{57}Fe was implanted into high purity Si (111) and (100) wafers and into $CoSi_2$. ^{57}Co was implanted in synthesized α-$FeSi_2$. The implantations were made at 80 keV energy in the Leuven isotope separator at room temperature. ^{57}Fe layers on Si samples were deposited by MBE. Since the measurements were made "ex situ", the surface of the deposited iron was covered with 50 Å layer of Ag deposited in the MBE chamber. The base pressure here was 3×10^{-11} Torr. The annealings of the samples were made in vacuum below 1×10^{-6} Torr. The annealing temperatures are shown in the figures. $FeSi_2$ and the ternary silicides were prepared by using the metal constituents of 99.99 purity and very high purity Si. The samples were melted in an induction oven in vacuum of 1×10^{-6} Torr and then thermally treated at temperatures necessary for the given phases. The Si_xGe_{1-x} layers were grown by CVD technique at IMEC (Leuven), and characterized by RBS and electronmicroscopy. The grown layer thickness is 300 nm with linearly graded Ge, starting at 0 at % and increasing up to 50 at % Ge. Mössbauer spectra of the layer samples were measured with CEMS, the spectra of bulk samples were measured in conventional transmission arrangement. The measurements were made at room temperature. The isomer shift values are given relative to α-Fe. Analysis of the spectra was performed by using a least squares fitting method allowing distributions in the spectral parameters.

3. Results and Discussion

The CEMS spectra of ^{57}Fe implanted in Si and deposited on Si are shown in Fig. 1/a-b. The spectrum of the as-implanted sample shows a quadrupole split doublet characteristic of the amorph-

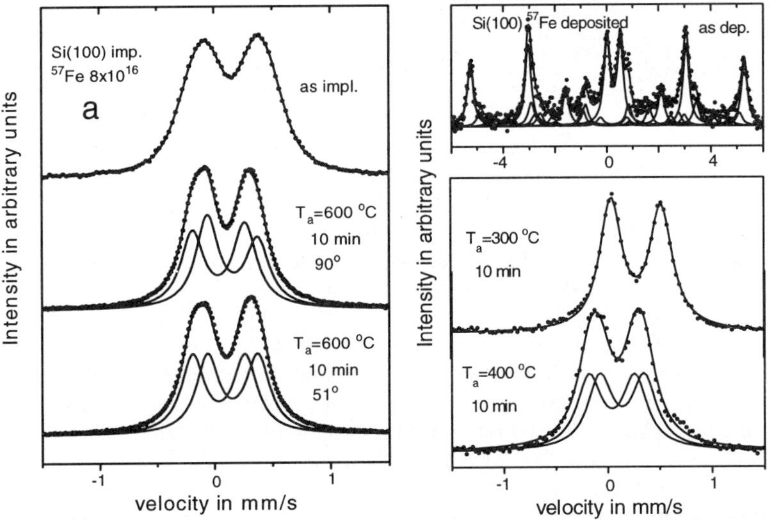

Fig. 1 The Mössbauer spectra of ^{57}Fe. a: implanted and annealed , b: MBE deposited at 200 °C and annealed on Si

ous phase formed after high dose implantation [10]. On annealing at 300 °C, the parameter changes indicate the formation of β-$FeSi_2$. $FeSi_2$ has two modifications, α and β. The α phase is tetragonal, stable between 937 and 1223 °C; the orthorhombic β phase is stable below 937 °C. Both structures can be viewed as distorted fluorite structures. However, because of the vacant iron sites in the α-phase (13-23 at %), the local symmetry around iron atoms are different. Therefore, the crystallographic equivalence is broken. This manifests itself in the complex iron Mössbauer spectrum that can be fitted with a distribution in the parameter values. The two spectral components in the Mössbauer spectrum of the β-phases correspond to the existence of the two iron atoms in crystallographic non-equivalent positions. The isomer shift (δ), quadrupole splitting (Q.S.), linewidth (Γ) and the relative intensities of the observed spectra are compiled in Table 1. The parameter values agree well with those measured earlier for stoichiometric pure silicide phases [7]. New parameters for α-FeSi$_2$ are included, because earlier spectra were only approximated by two quadrupole split spectral components.

Table 1. Mössbauer parameters of the various silicides.

Silicide	δ(mm/s)	Q.S.(mm/s)	Γ(mm/s)	Intensity(%)
α-FeSi$_2$	<0.230>	<0.53>	0.25	100
β-FeSi$_2$	0.037	0.457(5)	0.24(1)	50
	0.157	0.434(5)	0.24(0)	50
ϵ-FeSi	0.279	0.48(1)	0.27(1)	100
^{57}Fe in as imp. Si	<0.15>	<0.51>	0.25	100
(^{57}Co)CoSi$_2$ ($\ast$)	0.08(1)	-	0.50(1)	85(1)
	-0.37(1)	-	0.45(1)	15(1)
^{57}Co as impl. ($\ast$)	<-0.15>	<0.67>	0.50	100

$\ast$: measured in source experiment in SiGe. < >: average values obtained after fitting with parameter distribution

The MBE deposited sample shows a magnetic split spectrum of α-Fe with other resonance absorption lines (characteristic to Fe$_3$Si and ϵ-FeSi, for the latter in the central region of the spectrum). The appearance of these lines indicates the reaction of Fe and Si during deposition. The reaction was studied in detail in [8] and the authors concluded on efficient Fe diffusion into Si. The reaction between Fe and Si is complete at 300 °C, the parameter values indicate the formation of ϵ-FeSi. Although the structure of this monosilicide is cubic, the local symmetry around iron is lower than cubic. Therefore, the nuclear levels are split and the doublet spectrum appears. This phase gradually transforms to β-FeSi$_2$ after annealing at 600 °C. In single crystalline samples the relative intensities of the quadrupole doublet transition lines I_π and I_σ are angular dependent. In the case of axial symmetry and the isotropic Debye-Waller factor, the ratio is given as:

$$\frac{I_\pi}{I_\sigma} = \frac{3(1+\cos^2\theta)}{5-3\cos^2\theta},$$

where θ is the angle between the direction of the emission γ-ray and the main axis of the field gradient. The relative line intensities measured at the different angles of γ-ray incidence relative to the surface plain showed a small degree of orientation of the formed β-FeSi$_2$. The spectrum was fitted by considering the two existing models for the forming of the two doublets. Either the 1,3. and 2,4 or the 1,4 and 2,3 lines would form the doublets. In both cases, we obtained fits with the same quality.

The presence of Co strongly influences the formation of the iron containing phases, as can be seen in Fig. 2/a-b. In Fig. 2/a the spectra of ^{57}Fe implanted in CoSi$_2$ are shown. The as-implanted spectrum shows a quadrupole split doublet indicating the formation of a disordered structure in the implanted layer. After annealing the sample at 300 °C indicates the formation of α-FeSi$_2$, the spectrum obtained could be fitted with a distribution in the Q.S. and δ values because of the existence of the non-equivalent iron sites. The formation of the α-FeSi$_2$ is surprising since at this temperature the β phase is stable. The latter phase appears only after annealing the sample at higher temperature but the transformation is still not complete. The same behavior was observed for the MBE deposited samples, where the reaction of the iron and cobalt with Si also results in the α-phase.

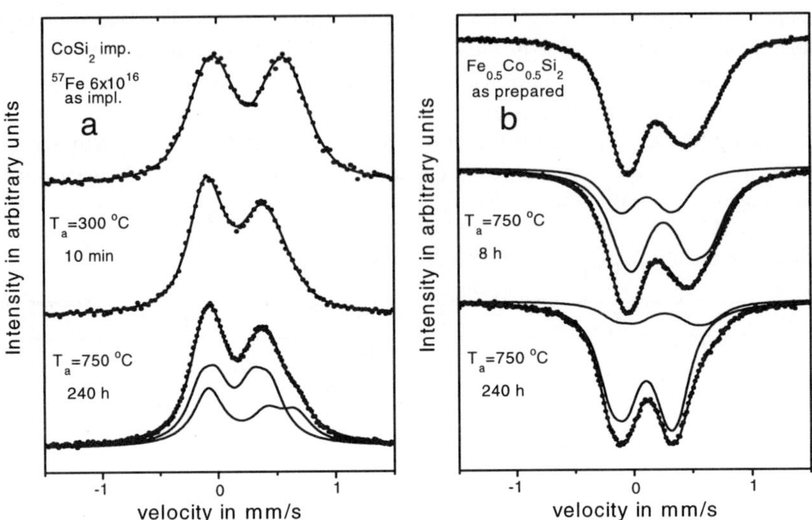

Fig. 2 The Mössbauer spectra of ^{57}Fe. a: implanted and annealed in $CoSi_2$, b: in synthesized and annealed $Co_{0.5}Fe_{0.5}Si_2$ ternary silicide phases deposited samples.

The α-phase transforms again to the β-phase at 750 °C after 240 hours annealing.
We performed a series of measurement on synthesized bulk $Fe_xCo_{1-x}Si_2$ samples in order to know which phase forms. We show here only the result obtained for the sample with x = 0.5 (Fig. 2/b). The Mössbauer parameters obtained for the as-prepared sample show the presence of the α-phase frozen in after cooling.
The β-phase appears only after long period of annealing at 740 °C. This shows that the cobalt and iron form separate phases at this concentration. This was also proved by the X-ray diffraction pattern of the sample, where the lines of both $CoSi_2$ and α-$FeSi_2$ were present after heat treatment. The phase separation could be observed down to 1.5 at % Fe. Below this value the iron dissolves in $CoSi_2$, showing two single lines indicating two iron atoms populating different lattice positions.
The phases of large x were studied by implanting and diffusing ^{57}Co in $FeSi_2$ samples. This concentration range is interesting because of the doping of β-$FeSi_2$, to make it n-type and to change its band gap energy. In order to ensure that Co finds the energetically preferred position the ^{57}Co was implanted and diffused in α-$FeSi_2$ and the annealing was made first at 1000 °C, then at 750 °C for 240 hours. The measured spectra are shown in Fig. 3/a-b. After annealing at 750 °C, the whole volume of the bulk samples was transformed into β-$FeSi_2$ This was proved by CEMS measurements.

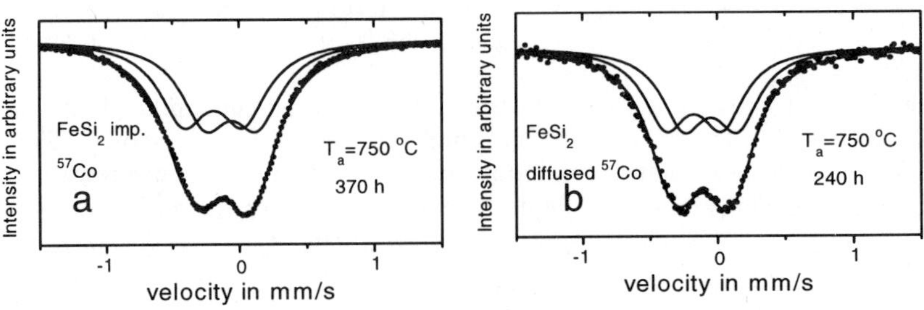

Fig. 3 The Mössbauer spectra of ^{57}Co. a: implanted and annealed, b: diffused in $FeSi_2$

Broader line widths and slightly different δ and Q.S. were found for the [57]Co spectra than for the [57]Fe in the undoped β-FeSi$_2$, probably because of the local distortions introduced by the Co atoms. The parameter values obtained where: δ_I = -0.047(4), δ_{II} = -0.17(1), Q.S.$_I$ = 0.39(1), Q.S.$_{II}$ = 0.42(2), Γ = 0.40(4). With these parameters, the intensities of the two spectral components of the iron were found to be equal. Assuming the same Debye-Waller factor for the different iron atoms, this result indicates no preferential location of iron on one crystallographic site, in contrast to that reported in ref. [11]. Present result is in agreement with that reported recently in [12], where for the preference a maximum of 55:45 value was given. The measurements made on the samples implanted with high dose values showed that the maximum equilibrium solubility of Co in the β-FeSi$_2$ lattice is close to 10 at. %.

The silicide phase formation in Si$_x$Ge$_{1-x}$ was studied by implanting [57]Co and [57]Fe in the samples. In the as-implanted samples a quadrupole split doublet appeared, indicating the disordered structure in the implanted layer (Fig. 4). The parameter values are shown in Table1.

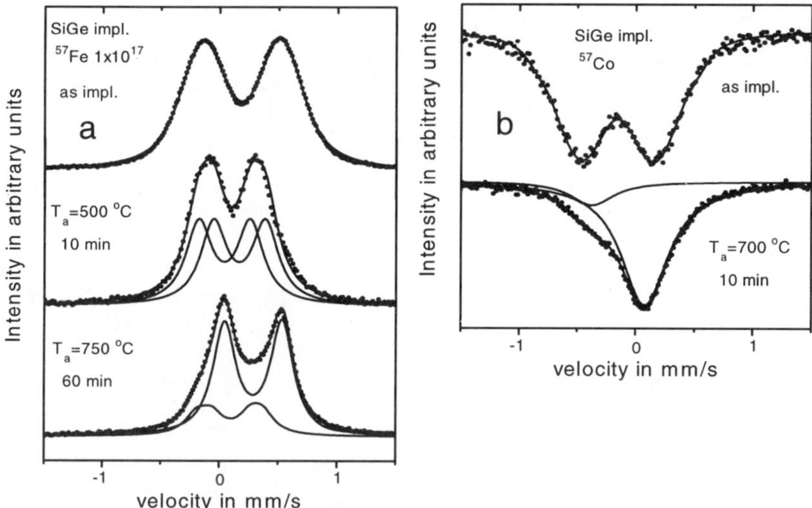

Fig. 4 The Mössbauer spectra a: [57]Fe implanted and annealed, b: [57]Co implanted and annealed in SiGe layer

In the spectrum of the annealed [57]Co implanted sample a single line and also a low intensity absorption line can be seen. The spectrum shape is characteristic to [57]Co in CoSi$_2$. The "extra" emission line at -0.4 mm/s observed for CoSi$_2$ spectra [4] is apparent here, too. Therefore, the maximum relative fraction of CoGe$_2$ that could be formed can not be given, because of a possible overlap of this line with the characteristic single line of [57]Co in CoGe$_2$ with δ = -0.243 mm/s. The spectrum of the [57]Fe implanted SiGe sample annealed at 400 °C for 10 min showed the formation of the β-FeSi$_2$ phase. It is interesting, that a spectrum with large quadrupole splitting value appeared after 750 °C annealing. This change indicates the transformation of a large fraction of the β-FeSi$_2$ phase into ε-FeSi.

Acknowledgements

The authors express their thanks to Dr. Matty Caymax (IMEC) for the CVD grown Si-Ge samples. This work was supported by the grants T 4405 (OTKA) and ERBCIPDCT940021.

References

[1] I. Dézsi, R. Coussement, G. Langouche, B. Molnár, D. L. Nagy and M. DePotter, J. Physique **41,** C1-425 (1980).
[2] H. Ishiwara, S. Saitoh and H. Hikosaka, Jap. J. Appl. Phys. **20,** 843 (1980).
[3] P. C. Norem and G. K. Wertheim, J. Phys. Chem. Solids **23,** 1111 (1962).
[4] I. Dézsi, H Engelmann, U. Gonser and G. Langouche, Hyp. Int. **33,** 161 (1987).
[5] A. Vantomme, M. F. Wu, I. Dézsi, P. Q. Zhang and G. Langouche, Hyp. Int. **57,** 2133 (1990).
[6] F. Fanciulli, C. Rosenblad, G. Weyer, A. Svane and N. E. Christensen, Phys. Rev. Lett. **75,** 1642 (1995).
[7] G. Langouche in Properties of Metal Silicides Ed. by K. Maex and M. Van Rossum, INSPEC, U. K. 1995, p. 313.
[8] M. Fanciulli, C. Rosenblad, G. Weyer, H. van Kanël, N. Onda, V. Nevolin and A. Zenkevich, Mat. Res. Soc. Symp. Proc. **402,** 319 (1996).
[9] M. Setton, ibid. [7] p. 129.
[10] P. Zhang, L. Urhahn, I. Dézsi, A. Vantomme and G. Langouche, Hyp. Int. **56,** 1667 (1990).
[11] S. Kondo, M. Hasaka, T. Morimura and Y. Miyajima, Physica B **198,** 322 (1994).
[12] K. Szymanski, L. Dobrzynski, A. Andrejczuk, R. Chrenowicz, A. Vantomme, , S. Degroote and G. Langouche, J. Phys: Condens. Matter **8,** 5317 (1996).

Materials Science Forum Vols. 248-249 (1997) pp. 345-350
© *1997 Trans Tech Publications, Switzerland*

Composition and Structure of Codeposited Layers on Plasma Facing Components in Controlled Fusion Devices: Ion Beam and Microscopy Studies

M. Rubel[1] and P. Wienhold[2]

[1] Royal Institute of Technology, Physics Department - Frescati, Association EURATOM - NFR, Frescativägen 24, S-104 05 Stockholm, Schweden

[2] Institute of Plasma Physics, Forschungszentrum Jülich, Association EURATOM - KFA, D-52425 Jülich, Germany

Keywords: Deuterium, Deposition, Plasma Facing Components, Ion Induced Detrapping

ABSTRACT

Ion beam analysis techniques (RBS, NRA) complemented by microscopy methods and surface sensitive spectroscopies were used in studies of layers deposited on carbon - based substrates during their exposure to the deuterium plasma in controlled fusion devices. The properties of the layers were determined and the ion-induced detrapping of deuterium from the deposits was measured under MeV helium irradiation. The rate and efficiency of D release was found dependent on the elemental composition of the layers.

INTRODUCTION

Controlled fusion in the deuterium - tritium mixture is extensively studied as a potential process for energy production. Two basic concepts have been developed to fuse light nuclei: an inertial confinement by using high power laser or ion beams to heat a solidified D-T pellet or, a magnetic confinement of the D-T plasma accomplished in toroidal vessels. In the latter case, plasma interacts - to some extent - with surrounding surfaces causing serious modification of in-vessel components by impact of particles escaping the plasma. The process can not be eliminated entirely, but it must be deeply studied and controlled when the construction of a thermonuclear reactor is considered [1]. As a result of particle interactions with the first wall, the plasma becomes contaminated by heavy atoms removed from the wall and the formation of co-deposited layers occurs in the so-called deposition zones of the reactor; i.e. certain areas of the divertor channel, rf antenna protection screens and limiters. Co-deposits consist of a mixture of hydrogen isotopes and plasma impurity atoms originating from the wall - mainly C, Be, B and metals - dependently on composition of the wall and its protective layers [2, 3]. The co-deposition, being a net effect of two competitive processes (erosion and deposition), leads to the accumulation of fuel atoms on plasma facing components (PFC) influencing the plasma density control and, in the case of radioactive tritium, raises the radioactivity level in the reactor [4]. Therefore, studies of co-deposits comprise both their detailed characterization and determination of conditions leading to the layer decomposition and release of H isotopes [5, 6].

Ion beam techniques, such as RBS, NRA, ERD, PIXE and other variants, due to their sensitivity and selectivity are widely applied for analyses of plasma facing components [7 - 9]. Broad information range, useful in depth profiling of species distributed in the surface and subsurface layers, makes these methods superior in studies of thick co-deposits: many microns in thickness. Accelerator based methods when combined with surface sensitive spectroscopies and microscopy techniques provide deep insight into the morphology of the layers.

This contribution is focused on properties of deuterium containing co-deposits formed on carbon-based substrates exposed to the scrape-off layer (SOL) plasma in the TEXTOR tokamak at the Research Centre - Jülich. The aim of the study was to determine: i) the structure and quantitative composition of co-deposits, and ii) the rate of deuterium release from co-deposits by MeV helium ion bombardment.

EXPERIMENTAL

The investigation was carried out with graphite and carbon fibre composites (CFC) used in fusion technology for PFC. The substrates were exposed to the plasma during ohmic or auxiliary heated pulses (by neutral beams or ion cyclotron resonance) fuelled with deuterium. The exposures were made in the presence of tokamak walls covered by carbonized, boronized or siliconized protective layers [10]in order to recognize the influence of C, B, Si on properties of co-deposits, especially on the retention and ion induced release of deuterium.

The morphology of exposed surfaces was examined with a number of methods including nuclear reaction analysis (NRA) and Rutherford backscattering spectroscopy (RBS). The deuterium content was determined with NRA using a 770 keV or 1500 keV ^{3}He$^+$ beam and detecting protons emerging from the ^{3}He(d,p)^{4}He process. The amount of D retained was also controlled during the prolonged irradiation (flux density 0.8 - 1.5 x 10 ^{13} cm^{-2} s^{-1} and the total ^{3}He$^+$ fluence up to 1 x 10^{17} cm^{-2}) accompanied by the deuterium detrapping. RBS, with a 1800 keV ^{4}He$^+$ beam, was applied to determine the quantity of heavy plasma impurity atoms in the co-deposits (Si, metals) and nuclear reaction [^{11}B(p, α)^{8}Be]for boron analysis. Atomic force (AFM), scanning electron (SEM) and optical microscopies were used to study the topography of the substrates prior and after their exposure, whereas sputter-assisted Auger electron spectroscopy was applied to determine the thickness of thin co-deposited layers.

RESULTS AND DISCUSSION

MORPHOLOGY OF CO-DEPOSITS

The co-deposition of species on the plasma facing surfaces occurs during the whole tokamak pulse, as it has been proven by collector probe measurements of ion fluxes in the scrape-off layer (SOL) [11]. An AFM image in Figure 1 shows the topography of graphite surface exposed to the SOL plasma during ohmic pulse lasting 6.9 s. Already after a short exposure the graphite surface - originally rough with sharp edges and terrases [12] - becomes tightly covered with a deposit, but the features of the underlying substrate can still be distinguished. The thickness of the deposited layer is approximately 15 nm, as determined with sputter-assisted AES and colorimetry technique [13], and the deuterium content measured with NRA amounts to 7 x 10^{16} cm^{-2}. Other major constituents are carbon and boron because the TEXTOR inner wall and limiter plates are furnished with graphite tiles which are frequently boronized [10]. The presence of other species in the co-deposits is shown in the spectra (Figure 2) recorded for two graphite substrates exposed to single plasma pulses lasting 6 s each: an ohmic pulse and another one auxiliary heated by neutral beam injection (NBI). The exposures were performed during tests of a tungsten limiter immersed in the plasma and in both spectra the W feature is pronounced [14]. The oxygen feature is related not only to the impurity deposition but also to the sorption of atmospheric oxygen by the layers subjected to air during their transportation from the fusion device to the surface analysis station [15]. Inconel components - Ni, Cr, Fe - originate from the machine liner. Fluxes of those metals are very low due to the efficient wall conditionig. The presence of Si is connected with the former siliconization of TEXTOR, whereas Mo with the former tests of a molybdenum limiter. Therefore, ion beam analyses of co-deposits sensitively reflect wall conditions in the machine and corresponding ion fluxes in the plasma edge.

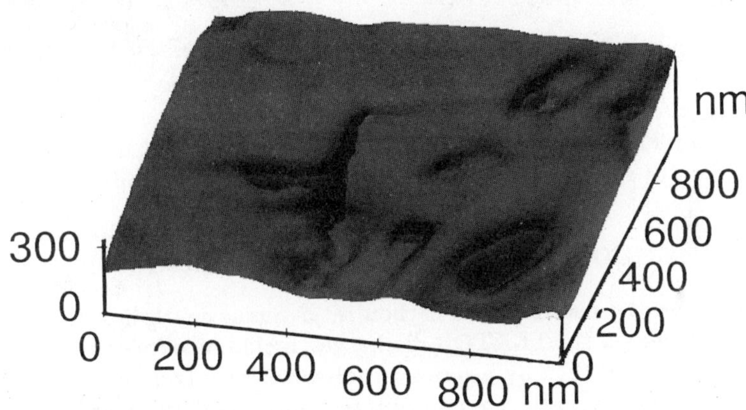

Figure 1.   AFM image of graphite surface exposed for 6.9 s to the TEXTOR plasma

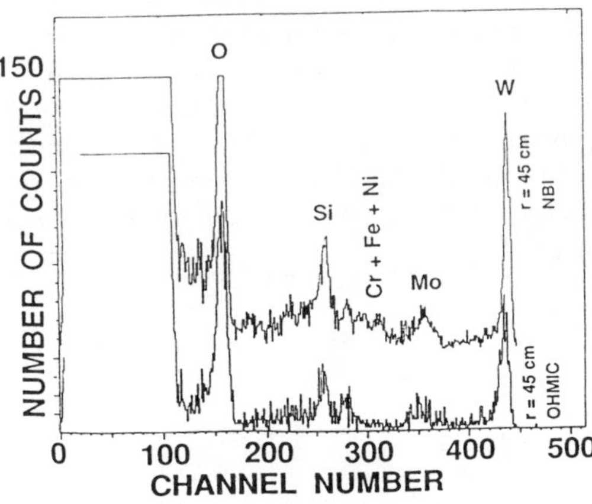

Figure 2.   Rutherford backscattering spectra of plasma impurity atoms in deposits collected on graphite during two pulses: ohmically heated (lower spectrum) and auxiliary heated with NBI (upper spectrum).

The correlation has been found between the concentration of species deposited, the layer thickness and the surface structure recorded with AFM and described in terms of fractal dimension [12, 16]. The growth rate of co-deposits is in the range from a few to approx. 100 nm s^{-1} dependently on the exposure conditions, i.e. the machine operation mode and the substrate position in the tokamak. Figures 3 and 4 exemplify the topography of CFC and graphite after prolonged exposures to the plasma. The thickness of the deposits ranges from 130 nm (in Fig. 3) to over 4 μm (in Fig. 4). In the case of thicker deposits, exceeding 8 - 10 μm, there are serious analytical and technical problems to determine simultaneously the content and depth distribution of deuterium through the layer, as well as to remove efficiently the deposited fuel [17].

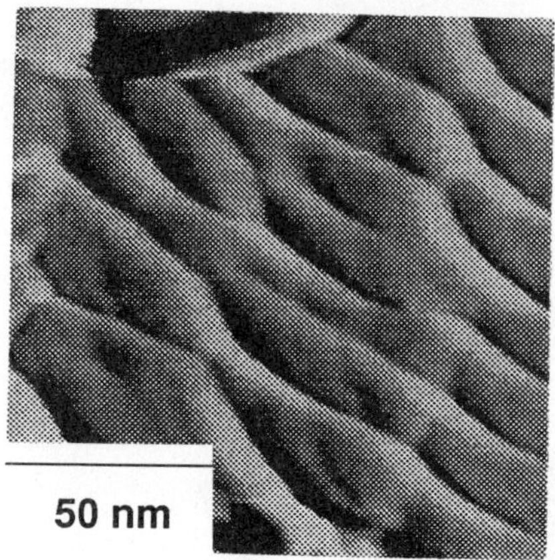

Figure 3.   AFM image of CFC exposed for 114 s to the plasma.

Figure 4.   Optical micrograph of a thick co-deposit on graphite facing hundreds of tokamak pulses.

ION - INDUCED DETRAPPING OF DEUTERIUM

Figure 5 shows the deuterium to carbon ratio *versus* depth in a thick deposit collected during a sequence of the TEXTOR discharges, including ohmically heated ones, followed by a disruption and then by a series of pulses with neutral beam injection. During the last two discharges with the co- and counter- injectors the exposed substrate was heated to the temperature exceeding 550 °C. The two plots represent the C_D/C_C ratio at the initial stage of irradiation with a 1.5 MeV ^{3}He$^+$ beam (plot a) and after the irradiation with a dose of 3.9 x 10^{16} cm^{-2} (plot b). There is observed a significant variation of the C_D/C_C ratio through the layer, reflecting the stratified structure of the deposit and - to large extent - the history of the exposure. Secondly, the measured C_D/C_C ratio is distinctly below the value 0.4 frequently reported for tokamak co-deposits. The initial D content is reduced by nearly 50 % due to the irradiation and the comparison of two plots indicates that the deuterium is more effectively detrapped from the deeeper region (2.5 - 4.0 μm) than from the near surface layer.

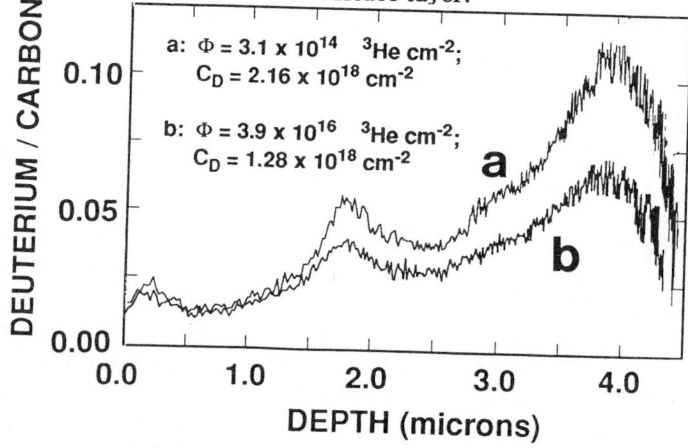

Figure 5.   The influence of 1.5 MeV ^{3}He$^+$ irradiation on the deuterium depth profiles in a deposit.

The results obtained can be interpreted in terms of factors governing the deposition and detrapping processes. Detrapping is mostly induced by electronic excitations [18]and the electronic stopping power of 1.5 MeV ^{3}He ions in carbon reaches its maximum at the depth between 3 and 4 μm. Moreover, the major components of co-deposits (C and D) do not form a homogeneous mixture but they occur in different states (traps) characterized by binding energy of up to 4.3 eV per hydrogen atom [19]and the states of low binding energy become easily decomposed [20]. The co-existance of various binding states in the deposit under investigation could be inferred from the detrapping characteristics. Upon the ^{3}He irradiation the most rapid decrease in the D content occurs within the initial stage with the fluence up to 1 x 10^{15} cm^{-2}; effective cross-section approx. 1 x 10^{-16} cm^2. At higher doses the rate decreases and the cross-sections are in the range of 0.7 - 1.0 x 10^{-17} cm^{-2}. Secondly, taking into account that the substrate was heated to above 500 °C during the exposure (two last plasma pulses), the loosely bound deuterium could be removed from the surface layer already at that time. The strong influence of preheating (300 or 550 °C) on the characteristics of the ion - induced detrapping was found in dedicated series of experiments [21]. The factors mentioned above may explain the non-uniform distribution of D in the co-deposit and the fact, that deuterium is more efficiently released from the depth of the deposit than from the surface layer containing small amout of this species.

Figure 6 summarizes the results of comparative detrapping measurements performed for thin films, 40 - 60 nm, containing 0.6 - 1.0 x 10^{17} D atoms cm^{-2} collected during ohmic pulses in the presence of carbonized, boronized and siliconized TEXTOR walls. Only very modest temperature increase (not exceeding 50 °C) of the collecting substrates was observed during the exposures.

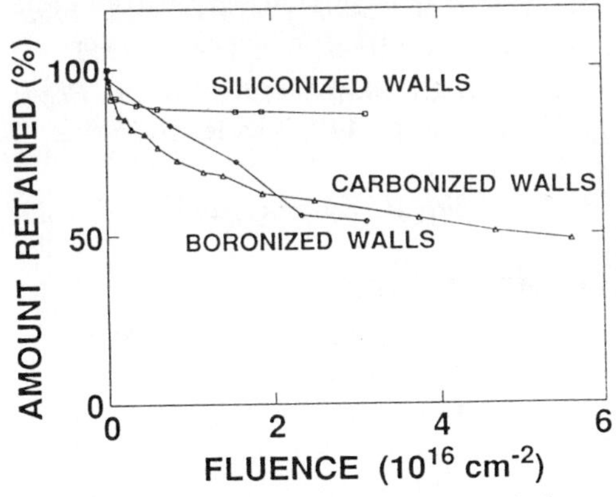

Figure 6.   Detrapping characteristics of D from co-deposits formed under different wall conditions at TEXTOR.

As one may notice, the efficiency and the rate of the deuterium release from films formed in the presence of the carbonized and boronized walls are nearly the same and up to 55 % of deuterium is detrapped with the ^{3}He$^+$ fluence of 6 x 10^{16} cm^{-2}. In contrary, only 10 - 12 % of D is released from the co-deposits formed very shortly after the siliconization and the release occurs only within the initial phase of irradiation with the dose not exceeding 5 x 10^{15} cm^{-2}. This result confirms also high stability of hydrogen in the Si-containing layers.

SUMMARY

The data gathered for tokamak co-deposits prove a great variety of their properties dependent on the operation conditions in the machine. They also indicate that the efficiency of ion-induced deuterium detrapping is strongly related by the layer composition. While only 10 % of D can be detrapped from the Si-containing films, up to 55 % is released from the C- and B-containing ones by irradiation with the helium ion fluence of 1×10^{17} cm^{-2}. One may tentatively expect that in the D-T reactor even low fluxes of unconfined α particles heating the wall (5 % or less [22]) will stimulate the release of fuel atoms and prevent their massive accumulation in plasma facing components.

ACKNOWLEDGEMENTS   The support of the studies by Grants F/AC - FF - 06571-307 and 312 from the Swedish Natural Science Research Council (NFR) is highly acknowledged. The authors are very greatful to the TEXTOR Team for kind and fruitful co-operation and to Dr. N. Almqvist for AFM images of plasma exposed materials.

REFERENCES

[1]   J.K. Ehrenberg, in *Physical Processes of the Interaction of Fusion Plasmas with Solids*, W.O. Hofer and J. Roth (Eds), Chapter 2, Academic Press, San Diego, 1996.

[2]   J.P. Coad, M. Rubel and C.H. Wu, J. Nucl. Mater., in press.

[3]   M. Rubel, P. Wienhold, N. Almqvist, B. Emmoth, H.G. Esser, L. Könen, J. Von Seggern and J. Winter, J. Nucl. Mater. **220 - 222**, 536 (1995).

[4]   C.H. Skinner *et al.*J. Nucl. Mater., in press.

[5]   R. Causey, W.R. Wampler and D. Walsh, J. Nucl. Mater. **176 & 177**, 987 (1990).

[6]   M. Rubel, H. Bergsåker and P. Wienhold, J. Nucl. Mater., in press.

[7]   J.P. Coad and B. Farmery, Vacuum **45**, 439 (1994).

[8]   A. Martinelli, A. Peacock and R. Behrisch, J. Nucl. Mater. **196 - 198**, 729 (1992).

[9]   P. Boergessen, L.G. Svedsen and J. Ehrenberg, J. Nucl. Mater. **144**, 181 (1987).

[10]   J. Winter, in *Physical Processes of the Interaction of Fusion Plasmas with Solids*, W.O. Hofer and J. Roth (Eds), Chapter 6, Academic Press, San Diego, 1996.

[11]   M. Rubel and P. Wienhold, in *Contributions to High Temperature Plasma Physics*, K.H. Spatschek and J. Uhlenbusch (Eds), Chapter 5, p. 445, Akademie Verlag, Berlin, 1994.

[12]   N. Almqvist, M. Rubel, S. Fredriksson, P. Wienhold, B. Emmoth and L. Ilyinski, J. Nucl. Mater. **220 - 222**, 917 (1995).

[13]   P. Wienhold and F. Weschenfelder, Vacuum **47**, 919 (1996).

[14]   V. Philipps *et al.*, Proc. 15th Int. Conf. on Plasma Physics and Controlled Nuclear Fusion Research, vol. **2**, 149 (1995), IAEA, Vienna.

[15]   P. Wienhold *et al.*, *J. Nucl. Mater.* **196 - 198**, 647 (1992).

[16]   N. Almqvist, M. Rubel, P. Wienhold and S. Fredriksson, Thin Solid Films **270**, 426 (1995).

[17]   Proc. 3rd Int. Workshop on Tritium Effects in Plasma Facing Components, Ispra, Italy, 1996.

[18]   H. Bergsåker, S. Nagata, M. Rubel and B. Emmoth, Proc. Symp on Amorphous Hydrogenated Carbon Films, E-MRS **17**, 433 (1987).

[19]   R. Causey, J. Nucl. Mater. **162 - 164**, 151 (1989).

[20]   J. Roth *et al.*, J. Nucl. Mater. **93/94**, 601 (1980).

[21]   M. Rubel *et al.*, Vacuum **45**, 429 (1994).

[22]   J.K. Ehrenberg, private communication.

Materials Science Forum Vols. 248-249 (1997) pp. 351-356
© *1997 Trans Tech Publications, Switzerland*

Ion-Beam Analysis of Solar-Cell Materials

H. Metzner

II. Physikalisches Institut, Universität Göttingen, Bunsenstr. 7-9, D-37073 Göttingen, Germany

Keywords: Solar-Cell Materials, Rutherford Backscattering, Nuclear Reaction Analysis, Perturbed Angular Correlations

A review of typical applications of ion-beam analysis techniques in the field of thin-film solar-cell materials is presented. Special emphasis is put on the chalcopyrite semiconductor $CuInS_2$ and its important production processes, which include: physical vapour deposition, reactive annealing of metallic precursors, and molecular beam epitaxy. Rutherford backscattering with α particles and heavy ions, hydrogen profiling with the ^{15}N method, ion channelling, and $\gamma\gamma$ perturbed angular correlations yield important information about the produced light-absorbing $CuInS_2$ layers and their precursors. Additionally, implanted nuclear probes are employed in order to characterize indium-tin-oxide films which serve as contact layers and anti-reflection coatings of solar cells.

1. Introduction

In principle, all semiconductors with energy band-gaps between 1.0 and 2.0 eV can be regarded as solar-cell materials for applications in single-junction solar cells. To date, however, only semiconductors with highly developed technologies such as single-crystalline Si, GaAs, or InP allow the production of photovoltaic devices with conversion efficiencies exceeding 20 %. Still, it is tempting to investigate alternative light-absorbing materials in view of future low-cost production processes. Recently, $CuInS_2$ (CIS) has attracted considerable interest in this field since polycrystalline CIS hetero-junction cells of only a few μm thickness with conversion efficiencies above 10 % could be produced on a laboratory scale [1]. CIS is a chalcopyrite semiconductor with a direct bandgap of 1.5 eV and can be made n- or p-type depending on slight deviations from the 1-1-2 composition of the stoichiometric compound. Fig. 1 (left) shows the cross-section of a typical CIS solar cell on a glass substrate with a Mo back-contact, the CIS layer for light absorption, a CdS buffer layer for the p-n-junction, a ZnO contact- and anti-reflection coating, and Al fingers of the front-contact grid. Instead of ZnO, also indium-tin-oxide (ITO) can be used as a transparent front-contact. Fig. 1 (right) shows the current-voltage curve of such a device with a conversion efficiency of 10.9 %. Despite the recent encouraging results, there remain many open problems in the CIS system, such as controlled production and electronic doping processes and the regulation of the electronic properties of grain boundaries and interfaces. It would be highly desirable, for example, to substitute the CdS buffer layer in CIS devices by an n-type CIS layer. The material development of CIS is intimately related to the state-of-the-art of the important production processes of this compound. In this paper, I discuss physical vapour deposition (PVD), reactive annealing of metallic precursors (RA), and molecular beam epitaxy (MBE) and put special emphasis on the film characterization by means of ion-beam methods. It should be noted, however, that the successful material development depends on a whole spectrum of experimental techniques and ion-beam methods. The full power of these techniques can only be exploited when applied in well-chosen combinations. As also in many other areas of thin-film technology, Rutherford backscattering (RBS) turns out to be extremely useful for the characterization of CIS layers. Due to the very different masses of the constituting

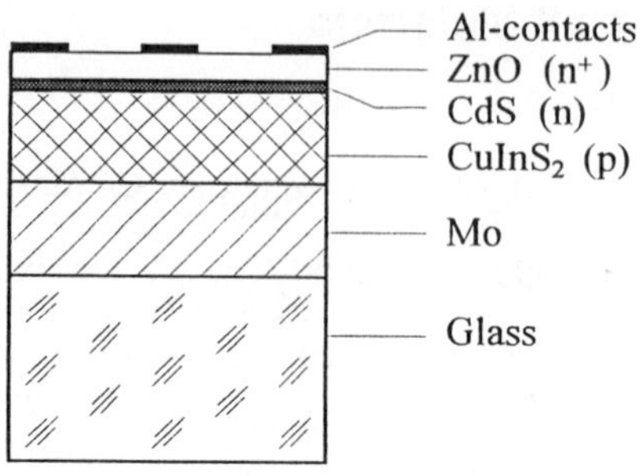

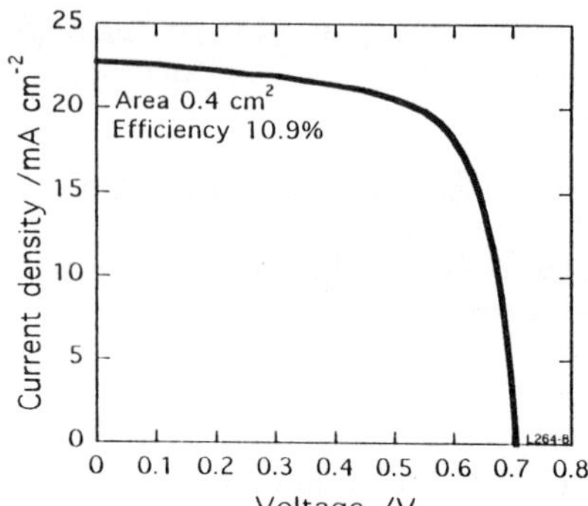

Figure 1: Cross-section of a typical CIS thin-film solar cell (left). The efficiencies of such devices meanwhile exceed 10% as demostrated by the current-voltage characteristics (right).

elements, these are well separated in an RBS spectrum of a thin CIS film even if the energy of the analyzing α beam is unusually low (here: 1 MeV). This effect is demonstrated by means of Fig. 2. RBS-spectra will be discussed in connection with all three CIS deposition techniques and so RBS is the leitmotif of the present paper. Resonant nuclear reaction analysis (RNRA) using the ^{15}N isotope, RBS in the channelling mode, and $\gamma\gamma$ perturbed angular correlations (PAC) using implanted ^{111}In nuclear tracers appear as additional ion-beam methods which are employed for the characterization of CIS and other solar-cell layers. The data presented in this paper have been obtained at the IONAS implanter in Göttingen and at the ISL facility of the Hahn-Meitner-Institut in Berlin.

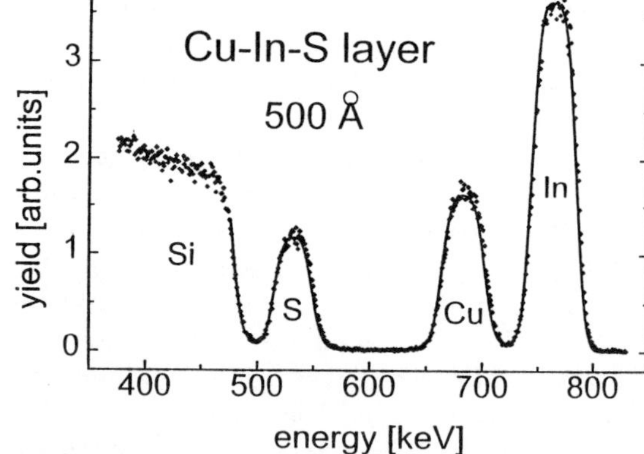

Figure 2: Cu-In-S is a quasi-ideal RBS system and the elements of a CIS thin-film on Si are nicely separated.

2. CuInS₂ Films by Physical Vapour Deposition

An optimized PVD process is employed for production of the CIS thin-film cells with the highest efficiencies to date [1]. A cross-section of the typical cell structure and representative current-voltage characteristics are given in Fig. 1. The photoactive p-type CIS layers of these high-efficiency cells are produced by means of a two-step process. In the first step, the elements Cu, In, and S are simultaneousely evaporated onto a Mo-coated glass substrate which is kept at elevated temperatures of up to 600 °C. It is essential to operate the deposition under moderate copper and extreme sulphur excess as compared to the 1-1-2 stoichiometry of CIS. This procedure does not lead to an homogeneous film but rather to the formation of a close-to-stoichiometric CIS layer on top of which the phase CuS segregates as an uncoherent layer. In the second step, the CuS secondary phase is removed by means of a chemical etching process in a KCN solution. Thus, CIS layers with sufficient p-type conductivity corresponding to hole concentrations of up to $10^{17}/cm^3$ are obtained. Here, I demonstrate that the features of this process can be studied by means of the RBS technique quite easily. It should be noted, that small amounts of the CuS phase are very hard to detect by means of X-ray diffraction (XRD) and the outlined description of the process is mainly based on photoelectron spectroscopy (X-PS) data [2]. Fig. 3 (left) shows the RBS spectrum of a thin Cu-In-S film which was deposited at Cu- and S-excess onto heated Si (200 °C) and subsequently annealed at 400 °C. Using the computer code RUMP [3], the spectrum was fitted and yielded two Cu-In-S layers of 230 Å (layer #1) and 630 Å (layer #2) thickness at the nominal compositions $Cu_{1.2}In_{0.8}S_{2.06}$ and $Cu_{0.99}In_{1.01}S_{2.13}$, respectively. The low energy tails in the spectrum are an effect of the surface roughnesses of the films which, as we show on this symposium, can be analyzed quantitatively by means of RBS [4].

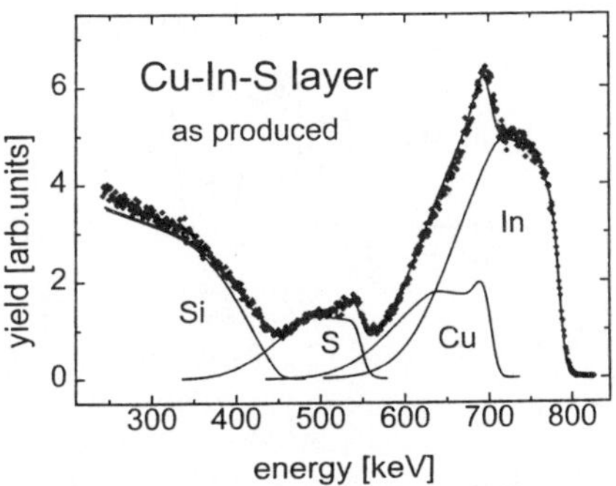

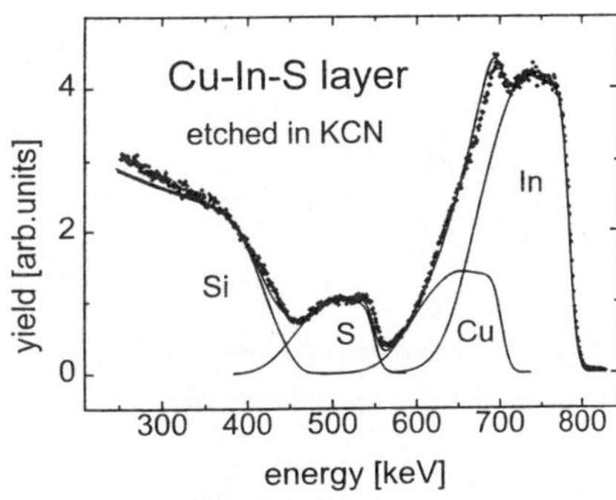

Figure 3: RBS spectra of a CIS film before (left) and after (right) a chemical treatment which removes Cu-S surface precipitates.

After an etching process in KCN, the RBS spectrum of Fig. 3 (right) is obtained. The analysis shows that layer #1 was essentially removed by the etching. The shown fit curve corresponds to layer #2 exhibiting the composition $Cu_{0.97}In_{1.03}S_{2.08}$ at the increased thickness of 690 Å. Additionally, the surface roughness has decreased. Since the applied type of chemical treatment only removes Cu-S phases, a straightforward interpretation of the RBS data is obtained. Layer #1 is laterally inhomogeneous and consists of close-to-stoichiometric CIS and a phase Cu_xS_{1-x}. The nominal composition of layer #1 suggests the following relation: $Cu_{1.2}In_{0.8}S_{2.0} \rightarrow 0.8\ CuInS_2 + 0.4\ CuS$. The selective removal of CuS leads to a nominally thicker layer #2 (pure CIS) and a decreased surface roughness.

3. $CuInS_2$ Films by Reactive Annealing

The production of CIS layers by means of RA of metallic Cu-In precursors in a H_2S atmosphere at typically 400 °C is a promising process for large-area CIS photovoltaic devices. Quite high conversion efficiencies have already been obtained for small CIS cells produced by this process [5]. Using PAC and XRD, we have investigated the phase formation in the metallic precursors [6] and the mechanisms of sulphurization in some detail [7]. Here, I shall confine myself to a brief outline of RBS and RNRA applications to this production process. As already stated, the Cu:In ratio plays a crucial role for the performance of CIS solar devices. Using conventional RBS with α particles of 1 to 3 MeV, the Cu and In-Yields are separated only for quite thin layers as demonstrated in Fig. 2. Fig. 4 (left) shows the RBS spectrum of a Cu-In precursor of 0.5 μm thickness which corresponds to a 1.3 μm thick CIS layer after sulphurization. 1 to 3 μm is the typical thickness of the CIS layer in a solar device. The Cu and In yields now have a considerable overlap and RBS is, therefore, not so well-suited for a precice precursor characterization in the technically relevant thickness range. The situation is changed completely, if high-energy heavy ions are utilized. As an example, Fig. 4 (right) shows the RBS spectrum obtained for the identical Cu-In precursor by means of ^{15}N ions of 18 MeV [8]. The Cu and In yields are clearly separated and details of the precursor morphology become visible. Consequently, heavy-ion RBS allows a very precise determination of the Cu:In ratio also for precursors of thicknesses relevant for CIS solar cells.

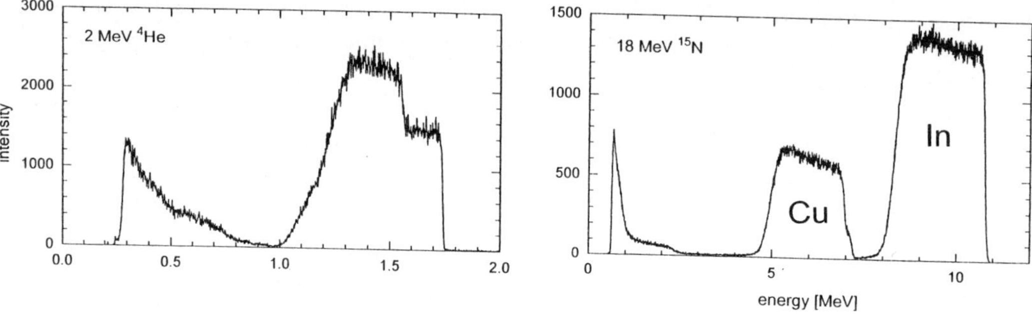

Figure 4: Conventional (left) and heavy-ion (right) RBS spectra of a Cu-In precursor for the production of CIS by means of reactive annealing in H_2S.

A characteristic feature of the RA process is the incorporation of hydrogen into the produced CIS layers. It is well known from other semiconductors that hydrogen may cause dramatic effects on the electronic properties. A powerful tool for the determination of hydrogen profiles is RNRA using ^{15}N ions. Fig. 5 shows typical hydrogen profiles of CIS films prepared by means of sulphurization at different temperatures. The decrease of the H-content with increasing tempertaure becomes apparent. We found that hydrogen leads to an increased conductivity of CIS layers [9]. This effect might help to increase conversion efficiencies of CIS solar devices and has tempted further research into this direction.

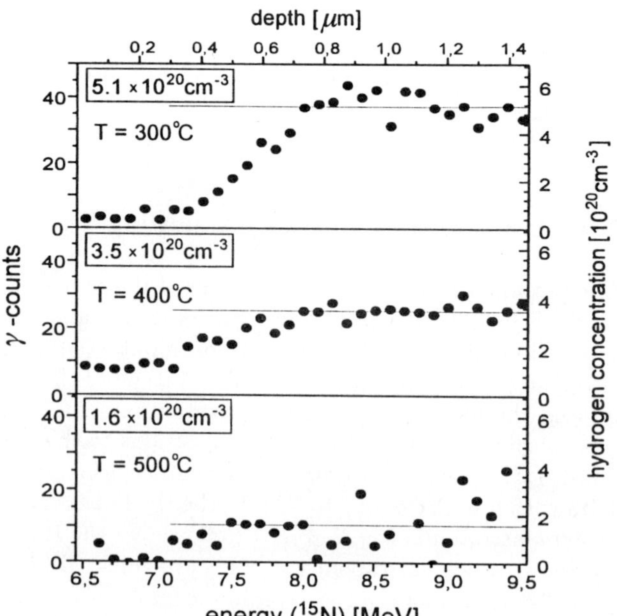

Figure 5: ^{15}N-RNRA hydrogen profiles of CIS films which were produced by means of reactive annealing at different temperatures.

4. CuInS$_2$ Films by Molecular Beam Epitaxy

The heteroepitaxial growth of CIS layers appears to be the only practicable way to obtain single crystalline CIS at present, since all attempts to prepare large CIS single crystals by means of conventional growth techniques essentially failed. Only recently, we were able to prepare CIS epilayers on sulphur terminated Si (111) surfaces by means of MBE [10]. We characterized these layers by various methods including XRD and RBS in the channelling mode. Fig. 6 (left) shows the RBS random and aligned spectra of an epitaxial CIS layer indicating a pronounced channelling effect. However, the minimum yield of 40 % appears to be somewhat high regarding the quality of the layer as it was indicated by XRD. XRD gave no signs of minor phases or polycrystalline fractions and revealed narrow rocking curves. A hint for the solution of this discrepancy is obtained by means of a closer inspection of the yield ratio of aligned and random spectrum which is plotted in Fig. 6 (right). These data indicate a significant dechannelling effect which is exclusively caused by Cu and S atoms. Hence, we have to postulate precipitates of a Cu-S phase which is practically invisible to XRD but which causes considerable dechannelling effects. We assume, that this minor phase has a major influence on the performance of our CIS/Si heterojunction devices.

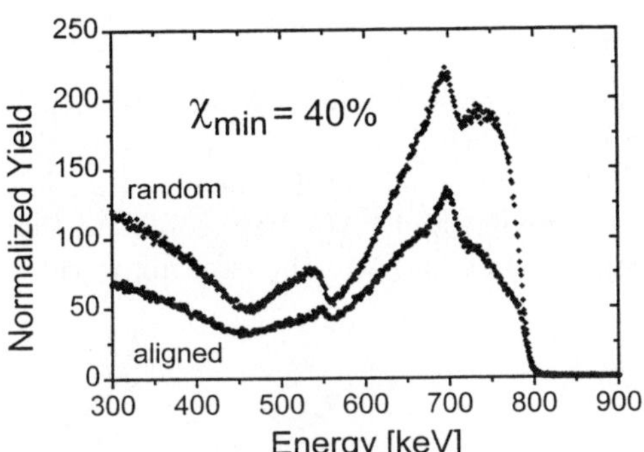

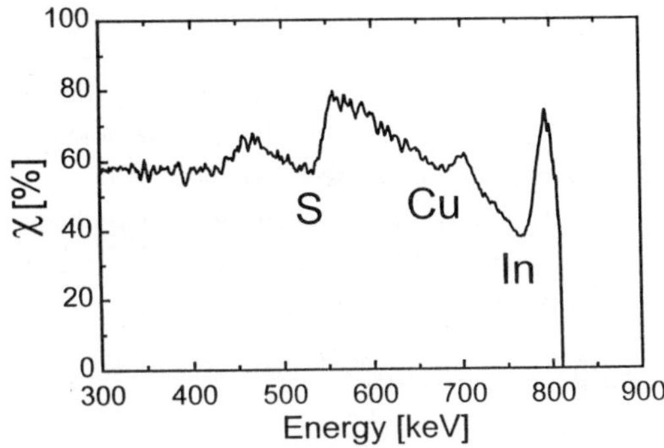

Figure 6: RBS-channelling effect of a CIS film which was epitaxially grown on a sulphur-terminated Si-(111) surface (left). The Yield ratio of aligned and random spectrum (right) indicates a dechannelling effect which is caused by copper and sulphur atoms exclusively.

5. Indium-Tin-Oxide Layers

ITO is a transparent conducting oxide of special interest for solar devices and we use it as a contact material for CIS/Si heterojunction devices. Fig. 7 (left) shows a typical RBS spectrum of such a device before the deposition of the metallic front-contact grid. It turns out that the thickness of the ITO layer can easily be determined by means of the In(Sn) yield in RBS. We also characterize ITO layers by PAC employing implanted nuclear ^{111}In probes [11]. Thus, we are able to study ITO layers from the local viewpoint of probes on lattice sites of the cation sublattice. This technique has the advantage that ITO layers can still be characterized even if they are already part of a device. Typical PAC spectra are shown in Fig. 7 (right). In the Fourier transform, the contributions of the two cation lattice sites #1 and #2 are indicated. Technical ITO layers have tin contents of typically 10 met. at.% and the tin atoms cause a considerable damping of the PAC spectrum of such films. It was recently shown that PAC is also directly sensitive to the concentration of conduction electrons in ITO samples [12].

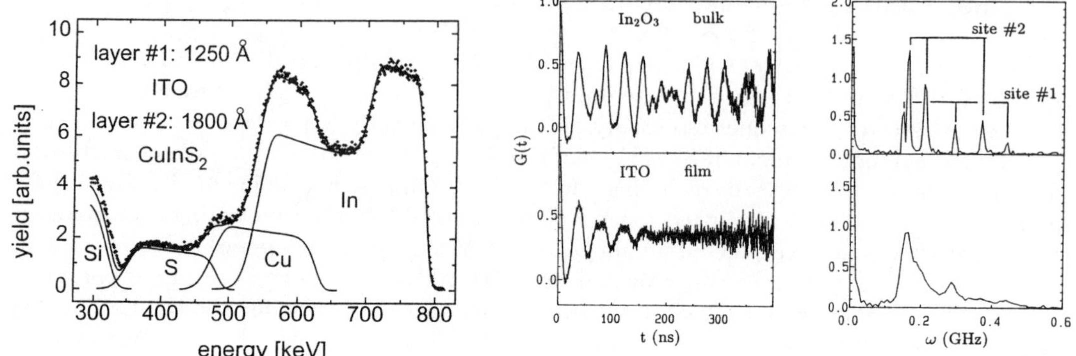

Figure 7: For the production of Si/CIS heteroepitaxial devices, an ITO layer is deposited on top (left). PAC allows us to investigate ITO from the viewpoint of nuclear tracers in bulk as well as in film material (right).

6. Conclusions

In this paper, I presented several examples of ion-beam applications to solar-cell materials. It turned out that especially RBS is a very usefull experimental technique for the characterization of layers which are relevant for thin-film solar cells, but the material development in this field also benefits from more sophisticated techniques such as RNRA and PAC. I am confident that this paper will tempt one or other of the ion-beam specialists to turn to the exciting world of solar materials.

Acknowledgements

Thanks are due to my colleagues, co-workers, and hosts in Berlin (Hahn-Meitner-Institut), Hagen (FernUniversität) and Göttingen (Universität) for fruitful cooperation, excellent collaboration, continuous support, and encouragement. The financial support of the Deutsche Forschungsgemeinschaft is gratefully acknowledged.

[1] R. Scheer et al., Appl. Phys. Lett. **63**, 3294 (1993).
[2] R. Scheer and H. J. Lewerenz, J. Vac. Sci. Technol. **A12**, 51 (1994).
[3] L. R. Doolittle, Nucl. Instr. Meth. **B9**, 344 (1985) and **B15**, 227 (1986).
[4] Th. Hahn et al., contribution to this symposium.
[5] Y. Ogawa et al., Jpn. J. Appl. Phys. **33**, L1775 (1994).
[6] C. Dzionk et al., J. Appl. Phys. **78**, 2392 (1995).
[7] C. Dzionk et al., Thin Solid Films, submitted.
[8] W. Bohne, S. Hessler et al., unpublished results.
[9] C. Dzionk et al., J. Vac. Sci. Technol. **A13**, 2275 (1995).
[10] H. Metzner et al., Appl. Phys. Lett., in press.
[11] S. Habenicht et al., J. Cryst. Res. Technol. **31**, 821 (1996).
[12] S. Habenicht et al., Z. Phys. B, in press.

Materials Science Forum Vols. 248-249 (1997) pp. 357-360
© *1997 Trans Tech Publications, Switzerland*

Evaluation of Coatings Produced by Low-Energy Ion Assisted Deposition of Co on Silicon

I.S. Tashlykov[1], G. Carter[2] and J.S. Colligon[2]

[1] Dept. of Physics, Belorussion State Technological University,
13-a Sverdlova Str., ,220630 Minsk, Belarus

[2] Dept. of Electronic & Electrical Engineering, University of Salford,
Salford M5 4WT, UK

Keywords: Cobalt, Ion-Assisted Deposition, Sputtering, Atom Mixing

Abstract

Ion assisted deposition (IAD) is a commonly used technique in modern surface modification treatments. Ever since the introduction of IAD (IBAD) methods it has been debated whether the low-energy ion beam, intended for the formation of dense coatings on surface, can also cause ion mixing, how it can influence incorporation of admixtures into thin films and on the growth of films under variable ion-to-atom ratios and deposition rates.
The purpose of this study is to determine the influence of 200 and 500 eV Ar^+ ion beams on the evolution of Co coatings formed on Si by means of IAD, when the ion-to-atom ratio and deposition rate were varied in the intervals of 0.1 - 0.9 and 3.4 - 10.2 x 10^{14} at/(cm^{2}s) respectively. The composition of the thin films produced by the IAD technique has been investigated using Rutherford back scattering of 2.0 MeV He^+ ions. The thickness of the Co layers deposited on Si exhibited major dependencies on ion-to-atom ratio. This decreases with increase of this ratio. The relative amount of argon captured into coatings was two to three times more when the energy of assisting ions increases from 200 to 500 eV. A minor mixing effect is observed in an interface region at both energies applied for IAD. A quantitative evaluation of ion irradiation effects during low-energy IAD of Cobalt on Silicon is given.

Introduction

There is much interest in the deposition of thin films using ion assistance techniques including directed ion beams [1] or plasma sources [2]. Generally such techniques involve simultaneous bombardment of a thin film during deposition with inert gas ions or chemically active ions to achieve a desired composition. Only in the case of self-ion assisted deposition [3] can the presence of a species, desired or not, different to the deposited species be avoided. Although inert gas occlusion in deposited films is well known [4] little quantitative information on its concentration or the effects on film properties as a function of the processing parameters exists [4-6]. The present study represents a step in this direction in which, in a directed ion beam system, the ratio of ion: atom arrival rate and the ion energy were varied when cobalt was deposited in the presence of Ar^+ ion irradiation and the rates of Co deposition and the Co:Ar ratio were measured as functions of these parameters.

Experimental Methods and Results

The system employed for the current studies has been described in detail elsewhere [7] and consists, essentially, of a high vacuum chamber pumped by diffusion pumps to which is attached two ion sources. Both of these sources are of the broad-beam Kaufman type and one delivers Ar^+ ions at 1 keV to sputter a plane Co target which provides the Co atom deposition flux. The second source provides an Ar^+ ion flux at energies from 200 eV to 500 eV to directly bombard the substrate (Si) upon which the Co is deposited.

The target and substrate are so positioned that the Co arrival flux varies with position across the Si substrate but the assisting Ar^+ ion flux varies only slightly over this area. The Ar^+ ion flux variation, as a function of source parameters (energy and total beam current) was measured carefully with a many position Faraday cup arrangement as was the Ar^+ ion current to the sputtered Co target. System pressures were measured with ion gauges and quadrupole mass spectrometers, target and substrate temperatures with embedded thermocouples and the Co deposition rate could be measured with a moveable quartz crystal oscillator.

The main method of determining both the quantity of Co deposited, and hence deposition rate from knowledge of deposition time, and Ar incorporated in films, however, was by Rutherford Back Scattering Spectroscopy [8] using 2 MeV He^+ ions as the probe, after a deposition cycle had been completed. This technique gives accurate and quantitative information on the depth profiles of concentrations of species throughout a film and, by integration, the average concentrations of all species. Such averaged concentrations are presented in the current work. Since the atomic masses of the substrate (Si), Ar and Co are significantly different there is relatively little overlap in the as-received RBS spectra of these components in the Si-film structure and so the atomic concentrations of the species could be determined rather accurately. Although not the main emphasis of the present work examination of the forms of the RBS spectra obtained from Co films deposited without or with congruent Ar^+ ion bombardment assistance revealed significant evidence of atomic mixing at the Si-film interface in the ion assisted case. This well-known, collisionally driven, process might be expected to improve, for example, film-substrate adhesion.

Using the above methods the rate of Co deposition, R_d, was first measured without simultaneous ion assistance at different positions on the substrate for defined target sputtering conditions. The rate of Co deposition, R_{Co}, was then measured for identical target sputtering conditions but with simultaneous application of Ar^+ ion assistance with energy $E_o = 200$ eV or 500 eV and with a range of ion assistance flux densities, J_i, achieved by varying both the total ion flux from the source and/or assessing the composition of the film at different regions on the substrate. The rate of incorporation of Ar, R_{ar}, was determined for each case studied.

The main results of these investigations are presented in Figs 1 and 2. Fig. 1 shows the ratio of

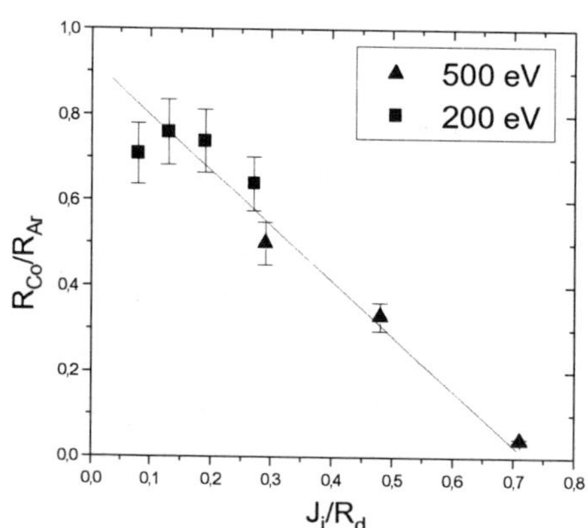

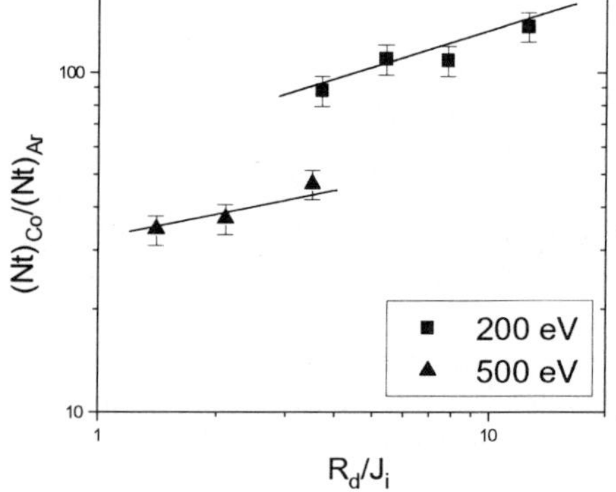

Fig. 1 The ratio of the Cobalt deposition rate with of ion bombardment to that without ion bombardment as a R_{Co}/R_d as a function of the Ar^+ ion flux: non ion Co assisted Co deposition rate ratio J_i/R_d

Fig. 2 The depth integrated ratio of the quantity Cobalt deposited to Ar trapped Nt(Co)/Nt(Ar) function of the ratio of the non ion assisted deposition rate to the Ar^+ ion flux R_d/J_i

the Co:Ar deposition rate, R_{Co}/R_{Ar}, as a function of the ion:atom arrival rate ratio J_i/R_d while Fig. 2 displays the ratio of the depth integrated concentrations of Co:Ar as a function of the atom: ion arrival rate ratio R_d/J_i. In Fig. 1 the data for 200 eV and 500 eV Ar^+ ions fall upon the same line but, in Fig. 2, the data for 200 eV and 500 eV Ar^+ ions lie upon different curves. It may be noted that, in both cases, the ion:atom arrival rates were varied by about one order of magnitude.

Theory and Discussion

The effects of ion bombardment are not only to cause atomic displacements and ensuing intermixing of substrate and film components (as noted earlier) and intermixing and spatial homogenisation of film components, but also to introduce, by implantation, bombarding Ar^+ ions into the surface and to simultaneously sputter erode the growing film and incorporated Ar. Following an initial incubation period where ions penetrate and trap in both the substrate and the earliest thickness of the film an equilibrium will be established in which the rate of Ar incorporation by implantation is balanced by sputtering of Ar from the near surface region of the growing film. Under these conditions, as the film continues to grow the Ar concentration, Nt(Ar), and the Co concentration Nt(Co) will become depth independent relative to the film surface.

If the sputtering yield, in equilibrium, of the film is Y and the Ar^+ ion incorporation probability is η, then the following steady-state equations apply:

$$R_{Co} = R_d - J_i Y N^{-1} \tag{1}$$

$$\frac{Nt(Co)}{Nt(Ar)} = \frac{R_{Co}}{R_{Ar}} = \frac{R_d - J_i Y N^{-1}}{J_i \eta N^{-1}} \tag{2}$$

where N is the effective atomic density of the Co:Ar composite film. Eq.1 shows that the net Co deposition rate is the balance between atomic Co deposition and removal, by sputtering of Co, while Eq. 2 simply reflects the requirement that the concentration ratio should equal the 'effective' species deposition ratio. Eq. 1 may be rewritten as:

$$\frac{R_{Co}}{R_d} = 1 - \left(\frac{J_i}{R_d}\right) Y N^{-1} \tag{3}$$

and Eq. 2 may be rewritten as:

$$\frac{Nt(Co)}{Nt(Ar)} = \left(\frac{R_d}{J_i}\right)\frac{N}{\eta} - \frac{Y}{\eta} \tag{4}$$

Clearly Eq. 1 predicts that R_{Co}/R_d should be a linear function of J_i/R_d with a gradient of $- YN^{-1}$ and Eq. 2 predicts that $Nt(Co)/Nt(Ar)$ should be a linear function of R_d/J_i with gradient N/η and intercept $-Y/\eta$. Apart from a small deviation for small J_i/R_d it is evident that Fig. 1 displays the type of behaviour predicted by Eq. 1. The deviation for small J_i/R_d may be attributed to larger errors in ion flux determination and/or to fast Ar neutral atoms which are backscattered from the sputtered Co target and strike the film but are unrecorded as ions. This would effectively increase J_i and move the data points closer to the linear fit. The fit of both 200 eV and 500 eV Ar^+ ion bombardment to a common line is somewhat unexpected since the sputtering yield for 500 eV might be expected to be substantially larger than for 200 eV ions [9]. It is, however, possible that more data points would modify this result and it should be noted that as ion flux increases the films become thinner so that deposition rates are less reliably evaluated.

In a similar manner Fig. 2 also corresponds well with the linear function prediction of Eq. 4. For both 200 eV and 500 eV ion assistance a linear fit is rather well obeyed and the gradient for the 200 eV data appears to be slightly larger than that for 500 eV data. This, from Eq. 4, implies that $\eta(200$ eV$) < \eta(500$ eV$)$ which is entirely expected since the penetration and incorporation probability would be expected to increase with increasing ion energy [10]. It should be pointed out that Eq. 1, (and hence Eq. 2→4), is only valid for positive R_{Co} where the cobalt deposition rate exceeds the sputtering rate. When these are equal R_{Co} and R_{Ar} become zero and the ratio Nt(Co)/Nt(Ar) is indeterminate. Consequently the apparent intercepts of the Nt(Co)/Nt(Ar) data as $R_d/J_i \to 0$ have no physical significance. Indeed the results of Fig. 1 suggest that no net film deposition occurred when $J_i/R_d \geq 0.7$, i.e. $R_d/J_i \leq 1.4$ and so, in Fig. 2 concentration ratios in this region of R_d/J_i should be regarded with some circumspection. Moreover when the films are very thin ion penetration into the substrate and species intermixing between film and substrate become increasingly important so that effective values of η, Y and N may vary. Consequently it is the data at larger values of R_d/J_i which is more reliable and, as indicated fits to Eq. 4.

Conclusions

Quantitative measurements have been made of both the net Cobalt deposition rate and the Cobalt: Argon concentration ratio when Co films were deposited with and without simultaneous Ar^+ ion irradiation for Ar^+ ion: Co atom arrival rate ratios ranging from 0.07 to 0.5 and for Ar^+ ion energies of 200 eV and 500 eV. As predicted by simple theory the Co deposition rate decreases with increasing Ar^+ ion flux as a result of sputtering of the growing film and the fractional Ar concentration decreases as the Ar^+ ion flux decreases (i.e. Nt(Co)/Nt(Ar) increases as R_d/J_i increases). This latter behaviour results from the fact that decreasing the Ar^+ ion flux both increases the net Co deposition rate and decreases the Ar^+ implantation rate. The clear conclusion is that, in ion assisted deposition, the assisting ion concentration in the growing film will be reduced by employing large atom: ion arrival rates.

References

[1] J.J. Cuomo, S.M. Rossnagel and H.R. Kaufman. (Eds). Handbook of Ion Beam Processing Technology (Park Ridge, Noyes, 1989).

[2] S.M. Rossnagel, J.J. Cuomo and W.D. Westwood. (Eds). Handbook of Plasma Processing Technology (Park Ridge, Noyes, 1990).

[3] I.S. Tashlykov, O.G. Bobrovich, V.Ch. Palchekh, V.V. Tuljev, N.V. Alov, V.S. Kulikauskas and G.K. Wolf, Surf. Coating Technol. **74-75**, 945 (1995).

[4] J.M.E. Harper, J.J. Cuomo, R.J. Gambino and H.R. Kaufman. In: Ion Bombardment Modification of Surfaces, Fundamentals and Applications. Eds. O. Auciello and R Kelly. (Amsterdam, Elsevier). Ch.4. (1984)

[5] M.J.W. Greuter, L. Niesen, A. van Veen, R.A. Hakvoort, M. Verwerft, J.Th. M. de Hosson, A.J.M. Berntsen and W.G. Sloof, J. Appl. Phys. 77, 3467, (1995).

[6] Z. He, S. Inoue, and G. Carter, J. Vac. Sci. Technol. **A14**, 1 (1996).

[7] J.S. Colligon, J. Vac. Sci. Technol. **A13**, 1649 (1995).

[8] W.K. Chu, J.W. Mayer and M.A. Nicolet. Backscattering Spectrometry (N.Y., Academic Press, 1978).

[9] H.H. Anderson and H.L. Bay. Topics in Applied Physics. Vol. 47. Sputtering by Particle Bombardment. I. Ed. R. Behrisch (Berlin, Springer, 1983). Ch. 4.

[10] G. Carter, D.G. Armour, S.E. Donnelly, D.C. Ingram and R.P. Webb, Rad. Effects, **53**, 143 (1980).

Materials Science Forum Vols. 248-249 (1997) pp. 361-364
© 1997 Trans Tech Publications, Switzerland

Application of the Combined Channeling Method at Higher Ion Energies on Cyclotron

F. Ditrói[1], J.D. Meyer[2], R.W. Michelmann[2] and K. Bethge[2]

[1] Institute of Nuclear Research, P.O. Box 51, H-4001 Debrecen, Hungary

[2] Institut für Kernphysik, J.W. Goethe University, August Euler Str. 6, Frankfurt/M, Germany

Keywords: Channeling, Cyclotron, High-Energy, Silicon, GaAs

Abstract

The method of ion channeling combined with other analytical methods is widely used for investigation of highly ordered materials by using the low energy linear accelerators. The energy of the bombarding particles (p, d and He ions) extends to several MeV/nucleon. The use of a cyclotron with the channeling method makes possible to apply higher bombarding energies (20 MeV maximal energy for protons in our case).

The scattering chamber to be used for the channeling measurements is situated after an analyzing magnet and a pair of slits. The beam path is calculated electron-optically so, that practically no collimating is necessary after the last slit of the analyzing magnet (approx. 5 m from the target position) to produce a spot of 1 mm^2 on the sample surface. This can lower the scattering-background and enhance the beam-quality for channeling purposes.

The detectors are placed on a ring around the center of the chamber, and their distance from the target is also variable. The detectors can be moved on the ring by remote-controlled stepping motors. For positioning and rotating the samples a FISONS two-axis goniometer* was used. For testing and calibrating the irradiation and measuring equipment simple Si crystals were used. Their surfaces were cut and polished perpendicular to main crystallographic directions (<110>, <100>).

1. Introduction

The combination of the channeling method with other prompt or delayed nuclear analytical methods is widely used in materials science especially in analysis of highly ordered materials (Si, GaAs, etc.)[1-4]. The theory of interaction of energetic ions with the crystal-atoms has been developed yet [6-8]. In the majority of laboratories dealing with the channeling phenomena the Van de Graaff type accelerators are used for ion bombardment within the energy range of 0.5-3 MeV[9]. This technique has been used only in a few cases at higher energies because of the severe requirements on the beam quality (parallel beam, angular spread,...).

The main purpose of extending the energy range of channeling towards the high energy range is to achieve a larger penetration depth into the crystal to investigate the properties of deeper layers and multilayers. The other possible goal of the method is a possibility the check the validity of different theories on ion-crystal(atom) interactions in an extended energy range.

2. Experimental set-up

A channeling facility has been developed in the Cyclotron Laboratory of the Institute of Nuclear Research of the Hungarian Academy of Sciences, Debrecen, by using a big scattering chamber, which has been used for angular distribution and RBS measurements. The goniometer (FISONS type, bought by financial support of the Volkswagen Foundation in co-operation with the Institute for Nuclear Physics, University of Frankfurt) was built on the top of the chamber allowing to change it back to the former sample holder.

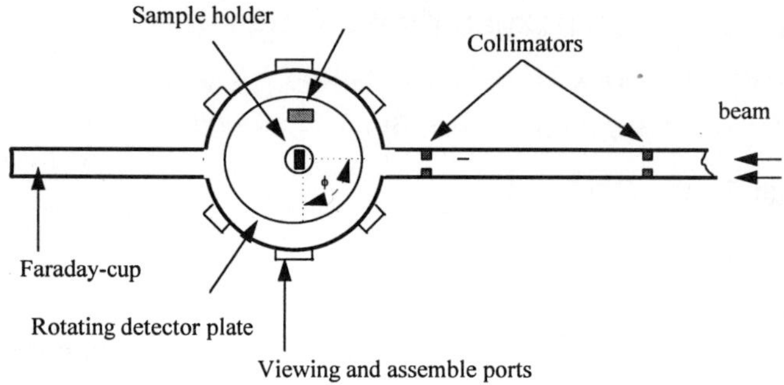

Figure 1   Top view of the scattering chamber with the goniometer

The goniometer has three translations (x, y, z), the z motion (vertical, range = 150 mm) is stepping motor driven to provide a remote control possibility for replacing the point of irradiation on the sample. Two stepping motor driven angular motion is provided, a primary rotation (ϕ = 0-360°, resolution = 0.01°) which is parallel to the horizontal plane (Fig.1.). A secondary rotation is also provided through the vertical axis (Fig. 2.), its plane is perpendicular to the that of the primary rotation. ($\Theta = \pm\,10°$). Both rotations are also remote controlled.

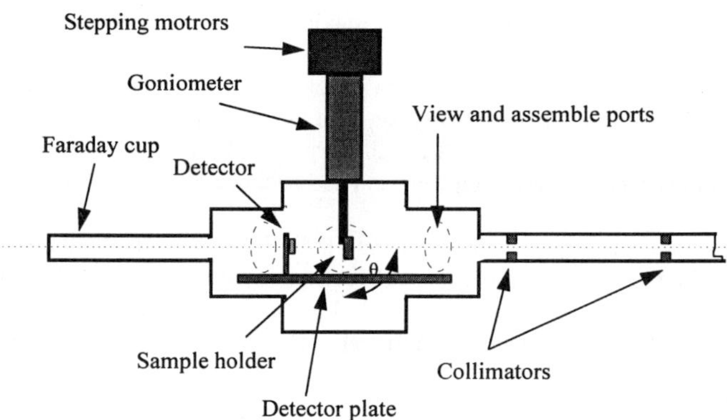

Figure 2   Side view of the scattering chamber with the goniometer on it.

There are two adjustable collimators in front of the chamber to ensure the necessary beam parallelity. An alternative method has also been developed by the operating staff of the cyclotron to produce a several mm^2 beam spot on the sample and highly parallel beam only by adjusting the lenses and magnets so, that the last slit(collimator) in the beam path is the exit-slit of the analyzing magnet, which is situated about 8 m from the sample position. This method can enormously reduce the background coming from the scattering on the collimator edges.

The sample is fastened on the sample holder of the FISONS goniometer so, that the irradiation point is always in the center of the angular motions (e.g. in the center of the motion sphere). The detector(s) can be placed on detector holders in different distances from the sample on a rotating plate which makes possible a 360° remote controlled rotation of the detectors. The signals are going through feedthrough into the pre-amplifiers and after that to the measuring room, which is situated outside of the irradiating area.

3. Theoretical considerations for high energy channeling

When using higher bombarding energies (e.g. above 2 MeV) the different parameters of the channeling change in such a direction, that the requirements for the beam and instrumentation parameters will be sharper.

The most important parameters describing the channeling phenomena are the critical angle, the minimum yield and the FWHM (full width at half maximum) of the channeling dip. These parameters can be calculated at a given energy range by using semi-empirical or theoretical formulas [5-6, 8].
For our calculations we have used the following terms.

$$a = 0.04685 \left(Z_1^{1/2} + Z_2^{1/2} \right) \tag{1}$$

is the Thomas-Fermi screening radius in units of nm, and Z_1 and Z_2 are the atomic numbers of the ion and the target atom. The critical angle Ψ_1, which describes the impact angle, under which the bombarding ion can not penetrate into the atomic rows (remains in the channel) is expressed as follows:

$$\Psi_1 = \left(2\, Z_1 Z_2 e^2 / E\, d \right)^{1/2} \tag{2}$$

Eq. 2 gives the critical angle in radians. E is the ion energy and the d is the atomic spacing along the axial direction. By using Eq. 2 one can calculate the energy-dependence of the critical angle.
From the critical angle the FWHM of the channeling dip can be calculated [8].

$$\Psi_{1/2} = 0.8\, F_{RS}(x')\Psi_1 \tag{3}$$

Eq. 3 gives the FWHM of the channeling dip in radian. F_{RS} is the square root of the Moliere string potential and x' is 1.2 u_1 /a where u_1 is the one dimensional vibrational amplitude. The values of F_{RS} can be calculated by using the Moliere potential or can be taken from tabulated form [8].
If one uses a potential other than the Moliere potential as string potential, the result can be slightly different, but this difference is not significant.
As an example the energy dependence of the FWHM of the channeling dip is calculated and plotted for the case of bombardment of Si in <110> axial direction with ^{4}He ions.
The most important channeling parameter is the χ minimum yield of the channeling dip, which represents the normalized ratio of the channeling yield at $0°$ impact to that of the random case. The minimum yield can also be estimated by using the following formula [8], which is deducted from a fit on experimental data:

$$\chi = 2\,\pi\, N\, d\, c u_1^2 \left(1 + \frac{\Psi_{1/2}^2 d^2}{k^2 u_1^2} \right)^{1/2} \tag{4}$$

In Eq. 4 the N is the atomic density/unit volume c and k are tabulated constants [8]. The example plot based on Eq. 4 is seen on Fig. 4. To measure the minimum yield as calculated the fulfillment of the requirements for the beam quality is necessary.

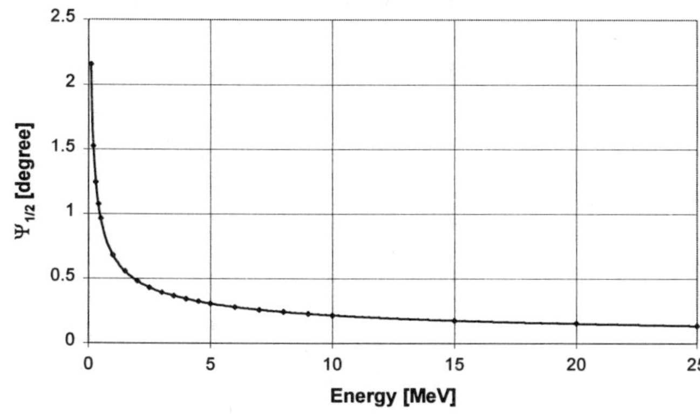

Figure 3   Full Width at Half Maximum   (FWHM)   of   the channeling   dip   for ^{4}He bombarding ions in the <110> axial direction in silicon.

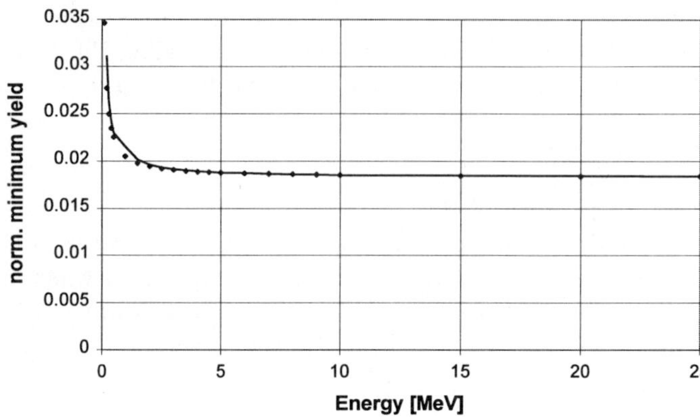

Figure 4   Energy dependence of the minimum yield of ^{4}He particles at the <110> axial channeling direction in silicon.

4. Conclusions

From Fig 3 and 4 it is obvious, that the parameters (i.e. critical angle, FWHM of the channeling dip and the minimum yield for ^{4}He in silicon) are changing most significantly in the 0-5 MeV energy range. Above 5 MeV the angles decrease very slightly and the minimum yield practically remains constant. It means, if the requirements for the beam quality are fulfilled for 5 MeV (in the case of ^{4}He in Si) , than this beam quality is suitable for the measurement in higher energy range.
The calculations (partly based on experimental data) have been proved, that the channeling method can be successfully used at higher energies, exploiting the advantages of high energy irradiations.

References

[1] F. Ditrói, J.D. Meyer, R. Michelmann, D. Kislat, K. Bethge, Nucl. Instr. Meth. B **89**, 164 (1994)
[2] J.D. Meyer, R.W. Michelmann, F. Ditrói, K. Bethge, Nucl. Instr. Meth. B **89**, 186 (1994)
[3] F.Ditrói, J.D. Meyer, R.W. Michelmann, K. Bethge, Nucl. Instr. Meth. B **99**, 182 (1995)
[4] J.D. Meyer, R.W. Michelmann, F. Ditrói, K. Bethge, Nucl. Instr. Meth. B **99**, 440 (1995)
[5] J. Lindhard, K. Winter, K. Dan, Vidensk. Selsk. Mat. Fys. Medd. **34**(4) (1964)
[6] J. Lindhard, K. Dan, Vidensk. Selsk. Mat. Fys. Medd. **28**(8) (1954)
[7] H.H. Anderson, J.F. Ziegler, Hydrogen Stopping Powers and Ranges in All Elements, Vol. 3. (Pergamon, New York, 1977)
[8] J.R. Tesmer, M. Nastasi, Handbook of Modern Ion Beam Materials Analysis, Materials Research Society, Pittsburgh, Pennsylvania, 1995
[9] K. Lenkeit, Ch. Trikalinos, L.L. Balashov, N.N. Kabachnik, V.I. Shulga, Phys. Status. Solidi B **161**, 513 (1990)

Materials Science Forum Vols. 248-249 (1997) pp. 365-368
© *1997 Trans Tech Publications, Switzerland*

Corrosion Depth Profiles by Rutherford Backscattering Spectrometry and Synchrotron X-Ray Reflectometry

E. Szilágyi[1], L. Bottyán[1], L. Deák[1], E. Gerdau[2], V.N. Gittsovich[3], A. Gróf[1], E. Kótai[1], O. Leupold[2], D.L. Nagy[1] and V.G. Semenov[3]

[1] KFKI Research Institute for Particle and Nuclear Physics, P.O.B. 49, H-1525 Budapest, Hungary

[2] II. Institut für Experimentalphysik, Universität Hamburg, Luruper Chaussee 149, D-22761 Hamburg, Germany

[3] Chemical Institute of Saint-Petersburg University, Universitetsky pr. 2, St. Petersburg 198904, Russia

Keywords: Depth Profiling, Rutherford Backscattering, X-Ray Reflectometry, Synchrotron Radiation, Corrosion, Thin Films

Abstract

Rutherford backscattering and synchrotron x-ray reflectometry was used to analyse the depth profile of elements in a sputtered iron thin film of originally 20 nm thickness following corrosion heat treatments. An "up to self-consistency" simultaneous evaluation of both kinds of spectra allowed for accurate determination both elemental composition and thickness of the sub-layers. Different iron oxide and oxi-hydroxide layers were identified on top of the iron layer depending on the treatment. An oxide layer of overall composition close to Fe_2O_3 was also observed at the iron/glass interface.

Introduction

Rutherford backscattering spectrometry (RBS) as well as grazing incidence scattering (reflectometry) of x-rays are established methods for depth profiling elements in stratified structures. By depth profiling we mean to determine *which elements* in *what concentration* can be found *at a certain depth* below the surface. From that point of view, x-ray reflectometry and RBS yield partly identical, partly complementary information.

In *identifying* layer compositions, RBS gives reliable results since this information is based on the kinematic factor of the scattering that is known to a high accuracy. Also the *concentration* of the elements can be extracted quite accurately from RBS spectra. The layer thickness (in units of at/cm^2) is a parameter less accessible to RBS. To extract this latter information, the stopping power (in eV/at/cm^2 units) is often not known to a high enough accuracy. Moreover, derivation of the *geometrical* layer thickness (in nm) is further obscured by the insufficient knowledge of the density of the individual layers.

In contrast to that, the geometrical layer thickness can be determined quite precisely by x-ray reflectometry. This technique is based on the analysis of the angular dependence of the intensity of grazing incident x-rays specularly reflected from a flat layered structure. The interference pattern is governed by the depth profile of the (complex) refraction index n, or the electronic susceptibility, χ, a small quantity for x-rays ($|\chi| \cong 10^{-5}$; $\chi \cong 2(n-1)$). Both the real and imaginary part of χ (see. Eq. 1) bear integral information on the elemental composition and concentration. However, since χ is small, there is almost no difference between the geometrical and optical path. Therefore, the beating in the reflectivity spectrum very well defines the *geometrical* thickness of the individual layers.

The complementary character of the two techniques allows for a quasi-simultaneous evaluation of the experimental data taken on the same sample with both methods. The composition of each layer as determined by RBS can be used for calculating the index of refraction for x-rays while the absolute layer thickness yielded by x-ray reflectometry refines the layer structure given by RBS. Such a way the sensitivity of RBS to the kind and concentration of elements is retained and the layer thicknesses are determined to an accuracy of a few tenths of nm. The efficiency of the combination of these techniques will be demonstrated on corrosion heat treated thin films of ^{57}Fe.

Experimental

In order to study the progress of corrosion of thin iron films a series of samples was prepared and exposed to corrosion under different conditions. In view of an extension of the study to synchrotron Mössbauer reflectometry [1], ^{57}Fe iron isotope (enrichment 95%) was used.

A nominally 20 nm thick layer (controlled by small angle x-ray scattering) of ^{57}Fe was laser ablated on a high quality float glass substrate of 80×40×5 mm and r.m.s. surface roughness of 0.3 nm. The radiation of a Nd^{3+} laser (λ=1.06 µm) of up to 3J pulse energy and 30 ns duration was brought into a vacuum chamber and focused on the ^{57}Fe metallic target mounted on a wolfram plate. The chamber pressure did not exceed 1×10^{-4} Pa. The center of the substrate was placed at 10 cm from the target. No facility was provided for the rotation of the substrate. The sample was then cut into four rectangular pieces. Sample H1011 was kept as prepared, the other three were annealed at various temperatures for 4 h in air (T_{ann} = 170, 250 and 285 °C for samples H1012, H1013 and H1014, respectively). Both RBS and x-ray reflectometry measurements were performed 9 months after preparation. During this period the samples were exposed to air at ambient pressure and temperature.

RBS measurements were performed with 1620 keV $^{4}He^{+}$ ions at scattering and tilt angles of 165° and 70°, respectively, by the EG–2R 5 MV Van de Graaff accelerator of the KFKI Research Institute for Particle and Nuclear Physics, Budapest. An ORTEC surface barrier Si detector of 2.7 msr solid angle was used. In order to check the samples for possible lateral inhomogeneity, RBS spectra were taken on each of them at five different spots; one at the centre, the others close to the middle points of the sides of the rectangular samples.

X-ray reflectometry measurements were done at the bending magnet beamline F4 of HASYLAB, Hamburg at wavelength λ = 0.86 nm (E = 14.4 keV). The width and height of the beam was 9 mm and 60 µm, resp. It was, thereby, ensured that the sample reflected the whole beam down to a glancing angle Θ well below the angle of total reflection. The specularly reflected beam was detected by an avalanche photodiode (APD). The Θ–2Θ geometry was ensured by simultaneously moving the sample goniometer and the APD. Special care was taken to avoid saturation effects both of detector and electronics.

Data Evaluation

The RBS spectra showed a slight variation of the total layer thickness from sample to sample. The lateral variation of the total layer thickness did not exceed 7% within one sample. For sample H1013, the x-ray reflectometry spectrum could only be described assuming a wedge-like change of the total thickness while RBS on the same sample showed a lateral inhomogeneity of the composition. In case of sample H1014, both methods clearly showed an enhanced inter-diffusion at the iron-glass interface. Therefore samples H1011 and H1012 were selected for detailed analysis.

Figs. 1 and 2 show RBS and x-ray reflectivity spectra of samples H1011 and H1012, respectively. In order to determine the best layer structure describing both types of measurements an iterative self-consistent approach was used. Starting with the evaluation of RBS spectra using the code RBX [2], the elements C, O, Si and Fe were identified along with traces of Sn in the float glass and traces of W in the deposited layer. To decrease the number of parameters, a model of *four* sub-layers was chosen. The composition of the sub-layers was adjusted to optimally describe the RBS spectra. First, the density ρ_i of each sub-layer was estimated from literature data of that of the probable constituent phases. This way the geometrical thickness d_i of sub-layer i (i=1...4) was also estimated. Furthermore, from the composition the indices of refraction $n_i = 1+\chi_i/2$ of the individual layers can be calculated by

$$\chi_i = \frac{4\pi}{k^2} \sum_j N_{ij}\left[-\left(Z_j + \Delta F_a\right)r_0 + \frac{i}{4\pi}\sigma_j k\right] \quad (1)$$

where N_{ij} is the density of the jth kind of atom in the ith layer, Z_j the atomic number, $r_0 = e^2/mc^2$ the classical electron radius and σ_j the total absorption cross section for element j [3] (at E = 14.4 keV for our case). The anomalous scattering correction ΔF_a ($\leq$ 1%) has been neglected.

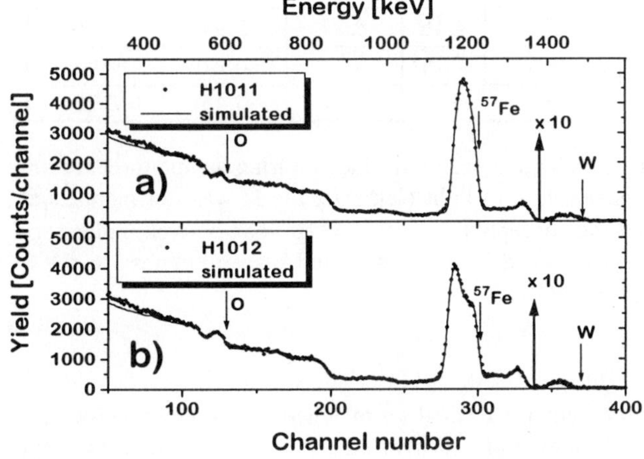

Fig. 1 RBS spectra taken on samples a) H1011 and b) H1012 at 1620 keV He⁺, Θ=165°, tilt angle 70°

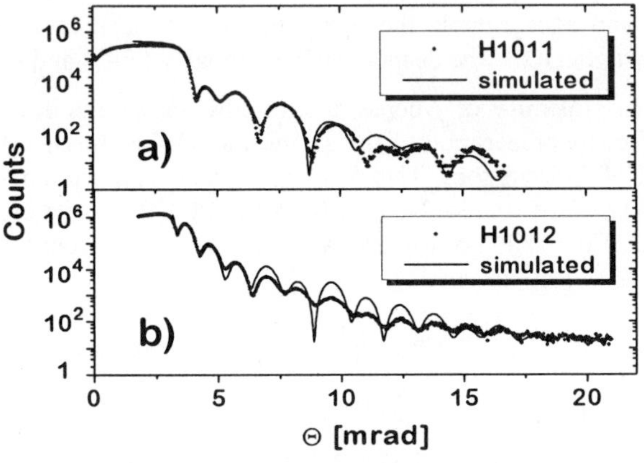

Fig. 2 X-ray reflectrometry spectra taken at λ=0.086 nm on samples a) H1011 and b) H1012

The reflectivity curves were calculated using the method of characteristic matrices [4] and accounting for roughness [5]. The values of d_i and $Re(\chi_i)$ were least squares fitted while, in view of the relative insensitivity of the reflectivity curves to the absorption coefficients, $Im(\chi_i)$ not. The new values of d_i and $Re(\chi_i)$ were used to recalculate the RBS spectra, also re-adjusting the densities. This procedure was repeated several times until self-consistency was achieved. The corresponding simulated curves are shown by solid lines on Figs. 1 and 2. The agreement with experiment is quite satisfactory although marked deviations in the x-ray reflectivity exist. These are most probably due to the neglected lateral inhomogeneity rather than the simplified layer structure or the way of accounting for roughness [5]. The parameters of the final layer structures are summarised in Table 1.

Table 1. Parameters of the final layer structure with layer 1 on top, layer 4 on the substrate

Sample	H1011				H1012			
Layer	1	2	3	4	1	2	3	4
d [nm]	3.7	3.6	8.7	5.5	13.3	5.2	7.6	4.3
layer composition	$Fe_2O_{2.9}$ $W_{0.0005}$	$Fe_1O_{0.94}$ $W_{0.0005}$	$Fe_1O_{0.12}C_{0.1}$ $W_{0.0005}$	$Fe_2O_3C_{0.1}$ $Sn_{0.07}W_{0.001}$	Fe_1O_2	$Fe_2O_{3.4}$ $C_{0.2}W_{0.001}$	$Fe_1O_{0.5}C_{0.1}$ $W_{0.005}$	$Fe_2O_3C_{0.2}$ $Sn_{0.15}W_{0.005}$
$\rho\,[10^{22}\text{ at/cm}^3]$	9.84	9.52	8.44	9.25	8.42	10.4	8.64	8.95
$\rho\,[\text{g/cm}^3]$	5.35	5.87	6.92	5.10	4.15	5.27	5.94	5.06
$Re(\chi)\,[10^{-6}]$	-10.0	-10.9	-12.7	-9.5	-7.8	-10.0	-11.0	-9.4
$Im(\chi)\,[10^{-6}]$	0.34	0.41	0.58	0.32	0.24	0.31	0.45	0.28

The C and the W contamination in the Fe layer may have been introduced during sample preparation. Since neither RBS nor x-ray reflectometry is sensitive to light elements, the H was not manifested. However, the nearly 1:2 ratio of Fe to O and the susceptibility $\chi \cong -7.8 \times 10^{-6}$ can be well understood assuming the presence of FeOOH. The further observed Fe to O ratios and susceptibilities are due to mixtures of C, Fe, Fe_3C, FeO, Fe_3O_4, Fe_2O_3 and FeOOH.

Conclusions

In conclusion, the up to 'self-consistency' simultaneous evaluation of RBS and x-ray reflectometric spectra allowed for revealing various iron oxides and oxi-hydroxides within the thin film and determining the corresponding layer thicknesses to a high accuracy. The as prepared Fe film aged for 9 months was covered by a thin Fe_2O_3 layer and, at the same time, the iron at the Fe/glass interface appeared to be oxidized. In the annealed and aged sample the upper half of the iron layer was converted, the top of the converted layer being FeOOH, the deeper regions probably Fe_3O_4 and FeO.

Thanks are due to Dr. N.N. Salashchenko (Institute of Microstructure Physics of the Russian Academy of Sciences, Nizhnii Novgorod) for the preparation of the samples as well as to Dr. M.A. Andreeva (Moscow State University) for helpful discussions. This work was partly supported by the Hungarian Scientific Research Fund (OTKA) under Contract Nos. T016667 and F022150 and by the Hungarian Academy of Sciences (Contract No. AKP 96–230/25) as well as by the Hungarian–Russian Intergovermental S&T Cooperation (Project No. 4/94).

References

[1] D.L. Nagy, V.V. Pasyuk, Hyp. Inter. **71**, 1349 (1992)

[2] E. Kótai, Nucl. Instr. and Meth. **B85**, 588 (1994)

[3] J.P. Hannon, N.V. Hung, G.T. Trammell, E. Gerdau, M. Mueller, R. Rüffer and H. Winkler, Phys. Rev. **B32**, 5068 (1984)

[4] M. Born, E. Wolf: *Principles of Optics*, 7th edition, Pergamon Press, New York, 1978.

[5] L. Névot and P. Crocé, Phys. Appl. **15**, 761 (1980), R.A. Cowley and T.W. Ryan, J. Phys. D: Appl. Phys. **20**, 61 (1987)

Materials Science Forum Vols. 248-249 (1997) pp. 369-372
© *1997 Trans Tech Publications, Switzerland*

Hydrogen Elastic Recoil Detection Depth Resolution and Sensitivity as a Function of Sample Composition

L.S. Wielunski[1], G.L. Harding[1] and A. Bendavid[2]

[1] Division of Applied Physics CSIRO, PO Box 218, Lindfield, NSW 2070, Australia

[2] Cooperative Research Centre for Molecular Engineering and Technology, CSIRO, Division of Applied Physics, PO Box 218, Lindfield, NSW 2070, Australia

Keywords: Thin Film Analysis, Hydrogen Detection, Hydrogen Depth Resolution, Elastic Recoil Detection (ERD), Sensitivity, Multiple Scattering, Rutherford Backscattering (RBS)

Abstract

The depth resolutions obtained in 4.5 MeV ^{4}He ion Elastic Recoil Detection (ERD) and 2 MeV ^{4}He Rutherford Backscattering (RBS) are compared in different materials. Observed substantial differences are interpreted in terms of multiple scattering effects in the ERD case. The differences are explained by kinematic effects due to substantial differences in the experimental geometry used. The ERD hydrogen detection sensitivity is strongly limited by double scattering process. Results presented show how the ERD depth resolution and sensitivity depend on sample composition.

Introduction

Elastic Recoil Detection (ERD) is used to profile hydrogen and other low mass elements when studying thin films, interfaces and solid surfaces [1-3]. Generally, ERD is used to detect and profile low mass elements in the same way as RBS is used to detect and profile heavy elements in thin film structures. Usually 2-5 MeV ^{4}He ions are used as a primary beam to detect hydrogen with ERD or heavy elements with RBS and it is often assumed that the depth resolutions and sensitivities are similar for these two techniques. However in contrast to RBS, the ERD depth resolution and sensitivity are substantially limited by double and multiple scattering effects [4].

In this paper, we experimentally demonstrate the limitations of the depth resolution and sensitivity of ERD analysis on practical thin metal film structures with intentionally introduced thin interfacial films containing hydrogen.

Experimental

Magnetron sputtered thin metal double layers of Al, Cu, Ag and Au with interfacial hydrogenated amorphous carbon (α-C:H) films [5] were deposited on typical semiconductor quality silicon wafers. The thicknesses of the different metal double films were chosen to be almost equivalent in ion energy losses (for 2 MeV ^{4}He ions used in RBS): Al - 2x400 nm, Cu - 2x180 nm, Ag - 2x171 nm and Au - 2x139 nm. The interfacial hydrogenated amorphous carbon film (α-C:H) deposited between identical metal films was about 50 nm thick in all the above cases. The layer thicknesses and the structure of each sample were analysed using 2 MeV ^{4}He RBS. The RBS results are shown on Fig. 1 and the ERD analysis of the same samples are shown on Fig 2.

In order to observe the hydrogen signal from a very thin interfacial organic film an Au double film with thicknesses 2x206 nm was used with about 4 nm thick, organic self-assembled monolayer (SAM) deposited between the gold layers [6]. Figs 3 and 4 show RBS and ERD results from this sample.

The RBS and ERD analysis were performed in the same vacuum chamber at a pressure level of about 0.05 mPa. The samples were first positioned normal to a 2 MeV ^{4}He beam from a tandem accelerator for RBS analysis and the scattered ^{4}He ions were detected by a standard silicon surface barrier detector positioned at a scattering angle of 160^0. In order to perform ERD analysis the ^{4}He

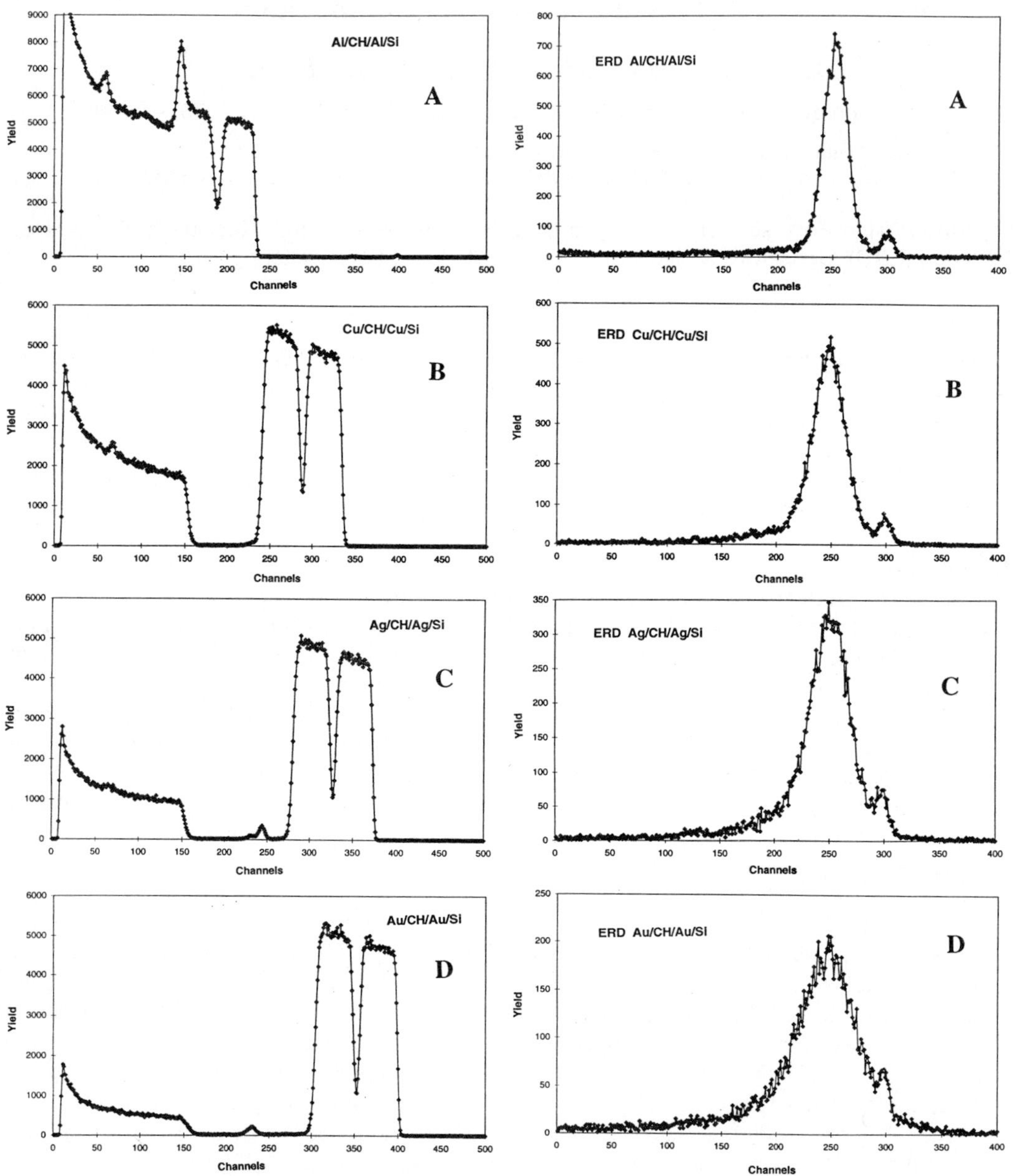

Fig. 1. 2 MeV ^{4}He RBS spectra from different metal double films with 50 nm of interfacial amorphous hydrogenated carbon layer (CH) between metal films. The single metal thicknesses used are 400nm of Al in A, 180nm of Cu in B, 171nm of Ag in C and 139nm of Au in D.

Fig. 2. 4.5 MeV ^{4}He ERD spectra from samples describe in Fig 1. The surface H peak is located in channel 300 and H signal from interfacial CH film is in form of broad peak around channel 250. The metals used are Al in A, Cu in B, Ag in C and Au in D.

beam energy was increased to 4.5 MeV and the samples were rotated 75^0 from the normal position. The recoiled H atoms were detected by a second silicon surface barrier detector positioned at a scattering angle of 30^0. In front of this detector a 24 µm mylar absorber foil was used to absorb scattered ^{4}He ions. The RBS and ERD spectra were each recorded using 512 channels with energy conversion of about 3.8 keV per channel using a standard ion beam analysis system (HYPRA).

Results and discussion

The results of RBS analysis shown on Fig 1 demonstrate that all four samples are composed of two identical uniform metal films (Al, Cu, Ag and Au) with the thicknesses of each film equivalent to about 150 keV energy width (in 2 MeV RBS) and the interfacial hydrogenated amorphous carbon film between two metal films with the thickness equivalent to about 30 keV (FWHM). The α-C:H film is observed as a sharp dip between RBS signals from the two metal films and as a small Carbon peak in the low energy part of the spectra. It is clear from this data that in each case there is no substantial mixing between α-C:H and metal films. The RBS depth resolution in the region of the α-C:H interface layer is about 1/5 of one metal film thickness or better. There is no noticeable difference in RBS depth resolution between different metals.

The ERD results from the same samples are shown on Fig. 2. The H signal from the α-CH interfacial film is in the form of a broad peak with the centre at about channel 250 and the surface hydrogen contamination is responsible for a small peak around channel 300. The energy width of the Hydrogen surface peak is about 50 keV (FWHM) and it is a measure of the energy resolution of the ERD detection system including the energy spread related to the detector solid angle and the energy straggling in the absorber foil. The interfacial H signal from the α-C:H film is much broader and its width (FWHM) varies from sample to sample: 91 keV in Al, 127 keV in Cu, 150 keV in Ag and 237 keV in Au. This depth resolution is low compared with a single metal film thickness equivalent to about 190 keV in the ERD case.

Fig 3 shows the RBS spectrum from the uniform Au double layer (each 206 nm thick) with a thin (4 nm) SAM organic layer deposited between Au layers. The ERD spectrum from this structure is shown in Fig. 4. The peak in channel 300 is a surface H signal and the very broad peak between channels 150 and the surface H peak corresponds to H in the SAM layer. The FWHM of this peak is about 300 keV and the single Au layer thickness corresponds to about 240 keV in this case. This sample is very close to the real technological organic contamination on the interface. It is clear from this data that ERD depth resolution is a strong function of depth and atomic number of the sample material. The depth resolution is rapidly reduced with increase of depth and atomic number of sample material.

Similar effects have been observed in H ion implanted materials and interpreted as consequences of multiple scattering in the sample material and the strong angular dependence of ERD energy [4]. The results presented in this work agree with predicted multiple scattering effects and clearly indicate that this reduced depth resolution limits ERD applications especially in the practical case of multilayer structures. It must be pointed out that this effect is related to the nature of the small angle scattering ERD spectroscopy and can be reduced in transmission ERD (0^0 recoils), however in most applications samples cannot be studied in this geometry [7]. On the other hand Nuclear Reaction Analysis (NRA) can offer better H depth resolution in the case of H located at depth [8-9].

Another experimental limitation of ERD relates to the long tails associated with H peaks which are clearly visible in the case of heavy element target materials such as Au. These tails are visible in the region of channels from 0 to 150 and in high energy in channels above the H surface peak (above 310) on Fig 2 b, c and d. It was suggested that these tails are generated by double scattering which includes ^{4}He or H small angle scattering on a heavy sample atom (strong Z dependence) and hydrogen forward recoil scattering [4]. The sum of the hydrogen recoil angle and the secondary scattering angle must be equal to the nominal ERD scattering angle. This may explain the high energy tails which are higher than the normal ERD signal from H on the surface. In the double scattering case ERD can occur with an angle substantially smaller than the nominal value and this will produce a higher recoil energy. The secondary scattering will not change the energy too much as it is a small angle scattering on a heavy target atom. The final energy can be substantially higher than from a single ERD process providing that the energy losses for all paths in the sample are small. Usually the extra path in the case of the double scattering is responsible for a reduction of the final energy and can explain the low energy tail as well. The level of this tail is a function of the total amount of hydrogen on the sample surface and in the near surface region, and of the atomic numbers of the major sample components, which is in general agreement with the double scattering model[4].

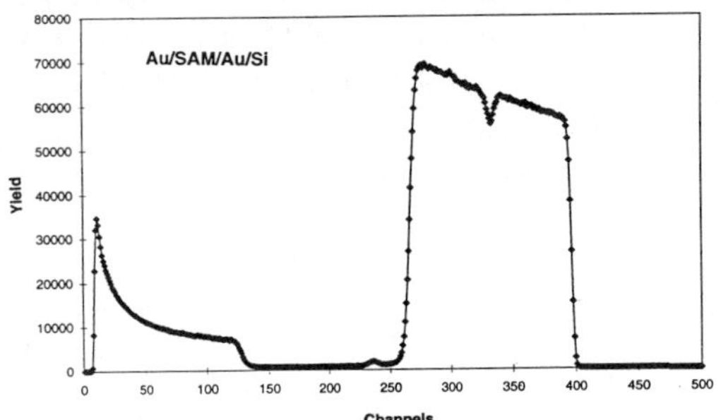

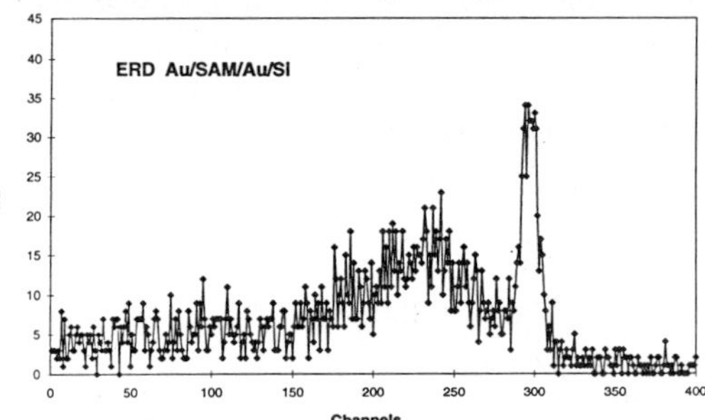

Fig. 3. 2 MeV RBS spectrum from double Au layer with SAM - organic layer 4 nm thick on the Au-Au interface.

Fig. 4. 4.5 MeV ERD spectrum from double Au layer with SAM - organic layer 4 nm thick on the Au-Au interface.

Conclusions

The comparison of the depth resolution of ERD and RBS analysis show that multiple scattering is responsible for relatively strong reduction of ERD depth resolution. The RBS resolution is not affected by this process in the same way due to the different geometry used. In the case of interfacial hydrogenated layers between heavy metal films, ERD depth resolution can be comparable to the metal film thickness and exact hydrogen location is not possible. The sensitivity of ERD hydrogen detection is limited by the long tails associated with hydrogen signals related to the double scattering process. The effects demonstrated in this work are related to the nature of low angle scattering ERD spectroscopy.

References:

[1] L.S. Wielunski, R.E. Benenson and W.A. Lanford, Nucl. Instr. and Meth. in Phys. Res. 218, 120 (1983).

[2] A. Turos and O. Meyer, Nucl. Instr. and Meth. in Phys. Res. B 4, 92 (1984).

[3] J. E.E. Baglin, A.J. Kellock, V.R. Deline, M.A. Crockett and A.H. Shih, Nucl. Instr. and Meth. in Phys. Res. B 64, 469 (1992).

[4] L.S. Wielunski, Nuclear Instr. and Meth. in Phys. Res. (IBA-95 conference in print).

[5] S. Craig and G.L. Harding, Thin Solid Films 97, 345 (1982).

[6] R.G. Nuzzo and D.L. Allara, J. Am. Chem. Soc. 105, 4481 (1983).

[7] L.Wielunski, R.E. Benenson, K. Horn and W.A. Lanford, Nucl. Instr. and Meth. in Phys. Res. B15 469 (1986).

[8] A.E. Jaworowski, L.S. Wielunski and G. Bambakidis in Mat. Res. Soc. Symp. Proc. 59, 501 (1986).

{9] A.E. Jaworowski and L.S. Wielunski in Mat. Res. Soc. Symp. Proc. 104, 305 (1988).

Materials Science Forum Vols. 248-249 (1997) pp. 373-376
© *1997 Trans Tech Publications, Switzerland*

Investigation of the Morphology of Porous Silicon by Rutherford Backscattering Spectrometry

E. Szilágyi[1], Z. Hajnal[2], F. Pászti[1], O. Buiu[3], G. Craciun[3], C. Cobianu[3], C. Savaniu[3] and É. Vázsonyi[2]

[1] KFKI - Research Institute for Particle and Nuclear Physics,
P.O.B. 49, H-1525 Budapest, Hungary

[2] KFKI - Research Institute for Materials Science, P.O.B. 49, H-1525 Budapest, Hungary

[3] Institute of Microtechnology, P.O.B. 38-160, RO-72225 Bucharest, Romania

__Keywords:__ Ion Beam Analysis, Rutherford Backscattering Spectrometry, Porous Silicon, Morphology

Abstract

The pore walls of a porous Si sample of columnar type were coated by SnO_2 using the sol-gel technique. The sample was characterised by Rutherford Backscattering Spectrometry (RBS). The Sn signal in the RBS spectra revealed that the coating was homogeneous in depth. The low energy edge and the total width of the Sn peak showed significant variations with sample tilt angle yielding information on the 3D morphology of the porous layer. These effects could be simulated by Monte Carlo type calculations of RBS measurements on 3D structures.

Introduction

Ion beam analysis methods, mainly Rutherford Backscattering Spectrometry (RBS), are widely used to determine the elemental composition of thin films. The samples are usually considered to have only one dimensional structure, i.e., the elemental concentrations vary only with depth. Possible lateral inhomogeneity over the beam spot appears just as an enhanced energy spread in the spectrum, or for larger scale variations the composition can be mapped by measuring at different impact points.

Recently, porous silicon (PS), a very promising material for optoelectronic devices and sensor applications, is widely studied by RBS and other ion beam analytical techniques [1, 2]. PS layers with a thickness in the μm range can be produced with two basically different morphologies. The columnar type structure contains long pores of typically 15 nm diameter which are surrounded by 5-10 nm thick silicon walls and run parallel to the <100> direction of the silicon lattice. In the case of sponge-like morphology, small pores are randomly distributed in the film, their diameter are about 2 nm and the wall thickness is around 2-5 nm depending on the preparation conditions. Porous layers have very high specific internal surface areas of up to 600 m^2/cm^3.

Since ions are not sensitive to holes but to matter only, two main questions might arise: i) is there any difference between RBS spectra recorded on compact and porous samples and ii) can we obtain any information on the 3-dimensional (3D) morphology of a PS layer. It will be shown that a structure induced energy spread arises from the variation of the amount of material crossed by ions for a given PS layer and this spread may strongly depend on the morphology. This dependence has been demonstrated recently in ref. 1, where the pore walls were coated with ^{15}N by high temperature nitridation and were studied by narrow resonance in the $^{15}N(p,\alpha\gamma)^{12}C$ nuclear reaction. In the present work, the internal surface of the pores of a columnar PS sample was covered by SnO_2 using a sol-gel technique. RBS spectra taken at various sample tilt angles will be presented and the observed variations in the energy spread at the PS-bulk interface will be discussed in details. The results will be interpreted using our recently developed simulation program RBS-MAST, that can model RBS measurements on samples with 3D structures on the nm-μm scale [3].

Experimental

A p-type Si wafer of <100> orientation was doped by boron diffusion to obtain a ~6 μm thick surface layer of 1.2×10^{-3} Ωcm resistivity. The PS layer was then formed by anodisation in dark condition at room temperature using 3 mA/cm^2 current density and 10 minutes etching time [4]. The electrolyte consisted of hydrofluoric acid (HF) 49 w% and absolute ethanol in 1:3 volume ratio. The resulted columnar-type PS layer was shiny and very homogenous with an expected porosity of ~55 % and thickness of ~1 μm.

The internal surface of this PS layer was covered by a thin SnO_2 film using the sol-gel technique. An alkoxide precursor of chemical formula of $Sn(OC_2H_5)_4$ - 1.5 C_2H_5OH was diluted by ethanol to a concentration of 0.046 M. Since no water was added to the mixture, the obtained sol contained only residual water from atmospheric humidity. The pH of the solutions was 1.41 and the sol viscosity of 1.5 cP was constant for more than 30 days. The same period of time can be assigned to the stability of the sols. The solution was filtered then spun onto the PS sample with 0.2 μm thickness resulting in a thin film of good adhesion and homogeneity. To remove the residual organic compounds the sample was dried and then pyrolysed at 500 °C in O_2 flow.

The sample composition was determined by RBS using 3555 keV $^4He^+$ ions obtained from the 5 MeV Van de Graaff accelerator of KFKI, Budapest. At this beam energy the cross section for carbon is about 7 times higher than the Rutherford value. The ion beam was collimated by two sets of four sector slits to the dimensions of 0.2×1 mm^2 and angular divergence of < 0.04°. The applied ion current of typically 10 nA was measured by a transmission Faraday cup [5]. In the experimental chamber the vacuum was kept better than 2×10^{-4} Pa. To minimize carbon deposition from the vacuum, the cold traps located along the beam path and the wall of the scattering chamber were filled with liquid nitrogen 10 minutes after starting the evacuation.

The RBS experiments were performed using a 25 mm^2 ORTEC surface barrier detector at 165° scattering angle with a solid angle of 2.67 msr and an energy resolution of 16 keV. The RBS spectra were taken on the same spot but at different tilt angles ranging from -2° to 30° including the channeling direction at tilt 0.3°.

Results and discussions

The average atomic composition of the PS layer was determined using the RBX code [6] by simulating the RBS spectrum taken at tilt 30° and it was found to be $Si_1O_{1.09}C_{0.28}Sn_{0.024}$ all over the PS layer. Typical measured and simulated spectra are shown in Fig. 1.

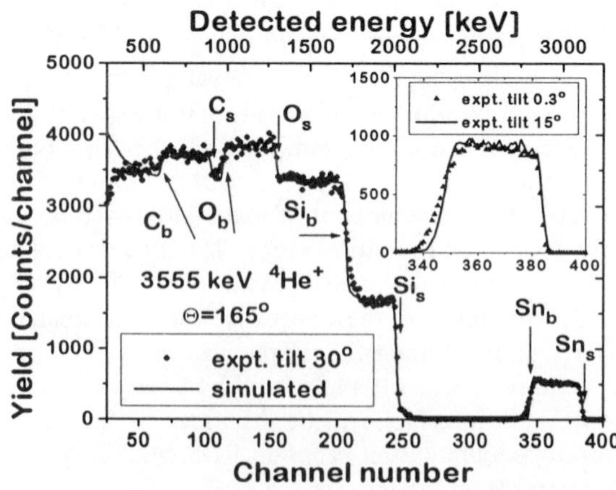

Fig. 1. Typical measured and simulated RBS spectra of a porous silicon layer treated by the sol-gel technique. The simulations are performed by RBX. The inset shows Sn peaks of channeled (tilt 0.3°) and random spectra (tilt 15°). "s" and "b" indexes stand for sample surface and back side positions.

The introduced Sn appears in the experimental spectra as a wide, flat peak, hence it is distributed on the internal surface probably as a uniform film of SnO_2. All the experimental back edges in Fig. 1. are smeared out much more than the simulated ones, in spite of the fact that during the simulations both Bohr energy straggling and detector energy resolution were taken into account. Sn regions of channeled and tilted spectra are compared in the inset of Fig. 1. Let us note that in spite of the probably decreased stopping power [7] and smaller tilt angle, the peak is widened in the channeled spectrum. Furthermore, its back edge is smeared out much more than in the tilted spectrum. To follow these phenomena, the Sn peaks in all the experimental spectra were fitted with error functions. Fig. 2 shows both the total width of the Sn signal and its energy spread at the interface as a function of the tilt angle. The energy spread shows two local maxima around $0°$ and $1°$ and a minimum between them at tilt $0.6°$ and it gradually decreases at higher tilt angles. The total width of the Sn peak shows a maximum at the same position where the local minimum in the energy spread was found.

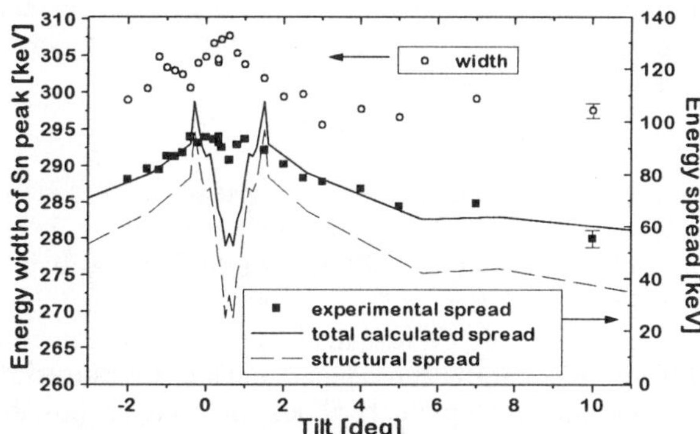

Fig. 2. Full width of Sn peaks and energy spread at their back edge (in FWHM sense) as a function of tilt angle. Symbols: extracted from the measured spectra by fitting error functions, dashed line: structural energy spread alone, solid line: calculated total energy spread (see text).

By taking into account all the classical contributions for an equivalent compact sample using the DEPTH code [8] we were unable to reproduce the measured energy spread. The obtained spread (~28 keV) was just a fraction of the experimental value (60-100 keV) and it was practically independent of the tilt angle, in marked contrast to the experimental trend.

To understand the above discrepancies we have to take into consideration an extra energy spread caused by the porous structure itself. It is easy to see for an extreme case when the incident beam is nearly parallel to the columnar pores that some ions reach the interface mainly travelling along pores, i. e., without considerable energy loss, while others penetrate mainly in walls losing almost as much energy as if the PS layer would be compact. This *structure induced energy spread* contribution should decrease if all those ions suffering backscattering from the pore walls at the interface cross more or less the same amount of wall material. There are two ways to reach this situation: i) for large tilt angles the beam turns away from the pore axis and the ions reach the interface after crossing numerous pores and walls; ii) at the tilt angle where the beam is exactly parallel to the pores ($0.6°$ in our case) the mentioned ions should always travel within the walls. These two mechanisms are responsible for the experimentally observed trends in the energy spread. Furthermore, in the latter case the total Sn peak width increases, because those ions suffering backscattering from the interface lose more energy before the scattering than in the tilted cases.

In the experiments both the peak broadening and the peculiar behaviour of structure induced energy spread appears (see Fig. 2.) clearly indicating that *the pores are well ordered and run along a direction deviating from the <100> axis of the substrate by 0.3°.* This clearly measurable deviation is probably caused by stress relaxation in the PS layer but it might also be introduced somehow during the sample preparation.

The structure induced energy spread was calculated using a Monte Carlo type simulation program called RBS-MAST [3] on a 3D structure model corresponding to the sol-gel treated film. The structure model reproduced the average material composition determined by RBX, assuming the different materials to be distributed along the pore walls in successive shells. The shells are characterised by the following material compositions and surface area ratios: void 19.3%, $CH_{0.36}$ 7.3%, $SiO_2Sn_{0.044}$ 52.9% and Si 20.5%. The assumed porous layer was 0.85 μm thick with average pore diameter and inter-pore distance of ~10 and 20 nm, respectively. There are no classical energy spread effects built in RBS-MAST, hence it gave only the structure induced energy spread at the interface. This contribution and the 28 keV spread given by DEPTH were considered to be independent Gauss distributions, and added in quadrature by their FWHM. To reproduce better the experimental energy spread, an extra contribution of 37 keV FWHM was also introduced, which might originate from inhomogeneities in layer thickness and porosity, surface roughness of the sample or the weakness of the model. The obtained energy spread curve is also shown in Fig. 2. The agreement with the experimental values is satisfactory except for the central part, where the calculation predicts an extremely deep minimum. This discrepancy is probably caused by some effects not considered yet in RBS-MAST (e.g., roughness of the pore walls, the ions do not fly along a straight path because of small angle multiple scattering, etc.).

Conclusions

It was demonstrated that, by determining the structure induced energy spread of a suitable decoration on the pore walls, RBS may characterize the morphology of the porous structures. In our case the sample was decorated by Sn using sol-gel technique, without destroying the porous structure. A deviation of 0.3° between the main pore direction and the <100> axis of the single crystal silicon substrate was clearly observed. Using Monte-Carlo calculations the experimental trends of the structure induced energy spread were satisfactorily reproduced. Further investigations of porous materials of different types and of various decoration methods, as well as possible applications, are in progress.

References

[1] G. Amsel, E. d'Artemare, G. Battistig, V. Morazzani and C. Ortega, in print in Nucl. Instr. and Meth. B.

[2] G. Battistig, V. Schiller, E. Szilágyi and É. Vázsonyi, Nucl. Instr. Meth. B **118**, 654 (1996).

[3] Z. Hajnal, E. Szilágyi, F. Pászti and G. Battistig, Nucl. Instr. and Meth. B **118**, 617 (1996).

[4] G. Craciun, A. Dafinei, C. Vranceanu, E. Vasile and C. Flueraru, Int. Semicond. Conf., 18th edition, Sinaia, Romania p. 331, (1995).

[5] F. Pászti, A. Manuaba, C. Hajdu, A. A. Melo and M. F. da Silva, Nucl. Instr. and Meth. B **47**, 187 (1990).

[6] E. Kótai, Nucl. Instr. and Meth. B **85**, 588 (1994).

[7] E. Kótai, N.Q. Khanh and J. Gyulai, Proc. of 9th Int. Conf. on Ion Beam Modification of Materials, p.823 (1996).

[8] E. Szilágyi, F. Pászti and G. Amsel, Nucl. Instr. and Meth. B **100**, 103 (1995).

Materials Science Forum Vols. 248-249 (1997) pp. 377-380
© *1997 Trans Tech Publications, Switzerland*

Combining RBS and FTIR Spectroscopy for the Ge Analysis in $Si_{1-x}Ge_x$ Single Crystals

A. Gerhardt[1], J. Donecker[1], B. Selle[2] and J. Wollweber[1]

[1] Institut für Kristallzüchtung, Rudower Chaussee 6, D-12489 Berlin, Germany

[2] Hahn-Meitner-Institut, Abt. Photovoltaik, Rudower Chaussee 5, D-12489 Berlin, Germany

Keywords: RBS, FTIR Spectroscopy, Element Analysis, $Si_{1-x}Ge_x$

Abstract Unstrained bulk $Si_{1-x}Ge_x$ single crystals with Ge contents up to x = 0.16 were grown by a modified float zone (FZ) technique. The alloy composition x of the crystals was obtained from RBS random spectra. The threshold energies of the phonon-assisted and no-phonon transitions at the indirect optical band gap were determined from the second derivative of the transmission spectra. The data were used to reexamine the correlation between the optical band gap energy and the Ge concentration. Channeling spectra indicate that in the range x < 0.16 the Ge atoms are incorporated substitutionally on regular Si sites.

Introduction Recently, the $Si_{1-x}Ge_x$ system has attracted renewed interest for potential use in silicon-based photovoltaics. This system allows to control the value of the band gap energy in a wide range by the alloy composition and, thus, appears to be a candidate for multiple band gap devices [1]. Furthermore, it has been shown that at a certain composition the shape of the band structure is highly favourable for the impact ionization generation and that, therefore, at high photon energies internal quantum efficiencies in excess of unity can be achieved [2,3].
The band gap variation of $Si_{1-x}Ge_x$ was first measured by Braunstein et al. [4] in 1958. However, in their publication the quality of the investigated samples, which partly were polycrystalline, and the methods used for the alloy composition analysis were not specified in much detail. The gap energies were derived by a one-phonon analysis of the absorption curves, a procedure which involves some degree of ambiguity. In this paper, we present a reexamination of Braunsteins results in the composition range of x = 0 to 0.16 where we had access to high-quality single crystals. We use Rutherford backscattering (RBS) measurements to determine the Ge concentration and Fourier transform infrared (FTIR) spectroscopy to study the band edge absorption. To circumvent problems inherent in absolute intensity measurements the relevant energy positions at the band gap were determined from differentiated transmission spectra.

Experimental Procedure $Si_{1-x}Ge_x$ single crystals with Ge concentrations up to 16 at% were obtained by a float zone technique [5]. An essential modification of this growth procedure was a continuous Ge supply of the melted zone. The crystals were about 20 cm long and 2 cm in diameter. Samples of about 0.5 to 1 mm thickness were cut perpendicular to the growth direction (mostly [111]) and polished by conventional techniques.
RBS spectra were measured with 1.4 MeV $^4He^+$ ions at a detector angle of 170°. After a background correction the random spectra were evaluated by an iterative computer simulation based on the algorithms given in [6] (RUBSODY, Friedrich-Schiller-Universität Jena, Institut für Festkörperphysik). For channeling measurements the samples were mounted on a high-precision two-axis goniometer. Optical transmission spectra were recorded in the wavenumber range from 11000 to 7000 cm^{-1} using a Fourier transform spectrometer (Bruker IFS 66v). The maximum spectral resolution was 0.1 cm^{-1}. All measurements were made at room temperature.

RBS Results Due to the fairly different masses of Ge and Si atoms and to the comparatively high backscattering cross section of Ge, RBS is a well-suited method for the element analysis of Ge in $Si_{1-x}Ge_x$. This is illustrated by Fig. 1 where random spectra of samples with different Ge concentrations are shown. Because of the low accessible depth of RBS ($\leq$ 1.4 μm) both sample surfaces were measured separately, then the spectra were added up to give a spectrum which was believed to be representative for the average composition of the whole sample. This approach is justified by

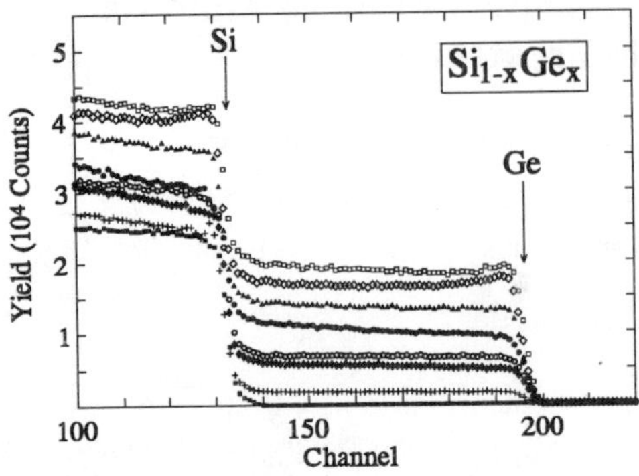

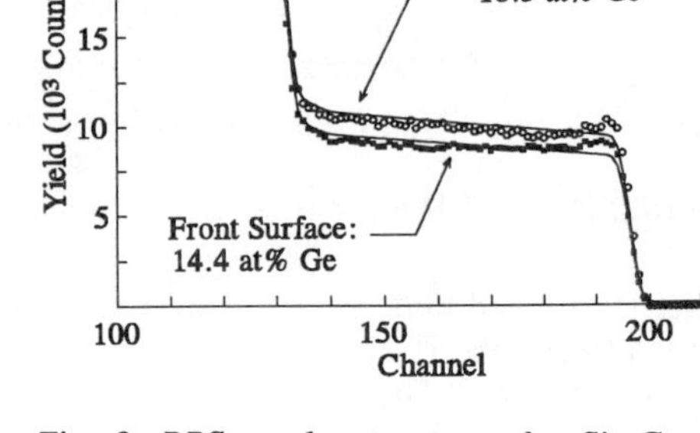

Fig. 1. RBS random spectra of Si$_{1-x}$Ge$_x$ single crystals with different Ge concentrations. These spectra were obtained by measuring the two sample surfaces separately and then adding the corresponding yields. Channel width: 5.23 keV.

Fig. 2. RBS random spectra of a Si$_{1-x}$Ge$_x$ single crystal with an axial concentration gradient. Solid lines are simulation fits yielding the indicated Ge concentrations.

the channeling experiments which did not reveal significant changes of composition and crystallinity in the near-surface region caused by the preceding surface preparation. Axial concentration gradients are assumed to be approximately linear over the sample thickness. The main source of the error in the Ge content determination is the uncertainty of the spectrum background correction ($\pm$ 0.15 at%). However, sample inhomogeneity might also contribute significantly to the statistical error. Fig. 2 presents RBS spectra of the front and back side of a selected sample with a strong

axial concentration gradient. The simulation reveals a difference in the Ge content of about 2 at% between the two sample surfaces (14.4 and 16.5 at%, respectively). In accordance with these single surface measurements an average concentration of 15.5 $\pm$ 0.4 at% Ge was obtained by evaluating the added-up spectrum.

The [111] channeling spectrum of a sample with 13.7 at% Ge is shown in Fig. 3. The surface peaks of Si and of Ge are clearly seen. The minimum yield values of only 3.3 % for the Ge signal and of 4.4 % for the Si(+Ge) signal, close to the theoretical limits [6], give evidence of a low level of disorder and a substitutional incorporation of the Ge atoms on regular lattice sites. Similar channeling spectra were found for all of our samples indicating their high crystalline quality in the range up to 16 at% Ge.

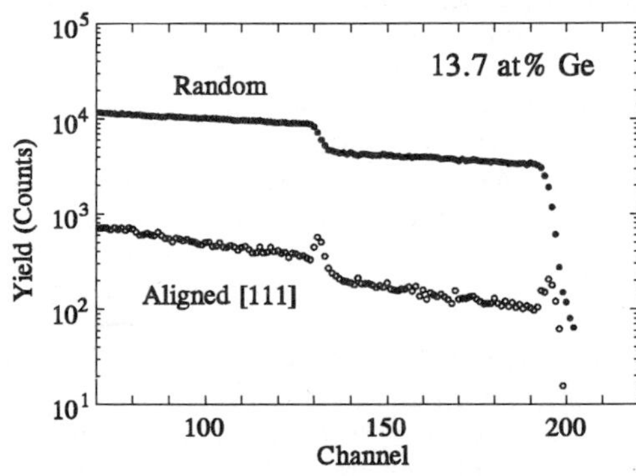

Fig. 3. Random and [111] aligned spectrum of a Si$_{1-x}$Ge$_x$ single crystal with 13.7 at% Ge.

Results of FTIR Spectroscopy The Si$_{1-x}$Ge$_x$ material is an indirect semiconductor, i.e. for a transition of an electron from the valence band maximum to the conduction band minimum phonon participation is needed to conserve momentum. Assuming parabolic energy bands the absorption coefficient α at the indirect energy gap E_g is given by the well-known quadratic dependence on the photon energy $\hbar\omega$:

$$\alpha_{el,i} = A_{el,i} \left[\hbar\omega - (E_g \pm E_{ph,i})\right]^2 \qquad \text{for} \quad \hbar\omega > (E_g \pm E_{ph,i}) \qquad (1)$$

$E_{ph,i}$: energy of a phonon i

Excitonic contributions can be described by a hydrogen-like series of states below the conduction band edge. For these transitions α follows the square-root law:

$$\alpha_{exc,i} = B_{exc,i} \left[\hbar\omega - (E_g \pm E_{ph,i} - E_{exc}) \right]^{1/2} \quad \text{for} \quad \hbar\omega > (E_g \pm E_{ph,i} - E_{exc}) \tag{2}$$

E_{exc} : exciton binding energy

The edge absorption is determined by phonon-assisted excitonic and band-band transitions as well. Interactions with different kinds of phonons (TO, LO, LA, TA) are involved. The ultimate shape of the indirect absorption edge is given by a superposition of all these contributions and is characterized by a number of weak changes of the slope which mark the onset of the distinct transitions described by Eqs. 1 and 2.

Previous work on the absorption edge was mainly based on the measurement of absolute transmission values with their very sensitive dependence on sample parameters such as effective reflectivity and thickness. In this case, it is difficult to find the appropriate threshold energies. In order to surmount these problems we determine the *positions* of the corresponding features by

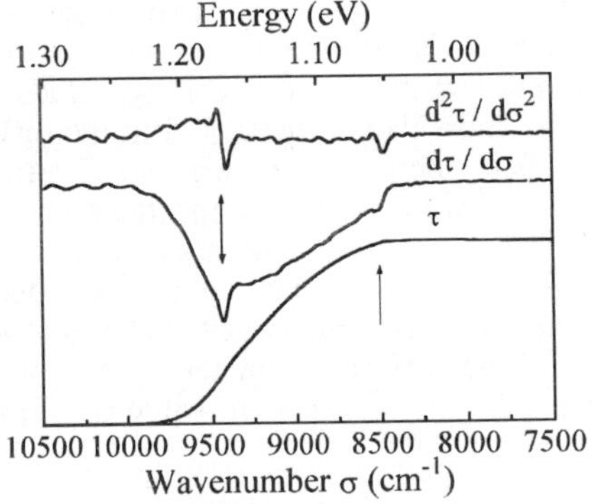

Fig. 4. Optical transmission τ of an FZ Si single crystal and the first ($d\tau/d\sigma$) and second ($d^2\tau/d\sigma^2$) derivatives. The energy positions of the transitions involving TO phonons become discernible.

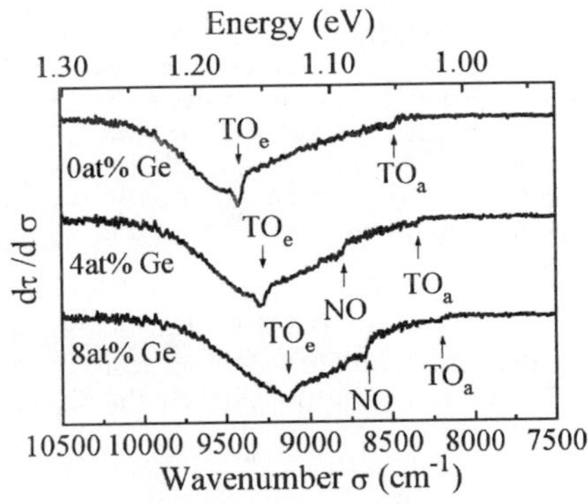

Fig. 5. Shift of the optical threshold energies with the alloy composition. With increasing Ge content the excitonic no-phonon transition appears, additionally.

spectrum differentiation. Fig. 4 shows that in the transmission spectrum τ the changes of the slope are only weakly pronounced whereas in the first and second derivatives of τ the threshold points of the relevant transitions become well visible. The exact position of a threshold can be determined by the ratio between the minimum of the first and second derivative [7]. In Fig. 4 one finds two thresholds in both derivatives being separated by an energy of 932 cm^{-1}. Since this value equals twice the TO-phonon energy of Si [8] the thresholds mark transitions with TO-phonon absorption and emission, respectively. In measurements with an improved signal-to-noise ratio the sensitivity allows to determine the position of the TA-phonon thresholds, additionally.

The first derivatives of three Si$_{1-x}$Ge$_x$ samples with different Ge concentrations are plotted in Fig.5. For an increasing Ge content both TO thresholds shift by the same amount thereby maintaining their energy interval constant. For the investigated compositions (x = 0 to 0.16) we did not observe any significant change of the measured phonon energies [7], in accordance with other reports [8]. For Ge contents above 0.2 at% a further feature appears positioned exactly at the half energy between the TO-phonon-assisted thresholds. This additional structure can be attributed to a no-phonon transition due to the perturbation of the translation symmetry of the lattice by the incorporated Ge atoms. From the no-phonon threshold position the excitonic gap $E_{gx} = E_g - E_{exc}$ can be found directly. E_{gx} also follows from the TO-phonon-assisted thresholds by taking account of the TO-phonon energy (Eq. 2).

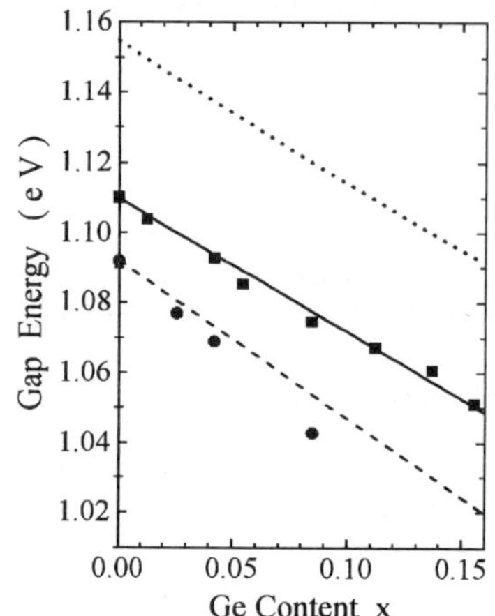

Fig. 6. *Gap energies of Si$_{1-x}$Ge$_x$ crystals as a function of the Ge content.* ■ *and solid line: room-temperature excitonic gap (our results),* ● *and dashed line: room-temperature indirect gap (Braunstein et al. [4]), dotted line: low-temperature excitonic gap (Weber and Alonso [8]).*

The gap energy of Si$_{1-x}$Ge$_x$ as a function of x is shown in Fig. 6. Our room temperature values of E_{gx} are plotted against the Ge concentrations measured by RBS. The solid line represents a linear least-squares fit expressed by

$$E_{gx}(x) = 1.110 - 0.382x \ \text{ eV} \tag{3}$$

For homogeneous samples the threshold energies can be fixed, as a rule, with an optimum accuracy of $\pm$ 10 cm^{-1} ($\sim$ $\pm$ 1.3 meV). Ge concentration gradients in the growth direction of the samples, i.e. along the optical path, may increase the experimental error to $\pm$ 2.5 meV by broadening the structures in the differentiated spectra. In Fig. 6 the scatter of our experimental points is caused predominantly by these axial composition gradients.

Discussion The comparison of our results with those of Braunstein et al. (see Fig. 6) reveals two points of disagreement: a minor deviation in the slope and, more pronounced, a difference in the absolute energy values. Both these discrepancies most likely are related to the different approach to the gap energy determination. Braunsteins data are room-temperature values of the indirect gap E_g which result from an inaccurate one-phonon approximation of the absorption curves. Our composition dependence of the room-temperature E_{gx} has the same slope as the more recent curve of Weber and Alonso [8] derived from low-temperature luminescence data. The difference on the energy scale is given by the temperature shift of the band gap and approaches the well-known value of Si [9] for x = 0.

Conclusion We have studied the composition dependence of the indirect band gap of high-quality Si$_{1-x}$Ge$_x$ single crystals in the range of x = 0 to 0.16. In order to improve the accuracy of the composition scale the Ge content x was analyzed by RBS. Furthermore, a differentiation method was applied to obtain the position of the excitonic band gap directly from the optical spectra. The data of Braunstein et al. [4], most probably, deviate from ours due to the fact that they result from intensity measurements and an inappropriate one-phonon approximation. For practical purposes our results, in turn, provide a reliable calibration curve for a rapid and rather simple composition analysis of Si$_{1-x}$Ge$_x$ bulk samples by an optical method [7].

References

[1] J. M. Ruiz, J. Casado, and A. Luque, *Proc. 12th European Photovolt. Solar Energy Conf.*, Amsterdam, 1994, p.572.
[2] J. H. Werner, S. Kolodinski, and H. J. Queisser, Phys. Rev. Lett. **72**, 3851 (1994).
[3] J. H. Werner, B. Winter, M. Wolf, S. Kolodinski, R. Brendel, M. Hirsch, H. J. Queisser, J. Wollweber, and W. Schröder, *Proc. 13th European Photovolt. Solar Energy Conf.*, Nice, 1995, p.111.
[4] R. Braunstein, A. R. Moore, and F. Herman, Phys. Rev. **109**, 695 (1958).
[5] J. Wollweber, D. Schulz, and W. Schröder, J. Crystal Growth **163**, 243 (1996).
[6] W.-K. Chu, J. W. Mayer, and M.-A. Nicolet, *Backscattering Spectrometry*, Academic Press, New York, 1978.
[7] J.Donecker, A.Gerhardt, and J.Wollweber, Mat. Sci. Engineerg. B **28**, (1994) 18.
[8] J. Weber and M. I. Alonso, Phys. Rev. B **40**, 5683 (1989).
[9] W. Bludau, A. Onton, and W. Heinke, J. Appl. Phys. **45**, 1846 (1974).

Materials Science Forum Vols. 248-249 (1997) pp. 381-384
© *1997 Trans Tech Publications, Switzerland*

Hydrogen and Nitrogen Loss during ERD Analysis of Silicon (Oxy)nitrides

S.G. Wallace[1], A.C. Kockelkoren[1,2], M.G. Boudreau[1], P. Mascher[1], J.S. Forster[1,3] and J.A. Davies[1]

[1] Department of Engineering Physics, McMaster University, Hamilton, Ontario, Canada

[2] Department of Applied Physics, Eindhoven University of Technology, Eindhoven, Netherlands

[3] AECL, Chalk River Laboratories, Chalk River, Ontario, Canada

Keywords: Elastic Recoil Detection, Silicon, Hydrogen, Loss, Damage, Release Cross Section

Elastic recoil detection (ERD) has been used to profile the near-surface region of silicon (oxy)nitride thin films. During analysis significant amounts of nitrogen and hydrogen were lost. Separate analysis with ^{79}Br, ^{127}I and ^{209}Bi beams at the same velocity have been performed to investigate the loss per recoiled particle, which should be larger for the lighter beams due to the Z_1^2 dependence of the cross section and only a Z_1 dependence of the electronic stopping power. The results of this study indicate that within the uncertainties, all three beams result in approximately the same loss of hydrogen and nitrogen per recoil.

I. Introduction

Ion beam analysis techniques such as Rutherford backscattering spectrometry (RBS) and elastic recoil detection (ERD) have been classified as non-destructive or semi-non-destructive. This description is accurate for the analysis of most materials, but should not be assumed to be universal. Various materials have been shown to be susceptible to ion beam induced losses, resulting from the beam energy loss as it penetrates the sample. Characterization of this loss with respect to the beam species and energy is at the present time incomplete. In this paper, a comparison of beam-induced damage caused by Br, I and Bi ions at the same velocity is presented and the conclusions based on these results discussed.

Elastic recoil detection has been used extensively over the last two decades, particularly for the study of hydrogen; an element that is difficult to study using other analysis techniques. ERD was developed by L'Ecuyer *et al.* [1] in 1975 for precisely this purpose, using a 30 MeV chlorine beam. In recent years with the increased access of large terminal voltage accelerators much heavier beams have been used that allow a much larger mass range of elements to be profiled [2], [3]. However, with the increased mass and energy of the analysing beam came an increase in the concern of target damage.

Estimates of the maximum sputtering yield, Y, based on the assumption that only the nuclear stopping power contributes to sputtering are often made [4] using

$$Y = \alpha \frac{0.042}{NU} \left. \frac{dE}{dx} \right|_n . \tag{1}$$

For typical heavy ion analysis fluences in the 10^{11} ions/cm^2 range the total sputtering yield is normally estimated to be much less than a monolayer ($\simeq 10^{15}$ atoms/cm^2). However in some cases sputtering yields have been measured to be up to three orders of magnitude greater than that estimated from nuclear stopping. This increase is caused by the electronic stopping power's contribution to the sputtering process.

II. Theoretical Aspects

The loss of target elements is a result of the transfer of energy from the energetic ion to either the target nuclei directly or indirectly to the target electrons. Once an element has been 'freed' it may contact a second 'freed' element forming a molecule that may then diffuse out of the target, as proposed by Adel *et al.* [5]. The absolute maximum fraction of atoms that may be displaced from a target, based on purely energetic considerations, is approximated by

$$F_d = \frac{\epsilon \times \Phi}{E_d} \qquad (2)$$

where ϵ is the stopping power of the ion in the target, Φ is the ion dose, and E_d is the displacement energy of the target element. E_d is normally about 30 eV for nuclear collisions, and about 5 eV for inelastic electronic energy transfers. From this, the maximum percentage of the target elements that can be freed in the analyzed volume can be estimated. Results for a total dose of 0.5×10^{12} ions/cm^2 230 MeV Bi ions, and for equivalent 'effective doses' [1] of 1.25×10^{12} ions/cm^2 140 MeV I ions and, 3.1×10^{12} ions/cm^2 87 MeV Br ions are given in Table 1. These values are maximum and are based on the assumption that only the minimum amount of energy is required to cause the displacement. It is clear that for the doses typically used in heavy ion

Beam	Stopping Power		Maximum Fractional Displacement	
	ϵ_n	ϵ_e	F_n	F_e
	[eV/10^{15}atoms/cm^2]			
Bi	29	3700	0.00048	0.37
I	13	2635	0.00055	0.65
Br	6	1800	0.00063	1.13

Table 1: Stopping powers and resultant maximum fractional displacement for Bi, I and Br in Si.

ERD, the nuclear energy loss is insufficient to account for the elemental loss that was observed.

III. Experimental Aspects

In our measurements three beams of equal velocities were used, namely 230 MeV $^{209}_{83}$Bi, 140 MeV $^{127}_{53}$I and 87 MeV $^{79}_{35}$Br. The ERD cross section in the Rutherford window (0.5-2MeV/u) is given by

$$\sigma(\xi) = \left(\frac{e^2}{8\pi\epsilon_0}\right)^2 \left(\frac{Z_1 Z_2(M_1 + M_2)}{M_2 E_0}\right)^2 \frac{1}{\cos^3 \xi} \qquad (3)$$

where ξ is the scattering angle, Z_1, M_1 and Z_2, M_2 are the atomic number and mass of the projectile and the scattering target atom respectively. For very heavy projectiles, $M_1 \gg M_2$ Equation 3 can be approximated by

$$\sigma(\xi) = \left(\frac{e^2}{8\pi\epsilon_0}\right) Z_1^2 \left(\frac{M_1}{E_0}\right)^2 \left(\frac{Z_2}{M_2}\right)^2 \frac{1}{\cos^3 \xi}. \qquad (4)$$

As shown above, for different beams with the same velocity (M_i/E_0=constant) the cross section increases as Z_1^2. The dependence of the electronic stopping power [6] (in silicon) on the atomic number of the projectile (at a velocity given by $\simeq 1$ MeV/u) is illustrated in Figure 1. An approximately linear relationship between the atomic number and the electronic stopping power is observed. Since the electronic stopping power is responsible for the beam induced loss

---

[1] Effective doses of I and Br describe the doses required to produce an approximately equivalent number of recoils to that from Bi.

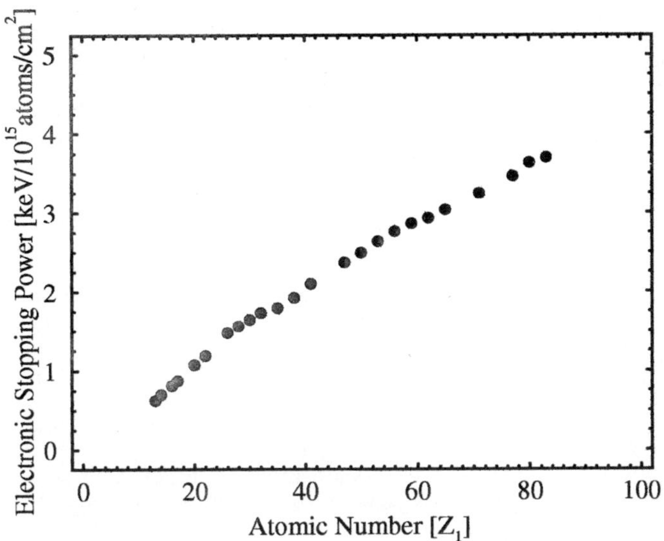

Figure 1: Electronic stopping power dependence on projectile atomic number (Z_1) for projectiles with velocity $\sim$1MeV/amu.

and the recoil scattering cross section determines the probability of a given scattering event, one would expect that heavier ion beams should not produce more damage per recoil, due to the Z_1 dependence of the electronic stopping power and the $Z_1{}^2$ dependence of the scattering cross section.

IV. Results

In this study two silicon nitride samples and one silicon oxynitride sample, deposited by Electron Cyclotron Resonance Plasma Enhanced Chemical Vapour Deposition (ECR-PECVD) [7], were studied with respect to ion beam induced loss. In all samples the release of both hydrogen and nitrogen was observed. The initial loss of these elements can be described by a simple exponential function

$$y_i = A_i \exp(-\sigma_{D_i}\Phi) \tag{5}$$

where y_i is the normalized yield of element i, σ_D is the release or damage cross section and Φ is the ion dose [ions/cm^2]. The damage or release cross sections obtained for these samples as a result of the three beams ranged from about 5×10^{-15} cm^2 to 4×10^{-13} cm^2.

For ERD analysis, the important parameter is not simply the damage cross section, but rather the ratio of the damage cross section to the recoil cross section, which is given by $\sigma_{D/R}$ and is named the normalized damage cross section. The beam with the smallest $\sigma_{D/R}$ will produce the least damage per recoil. The results for the three samples studied are given in Table 2. Sample #1 is a silicon oxynitride and samples #2 and #3 are silicon nitrides. These results indicate that the damage created per recoil is approximately independent of the beam type. This implies a quadratic relationship between the release cross section and the electronic stopping power, which is supported by the findings of Mareé *et al.* [8]. In the lower mass and energy regime (2MeV He - 78 MeV Ag) of their experiments, $\sigma_{D/R} \propto (dE/dx)^{1.8}$ was observed.

V. Conclusion

Our findings support the idea that the increased damage created by high energy heavy ions is offset by the increased recoil cross section. This makes heavy ion ERD an effective

Sample	Element	$\sigma_{D/R}$ [10^9]		
		Bi	I	Br
#1	H	3.3	6	5.9
	N	11	15	8.2
#2	H	2.5	4.8	4.1
	N	2	4	1.4
#3	H	3	3.6	3.7
	N	3.1	5.5	1.1

Table 2: Normalized damage cross sections for a silicon oxynitride and two silicon nitride thin films.

analysis technique even in situations where beam-induced target element loss is observed. The increased probing depth and recoil mass range available with heavy ion ERD makes it an even more attractive analysis technique.

References

[1] J. L'Ecuyer *et al.*, Journal of Applied Physics **47**, 381 (1976).

[2] R. Siegele, J. A. Davies, J. S. Forster, and H. R. Andrews, Journal of Applied Physics **76**, 4524 (1994).

[3] W. Assmann *et al.*, Nuclear Instruments and Methods in Physics Research **B89**, 131 (1994).

[4] L. C. Feldman and J. W. Mayer, *Fundamentals of Surface and Thin Film Analysis* (North Holland, New York, 1986).

[5] M. E. Adel, O. Amir, R. Kalish, and L. C. Feldman, Journal of Applied Physics **66**, 3248 (1989).

[6] J. F. Ziegeler, J. P. Biersack, and U. Littman, in *The Stopping and Ranges of Ions in Matter*, edited by J. F. Ziegeler (Permagon Press, New York, 1985), Vol. 1.

[7] M. Boudreau, M. Boumerzoug, P. Mascher, and P. E. Jessop, Applied Physics Letters **63**, 3014 (1993).

[8] C. H. M. Marée, A. Kleinpenning, A. M. Vredenberg, and F. H. P. M. Habraken, Ion beam analysis of electropolymerized porphyrin layers, Accepted for publication in, *Nuclear Instruments and Methods in Physics B.*

Materials Science Forum Vols. 248-249 (1997) pp. 385-388
© *1997 Trans Tech Publications, Switzerland*

Composition and Structural Evolution of Al-Me Alloys (Me = Fe, Cu, Sb) Prepared by Means of Ultrarapid Quenching from the Melt Studied by RBS Technique

V. Shepelevich and I. Tashlykova-Bushkevich

Department of Solid State Physics, Belorussian State University,
4 Skaryna Ave, 220080 Minsk, Belarus

Keywords: Ultrarapid Quenching, Aluminium Alloys, RBS Measurements, Hardness

Abstract

The hardness of thin ribbons and related composition in rapidly solidified Al-Sb, Al-Cu, Al-Fe alloys have been investigated by Vickers measurements and Rutherford backscattering spectroscopy. Optical microscopy was employed for microstructural analysis of alloys. Rapid solidification of Al-Cu and Al-Fe alloys leads to a substantial increase of the hardness of the Al-Fe foils. In the investigated range of the admixtures the hardness increases from 220 to 412 and 840 MPa in Al-Cu and Al-Fe alloys respectively. The hardness of Al-Sb alloy does not rise strongly and the dependence on dopant concentration is more complicated. A composition study of the alloys showed a many fold enhancement of the admixture concentration in a surface (10-20 nm) region of all the investigated Al-Me foils.

Introduction

Al-Me alloys have been studied extensively because of their importance to the transportation industry in general and to the aerospace industry in particular. Rapid solidification processing has been proved as a means of improving the behaviour of existing Al alloys systems and also developing novel alloy compositions [1]. Rapid solidification techniques have focused attention on finer microstructures and reduction of microsegregations in alloys and achievement of superior fracture related properties. Therefore the solid solution alloys with dispertion hardening additions (aluminium-iron) and the precipitation hardenable wrought compositions (based on the aluminium-copper) alloys are widely investigated [2-5]. However hardening phenomena and microstructure of rapidly solidified Al-Me alloys are investigated in systems with calculated average composition, but insufficient attention has been devoted to study of real concentration of components at least in a near surface region of samples. The present study thus involved the investigation of the composition in thin surface layers of aluminium alloys and the effects of additional metal (Me) elements (Fe,Cu,Sb) on hardening and microstructure in ultrarapid quenched Al-Me alloys.

Experimental

High purity Al and other metals (Fe,Cu,Sb) were melted in a quartz tube in a nitrogen atmosphere. Nominal alloy compositions were in a range from 0.2 to 4.2 at.% and are indicated in Table 1. Specimens from each alloy were rapidly solidified at a cooling rate of about 10^6 K/s via melt spinning. The resultant ribbons were approximately 30-60 mm in thickness and 5-10 mm in width. Some specimens of the resulting meltspun ribbons were then annealed for 30-60 min at temperatures in the range 250-500 C. Thick samples   (1-2 mm) in the form of cylinders (d=10 mm) cut from cust homogenized Al-Me alloys were used for comparison. Compositions of as-solidified and annealed ribbons were examined by means of Rutherford backscattering using He$^+$ ions with an energy of 1.4 and 2.0 MeV.

The relative concentration of admixture in Al-Me alloys was determined using the known [6] expression

$$\frac{N_{Me}^{Al-Me}}{N_{Al}^{Al-Me}} = \frac{H_{Me}^{Al-Me}}{H_{Al}^{Al-Me}} \cdot \frac{Z_2^{Al\,2}}{Z_2^{Me\,2}} \cdot \frac{[\varepsilon]_{Me}^{Al-Me}}{[\varepsilon]_{Al}^{Al-Me}},$$

where the H_X^{Al-Me} - yield from the X component of alloy in RBS spectrum; Z_2^X - atomic number of X element and $[\varepsilon]_X^{Al-Me}$ - factor of the stopping cross-section. Some of the RBS spectra were additionally compared with RUMP code computer simulation [7].

Microhardness measurements were made for all specimens using a 20 g load and 30 second dwell time with a Vickers indenter. For each hardness value at least ten indentations were made.

Results and Discussion

Directly after quenching (Al-Sb) and after short periods of ageing (a few weeks for Al-Cu and Al-Fe) the microhardness of the treated samples was measured. The Vicker hardness data of the Al-Me alloys and related concentration of the dopants are presented in Table 1. The measured hardness is a function of impurity concentration in the Al-Me alloy.

Table 1
Vickers hardness and concentration of dopants in Al-Me alloys

Me	Calculated concentr. of Me, at. %	Microhardness, 10^7 Pa	H_{Al}^{Al-Me}, rel. un.	H_{Me}^{Al-Me}, rel. un.	H_{Me}^{Al-Me}, max,rel. un.	Measured concentr. of Me, at. %
Al	100	22.0	-	-	-	-
Fe	0.25	29.4	92.5	4.0	24.0	0.12
	1.00	30.2	90.0	4.5	16.0	0.14
	2.00	32.6	101.0	5.5	28.0	0.15
	3.00	84.0	-	-	-	-
Cu	1.20	62.4	34.5	0.8	1.5	0.42
	2.10	66.4	33.8	2.9	5.5	1.55
	3.00	72.8	33.5	3.5	6.1	1.89
Sb	0.20	18.9	48.8	0.3	4.2	0.04
	0.80	19.7	49.0	1.0	4.0	0.013
	1.60	19.3	49.0	1.6	8.5	0.21
	2.40	20.9	52.0	2.1	10.5	0.28
	3.20	18.3	48.5	1.2	5.5	0.15

The observed behaviour can be explained in part by the influence of impurities in Al in the movement of dislocations during indenter loading. Decreasing the grain size to 2 μ and less in the alloys investigated also may cause their increased hardness as has been frequently observed in aluminium alloys [1].

An unexpected feature of all of the investigated Al-Me foils was the observation that the measured concentration of the dopants was usually less than that calculated from the initial composition ratio. Additionally a Me enrichment of a thin (10-15 nm) surface region was found. Fig.1 illustrates the more then 5 fold increase of the Fe yield in the maximum compared with the height of the Fe signal just after surface peak of iron in aluminium. This effect differs when various dopants are used and depends on the concentration of the admixture in the aluminium foils. Some of the observed during investigation of Al-Me alloy data are shown in Table

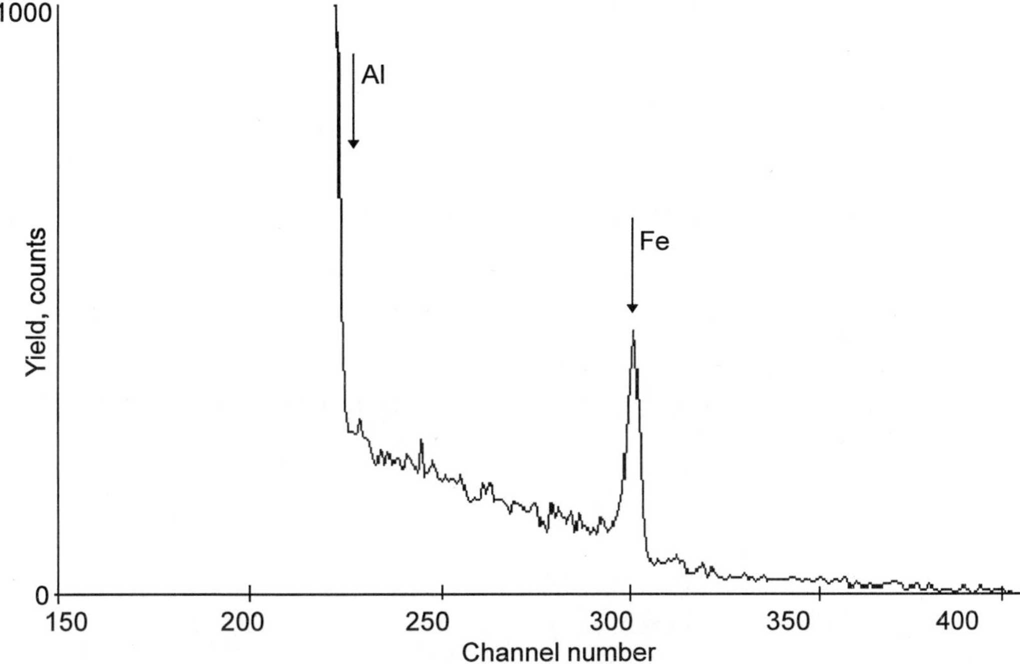

Fig.1 Backscattering energy spectrum of ^{4}He ions with E_0=2 MeV from Al-Fe alloy (2 at.%)

1. We consider that such a behaviour of a depth distribution indicates that iron, copper and antimony atoms diffuse during cooling of foils to the surface in quenched Al-Me alloys. An enlargement of the dopant atom concentration near grain boundaries, the relative amount of which is much bigger in foils than in samples prepared from cast homogenized aluminium alloys, is caused by decreasing size of polycrystals in foils. Nonhomogeneous distribution of components and their phases in Al-Cu rapidly solidified alloys is reported in [4]. The authors of this work consider that oscillatory instabilities in microstructural features may be connected with an effect of growth kinetics. An important role of grain boundaries on precipitation of the dopant included phases is estimated by Prasad et al [8] in their investigation Al-Li-Cu,Mg, Zr alloy. Similar decomposition of precipitates in Al-Fe alloy leads to attainment of the highest strength ever exhibited for an alloys [3]. The small influence of Sb on hardness of Al alloys is not fully understood because of a comparable solubility of Sb and Fe atoms in aluminium [1]. Therefore further investigation of this behaviour should be undertaken.

Conclusions

Composition and structural evolution of Al-Fe, Al-Cu and Al-Sb alloys has been studied. It was found that the concentration of admixtures in thin Al foils is much less than calculated. The second feature of such alloys is connected with enlargement of the admixture atom concentration in a thin surface layer. The last effect may be explained by precipitation of Me atoms and their diffusion in part to the surface of the samples.

Acknowledgements

One of the authors (I.T-B.) is grateful the International Soros Science Education Programm for financial support during the course of this work.

References

[1] E. J. Lavernia, J. D. Ayers and T.S. Srivatsan, Intern. Mater. Reviews **37**, no. 1, 1 (1992).

[2] D. H. Kim and B. Cantor, J. Mater. Science **29**, 2884 (1994).

[3] X. D. Zhang and M. H. Loretto, Philos. Mag. Letters **69**, no. 4, 205 (1994).

[4] S. C. Gill and W. Kurz, Acta Metall. Mater. **44**, no. 12, 3563 (1993).

[5] M. J. Starink and P. Van Mourik, J. Mater. Science **29**, 2835 (1994).

[6] W. K. Chu, J. W. Mayer and M.-A. Nickolet, Backscattering Spectroscopy (Academic Press, N. Y., 1978).

[7] L.N. Doolittle, Nucl. Instr. and Meth. **B9**, 344 (1985).

[8] K. S. Prasad, A. K. Mukhoradhyay, A. A. Gokhale, D. Banerjee and D. B. Goel, Scr. Metal. et Mater. **30**, 1299 (1994).

Materials Science Forum Vols. 248-249 (1997) pp. 389-393
© *1997 Trans Tech Publications, Switzerland*

Diffusion of Water into Quartz and Silica Glass

O. Dersch[1], A. Zouine[1], F. Rauch[1] and E. Arai[2]

[1] Institut für Kernphysik, Johann Wolfgang Goethe-Universität, D-60486 Frankfurt, Germany

[2] Tokyo Institute of Technology, 152 Tokyo-Meguroku, Japan

Keywords: Water Diffusion, Quartz, Silica Glass, Diffusion Rate Constant, Activation Energy, Hydrothermal Treatment, Ion Beam Techniques

Diffusion of water into quartz and silica glass is studied by measuring H, D and ^{18}O concentration profiles in samples treated hydrothermally in normal or isotopically labeled water. The temperature range covered is 125°C to 200°C. The concentration profiles are obtained using the nuclear reactions $^{1}H(^{15}N,\alpha\gamma)^{12}C$, $D(^{3}He,p)^{4}He$ and $^{18}O(p,\alpha)^{15}N$. The measured profiles have widths of up to 500 nm in quartz and 7 µm in silica glass. Diffusion rate constants derived from the profiles are lower for quartz than for silica glass by about 3 orders of magnitude. Based on the data gained for quartz, a tentative description of the diffusion mechanism is proposed.

1. INTRODUCTION

Silica (SiO_2) in its forms of quartz and silica glass is a material of great scientific and technical interest. For example, quartz is an important rock-forming mineral and has many industrial applications. Also for the amorphous forms, silica glass and thermally grown SiO_2 films, there are numerous uses. In a humid or aqueous environment, indiffusion of water can occur, affecting the material properties in the hydrated layer, for example conductivity, refractive index and viscosity. This makes it desirable to study the hydration process in its extent (thickness of affected layer, water concentration) and dependence on external variables, of which temperature is the most important. From a more basic point of view, also the underlying mechanism is of interest.

Most of such investigations have been done above 500 °C, especially in the case of quartz, so that there is a lack of data for low temperatures. There are two studies on silica glass which cover the range 200 °C to 700 °C [1] and 100 °C to 200 °C [2] and have yielded diffusion rate constants and activation energies. The diffusion mechanism is well understood [3]. Hydrogen and oxygen are transported as H_2O molecules over distances large compared to atomic spacings. They can react with the silica network, according to the reaction Si-O-Si + H_2O → 2 SiOH. The reverse reaction is insignificant below 200 °C but becomes important for higher temperatures. The movement of H_2O molecules is supposed to occur via "holes" of about 3 Å diameter [4]. Low temperature studies on quartz are more scarce [5,6,7], perhaps owing to the experimental difficulties which arise from the much lower diffusion rates compared to silica glass. A picture for the diffusion mechanism has not been advanced so far. The very fact that quartz does show hydration is somewhat unexpected since its ordered atomic structure does not provide interstitial space for water molecules.

We have an ongoing program to study water diffusion in quartz and silica glass in the range below 250 °C. One of the aims in the experiments on quartz is to gather information which may be helpful in clarifying the diffusion mechanism; the empirical data and the well-supported diffusion model for silica glass can be used as a reference frame. The present experiments on silica glass are intended to complement previous studies [2]; three types of silica glass are compared which are produced by different methods and, thus, show differences in water content, other impurities and short-range order. For these studies, we use ion beam analysis techniques as a tool. With the ^{15}N technique the indiffusion of water can be followed by measuring H depth profiles. By using $H_2{}^{18}O$ instead of normal water also oxygen incorporation can be observed, employing the reaction $^{18}O(p,\alpha)^{15}N$; this is crucial for deducing the mechanism [2]. Recently, we have also begun with hydrothermal treatment in D_2O. By utilizing the $D(^3He,p)^4He$ reaction, a larger analysis depth range than that of the ^{15}N technique will be provided. Furthermore, isotope effects can be examined and isotope exchange experiments become possible. This is expected to give additional information on the diffusion mechanism(s).

In this paper, the experimental procedures are described and some results are shown which illustrate how the indiffusion of water can be explored by ion beam techniques.

2. EXPERIMENTAL

Two sets of quartz samples were used, from a natural and from a synthetic quartz crystal. The sample surfaces were oriented normal to the c-axis; thus, diffusion parallel to the c-axis was studied. The experiments on silica glass were done with Suprasil, Herasil and Infrasil. Before hydrothermal treatment the samples were etched in 10 % HF to obtain fresh surfaces. They were then heated for predetermined times in closed steel cylinders containing an amount of water. The water used for the ^{18}O-tracer experiments had an enrichment of 20%, the D-enrichment of the deuterated water was 99.75%. After the treatment, the samples to be profiled for ^{18}O and D were stored in the hydration solution in a refrigerator until ion beam analysis to avoid isotope exchange caused by interaction with ambient humidity.

Hydrogen depth profiles were obtained using the ^{15}N technique which is based on the resonant nuclear reaction $^1H(^{15}N,\alpha\gamma)^{12}C$. The depth resolution in SiO_2 is about 7 nm at the surface and 26 nm in a depth of 500 nm. The measurements were performed with a set-up for high-sensitivity analysis of hydrogen [8].

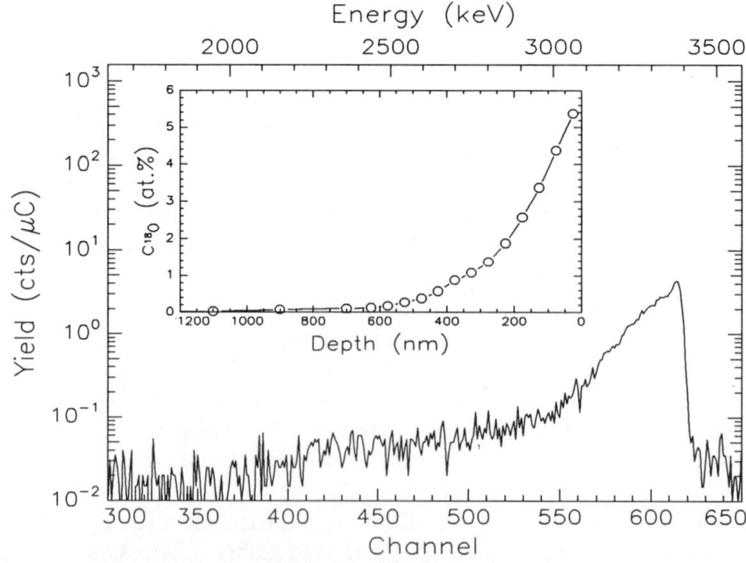

Fig. 1: α spectrum obtained with the reaction $^{18}O(p,\alpha)^{15}N$ on a quartz sample treated at 200 °C for 30 days. The inset shows the ^{18}O depth profile deduced from the spectrum.

The ^{18}O profiles were determined with the reaction $^{18}O(p,\alpha)^{15}N$ at a proton energy of 800 keV. The α particles were detected with a Si detector at an angle of 165°. The beam current was kept low enough that pile-up events from backscattered particles did not disturb the α particle spectrum. The samples were tilted to an angle of 30° between beam and sample surface. This results in a depth resolution of 20 nm at the surface and 35 nm at 500 nm depth. Conversion of the measured energy spectra into concentration profiles was done using a simulation program [9], taking into account the known cross section data. Fig. 1 shows the high-energy part of a spectrum containing the α particle range and a profile deduced from it.

D was profiled by detecting the protons produced in the $D(^3He,p)^4He$ reaction. The protons have typically energies greater than 11 MeV, requiring a thick (2 mm) surface barrier detector; it was placed at 165°. The reaction at backward angles has inverse kinematics, i. e. the lower the energy of the incident ^3He, the higher the energy of the protons. Since the Stopping Power of the protons is rather small, the depth scale is determined mainly by the Stopping Power of the ^3He ions. In order to optimize the analysis depth (7 μm), the samples were not tilted, so that the depth resolution is only about 300 nm. Conversion of the measured data was done with the same simulation program as used for ^{18}O analysis. As example, the spectra of silica glass samples hydrated for different times and the profiles deduced from them are shown in fig. 2.

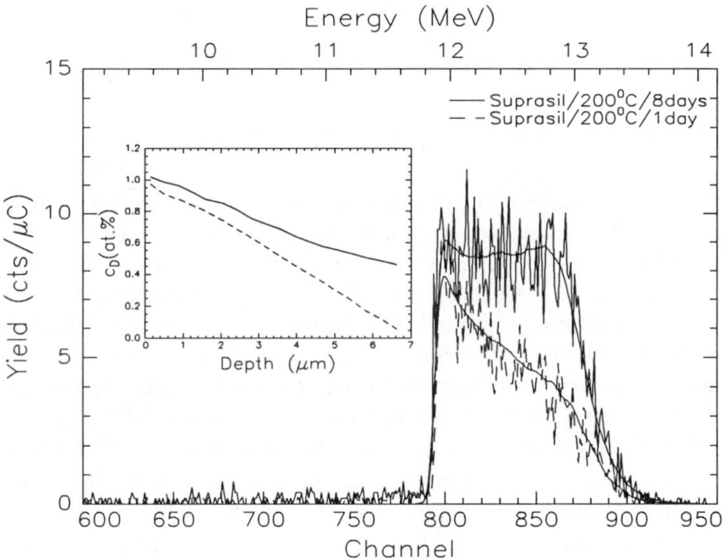

Fig. 2: Two proton spectra together with the corresponding simulations. The spectra were obtained with the reaction $D(^3He,p)^4He$ on two silica glass samples treated at 200 °C for 1 and 8 days. The inset shows the D depth profiles.

3. RESULTS

Experiments with D_2O were done so far only on silica glass. The hydration temperature was 200 °C, with treatment times between 1 and 8 days. The width of the D profiles obtained for Suprasil were found to agree with those of the corresponding H profiles [2]. This means that an isotope effect is zero or small and supports the model that water molecules are the diffusion species. If protons (deuterons) were the diffusing species, an isotope effect given by the square root m_D/m_H would be expected. Furthermore, it was found that Herasil shows a lower D concentration than Suprasil and Infrasil, indicating that this silica glass has a lower concentration of "holes".

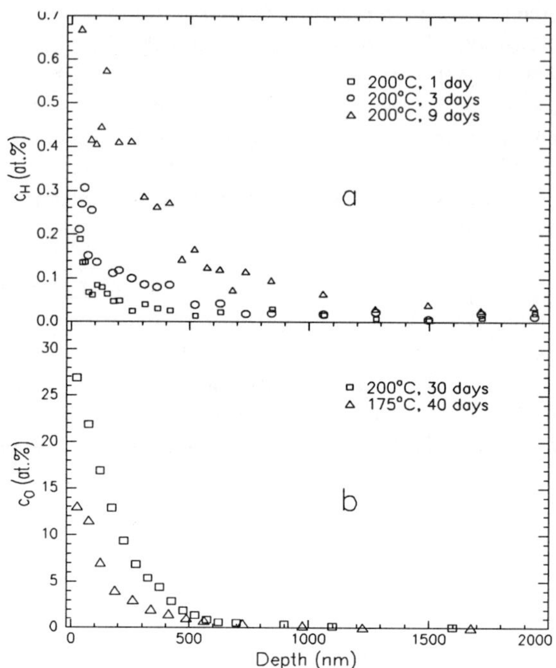

Fig. 3: H profiles (a) and ^{18}O profiles (b) from hydrothermally treated quartz samples.

From the measurements on quartz, three of the H profiles and the two ^{18}O profiles obtained so far are shown in fig. 3. The ^{18}O concentration values have been scaled by a factor 5 compared to that obtained in the hydrothermal experiments, accounting for the fact that the ^{18}O enrichment in the hydration solution was not 100%. As can be seen, H and ^{18}O concentration and the the penetration depth grow with increasing time. A remarkable finding is that the ^{18}O concentrations are much higher than those for H. This behaviour has been found in high-temperature experiments on silica glass where it is interpreted as arising from exchange of oxygen in water molecules with oxygen in the SiO_2 matrix ("network exchange"), involving the recombination of SiOH groups.
The width of the H profiles was found to increase with the square root of the treatment time. This behaviour is characteristic for a diffusion process. Using the depth σ when the concentration has decreased to half the value at depth 0 for parametrization, one can define a diffusion rate constant K, given by $K=\sigma^2/t$. (This rate constant is not identical to the diffusion coefficient for H_2O molecules, although it is closely related [3]).

In fig. 4 the K values for quartz deduced in this way are plotted in an Arrhenius diagram and compared with the K values for silica glass obtained in previous experiments [2] and in the experiments discussed above. The K values have errors between 10% and 30%. It is seen that diffusion is strongly retarded in quartz compared to silica glass. The quartz data points indicate a slight difference between the natural and synthetic material; also the $K(^{18}O)$ values are lower than the corresponding K(H) values. Further experiments are needed to substantiate these differences.

On the whole, the quartz data points have roughly the same temperature dependence as that for silica glass. This means that the activation energies are similar (58 kJ/mol [2]). From this observation a tentative picture for the diffusion process can be deduced. Since the activation energies for quartz and silica glass are similar, the large difference in the diffusion rate coefficient comes from the difference in the preexponential factor which contains the mean free path (jump distance). It can be concluded that it is much smaller in quartz than in silica glass, which means that water molecules reside longer at a given site. This is in accordance with the inference on network exchange discussed above: a long residence time will facilitate reaction with the SiO_2 matrix.

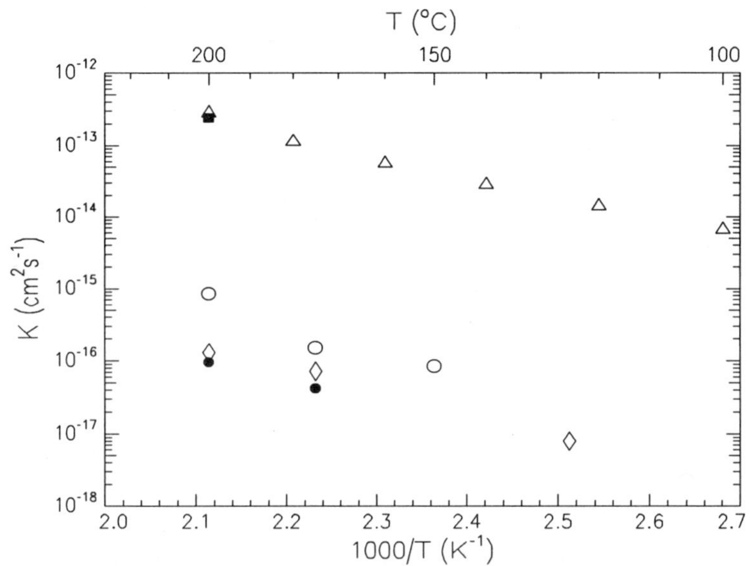

Fig. 4: Arrhenius plot of the H diffusion rate constants in natural quartz (O), in synthetic quartz (◊), and in silica glass (△) and of the ^{18}O diffusion rate constants in natural quartz (●). Also, the K(D) value obtained on silica glass is shown (■).

In conclusion, ion beam techniques have been shown to be a valuable tool for studying indiffusion of water in silica at relatively low temperatures. In particular, their ability to discriminate between isotopes of hydrogen and oxygen is of significance.

ACKNOWLEDGEMENTS

We want to thank K. Röller and J. E. Ericson for supplying quartz samples and for stimulating discussion. We are indebted to A. S. Clough for his support in the experiments with D_2O at Guildford.

REFERENCES

[1]   H. Wakabayashi, M. Tomozawa, J. Am. Ceram. Soc. 12, 1850 (1989)
[2]   M. Helmich, F. Rauch, Glastech. Ber. 66: 195-200 (1993)
[3]   R. H. Doremus, J. Mater. Res. 10: 2379 (1995)
[4]   R. H. Doremus, Glass Science, 2nd ed. (John Wiley, New York, 1994)
[5]   G. J. Clark, C. W. White, D. D. Allred, B. R. Appleton, S. T. Tsong, Phys. Chem. Minerals 3, 199 (1978)
[6]   D. R. Baer, L. R. Pederson, G. L. McVay, J. Vac. Sci. Technol. 12, 738 (1984)
[7]   J. E. Ericson, R. P. Livi, J. Y. Tang, C. R. Shi, T. A. Tombrello, Abstract, Society for Archaeological Sciences, Denver, CO, USA (1985)
[8]   D. Endisch, H. Sturm, F. Rauch NIM B, 84, 380 (1994)
[9]   G. Vizkelethy, NIM B 45 (1990)

Materials Science Forum Vols. 248-249 (1997) pp. 395-398
© *1997 Trans Tech Publications, Switzerland*

Lattice Location of Hf in Near-Stoichiometric LiNbO$_3$: RBS/Channeling and PAC Studies

J.G. Marques[1], A. Kling[1], L. Rebouta[2], M.F. Da Silva[3], A.A. Melo[1], J.C. Soares[1], M.D. Serrano[4], E. Diéguez[4] and F. Agulló-López[4]

[1] Centro de Física Nuclear, Univ. Lisboa, Av. Prof. Gama Pinto 2, P-1699 Lisboa Codex, Portugal

[2] Departamento de Física, Universidade do Minho, P-4719 Braga, Portugal

[3] Instituto Tecnológico e Nuclear, Estrada Nacional 10, P-2685 Sacavém, Portugal

[4] Departamento de Física de Materiales, Universidad Autónoma de Madrid, E-28049 Madrid, Spain

Keywords: Lithium Niobate, Lattice Location

The lattice site of Hf in near-stoichiometric LiNbO$_3$ was studied combining ion beam and hyperfine interaction methods. In contrast with congruent crystals, where Hf fully replaces Li, in near-stoichiometric crystals Hf replaces both Li and Nb, in relative fractions that depend on the amount of HfO$_2$ introduced in the melt. The amount of dopant affects also the defect structure of the crystals.

1. Introduction

Lithium niobate (LiNbO$_3$) is an important material for applications in optoelectronics in which a strong effort has been done to determine the lattice location of dopants [1]. In congruent crystals which are lithium deficient ([Li]/[Nb]=0.945), Hf occupies Li positions, while it moves to Nb sites in crystals co-doped with MgO [2,3]. Comparative investigations in congruent and stoichiometric LiNbO$_3$ became practical with the recent growth of near-stoichiometric ([Li]/[Nb]=0.979) crystals from melts containing K$_2$O [4]. Stoichiometric crystals are attractive for some applications, since they show a strong reduction of the photorefractive effect [5]. However, little is known on the lattice location of dopants in these crystals and on the effect on the structure of the crystal of even small amounts of dopants. In this work, for the first time, the lattice site of Hf was investigated in near-stoichiometric crystals doped with different amounts of Hf. The combination of RBS/channeling (RBS/C) with Perturbed Angular Correlations (PAC) enables the characterization of both the lattice site and the near surroundings of the Hf dopant.

2. Experimental Details

Near-stoichiometric LiNbO$_3$ single crystals were grown in Madrid, adding 4.6 wt% K$_2$O and 0.2 mol% HfO$_2$ (crystal HF02) or 1.0 mol% HfO$_2$ (crystal HF10) to the congruent melt. Plates were cut parallel and perpendicular to the c-axis and polished with 0.3 μm alumina powder. The RBS/C experiments were carried out at the 3.1 MeV Van de Graaff accelerator of ITN using a 1.6 MeV He$^+$ beam. The backscattered particles were detected at an angle near 180° using an annular silicon surface barrier detector with 18 keV resolution.

The crystals were irradiated with thermal neutrons to produce the ^{181}Hf activity necessary for the PAC experiments, through the ^{180}Hf(n,γ)^{181}Hf reaction, and then annealed in air at 873 K for 30 min to remove the damage. The γ–γ PAC measurements were done using a 4-detector spectrometer equipped with conical BaF$_2$ scintillators, with a time resolution of 0.6 ns (FWHM) for the well known 133-482 keV cascade from the decay of ^{181}Hf [6]. Two complementary geometries were used, with (i) the c-axis in the detectors plane, at 45° with two detectors, and (ii) the c-axis perpendicular to that plane. From the twelve recorded coincidence spectra N(θ,t), where θ is the angle between detectors and t is the time delay between events, the time differential anisotropy,

$$R(t)=2[N(180°,t)-N(90°,t)] / [N(180°,t)+2N(90°,t)] \qquad (1)$$

was calculated. Details on the evaluation of experiments in single crystals can be found elsewhere [3,5].

3. Results and Discussion

Figs. 1 a) and b) show the Hf and Nb angular scans for the $<01\bar{1}0>$, $<11\bar{2}0>$ and $<02\bar{2}1>$ axes in the HF10 crystal, and for the $<02\bar{2}1>$ axis in the HF02 crystal, respectively. In the latter case the rapid buildup of beam-induced damage, even for very low currents, prevented the recording of reliable angular scans for other axes. The Hf dips in the HF10 crystal show well defined symmetrical kinks, which are typical for Hf sitting on both Li and Nb sites [2]. The solid lines are Monte-Carlo simulations using a modified version of the FLUX code [7] considering 66(5)% of the Hf ions in Nb sites and the remaining 34(5)% in Li sites. In the case of the HF02 crystal, Monte Carlo simulations using the CASSIS code [8] show that at least 85(5)% of the Hf ions occupy Li sites. The fraction of Hf residing on Nb sites is therefore limited to a maximum of 15%.

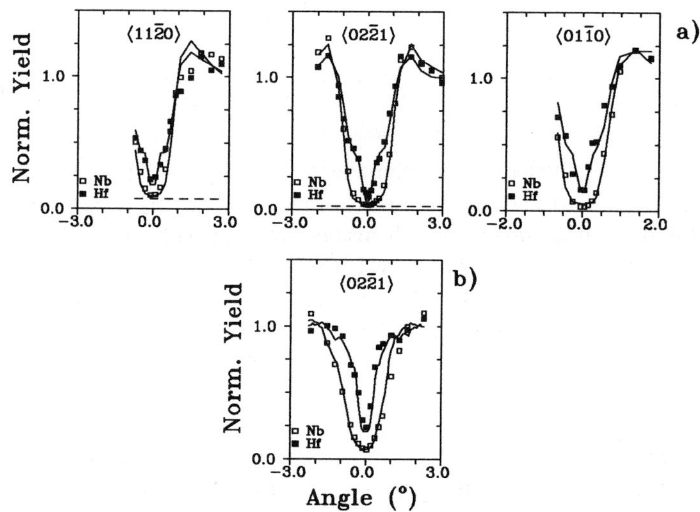

Fig. 1. Experimental and simulated angular scans of the backscattered α particles from Hf and Nb for (a) the $<01\bar{1}0>$, $<11\bar{2}0>$ and $<02\bar{2}1>$ axes in the near-stoichiometric crystal doped with 1.0 mol% HfO₂; (b) the $<02\bar{2}1>$ axis in the near-stoichiometric crystal doped with 0.2 mol% HfO₂.

Figs. 2a) and b) show the PAC spectra and respective Fourier analyses obtained in the HF10 and HF02 crystals, respectively. The c-axis was perpendicular to the detectors plane. Two peaks are visible in the Fourier analyses, at ~350 and ~1200 Mradians/s, corresponding to Hf probes ocupying Nb and Li sites, respectively [3]. From their heights it can be estimated that about 60% of the Hf probes occupy Nb sites in the HF10 crystal, while this fraction is only about 10% in the HF02 crystal, in good agreement with the RBS/C data. However, a detailed analysis shows that the spectra cannot be simulated considering only two fractions of Hf probes, as shown in Table 1.

In the HF02 crystal, the probes occupying Nb sites are well described by one quadrupole interaction frequency, $\nu_Q(Nb)=353(4)$ MHz, but two frequencies have to be considered for the probes occupying Li sites, $\nu_Q(Li^I)=1154(11)$ MHz and $\nu_Q(Li^{II})=1213(11)$ MHz, with non-zero η asymmetry

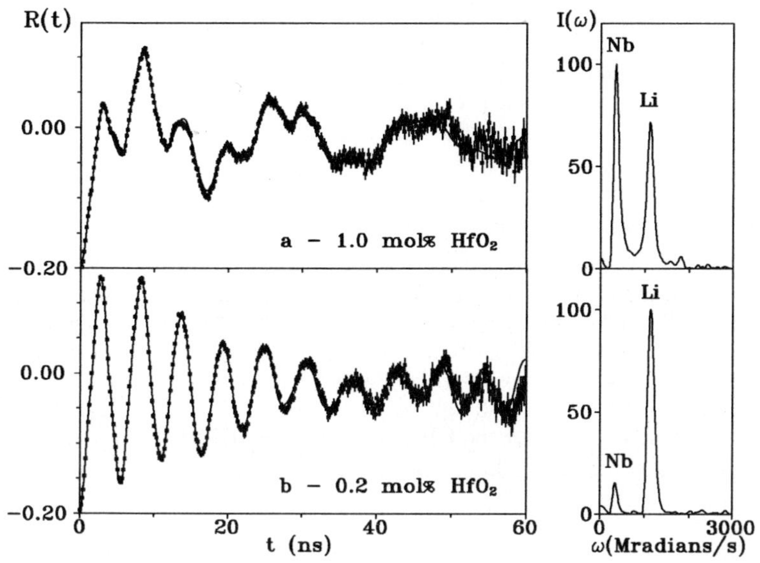

Fig. 2. PAC spectra and respective Fourier analyses for near-stoichiometric LiNbO₃ single-crystals doped with (a) 1.0 mol% HfO₂; (b) 0.2 mol% HfO₂. The c-axis of the single crystals was perpendicular to the detectors plane.

parameters, identical to those obtained in a congruent crystal doped with 1.0 mol% HfO_2 [3]. The frequency $v_Q(Li^I)$ corresponds to undisturbed Li sites in the crystal and $v_Q(Li^{II})$ is related with the defect structure of $LiNbO_3$, by comparison with results obtained with the ^{111}Cd probe, where two frequencies were observed in congruent crystals, $v_Q(Li^I)=192(2)$ MHz and $v_Q(Li^{II})=205(2)$ MHz, also with non-zero η values, while in near-stoichiometric crystals only $v_Q(Li^I)=192$ MHz with $\eta=0$ is observed [4]. For the HF10 crystal, only $v_Q(Li^I)$ is found, but an additional frequency had to be considered for the probes in Nb sites, $v_Q(Nb^{II})=494(10)$ MHz, with a large distribution width, $\delta_L=0.24$.

Table 1. Values of the quadrupole interaction of ^{181}Ta in LiNbO₃ single crystals grown from different melts. The data for the congruent crystal was taken from ref. [3].

sample	site	v_Q (MHz)	η	δ_L	fraction
congruent	Li site I	1154(11)	0.21(1)	0.04(1)	0.68(5)
1.0 mol% HfO₂	Li site II	1213(11)	0.28(1)	0.03(1)	0.32(5)
near-	Li site I	1154(11)	0.15(1)	0.02(1)	0.55(5)
stoichiometric	Li site II	1213(11)	0.23(1)	0.02(1)	0.35(5)
0.2 mol% HfO₂	Nb site I	353(4)	0.15(1)	0.05(1)	0.10(5)
near-	Li site I	1154(11)	0.16(1)	0.04(1)	0.40(5)
stoichiometric	Nb site I	353(4)	0.15(1)	0.05(1)	0.40(5)
1.0 mol% HfO₂	Nb site II	494(10)	0.00	0.24(1)	0.20(5)

In an ideal stoichiometric crystal doped with Hf^{4+} ions, charge compensation would be achieved distributing the Hf ions in a ratio of 1:3 in Li and Nb sites, respectively. This was clearly not observed. In the crystal doped with 0.2 mol% HfO_2, Hf is found mostly in Li sites. Additionally, this crystal presents a defect structure similar to a congruent one. A similar effect has been reported for Fe doping. Sharp EPR lines were obtained in crystals doped with a negligible amount of Fe, but doping at a 500 ppm concentration was found to produce larger EPR linewidths, equivalent to a strong decrease of the Li content [9]. In the case of the crystal doped with 1.0 mol% HfO_2, the distribution of the Hf dopant by Li and Nb sites is closer to what could be expected for charge compensation. The additional frequency observed for probes in Nb sites can be explained by small displacements of the probes from the ideal Nb site. Monte-Carlo simulations show that displacements of the order of 0.1 Å are compatible with the data. Such displacements were previously observed for rare-earth dopants [7]. These results show that the stoichiometry has a direct effect on the lattice location of Hf, but also that the dopant itself changes the stoichiometry of the crystal.

The RBS data show also that the amount of Hf found in the crystal depends on the amount of HfO_2 introduced in the melt. While for the HF10 crystal the actual amount of Hf was found to be 1.7 mol%, in good agreement with the value of the distribution coefficient in congruent crystals [2], in the HF02 crystal it was 0.6 mol%.

4. Conclusions

Ion beam and hyperfine interaction methods were used to characterize the lattice location of Hf in near stoichiometric $LiNbO_3$ crystals. In contrast with congruent crystals, where Hf fully replaces Li, in near-stoichiometric crystals Hf replaces both Li and Nb, in relative fractions that depend on the amount introduced in the melt. For crystals doped with 0.2 mol% HfO_2, 90% of Hf ions are found in Li sites, and 10% in Nb sites, while for crystals doped with 1.0 mol% HfO_2 these fractions are 40% and 60%, respectively. The amount of dopant affects also the defect structure of the crystals.

Acknowledgments

This work was partially funded by JNICT, Portugal, through project CERN/CA/137/94 and a Post-Doctoral Fellowship (J.G.M.).

References

[1] A. Kling, J.C. Soares, M.F. da Silva, in *Insulating Materials for Optoelectronics: New Developments*, edited by F. Agulló-López (World Scientific, Singapore, 1995) p. 175

[2] L. Rebouta, M.F. da Silva, J.C. Soares, M.T. Santos, E. Diéguez, F. Agulló-López, Opt. Mat. **4**, 174 (1995)

[3] J.G. Marques, C.M. Jesus, A.A. Melo, J.C. Soares, E. Diéguez, F. Agulló-López, *Proc. 10th Int. Conf. on Hyperfine Interactions*, edited by M. Rots, A. Vantomme, J. Dekoster, R. Coussement and G. Langouche, Hyp. Int. (C) **1**, 348 (1996)

[4] A. Kling, J.G. Marques, J.G. Correia, M.F. da Silva, E. Diéguez, F. Agulló-López, J.C. Soares, Nucl. Instr. and Meth. B **113**, 293 (1996)

[5] G.I. Malovichko, V.G. Grachev, E.P. Kokanyan, O.F. Schirmer, K. Betzler, B. Gather, F. Jermann, S. Klauer, U. Schlarb, M. Wöhlecke, Appl. Phys. A **56**, 103 (1993)

[6] G. Schatz, A. Weidinger, *Nuclear Condensed Matter Physics* (Wiley, Chichester, 1996) p. 78

[7] L. Rebouta, P.J.M. Smulders, D.O. Boerma, F. Agulló-López, M.F. da Silva, J.C. Soares, Phys. Rev. B **48**, 3600 (1993)

[8] A. Kling, Nucl. Instr. and Meth. B **102**, 141(1995)

[9] G.I. Malovichko, V.G. Grachev, O.F. Schirmer, Sol. State Comm. **89**, 195 (1994)

Materials Science Forum Vols. 248-249 (1997) pp. 399-403
© *1997 Trans Tech Publications, Switzerland*

(1 - 3) MeV / amu Heavy Ion Irradiation Effects on Optical Properties of Al_2O_3

V.A. Skuratov[1], S.M. Abu-Al Azm[2] and V.A. Altynov[1]

[1] Joint Institute for Nuclear Research, Flerov Laboratory of Nuclear Reactions,
141980, Dubna, Russia

[2] On leave from the Solid State Laboratory, Nuclear Research Center
AEA, P.C. 13759, Cairo, Egypt

Keywords: Aluminium Oxide, Ionoluminescence, Cathodoluminescence, Color Centers, Heavy Ions, Ion Tracks, Radiation Damage

Abstract The effects of $(1 \div 3)$ MeV/amu ion irradiation on the luminescence spectra of single crystalline α-Al_2O_3 in visible and near ultraviolet range have been studied. The peculiarities of high-energy ionoluminescence are discussed in comparison with cathodoluminescence of samples, pre-irradiated with high-energy ions and low-energy ionoluminescence.

Introduction Ion irradiation of many solids are accompanied by luminescence generated in conditions of formation and evolution of radiation defects. Therefore, studies of ionoluminescence (IL) and mutual relation between IL and radiation damage processes in insulators are interesting for radiation solid state physics and several practical applications.

Net of experimental results of the IL investigations of α-Al_2O_3 have been reported last few years [1-4] for low-energy ions. High-energy heavy ion irradiation effects on optical properties of sapphire have been considered in works [5, 6]. Some preliminary data related to high-energy IL of sapphire were published in our work [7]. The current paper continues the presentation and discussion of experimental studies of high density excitation effects on luminescence spectral structure from $\alpha-Al_2O_3$ under 1 MeV/amu B, Ne, Ar and 3 MeV/amu Kr ion irradiation.

Experiment Experiments have been carried out with the IC-100 and U-400 cyclotrons beams. The ion beam induced luminescence spectra has been recorded in the wavelength range of 225-550 nm in the temperature range 80-350 K and corrected for system response. The flux density of B, Ne, Ar and Kr ions was about of $10^9 cm^{-2}s^{-1}$. We studied unpolarized luminescence from nominally pure crystals. The impurity analysis for these samples did not show that V, Mn, Cr and Fe were present at levels of a few tens of ppm.

Results and discussion The typical luminescence spectrum measured under 1.2 MeV/amu Ne ion irradiation at 80 K and damage dose $\approx 2.6 \times 10^{-5}$ dpa, decomposed by three Gaussians is given in Fig. 1. As can be seen, spectral structure of IL at low ion fluences (for 1 MeV/amu ions it corresponds to damage dose less than 10^{-4} dpa) besides well-known broad bands peaked around 3.0 and 3.8 eV and associated with F and F^+-centers, consists of band near 4.35 eV , the nature of which is not uniquely determined. Similar band for low-energy ion was observed at 4.0 eV [4].

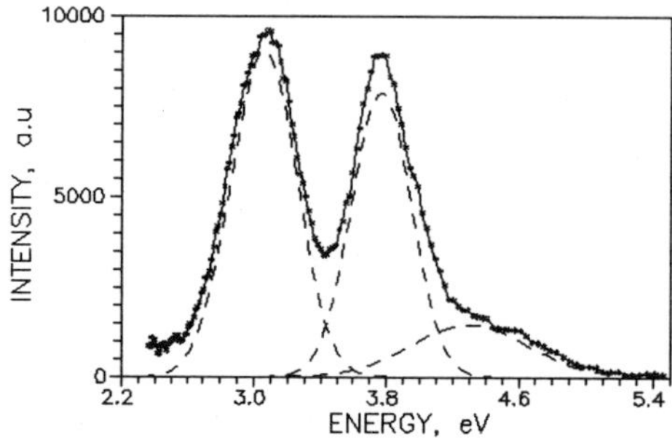

Fig. 1. The typical luminescence spectrum of α-Al$_2$O$_3$ under Ne ion irradiation

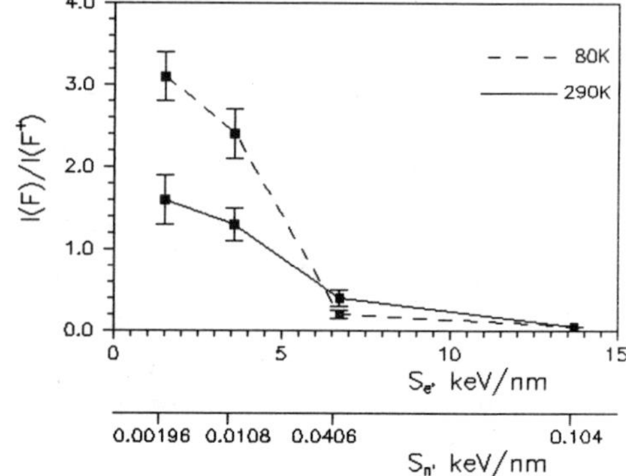

Fig. 2. The I(F)/I(F$^+$)- centers luminescence intensity ratio in dependence on ionizing and nuclear energy losses.

With damage dose increasing the luminescence yield in these ultraviolet bands rapidly decreases, whereas the intensity of the band 3.0 and 3.8 eV increases and reaches saturation with following fast decreasing for low-energy ions.

One of the most interesting futures of ionoluminescence spectra is the supression of F-centers emission in comparison with F$^+$-centers for Ar and more heavier ions, regardless to ion energy. These data collected for high-energy ions are presented in Fig. 2. as dependence of intensities ratio in these bands at the saturation stage versus the excitation density S_e defined as total ion ionization energy losses, normalized per path length unit. The same values for nuclear energy losses S_n are also presented. The data given in this figure are averaged for a three series of samples. Four values of S_e and S_n correspond to the irradiation by the B, Ne, Ar and Kr ions. The similar behavior of the I(F)/I(F$^+$) ratio is observed and for low-energy IL. Therefore, one can state that physical reasons, enhanced F-centers luminescence suppression are similar. It is very important result, especially taking into account, that I(F)/I(F$^+$) in absorption and luminescence spectra from $\alpha-$Al$_2$O$_3$ is a parameter sensitive to the damage production rate, the ratio of ionizing and damaging components of ion irradiation spectrum.

As known, the supression of F-centers emission can be attributed to the concentration quenching, due to a large number of the F-centers, formed in cascade atomic collisions regions. Really, the number of Frenkel pairs, created by high-energy primary knock-on atoms (PKA), grows with the increase of mass and energy of ions and this correlates with the behavior of I(F)/I(F$^+$) dependence presented in Fig. 2. Our calculations, based on work [8], show that part of point defects, formed by PKA with energy more than 5 keV, for example, for 1 MeV/amu ions differ by not more than 10%, however, the difference in the density of energy losses on elastic scattering, leading to formation of displacements, is much higher. So, the number of Frenkel pairs per track length unit (per micron) for B, Ne, Ar and Kr ions equals 45, 180, 480 and 790, respectively. The same value for 200 keV Ar ions is 6520 and this explains why F-centers, as well as F$^+$-centers luminescence yield reduces immediately after beginning of irradiation. Thus, one of the possible reasons of I(F)/I(F$^+$) decrease with the growing S_n may be the increase in the local defect density.

In the analysis of possible effects of high density ionizing energy losses on luminescence spectra of $\alpha-$Al$_2$O$_3$ it is appropriate to start with a question - to what extent does the IL

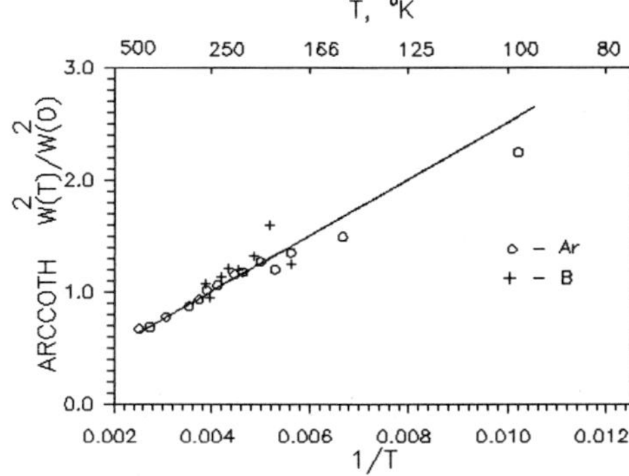

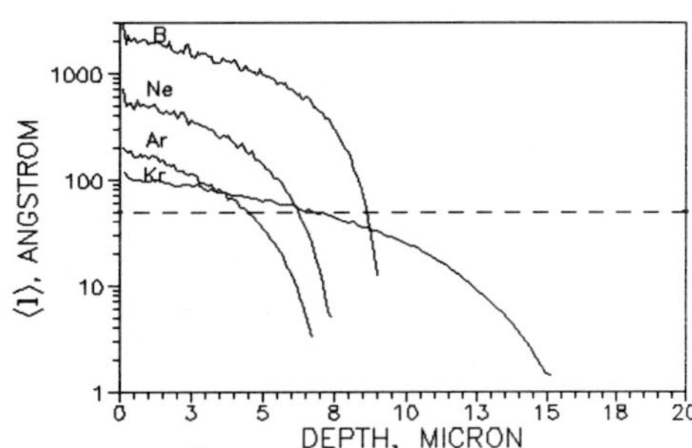

Fig. 3. Temperature dependence on the FWHM of F^+ band for B and Ar ion excitation.

Fig. 4. Mean free length of B, Ne, Ar and Kr ions as function of ion penetration depth, (- -) dashed line is the value of charge carriers thermalization length.

spectra reflect the information on intermediate stages energy dissipation in ion tracks? By our estimates, the maximum duration of the process of equalizing the temperature of the lattice in the track region in sapphire at room temperature does not exceed 10^{-10} s. Taking into account the life-times of excited states of $F^+(\approx 4$ ns) and, especially, of F-centers (30 ms), it is difficult to expect that IL spectral structure will be sensitive to the temperature transient processes in ion tracks, although there are numerous experimental data showing strong dependence of thermal conductivity on radiation damage level. To check it, we compared the FWHM of F^+ bands, as parameter sensitive to the crystal's temperature, for several excitation density values. As known, for a Gaussian-shaped band and linear electron-lattice interaction, the width of the band at some temperature T, called $W(T)$, can be related to the width at absolute zero, $W(0)$, by the equation [9]:

$$W^2(T)/W^2(0) = coth(h\nu/2kT) \qquad (1)$$

where, k - is the Boltzman constant and ν is a weighted average over frequencies of the defect-lattice modes that couple to the excited state of the F^+-center. In the case of their difference one could say about different temperatures in the ion tracks during the time compared, at least, with the life-time of the excited state of F^+-centers. In Fig. 3 the data on $W(T)$ are presented as a function of the reciprocal temperature. A good fit of the $W(T)$ dependence to the experimental data is shown by a straight line in this figure and yields $h\nu$=0.044 eV. It is possible to say that practically no difference is observed for the F^+ band FWHM values under different ion irradiation. So, the luminescence spectral distribution is not sensitive to thermal relaxation processes in ion tracks and is most likely reflects the degree of anion crystal sublattice disorder by the time of measurement. The observed differences in the spectra for high-energy ions is evidently caused by different evolution of defect structure. Spectra dependence on the prehistory of a sample is supported by following examples. The ratio I(F)/I(F^+) obtained for a sample irradiated at 300 K does not change at the subsequent decrease of temperature up to 80 K, i.e. the spectra registered at 80 K are determined by defects formed at 300 K.

Mean free lengths of B, Ne, Ar and Kr ions versus ion penetration depth in the target are presented in Fig. 4. The length of charge carriers thermalization l_{th}=50 Å, is also shown in the figure, which has been found on the condition of equality of potential energy of interaction

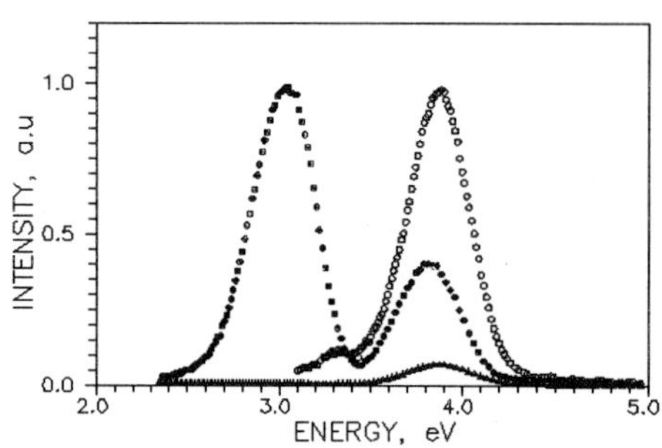

Fig. 5. Cathodoluminescence (10 keV, 10μA, 80K, o–Ne ion irradiated sample, D $\approx 10^{-4}$ dpa, ▷-unirradiated sample) and ionoluminescence spectra from α-Al$_2$O$_3$ (• – Ne ion irradiation, 80K)

between the electron-hole pair and kinetic energy of their thermal motion [10]. As it is seen from the figure, the value l_{th}, which determines the size of the ion track's luminescencing region, exceeds $< l >$, only in the case of Kr ions. Thus, charge carriers recombination occur in the zone of mainly structural defects which previously existed.

Although IL spectra are determined by pre-existing defect structure, some data could not be explained without suggestion about complicated transient processes in ion tracks. As was shown, cathodoluminescence (CL) spectra of α-Al$_2$O$_3$ samples pre-irradiated with B, Ne, Ar and Kr ions are characterized by absense of F band emission, as can be seen, for example, from Fig. 5, where the CL and IL spectra, measured from the same samples are given. High-energy ion excitation of luminescence differs in principle from the excitation by low-energy electrons by a space separation of the charge carriers and, probably, also of the electron and hole centers. Such a state is not evidently an equilibrium one, and the concentrations of positively and negatively charged centers will be become gradually equal. This assumes a relaxation stage like as observed in work [3], where $I(F)/I(F^+)$ ratio reduced at registration after the interruption of the D^+, 25 keV ion irradiation. At the same time, our experiments did not show any changes in spectra after the interruption of irradiation. So, the nature of the difference in IL and CL spectra could not be attributed with the post-irradiation relaxation processes in α-Al$_2$O$_3$ crystals and it is require further detail experiments.

Conclusion The ratio of luminescence intensities of F and F^+-centers depends on the density of ionizing and nuclear energy losses of heavy ions and decreases with increase of total energy losses. The suppression of the F-centers luminescence with increase of the S_e and S_n is associated with higher local damage density. the CL spectra of α-Al$_2$O$_3$ samples previously irradiated with high-energy heavy ions are characterized by presence of F^+ band only.

Acknowledgements This work was accomplished with the partial support of the Russian Foundation of Basic Research under grant No 95-02-4885.

References

[1] A.AlGhamdi and P.D.Townsend, J. Nucl. Instr. and Meth. B46, 133 (1990).

[2] A. AlGhamdi and P.D.Townsend, Rad. Eff. vol.115, 73 (1990).

[3] T.Tanabe, S.Tanaka, K.Yamagushi, N.Otsuki, T.Iida, M.Yamawaki, J. Nucl. Mater. vol.212-215, 1050 (1994).

[4] Y.Aoki, Nguyen T.My, S.Yamamoto, and H.Naramoto, J. Nucl. Instr. and Meth. B114, 276 (1996).

[5] B.Canut et al., J. Nucl. Instr. and Meth. B80/81, 1114 (1993).

[6] B.Canut et al., Phys. Rev. B51, 12194 (1995).

[7] V. A. Skuratov, V. A. Altynov and S. M. Abu AlAzm, J. Nucl. Instr. and Meth. **B107**, 263 (1996).
[8] G.Bardos, Rad. Eff. **105**, 203 (1988).
[9] W. B. Fowler, ed. by Fowler B., Academic, New-York, (1968).
[10] A.S.Portnyagin, I.I.Milman and V.S.Kortov, Sov. Phys. Solid State. **33**, 1272 (1991).

Materials Science Forum Vols. 248-249 (1997) pp. 405-408
© *1997 Trans Tech Publications, Switzerland*

Ion Beam Analysis and Applications in On-Line Monitoring of Ion Induced Modifications of Materials

D.K. Avasthi

Nuclear Science Centre, Aruna Asaf Ali Marg, Post Box 10502, New Delhi - 110 067, India

Keywords: ERDA, RBS, Electronic Energy Loss, Sputtering, Detector Telescope

Abstract: The developments in ion beam analysis at Nuclear Science Centre (NSC) are briefed with a few examples. Possibilities of on-line monitoring of ion induced modifications at interface, in the bulk and sputtering from the surface are discussed.

1. Introduction:

KeV ions are being exploited for materials modification and MeV energy α ions have been in use for the analysis of materials by Rurtherford backscattering (RBS) [1] since early sixties. The technique of RBS has poor sensitivity for light mass elements. To overcome this limitation, another technique known as elastic recoil detection analysis (ERDA) was demonstrated by L'Ecuyer et al [2] in mid seventies. The availability of heavy ion accelerators in seventies proved to be a boon for ERDA. The use of particle identification techniques such as time of flight (TOF) [3], detector telescope[4], Bragg curve spectrometer[5] etc. with ERDA made it possible to carry out the depth profiling of several elements ranging from light mass to middle mass (A=130 or so) region. There have been studies on materials modifications with heavy ions of a few MeV to a few hundreds of MeV. High electronic excitations of target atoms by the incident ions have been responsible for such changes in the materials. The MeV (10 to a few hundreds) energy ions have also been used widely for materials characterization especially by ERDA. GeV ion accelerators too have been in use to provide extremely high degree of electronic excitation of material causing exotic changes in it. Out of these three energy regions (viz. keV, MeV and GeV), the MeV ions have been used for materials modification as well as materials analysis.

The present article reports on some of the ion beam analysis (ERDA) work carried out at NSC[6] and the possibilities of the application of ion beam techniques in monitoring the ion induced modifications at surface, bulk and interface are discussed.

2.0 Elastic recoil detection analysis(ERDA):

Backscattering technique has limitation of poor sensitivity for light mass elements (say for atomic numbers <14 or so). The use of heavy ions can improve the sensitivity due to larger cross section but this also meets a limitation, which is due to the fact that high mass ions cannot scatter back from a lighter mass and hence do not provide any signal. The lighter elements in a sample however can be detected by ERDA technique. In this technique, when ions are incident on a tilted sample, they impart energy to the atoms (in the sample) in the collision as a result of which the atoms recoil in forward direction. The detector may have a stopper foil of an appropriate thickness to stop the unwanted scattered ions and heavier recoils so that only light mass recoils can be detected. The use of stopper foil can be avoided in some specific cases when the incident ion is heavier than the atom of the sample and the detector is kept beyond the maximum scattering angle. The principle of ERDA, the mathematical formulation alongwith the examples are described elsewhere[7]. However, some examples are given here.

2.1   ERDA in transmission geometry: The analysis of thin self supporting films in ERDA is quite simple. Thin self supporting film is kept perpendicular with respect to the incident ion beam and detector is kept behind in transmission geometry. One such example [8] is the analysis of a thin self supporting carbon film. 100 MeV I ions were incident on a thin carbon film and the recoils were detected in a solid state surface barrier detector kept at an angle of 30°. The detection angle was more than the maximum scattering angle so that unwanted elastically scattered ions do not reach the detector. The experiment indicated the capability of analyzing the light elements C,N,O, etc. of adjacent masses in a thin self supporting film.

2.2   ERDA in reflection geometry: This is applicable in the cases when the sample is thick or it is a thin film on thick substrate. Most often the samples to be studied fall in this category. Some of the examples are quoted here.

2.21   Study of H in NEG strip: Non evaporable getter (NEG) is an alloy of Zr, V and Fe which is used in vacuum applications for pumping [9]. A saturated NEG strip and a virgin NEG strip were taken for the study of H intake in it during vacuum pumping. Figure 1 shows the recoil spectra in the two cases. It clearly indicated that H was absorbed by the strip in appreciable amount. Such analytical   studies using ion beam will reveal information on the process of adsorption and diffusion of H.

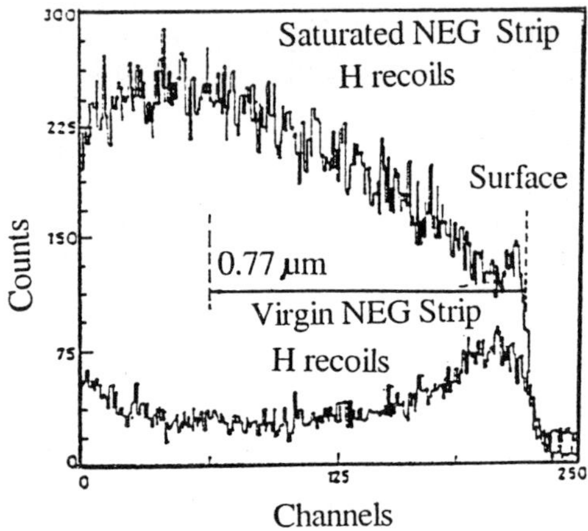

Fig. 1 ERDA spectra of a saturated and a virgin NEG strip.

2.22   Simultaneous detection of light elements: We noticed in the example in the section 2.1 that the simultaneous detection of neighbouring mass light elements is possible by conventional ERDA in case of a thin self supporting foil. In case of a thick film , the recoil energies of the neighbouring elements overlap which results in loss of mass resolution. In the case of a thin film on a substrate, one needs to have a stopper foil to stop unwanted recoils and scattered ions. The use of stopper foil produces straggling resulting in poor mass resolution. The use of a detector telescope is an answer to such cases. The detector telescope consists of two detectors ΔE and E. The details of these are available elsewhere [4] and references therein. Detector telescope with ERD technique has been exploited at NSC for simultaneous detection of H,C,N and O in a DLC film [10] and depth profiling of B and C in B implanted stainless steel [11].

3. Dynamic monitoring during ion irradiation: The ion beam techniques described in previous sections are exploited to monitor the changes in concentration of elements during ion irradiation. It has been noticed that evolution of H takes places in the process of irradiation [12]. The changes in H concentration were observed in the diamond, diamond like carbon (DLC), porous Si, a-Si films etc. in the work at NSC. Therefore, the ion irradiation offers a possibility to alter the hydrogen concentration to desired level in hydrogenous systems. This process, however, will always be along with structural changes and, therefore, requires detailed investigation. Figure 2 shows a typical example   of loss of H   during   ion irradiation   in case of a porous Si layer   [13].

Four spectra are recorded for the same incident ion charge at different fluences. The spectrum at the top is at the lowest fluence.

4. On-line monitoring of transport of H atoms:

H being the lightest element has high diffusivity. The transport of H atoms under heating can be monitored on-line by ERDA. An investigation was carried out to study the transport of H atoms in GaAs. H implanted GaAs and the same treated at 300° C for 15 minutes were studied by ERDA. The recorded spectra are shown in figure 3. The peak at higher channels represent H at the surface and that at lower channels represents the implanted H. The implanted H peak got broadened and reduced in the height in heat treatment which is due to diffusion of H. It clearly indicates that with a set up of on line heating of H implanted samples in the ERD analysis vacuum chamber, one can monitor the diffusive motion of H atom by recording the H recoil spectra at different times. The broadening in H implanted peak will provide a measurement of diffusion co-efficient. Such measurements at different temperature will enable one to determine the activation energy.

5.0 Discussions on possibilities of on-line monitoring of ion induced modifications:

5. 1 On-line monitoring of mixing at interface:

Mixing at interface and formation of new compounds is of wide interest. Thin film deposited on Si can be heated in situ in the ERD analysis vacuum chamber and at the same time ion beam analysis can be performed to investigate the transport of atoms at the interface. There is enormous work in the field of ion beam mixing at the interface induced by elastic collisions of the incident ion with the atoms. KeV energy ions having high value of nuclear stopping cross sections are normally used for it. In the development of recent years, an affective mixing at interface has been reported with GeV ions [14] at low dose. In such cases, the atomic motions are believed to be due to extremely high electronic excitations of target atoms. There have been developments [15] in which it has been shown that the depth resolution of a few nm can be achieved by using the high energy heavy ion beam at glancing angle on the sample along with the use of gaseous detector telescope. It will, therefore, be possible to investigate the electronic excitation induced mixing at the interface.

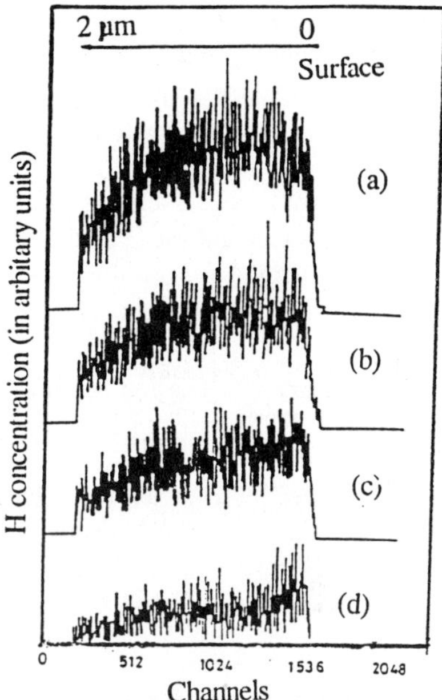

Fig. 2 H recoil spectra indicating loss of H under 100 MeV Ag ion irra-adition.

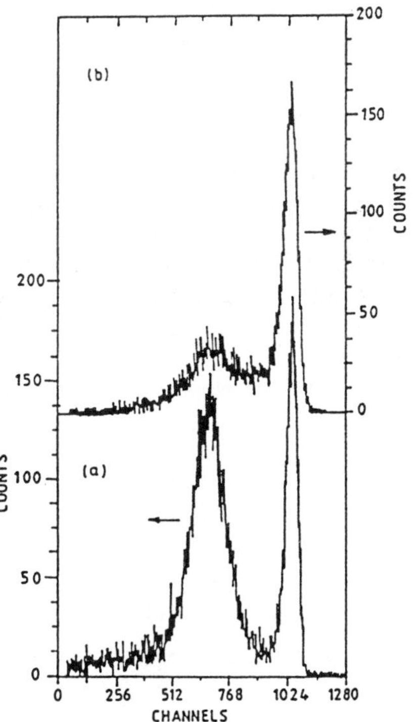

Fig. 3 H depth profiling of (a) H implanted GaAs and (b) the same treated at 300° C for 15 minutes.

5.2 On-line monitoring of sputtering induced by electronic excitation: Sputtering of insulators and semiconducting materials by high energy heavy ions have attracted many researchers. The mechanism of the sputtering process by electronic excitation is yet to reach a complete understanding. It is possible to measure sputtering yield by recording the recoil yield for a fixed dose at different fluences. The decrease in the ratio of recoil yield to the dose if observed at different ion fluences will indicate sputtering. Such a study, however, is possible with the use of heavy ion of high Z (such as Ag, Au ,U, etc.) and the Z of the sputtering material has to be much lower than that of the incident ion to employ ERDA for this purpose. In the case of light Z incident ion and high Z sputtering element, the backscattering technique can be employed to determine the sputtering yield.

6.Conclusion: The ion beam technique (ERDA) utilized at NSC Pelletron is discussed with a few examples. High energy heavy ions and the use of detector telescope make it possible to analyze and depth profile the neighbouring mass light elements such as B,C,N,O, etc. which, in general, is difficult by other techniques. These techniques allow monitoring of the changes in concentration of light gaseous elements like H during ion irradiation. This is likely to provide better understanding of materials modification caused by ion beams. The possibilities of on-line monitoring of the mixing at the interface and sputtering from the surface are discussed.

Acknowledgement: The financial support by Department of Science and Technology to set up the facilities for materials science research with high energy heavy ions is gratefully acknowledged.

References:

[1] W.K.Chu, J.W.Mayer and M.A.Nicolet, Backscattering Spectrometry, Academic Press, New York (1978).

[2] J.L'Ecuyer, C.Brassard, C.Cardinal and B.Terrault, Nucl. Instr. Meth. 149,271(1978).

[3] H.A.Rijken,S.SKlein and M.J.A. de Voigt, Nucl. Instr. and Meth. B64,395(1992).

[4] W.M.Arnold Bik and F.H.P.M.Habraken, Rep. Progr. Phys. 56,859(1993).

[5] R.Behrisch, R. Grotzschel, E.Hentschel and W. Assmann, Nucl. Instr. and Meth. B68,245(1992).

[6] G.K.Mehta and A.P.Patro, Nucl. Instr. Meth. A268,334(1988).

[7] D.K.Avasthi,Bull.ofMats.Sci.19,3(1996)

[8] Jaipal,D.Kabiraj and D.K.Avasthi, Nucl. Instr. Meth. A334,196(1993).

[9] A.Tripathi,L.Senapati and D.K.Avasthi, Vacuum 45,893(1994).

[10] D.K.Avasthi, D.Kabiraj, A.Bhagwat, G.K.Mehta, V.D.Vankar and S.B.Ogale, Nucl. Instr. Meth. B93,480(1994).

[11] D.K.Avasthi,D.Kabiraj,G.K.Mehta,A.Jain, Rev. Sci. Instr. (In press).

[12] D.C.Ingram and A.McCormick, Nucl. Instr. Meth. B34,68(1988).

[13] B.R.Mehta,L.K.Malhotra, R.K.Soni and D.K.Avasthi, Thin Solid Films (In press).

[14] C.Dufour, Ph.Bauer,G.Marchal,J.Grilhe,C.Jaouen,J.Pacaua and J.C.Jousset, Euro Phys. Lett. 21,671(1993).

[15] W.Assmann, P.Hartung, H.Huber, P.Staat, H.Steffens, Ch. Steinhausen, Nucl. Instr. Meth. B85,126(1994).

Materials Science Forum Vols. 248-249 (1997) pp. 409-412
© *1997 Trans Tech Publications, Switzerland*

Thin-Film Morphology and Rutherford Backscattering Spectrometry

Th. Hahn[1], H. Metzner[1], M. Gossla[2] and J. Conrad[1]

[1] II. Physikalisches Institut der Universität Göttingen, Bunsenstr. 7-9, D-37073 Göttingen, Germany

[2] Hahn-Meitner-Institut, Abteilung FD, Glienicker Str. 100, D-14109 Berlin, Germany

Keywords: Rutherford Backscattering, Surface Morphology , Scanning Tunneling Microscopy

In this paper we present a new approach for the analysis of Rutherford backscattering spectra of thin films. The developed formalism allows, under some simplifying assumptions, the *direct* derivation of the height distribution of rough films. The advantages and limitations of this method compared to conventional methods of measuring the surface roughness are discussed.

1. Introduction

It is commonly accepted, that Rutherford backscattering spectrometry (RBS) is a powerful method for the quantitative analysis of thin films, yielding such valuable information as film composition, film thickness, or diffusion profiles. However, the analysis of RBS spectra is regarded to be restricted to laterally homogeneous and smooth samples. In case of rough samples, even simple parameters such as the film thicknesses are difficult to obtain from RBS measurements due to the typical low energy tails in the corresponding spectra. Although such spectra can be described phenomenologically by assuming an interdiffusion between a smooth film and its substrate, the parameters of such a distribution are of low significance, since they are obtained under wrong assumptions. In case of a smooth, interdiffused sample, there is a region of mixed substrate and film material, and the stopping powers in this region are calculated according to Bragg's rule. In case of a rough sample without interdiffusion, however, there is a sharp interface between film and substrate and the stopping power acting on an α particle will either be that of the film material or that of the substrate material.

In this paper, we present a new formalism for the evaluation of RBS spectra of rough thin films, which allows the derivation of the *absolute* (i.e. scaled to the *substrate* surface) height distribution $p(h)$ of the sample. In the first section of the paper, we will give a short description of this formalism. In the second part, we will present its application to two examples, where conventional methods of surface imaging failed to give correct height distributions.

2. Theoretical Background

First, we will give a brief sketch of the analytical description of RBS spectra of smooth films under simplifying assumptions. More elaborated descriptions can be found elsewhere [1,2]. Then, we will expand this formalism to rough films on smooth substrates.

In the following, we assume α particles with an initial energy E_0 (typically, E_0 lies between 1 and 3 MeV), which hit the film in normal incidence and are backscattered at an angle of $\theta = 180°$. For simplicity, the energy loss $g_{in,out} := (dE/dz)$ in the film is assumed to be constant for the inward (g_{in}) as well as for the outward (g_{out}) path. The RBS-Yield Y_{smooth} of a smooth elemental film is then given by

$$Y_{smooth}(E_z) = Y_0 \cdot \left(\frac{E_0}{E_z} \right)^2 . \tag{1}$$

Here Y_0 is a normalization factor containing constants of the utilized experimental setup as well as parameters of the measured sample [1]. E_z denotes the energy of the α particle before scattering at a depth z. Eq.(1) can be rewritten in order to make use of the detected energy of the backscattered α particle, E_1, if one considers that $E_z = E_0 - g_{in} \cdot z$ holds for the inward path and $E_1 = k \cdot E_z - g_{out} \cdot z$ for the outward path, where $k = (M_f - M_\alpha)^2/(M_f + M_\alpha)^2$ is the kinematic factor for the film atoms (mass M_f; M_α denotes the mass of the α particles):

$$Y_{smooth}(u) = Y_0 \cdot \left(\frac{(g_{in}/g_{out}) \cdot k + 1}{(g_{in}/g_{out}) \cdot u + 1} \right)^2 . \tag{2}$$

Here we have introduced a normalized energy $u := E_1/E_0$. Eq. (2) holds as long as u is in the range $(k - b) \leq u \leq k$, where b denotes the maximum possible energy loss for an α particle backscattered from a film atom. It is related to the film thickness h via $b := (h/E_0) \cdot (kg_{in} + g_{out})$.

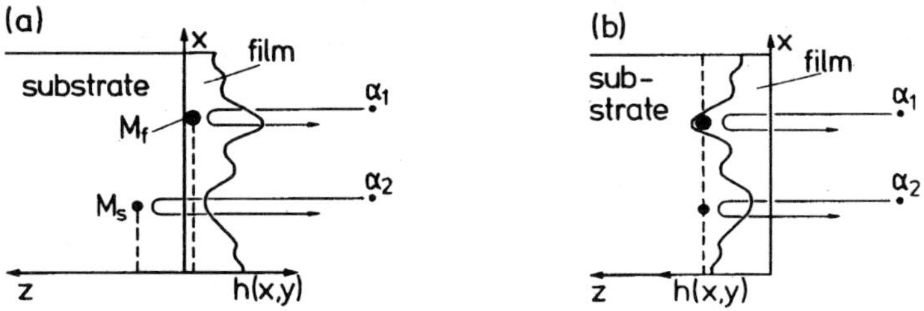

Figure 1: (a) Backscattering of α particles from a rough film. Two backscattering events are shown, one from a film atom (M_f, α_1) and one from a substrate atom (M_s, α_2). Both events take place at the same distance from the film surface, but have different z coordinates. (b) Inversion of the height axis $h(x, y)$ leads to a completely equivalent situation, but now both events occur in the same depth z.

Now we want to consider a rough film on a smooth substrate. Fig. 1(a) depicts the corresponding situation. The assumption of normal incidence and ideal backscattering ($\theta = 180°$) allows us to transform this situation into the equivalent one shown in Fig. 1(b).
From Fig. 1(b) it is easy to see, that the resulting RBS spectrum can no longer simply be described by Eq. (2), since different lateral regions of the film produce different backscattering yields according to their individual heights $h(x, y)$. Thus, the resulting RBS spectrum will be a linear superposition of RBS spectra belonging to different height intervals $h + \Delta h$. In this summation, each of these individual spectra are weighted with the area for which a height in the respective height interval occures. This weighting is done by a suitable integration over the height distribution $p(h)$ of the film. Thus, we obtain the following expression for the RBS Yield of a rough film:

$$Y_{rough}(u) = Y_0 \cdot \left(\frac{(g_{in}/g_{out}) \cdot k + 1}{(g_{in}/g_{out}) \cdot u + 1} \right)^2 \cdot \left(1 - \int_0^z p(h)dh \right) . \tag{3}$$

A first glance at Eq. (3) suggests that the RBS Yield of the film has become not only a function of the detected α energy u, but also of the depth variable z. However, one must remember, that the above-mentioned relations between the α energies E_1, E_z, E_0, and the depth variable z are still valid. These relations can be utilized to relate z to u and one obtains

$$z = \frac{(k - u) \cdot (E_0/g_{out})}{k \cdot (g_{in}/g_{out}) + 1} . \tag{4}$$

According to Eq. (3) and Eq. (2), the height distribution of a rough film is derived from its RBS spectrum by simply calculating the derivative of the spectrum, normalized by that of an ideal, e.g. smooth, film. The thus obtained graph gives a distribution of the occurring thicknesses in the sample. However, this method can not give a microscopic picture of the sample morphology, unlike most topographic methods. On the other hand, these methods usually are able only to detect a height distribution relative to an arbitrary surface level, whereas, in the presented analysis of the RBS spectra, the information of the film thickness is retained.

3. Experimental Examples

Fig. 2(a) shows the RBS spectrum of a thin In film on a Si substrate, taken at an incident α energy of $E_0 = 900$ keV. The spectrum exhibits typical washed out edges for the film as well as for the substrate. However, it is well known, that there is no interdiffusion between Si and In. Therefore, the only plausible explanation for this effect is islanding of the evaporated In film. Measurements of the roughness of the film by conventional methods failed, mainly due to the occurrence of a thin, insulating oxide layer and the softness of the deposited In. Fig. 2(b) shows the height distribution as it was directly derived from the RBS measurement. For this analysis, a thick (500 nm), smooth In film was simulated using the computer code RUMP [5,6]. In the simulation, the detector resolution and Bohr straggling were taken into account. The obtained ideal yield was used to normalize that of the experimental spectrum and the resulting graph was differentiated with respect to u. The depth scale for the thus calculated height distribution was then obtained by relating the energy approximately to the depth via the energy loss. This analysis reveals a film with an average thickness of $< h >= \int_0^\infty h \cdot p(h)dh = 134$ nm and a surface roughness of 50 nm (fwhm of the fitted Gauss distribution).

A second example is given in Fig. 3. It shows STM and RBS measurements of a thin epitaxial film of the ternary semiconductor $CuInS_2$ on a (111)Si substrate. Fig. 3(a) depicts the height distribution for the sample as it was derived from the STM image, which is given as an inset. It reveals two maxima and a roughness of 160 nm (fwhm of the experimental spectrum). Since the thickness of the whole film is only 100 nm as adjusted and controlled via a thickness monitor during preparation, such a large value for the surface roughness, as measured by STM, is quite surprising. Previous work on thin films of this chalcopyrite has already shown, that slightly Cu rich samples tend to develop CuS_x precipitates on their surface [3,4]. These phases are known to have a high conductivity as compared to the ternary semiconductor below. So, in this case, the STM measurement reproduces the electronic conditions on the surface rather than the roughness of the sample. The RBS spectrum of this sample is shown in Fig. 3(b). Here, contrary to the STM measurement, the CuS_x precipitates are not visible. Thus, we must assume that their thickness is well below 5 nm, which is the detection limit of the experimental

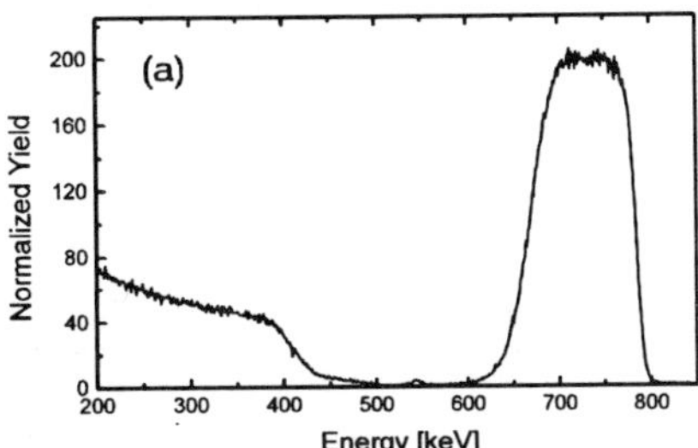

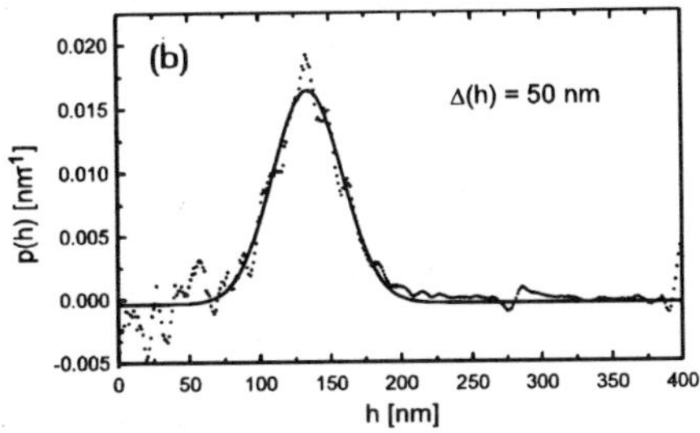

Figure 2: RBS spectrum (a) and derived height distribution (b) of an 134 nm thick, rough In film on Si.

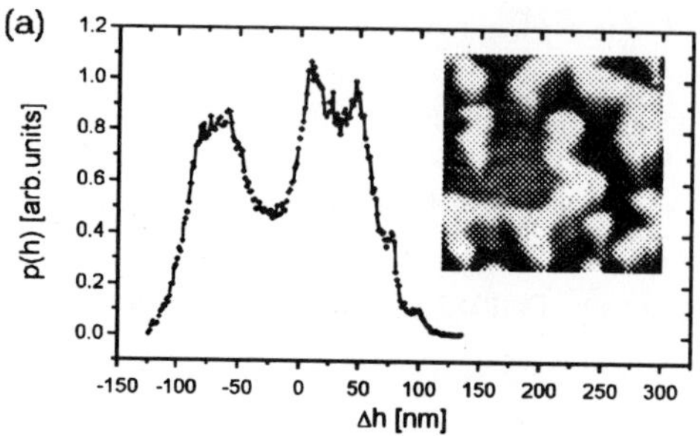

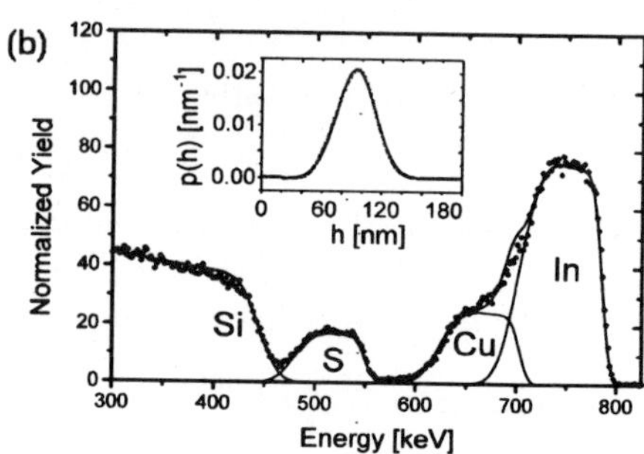

Figure 3: STM (a) and RBS (b) measurements of an epitaxial $CuInS_2$ thin film on a (111)Si substrate with the derived height distributions. The STM image (inset of (a)) covers a lateral scale of 2000×2000 nm^2, while the greyscale covers 200 nm. In the RBS measurement (b) the contribution of each element of the film is depicted.

setup used. However, since the RBS method is insensitive to the electronic inhomogeneities of the sample surface, we were able to obtain the film roughness after a suitable analysis of the RBS measurement. For this purpose, the RBS spectrum of a $CuInS_2$ film with a diffusion profile of the film material into the Si substrate below was fitted, using RUMP [5,6]. The fitted curve was then used to extract the yield for each element separately and these data were subjected to our roughness analysis. The inset in Fig. 3(b) shows the obtained height distribution, which is the same for all three elements, since all of them are described by an identical diffusion profile. Taking into account the detector resolution and straggling, we calculated a roughness of $\Delta h = 36$ nm and a total film thickness of 93 nm for the sample.

4. Conclusion

We have presented a new algorithm which allows a *quantitative* analysis of the roughness of thin films by means of conventional RBS spectra. Using this method, one can derive an height distribution function $p(h)$ of the film, which gives absolute values of the thicknesses occurring in the film. The application of this method in combination with supplementing methods of surface imaging, such as STM, were demonstrated.

5. Acknowledgements

We thank Prof. K. P. Lieb (II. Phys. Inst. d. Univ. Göttingen) for hospitality and support and Dr. W. Bolse (II. Phys. Inst. d. Univ. Göttingen) for making his RBS chamber available. M. Gimbel (I. Phys. Inst. d. Univ. Göttingen) is thanked for the STM images of our samples. We gratefully acknowledge the support of the Deutsche Forschungsgemeinschaft.

References

[1] W.-K. Chu, J. W. Mayer and M.-A. Nicolet, Backscattering Spectrometry (Academic Press, Orlando, 1978).

[2] H. Metzner, Th. Hahn, J.-H. Bremer and J. Conrad, submitted to Nucl. Instr. Meth. B.

[3] M. Gossla, Th. Hahn, H. Metzner, J. Conrad and U. Geyer, Thin Solid Films **268**, 39 (1995).

[4] H. Metzner, Th. Hahn, J.-H. Bremer and J. Conrad, Appl. Phys. Lett., in press.

[5] L. R. Doolittle, Nucl. Instr. Meth. B **9**, 344 (1985).

[6] L. R. Doolittle, Nucl. Instr. Meth. B **15**, 227 (1986).

Materials Science Forum Vols. 248-249 (1997) pp. 413-417
© *1997 Trans Tech Publications, Switzerland*

Electronic Sputtering and Desorption Effects in TOF-SIMS Studies Using Slow Highly Charged Ions like Au^{69+}

T. Schenkel[1,2], M.A. Briere[1,3], H. Schmidt-Böcking[2],
K. Bethge[2] and D. Schneider[1]

[1] Physics and Space Technology Directorate, Lawrence Livermore National Laboratory, Livermore, CA 94550, USA

[2] Institut für Kernphysik der J.W. Goethe University, D-60486 Frankfurt, Germany

[3] Present Address: Physics Department, University of Rhode Island, S. Kingston, RI

Keywords: Time of Flight-SIMS, Graphite, SiO$_2$, Highly Charged Ions, Electronic Sputtering

Secondary ion yields from highly oriented pyrolytic graphite (HOPG) and SiO$_2$ (native oxide on float zone silicon) targets at impact of slow (v $\approx$ 0.3 v$_{Bohr}$) highly charged ions like Th^{70+} have been measured by Time-of-Flight Secondary Ion Mass Spectrometry (TOF-SIMS). A direct comparison of collisional and electronic effects in secondary ion production using a beam of charge state equilibrated 300 keV Xe^{1+} shows a secondary ion yield increase by a factor of ≥ 100.

1. Introduction

Secondary ion mass spectrometry is among the most widely used surface analysis techniques [1]. In standard SIMS, primary ion beams consist of singly charged, keV ions. Secondary ion production results from momentum transfer along a collision cascaded caused by the incident ion in the target material; a process in which useful yield generation is accompanied by extensive sample damage [2]. Collisional sputter yields scale linearly with the nuclear stopping power, S$_n$, and amount typically to ~ 2-10 target atoms per incident ion, while secondary ion yields per incident ion are often less than 10^{-2}. It is the useful yield, the number of secondary ion counts per unit of sample consumption that primarily determines ultimate sensitivity limits of SIMS [3]. In the case of surface analysis by static SIMS, values for sensitivity limits are mostly in the order of 10^9 atoms/cm^2 e. g. for the detection of transition metal contaminations on silicon wafers [1], a problem crucial for the semiconductor industry. In the following we address this problem of secondary particle generation in static SIMS by use of slow, very highly charged primary ions, like Xe^{44+} or Au^{69+}.

Ions traveling in solids develop an equilibrium charge state distribution according to their velocity, atomic mass number and the Fermi velocity of electrons in the target material (v$_{Fermi}$ $\approx$ v$_{Bohr}$ = 2.2 $\times$ 10^6 m/s). Initial charge states of highly charged ions used in this study are much larger than the mean equilibrium charge states that correspond to their low velocities (v $\approx$ 0.3 v$_{Bohr}$, E$_{kin}$ $\leq$ 3 keV/amu). Calling these highly charged ions "slow" sets them apart from "fast" ions of similar high charge states with velocities v $>>$ v$_{Bohr}$, which are produced by charge state equilibration in gaseous or solid targets at relativistic energies (E$_{kin}$ >100 MeV /amu).

Studies of the interaction of slow highly charged ions with surfaces have drawn considerable attention over the last decade [4, 5]. Having reached a critical distance from a metal surface, incoming ions begin to resonantly capture electrons from the target conduction band into highly excited Rydberg states, with binding energies approximately equal to the metal workfunction. Corresponding principal quantum numbers of n $\approx$ 60 can be estimated from the "classical-over-the-barrier-model" [4] for the interaction of Au^{69+} with amorphous carbon. While the ion is approaching the surface, highly excited states decay predominantly by autoionization, giving rise to electron emission into the vacuum [6]. Typical transition times for Auger- and radiative transitions are by far too long for the ion to de-excite completely above the surface. With most of it's electrons populating high n Rydberg states, a highly excited, but neutral "hollow atom" finally hits the surface. Now electrons in Rydberg states with radii in excess of a characteristic surface screening length are peeled off. De-excitation continues below the surface via rapid side-feeding processes of target electrons into energetically favorable ion vacancies, accompanied by a multitude of radiative and Auger transitions. Total neutralization times are in the order of tens of femtoseconds (see below). During this time highly charged ion potential energies of several hundreds of keV are dissipated. The quantitative differentiation of the various energy dissipation channels (target lattice excitations, secondary photons, x-rays, Auger-electrons, secondary ions and neutrals) is currently a matter of intense research [5]. Secondary electron yields amount to ~180 electrons per incident ion for Th^{75+} impact at $v = 6\times10^5$ m/s on gold surfaces and to ~90 electrons/ion from thin SiO_2 films (50 nm on Si) [7]. In a model for electronic sputtering, Parilis et al. have suggested a "Coulomb explosion" mechanism for the description of high secondary particle yields from insulators at highly charged ion impact [8]. The model assumes that electron emission leads to the formation of a charge depleted region on insulator surfaces. Coulomb repulsion between ionized target atoms results in an explosive lattice relaxation before charge neutrality can be reestablished.

2. Experimental

Beams of multiply and highly charged ions were extracted from the electron beam ion trap (EBIT) at Lawrence Livermore National Laboratory [9]. Ions were analyzed according to their mass to charge ratios with a 90° bending magnet. Residual gas pressures in the beam transport system and the target chamber were 10^{-8} torr and 2×10^{-10} torr, respectively. For the neutralization and TOF-SIMS measurements, a highly charged ion flux of ~ 1000 ions/s was used. The beam spot size was 1 mm^2. A schematic representation of the experimental setup is shown in Fig. 1. A position sensitive detector was used for measurements of ion charge states after transmission through a 10 nm thick carbon foil. Electrostatic charge state analysis of transmitted ions was accomplished by a pair of parallel plates. A TOF coincidence between secondary electrons (start) and transmitted ions (stop) was used for energy loss measurements [10]. Each TOF-SIMS-cycle is triggered by secondary particles that are emitted from the target at ion impact. High yields of secondary electrons and protons were used in TOF-SIMS of negative and positive secondary ions. Start efficiencies were 100% for electron- and 10-80% proton starts. Start signals and secondary ion stop signals were both detected by an annular micro channelplate detector. Secondary ion acceleration is accomplished by biasing the target to + or - 3000 V. Typical secondary ion flight times are ~ 0.1 - 5 µs, while the electronic setup of detectors and amplifiers had a time resolution of ~ 2 ns. TOF-SIMS spectra where recorded by a high resolution ($\leq$ 1 ns) one-start-one-stop TAC-ADC system and a multi-stop multichannel scaler (time resolution: 5 ns). For preparation of a 300 keV Xe-ion beam in charge state equilibrium, a 10 nm thick carbon foil was placed into the beam line in front of the annular micro channelplate detector. TOF-SIMS targets consisted of as received highly oriented pyrolytic graphite (HOPG) and float zone silicon (native oxide).

3. Results

Fig. 2 shows charge state distributions of incident Th^{65+} ions after transmission through a 10 nm thick carbon foil. The incident ion velocity was 6.4×10^5 m/s. The potential energy that corresponds to the initial charge state is 113 keV. The thorium ions have developed a mean final charge state of 1.6+ in the foil, a result that agrees well with values for mean equilibrium charge

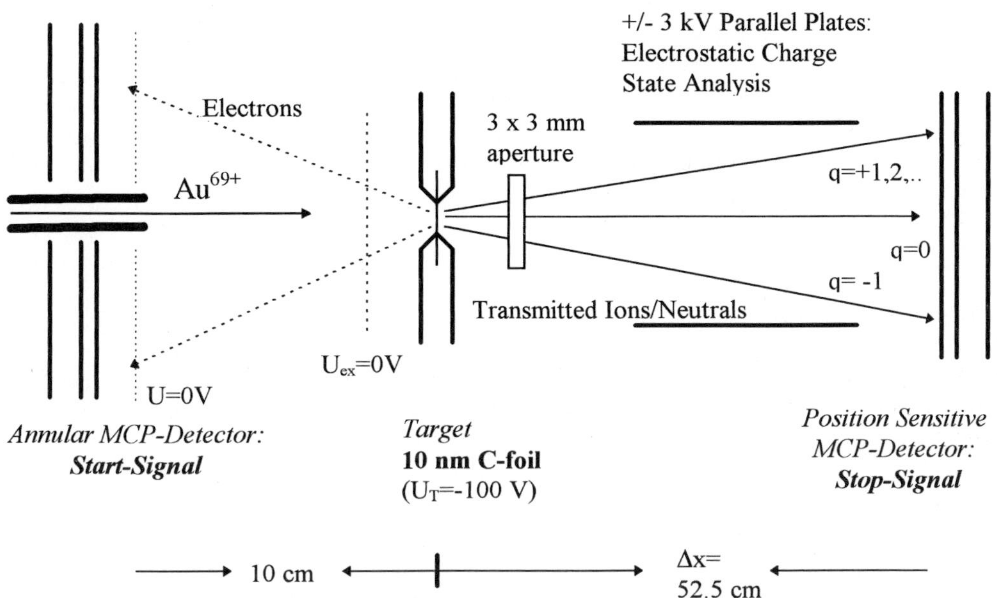

Fig. 1: Schematic of the experimental setup for studies of highly charged ion neutralization in 10 nm thick carbon films, and highly charged ion based TOF-SIMS.

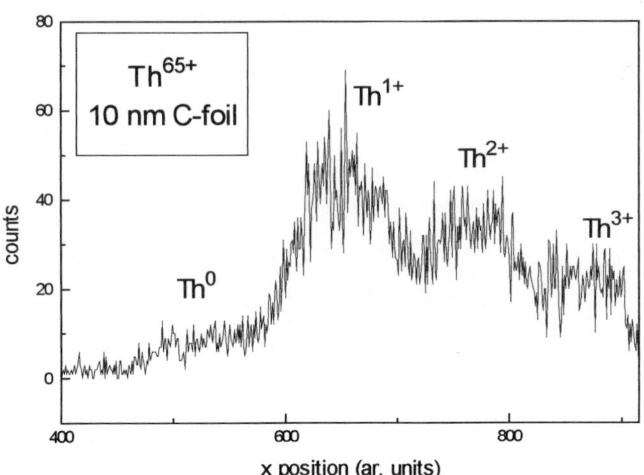

Fig. 2: Charge state distribution of highly charged thorium ions after transmission through a 10 nm thick carbon foil. Data represent projections of two dimensional position distributions on the axis perpendicular to the direction of electrostatic charge state separation.

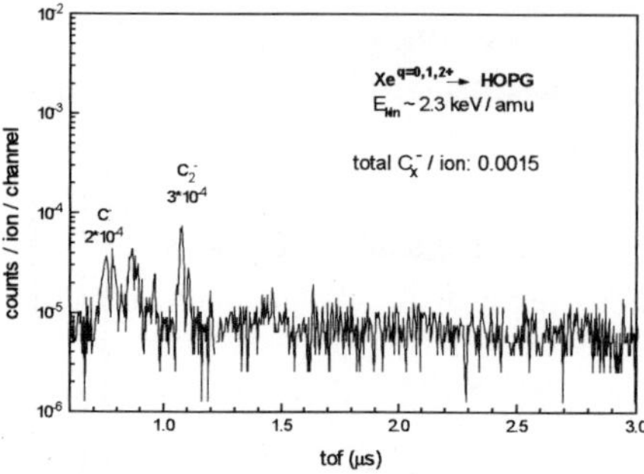
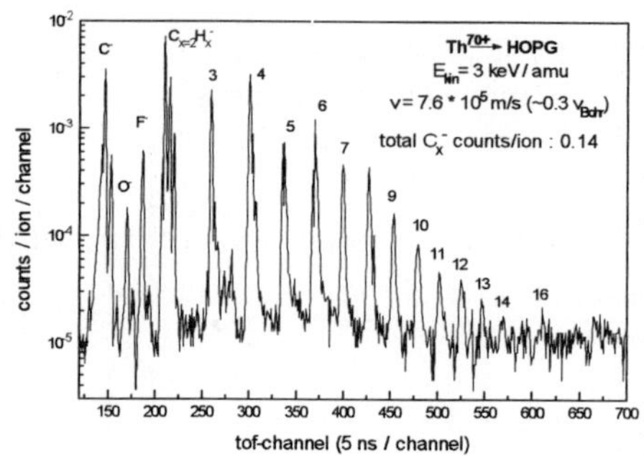

Fig. 3: Negative secondary ion production from graphite (HOPG) at impact of Xe^{1+} (top) and Th^{70+} (bottom) ions at kinetic energies of 2.3 and 3 keV/amu.

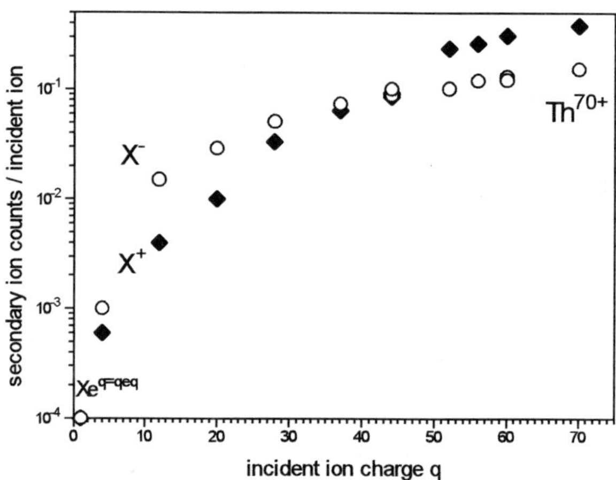

Fig. 4: Positive and negative secondary ion production from a FZ-Si native oxide target as a function of incident ion charge state. Ion energies are 4 (positive) and 10 keV×q (negative).

states of corresponding low charge state ions as determined by beam foil experiments [11]. We conclude that highly charged thorium ions reach charge state equilibrium in the 10 nm thick target, within a time of $\leq$ 17 fs (Th^{65+}).

In order to be able to directly compare contributions to secondary ion yields from collisional and electronic, i. e. ion charge state dependent, processes, we used the 10 nm thick carbon foil as a highly charged ion neutralizer. Incident ions loose approximately 10% of their initial energy in the foil [10]. A beam of ions exiting the foil in charge state equilibrium was used as a reference beam for in situ assessment of collisional contributions to observed secondary ion yields. Fig. 3 shows

TOF-SIMS negative secondary ion spectra from a HOPG target at impact of $Xe^{q=q_{eq.}}$ and Th^{70+} at 2.3 and 3 keV/amu. The energy density in highly charged ion neutralization is much higher than the corresponding near surface nuclear energy loss of singly charged ions at similar kinetic energies ($S_n \leq 5$ keV/nm). The ion charge (ion potential energy) based yield increase amounts to a factor of ~100. Significant is the observation of high yields of carbon clusters up to C_{16}^{-} at highly charged ion impact. Fig. 4 shows the incident ion charge state dependency of positive and negative secondary ion production from a float zone silicon native oxide target. Incident ions were C^{4+}, $Xe^{12, 20, 28, 37, 44+}$, $Au^{44, 52, 56, 60+}$ and $Th^{44, 60, 70+}$. The sum of positive and negative secondary ion counts per incident Th^{70+} amounts to 0.6, an increase of over three order of magnitude as compared to collisional secondary ion production from low charge state, 300 keV Xe ions. The detection efficiency of the setup of ~10% is not included. Corresponding total ablation rates in collisional sputtering (i. e. secondary ions and neutrals) are ~ 5 atoms/ion [12]. The observed yield increase is thought to be a convolution of an increase in total ablation rate and higher ionization probabilities. Measurements of ion charge dependent total sputtering yields are being performed.

4. Conclusion

Secondary ion production rates in highly charged ion solid interactions show substantial increases over corresponding rates in collisional sputtering. The very efficient electronic sputtering and desorption processes can be utilized for the development of highly charged ion based static SIMS. Given the compactness and availability of the basic ion source, EBIT, slow highly charged ions are promising candidates for the development of versatile surface analysis techniques with sensitivity limits $\ll 10^9$ atoms/cm^2.

This work was performed under the auspices of the U. S. Department of Energy by Lawrence Livermore National Laboratory under contract No. W-7405-ENG-48.

References

[1] A. Benninghoven, Angew. Chem. Int. Ed. Engl. 1994, 33, 1023, and references therein
[2] J. P. Biersack, Nucl. Instr. and Meth. **B 27** (1987) 21
[3] J. Schwieters, et al., J. Vac. Sci. Technol. **A 9** (1991) 2864
[4] J. Burgdörfer, et al., Phys. Rev. **A 44**, 5674 (1991); ibid., Aust. J. Phys. **49** (1996) 527
[5] D.H. Schneider, M. A. Briere, Physica Scripta, **53**, 228 (1996), and references therein
[6] F. Aumayr, et al., Phys. Rev. Lett. **71**, 1943 (1993); ibid., Nucl. Instr. Meth. B **90** (1994) 523
[7] T. Schenkel, A. V. Barnes, K. Bethge, H. Schmidt-Böcking, D. Schneider, UCRL-ID-124429, LLNL-EBIT Annual Report 1996; to appear in Nucl. Instr. and Meth. B
[8] Atomic Collisions on Surfaces, E. S. Parilis et al., (Elsevier, North Holland, 1993)
[9] D. H. Schneider et. al., Phys. Rev. A 44, 3119 (1991)
[10] T. Schenkel et al., to be published
[11] K. Shima, T. Mikumo and H.. Tawara, At. Data Nucl. Data Tables 34, 357 (1986)
[12] Y. Yamamura, H. Tawara, At. Data and Nucl. Data Tables **62** (1996) 149

Materials Science Forum Vols. 248-249 (1997) pp. 419-423
© *1997 Trans Tech Publications, Switzerland*

Ion Channeling Study of GaN Single Crystals

A. Turos[1,2], L. Nowicki[1], A. Stonert[1], M. Leszczyński[3],
I. Grzegory[3] and S. Porowski[3]

[1] Soltan Institute for Nuclear Studies, ul. Hoża 69, PL-00-681 Warsaw, Poland

[2] Institute of Electronic Materials Technology, ul. Wólczyńska 133, PL-01-919 Warsaw, Poland

[3] High Pressure Research Center, Polish Academy of Sciences,
ul.Sokołowska 29/37, PL-01-142 Warsaw, Poland

Keywords: Wide-Bandgap Semiconductors, Gallium Nitride, Crystal Growth, Ion Beam Analysis

Abstract

Gallium nitride single crystals grown at 1800K and under the pressure of 1 - 2 GPa were analyzed using the channeling technique with 2 MeV ^{4}He ions. These crystals of the wurzite structure are hexagonal platelets approximately 5 mm long and 0.1 mm thick. The uppermost layer of the top side of platelet consists of gallium atoms whereas the opposite one consists of nitrogen atoms. In order to study channeling properties of GaN single crystals angular scans across the principal axial and planar directions were performed and corresponding aligned spectra were measured. The properties of scans and spectra were correlated with the structure of atomic rows forming a given channel. Surface perfection of as-grown crystals was investigated on both platelet sides. Surface peaks for both sides of the same size were observed. A small disorder in the subsurface region of the N-side was detected that was attributed to the presence of Ga precipitates.

1. Introduction

The demand for increased data storage capabilities for optical recordings is the primary driving force behind the efforts to develop blue and ultraviolet lasers. Two semiconductor compound families are considered as the most promising candidates for constructing an efficient laser diode. Although ZnSe-based alloys have been successfully used for lasers operating in the continuous wave mode their lifetime of about 100 h. is too short for practical applications. It is limited by low energy of defect formation in these compounds. Demonstration of superbright high-efficient blue and green light-emitting diodes based on III-V nitrides [1] has attracted growing attention for this group of wide-gap semiconductors. It is generally accepted that the poor structural quality of III-V nitride epitaxial films is one of the main limitation here. It is principally due to the large lattice mismatch between GaN and commonly used substrate materials. Unfortunately, well matching substrates ($\Delta a/a \cong 0$) for GaN do not exist. Currently used are saphire ($\Delta a/a = 16.0\%$) and SiC ($\Delta a/a = 3.4\%$). Recently, two different approaches to the solution of this problem have been proposed. The first one is based on the use of new substrates like $LiGaO_2$, $LiAlO_2$ or $AlMg_2O_4$ that match better the GaN lattice. However, the use of GaN single crystals as substrates for the growth of homoepitaxial layers represents the most natural way to overcome difficulties caused by a large lattice mismatch and differences in thermal expansion between the substrate and an epitaxial film. Homoepitaxy has become possible owing to the ability to grow GaN single crystals. This has been achieved by means of the high-pressure, high-temperature method.

This paper reports on the preliminary results of the study on as grown GaN single crystals using the ion channeling technique [2].

2. Growth and Properties of GaN Single Crystals

Because of their very high melting temperature of approximately 2800K at a corresponding equilibrium pressure of nitrogen of about 4 GPa GaN single crystals cannot be obtained by conventional Czochralski or Bridgman methods in which stoichiometric melts are used. The Solid Diffusion Synthesis method was chosen where crystal growth has been carried out under conditions where GaN is still stable [3]. At high temperatures only the application of high pressure of nitrogen ensures a maximum solubility of nitrogen in the liquid gallium. Thus, the temperature and pressure of N_2 should be between 1500-1900K and 1-2 GPa, respectively.

Crystals grown by the high-pressure method have the wurtzite structure (hcp) with the [0001] axis perpendicular to the surface. They are hexagonal platelets about 5-10 mm long and 0.1 mm thick. Since the GaN semiconductor is ionic, the plates have a polar character i.e. on the one side the uppermost layer consists of Ga atoms, whereas the opposite one consists of N atoms. In contrast to the Ga side which is very smooth (1 - 4 nm roughness) the N side is more rough (30 - 50 nm). The obtained samples are gallium-rich and contain a high concentration of gallium antisites and interstitials as well as nitrogen vacancies [4].

3.Experimental

Gallium nitride single crystals of [0001] orientation were prepared at High Pressure Research Center Warsaw. The channeling experiments were performed at SINS Warsaw using the LECH Van de Graaff accelerator. Prior to the analysis the samples were not specially prepared. The samples were mounted on a 3-axis goniometer and analyzed with a 2 MeV ^{4}He beam. In order to avoid egde effects due to small dimenssions of the samples the beam spot was reduced to 0.5 mm. Backscattered particles were registered in a surface barrier detector at 165°.

4. Results and Discussion

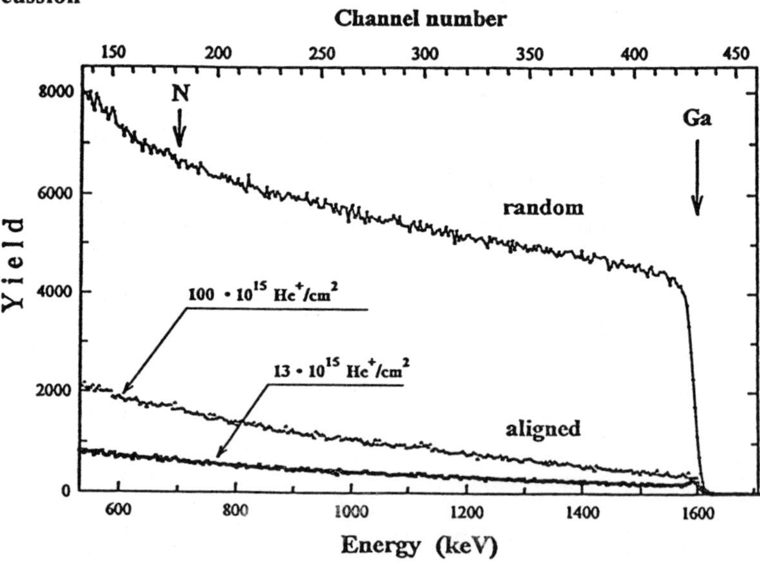

Fig. 1. Random and [0001] aligned spectra for the Ga-side of GaN single crystal.

Fig.1 shows the random and [0001] axial aligned spectra obtained for the Ga-side of a GaN single crystal using the 2 MeV ^{4}He analyzing beam. The good crystal quality resulted in the axial minimum yield χ_{min} = 0.04 as measured directly behind the Ga surface peak. One notes that neither in the random nor in the aligned spectra the N signal can be resolved from the Ga background. This is due to the much smaller scattering cross section of nitrogen being about 20 times lower than that for gallium. As the majority of III-V compound semiconductors GaN is sensitive to the ionisation effects. An increase of the dechanneling level in the aligned spectrum upon 2 MeV ^{4}He-ion bombardment was observed (cf. Fig.1). The minimum yield increases by a factor of 2 as a consequence of the damage accumulation due to the bombardment with 1×10^{17} He at./cm^2. Such a dose is almost 10 times higher than that sufficient to record a spectrum with a reasonable statistics.

Fig.2 shows the detailed structure of the [0001] channel and the structure of atomic rows. The lattice constants are: $a = 0.3189$ nm, $c = 0.5186$ nm and the structure paramenter $u = 0.375$ [5]. First of all it should be noted that this channel is formed by mixed rows. The interatomic distances are the same along rows but every second row is displaced by $c/2$.

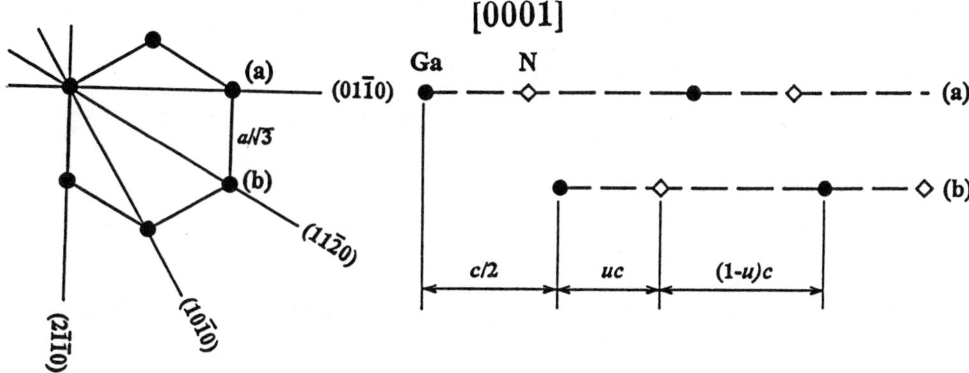

Fig.2. View along the [0001] direction of of the GaN single crystal
and the structure of atomic rows forming this channel.

The structure and location of atomic rows forming axial channels in hcp crystals are too complex to permit the simple analysis of ion channeling applying analytical models commonly used in the case of monoelemental crystals [2]. A detailed information can only be provided by the Monte Carlo simulation of channeled ions trajectories [6]. Nevertheless some quantitative observations can be made for the [0001] channel basing on its structure as shown in Fig. 2.

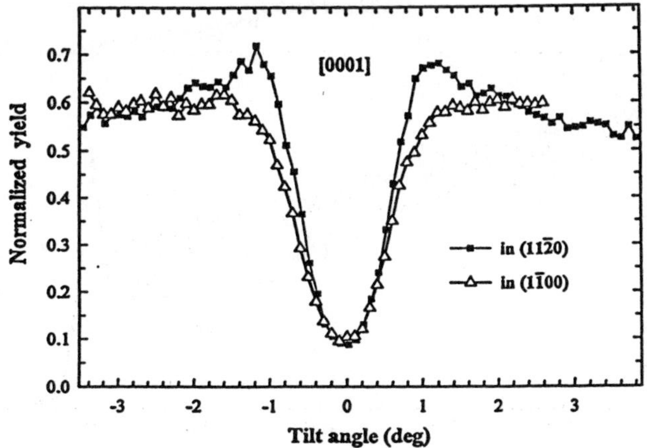

Fig.3. Angular scans across the principal [0001] axis in two different planes

Scans shown in Fig.3 are recorded for different tilt planes i.e. $(11\bar{2}0)$ and $(1\bar{1}00)$. It should be pointed out that the planes $(1\bar{1}00)$, $(01\bar{1}0)$ and $(10\bar{1}0)$. are equivalent. The scan for the $(11\bar{2}0)$ plane is a typical one with sharp edges and well developed shoulders. The shoulders disappear for the $(1\bar{1}00)$ tilt plane, instead a slow increase of the yield is observed in this region. This effect can be attributed to the structure of the planes in question. The $(11\bar{2}0)$ planes are regular with the interplanar distance $a/2$. The other planar channel is limited by two „walls" ($d = a/2$) each of them composed of two atomic planes distant by $a\sqrt{3}/2$.   Ion channeling in such a structure is much more complicated.

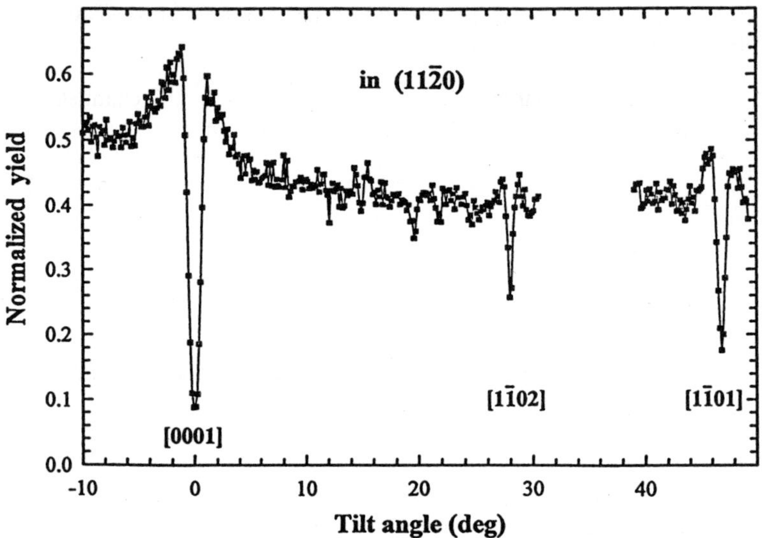

Fig.4. Angular scan along the $(11\bar{2}0)$ plane.

In most cases of lattice location studies an unambiguous determination of impurity atom positions requires scans across different axial and planar directions. Fig.4. shows the scan along the $(11\bar{2}0)$ plane. Besides of the principal [0001] axis two other axial minima can be seen. The first one appears too weak to be useful, however, the [1101] axis with $\chi_{min} = 0.18$ can be applied for lattice location experiments.

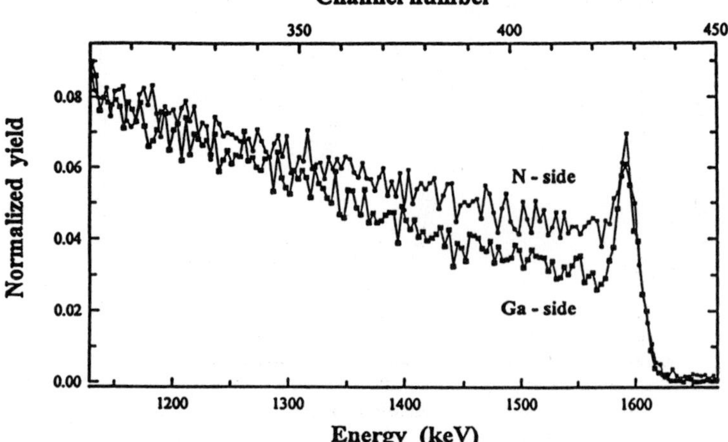

Fig.5. Comparison of axial channeling spectra for Ga - and N - sides of the GaN single crystal.

One of the most important applications of the ion channeling technique is the defect analysis. With this respect two kind of data can be obtained: the analysis of surface peak allows to study the surface perfection and the shape of dechannelig supplies information about the amount of disorder and defect depth distributions. An example of such an approach for GaN single crystals is demonstrated in Fig. 5 where aligned spectra recorded for two sides of the same crystal are shown. Comparing these spectra one can conclude that: (i) the surface peak for the N-side is somewhat higher that may be caused by the greater roughness of this side, (ii) the dechanneling level in the near surface regiion of the N-side exceeds substantially that for the other side. One can speculate that this is due to the presence of Ga precipitates in this region.

4. Conclusions

The presented paper summarized the preliminary results on the channeling properties of GaN single crystals. The minimum yield of 4% found for c-axis oriented crystals demonstrates the high quality of the single-crystalline platelets. It has been demonstrated that the behavior of surface peaks and dechanneling for as grown crystals differs slightly between the nearly perfect Ga-side and the somewhat poorer N-side. In addition, it was observed that during the analysis the damage accumulation in the sample occurs. The shape of the channeling dips mesured across the c-axisis in different planes can be attributed to the strycture of the respective atomic planes. Channeling along higher index directions was found to be much weaker than along the c-axis, but in some cases might be sufficient for lattice location studies. Extended study on the defect formation and transformation mechanism as well as on the lattice location of implanted impurity atoms is in progress.

References

1. S.Nakamura, T.Mukai, and M.Senoh, Appl.Phys.Lett. **64**, 1687 (1994).
2. L.C.Feldman, J.W.Mayer and S.T.Picraux, Materials Analysis by Ion Channeling, Academic Press, New York, 1982
3. I.Grzegory, S.Krukowski, Physica Scripta, **T39**,242 (1991).
4. M.Leszczynski, I.Grzegory, H.Teissere, T.Suski, M.Bockowski, J.Jun, J.M.Baranowski, S. Porowski and J.Domagala, J.Crystal Growth in press (1996).
5. H.Schulz and K.H.Thiemann, Solid State Comm. **23**, 815 (1977).
6. A. Dygo and A.Turos, Phys.Rev. **B40**, 7704 (1989).

New Technical Developments
and
Industrial Applications

Materials Science Forum Vols. 248-249 (1997) pp. 427-432
© *1997 Trans Tech Publications, Switzerland*

Highly Focused Ion Beams in Integrated Circuit Testing

K.M. Horn, P.E. Dodd and B.L. Doyle

Sandia National Laboratories, Albuquerque, NM 87185-1167, USA

Keywords: Single Event Upset, Ion Beam-Induced Charge Collection, Ion Microbeam, Radiation Testing

Abstract

The nuclear microprobe has proven to be a useful tool in radiation testing of integrated circuits. This paper reviews single event upset and ion beam induced charge collection imaging techniques, with special attention to damage-dependent effects. Comparisons of charge collection measurements with three-dimensional charge transport simulations of charge collection are then presented for isolated p-channel field effect transistors under conducting and non-conducting bias conditions.

1. Introduction

Single event upset (SEU) is the disruption of a circuit's normal operation, caused by the passage of an ionizing particle through an integrated circuit (IC). In space-based systems, the sources of ionizing radiation can be galactic cosmic rays, particles emitted by solar activity, or protons trapped in the earth's magnetic fields. Terrestrially, energetic ions arise from naturally occurring, alpha-emitting transuranic impurities in IC manufacturing materials, products of cosmic ray showers and neutron-induced Si-recoils. When such particles lose energy transiting an IC, electron-hole pairs are created. The magnitude of this electrical charge is referred to as the particle's linear energy transfer (LET); it is the electronic portion of the stopping power (dE/dx) measured in units of $MeV/mg/cm^2$. High electric field gradients present in a circuit can enhance separation and collection of this electrical charge before recombination occurs. If the magnitude and duration of the charge collection transient is greater than a critical value for the circuit, upset may occur.

In order to probe circuits with higher resolution than is possible with apertured ion beams ($\cong 2\mu m$), nuclear microprobes, which utilize magnetic and/or electrostatic lenses to focus high-energy ion beams, were applied to SEU testing and charge collection imaging of integrated circuits in 1992 [1,2]. Using this means of controlled, highly localized ion delivery, the upset susceptibility and charge collection characteristics of individual transistor components within a circuit can be directly measured. Additionally, calibrated charge collection spectra from isolated circuit structures can be compared with charge transport simulation results in order to test the predictive capability of simulation codes for future IC designs.

2. Nuclear Microprobe System

The Sandia Nuclear Microprobe is installed on a 6 MV Pelletron-upgraded EN Tandem Van de Graaff accelerator. The microprobe system is described in detail in reference [1]; it routinely produces beam spot sizes of 1.0 micron diameter. The scanning of the incident microbeam, and therefore the impact point of the ions on the circuit, is controlled by the data acquisition computer. Registry is thereby maintained between the position of the incident ion beam on the target and the signals arising from that exposure. The rastered scan of the normally incident beam consists of a 512x512 position grid. Owing to the highly damage-dependent effects on charge collection within ICs, event-by-event recording of the pulses arising from the ion exposures has been implemented. The resulting data file consists of a sequence of (x,y,z,t) entries which record the x and y position of the microbeam at the time the event occurred, the digitized height, z, of the event pulse, and the time, t, at which the event occurred (as derived from the computer's system clock).

The occurrence of SEU is detected by an external computer which is programmed to repetitively exercise the target circuit and issue a +5V pulse to the data acquisition computer if a malfunction is detected.

Measurement of the charge collected from the passage of an energetic ion through the circuit (IBICC) is accomplished by directly recording the charge pulse height resulting from each ion strike. In direct analogy to the configuring of a surface barrier detector, a charge-sensitive pre-amp is connected to the bias pin (V_{dd}) of the IC package, V_{ss} is grounded, and normal operating bias is applied to the IC through the pre-amplifier.

The duration of the ion beam exposure at each point in the scan is controlled by the data acquisition computer. All pixels in the scan are exposed for an equal period of time, and therefore to an equal average incident dose. The practical lower limit for exposure of a single pixel is dictated by the data conversion and storage time needed by the data acquisition system to process a single event; this value is 20 microseconds. Therefore, the average ion arrival rate must be less than 50,000 ions/sec, if single ion resolution is expected. Typically beam currents of only 500 to 2000 ions per second ($\cong 0.1$ femtoamp) are used, providing on average a 500 microsecond period between ion arrivals. To improve display statistics, the data is normally binned to a resolution of 64x64x64 pixels, though the original data file retains the original 512x512x4096 resolution.

The SEU, ion beam-induced charge collection (IBICC), and circuit simulation results presented here are all based on the Sandia-designed TA670 16K SRAM test chip. The TA670 was designed in 1987 as a total dose and SEU rad-hard test chip using 2 μm design rules and fabricated on a 1 μm process line. Since we have access to the complete circuit design masks, vertical layout, doping concentrations, and other circuit parameters, the TA670 has been a valuable test-bed for evaluating microbeam radiation testing techniques.

3. Damage Effects in Single Event Upset (SEU) Imaging

Using a small number of TA670s which had been specially fabricated with reduced feedback resistance in order to increase their otherwise low upset susceptibility, SEU-images were recorded using 33 MeV Cu ions which have an LET of 27 MeV/mg/cm^2. The circuit layout of an individual memory cell being scanned in this measurement is shown in figure 1a. The dimensions of the memory cell are 30μm x 37μm. Alignment of the circuit design mask with the accompanying upset images is done through comparison to subsequently measured charge collection images. For the upset measurements in figures 1b and 1c, the memory cell contains a '1' logic state. In figure 1b, with no accumulated damage, upsets are produced only in the p-drain region; no upsets are observed in the n-drain. After a dose of approximately 500 ions/μm^2 [20Mrad(Si)], the n-drain begins to exhibit upset. After a dose of roughly 1200 ions/μm^2 [52 Mrad(Si)], both the n- and p-drain upset cross-sections saturate, as shown in figure 1c. A full description of this dose-dependence is given in reference [3], where it is proposed that the effect of the accumulated ion-induced damage is to decrease the restoring drive in the "on" transistors, and thereby slow charge dissipation and effectively reduce the LET-threshold of the n-drain to below the ~27 MeV/mg/cm^2 LET of the Cu beam.

When the measurement is repeated on a new, unexposed memory cell containing the opposite logic state, the same dose-dependence is observed, but with upsets occurring in the complementary pair of p- and n-drains. The resulting SEU-image, after an exposure of 1200 ions/micron^2, is shown in figure 1d.

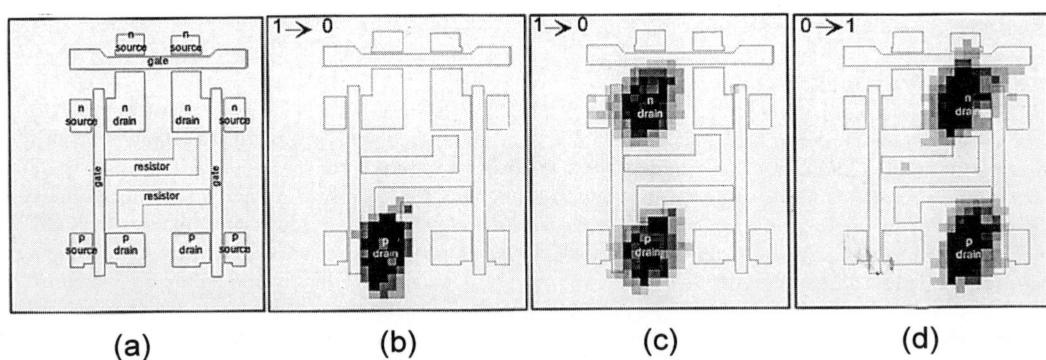

Figure 1. Dose and logic state dependence of SEU-Imaging. (a) Circuit layout of a single TA670 memory cell, scaled and oriented to the accompanying upset images; the cell size is 30 x 37 μm^2. (b) Initial upset image measured with 33 MeV Cu ions. (c) Upset image recorded after a dose of 1200 ions/μm^2. (d) Upset image for the opposite logic state after a dose of 1200 ions/μm ^{2}. Darker pixels reflect higher numbers of upsets.

The fact that such high doses (Mrad) were necessary to change the upset susceptibility of this part is a consequence of this device's initial radiation hardness. In 'softer' commercial devices, which often have upset thresholds of 2 to 4 MeV/mg/cm^2, we have observed changes in device response with dose much more quickly. Thus, damage effects play a dominant role in microbeam-based radiation testing, even at beam intensities of less than a thousand ions/second. Unlike analytical applications of the nuclear microprobe, in which samples are routinely exposed to picoamp beam currents for many minutes during ion beam analysis, both the SEU and IBICC response of a device can degrade after only hundreds of ions per exposure location. This has been a prime motivation for implementing event-by-event recording and optical targeting of the ion beam using a front-viewing microscope.

4. Upset Effects in Ion Beam Induced Charge Collection (IBICC) Imaging

An SEU-image displays only those device structures which cause circuit malfunction under irradiation. Alternatively, the continuum of ion-induced charge collection within the integrated circuit can be measured, thus producing a detailed picture of the circuit layout, as shown in figure 2.

Since the generation of charge within an IC is the underlying physical process which causes SEU, the amount of charge generated and collected at specific circuit nodes is an important point of comparison between experiment and computer simulations of circuit response to ionizing radiation. The complementary nature of the IBICC- and SEU-images also allows the two related phenomena to be examined independently. This is especially important since not all structures exhibiting high charge collection are necessarily sources of upset. Simulation results, in fact, have

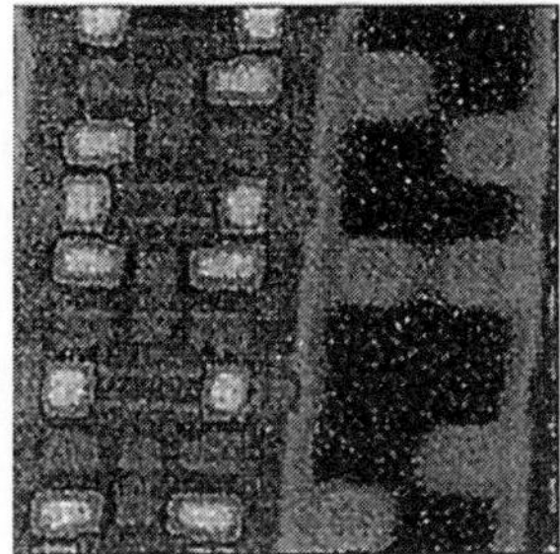

Figure 2. IBICC image of TA670.

indicated that some regions that exhibit lower charge collection can in some circumstances induce upset [4].

The effect of accumulated dose on the pulse height of IBICC signals has previously been examined and been shown to play a significant role in IBICC measurements [3, 5]. We will comment on the effects of dose on the lateral collection of charge in section 6. In this section, the effect of the circuit's normal operation on the resulting IBICC image will be examined. This will require a slightly more detailed description of the circuit's operation and especially the conduction paths existing in the circuit for a given logic state.

The six transistor memory cell shown in figure 3a can be thought of as consisting of two cross-coupled inverters. Each inverter is formed by the serial connection of one n-channel and one p-channel field effect transistor. The overall logic state held in each inverter can be accessed through a transistor (either #5 or #6) which controls access to the "bit" or "not-bit" outputs of the memory cell. One inverter in figure 3a is composed of transistors #1 and #2; these two transistors share a common gate as well as an electrically conducting path between their drains. The second inverter is similarly formed out of transistors #3 and #4. Owing to the complementary nature of the memory cell's operation, when n-channel transistor #1 is "on", p-channel transistor #2 is "off", and vice versa. Additionally, when n-channel transistor #1 is "on", n-channel

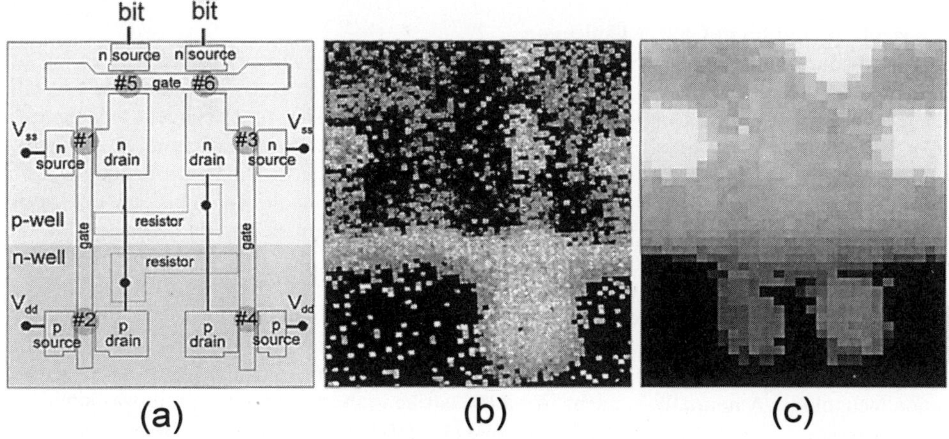

Figure 3. Upset effects in IBICC Images. (a) Circuit layout of TA670 memory cell. (b) IBICC-image of SRAM cell using 2.5 MeV He, (c) IBICC-image of SRAM cell using 12 MeV C while softened TA670 is biased at 2.5V. Greater charge collection is indicated by greater brightness in the IBICC images

transistor #3 is "off". Thus, for this condition, (i.e. #1-on, #2-off, #3-off, and #4-on), charge collected from the p-drain of transistor #4 and n-drain of transistor #3 can be measured with a pre-amp connected to the IC bias line, V_{dd}; charge collected into the n-off and p-on drains of transistors #1 and #2 cannot be measured since they have no conducting path to V_{dd} in this logic state. Conversely, when the memory cell contains the opposite logic state, the state of each transistor is likewise reversed, (i.e. #1-off, #2-on, #3-on, and #4-off) and charge collected into the n-off and p-on drains of transistors #1 and #2 can be measured, but not the n-on and p-off drains of transistors #3 and #4. Thus, as corresponding pairs of transistors turn on or off, two separate charge collection paths become connected to the V_{dd} bias line from which the IBICC measurement is being made. Since not all regions of the memory cell can, by definition, be measured and imaged simultaneously for a given logic state, any IBICC image of a static device will display only those circuit structures with a conducting path to the external charge-sensitive pre-amp. This is the condition displayed in figure 3b, in which a 2.5 MeV He ion beam has been used to image a single memory cell in the TA670. If, however, a higher LET beam is used, such that the memory cell is caused to upset by the beam's exposure, BOTH available conduction paths to the pre-amp will alternately become available as the memory cell successively 'flips' from one logic state to another. Thus in figure 3c, an IBICC image measured with a 12 MeV carbon beam, with 2.5V biasing of the device, depicts both p-drains in the memory cell, whereas in the 2.5 MeV He IBICC image of figure 3b, only one p-drain can be seen.

Charge collection measurements made from a transistor imbedded in surrounding circuitry can be quite different from the results obtained when the same transistor is irradiated in isolation. Differences arise from the introduction of external loading, capacitive coupling, and additional charge conduction paths or biasing conditions enforced by neighboring circuit structures. Without explicit control or measurement of this chorus of external influences, the conditions of the measurement can differ significantly from those assumed in the simulation. Therefore, as a first step toward validation of the simulation code for the entire circuit, we have begun by comparing the charge collection predicted by simulation, and measured by experiment, for an isolated, p-channel field effect transistor (FET) fabricated on the same die as the TA670 SRAM. A cross-sectional view of this p-channel FET is shown in figure 4, along with biasing conditions for the conducting ('on') and non-conducting ('off') states.

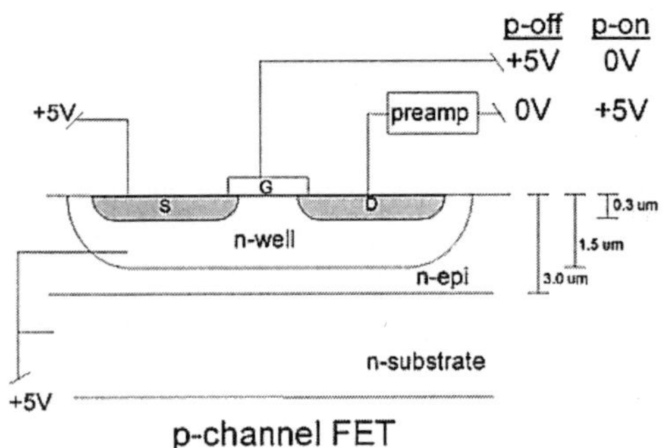

Figure 4. Biasing and vertical structure of the isolated p-channel FET.

5. Numerical Simulation of Charge Collection

The charge-collection characteristics of Si junction diodes and transistors in response to an ion strike are often computed using semiconductor device simulation codes [6]. Such codes solve Poisson's equation and the current continuity equations in one, two, or three dimensions, using finite element techniques. These equations describe the physics of carrier transport under potential and carrier concentration gradients, giving detailed microscopic information such as transient electron and hole densities throughout the simulated device. In this work, we have used DAVINCI, a commercially-available three-dimensional device simulator [7]. The simulations reported here typically require 6-8 CPU hours to execute on a Hewlett-Packard J210 workstation.

Inputs required for the simulations include structural information and doping profiles for the TA670. For the p-channel transistors simulated here, we have modeled only the drain junction diode. Simulations of the entire transistor produce identical results, because the charge collection is dominated by the drain junction. Doping profiles were taken from process simulations and spreading resistance measurements of the TA670 fabrication technology. A normally-incident 14 MeV C strike to the center of the drain was simulated using the charge generation profile (LET vs. depth) computed by TRIM [8], and a uniform track radius of 0.1 μm.

Following electron-hole-pair generation in the device, 3D drift and diffusion of the free carriers is computed by DAVINCI. The resulting terminal currents, (lower curve in figure 5), are integrated to give the simulated circuit node's charge collection characteristics, (upper curve). The transition from drift-dominated charge transport to diffusion-dominated transport is visible in the charge collection curve as the region of inflection occurring approximately 0.5 nanosecond after the ion strike.

6. Comparison of 3-D Simulations and IBICC Experiment

Calibrated IBICC measurements were performed on a series of isolated p-channel FETs. Since degradation of IBICC pulses can occur after exposures of only hundreds of ions/μm^2, the target IC was replaced with an unexposed part upon exceeding ion exposures of 200 ions/μm^2. Furthermore, the data files were examined after the measurement for evidence of any signal degradation during the irradiation. The laboratory IBICC measurements were performed using a 14 MeV carbon beam, focused to 1 μm diameter as determined from STIM (scanned transmission ion microscopy) images of a 1000 mesh grid. The size of the beam scan was calibrated using the pitch of the 1000

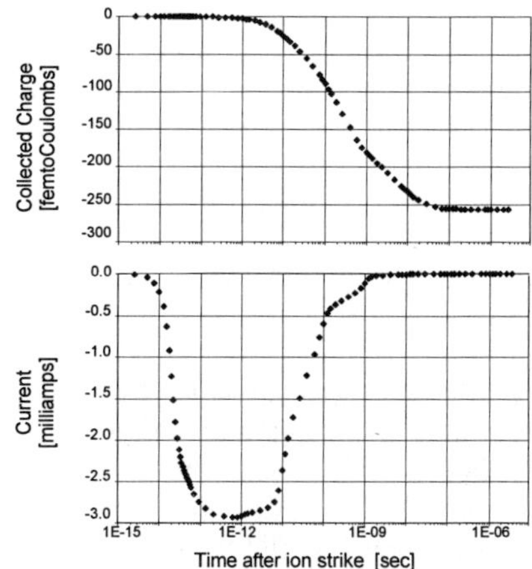

Figure 5. The DAVINCI simulation calculates the collected charge during the first microsecond after a 14 MeV C ion strikes the 'off' p-channel FET drain.

mesh grid in the STIM image; the IBICC images shown in figure 6 are 40 x 40 μm.

The dwell time used in the calibrated IBICC measurements was 50 microseconds per exposed pixel. The original 512 x 512 resolution of the scanned beam has been binned to produce a 64 x 64 pixel image; thus each image pixel has a cumulative dwell time of 3.2 milliseconds, (64 exposure pixels * 50 microseconds/exposure pixel). The ion arrival rate was measured in a PIN diode four times during the IBICC measurements, yielding an average of 775 ±144 ions/sec. This results in an average ion exposure per image pixel per scan of 2.5 ±0.5 ions.

The range of 14 MeV carbon in silicon is 14 μm, with an LET of 4.2 MeV/mg/cm^2 at the surface, a maximum LET of 5.5 MeV/mg/cm^2 at a depth of 10 μm, and dropping below 4.2 MeV/mg/cm^2 at a depth of 11.5 μm. The vertical structure of the FET is shown in figure 4; the distance from the top of the encapsulation layer (not shown) to the underlying silicon substrate is 7.3 μm. Thus the 14 MeV carbon beam has an LET in excess of 4.2 MeV/mg/cm^2 throughout the entire active device and 4 μm into the underlying substrate. The data collection electronics were calibrated by measuring charge collection spectra of this ion in a previously unexposed PIN diode, biased to form a depletion depth of 30 μm (twice the calculated range of the beam). The full charge collection peak of the resulting spectrum, shown in figure 6, was equated to the calculated maximum charge production of this ion in silicon, 622 femtoCoulombs

In order to avoid damage effects such as discussed in section 3 and 4, the p-drain structures were optically targeted using a front-viewing microscope and only then exposed to the scanned 14 MeV carbon beam. In this way, the effects of any initial beam-induced damage on the IBICC pulse height could be detected if present. From examination of the time-evolution of the IBICC pulse heights measured from different regions of the scanned image, it was seen that the collection of diffusive charge from outside of the p-drain was quickly suppressed, (presumably by the introduction of defect sites acting as recombination centers), but that for doses up to 200 ions/image pixel, charge collection pulses from the drain itself did not show any measurable degradation. The images and charge collection spectra shown in figure 6 were generated using events collected in the first 20 scans of the beam across the test structure, or an equivalent ion exposure of 50 ions/image pixel. The charge collection from the p-drain is therefore taken to be unaffected by damage for the ion exposures used, and therefore representative of the p-drain charge collection under 14 MeV C irradiation. The boundaries of the p-drain, as specified in the design mask, are outlined in white dashed lines in the IBICC-images inset in figure 6.

The charge collection spectra recorded from the outlined region of the p-on and p-off drains are shown in figure 6. Charge collection measurements from the p-on drain display a peak at 245 femtoCoulombs (fC), while the DAVINCI simulation predicts a charge collection magnitude of 242 fC. Charge collection from the p-off drain is measured to be 262 fC, while the simulation again predicts a slightly smaller value of 257 fC. Thus, the magnitudes of the charge collection determined from simulation and experiment for both biasing conditions are found to agree to within 2%. Additionally, the roughly 15 fC difference in charge collection from the p-on and p-off drains is clearly resolved experimentally and predicted from simulation. In both biasing

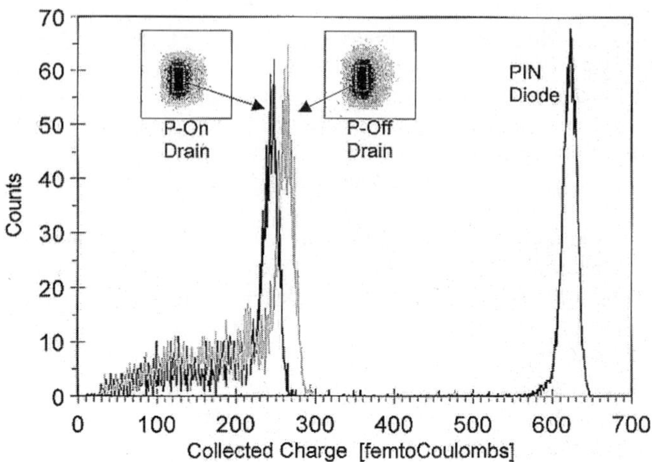

Figure 6. Charge collection spectra and IBICC images (inset) of isolated p-on and p-off drains for TA670 test FETs. The full charge generation of the 12 MeV carbon ions is indicated by the PIN diode measurement.

conditions, the simulation predicts charge collection which is slightly lower than the experimentally measured result. Though this is quite good agreement, we note that the simulation results can be extremely sensitive to inaccuracies in the device parameters supplied to it. In performing these simulations, device design parameters such as the dopant concentrations and thicknesses have, where possible, been subjected to independent verification using techniques such as spreading resistance measurements.

7. Conclusions

One of the most beneficial applications of microbeam-based radiation testing may lie in the verification of charge transport results from three-dimensional device simulations. However, unlike many analytical applications of ion microbeams, in which sub-picoamp beam currents cause little degradation of most samples, great care must be taken to measure and control ion dose when performing charge collection measurements from circuit structures that are susceptible to ion dose damage effects. This has been shown with explicit examples from both SEU- and IBICC-imaging measurements. Calibrated charge collection measurements from p-drains of isolated field effect transistors have been compared to DAVINCI device simulations. With measurements made under very controlled dose conditions, and inspected for the presence of any damage effects, microbeam-based calibrated charge collection measurements are found to agree with computer modeling to within 2%. The relative differences in charge collection from the p-off and p-on drains also conform to simulation predictions. Validation of device simulation codes is a necessary first step in establishing the capacity of such codes to predict the radiation tolerance of future circuit designs.

References

[1] K.M. Horn, B. L. Doyle and F.W. Sexton, IEEE Trans. Nucl. Sci., **39**, no. 1, 7 (1992).

[2] M.B.H. Breese, P.J.C. King, G.W. Grime and F. Watt, J. Appl. Phys., **72**, no. 6, 2097 (1992).

[3] F.W. Sexton, K.M. Horn, B.L. Doyle, M.R. Shaneyfelt and T.L. Meisenheimer, IEEE Trans. Nucl. Sci., **42**, no. 6, 1940 (1995).

[4] P.E. Dodd and F.W. Sexton, IEEE Trans. Nucl. Sci., **42**, no. 6, 1764 (1995).

[5] M.B.H. Breese, J. Appl. Phys., **74**, no. 6, 3789 (1993).

[6] P. E. Dodd, IEEE Trans. Nucl. Sci., **43**, no. 2, 561 (1996).

[7] DAVINCI 3.1 (Technology Modeling Associates, Inc. 1995).

[8] J.F. Ziegler, J. P. Biersack, and U. Littmark, The Stopping and Range of Ions in Solids, New York: Pergamon Press, 1985.

Materials Science Forum Vols. 248-249 (1997) pp. 433-438
© *1997 Trans Tech Publications, Switzerland*

Atomic Structure and Electrical Properties of a Supertip Gas Field-Ion Source

Th. Miller, A. Knoblauch and S. Kalbitzer

Max-Planck-Institut für Kernphysik, Postfach 103980, D-69029 Heidelberg, Germany

Keywords: Field-Ion Microscopy, Supertip, Field Ionization

Abstract: By using field-ion microscopy the atomic structure of tungsten supertips was revealed. They are small protrusions of only a few nm of height and width. Surface and volume contain some 10 to some 100 atoms respectively, arranged in bcc orientation. The polyhedral shape leads to local emission patterns with overall emission angles of less than 1°. By performing electrical field calculations the virtual source size was quantified to be of atomic size corresponding to a brightness in the range of 10^{10}-10^{12} A/cm^2 sr. The geometry of the supertip appears not to be influenced by formation temperature or basetip geometry. Modification of supertips by field evaporation or other postformation treatment will be described.

1 Introduction

Focused ion beams have already become a powerful tool for sub-μ applications. Due to their multiple ways of application they are irreplaceable, e.g., in integrated circuit fabrication or photo mask repair [1]. In addition to surface modification, focused ion beams can also be used for surface analysis. With focused ion beams secondary ion mass spectroscopy (SIMS), proton induced x-ray emission (PIXE) or scanning ion microscopy (SIM) can be performed with very high spatial resolution. For some applications SIM appears to be more suitable than scanning electron microscopy (SEM) because of better contrast in the secondary particle signal.

Focused ion beam systems contain three main parts: ion source, optical column and target chamber. For some time ion micro-beam systems equipped with a liquid metal ion source (LMIS) [2] are used for several steps of industrial integrated circuit fabrication. Because of their limitation to metallic ions, LMIS are not so well suited for repair of chips or photomasks because of the concomitant change of electrical or optical properties by the implanted metal ions. Gas field ion sources (GFIS) [3] use hydrogen and rare gases which produce much less disturbing effects in the processed target material.

A GFIS with a supertip [4] produces target-current densities up to 2 orders of magnitude higher than a LMIS. For better understanding of the supertip formation process and the resulting emission features it is at first necessary to know its atomic stucture.

2 The supertip GFIS

Regular field emitter tips produced by an electrochemical etch process exhibit curvatures of 100 - 500 nm. At an applied electrical field of more than 10 V/nm gas field ioniziation occurs at a distance of less than 1 nm above the emitter surface. Gas atoms are ionized by electron tunneling into the emitter conduction band. Fig.1 a) shows the emission angle of a regular tip. A supertip, a small protrusion on a <111>-oriented tungsten base tip, emits into much narrower solid angles (see Fig.1 b) . Hence beam intensity losses caused by the small acceptance

Figure 1: a: A regular basetip with a typical emission angle of about 30°. b: A supertip on a basetip with a typical emission angle of about 1°.

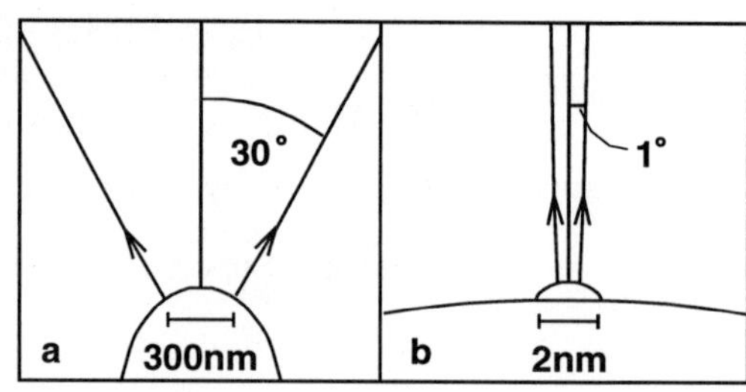

angles of ion optical columns of highly focussing systems are considerably reduced. Owing to its smaller radius of curvature the electrical field at the supertip is correspondingly higher than at the basetip. A typical field enhancement factor amounts to $\sim$1.5. Thus, by appropriate choice of the applied voltage ion emission can be confined to the supertip area . By taking over nearly the total emission current of the basetip supertips reach an approximately 1000 times higher angular emission density than a regular emitter. Due to these special features a supertip GFIS is also predestinated for fine-focus applications where a certain current intensity is a prerequisite. Tab.1 compares emission data of a supertip and of a LMIS [5] .

3   Experimental setup

In order to image supertip structures a field-ion microscope (FIM) after Müller [6] was built up as shown in Fig.2. At the top the ions are formed near above the supertip surface. In emission mode the tip is ideally cooled down close to the condensation point of the respective gas. For field-ion imaging a gas pressure of 10^{-5} - 10^{-4} mbar is maintained. Depending on supertip geometry voltages in the range of 10 - 40 kV are applied to generate field ion emission. The ions, accelerated through an aperture, drift towards the multichannel plates (MCP). The total emission current can be measured with a retractable screen. Because of the high amplification of the Chevron-type MCP the ion image of the supertip is directly visible on the screen and photos can be taken without any special equipment. With a drift length of 234 mm between tip and MCP the magnification of this assembly ranges from about 10^6 - 10^7 depending on the tip shape. With He as an imaging gas a resolution of about 0.2 nm is achieved.

4   Supertip structure

Two different W supertips with atomic resolution are shown in Fig.2. Each visible bright point corresponds to a surface atom. The emitting supertip volume in Fig.2 (middle) amounts to

Table 1: Emission data GFIS and LMIS systems.

source type		GFIS			LMIS
ion species		H_2^+	He^+	Ne^+	Ga^+
current into $\pm 0.5°$ $[nA]$		8	5	5	5
angular density $[\mu A/sr]$		35	20	20	20
emission half angle $[°]$		≤ 1	≤ 1	≤ 1	30
energy spread $[eV]$		$\simeq 1$	$\simeq 1$	$\simeq 1$	5 - 20

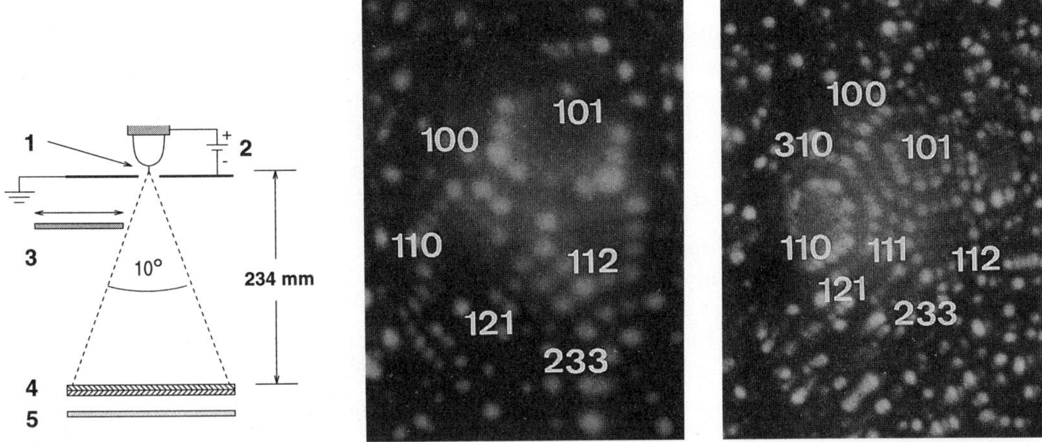

Figure 2: Left: Scheme of field-ion setup for imaging: (1) Cryo-cooled tip on a hairpin support; (2) high-voltage supply; (3) retractable screen; (4) multichannel plate image intensifier; (5) fluorescent screen. Right: Field-ion micrographs of two different supertips on <111> W basetips. The emitting atoms are arranged on polyhedral surface planes characterized by their Miller-indices.

about 300 atoms (~ 7.5 nm^3), the surface to about 85 atoms (~ 7.5 nm^2). The mean diameter is about ~ 12 atoms (3.7 nm).

All 26 supertips inspected so far exhibit regular atomic arrangements. They constitute single crystals in a bcc array formed by an epitaxial growth process on the single crystal beneath. Determining the Miller indices of the planes we find that about 60% of the supertip surfaces are composed of (110) planes. Presumably the (110)-planes are preferred as they are the closest packed planes in a bcc lattice.

Due to this polyhedral shape it is not possible to define a supertip radius in the usual geometric sense. But via the relation M=l/$\beta\cdot$r which gives the magnification of a FIM [3, 8] (M: magnification, l: distance from emitter tip to MCP, r: tip radius, β: compression factor) an effective radius r_t^* can be obtained by determining M experimentally. The actual magnification obtained in a measurement is derived from the known interatomic distances of the W crystal and the corresponding image at the MCP-screen. This leads to a magnification of $\sim 4\cdot10^6$ for the supertip in the middle of Fig.2. Using a compression factor of ~ 11 [7] this approach yields an effective radius of $r_t^* \sim 5$ nm.

In order to evaluate the field strength at the emitter surface commonly the relation F=U/kr [9] (F: field strength, U: emitter voltage, r: tip radius, k: shape factor) is applied. From field-electron characteristics using the Fowler-Nordheim equation ([10], [12]) a kr of 510 nm was calculated for the supertip in the middle of Fig.2. With $r_t^* \sim 5$ nm one derives a supertip shape factor of about 100. This figure is much larger than of a basetip (k$\sim$4) or of a sphere (k=1). Assuming a value of k$\sim$100 for other supertips one can approximate a radius for an unknown supertip just by determining the respective Fowler-Nordheim-characteristics which can be done with relatively little effort.

By using the geometrical information from the FIM measurements a typical supertip was modelled in order to perform calculations on electrical field strength and virtual source size. A computer code [11] was used to solve Laplace's equation numerically. Fig.3 shows the shape of the supertip model. The lateral angle of 35° corresponds to the inclination between the (111) and (110) planes in a bcc-lattice. The height amounts to about 4 atomic layers and the flat part at the top to about 0.3 nm. After determination of ion trajectories, their asymptotes were extrapolated to the region of smallest spatial extent which determines the virtual source size.

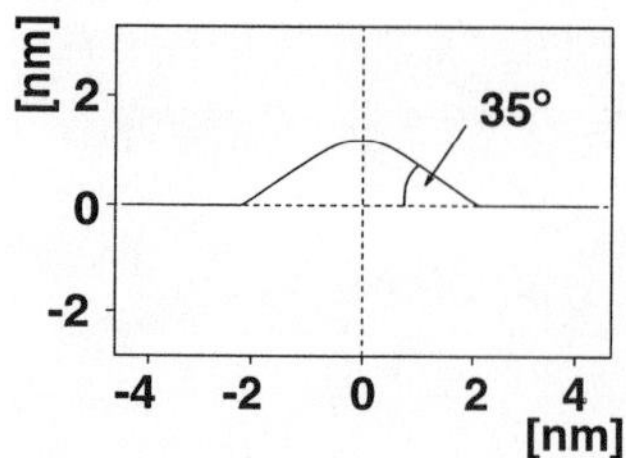

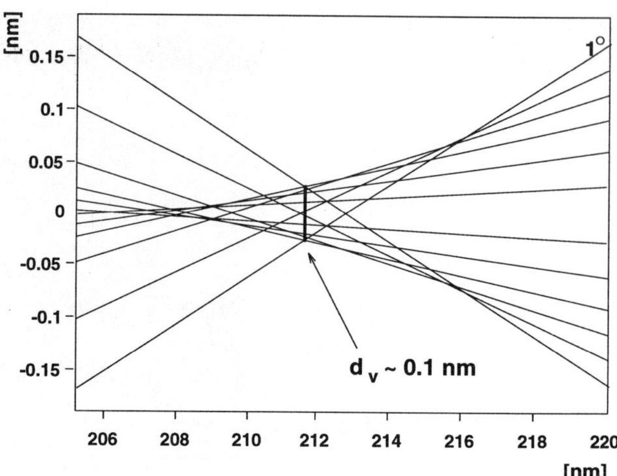

Figure 3: Left: Geometrical shape of the supertip model. Right: Calculated asymptotes of the trajectories of the assumed supertip structure. Emission angle $\alpha_{1/2} \sim 1°$, virtual source size $d_v \sim$ 0.1 nm.

For an emission half angle of up to 1° a virtual source of about the size of an atom is obtained (Fig.3).

The total emission current of the left supertip of Fig.2 amounts to $\sim$ 0.2 nA. With a virtual source size of 0.1 nm and an emission angle of $\pm 1.5°$ one calculates a brightness of $\sim 10^9 \text{A/cm}^2$ sr for this supertip which is on the low side because temperature and gas pressure have not been at the optimal working point. Emission currents of some nA can be obtained under optimum conditions (see Tab. 1), which leads to brightness figures of 10^{10} - 10^{12} A/cm^2 sr. In Fig.4 the different source types are compared due to their spectral brightness and target current density. A LMIS reaches a much lower brightness of 10^6 - 10^7A/cm^2 sr due to its larger virtual source size of $\sim$ 50 nm [13]. With a supertip GFIS target current densities can be obtained which are 2 orders of magnitude higher than of LMIS.

During ionization a flickering of emission of single supertip atoms can be observed. The flickering frequency can be estimated to about 1 Hz. Presumably this is due to inhomogeneities in the gas supply.

Tab. 2 summarizes the data of the supertips shown in Fig.2.

Table 2: Features of the supertips shown in Fig.2.

supertip of figure 2	#1 (left)	#2 (right)
volume	$\sim$300 atoms ($\sim$7.5nm^3)	$\sim$450 atoms ($\sim$9nm^3)
surface	$\sim$85 atoms ($\sim$7.5nm^2)	$\sim$130 atoms ($\sim$12.5nm^2)
effective radius r_t^*	$\sim$5 nm	$\sim$9 nm
shape factor k	$\sim$100	$\sim$90
max. emission current	0.2 nA	0.1 nA
total emission angle	$\pm$1.4°	$\pm$1.5°
virtual source size	$\sim$0.1 nm	$\sim$0.2 nm
brightness	10^9 A/cm^2 sr	10^8 A/cm^2 sr

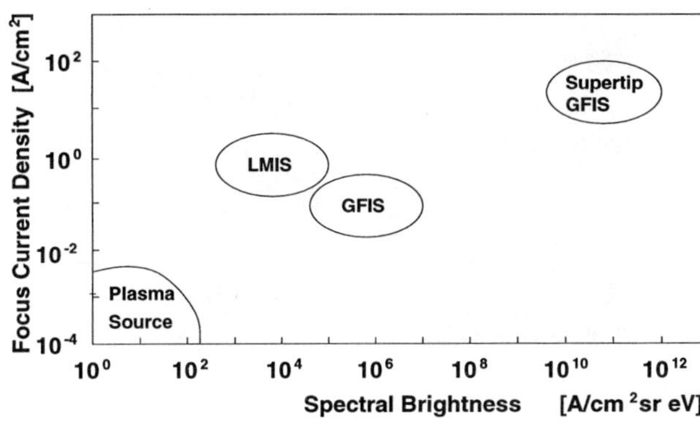

Figure 4: Features of the different ion source types.

5 Supertip formation

The generation procedure of a supertip is as follows: The base tip is heated up to a temperature in the range of 600 - 1000 K. In a He atmosphere of about 10^{-5} mbar electron emission of some 100 nA is started from the tip.

During this formation procedure the emitted electrons are ionizing helium atoms which are accelerated towards the tip. The deposited energy serves to displace W atoms from their lattice positions. They undergo surface-diffusion and drift in the electrical polarization field. At certain surface positions on the basetip the diffusing atoms can be trapped to form small aggregate centers growing with irradiation time. Preferred places could be the edges of lattice-planes or lattice defects possibly introduced during irradiation. These small protrusions generate local field enhancements from where an intensified electron emission occurs. This is followed by an amplified helium-ionization rate above this site which leads to a self-intensifying formation process. Finally, in order to protect the growing protrusion from self-destruction the emission is cut-off at a preselected current level of about 1 μA.

6 Modification of supertip properties

Fabrication of supertips with predefined features would be a great advantage for all applications. In order to investigate this possibility the supertip formation conditions were varied. 26 supertips were produced at different temperatures and basetip radii ranging from 600 - 1000 K and 150 - 400 nm, respectively. Surprisingly, little or no influence on volume, surface, emission angle and supertip formation positions has been observed so far. The supertip formation seems to have a statistical origin. That means that a predefinition of supertip properties by using special production conditions has not been successful.

Supertips produced below the above temperature range appear to be not suited for source applications: If produced at cryogenic temperatures the supertips show structural instabilities which can lead to a complete removal of the supertip at high electrical fields.

It is possible, however, to modify properties of already existing supertips. In Fig.5 a) it is shown how an existing supertip is sharpened. The supertip on the left side has an effective radius of $r_t^* \sim 8$ nm. Application of the supertip production procedure leads to a modified supertip shown in the other picture. Its effective radius has changed to $r_t^* \sim 4$ nm which means that the effective curvature has doubled. This modification leads to a $\sim 60\%$ higher emission current and lower working voltages (see Fig.5 b)). A blunting of supertips can be achieved by field desorption where surface atoms and planes can be removed.

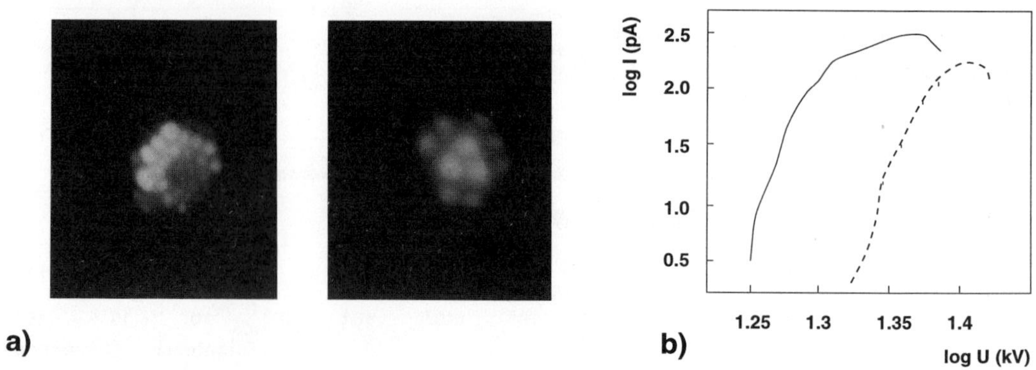

Figure 5: a) Resharpening of a supertip. Left: before, right: after. b) Measured current-voltage-characteristics of the supertip before (dotted) and after (line) the sharpening procedure.

7 Conclusion

With our FIM setup it is possible to investigate supertips at atomic scale. Exact geometrical and electrical features have been determined and modification of supertips has been performed. For application purposes it is further of high interest how supertips behave under special working conditions, e.g, the influence of adsorbates to emission behaviour or geometrical stability during long-time ionization.

References

[1] J. Melngailis, J. Vac. Sci. Technol. B5(2), 469-495 (1987)

[2] L. W. Swanson, Appl. Surf. Sci. 76/77, 80-88 (1993)

[3] T. T. Tsong, *Atom probe field ion microscopy*, Cambridge Univ. Press (1990)

[4] K. Jousten, K. Böhringer, R. Börret, S. Kalbitzer, Ultramicroscopy **26**, 301-312 (1988)

[5] J. W .Ward, J. Vac. Sci. Technol. B3(1), 207-213 (1985)

[6] E. W. Müller, Zeitschrift der Physik, **131**, 136 (1951)

[7] Th. Miller, PhD Thesis, MPI für Kernphysik Heidelberg (1996)

[8] M. K. Miller, *Atom probe microanalysis*, Materials Research Society (1989)

[9] R. Gomer, *Field emission and field ionization*, Harvard Univ. Pres Cambridge (1961)

[10] R. Fowler, L. Nordheim, L. Proc. Royal Soc. (London), A **119**, 173 (1928)

[11] E. Kasper, private communication, Universität Tübingen

[12] A.Knoblauch, Ch. Wilbertz, Th. Miller, S. Kalbitzer, J. Phys. D: Appl. Phys. **29**, 470-473 (1996)

[13] P.D. Prewett, G.L.R. Mair, *Focused Ion Beams from LMIS*, Research Studies Press LTD, Tanton Somerset, England (1991)

Materials Science Forum Vols. 248-249 (1997) pp. 439-443
© *1997 Trans Tech Publications, Switzerland*

High Energy Ion Microprobes: Where are we going ?

T. Butz and R.-H. Flagmeyer

Fakultät für Physik und Geowissenschaften, Universität Leipzig,

Linnéstr. 5, D-04103 Leipzig, Germany

Keywords: Ion Beam Analysis, Nuclear Microprobe

Abstract: In this contribution the compilation of data on high energy (> 1 MeV) ion microprobes (lateral resolution < 50 μm) prepared for ECAART-4, 1995, is analyzed statistically in order to provide an overview over existing installations, technical details, and research activities, possibly indicating future trends.

Where are we going? This question appears on the famous canvas by Paul Gauguin - now at the Museum of Fine Arts at Boston - entitled "Where do we come from, where are we, where are we going" which he painted in a rush of fever shortly before his death. There is no convincing answer to this question and we therefore restrict ourselves to the present status of high energy (> 1 MeV) ion microprobes with lateral resolutions below 50 μm. The statistical analysis is based on the compilation for the fourth European Conference on the Application of Accelerators in Research and Technolgy (ECAART-4), Zurich, 1995 [1]. Although some 20 institutions not contained in this compilation and supposed to have microprobes were brought to our attention since august 1995 and contacted, most of them either did not respond, did not have microprobe facilities, or had facilities under construction without relevant performance parameters yet being available. Therefore the data are based on some 30 systems, whereas around 50 installations are suspected to exist worldwide. Nevertheless, we hope that the data are representative for the present time and future trends.

Accelerators

Since for high resolution high energy ion microprobe work beams with extraordinary brightness on the order of some 10 A/m^{2}rad^{2}eV and low energy spread $\Delta E/E < 2.10^{-5}$ are highly desirable, the choice of accelerator and ion source is of crucial importance. In this respect, single ended machines should be favoured over tandem accelerators both because of the brighter (positive) ion source as well as because losses in brightness and the deterioration of $\Delta E/E$ in the stripper can be avoided. Nevertheless, about 35% of all accelerators used for microprobes are tandems, about 65% only are single ended machines (there are outlyers like the Eindhoven cyclotron or the Darmstadt LINAC). This certainly reflects the fact that microprobe applications are frequently added to the research program to installations which formerly were constructed for nuclear research. This is particularly evident for machines with terminal voltages largely exceeding 5 MV. Nonetheless, such systems can apparently be combined with microprobes and exhibit excellent performance. On the average, terminal voltages in the range from 2 to 4 MV dominate by far. The large majority of voltage generating systems is based on the van de Graaff principle using a rotating belt or segmented belt (Pelletron), whereas Cockroft-Walton (all solid state) power supplies without moveable parts are progressively entering the market in the 3-4 MV range. These power supplies avoid vibrations which can deteriorate the spatial resolution of the microprobes and show promising performance as far as energy stability and ripple is concerned

(cf. [2]). There is a strong preference for RF ion sources over other types like duoplasmatron, sputter, and penning sources, and the choice is frequently based on a compromise between versatility of the system (e.g. ions available, high currents - but not brightness) and optimal performance of the microprobe regarding spatial resolution. It should be mentioned that the energy spread in the ion source tends to be the limiting factor of the energy spread of the ion beam because terminal voltages around 1 MV with $\Delta E/E \sim 10^{-6}$ have been achieved already for electron microprobes [3] and are certainly not out of reach for 3 MV supplies. New developments like the field ionization super-tip [4] for MV-applications at terminal voltage potential are anxiously awaited.

Microprobes

<u>Total Length</u>
There is a large variation of the total length of the microprobe systems ranging from 1.8 m only up to 30 m. However, the vast majority is in the range from 2 - 10 m (see Fig. 1a). It seems that spatial limitations in the laboratories dominate over ion optical considerations. However, considering the large variety of slit/diaphragm arrangements in use, the variety of lens systems (see below) as well as beam deflection systems (~2/3 use magnetic deflection, ~1/3 electrostatic deflection; some use no deflection at all and translate the specimen) it is hard to suggest an "optimum" length at all. The concept for a compact, low-cost microprobe was recently presented by Grodzins et al. [5].

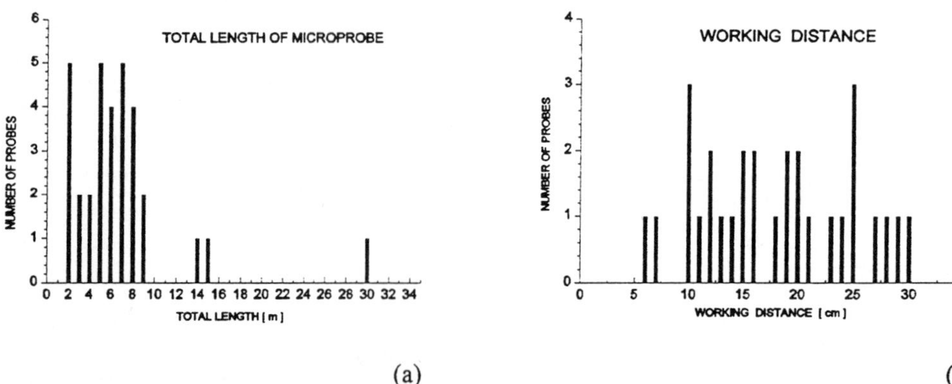

(a)　　　　　　　　　　　　　　　　　　　　　(b)

<u>Fig.1</u>
Total lengths (a) and working distances (b) of nuclear microprobes. Further 4 installations use working distances between 35 and 60 cm.

<u>Lens systems</u>
Almost all lens systems are based on magnetic quadrupoles with roughly equal shares of doublets, triplets (many of them the "classical" Oxford triplet), and quadruplets (most of them as "Russian quadruplets"). Exceptions are the superconducting solenoid at Bochum/Germany, the Martin-type lens at Guelph/Canada, and the Harwell-lens at Namur/Belgium. For high-resolution work it is essentially chromatic and parasitic aberrations which have to be minimized. There is no simple "unique solution" to the problem and mechanical precision as well as "intrinsic" or "extrinsic" compensations of aberrations seem to point to the believe that there is not too much of an improvement to be expected from lens designs as far as resolution is concerned. This partly

reflects the fact that the chromatic aberration coefficients enters the resolution in a low power only and that parasitic aberration can be kept sufficiently low. This situation might change with ultrastable ($\Delta E/E=10^{-6}$) beams.

<u>Working distance</u>
The working distance, although a rather important parameter, seems to be uniformly distributed between 6 to 30 cm, with values up to 60 cm (see Fig. 1b). It is clear that a lower limit is set by constructive considerations of the specimen chamber and detector arrangements. The actual working distance appears to be dictated by the lens parameters (magnetic rigidity, demagnification factor) or by beam deflection devices (sometimes integrated into the lens) rather than by ion optical optimization. For a high demagnification and not too small object slits, of course, the working distance (as the focal length) should be as short as possible.

<u>Lateral resolution</u>
Fig. 2 illustrates the lateral resolutions on about 30 microprobes where the "best" values (usually with the lowest currents) are reported. These numbers are based essentially on H^+ (and some $^4He^+$) data. There is a clear dominance of systems with lateral resolutions in the range from 1 - 10 µm. Thus far, only 5 systems reported values below 1 µm with the present lower limit at 50 nm. Several systems under construction are aiming at resolutions in the submicrometer regime ("nanoprobes"), e.g. Denton/TX, Leipzig/Germany, or Sumy/Ukraine, to mention but a few.

<u>Fig. 2</u>
L a t e r a l
resolutions

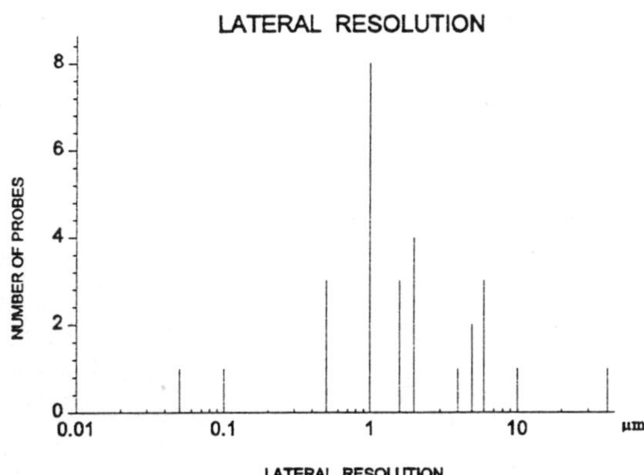

Techniques used

The list of techniques used is rather large and often it is the *combination* of techniques which really pushes a field. Nevertheless, Fig. 3a shows that PIXE (i.e. µ-PIXE including PIXE-Tomography) and RBS (including Channeling) are the most frequently employed techniques (i.e. the working horses). µ-NRA and STIM (including STIM-Tomography and Channneling-STIM) both comprise about 10% of all research effort. Secondary electron imaging (SE), IBIC, PIGE as well as other techniques like IL, SEU, RFS, and ERDA are surprisingly seldom mentioned, but their potential is expected to be large. Since the future of a technique strongly depends on technical developement (e.g. detectors) and competition with other techniques - apart from bright researchers - it is difficult to make any prognoses. Alas, it is good to have such a plethora of techniques available with ion beams to face future developments with optimism.

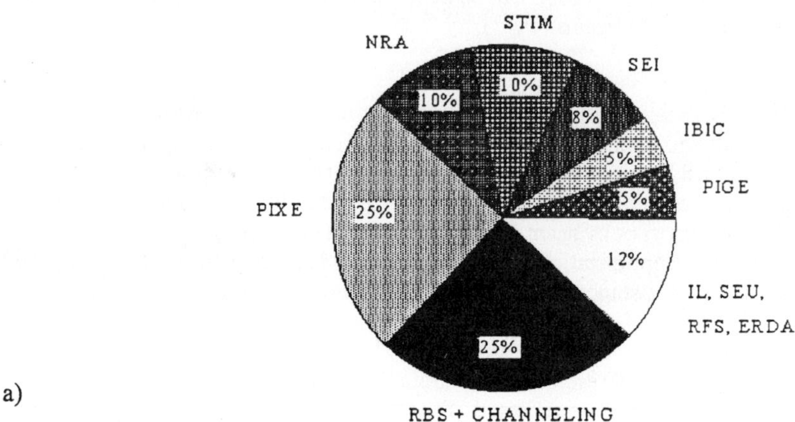

a)

Fig. 3 Techniques used (a) and research areas (b) in nuclear microscopy

b)

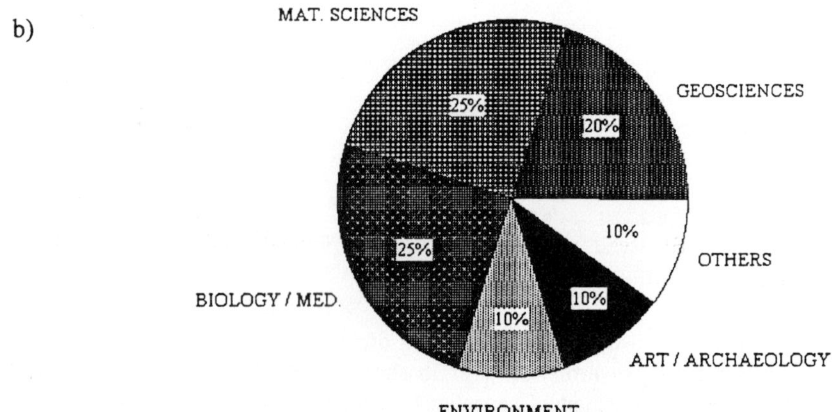

Research areas

It is difficult to derive statistics on research areas based on nominations in a compilation only, but we believe that data based on the number of publications or research funds would not look very different. By far the most important areas are materials science (including semiconductor devices, material modification, etc.), bio-medical research, and geosciences. To our surprise, environmental studies comprise about 10% of all research efforts only, as does art and archaeology (see Fig. 3b).

It is a pleasure to thank G. J. F. Legge and D. J. W. Mous for fruitful discussions and collaboration.

References

[1] High Energy Ion Microprobes. A compilation of data on high energy (≥ 1 MeV) ion microprobes (lateral resolution ≤ 50 µm) prepared for ECAART-4, Zurich 1995.
T. Butz and R.-H. Flagmeyer
Supplement to Proc. 4th European Conference on Accelerators in Applied Research and Technology (ECAART-4), August 29- September 2, 1995, Zurich (Ed. M. Suter)

[2] A novel ultrastable 3 MV Singletron^{TM} accelerator for nanoprobe applications.
D. J. W. Mous, R. G. Haitsma, T. Butz, R.-H. Flagmeyer, D. Lehmann, and J. Vogt
Proc. 5th Int. Conf. on Nuclear Microprobe Technology and Applications, Santa Fe, Nov. 10-15, 1996 (submitted)

[3] Applications of Electron Beams.
W. Sigle, *this conference*

[4] Development of a High-Brightness Gas Field-Ionization Source.
K. Böhringer, K. Jousten , and S. Kalbitzer,
Nucl. Instrum. Meth. Phys. Res. B **30** (1988) 289

[5] Holistic design of a scanning nuclear Microprobe.
L. Grodzins, P. Boisseau, H. Glavish, R. Klinkowstein, W. Nett, and R. Shefer,
Nucl. Instrum. Meth. Phys. Res. B **104** (1995) 1

Materials Science Forum Vols. 248-249 (1997) pp. 445-450

Application of Highly Focused Ion Beams

L. Bischoff and J. Teichert

Research Center Rossendorf Inc., Institute of Ion Beam Physics and Materials Research,
P.O. Box 510119, D-01314 Dresden, Germany

Keywords: Focused Ion Beam, Liquid Metal Ion Source, Ion Optics, Writing Implantation, Ion Milling, $CoSi_2$, Etch Stop, 3D Structures

Abstract: The impressive development of focused ion beam (FIB) systems from the laboratory level to high performance industrial machines during the last twenty years is briefly reported. The design and the functional principle of a liquid metal ion source as well as a FIB column are described. The main application fields of the FIB technology are discussed. More in detail the Rossendorf FIB activities, like writing Co^+ implantation in silicon to fabricate $CoSi_2$ microstructures by ion beam synthesis and the writing Ga^+ FIB implantation to create a patterned etch stop during a following anisotropic etching for micromechanical 3D structure fabrication are presented.

1. INTRODUCTION

In the last two decades focused ion beams (FIB) have become a very useful tool for many tasks in micron and submicron technology [1,2]. After the invention of the liquid metal ion source (LMIS) in the early 1960′s about ten years later a vehement development of FIB systems started from the laboratory level to high performance industrial machines. Probe sizes of less than 50 nm and current densities of more than 10 A/cm^2 are now available and allow to use these beams for many applications. Integrated circuit repair and modification [3], failure analysis [4], lithographic mask repair [5], FIB lithography [6] are the main applications in microelectronic industry. Especially during the R&D phase the FIB is very advantageous because of its high spatial resolution and its flexibility in variation of dose, energy and pattern design on one chip, or even in one structure detail. If the FIB equipment contains a mass separator and an alloy LMIS is used different needed ion species can be extracted. This is very important for writing FIB implantation purposes [7]. FIBs are applied for direct patterning by ion milling to fabricate micromechanical components [8] or to prepare specimen for SEM or TEM investigations [9]. Utilizing a precursor gas submicron conducting pattern can be deposited [10] or selectively enhanced etching is possible [11]. Furthermore the FIB can be used for surface analysis and as a scanning ion microscope [12].

2. FOCUSED ION BEAM EQUIPMENT

2.1. Liquid Metal Ion Sources

In order to focus an ion beam into a spot size of a diameter lower than one micron a source is needed which emits the ions from a very small area (point source) into a limited solid angle. LMISs as well as gaseous field ionization sources approach these demands [13]. Because of their broad and versatile applications the following discussion is concentrated on the LMIS only. This source type usually consists of an emitter needle with a tip radius of about 10 μm, which is covered with a pure metal, (i.e. Ga, In, ...) or an alloy (i.e. Au/Si/Be, Co/Nd,...) and can be directly or indirectly heated. The source material should have a high surface tension and a low vapour pressure at its melting point. A counter electrode is placed in front of the needle where a high voltage (extraction voltage) in the range of 2 to 10 kV is applied. At a critical voltage a liquid metal cone with a half angle of 49.3 deg forms (Taylor-cone [14]) and ion emission occurs from a liquid jet based on this cone, mainly due to a field evaporation process [15]. Fig. 1 shows a Co-Nd alloy wetted emitter of the directly heated needle type and an indirectly heated unwetted capillary emitter (Research Center Rossendorf). An LMIS is characterized by its brightness $B = (d^2I/ dA\ d\Omega)$ where dI is the differential current element emitted from an infinitesimal source area dA into the solid angle element dΩ. A typical value for an LMIS is 10^6 A/cm^2 sr. Another source parameter which is important for the spatial resolution of the FIB is the width of the energy distribution. It is caused by the ionization process itself and amounts from 5 to 25 eV depending on the source material and the emission current. The influence on the spot size is explained below.

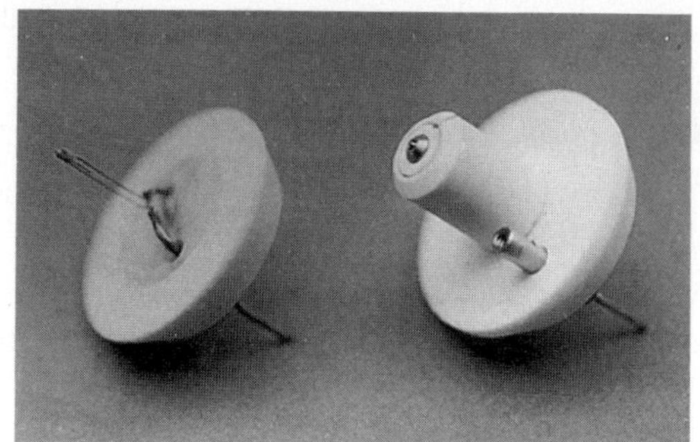

Fig.1:   Liquid metal ion source emitters, needle type (left) capillary type (right).

2.2. Ion Optical Column

Modern FIB columns are rather similar and so the design is explained using as an example the IMSA-100 FIB system (Research Center Rossendorf). This equipment was designed to achieve current densities of more than 10 A/cm^2 [16]. The column consists of an LMIS, two electrostatic lenses, a pre-lens octopole double deflector, the ExB mass separation system, two beam blankers, two stigmators, and a secondary electron/ion detector for imaging. A schematic presentation of this column is given in Fig. 2. The upper (objective) lens accelerates the ions from the LMIS extraction energy of about 7 keV to the final kinetic energy, adjustable between 25 and 50 keV for singly charged ions. The lower (projective) lens is an asymmetrical einzel lens. The source is placed in the object-side focal plane of the objective lens and the target plane corresponds with the image-side focal plane of the projective lens. Thus the beam is collimated between the two lenses. The lack of a cross-over minimizes the beam broadening due to Coulomb interaction. The beam spot size can be estimated by

$$d^2 = (Md_q)^2 + d_s^2 + d_c^2 \tag{1}$$

where d is the beam spot diameter at the sample, d_q is the virtual source size and M denotes the magnification of the column. The quantities d_s and d_c represent the contribution due to the spherical and axial chromatic aberrations, respectively, which are defined as

$$d_s = \frac{1}{2} C_s \alpha^3 \ ; \ d_c = C_c (\frac{\Delta E}{E}) \alpha. \tag{2}$$

C_s and C_c are the spherical and chromatic aberration coefficients of the ion optical column, respectively, α is the acceptance half-angle on the target, ΔE is the energy spread of the LMIS and E is the final ion energy. For the above described column, operated with an ion current in the range of 1 nA, the chromatic aberration becomes the dominant contribution to the spatial resolution d [2]. To minimize this influence a high final ion energy is required, but there are constructive limits, e.g. the breakdown voltage of ion optical components or the LMIS has to be operated near the onset current with its lowest energy spread.

The ion optical elements of the IMSA-100 column are accomodated in an UHV chamber with two integrated coaxial getter pumps, which guarantee a base pressure better than 10^{-5} Pa. The column is mounted on a vibration isolated target chamber containing a laser interferometer controlled x-y table where wafers up to 6 in. and masks up to 7 in. can be processed. For *in situ* electrical measurements or alternatively for target heating up to 500 °C an integrated 5 pin sample holder connection is available. The operation of the FIB system is completely controlled by computer.

3. APPLICATIONS

Today highly focused ion beam systems are used for many purposes in the production of microelectronic or micromechanical devices in industry as well as in research laboratories. The broad spectrum of applications has been mentioned in the introduction. This paper will focus on examples of recent activities on the IMSA-100 FIB equipment. Writing Co^+ implantation in silicon to fabricate $CoSi_2$ microstructures by ion beam synthesis (IBS) and the writing Ga^+ FIB implantation to create a patterned etch stop during a following anisotropic etching for micromechanical 3D structure fabrication will be described in more detail.

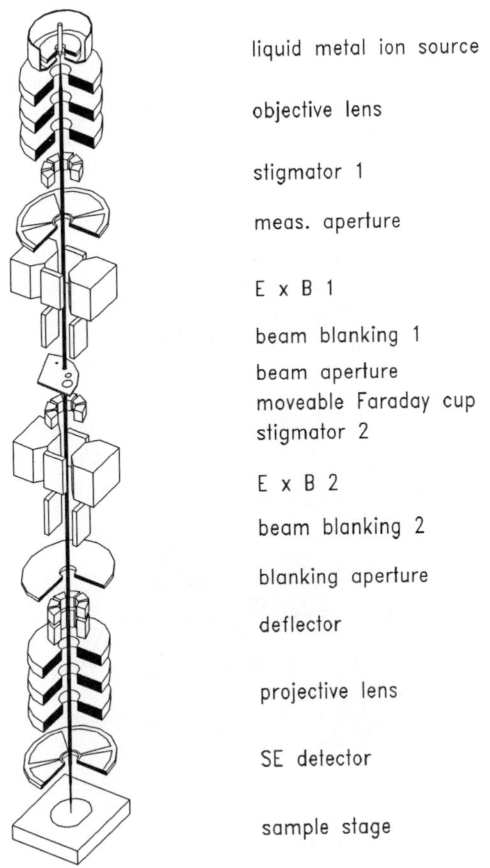

Fig. 2: Schematic drawing of the ion optical column of the IMSA-100 FIB system.

Creating microstructures by direct writing, maskless FIB implantation requires a suitable LMIS and a highly focused mass separated beam. The IBS of $CoSi_2$ as well as the Ga^+ etch-stop implantation need doses in the range of 10^{16} to 10^{17} ions / cm². For the fabrication of submicron features it is important to know the amount of real implanted ions and their depth profile. So the sputtering effect has to be taken into consideration. Instead of the usual approximation of a Gaussian depth distribution a saturation distribution is obtained, which depends on the dose $D = j\,t$ and can be described by [17]

$$N(z,j,t) = \frac{N_t}{2Y}\left(erf\ \frac{z-R_p+jt\ Y/N_t}{\sqrt{2}\Delta R_p}\ +\ erf\ \frac{z}{\sqrt{2}}\right) \tag{3}$$

where z is the depth coordinate, j the particle flux density, t the implantation time, N_t the sample density, and Y the sputtering coefficient. R_p and ΔR_p are the mean projected range and the straggling, respectively. Integrating Eq. 3 for z > 0 gives the retained dose of the implanted ions. By comparing this value with the irradiation dose D one finds a dose loss due to sputtering [18]. The IBS of $CoSi_2$ films with a broad beam implantation is well investigated [19]. A similar technology has been applied to produce microstructures using the FIB technique [7]. For $CoSi_2$ micropatterns fabricated on crystalline Si substrates a resistivity of about 20 $\mu\Omega$cm and a temperature stability up to 1000°C were obtained. On poly Si (high bulk resistivity) these values were 60 $\mu\Omega$cm and 700 °C, respectively [18]. Submicron $CoSi_2$ structures are now under investigation to use them as gate electrode for permeable base transistors and for lateral metal semiconductor metal (MSM) photodetectors (Fig. 3).

The application of the dynamical focus control (DFC) method [20] allows to write patterned $CoSi_2$ structures following the surface topography which is useful for contacting components of micromechanical 3D devices. Fig. 4 shows a test pattern with $CoSi_2$ interconnects between the top surface and the bottom of a 280 μm deep anisotropically etched groove.

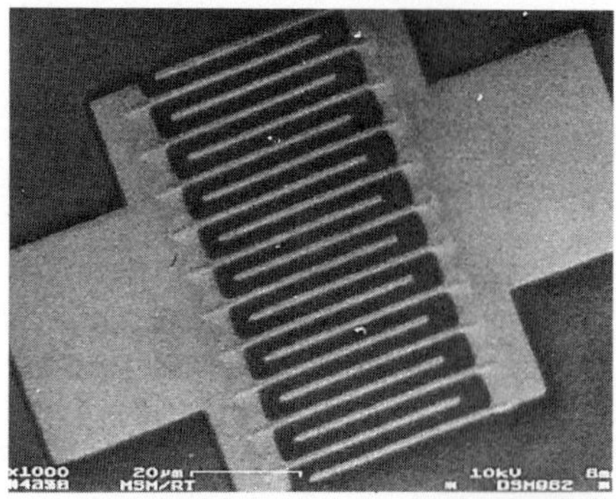

Fig. 3: SEM image of a MSM photodetector test structure, $CoSi_2$ on Si, fabricated by writing IBS.

A modern technique to fabricate three-dimensional micromechanical structures is doping of Si by high concentrations of p^+-type ions using a focused ion beam. In this case the p^+-type ions act as an etch stop. A subsequent anisotropic and selective wet chemical etching makes it possible to fabricate 3D structures. Micromechanical test structures were defined by implantation into (100) oriented silicon wafers at room temperature with a 35 keV $^{69}Ga^+$ beam. The ion beam current was 2.5 nA and the implantation dose was varied between $1 \cdot 10^{14}$ cm^{-2} and $2 \cdot 10^{17}$ cm^{-2}. In order to obtain different structure shapes, which depend on the orientation of the crystal planes at the edges and corners, two orientations for the designed structure were chosen. One was orientated along the [110] direction in the wafer surface, a second one under an angle of 45° to the [110] direction. The etching was carried out in KOH/H_2O etch solution of 30% concentration at 80 °C for a time of about 5 minutes. The etch selectivity ratio for the Si-planes (100) /(111) amounts to about 40. An SEM micrograph of a free standing grid pattern, 3.5 μm over the substrate plane is shown in Fig. 5. The wires are 1 - 2 μm wide, 20 μm long and 50 nm thick, corresponding to the data of the implantation profile.

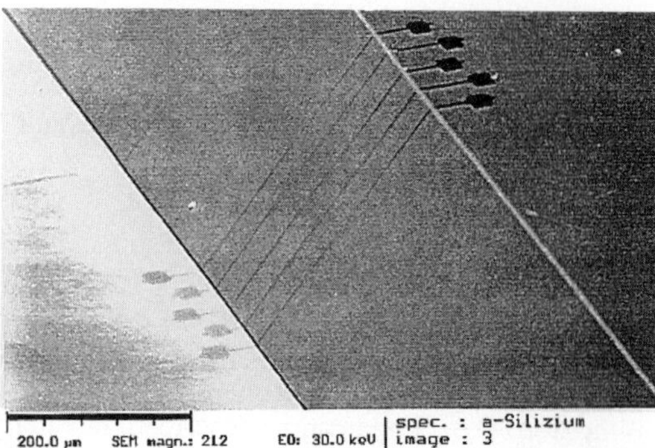

Fig. 4: SEM image of $CoSi_2$ wires over a slope of 280 μm height in poly Si.

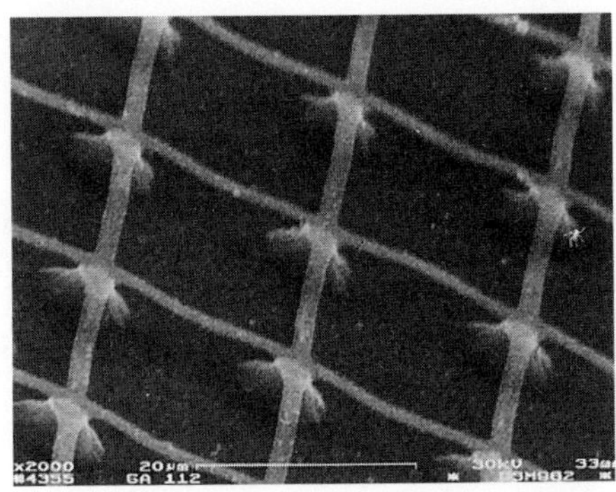

Fig. 5: SEM image of a free standing grid pattern, fabricated by writing Ga^+ FIB implantation and anisotropical etching.

4. SUMMARY

The focused ion beam technology has passed a remarkable development from the laboratory level to high performance industrial machines during the last two decades. Especially during the research and development phase of micromechanical or microelectronical devices FIBs are very useful as preparation as well as analytical tools, because of their high spatial resolution and their flexibility in variation of dose, energy and pattern design. Modern applications are the writing ion beam synthesis of $CoSi_2$ microstructures and writing Ga implantation as an etch stop patterning for 3D structure fabrication.

Acknowledgements

The authors would like to thank Dr. B. Schmidt for carrying out the anisotropical etching and to Dr. R. Müller and Mrs. E. Christalle for SEM imaging.

References

[1]   J. Orloff, Rev. Sci. Instr. **64**, 1105 (1993).
[2]   P.D. Prewett and G.L.R. Mair, "Focused Ion Beams from Liquid Metal Ion Sources",
        Taunton: Research Study Press, 1991.
[3]   T. Tao, W. Wilkonson, and J. Mengailis, J. Vac. Sci. Technol. B **9**, 162 (1991).
[4]   R. Boylan, M. Ward, and D. Tuggle, Proc. Int. Sym. for Testing and Failure Analysis (ASM
        International, Los Angeles, 1989), p. 249.
[5]   P.D. Prewett and P.J. Heard, J. Phys D: Appl.Phys. **20**, 1207 (1987).
[6]   R.L. Kubena, F.P. Stratton, J.W.Ward, G.M. Atkinson, and R.J. Joyce,
        J. Vac. Sci. Technol. B **7**, 1798 (1989).
[7]   L. Bischoff, J. Teichert, E. Hesse, D. Panknin, and W. Skorupa,
        J. Vac. Sci Technol. B **12**, 3523 (1994).
[8]   M.J. Vasile, C. Biddick, S.A. Schwalm, J. Vac. Sci Technol. B **12**, 2388 (1994).
[9]   A. Yamaguchi, M. Shibata, and T. Hashinaga, J. Vac. Sci Technol. B **11**, 2016 (1993).
[10]   P.G. Plauner, J.S. Ro, Y. Butt, and J. Melngailis, J. Vac. Sci Technol. B **7** (4), 609 (1989).
[11]   R.J. Joung,J.R.A. Cleaver, and H Ahmed, J. Vac. Sci Technol. B **11**, 3523 (1993).

[12]   R. Levi-Setti, J.M. Chabala, and Y.L. Wang, Ultramicroscopy **24**, 97 (1988).

[13]   J. Melngailis, J. Vac. Sci. Technol. B **5**, 469 (1987).

[14]   G. Taylor, Proc. R. Soc. A **133**, 383 (1964).

[15]   D.R. Kingham and L.W. Swanson, Appl. Phys. A **34**, 123 (1984).

[16]   L. Bischoff, E. Hesse, D. Janssen, F.K. Naehring, F. Nötzold, G.Schmidt, and J. Teichert, Microelectronic Engineering **13**, 367 (1991).

[17]   E.F. Krimmel and H. Pfleiderer, Rad. Eff. **19**, 83 (1973).

[18]   J. Teichert, L. Bischoff, E. Hesse, P. Schneider, D. Panknin, T. Gessner, B. Löbner, and N. Zichner, J. Micromech. Microeng. **6**, 272 (1996).

[19]   A.E. White, K.T. Short, R.C. Dynes, J.P. Garno, and J.M. Gibson, Appl. Phys. Lett. **50**, 95 (1987).

[20]   L. Bischoff, J. Teichert, E. Hesse, Microelectronic Engineering **27**, 351 (1995).

[21]   P.H. LaMarche, R. Levi-Setti, and Y.L. Wang, J. Vac. Sci. Technol. B1 1056 (1983).

[22]   B. Schmidt, L. Bischoff, and J. Teichert, Proc. of the 10th Eur. Conf. on Solid-State-Transducers, 9. -12. 9. 1996, Leuwen, Belgium.

Materials Science Forum Vols. 248-249 (1997) pp. 451-458

Use of an Ultra-High Resolution Magnetic Spectrograph for Materials Research

D.O. Boerma[1,2], W.M. Arnoldbik[1] and W. Wolfswinkel[1]

[1] Dept. of Atomic and Interface Physics, Debye Institute, Utrecht University,
P.O. Box 80.000, NL-3508 TA, Utrecht, The Netherlands

[2] Dept. of Nuclear Solid State Physics and Materials Science Centre, University of Groningen,
Nijenborgh 4, NL-9747 AG, Groningen, The Netherlands

Keywords: Magnetic Spectrograph, Ion Beam Analysis, High Resolution

Abstract

A brief description is given of a magnetic spectrograph for RBS and ERD analysis with MeV beams, delivered by a Tandem accelerator. With a number of examples of thin layer analysis it is shown that the spectrograph is uniquely suited for the measurement of concentration depth profiles up to a depth of about 50 nm. The positions of light and heavy atoms in epitaxial layers can be determined by using blocking, or channeling/blocking geometries in ERD or RBS measurements. Such determinations are especially interesting for samples with shallow interfaces and δ-doped layers. It is also shown that the impact-parameter dependence of energy loss and charge exchange in ion-solid interactions can be determined from measurements at samples with sub-monolayer coverage of elements at the surface, single crystalline samples, or (in transmission geometries) very thin layers. Some features of this spectrograph are compared with those of a selection of other high- resolution analyzers.

1. Introduction

A magnetic spectrograph, designed and constructed at the University of Groningen for interface and thin-film analysis, has been installed at the 6.5 MV tandem accelerator of the Utrecht University. The instrument (see fig. 1) has been described in an earlier publication[1].

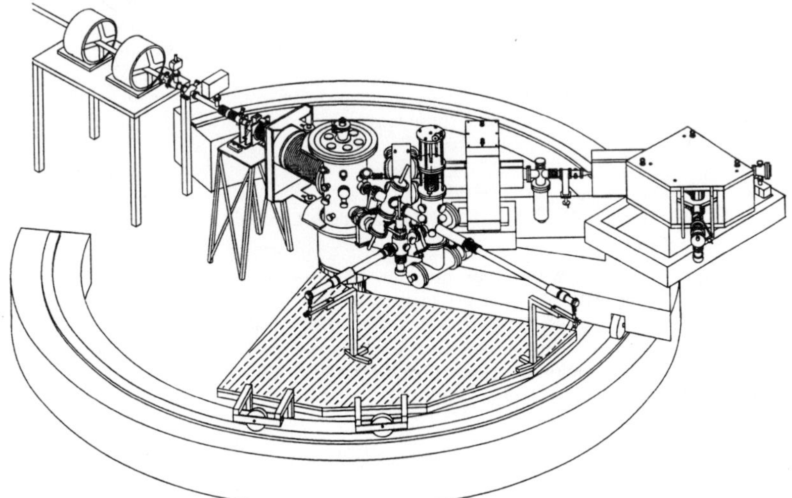

FIG. 1. 3-D image of the magnetic spectrograph

It consists of a quadrupole lens, a Wien filter and a magnetic dipole provided with higher-order correcting fields. With this combination of elements one unique charge-to-mass ratio of ions with energies within a chosen range can be selected for detection with a two-dimensional position sensitive detector placed behind the dipole. On this detector an energy range with a width of 13% of the median transmitted energy is imaged with a resolution $\Delta E/E$ of 3×10^{-4}. At the same time an angular range of 5° is imaged with a resolution of 0.15°. The energy dispersion is in the horizontal plane; the angular dispersion is in a vertical plane (see fig. 1). The spectrograph is coupled to an ultra-high vacuum scattering chamber provided with a three-axes goniometer and a chamber for on-line sample preparation. These vacuum chambers, together with the spectrograph, can be rotated over an angular range of 120°, while maintaining the vacuum. For this pupose the whole arrangement is coupled to the beamline with a combination of three bellows, forming an "S"-shape, through which the beam can pass, when the system is in a rotated position. The specifications of the spectrograph are summarized in table I.

Bending radius	0.5 m
Magnetic rigidity	0.5 Tm (6 MeV He^{++}, 1.7 MeV Si^{++})
Range of rotation	$20°$–$140°$ or $-10°$–$110°$
Acceptable beamspot	1x3 mm
Energy range on detector	13 %
Angular resolution	$0.1°$
Angular acceptance	$0.3° \times 5° = 0.5$ msr
Angular range	$5°$
Wien Filter	selection q/m ratio
Energy resolution ($\Delta E/E$)	
Spectrograph	3×10^{-4}
Ion beam	8×10^{-4}
2 D detector	4×10^{-4}

Table I. Specifications of the magnetic spectrograph.

In this paper we discuss the possibilities and the limitations of this spectrograph for its use in materials science, and for the study of ion-solid interactions. Some examples of thin-layer analysis with Rutherford backscattering (RBS) and elastic recoil detection (ERD) are given. With these examples it is illustrated that the spectrograph is uniquely suited for studying the composition and the structure of near-surface layers, with a depth below the surface up to a few tens of nm's. The advantages of the use of the spectrograph over the application of more conventional means of ion-beam analysis disappear when this maximum depth is surpassed. The study of near-surface layers requires on-line UHV sample preparation facilities as are installed in our apparatus.

The spectrograph is also suitable for the study of ion-solid interactions. We will describe three different approaches that can be used to study as a function of impact parameter the charge exchange and energy loss of ions scattered or recoiled from a solid. Since the cleanliness and the structure of the surface are influencing the charge exchange, also these measurements require surface preparation facilities as well as ultra-high vacuum conditions. At the moment the experimental and theoretical knowledge of charge exchange processes of high-energy ions in solids is very incomplete. Such knowledge is needed for a quantitative interpretation of a part of the data obtained with the spectrograph.

In the last section the features of the spectrograph are briefly discussed in relation with the requirements for both mentioned types of experiments. Then these features are compared with those of some other high-resolution instruments.

2. Near-surface analysis

The spectrograph will mainly be used for concentration depth profiling of elements (isotopes) in thin (multi-)layers, where compositional changes near the surface or shallow (< 50 nm) interfaces are of particular interest. Such measurements are important in the study of, for instance, early stages of gas-solid reactions, shallow interfaces or δ-doped layers in semiconductor structures, or interfaces in magnetic multilayers. Also interdiffusion in bi-layers of different species, like deuterated and non-deuterated block-copolymers can be determined from measured H and D depth profiles.

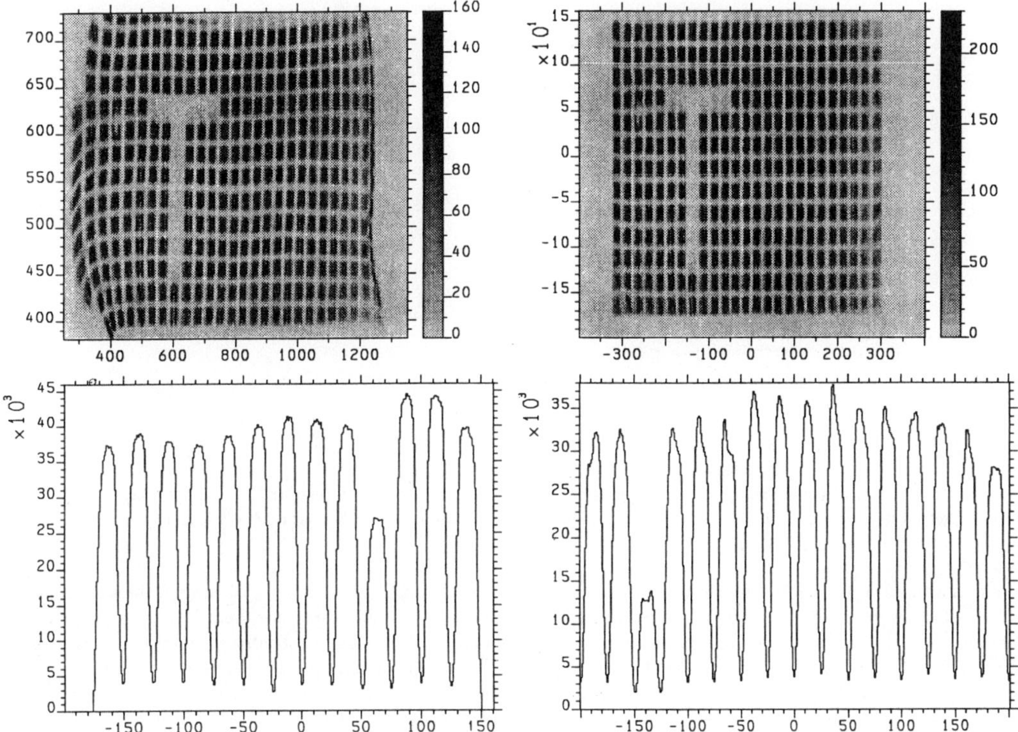

FIG. 2. Image of a calibration grid measured under radiation of 2 MeV He ions. Above: before and after the transformation procedure. Below: Projection of the image on the vertical and horizontal axis, respectively. The channel width after transformation is 0.1 mm.

For depth profiling 2-dimensional (energy-angle) spectra (7×5 cm) are taken with a detector consisting of three channelplates and an anode with a wedge-and-strip pattern [2,3]. Due to various effects, like incomplete charge collection near the edges of the detector, the 2-D spectra are distorted. This distortion is removed by an affine transformation as illustrated in fig. 2. The input for the transformation consists of the proper position of grid points derived from a measured spectrum with a square calibration grid pattern in front of the detector. During the on-line correction for distortion it is possible to correct also for the small kinematic shifts inherent to the opening angle of 5° perpendicular to the scattering plane. The kinematic spread over the opening angle of 0.2° - 0.4° in the scattering plane can be partly removed by positioning the detector at a calculated distance from the image plane[1]. Finally the 2-D spectra are projected on the energy axis and the energy scale is converted into a depth scale.

The program DEPTH developed by Szilágyi and Pászti [4,5] is used to estimate the depth resolution as a function of depth for the relevant measuring geometry, sample composition, and ion species and energy. This can be done both for RBS and ERD. In the calculations the following contributions are taken into account: electronic and nuclear straggling, multiple scattering, energy spread in the beam, detector resolution. Also geometrical spread due to finite opening angles, beam size and angular spread in the beam are taken into account. As reported earlier [1] we found that the overall energy resolution of our set-up is $\Delta E/E = 9 \times 10^{-4}$. This resolution is mainly determined by the energy spread in the applied ion beam from the tandem accelerator. The resolution is still sufficient to obtain near-to monolayer resolution at the surface both for RBS and ERD, when glancing-angle geometries are used. The depth resolution decreases with increasing depth so that a depth range can be defined for which the depth resolution obtainable with the spectrograph is superior to what is achievable with more conventional methods. In table 1 the various contributions to the resolution are given. For ERD the contribution due to the kinematic spread is strongly dependent on the measuring geometry. For an opening angle $\Delta\phi = 0.2°$ the spread $\Delta E/E$ amounts to 10^{-3} for a recoil angle $\phi = 10°$ and 5×10^{-3} for a recoil angle $\phi = 40°$. For RBS this contribution is negligible in most practical cases. Roughly speaking, the useful depth ranges as defined above are 0-40 nm for RBS and 0-50 nm for ERD.

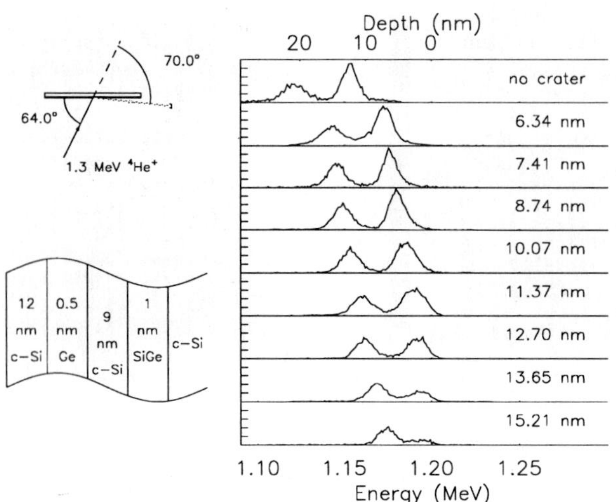

FIG. 3. RBS spectra of epitaxially grown Ge δ-layers in c-Si. The spectra are recorded at various stages of a 2 keV O_2^+ sputter process as used for depth profiling with SIMS. Configuration and primary beam are indicated, as well as the crater depths.

In fig. 3 an example of RBS depth profiling is presented for a multi-layer structure of Si and Ge as indicated. Depth profiles were taken at various stages of a sputter process to deduce the ion-beam mixing and surface roughening induced by a 2 keV O_2^+ beam used in high-resolution SIMS. The preliminary conclusions from this work are that the δ-doped layers are broadened due to diffusion of Ge mainly into the Si grown on top, and that ion beam mixing is of influence to the depth resolution obtainable with SIMS. This resolution of 1.5 nm FWHM was slightly worse than that obtained with the spectrograph. Near the surface the resolution of the spectrograph would be 0.3 nm in this case.

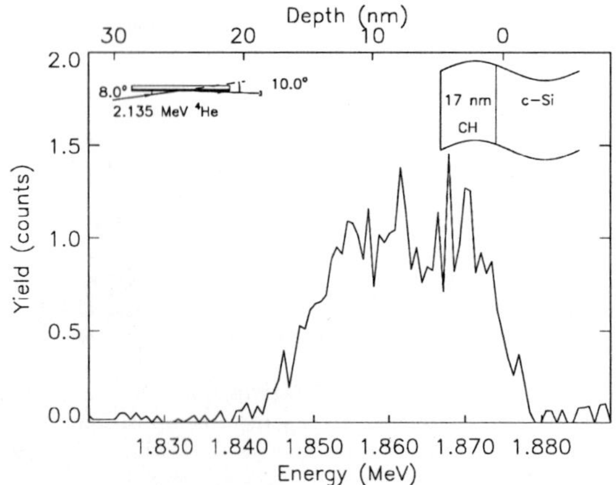

FIG. 4. Deuterium profile of a 17 nm thick deuterated polystyrene layer on c-Si. The geometry and primary beam used in this measurement are indicated in the figure. The spectrum reveals a resolution at the interface of 5 nm. The width of the surface edge amounts to 4 keV (FWHM) corresponding to 3 nm. Here the resolution is restricted by the kinematic spread resulting from the 0.25° opening angle in the scattering plane.

In fig. 4 we present an example of an ERD depth profile of deuterium in a 17 nm thick deuterated polystyrene layer on c-Si. The geometry of the measurement was optimized for depth resolution at the interface with Si, with the position of the detector in the image plane (i.e. without correction for kinematic spread). The resolution of 5 nm at the interface is "ideal", i.e. dominated by straggling in the sample. The beam energy and ion species were chosen to take advantage of a resonance in He-D scattering at 2.135 MeV, which is strongly peaked at 0° in ERD, with a cross section of 6.6 barn at 10° [6]. From calculations with DEPTH it followed that it would be advantageous to apply an even more glancing geometry for a slightly better depth resolution. Also the cross section would be higher at more forward angles. However, at more forward angles a strong background appears in the ERD spectra, due to multiple scattering in the spectrograph of the abundant He ions scattered from the sample. We hope to solve this problem by applying a new type of 2-D detector (see section 4).

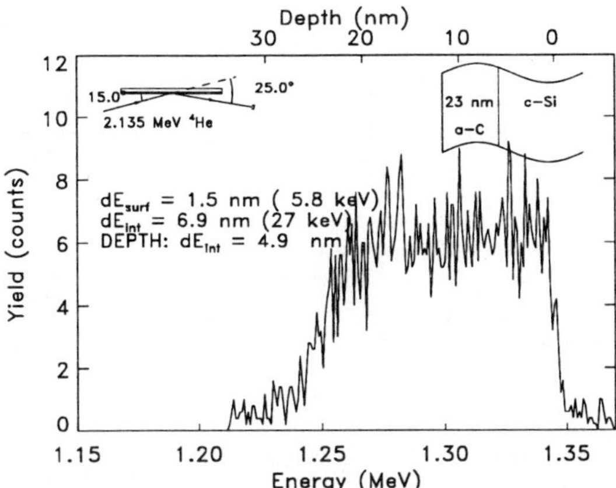

FIG. 5. Carbon profile of a 23 nm thick amorphous C:H layer on c-Si. The geometry and primary beam used for this measurement are indicated in the figure.

In fig. 5 a depth profile of C is given for a similar sample. Here the depth resolution both at the surface and at the interface is very good in comparison with the 15 nm obtainable with a more conventional method, in which a dE-E detector is used [7].

The spectrograph can be used efficiently to determine the structure of a thin epitaxial layer from blocking patterns measured with the spectrograph positioned in an appropriate crystalline direction of the sample material. This can be done both in RBS as well as in ERD mode. The structure information can be obtained by comparing measured blocking patterns with results obtained by computer simulations of ion beam trajectories. In such simulations a model for charge exchange as a function of impact parameter should be applied to be able to predict the blocking patterns for the various charge states observed. In a recent publication [8] we showed for the case of H scattering from NiSi$_2$ that the ratio of intensities in blocking patterns observed for different charge states is very sensitive to the near-surface structure. At the moment we are developing this method further for structure determinations from RBS or ERD blocking patterns. As an example we show a blocking pattern observed for He^{++} scattering from a single crystal of Ag in fig. 6.

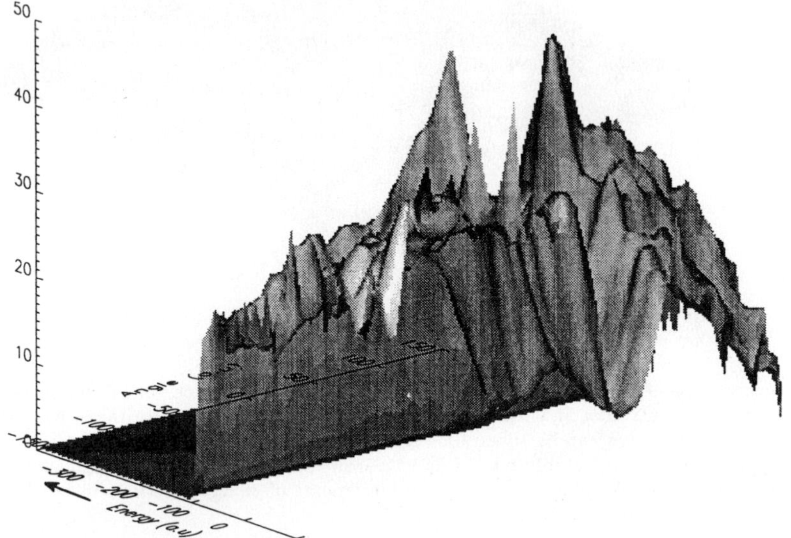

FIG. 6. 3 D representation of a spectrum as recorded during bombardment of a Ag(110) crystal with 2.135 MeV He ions. Here ϕ=70° and α=14°.

3. Measurements on charge exchange of ions propagating in solids

With the spectrograph certain types of measurements are possible in order to get information on the charge exchange (and energy loss) as a function of impact parameter of MeV ions which undergo a sequence of collisions with atoms while penetrating through a solid. Such experimental information is scarce at the moment, and also a proper theoretical description of charge exchange as a function of impact parameter is not available for ions with Z>2. A more thorough understanding of charge exchange in solids would be desirable, not only for the interpretation of ERD spectra measured for particular charge states, but also to better predict the interaction of implanted ions with solids and the interaction of energetic ions in plasmas with the walls of fusion reactors. It is our aim to contribute to a description of charge exchange in solids as a function of impact parameter, based on measurements and calculations applying a quantum- mechanical description of electron capture and loss. This program is in collaboration with a group of the Moscow State University. Here we mention three types of experiments.

1. By scattering ions from atoms present at the surface of a sample with a sub-monolayer coverage, and observing the various charged fractions, it is possible to get information on the charge exchange in a single collision at very small impact parameters. From such measurements for the scattering of 3 MeV C ions from Ag and Au, and a comparison with calculations of the capture and loss processes, the following picture emerged [9]: The C ions acquire a certain charge state distribution during the change of momentum taking place close to the target atom nucleus, which is only very weakly dependent on the charge distribution before this collision. The distribution is altered towards the equilibrium distribution by electron capture and loss of the ion on its way out through this target atom.

2. By measuring the charge exchange and energy loss of ions penetrating through a very thin foil at scattering angles near 0°, a range of small impact parameters is probed.

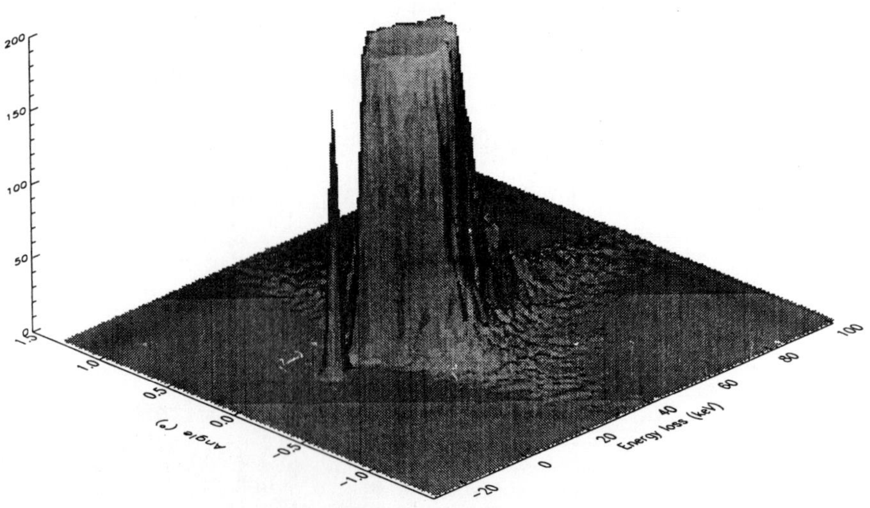

FIG. 7. Plot of intensity versus energy loss and scattering angle of Li^{3+} particles transmitted through a 20 nm thick gold foil. The measurement of the primary beam is added to this figure and visible at 0° and $\Delta E{=}0$.

In fig. 7 the energy loss of 2 Mev Li^{3+} ions after penetrating a 20 nm thick Au foil without charge exchange is shown as a function of scattering angle in the form of a corrected 2-D spectrum taken with the spectrograph positioned at 0°. Similar spectra can be taken for cases where a defined charge exchange is occuring.

3. In blocking geometries a predictable set of impact parameters is implicated in establishing the final charge distribution and energy loss. We plan to perform measurements of charge fractions in blocking geometries as soon as we have the surface preparation under control.

4. Conclusions

The spectrograph is well suited for near-surface analysis of solids with RBS and ERD (composition and structure as a function of depth) and for the study of ion-solid interactions. The main requirements for these studies as realised in our apparatus are:

- Energy resolution $\Delta E/E < 10^{-3}$; angular resolution $< 0.2°$.
- Option to select one type of ion to be detected.
- Spectrograph rotatable over a wide range of angles, including forward angles.
- Two-dimensional position-sensitive detector to measure simultaneously a reasonable range of energies at a range of angles.
- Ultra-high vacuum conditions, combined with surface preparation facilities.
- Samples mounted on a two- or three-axes goniometer.

This spectrograph has a solid angle close to 0.5 msr, which is smaller than desired. This can be compensated to a certain extent by selecting the ion species and energy and the detection angle for optimal yield within the limits imposed by the requirements for optimal depth resolution, minimal ion- induced damage, etc. In a number of cases use can be made of resonances in the cross section near $0°$ for ERD. It was found that detection of relatively small recoil yields at angles $< 10°$ is impossible, because a strong background of scattered ions can reach the detector after multiple reflection in the spectrograph. To avoid this problem we are developing a position- and energy dispersive Si detector to replace the channelplate detector. With such a detector a better separation of different ion species to be detected can be made on the basis of their difference in energy.

We may compare the characteristics of this spectrograph with some other high-resolution machines. These spectrographs are not provided with (the equivalent of) a Wien filter, so that particle identification must be done at the detector side of the instrument, making use of differences in specific energy loss or flight time.

A magnetic spectrograph for thin layer analysis is installed in München[10], where it is applied for ERD measurements with heavy projectiles. This instrument exhibits a comparable energy resolution, but has a larger applicable acceptance angle of 2 msr. The angular distribution however, is not resolved, so that the instrument can not be used for structure determination. This instrument is suited to correct for kinematic spread.

A few electrostatic analyzers for energies in the MeV range have been designed[11,12]. The energy resolution of the instrument described in [11] is slightly better than that of our magnetic spectrograph. The angular distribution is not resolved. The instrument has a smaller energy acceptance and solid angle.

A number of electrostatic analyzers are in use for medium-energy ion scattering with H or He beams of energies around 100 keV. At these low energies the scattering cross sections and the stopping powers are relatively high, making these analyzers very well suited for RBS depth profiling of heavy elements near the surface. The commercial versions of these instruments, produced by High Voltage Engineering Europe[13], have a solid detection angle close to 1 msr, and are combined mostly with a goniometer, an ultra-high vacuum chamber and surface preparation facilities. They are commonly used for determining the structure and composition of the outermost layers of solids with RBS [13–18] or ERD of H or D[19]. For these purposes they are preferable above our spectrograph, which, in its turn, has as advantages a larger depth range for RBS depth profiling and its suitability for ERD of all elements with A < 30, in combination with structure determinations.

5. Acknowledgements

This work is performed with financial support from the Foundation for Fundamental Research on Matter (FOM) and the Foundation for Technical Research (STW). The mentioned cooperation on charge-exchange studies with a group of the Moscow State University is supported by the Dutch Organization for the Advancement of Research (NWO).

6. References

[1] W. M. Arnoldbik, W. Wolfswinkel, D. Inia, V. C. G. Verleun, S. Lobner, J. A. Reinders, F. Labohm, and D. O. Boerma, Nucl. Instrum. Methods Phys. Res. B **118**, 566 (1996).
[2] C. Martin, M. Lampton, R. F. Malina, and H. O. Anger, Rev. Sci. Instr. **52**, 1067 (1981).

[3] J. S. Lapington, A. D. Smith, D. M. Walton, and H. E. Schwarz, IEEE Trans. Nucl. Sci. **NS-34**, 431 (1987).

[4] E. Szilágyi and F. Pászti, Nucl. Instrum. Methods Phys. Res. B **85**, 616 (1994).

[5] E. Szilágyi, F. Pászti, and G. Amsel, Nucl. Instrum. Methods Phys. Res. B **100**, 103 (1995).

[6] F. Besenbacher, I. Stensgaard, and P. Vase, Nucl. Instrum. Methods Phys. Res. B **15**, 459 (1986).

[7] W. M. Arnold Bik and F. H. P. M. Habraken, Rep. Prog. Phys. **56**, 859 (1993).

[8] V. A. Khodyrev and D. O. Boerma, in *proceedings COSIRES '96* (in press).

[9] D. O. Boerma, W. M. Arnoldbik, N. M. Kabachnik, and V. A. Khodyrev, Nucl. Instrum. Methods Phys. Res. B (1996), accepted for publication.

[10] G. Dollinger, T. Faestermann, and P. Maier-Komor, Nucl. Instrum. Methods Phys. Res. B **64**, 422 (1992).

[11] O. Kruse and H. D. Carstanjen, Nucl. Instrum. Methods Phys. Res. B **89**, 191 (1994).

[12] D. J. O'Connor, J. Phys. B **20**, 437 (1987).

[13] R. M. Tromp, M. Copel, M. C. Reuter, and M. Horn von Hoegen, Rev. Sci. Instr. **62**, 2679 (1991).

[14] J. Vrijmoeth, P. M. Zagwijn, E. Vlieg, and J. W. M. Frenken, Surf. Sci. **290**, 255 (1993).

[15] J. C. Lee, C. S. Chung, H. J. Kang, Y. P. Kim, H. K. Kim, and W. Moon, J. Vac. Sci. Technol. A **13**, 1325 (1995).

[16] K. Sumitomo, T. Nishioka, and T. Ogino, J. Vac. Sci. Technol. A **13**, 289 (1995).

[17] E. P. Gusev, H. C. Lu, T. Gustafsson, and E. Garfunkel, Phys. Rev. B **3**, 1759 (1995).

[18] T. Tappe, A. Chatziparaskewas, J. Schäffer, J. Schlosser, J. Schmalhorst, and B. Schmiedeskamp, Appl. Surf. Sci. **104/105**, 661 (1996).

[19] M. Copel and R. M. Tromp, Rev. Sci. Instrum. **64**, 3147 (1993).

Materials Science Forum Vols. 248-249 (1997) pp. 459-466
© *1997 Trans Tech Publications, Switzerland*

Ion Beam Analysis with Monolayer Depth Resolution

H.D. Carstanjen

MPI für Metallforschung, Heisenbergstr. 1, D-70569 Stuttgart

Keywords: Ion Beam Analysis, High Resolution ERD, High Resolution RBS, ERD of Hydrogen, ERD of Oxygen, High Depth Resolution, Electrostatic Spectrometer, Polymer Multilayer, X-Ray Mirror, Oxidation

Abstract

The paper is concerned with the analysis of surfaces and near-surface layers with monolayer depth resolution by means of high resolution Rutherford backscattering (HRBS) and elastic recoil detection (HERDA) of MeV ions in combination with an electrostatic spectrometer. With this instrument which has recently been set up at the 6 MV Pelletron accelerator of the Max-Planck-Institut für Metallforschung in Stuttgart depth resolutions of 0.1 nm are obtained in HRBS and 0.3 nm in HERDA experiments. After an introduction the paper gives a short outline of the design and performance of the spectrometer followed by various examples of applications. These comprise - besides examples for the analyzing power of the instrument - the analysis of an X-ray mirror by HRBS, the analysis of partially deuterided polymer multilayers and the study of the initial oxidation of surfaces of aluminum single crystals by HERDA.

1. Introduction

During the last twenty years the analysis of surfaces and near-surface layers has gained an increasing interest. Meanwhile there exists a wide range of efficient and well established methods for the analysis of surfaces, such as LEED, AES, XPS, culminating in recent years by the development of the STM by Binnig and Rohrer and the SFM. While the techniques for surface analysis and also for the analysis of thicker layers (we mention SIMS, RBS or ERDA and the various partitioning techniques) seem to be well developed, there is only a small number of methods available which allow to analyze near-surface layers with thicknesses of less than e.g. 20 nm with reasonable depth resolution. Most surface analysis techniques fail, because of the limited exit depth (at most three monlayers) of the particles used in the analysis. The now widely used scanning microscopy techniques are even limited to the very surface only.

From the small number of methods for the analysis of near-surface layers we would like to mention SIMS and related techniques, the transmission electron microscopy, X-ray surface diffraction techniques, RHEED and ellipsometry. All these methods have their merits, but also their disadvantages. SIMS is destructive and has only a limited depth resolution (generally 2 to 3 nm, in best cases also 1 nm), the sample preparation for TEM is rather complicated and tiresome, X-ray diffraction techniques and RHEED require very flat surfaces, both as well as ellipsometry are not element specific. So other techniques which are able to provide additional information are highly welcome, since there is a variety of problems in near-surface layers to be solved: the measurement of shallow diffusion profiles or segregation to surfaces, oxidation problems such as preferential segregation, interdiffusion in multilayer samples, surface phase transitions and others more.

In principle ion beam techniques such as RBS and ERD should allow depth profiling of even very thin surface layers (below 1nm) with reasonable depth resolutions (of the order of atomic monolayers), since the mean energy loss of the particles in a material per unit length, which provides the depth scale, is a statistical quantity and a well defined function of depth. The limiting fact, indeed, has been the limited energy resolution of the commonly used silicon surface barrier detectors which for standard detectors is of the order of 14 keV for light ions such as protons and alphas. An estimate for 1 MeV alphas in Au shows that the depth resolution obtainable for normal incidence is of the order of 10 nm; for grazing incidence (e.g. 12°) this improves by about a factor of 5, yielding a depth resolution of 2 nm which, however, is still off from a monolayer resolution by a factor of 10.

A breakthrough has certainly been the development of the MEIS technique for high resolution RBS measurements by Smeenk et al. [1] which uses medium energy light ions (in the few hundred keV range) and a small electrostatic spectrometer for analysis. With this technique depth resolutions of somewhat better than 1 nm were possible which was recently improved by roughly a factor of three. But due to the use of light ions and medium ion energies the mass resolution in RBS is rather limited and ERD measurements which require heavy ions are hampered considerably.

In this contribution we would like to present an electrostatic spectrometer for MeV particles which allows depth profiling by RBS and ERD with monolayer depth resolution. The instrument has been set up recently at the 6 MV Pelletron accelerator of the Max-Planck-Institut für Metallforschung in Stuttgart. The instrument was at first used for high resolution RBS (= HRBS) purposes. Recently various techniques were adopted to allow for the use of the instrument for high resolution, background-free ERD measurements. The following section gives a short outline of the instrument; a more detailed description can be found in ref. [2]. Examples for HRBS experiments are given in section 3 and can be found in refs. [2] and [3]. The use of the instrument for the high resolution ERD analysis (= HERDA) of hydrogen and deuterium is described in section 4 and ref. [4]. Section 5 finally describes the-set up for the high resolution ERD analysis of medium mass elements such as oxygen [5] with some examples.

2.   The high resolution electrostatic spectrometer

The spectrometer consists essentially of three parts (see Fig. 1): i) an electrostatic lens system, ii) the analyzer and iii) a 1-dimensional position sensitive detector. i) The lens system (four quadrupole lenses and one hexapole) focusses particles emitted parallel to the optical axis of the instrument onto the entrence slit of the analyzer. This design allows in particular for extended samples and beam spots without spoiling the energy resolution of the instrument due to kinematic errors. The hexapole corrects for curved imaging of the slit by the analyzer. ii) The analyzer is a 100° cylindrical condensor (cylindrical sector field) of 700 mm radius and 19.8 mm gap width. With the maximum voltage of ± 60 kV applied to the condensor plates 2 MeV particles can be analyzed, which doubles with double charge state. Since these numbers do not depend on the particle's mass, also heavy particles of the same energy can be analyzed. The simultaneously available energy window amounts to 2.8 % of the analyzed energy; the energy resolution of the instrument is better than 3×10^{-4}. iii)

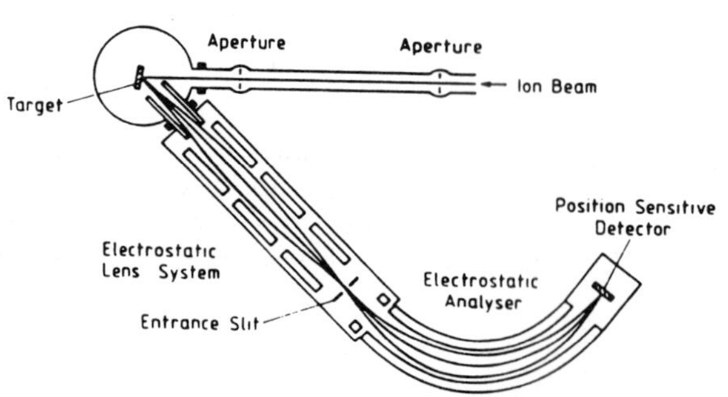

Fig. 1   Schematic drawing of the electrostatic spectrometer and the scattering chamber set up at the Pelletron accelerator of the Max Planck Institut für Metallforschung, Stuttgart.

For the detection of the particles, 1-dimensional position sensitive silicon-surface barrier detectors are used with lengths of typically 30 mm. The spatial resolution of these detectors is typically 1 % of the total length - for 5.5 MeV α-particles. Since the resolution drops almost linearly with 1/E, the spatial resolution for 1 MeV α-particles is about 1.65 mm. Cooling to liquid nitrogen temperature somewhat improves the resolution. If very high resolution is required, a 10 mm long detector is used. In many cases it is the finite spatial resolution of these detectors which limits the energy resolution of the spectrometer. On the other hand these detectors allow a rough energy analysis which is very usefull for the discrimination of background, e.g. of particles with different charge states.

With this set-up energy resolutions of 1.9 keV (30 mm detector) and 1.4 keV (10 mm detector) have been obtained in actual HRBS measurements. Typical depth resolutions are in the 2 to 3 monolayer range; this holds also for the HERD analysis of hydrogen and oxygen (see below). Because of the limited spatial resolution of these surface barrier detectors, recently 1-dimensional position-sensitive channel plates were used instead - with great success (see below). With these detectors energy resolutions of better than 0.7 keV have been obtained in actual HRBS measurements. By the use of a 1.5 MeV Ne beam a depth resolution of 0.1 nm was obtained on a gold target [6].

3.   High Resolution Rutherford Backerscattering Spectroscopy (HRBS)

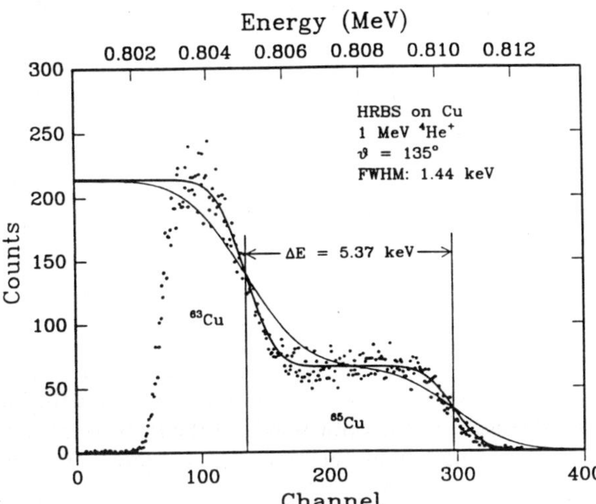

Fig. 2 shows as a first example the high energy edge of a HRBS spectrum of a clean Cu sample [2]. In the figure two distinct steps are visible. They are due to the two Cu isotopes ^{63}Cu and ^{65}Cu which are present in Cu with concentrations of 69% and 31%, respectively. From the scattering kinematics (1 MeV α-particles, scattering angle: 135°) the distance of the two steps is calculated to 5.37 keV which serves as energy reference. From the fit curve to the data as calculated by RUMP an energy resolution of 1.44 keV is obtained (for this experiment the 10 mm detector was used) which is by about 1 order of magnitude better than with ordinary silicon-surface barrier detectors. At grazing incidence (and exit, 12°) this depth resolution is sufficient to obtain a depth resolution of 0.2 nm, i.e. of 1 monolayer in Au. The figure further shows that with a by only a factor of 2 lower depth resolution no flat region is seen between the steps.

Fig. 2 High-energy edge of a high-resolution RBS spectrum (HRBS spectrum) as obtained with the electostatic spectrometer from a clean Cu single crystal (1 MeV He ions, 10 mm Si detector) [2]. Due to the high energy resolution of the analyzer the edge is resolved into two substages corresponding to scattering by ^{63}Cu (69%) and ^{65}Cu (31%). The thick line corresponds to a best computer fit to the experimental data using RUMP and provides an energy resolution of 1.44 keV in this measurement.

Fig. 3 shows an application of this HRBS technique to the analysis of X-ray mirrors as prepared by the group of W. Pompe in Dresden. The mirror consists of layers of ca 3 nm thick layers of carbon and nickel. The figure shows the HRBS spectrum from a mirror which is terminated by a Ni layer at the surface (only the scattering by the Ni atoms is shown). Different from an investigation by X-ray diffraction one is here able to determine the average thicknesses of individual layers: the first Ni layer has a thickness of 3.3 nm. It is followed by a 3.0 nm thick C layer (gap in the spectrum) which is again followed by an other Ni layer of 3.6 nm in thickness. By comparison with a simulation (full line, three layers only) also the thickness fluctions can be determined; in case of the first layer it amounts to ± 0.5 nm.

Fig. 4 finally shows two HRBS spectra (high-energy edge) from a clean gold surface (primary beam: 1 MeV Ne ions). In this case the silicon-surface barrier detector was replaced by a 1-dimensional position sensitive channelplate detector and the entrance slit of the analyzer set to 1 mm and 0.5 mm, respectively. Due to the very high spatial resolution of the detector (better than 0.2 mm) the energy and, hence, the depth resolution is now by far better: for a slit width of 0.5 mm an energy resolution of 0.78 keV can be derived from the fit to the edge (full curve) which corresponds in this case (incidence and exit angle: 37.5°) to a depth resolution of 0.1 nm, i.e. half a monolayer. Also the influence of different slit settings can now be seen clearly.

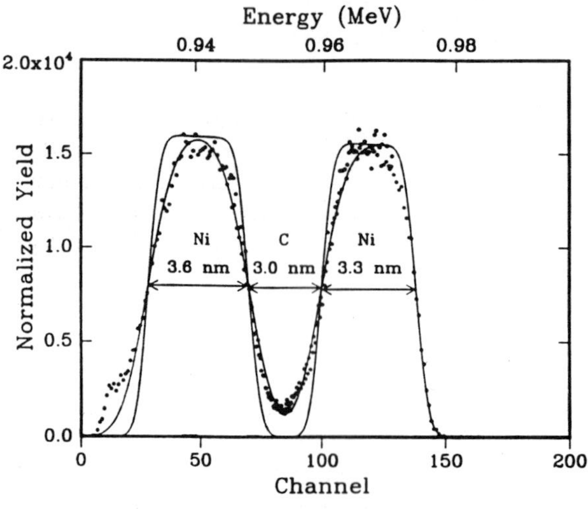

Fig. 3 HRBS spectrum of the first two Ni layers of an X-ray mirror (1 MeV He ions). The mirror consists of thin layers (ca 3 nm thick) of alternately Ni and C. The layer package is terminated by a Ni layer at the surface. The full line is a computer fit using RUMP which provides the layer thicknesses and the thickness straggling mentioned in the text.

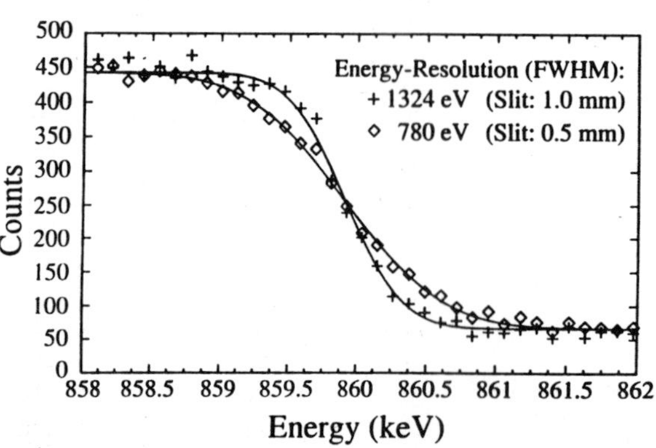

Fig. 4 High-energy part of HRBS spectra from a clean Au sample [6]. The spectra have been taken with the electrostatic spectrometer by employing a 1-dimensional position sensitive channel plate as detector and a primary beam of 1 MeV Ne ions. The two spectra correspond to two different settings of the entrance slit of the analyzer. Due to the excellent spatial resolution of the channel plates an energy resolution of 0.78 keV could be obtained for a slit of 0.5 mm which corresponds to a depth resolution of 0.10 nm in Au.

4. High Resolution Elastic Recoil Detection Analysis (HERDA) of Hydrogen

In conventional ERD measurements of hydrogen the energy and depth resolution is - besides the finite energy resolution of the detector - hampered by two effects: kinematic errors due to the finite opening of the aperture in front of the detector and the energy straggling of the recoil hydrogen atoms in the shielding foil, which is commonly used to stop primary particles, backscattered by the matrix. If the electrostatic spectrometer described above is used, all three sources are reduced considerably. First of all, the spectrometer has an excellent energy resolution which is by far better than that of common silicon-surface barrier detectors. Second, the problem of the energy straggling in the shielding foil is ruled out by a feature typical for spectrometers: the problem of energy resolution is converted into a problem of position resolution. If in a spectrometer the shielding foil is mounted directly in front of the position sensitive detector, the energy straggling in the foil can not influence the energy resolution any longer. The small amount of angular straggling by the foil does not influence the position resolution noticeably. Finally, the particular design of the spectrometer to focus only particles which are emitted parallel to the optical axis of the instrument onto the entrance slit eliminates the problem of kinematic errors due to a finite aperture of the instrument completely.

Thus, in order to be able to perform high resolution elastic recoil experiments the instrument was turned into the forward direction (38°) and a shielding foil put directly in front of the position sensitive detector. Details of this set-up, in particular also considerations about the energy resolution of the set-up can be found in ref. [4]. In the rest of this section various examples for applications will be given.

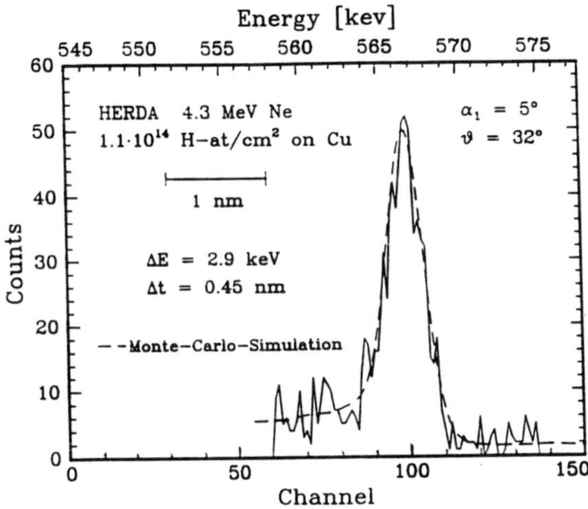

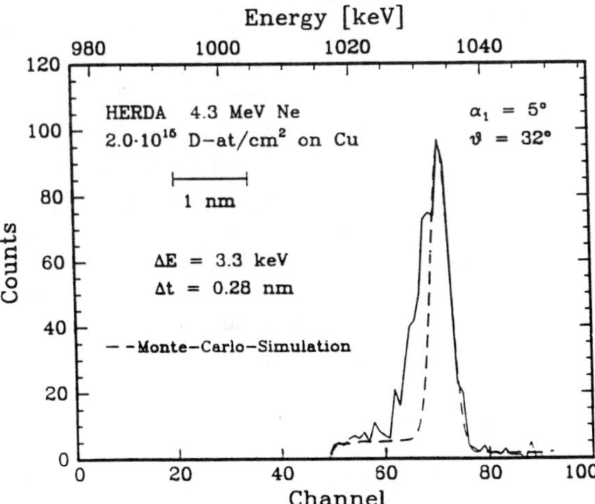

Fig. 5 High-resolution ERDA (HERDA) spectra of small amounts of H_2O and D_2O adsorbed on a clean (111) surface of a Cu sample [7]. Energy resolutions of 2.9 keV (H) and 3.3 keV (D) have been obtained which correspond to depth resolutions in Cu of 0.45 nm and 0.28 nm, respectively. (4.3 MeV Ne, scattering angle: 31.5°, incidence angle: 5°).

Fig. 5 shows HERDA spectra of small amounts of adsorbed water (ca 0.5· 10^{14} H_2O molecules/cm² and ca 1· 10^{15} D_2O molecules/cm², respectively) on a clean (111) Cu surface to demonstrate the excellent energy and depth resolution obtainable with this set-up [7]. Very narrow peaks are observed in both cases. For H the peak width is 2.9 keV which corresponds to a depth resolution of 0.45 nm (in Cu). In case of D the depth resolution is even better; from a fit to the data an energy resolution of 3.3 keV is derived which corresponds to a depth resolution of 0.28 nm (in Cu). Besides, one recognizes a slight broadening of the D spectrum towards lower energies. This is probably due to sub-surface D: the (111) surface in fcc metals is a 3-layer surface and with the relatively high amount of ca 1·10^{15} D_2O molecules/cm² as encountered here probably also subsurface sites have to be occupied by the D atoms. It should be noted that with this high energy resolution also the Doppler-broadening due to the motion of the hydrogen atoms adds a noticable contribution to the energy resolution. On the other hand, the influence of this effect is less than in the case of the well known nuclear resonance reaction for the analysis of H which uses 6.4 MeV ^{15}N ions [8]. Because of this reason the depth resolution with HERDA is in principal (and also experimentally) better than with the nuclear resonance reaction (for further details on this subject the reader is refered to refs. [4] and [9]. For the experiments presented in Fig. 5 a primary ion beam of 4.3 MeV Ne ions was used with a scattering angle of 31.5° and an incidence angle of 5°.

Fig. 6 finally shows as an example for application the HERDA spectra of H (top) and D (bottom) of a partially deuterided di-block copolymer multilayer sample [7]. The di-block copolymer molecules consist in this case of two long polymer chains A and B (each chain ca 1000 monomers) which are linked together to form the so called di-block copolymer A+B. In our case A is polystyrene and B polyparamethylstyrene; the latter has been deuterided for analysis purposes. A thin film of this material starts ordering when heated above a certain temperature, into multilayers by segregation of the molecules A and B, which, however, are linked together and, therefore, cannot separate completely. The ordering of such systems has been studied by F. Rauch from the university of Frankfurt by the nuclear resonance reaction with ^{15}N mentioned above [10]. Since this method can only see the hydrogen in the sample (in molecule A), it was of interest to investigate such samples with our HERDA technique which can see both isotopes, H an D. Fig. 6 shows the HERDA spectra of one of these samples. Both spectra show strong oscillations of a wave length of ca 17 nm which are out of phase by a half wave length. As the figure shows further, the H and D concentrations are complementary. A comparison with a computer simulation (broken line) for complete segregation of the two polymer species A and B shows that apparently strong segregation has occured in the sample. The simulation is even able to provide the pronounced damping of the oscillations in the H spectrum which is caused by the straggling of the low energy protons, but not by an incomplete segregation. For further details the reader is refered to refs. [7].

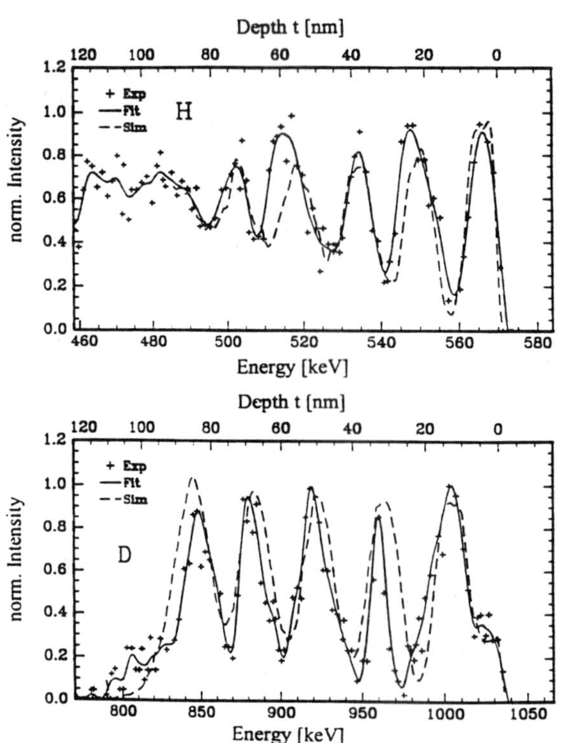

Fig. 6 HERDA spectra of H (top) and D (bottom) from a polymer multilayer sample [7]. Both profiles show strong oscillations with a wave length of ca 17 nm. The H and D concentrations are complementary. The strong damping of the H profile is due to straggling of the low energy protons.

5.　High Resolution Elastic Recoil Detection Analysis (HERDA) of Oxygen

If one is to analyze light atoms like carbon, nitrogen or oxygen by ERD, one encounters the problem, that is impossible to discriminate the backscattered particles of the primary beam (e.g. 1.5 MeV Ar ions) from the recoil atoms (e.g. 1 MeV O atoms) by a shielding foil, since the ranges of both particles in such a foil are almost equal. In order to separate the backscattered from recoil particles, the time of flight of each particle on the way through the spectrometer was measured in addition to its position on the detector, i.e. its energy. For this purpose a chopped primary ion beam (by sweeping the beam over an aperture with a frequency of 300 kcps) was used. A start signal was derived from the position sensitive detector and a stop signal from the high frequency oscillator which was used for sweeping. Typical times of flight are in the 1 μs range and, hence, easy to measure. In the HERDA experiment then 2-dimensional energy-time of flight spectra were measured with a 2-paramater data acquisition device. Further details of this set-up can be found in ref. [5].

Fig. 7 shows the result of such a 2-parameter measurement in a 3-dimensional presentation [11]. The sample consists in an oxidized Al sample, as primary ion beam 1.5 MeV Ar ions were used. As the figure shows, the backscattered Ar ions as well as the recoil Al atoms are well separated from the recoil O atoms.

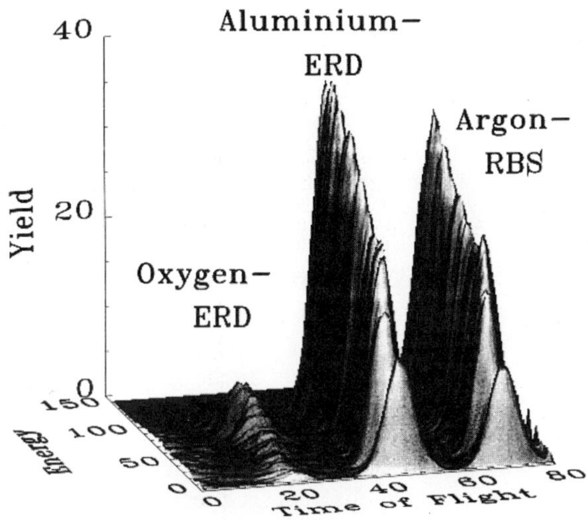

Fig. 7 Intensity versus time of flight and energy plots of particles ejected from an oxidized aluminum sample during an HERDA experiment (1 channel corresponds to 13 ns and 0.26 keV, respectively) [11]. The backscattered Ar^{++} ions are well separated from the Al^{++} and O^{++} HERDA particles. Primary beam: 1.5 MeV Ar^+ ions.

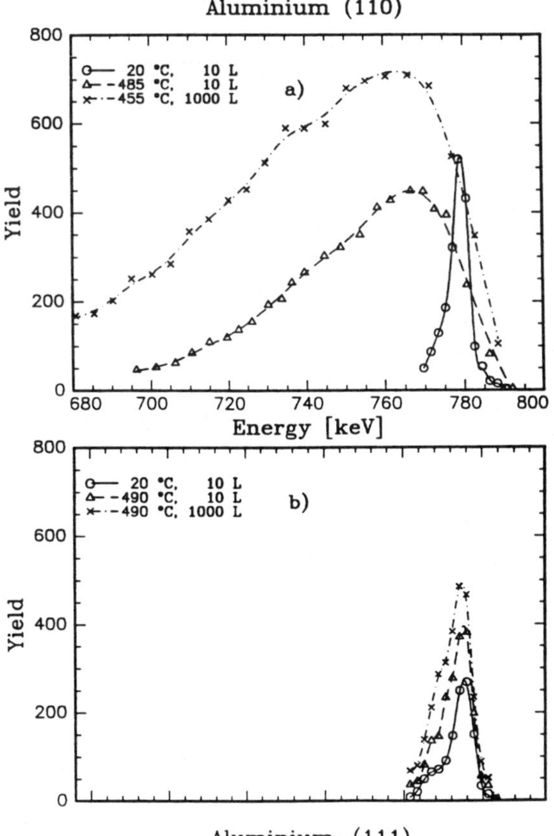

Fig. 8 Oxygen HERDA spectra of aluminum surfaces after different exposures to oxygen at temperatures around 500°C [11]. For comparison the depth profile after an exposure to 10 Langmuir at room temperature is shown. While a thin oxide layer is formed on the (111)-surface, which slightly increases in thickness with increasing temperature and exposure (lower part of figure), the oxygen diffuses into the bulk of the sample in the case of the (110)-surface, forming a thick oxide layer (upper part of figure). 10 keV on the abscissa correspond to a depth interval of 1 nm.

Fig. 8 finally shows, as an example for application, HERDA spectra of O atoms which were obtained in the way described above from Al single crystals of different surface orientations and at different stages of oxidation [11]. After the exposure of the samples to an amount of 10 Langmuir of O_2 gas at room temperature thin oxygen layers are formed an both surfaces ((111)-surface: ca $2.5 \cdot 10^{14}$ O-atoms/cm², (110)-surface: ca $4 \cdot 10^{14}$ O-atoms/cm²). The HERDA spectra show narrow peaks whose widths are fully determined by the depth resolution of the spectrometer which is ca 0.3 nm in this particular case. The formation of a thin, thermally stable oxide is indeed expected on Al surfaces. It is this oxide which protects Al samples against further oxidation. If the same procedure is performed at temperatures above ca 400°C, the picture changes completely. While the oxide layer on the (111)-surface still remains stable (only a slight increase of the amount of oxygen can be noticed), the oxygen starts diffusing into the sample bulk for the (110)-surface. The oxygen profile there extends to a mean depth of ca 3 nm; the total amount of oxygen is now ca $2.8 \cdot 10^{15}$ O-atoms/cm². After the exposure to 1000 Langmuir of oxygen at the same temperature (490°C) the oxide has finally grown to a total amount of ca $7.1 \cdot 10^{15}$ O-atoms/cm² on this surface and extends far into the sample. In contrast, the oxide layer on the (111)-surface only slightly increases to a total amount of ca $9.5 \cdot 10^{14}$ O-atoms/cm², while the thickness remains unchanged.

This anomalous oxidation behaviour of (110) Al surfaces is most probably due to premelting of the surface. Premelting of (110)-surfaces has been observed on aluminum down to ca 150 K below the melting temperature of 640°C [12]. At this temperature ca 1 monlayer of the material is disordered; but due to the logarithmic temperature dependence, some disorder is present at even lower temperatures. The effect of the premelting on the oxygen diffusion probably consists of an increased mobility of the oxygen due to the presence of defects in the partially disordered Al host lattice.

The author would like to thank his collaborators and the staff of the Pelletron accelerator at the MPI für Metallforschung in Stuttgart for kind cooperation.

References

[1] R.G. Smeenk, R.M. Tromp, H.H. Kersten, A.J.H. Boerboom, and F.W. Saris, Nucl. Instr. Methods **195**, 581 (1982)

[2] Th. Enders, M. Rilli, and H.D. Carstanjen, Nucl. Instr. Methods B **64**, 817 (1992)

[3] H.D. Carstanjen, W. Decker, and H. Stoll, Z. Metallkunde **84**, 368 (1993)

[4] O. Kruse and H.D. Carstanjen, Nucl. Instr. Methods B **89**, 191 (1994)

[5] R. Plieninger and H.D. Carstanjen, Nucl. Instr. Methods B, in press

[6] Th. Fischer, Diplom-Thesis Univ. Stuttgart 1996

[7] O. Kruse, Dissertation, Univ. Stuttgart 1995; O. Kruse and H.D. Carstanjen, to be published

[8] M. Zinke-Allmang, S. Kalbitzer, and M. Weiser, Z. Physik A **320**, 697 (1985)

[9] M. Zinke-Allmang and O. Kruse, Nucl. Instr. Methods B **90**, 579 (1994)

[10] K.-H. Gießler, F. Rauch, and M. Stamm, Europhysics Letters 27, 605 (1994)

[11] S. Jamecsny, D. Plachke, and H.D. Carstanjen, subm. to Phys. Rev. Letters

[12] A.W.D. van der Gon, R.J. Smith, J.M. Gay, D.J. O'Connor, J.F. van der Veen, Surface Science **227**, 143 (1990)

Materials Science Forum Vols. 248-249 (1997) pp. 467-470
© 1997 Trans Tech Publications, Switzerland

Applications of Ion Track Filters

S. Amrita Kaur[1], H.S. Virk[1] and S.K. Chakarvarti[2]

[1] Department of Physics, Guru Nanak Dev University, Amritsar - 143005, India

[2] Department of Applied Physics, Regional Engineering College, Kurukshetra, India

Keywords: Ion Track Filters, Makrofol-KG, Microstructures, Single Pore, Multipore

Abstract

We have developed ion track filters/polycarbonate sieves of Makrofol-KG of thickness 30μm, 60μm and 100μm by ^{132}Xe and ^{238}U ion beam irradiation at 5.9 MeV/u and 14.0 MeV/u energy, respectively from UNILAC, GSI, Darmstadt. Both single pore and multipore sieves were prepared by etching the plastic foils in 6.25 N NaOH at 70°C for different intervals of time. E.Coli and C. Bacillus (dia. 1μm) and malignant blood cells (diameter 7-15μm) were filtered using a conductivity cell designed in our laboratory. Commercial exploitation is possible in developing countries for getting bacteria free water supply. Solute concentration of different solutions is determined using ion track filters. The next on our agenda is the template growth of micro/nano ensembles which will have an extensive range of applications in a variety of fields of science and technology.

Introduction

Solid State Nuclear Track Detector (SSNTD) technique is one of the most developed tools of research in a number of scientific and technological areas due to its wide range of applications in various interdisciplinary fields[1]. The first application that found widespread acceptance is the ion track filters[2,3]. The so produced ion track filters or membranes can be used in wide range of applications in various fields, viz. Biophysics, Nuclear physics, Health physics, Material Science and many others. The main advantage of these track filters over conventional filters is their well-defined pore size, and uniform density[4]. Due to possibility of reuse, these filters will prove to be very economical.

Preparation of Ion Track Filters

Makrofol-KG ($C_{16}H_{14}O_3$) was used for the preparation of ion track filters in the present experiment. Heavy ion beams of ^{132}Xe and ^{238}U were used for irradiation of 30μm, 60μm and 100μm thick makrofol polycarbonate foils from UNILAC, GSI, Darmstdt, Germany at an angle of 90° with fluences of 10^4 and 10^5 ions/cm^2 having energy 5.9 MeV/u and 14.0 MeV/u, respectively. These irradiated foils have been etched in 6 N NaOH at 70°C for varying intervals of time for production of single and multipore filters. After washing and drying of these samples, the pores were observed under Carl Zeiss binocular optical microscope. The pore diameters were measured using a calibrated graticule in the eye-piece.

Applications of Ion Track Filters

There are growing number of applications of nuclear track- etched filters/membranes in various branches of science and technology like ultrafiltration, separation of emulsions, as model pores for biological membranes, fabrication of metal micro tubes using ion track filters ,etc. In the present study, we describe some of these applications:

(i) Conduction of bacteria and malignant cells through polycarbonate filters

The conduction effects through these filters have been observed in various liquids viz., polluted water, normal water, contaminated blood, beer, sugarcane juice etc., by using the conductivity cell. The cell is designed in such a way so that these filters act as a partition between the two chambers, serve as a barrier to bacteria and the change in current shows the resistance to the flow of the contaminants through the pores. The resistance through the pores of different diameters is measured

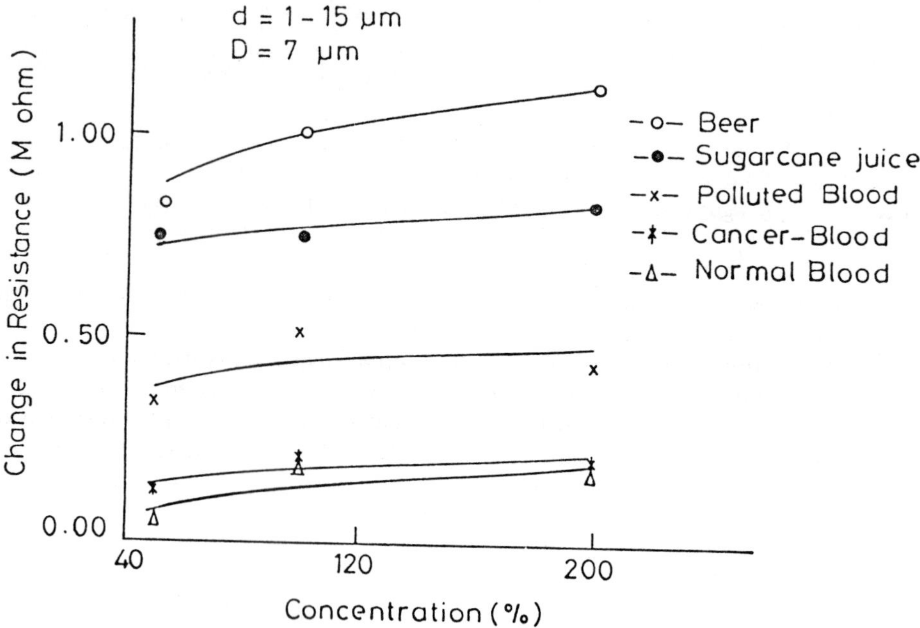

Fig. 1. Variation of change in resistance with concentration using single-pore filter for various liquids.

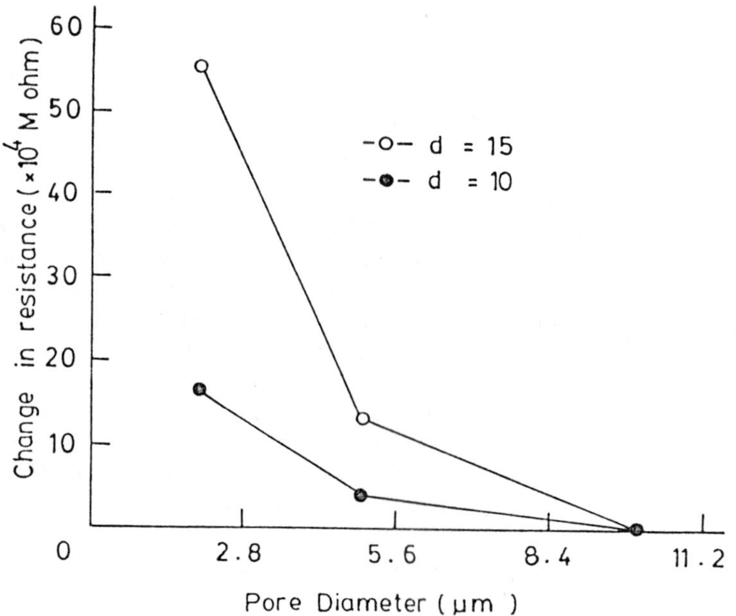

Fig. 2. Variation of change in resistance with pore diameter for blood sample.

using Ohm's law, and the change in resistance ΔR with respect to pore diameter is determined by using formula[1], $\Delta R = 4\rho d^3/\pi D^4$, where d is the diameter of the particle and D is the diameter of the pore and the corresponding resistivity of the solution is found by $\rho = RA/l$, where R is the resistance, A is the area of the cylindrical shaped conductivity cell and l is the length of the hole inside the cell. It has been observed that conduction is reduced progressively with increasing concentration of contaminants provided one side of the cell is filled with the media (i.e. polluted water, normal blood, cancerous blood, beer, infected sugarcane juice) and the other side is filled with unpolluted water or normal saline solution. Thus the change in resistance relative to concentration increases only if negative polarity is given to the media and positive polarity to the unpolluted water or normal saline solution. it is also observed that as the pore diameter of the filter increases, the change in resistance decreases (figs. 1 and 2 for blood).

(ii) Microhydrodynamical flow studies in various liquids using ion track filters

We have made microhydrodynamical flow studies on various liquids (water, alcohol and acetone) using these filters. For this study, a self-designed apparatus is used for the estimation of solute concentrations in different mixtures with water as a base. It has been observed that the rate of flow, dV/dt, decreases with the increase of solute concentration. The variation of the dV/dt versus concentration of the solutes in water at a constant pressure difference is studied (fig. 3).

(iii) Characterisation of metal-semiconductor microstructures grown using ion track filters

There has been an ever increasing interest exhibited towards fabrication of nano/micro structures (metallic as well as non-metallic) mainly because of the many possible potential applications. Microstructures comprising either metallic microdimensional devices- dots, fibres, wire, cones, tubules and whiskers have invited attention for use as horizontal technology in multidisplinary areas such as microelectronic mechanics, field emission electrodes[6], electrochemistry, conductive polymeric fibrils, transparent metal structures, biological substitutes for microtubles nano and microstructures, etc. There exist a variety of techniques like electron/ion beam lithography, photolithography (optical, UV and X-rays), selective electroless metallic deposition technique etc. We have used the technique of template growth of microstructures of metal-semiconductor (Cu-Se) through the etched pores of ion track filters of Makrofol-KG.

In the present work, we describe the simple method of electrodeposition of copper metal into the etched pores of polymeric ITFs by using an electrochemical cell. The basic principal underlying this process is an electrochemical process in which metallic ions in a supporting solution are reduced to the metallic state at the cathode which if closely covered by an ITF, would lead to the growth of an electroplated film. The electrolyte used here is $CuSO_4.5H_2O$ acid solution in $10M\Omega$ milli Q water + 25% vol H_2SO_4 and $Na_2SeO_3.5H_2O$ in 10 $M\Omega$ milli Q water. A current of 0.05 A/cm^2 was applied for 20 and 30 mins., respectively. After the plating process is over, the ITF is carefully removed, washed with distilled water and dried and is dissolved chemically in chloroform ($CHCl_3$). After drying, the thin film was carefully peeled off from ITF and sent for scanning. Two visible layers of Cu & Se were found to be deposited over the surface of ITF sandwiching the foil. In order to characterise these microstructures we have used AFM for finding out morphological and structural parameters (fig.4).

References

1. R.L. Fleischer, P.B. Price and R.M. Walker, Nuclear Tracks in Solids: Principles and Applications, University of California Press, Berkeley (1975).
2. B.E. Fischer and R. Spohr. Workshop on Industrial Applications of Energetic Heavy Ions, ATLE 89, C(1989).
3. R. Spohr, Ion Tracks and Microtechnology, Principle and Applications (Vieweg Verlag, 1990).
4. B.E. Fisher and R. Spohr, Rev. Mod. Phys. **55**, 907 (1983).
5. S. Amrita Kaur and H.S. Virk, Ind. J. Pure and Appl. Phys. **33**, 350 (1995).
6. Clampit R., Nucl. Instrum. Methods **62**, 109 (1981).

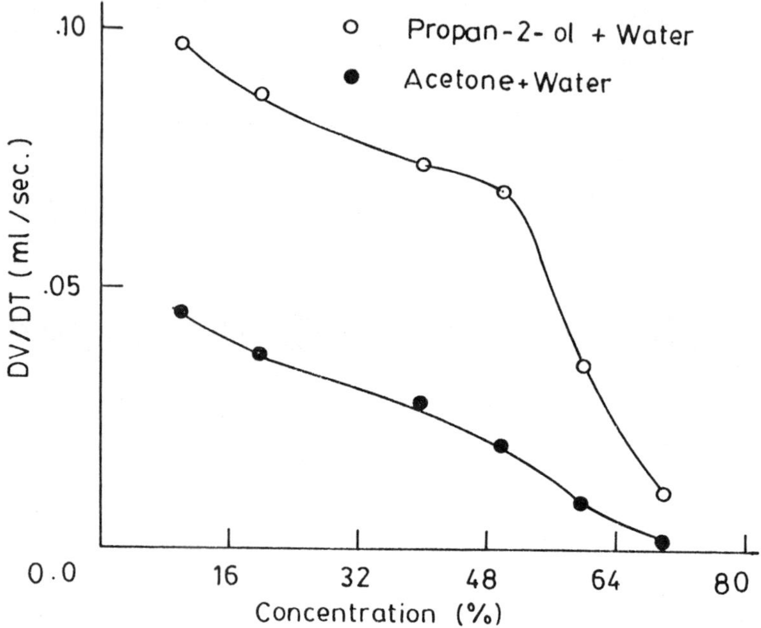

Fig.3. Variation of flow rate, dV/dt, with concentration of the solutes in water at constant pressure difference.

Fig 4 Microstructure ensemble (3-d)of Cu-Se grown electrochemically using ion track filters.

Materials Science Forum Vols. 248-249 (1997) pp. 471-475

Theoretical and Experimental Study of the Stationary and Kinetic Characteristics of the Thermo-Field Ion Sources

N.V. Egorov and B.V. Yakovlev

Faculty for Applied Mathematics of St-Petersburg State University,
Bibliotechnaya pl. 2, Petrodvorets, St-Petersburg 198904, Russia

Keywords: Ion Emission, Ion Source

Abstract: We investigated the thermo-field ion sources constructed as a low-melting metal layer deposited on the high-melting metal substrate with a sharp tip. The theoretical analysis was performed on the basis of the crater evolution model of its surface in the high electric field. We have calculated not only the stationary current-voltage plots but analysed the transitional processes, preceding the stabilisation of the ion emission and calculated the kinetic characteristics as well. Our calculation predicts an oscillative processes in ion emission. We performed experiments to examine the validity of the theory, and the agreement was found to be excellent.

1. INTRODUCTION

Nowadays a great attention is paid to thermo-field ion emitters (TFIE) as one of the promising elements of modern microelectronics. We investigated liquid-metal TFIE, which are consisting of a high-melting metal substrate (tungsten tip) and a low-melting metal film - gold, indium *etc.* deposited onto it [1]. Elevated operation temperature is obtained with ohmical heating. The operation electrical field strength is dependent on the bias between the TFIE and extractor, the special shape of extractor can lead to a one or two orders magnitude growth of the emission [2].

Usually, two models are used to explain the experimental results of the thermo-field ion emission investigations. In the first, the ions are emitted from the smoothed apex of the Taylor cone liquid metal surface with a balanced action of the electric field and surface tension in every point [3,4]. In the second, it is supposed that ions are emitted not from a region near the Taylor cone apex, but from a thin protrusion apex liquid metal jet [5]. Both models can explain only a part of the experimental results. There exist some results which can be explained only with some new ideas: A characteristic current leap at the first section of the current-voltage function [6], glowing at the TFIE apex [1,7], field ion images of the TFIE that are a number of the consecutive concentric rings with a dark spot in the centre [1,7], current increase if extractor has some specific shape [8,9]. The present work has the aim to propose a new model of TFIE that could explain these facts as well as the results which are explained on the basis of existing models.

2. THEORETICAL

2.1. Crater model

We suppose that for a proper interpretation of the results the electron beam action onto the liquid metal surface evolution is to be taken into account. This beam is generated simultaneously with the ion one and in the same area of the field ionisation of the atoms.

According to our model at the first stage as the bias U between extractor and emitter is increased to some extraction value U_0, the liquid metal surface attains the specific shape of the Taylor cone with a narrow protrusion. The formation of local areas with curvature radii of $r=10^{-7}$ cm is a necessary condition for onset of the emission at a bias that provides a local field strength of $E=2.10^{10}$ V/cm that is sufficient for liquid metal atom field ionisation [4].

We suppose that the second stage of the quasi-stationary liquid metal formation is the destruction of the Taylor cone apex as a result of liquid metal atoms field ionisation and the generation of two opposite beams - ion and electron, the initial leap at the current-voltage function being related to this process. So, the destruction of the Taylor cone apex leads to the crater appearance, and the constant pressure of the electron beam supports this shape in a quasi-stationary state. So, the quasi-stationary shape of the TFIE apex is not that of a Taylor cone, and even not of the Taylor cone with a protrusion, but a truncated cone with sharp edges and a crater in the centre - fig. 1.

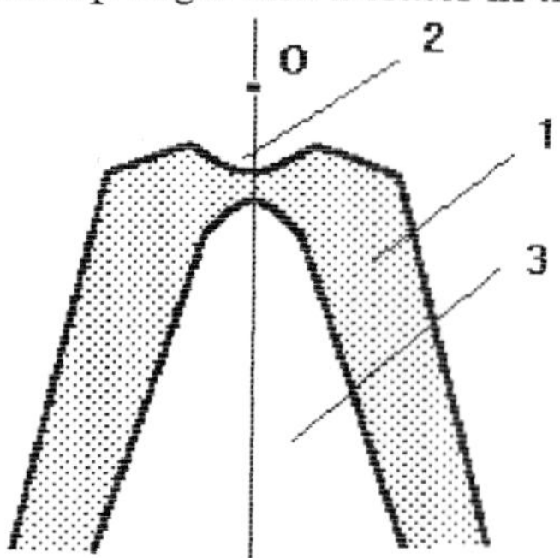

Fig. 1. Crater model of TFIE. 1-low-melting metal, 2-crater, 3-high-melting metal tip, O-field ionisation point.

It is worthy to remark that the destruction of the cone apex is not irreversible, as the field is switched off the generation process is stopped and electron pressure onto the central area of the cone disappears. As a result the cone appears again. So, the processes of cone formation, destruction and re-generation are some transitive processes and can be characterised with a relaxation time constant. We would like to underline once more, that the Taylor cone is a pre-emissive state of TFIE, emitting TFIE has a quasi-stationary shape of truncated cone with crater.

2.2. TFIE emission parameters

We performed a detailed mathematical analysis of the emission process for crater model of TFIE. We supposed that ion current is a superposition of: ions emitted from the edges of crater, field ionised atoms (near point O, see fig.1) evaporated from the crater surface, surface ionised atoms of the crater itself, the latter process is possible with the help of electrons generated in the same point O. The detailed analysis will be published elsewhere.

The transition processes time constant in this model can be assumed as time interval from the switch on of the field until the destruction of Taylor cone.

In this model the ion current can be presented as

$$I_i = (1 + exp[-\alpha / (U / U_0 - 1)]) \cdot AU$$

where $A = (\eta \cdot Y / E_o)\sqrt{e / 2m_a W}$, e - electron charge, ρ - liquid metal density, m_a - atomic mass, U-bias, U_0- extraction bias, $\alpha = (W-\Phi)/(eW)$, Φ-work function and W - ionisation potential, R - tip curvature radius. The time lag is

$$\tau = [e\rho R^3 / (m_a I_0)] exp[\alpha / (U / U_0 - 1)])$$

3. EXPERIMENTAL

3.1. Experimental set-up

The study was performed with indium, gallium and tin ions in the super-high vacuum with residual gases pressure about 1.10^{-9} Torr. The low-melting metal was deposited onto the tungsten tip substrate in a specialized preparation chamber, The tip substrate was welded to a wire that was used for heating the emitter. It is worthy to remark that the melting temperature of the working substance differed from that characteristic for the bulk metal and was about 2/3 of the value.

The process was investigated both in stationary and pulsed modes. The extractor was located 2-5 mm far from the emitter. We used two kinds of extractor shapes- that of diaphragm and of truncated cone with a sharpened edge directed to emitter. With the latter configuration we have had an ion current enhancement effect, described in [2]. We found out that in this case ion current enhancement is more than one order of magnitude.

All experiments based on the information-expert system have been described earlier[10,11].

3.2. Current-voltage function of TFIE

The results of TFIE investigation in the stationary mode can be summed as follows:

Current-voltage function is non-linear and has three characteristic sections: I-exponential growth, II-slow growth (saturation), III-sharp rise of the ion current. The ion emission starts only as some value of bias is achieved - extraction bias U_0. Depending on the tip substrate curvature radius its value varied from 4 to 15 kV for gallium TFIE and is approximately proportional to the curvature radius.

But if we applied a longitudinal magnetic field, so that the beam had no losses on its path through the extractor, collector and extractor current-voltage functions became qualitatively similar. It is important that as the extracting bias was attained the stabilisation of the stationary emission was preceded with a leap of current, its value being a constant for every specific TFIE. It is quite evident that this effect is a result of some transition process during formation of the TFIE emitting surface. It can be very important for practice that when some quantity of nickel dust was added to gallium (before the experiment) and the emitting cone was formed in the longitudinal magnetic field, the extracting bias could be reduced to 20% of its usual value. After the rapid

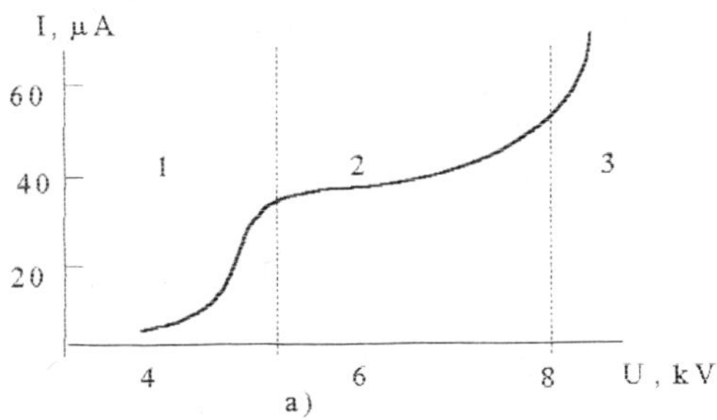

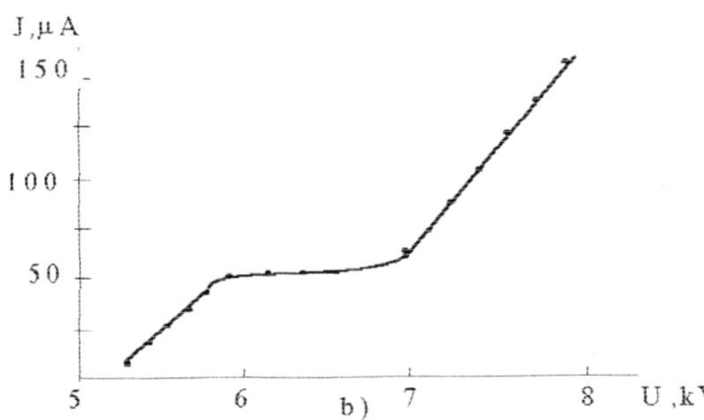

Fig. 2. Ion current as function of voltage, a)-theory, b)-experiment

increase of current the second section of the slow current growth begins at $U=1.3U_0$, which at $U>2U_0$ converts to section 3 of rapid growth of ion current again-fig. 2. Qualitatively, the current-voltage functions are reproducible without any dependence on curvature radius, distance, *etc*.

3.3. Kinetic characteristics and transition processes

Much attention was paid to the kinetic characteristics of the TFIE. These parameters were measured in the pulsed mode with high-tension rectangular pulses generator with amplitudes from 0 to 30 kV and duration from 10 µs to 1 ms. Custom designed device was used for pulse current measurement from 10 µA to 1 A. We observed the following experimental facts:

3.3.1. Current time lag. The investigation shows that even if $U>U_0$ (here U_0 is the extraction voltage in the continuous mode) there is some time lag between the field application and current advent. So, for one of the gallium samples time lag decreased exponentially from 350 µs to 10 µs as the bias was increased from 11.7 to 16 kV -see fig.3.

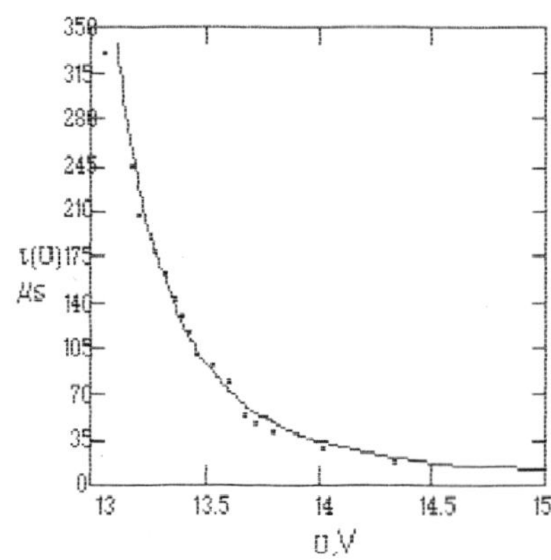

Fig. 3. TFIE current time lag, solid line-theory

3.3.2. The shape of the TFIE oscillogrammes. The oscillogrammes of the current pulses have a specific shape with maximum at the beginning of the transition process, i.e. at the initial stage of the transition process a current leap is observed. The value of the leap was constant for various pulse amplitudes, so, for gallium TFIE with curvature radius $=0.3$ µm it was 23 µA. After this, current decreases to some quasi-stationary value with time constant of some µs. It can be approximately described as $I(t)=f(\sqrt{t})$.

3.3.3. Ion current oscillations. At high ion currents (60-100 µA) we observed the oscillations of the ion current with 1 MHz frequency, amplitude vary from some µAmps to tens of µAmps, i.e. the amplitude is comparable to the main current value.

4. CONCLUSION

So, the crater model predicts, or at least, explains the following experimental facts:
1. The appearance of the ion current at some threshold voltage.
2. Current jump at the initial stage.
3. The emission images as light ring at the edge and a dark spot in the centre.
4. Ion current enhancement for the sharpened edge extractor.
5. Glowing near emitter.
6. Three characteristic sections of the current-voltage plot.
7. Time lag.
8. Parabolic law $I(t)=f(\sqrt{t})$.
9. TFIE current oscillations.

So, we can suppose, that the crater model adequately reflects the basic physical processes of TFIE mechanism.

REFERENCES

[1] M.D.Gabovich, Uspekhi Fisicheskikh Nauk, **140**, 137 (1983)
[2] N.V.Egorov and V.M.Zhukov, USSR Patent No. 1748571, 15.03.92
[3] G.Taylor, Proc.Roy.Soc. Ser.A **280**, 383 (1964)
[4] R.Gomer, J.Appl.Phys. **19**, 365 (1979)
[5] D.R.Kingham and L.W.Swanson, J.Appl.Phys. **41**, 157 (1986)
[6] A.Wagner and T.M.Hall, J.Vac.Sci.Techn. **16**, 1871 (1979)
[7] T.Sakurai, R.J.Culbertson and G.H.Robertson, Appl.Phys.Lett. **34**, 11 (1979)
[8] M.D.Gabovich and V.N.Starkov, Zhurnal Tekhnicheskoi Fisiki **52**, 1249 (1982)
[9] N.V.Egorov and A.E.Korol'kov, Zhurnal Tekhnicheskoi Fisiki **59**, 129 (1989)
[10] N.V.Egorov, K.O.Killistov and O.A.Kharitonov, Pribory i Tekhnika Eksperimenta, No.2, 165 (1985)
[11] N.V.Egorov, A.G.Karpov and B.V.Yakovlev *in* Abstracts of the 9-th Russian Conference on Photometry and Its Metrology, p.104, Moscow, 1992.

Materials Science Forum Vols. 248-249 (1997) pp. 477-480
© 1997 Trans Tech Publications, Switzerland

Status of a New Analytical Facility Based on the 2 MeV Electrostatic Accelerator in the Institute of Applied Physics

I. Karnaukhov[1], O. Ehichev[1], P. Gladkikh[2], S. Lebed, V. Miroshnichenko[1], A. Mytsykov[2], A. Shcherbakov[2] and V. Storizhko[1]

[1] Institute of Applied Physics, 244030 Petropavlovskaya St., 58, Sumy, Ukraine

[2] National Scientific Centre "Kharkov Institute of Physics and Technology", 310108 Academicheskaya St.,1, Kharkov, Ukraine

Keywords: Microprobe, Computer Simulation, Submicrometer Spatial Resolution, Elecrostatic Accelerator, Scanning Ion Microprobe

Abstract. Presented are state and progress in construction of the first analytical facility 'microprobe' with the energy of 2 MeV and the submicrometer spatial resolution in the Ukraine. Result of computer simulation of sensitivity of the beamline elements to the mechanical imperfections is also presented as well as acceptable errors in alignment of the elements.

1. Introduction

High technology and modern investigations in the fields of the microelectronics, material studies, mineralogy, metallurgy, ecology, biology requires the high energy probes with the micrometer resolution[1-4]. The first in the Ukraine analytical accelerator facility. 'The high energy ion microprobe' with 2 MeV beam energy and the submicrometer spatial resolution has been designed and is under construction now at the Institute of applied physics of the Ukrainian academy of science. This report presents the project of the facility and results of computer simulations on the sensitivity of the optical line to the mechanical imperfections.

2. The design of analytical facility

The accelerator based analytical facility is intended to provide both the high dose lines (6) and the beam line of the submicrometer resolution probe with a beam of high energy hydrogen or helium ions. The electrostatic accelerator and the magnets have been mounted already; the vibration resisted basements for the microprobe line equipment are constructed as well. The quadrupole lenses for this line are under tests now.

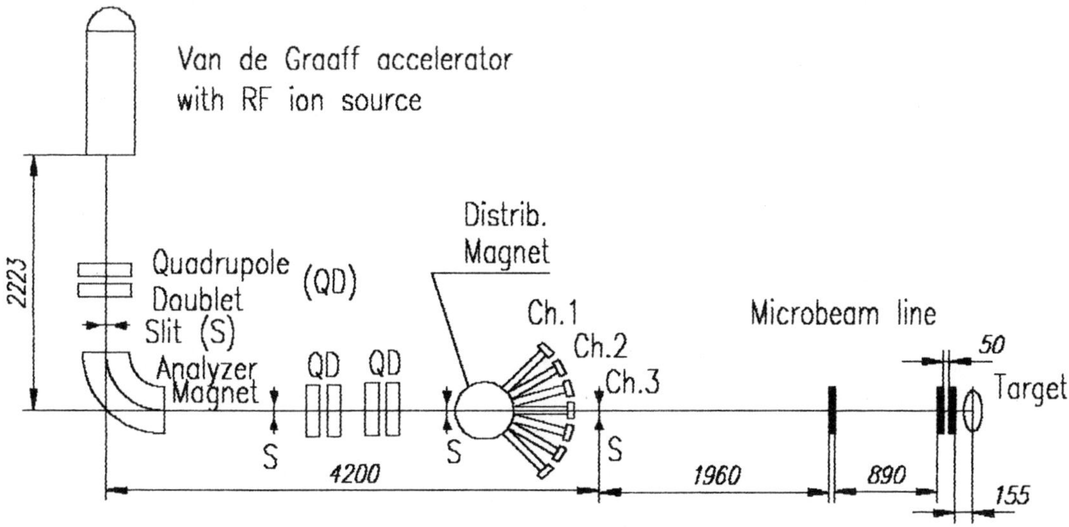

Fig. 1 The general scheme of the accelerator based analytical facility.

The general scheme of the facility is depicted in Fig.1.

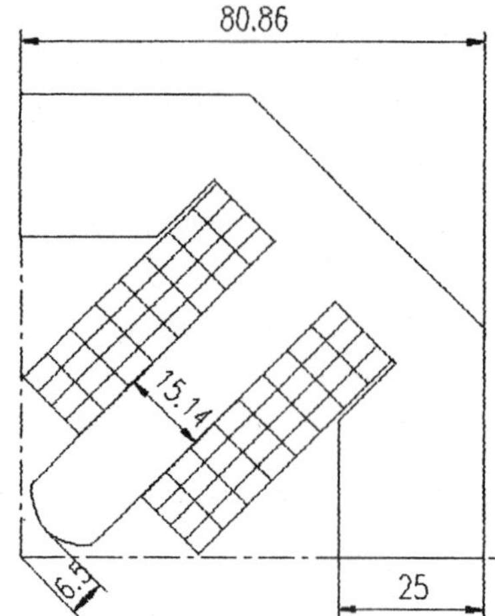

Fig.2. Quadrant of the quadrupole electromagnetic lens for the microprobe.

The analytical lines attached to the distributing magnet are dedicated for the elemental analysis, the express analysis of the rock samples, the environment monitoring, the ion-channeling studies of the monocrystal samples.

The line for express analysis (Ch.1 in Fig.1) is intended for the express multi elemental analysis of solid substances, materials, and manufactured articles used in the microelectronics, material science, biology, and medicine. The line for channeling and back scattering (Ch. 2 in Fig. 1) is intended for carrying out studies on the structure and elemental composition of the near-to-surface layers of solids, crystalline solids and materials. The converter line (Ch. 3 in Fig. 1) is intended to produce the characteristic X-ray radiation as a result of hitting of the converters by the accelerated ions.

Any line designed to be attached to the chamber of the distributing magnet consists of the vacuum valve, the ion line, the double gap beam collimator, the vacuum station, the chamber for interactions, and the set of detectors for registration, storing and processing of the data.

Table 1

Parameter	Value
Aperture radius, mm	6.5
Nominal gradient, T/m	64
Effective length, cm	5.
Amperes*turns, ÅW	1100
Accept. imperfections for the pole shape, mm	0.02
Accept. imperfections for the yoke assembling, mm	0.05
Accept. imperfections for the wedge assembling, mm	0.2
a3 - hexapolar component %G_{nom}	0.2
a4 - octupolar component %G_{nom}	0.15
a5 - decapolar component %G_{nom}	0.13
a6 - dodecapolar component %G_{nom}	0.13
Field at the pole face, T	
B1	0.281
B2	0.281
B3	0.395

The microbeam line is intended to form the ion beam of micrometer and submicrometer transverse dimensions for both the mode of the scanning ion microprobe and the mode of the transparent microscope. The focusing system of the microprobe is of the quadrupole triplet type (Fig.2). Main parameters of the quadrupole lenses are presented in Table 1.

Main parameters of the ion microprobe [5] are listed in Table 2.

Table 2

Parameter	Value
Beam energy, MeV	$0.2 \div 2.0$
Energy spread, %	0.01
Ion current, μA	$10 \div 100$
Source brightness, $A \cdot rad^{-2} \cdot m^{-2} \cdot (eV)$	$2 \div 5$
Length of the microprobe line, m	$7.2 \div 9$
Beam diameter on the target, μm	$1.0 \div 0.02$
Scanning area, mm·mm	$2.0 \cdot 2.0$
Total consuming power, kW	30.0

3. Calculations of ion optics

Optical characteristics of the microprobe line were computed with the codes TRANSPORT [6] and RAYTRACE [7]. Results of the calculations are summarized in Table 3.

Table 3

Parameter	Value
Chromatic aberrations, $\mu m/mrad \cdot \%$	
$\langle x/\theta \cdot \delta \rangle$	-200
$\langle y/\phi \cdot \delta \rangle$	1223
Spherical aberrations, $\mu m/mrad3$	
$\langle x/\theta^3 \rangle$	1712
$\langle y/\theta \cdot \phi^2 \rangle$	1619
$\langle y/\theta^2 \cdot \phi \rangle$	-5077
$\langle y/\phi^3 \rangle$	-1705

Sensitivity of the objective to transverse shifts of the lenses

Effects of the turns and shifts of the lenses were estimated in the coordinate system used by the RAYTRACE code[7].

Position of the beam centre at the crossover point as dependence on the i-th lens shift size Δx, Δy along X- or Y-axis can be estimated as:

$$s_x = \Delta x \cdot ki_x \quad s_y = \Delta y \cdot ki_y$$

where $k1_x$=0.125; $k1_y$=0.445; $k2_x$=0.88; $k2_y$=3.91; $k3_x$=1.98; $k3_y$=2.45, the coefficients were obtained for 10 μm shift size of the lenses along the corresponding coordinates.

Sensitivity of the objective to turn around the coordinate axes

Response of transverse position of the beam centre at the crossover point on the i-th lens turn around $X(\Phi)$ or $Y(\theta)$ axes can be described as follows:

$$s_x = \theta \cdot ki_x \quad s_y = \phi \cdot ki_y$$

where $k1_x$=3.92; $k1_y$=16.79; $k2_x$=37.73; $k2_y$=107.29; $k3_x$=0.215; $k3_y$=148, the coefficients were obtained for the turns of 0.1 mrad around the corresponding axis.

Based on the simulations carried out the acceptable misalignments of the lenses were defined, which are summarized in Table 4.

The cases of the most severe deviation of the microprobe objective parameter are depicted in Fig.3,4

Table 4

Coordinate or parameter	Z	X	Y	Φ	θ	Φz	I	Amplitude of vertical vibrations of a lens around the beam axis
Acceptable deviation	100 μm	10 μm	10 μm	0.5 mrad	0.5 mrad	0.02 mrad	2.5·10-5	0.01 μm

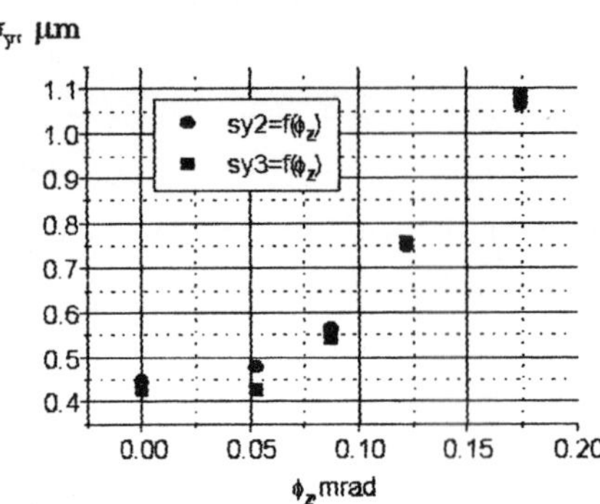

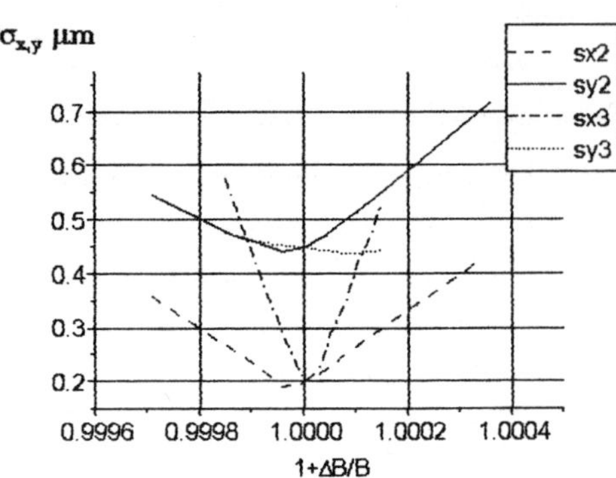

Fig.3. Response of the vertical spot size to the turn of second and third lenses around X axis

Fig.4. Response of the vertical and horizontal spot size to the pole face field of second and third lens.

4. Summary

The accelerator based analytical facility possesses the following advantages over the existing ones: satisfactory brightness of the beam, the upgraded optical system; submicrometer spatial resolution; compactness of the installation.

Computer simulation of sensitivity of the beam line elements upon the mechanical imperfections enables us to define the acceptable misalignments of these elements. Existing imperfections in manufacturing and allignments of the beam line elements enable us to get the target spot size of $2\sigma_x*2\sigma_y=0.5*0.8$ μm.

Commissioning of the 2 MeV electrostatic ion accelerator and the analytical lines is scheduled to the end of 1996. Operation of the microprobe will start by the end of 1997.

References

[1] F.Watt, G.W.Grime, D.G.Frazer, in: Principles and Applications of High Energy for Microbeams eds. F.Watt, G.W.Grime, (Adam Hilger, Bristol, 1987)

[2] R.D.Vis, The Proton Microprobe: Applications in the Biomedical Field (CRC Press, Boca Raton, 1985), 1-197

[3] F.Watt, G.W.Grime, Nucl. Instr. and Meth., B30 (1988), 356-366

[4] L.J.Cabri, Nucl. Instr. and Meth., B30 (1988), 459-465

[5] V.Braznik, V.Khomenko, S.Lebed, A. Ponomarev Nucl. Instr. and Meth., B104 (1995) 69-76

[6] K.L.Brown et. al., "TRANSPORT: a Computer Program for Designing Charged Particle Beam Transport Systems" SLAC-91 (1977).

[7] S.Kovalski and H.AEnge, "RAYTRACE", LNS Technical Report

KEYWORD INDEX